AUDIO VIDEO SYSTEMS

Principles, Practices and Troubleshooting

As per AICTE Model Curriculum 2018

S.P. Bali
Rajeev Bali

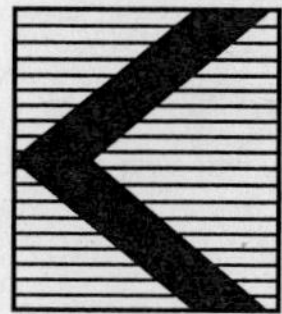

KHANNA BOOK PUBLISHING CO. (P) LTD.

Publisher of Engineering and Computer Books

4C/4344, Ansari Road, DaryaGanj, New Delhi-110002
Phone : 011-23244447-48 **Mobile :** +91-9910909320
E-mail : contact@khannabooks.com
Website : www.khannabooks.com

Dear Readers,

To prevent the piracy, this book is secured with HIGH SECURITY HOLOGRAM on the front title cover. In case you don't find the hologram on the front cover title, please write us to at contact@khannabooks.com or whatsapp us at +91-99109 09320 and avail special gift voucher for yourself.

Specimen of Hologram on front Cover title:

Moreover, there is a SPECIAL DISCOUNT COUPON for you with EVERY HOLOGRAM.

How to avail this SPECIAL DISCOUNT:

Step 1: Scratch the hologram
Step 2: Under the scratch area, your "coupon code" is available
Step 3: Logon to www.khannabooks.com
Step 4: Use your "coupon code" in the shopping cart and get your copy at a special discount
Step 5: Enjoy your reading!

Price: ₹ 650.00

ISBN: 978-81-87522-05-8

Audio Video Systems
by Bali & Bali

Edition: 2025

To view complete list of books, Please scan the QR Code:

Published by:
Khanna Book Publishing Co. (P) Ltd.
CIN: U22110DL1998PTC095547

Visit us at: **www.khannabooks.com**
Write us at: **contact@khannabooks.com**

Printed in Bharat.

This book is dedicated to

All those scientists, technocrats and engineers who have endeavoured in the field of electronics to bring it to the present-state-of-the-art and to make the short span of human life all more enjoyable.

About the Authors

Surinder Pal Bali has a long association, of over four decades with the field of electronics. He has worked in different capacities as a technician (repairing radio and line equipment) as an inspector (carrying out "in inspection and out inspection" of radio and line equipment) and as a lecturer (teaching diploma course in MCEME, Secunderabad). He is a professional trainer and has been conducting job oriented courses on "audio video" appliances. As a free lance technical journalist he has contributed articles to technical journals and authored about ten books.

Rajeev Bali is a graduate engineer with BTNL and has co-authored Elements of Electronic Instrument.

Preface

Audio Video Systems covers the entire syllabus laid down by the All India Council of Technical Education for *polytechnics* for the relevant subject. Some readers may have a good technical background, others will have a nodding acquaintance with the subject, while yet others will have no technical knowledge at all. The book has been written with this in mind and so the descriptions and explanations in the earlier chapters are selfcontained.

The greatest part of energy from the sun comes as a Wave Motion. Radio signals are waves that are used to transimit music or speech, as explained in Chapler 1.

Vibrating strings act as a source of sound. A column of air may also be made to vibrate. Wind instruments depend on the resonance of air coloumns. This is explained in Chapter 2 on *Vibrations of Strings and Columns.*

Hearing (Chapter 3) is the natural method for the reception of speech, music and noise and in particular is the ultimate significant termination of most acoustic systems involving the artificial reproduction of sound.

Speech, Music and Noise, dealt with in Chapter 4, are the three main categories into which practically all important natural sounds may be grouped.

Microphones serve two principal purposes: they are used for converting music or speech into electrical signals; they also serve as transducers which actuate indicating meters. These are covered in Chapter 5.

Headphones (Chapter 6), are a forgotten piece of equipment. When it breaks down especially during a program, it instantly causes the biggest headache any production person can ever have.

The science of *Loudspeakers* has been refined to the point where the response of a drive unit can be predicted with great accuracy. Just as the the choice of a microphone dictates what goes into the recording, the loudspeaker has the last word over the quality of what comes out. These are explained in Chapter 7.

An essential part of any loudspeakers system is the enclosure in which the loudspeaker is used. An enclosure for a loudspeaker is often called a baffle. *Speaker Baffles and Enclosures* are discussed in Chapter 8.

Monophony signifies one-source sound, one amplifier channel and one-speaker system one channel from start to finish. Stereophony denotes solid, or three dimensional sound-sound coming from different sources at different locations with different volume levels (two or more speakers, channels and speaker systems). Quadraphony is an extension of stereo. Quad denotes four-channel sound. *Mono Stereo and Quad* are discussed in Chapter 9.

If we can gain a basic understanding of what it takes to make a good listening room, we will find that there is much we can do to ensure that the money invested in hi-fi equipment doesn't go down the drain. This aspect of audio is discussed in Chapter 10 on *Acoustics.*

Disc Recording and Reproduction became firmly established with the invention of the gramophone in 1887. Ever since then, tremendous inprovements in the quality of reproduced sound have taken place. This is discussed in Chapter 11.

Some distortion is always present in the sound reproduced from a gramophone record. This is discussed in *Sources of Distortion In Disc Equipment* in Chapter 12.

Although magnetic tape offers some notable advantages in the recording and reproduction of sound, it features infrequently in A/V systems. *Magnetic Recording and Reproduction* is discussed in Chapter 13.

There are *Sources of Distortion in Tape Equipment* also. These are discussed in Chapter 14.

The use of solid state devices in audio systems has brought about a radical change in the appearance of the equipment. Never before have so many watts been packed into so small a volume. *Amplifying Systems* are discussed in Chapter 15.

Noise originates mainly in th recording process. *Noise Reduction Systems* are discussed in Chapter 16.

Typical *Recorder/Amplifier Circuits* are discussed in Chapter 17.

Electronic music was not designed to replace humans; rather it brings a powerful new tool to the hands of the musician. *Electronic Music Synthesizers* are discussed in Chapter 18.

Compact Disc Players are a very specialised form of record players. The information is present in the track in digital form. *Optical Recording and Reproduction* is discussed in Chapter 19.

Sound is commerical to the extent that a segment of the public has a preference for the same audio qualities consciously or subconsciously. *Commerical Sound* is discussed in Chapter 20.

The traditional aim of a hi fi system is to reproduce reorded sound with the highest possible fidelity to the original. Audio systems, both component and integrated, are discussed Chapter 21.

In the early 20th century, TV systems were mechanical. Today TV equipment is all electronic. The replacement of mechanical by electronic television has made it possible to reproduce a high-quality picture approaching the original scene in quality. *Monochrome TV Standards and Systems* are discused in Chapter 22. Chapter 23 deals with *Colour TV Standard and Systems*. The controls associated with monochrome and colour televisions are discussed in Chapter 24 on *Monochrome and Colour Television Controls*.

Before video recording sequences had to be shot on film which was expensive. They had to be transmitted live. *Video Tape Recording and Reproduction* is discussed in Chapter 25.

The possibility of recording video signals on a disc is discused in Chapter 26 on *Video Disc Recording and Playback*.

Theater Sound Systems Chapter 27, differ considerably from audio systems. The chief distinction is in the method of sound recording and reproduction.

Video Systems, both tape-based systems and disc-based systems, are discussed in Chapter 28.

Troubleshooting is the process of determining why a device or circuit does not operate as it should and is the most important aspect of AV Systems. The *Philosophy of Troubleshooting* is discussed in Chapter 29.

Preventive Maintenance (Chapter 30), is basically concerned with the correction of a potential trouble and thereby aviods repair costs and down time.

A scientific field advances at a speed determined by the availability of measuring instruments. *Acoustical Measurements* are discussed in Chapter 31.

Performance Tests on discrete components of AV Systems are explained in detail in Chapter 32.

Troubleshooting Procedure, with and without service data, is discussed in Chapter33. This is further supplemented by troubleshooting Charts.

Much useful reference information is tabulated in charts and tables and it is hoped that the complete work will of be continuing interest and assistance, both new and in the future, to all who, for either business or pleasure, set AV Systems in action.

—Surinder Pal Bali
—Rajeev Bali

March, 1999

Contents

1

WAVE MOTION

In all our activities the availability of energy is a continuous problem. The conversion of energy into the form that is useful for a particular purpose and the transportation of this energy from the place at which it is converted to the place at which it is utilised are problems of immediate concern. Developing means of communication has always been a challenge to man. In the different communication procedures we are attempting to carry our message through the motion of waves or through the motion of matter (particles). Waves provide a mechanism for the transfer of energy from one point to another without the physical transfer of matter between the points. The greatest part of energy from the sun comes as a wave motion. Sounds reach our ears by means of waves in the air around us. Radio signals are waves that are used to transmit music or speech. All waves have certain properties in common that will now be considered.

1.1. MECHANICAL WAVES

Energy and momentum can be transported from one place to another through the motion of particles. Wave motion provides an alternative way for *transfer of energy and momentum without the material particles making that journey.*

Water waves and sound waves, for example are mechanical waves, that travel through a *deformable or elastic medium.* They originate when some portion of the medium is displaced from its normal position and released.

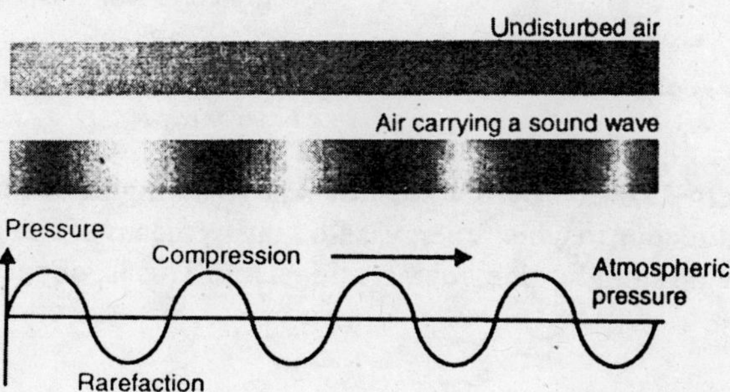

Fig. 1.1. Along the path of a sound wave in air, the longitudinal to and fro movement of the particles produces alternate compressions and rarefactions.

As a wave reaches a particle in the medium, Fig. 1.1, it sets that particle into motion and *displaces* it, thus *transferring* both kinetic and potential energy to it. We can regard the particles of the medium as moving only by small distances about their previous position, ***without undergoing any net overall displacement*** in the direction of travel of the wave.

1.2 TYPES OF WAVES

Waves can be *classified* according to their broad physical properties; according to how the direction of motion of the particles is related to the direction of propagation of the wave; according to the number of dimensions in which they propagate energy and also according to how the particles of the medium move in time.

Imagine a stone dropped in a still lake. Circular ripples spread out from the point where the stone entered the water. Along a given circular ripple, all points are in the same *state of motion.* These points define a surface called *wavefront.* If the medium is of uniform density, the *direction of motion* of the wave is at right angles to the wavefront. A line normal to the wavefront, indicating the direction of the motion of the wave is called a *ray*, Fig. 1.2.

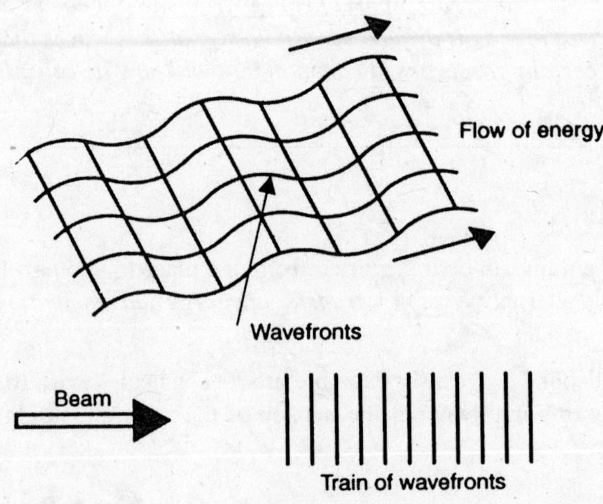

Fig. 1.2. A wavefront is a line on a wave joining points of equal phase. Waves may be represented by a train of wavefronts, or more simply by a beam or ray.

As a rule the energy exerted by a wave flows at right angles to the wavefront, and it is sometimes simpler to represent the direction in which energy is flowing by means of straight lines bounding the wavefronts, that is, by means of a single straight line, representing a ray. This is, of course, a simplified representation of wave motion.

1.3. LONGITUDINAL WAVES

The different kinds of waves have many *features* in common. Having considered one type of wave, we can apply the same methods to describe the behaviour of other waves, such as sound waves, light waves.

Waves in which the vibration of particles is in the *same direction* as the wave travels are called *longitudinal waves.* Waves in which the movement of particles is at *right angles* to the direction in which the wave travels are called *transverse waves.*

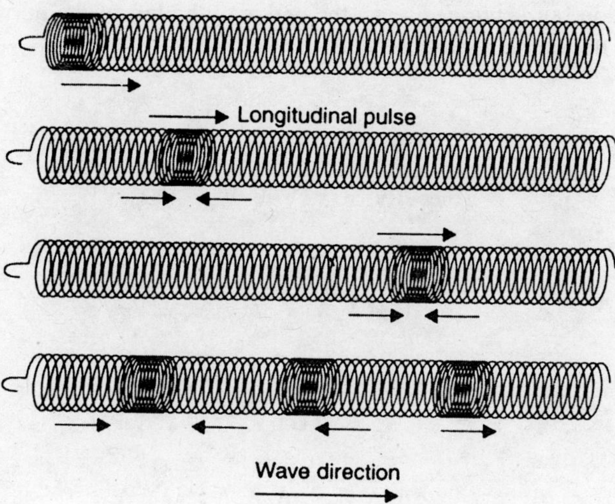

Fig. 1.3. A longitudinal pulse and a longitudinal wave are shown moving along a spiral spring. Individual coils move back and forth about their undisturbed positions.

Each particle on the path of a longitudinal wave performs simple harmonic motion about its *average position* as the wave moves through it. Although both types of wave may pass through the same material, their *velocities* are generally different because of the different *physical properties* of the material involved in propagating them.

If we plot the displacement as ordinate against the horizontal equilibrium position of the particles we obtain the familiar sinusoidal graph, Fig. 1.4.

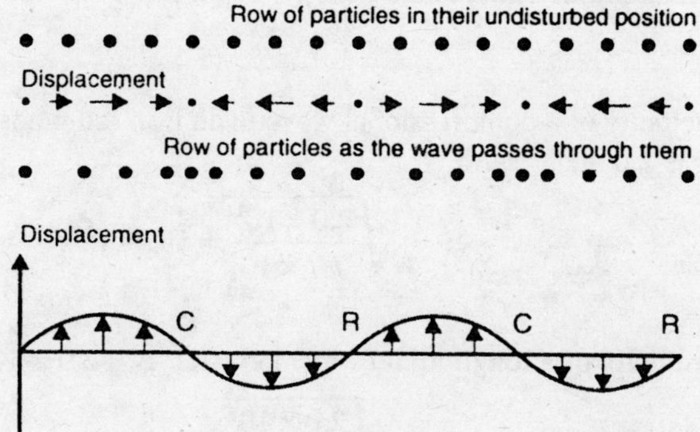

Fig. 1.4. Representing a longitudinal wave by a graph.

A complete wavelength includes a compression and a rarefaction. The instantaneously disturbed particles are at the *centre* of the compressions and rarefactions.

Considering a stretched spring. In Fig. 1.5 (*a*) the lines represent the position of the coils of the spring when the spring is undisturbed. A *harmonic longitudinal wave* can be set up in the spring if the end is moved with simple harmonic motion in a line parallel to the spring. The spring is *compressed* when the end is to the right of its normal position but *expanded* when the end is to the left of the normal. The result is that, as the end is moved in SHM, a harmonic succession of compressions and following expansions will move along the spring.

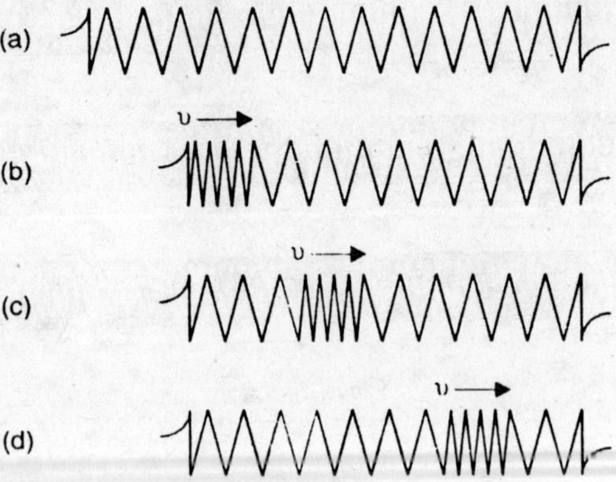

Fig. 1.5. Compression pulse in a spring. (a) The spring is undisturbed; (b) the pulse is started by moving the coils to the right; (c) and (d) the pulse has continued to move along the spring.

Longitudinal waves do not require shearing stress and hence may be transmitted through an *elastic medium.* The derivation of the relation between the speed of longitudinal waves and the properties of the medium is somewhat involved, only the end result is given.

$$v = \sqrt{\frac{E}{d}} \qquad \text{...(1.1)}$$

where E is the appropriate modulus and d is the density. For a rod, F is the Young's modulus; for a fluid, E is the bulk modulus B.

Example 1.1 : Find the velocity of a compressional wave in an iron rod whose density is 7.7×10^3 kg/m^3 and whose Young's modulus is 2.0×10^{11} N/m^2.

Solution. $$v = \sqrt{\frac{2.0 \times 10^{11}}{7.7 \times 10^3}} = \mathbf{5095 \ m/s}$$

Example 1.2 : Find the velocity of a longitudinal wave in water. B = 2.0×10^9 N/m and d = 1.0×10^3 kg/m^3.

Solution. $$v = \sqrt{\frac{2.0 \times 10^9}{1.0 \times 10^3}} = \mathbf{1414 \ m/s}$$

1.4. WAVE PROPERTIES

In observing waves we note several of their characteristics. Waves travel with a *definite speed* through a uniform medium. Also, if we watch a single spot, we see that the waves pass the spot at *regular intervals* of time. In wave motion such factors as wave speed, frequency, phase, wavelength and amplitude must be considered.

The *speed* of a wave is the distance it advances per unit time. The number of waves that pass a point per unit time is the *frequency* of the wave motion. The time required for a single wave to pass is called the *time period T* of the wave motion. The *wave length* λ is the distance between two adjacent particles that are in the same phase. The *amplitude* A of the wave is the maximum displacement of the particle from its equilibrium position. A *wavefront* is a surface that passes through all points in the wave that are in the same phase. In a medium in which the speed is the same in all directions, the wavefront is perpendicular to the direction that the wave travels. *If the wave passes from one medium into another, the speed changes. In this process the frequency remains the same but the wavelength will change in proportion to the speed.* If v increases, λ also increases.

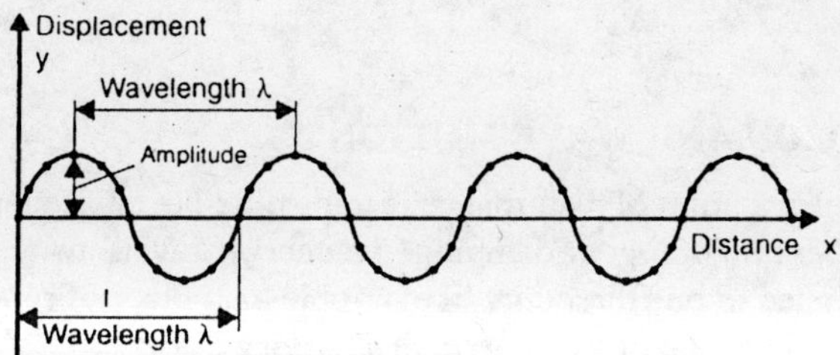

Fig. 1.6. A picture of a wave frozen at an instant in time showing the displacement of particles along its path.

If the energy which causes the disturbance carries on producing *recurrent* complete vibrations, then a *continuous periodic wave* is generated. We can form an accurate picture of such a wave by plotting the position, at one instant in time, of each particle in its path, Fig. 1.6. The distance between the wave crests or between the corresponding points on the curve is the wavelength.

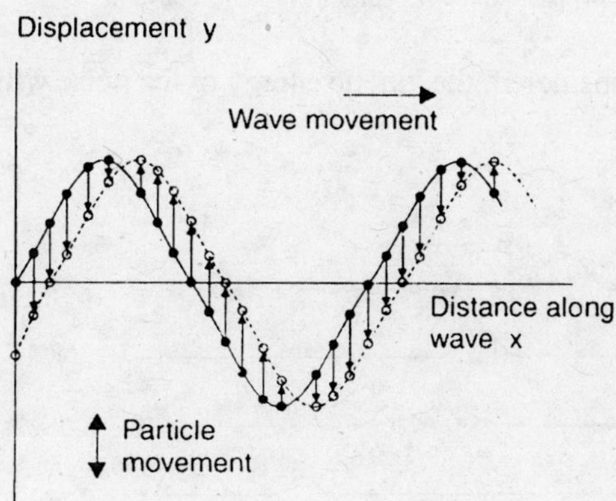

Fig. 1.7. Picture taken of a wave at two close instants of time shows that the particles do not move in the direction of the wave but at right angles to it.

As the wave profile moves, it displaces the position of each particle on its path. All the particles on the wave vibrate with S.H.M. but each one *lags* slightly behind the one on its left because of its *inertia.* Fig. 1.7 shows two positions of a wave separated by a short time interval. The wave profile moves from left to right but note that each particle moves up and down about its equilibrium position.

$$\text{Velocity of wave} = \frac{\text{distance travelled by wave}}{\text{time taken}}$$

$$= \frac{\text{number of waves} \times \text{wave length}}{\text{time}}$$

$$i.e.\ \text{velocity} = \text{frequency} \times \text{wavelength}$$

$$v = n\lambda \qquad \ldots(1.2)$$

1.5 ENERGY DISSIPATION AND WAVE MOTION

The energy supplied at a point of disturbance is transmitted to adjacent particles by the forces between the particles. If there is *internal friction* or damping, the energy travels away from the point in the form of a pulse. If the pulse is continued to one direction, like a ripple on a piece of rope, each particle in its path will, in turn, assume the movement of the original particle as shown in Fig. 1.8.

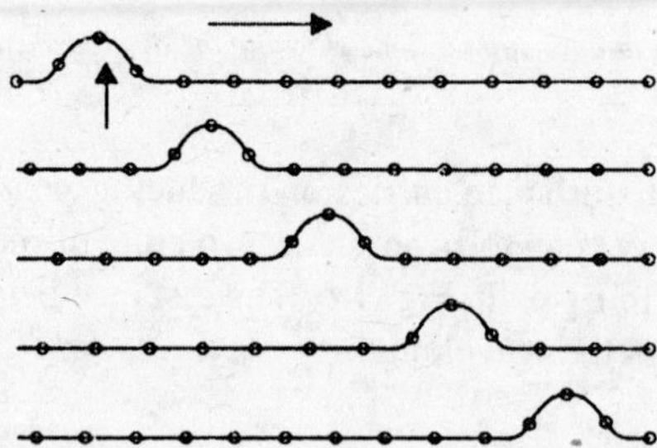

Fig 1.8. A single pulse travels away from a source of energy. The pulse shown is only half a complete cycle, and in an elastic medium the amplitude remains constant.

If the *external friction* is considered, the kinetic energy of the pulse will be converted to heat in a short distance, Fig. 1.9.

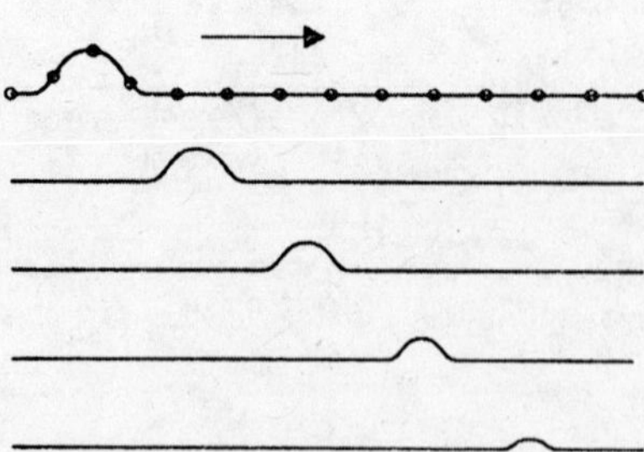

Fig. 1.9. A pulse is shown travelling in a non-elastic medium. The shape is preserved as before but the amplitude gradually reduces because of internal friction.

Example 1.3. : One end of a rope is vibrated at 8 Hz and the waves generated measure 60 cm from crest to crest. How fast do they travel and what is the period of vibration of any part of the rope ?

Solution
$$v = n\,\lambda = 8 \times 0.6$$
$$= 48 \text{ m/s}$$

Period
$$= \frac{1}{8} \text{ sec} = 125 \text{ ms}$$

If v_t and v_o be the velocities of sound at t°C and 0°C respectively, then

$$\frac{v_t}{v_0} = \sqrt{\frac{T_t}{T_o}} \qquad \text{...(1.3)}$$

where T_t and T_o are absolute temperatures.

Thus *the velocity of sound is directly proportional to the square root of the absolute temperature.*

Example 1.4 : Calculate the increases in the velocity of sound for 1°C rise of temperature, if the velocity of sound at 0°C is 332 m/s.

Solution.
$$\frac{v_t}{v_o} = \sqrt{\frac{T_t}{T_o}} = \sqrt{\frac{273+t}{273}}$$
$$v_t = v_o\left(1+\frac{t}{273}\right)^{1/2} = v_o\left(1+\frac{t}{2\times 273}\right)$$
$$= v_o\left(1+\frac{t}{546}\right) = v_o + v_o\left(\frac{t}{546}\right)$$
$$v_t - v_o = v_o\left(\frac{t}{546}\right) = \frac{332\times 1}{546} = 0.61 \text{ m/s}$$

The speed of sound in different mediums is given in Table 1.1.

Table 1.1 The Speed of Sound

Medium		Speed (m/s)
Gases	Air (0°C)	331
	Air (20°C)	343
	Helium	965
	Hydrogen	1284
Liquids	Water (0°C)	1402
	Water (20°C)	1482
	Sea water	1522
Solids	Steel	5941
	Granite	6000
	Aluminium	6420

1.6 INTENSITY OF A WAVE

The intensity I or *rate of flow of energy* by wave motion through unit area of the medium in a direction perpendicular to the direction of propagation of the wave is given by

$$I = 2\pi^2 f^2 a^2 vd \qquad \text{...(1.4)}$$

where f = frequency ; a = amplitude
v = velocity of the wave, and d = density of the medium.

If P is the power of an isotopic source of sound at a distance r, the intensity or *loudness* of sound.

$$I = \frac{P}{4\pi r^2} \qquad \text{...(1.5)}$$

Since $$P \propto a^2 \qquad \therefore \quad I \propto \frac{a^2}{r^2} \qquad \text{...(1.6)}$$

If a wave diverges in all directions from a source, *the intensity reduces with the distance from the source according to an inverse square law.*

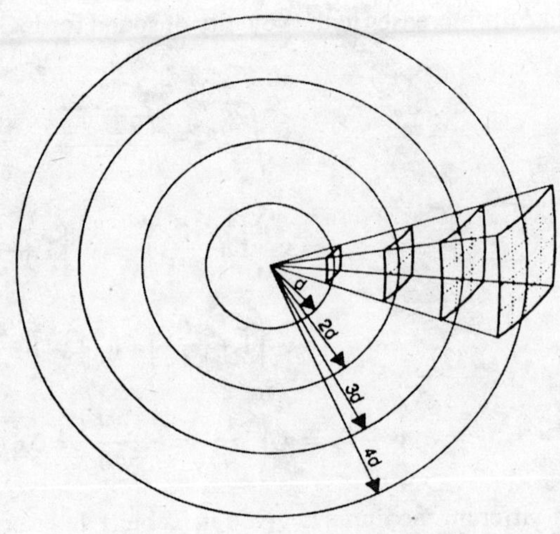

Fig. 1.10. In the solid angle shown, the same energy is distributed over spherical surfaces of increasing area as d is increased. The intensity of sound is inversely proportional to the square of the distance from the source.

It can be deduced from the above two relationships that *the amplitude of a divergent ray falls off linearly with distance.*

$$a \propto \frac{1}{r} \qquad \text{...(1.7)}$$

The intensity level is defined by an arbitrary scale that corresponds roughly to the sensation of loudness. Its unit is bel. The decibel is 1/10 of a bel. This is taken as the standard unit.

$$dB = 10 \log (I/I_o) \qquad \text{...(1.8)}$$

where I = intensity of sound and
I_0 = the weakest audible sound (10^{-12} W/m^2)

1.7. SOUND LEVEL

The reference intensity level of 10^{-12} W/m^2 is a typical value for the *threshold of human hearing.* A sound of intensity I_0 has a sound level of 0 dB, while a sound at the upper range of human hearing, called the *threshold of pain,* has an intensity of 1 W/m^2 and a sound level of 120 dB. Thus the human ear is capable of responding to intensities ranging over 12 orders of magnitude on a *logarithmic scale* of intensity called *sound level,* SL.

Example 1.5 : The intensity of waves at a distance of 20 m from their source is 25 W/m^2. What is the intensity at a distance of 50 *m* and how do the amplitudes of the waves compare at the two places ?

Solution. Intensity $\propto 1/d^2$

$$I_2/I_1 = d_1^2/d_2^2$$

$$I_2 = \frac{25 \times 400}{2500} = 4\text{W/m}^2$$

The amplitudes are in the *inverse ratio* of the distances from the source.

$$\frac{\text{amplitude of 20 cm}}{\text{amplitude of 50 cm}} = \frac{50}{20} = \frac{5}{2}$$

Each increase in intensity I by a multiplication factor 10 corresponds to adding 10 dB to the SL.

We can also use dB as a *relative measure* to compare different sounds with one another rather than with the reference intensity. Suppose we wish to compare two sounds of intensities I_1 and I_2.

$$SL_1 - SL_2 = 10\log I_1/I_0 - 10\log I_2/I_0$$

$$= 10 \log I_1/I_2$$

For example two sounds whose intensity ratio is 2 differ in SL by 10 log^2 = 3 dB.

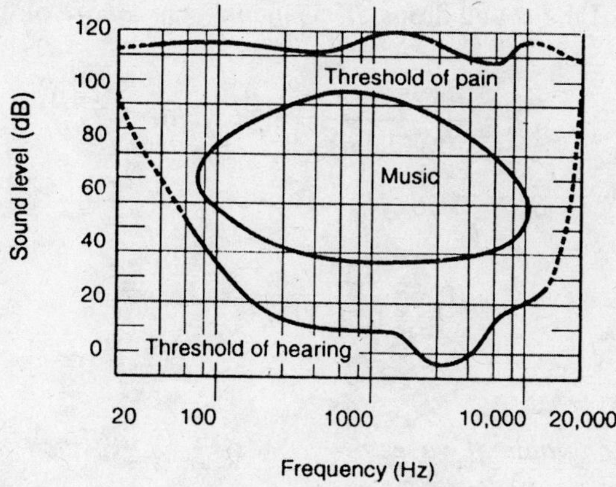

Fig. 1.11. The average range of sound levels for human hearing. Note the dependence of the threshold levels on frequency. A sound that we can just hear at 100 Hz must be 1000 times the acoustic power (30 dB greater sound level) than one we can just hear at 1000 Hz, because our ear is much less sensitive at 100 Hz.

The sensitivity of the human ear varies with frequency. The threshold of 10^{-12} W/m^2 applies only to the *midrange frequencies* around 1000 Hz. At the *higher frequencies* say 10,000 Hz, the threshold rises to about 10dB (10^{-11} W/m^2), while at *lower frequencies*, say 100 Hz the threshold is about 30dB (10^{-9}W/m^2). It takes 1000 times the sound intensity at 100 Hz to produce the same physiological response as a given sound intensity at 1000 Hz. Fig 1.11 shows the variation with frequency of the thresholds of hearing and pain and Table 1.2 shows some representative sound levels and their corresponding intensities.

Table 1.2. Some Intensities and Sound Levels

Sound	Intensity (W/m^2)	Relative Intensity (I/I_0)	Sound Level (dB)
Threshold of hearing	1×10^{-12}	10^0	0
Rustle of leaves	1×10^{-11}	10^1	10
Whisper (at 1 *m*)	1×10^{-10}	10^2	20
City street, no traffic	1×10^{-9}	10^3	30
Office, classroom	1×10^{-7}	10^5	50
Normal conversation (at 1 *m*)	1×10^{-6}	10^6	60
Jackhammer (at 1 *m*)	1×10^{-3}	10^9	90
Rock group	1×10^{-1}	10^{11}	110
Threshold of pain	1	10^{12}	120
Jet engine (at 50 *m*)	10	10^{13}	130
Saturn rocket (at 50 *m*)	1×10^8	10^{20}	200

Example 1.6 : Spherical sound waves are emitted uniformly in all directions from a point source, the radiated power P being 25 W. What are the intensity and the sound level of the sound wave at distance $r = 2.5$ m from the source?

Solution : All of the radiated power must pass through a sphere of radius r centered on the source. Thus

$$I = P/4\pi r^2$$

We see that the *intensity* of the sound drops off as the *inverse square* of the distance from the source.

Thus

$$I = \frac{25W}{(4\pi)\,(2.5m)^2} = \mathbf{0.32\ W/m^2}$$

and

$$SL = 10 \log I/I_0$$

$$= 10 \log \frac{0.32 W/m^2}{10^{-12}\ W/m^2} = \mathbf{115\ dB}$$

1.8 PROPERTIES OF SOUND WAVES

(*i*) Sound waves are *longitudinal waves.*

(*ii*) Sound travels through *all states of matter*

(*iii*) The velocity of sound is *maximum* in solids and *least* in gases.

(*iv*) Sounds which can be heard by human beings are known as *audible sounds*. Their frequency range is from 20 to 20,000 Hz.

(v) Sounds of frequencies greater than 20,000 Hz are called *ultrasonic sounds.* Ultrasonic sounds can be received and emitted by bats.

(vi) Sounds of frequencies less than 20 Hz are called *infrasonic sounds.* Infrasonic sounds can be received by snakes.

(vii) Sound waves can be *reflected.* In the phenomenon of reflection, the speed, frequency and wavelength of the sound waves do not change.

(viii) Sound waves can be *refracted.* In the phenomenon of refraction, the frequency of the sound waves does not change but their velocity and wavelength change.

(ix) Sound waves exhibit the phenomenon of *diffraction.*

(x) Sound waves do not show the phenomenon of *polarisation.*

(xi) The *normal level* of sound is 60 dB to 120 dB.

(xii) Sound waves cannot travel through *vacuum.*

1.9. TRANSVERSE WAVES

We are familiar with the way in which light energy travels from a source to arrive essentially at some other point which it illuminates. What is happening in the region through which the light is passing ? Were it possible to stop such a light wave and take measurements along its path, we would detect two influences; *a fluctuating electric field at right angles to the direction of the beam and a fluctuating magnetic field at right angles both to the beam and the electric field.*

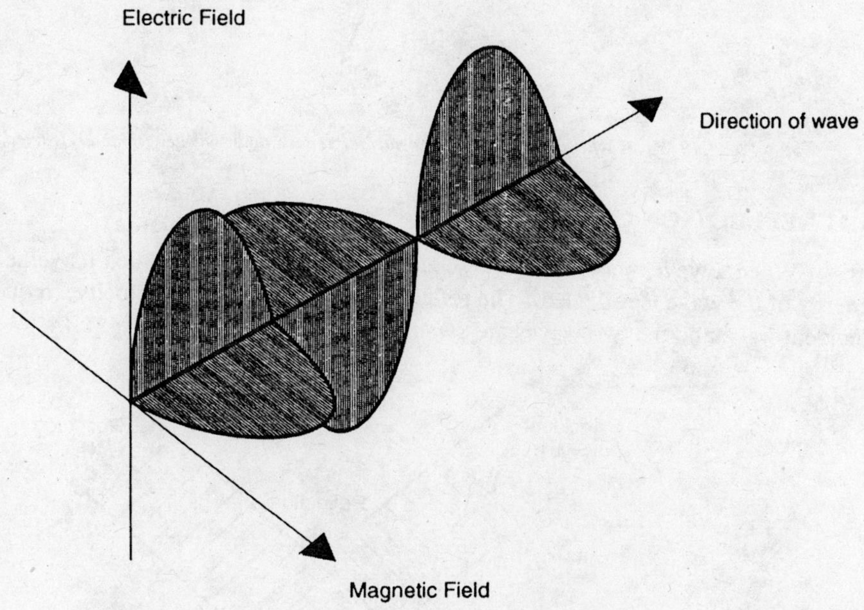

Fig 1.12. Field variations along an electromagnetic wave.

The electric field (E) and the magnetic field (H) have the *same wavelength* and they *keep in phase.* The energy of light is *stored* in these two fluctuating fields, the pattern of which is preserved as it travels along. In fact, *electromagnetic waves may be produced by a vibrating charged particle, such as an electron, which creates a varying electric field.*

1.10 REFRACTION OF WAVES

A change in medium or a change in conditions may cause a wave to *slow down.* This slowing down has the effect of shortening the wavelength, although the frequency of the wave remains the same. This is shown for normal incidence in Fig 1.13 which also shows what happens when the wavefronts meet the surface obliquely. One end of each wavefront, the end which touches the obstacle first, slows up *before* the other and that whole wavefront is made to *swing round.* The retardation of the wave causes deviation *towards the normal* at the surface. Conversely, if the waves travel faster in the second medium, they deviate *away from the normal.* This phenomenon is called *refraction.* It may occur *sharply* or it may occur *gradually.*

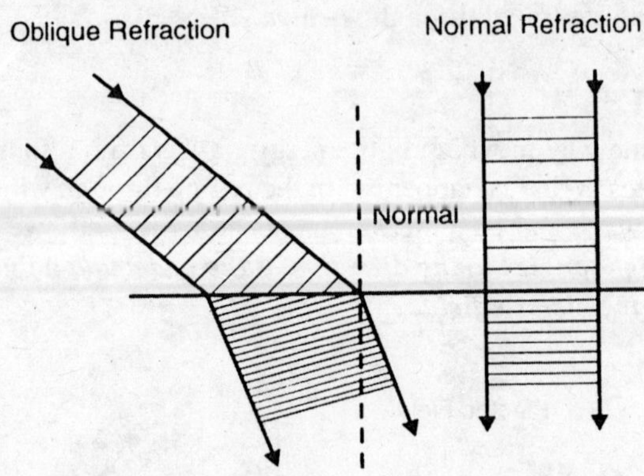

Fig. 1.13 Waves which are retarded at a surface reduce their wavelength and refracted towards the normal.

1.11 REFLECTION OF WAVES

When a wave meets a surface separating two materials in which its velocity is different, some of the energy of the wave is reflected. The reflected wave has the same velocity, frequency and wavelength as the incident wave and the two waves are *symmetrical* with the surface, Fig. 1.14.

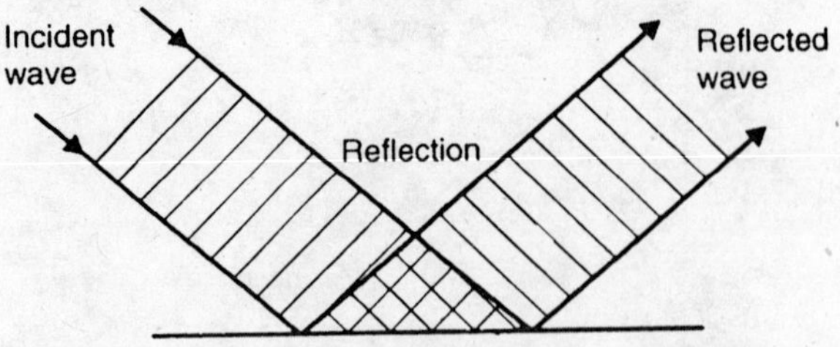

Fig. 1.14. The reflected wave and the incident wave are symmetrical with the surface.

In some cases of reflection, a *change of phase* occurs which is of importance when considering the details of sound and light waves. In *external reflection*, the phase of a wave changes by 180°. In the case of *internal reflection,* the incident and reflected waves are in phase. This is shown in Fig. 1.15.

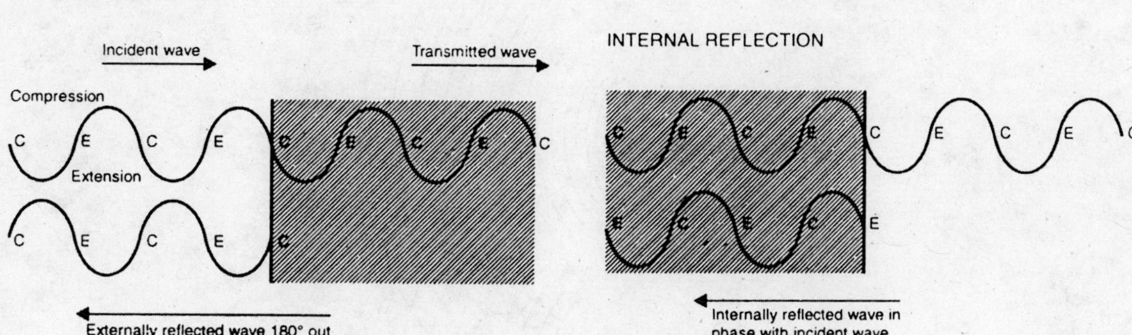

Fig. 1.15. A phase change occurs on external reflection but no change occurs on internal reflection.

The greatest the change in conditions at a surface the greater is the fraction of energy reflected. Where we want a wave to cross a boundary with minimum reflection, we try to reduce the *abruptness* of change. We do this in several ways, Fig 1.16, depending on the kind of wave.

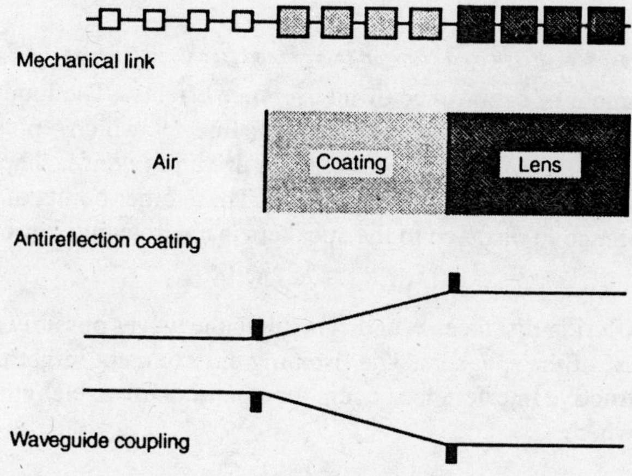

Fig. 1.16. The transfer of energy across a boundary is increased by interposing a section having intermediate properties which reduces reflection.

1.12 INTERFERENCE OF WAVES

When two or more waves combine at a particular point, they are said to interfere and the phenomenon is called interference. The resultant waveform is strongly dependent on the *relative phases* of the interfering waves.

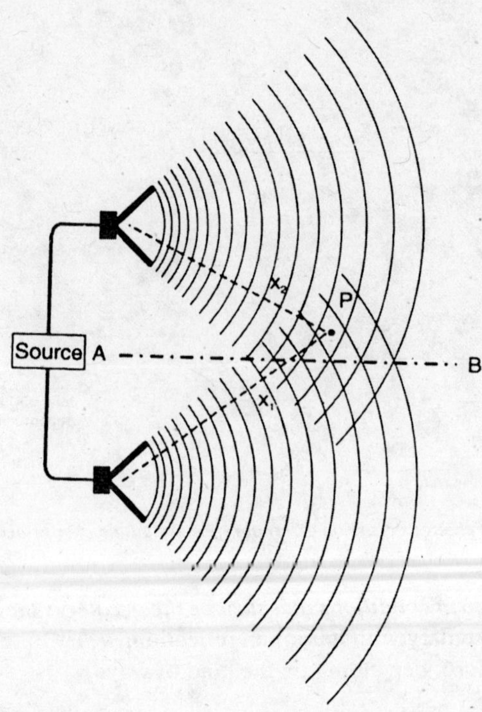

Fig. 1.17. Two loudspeakers driven by a common source send signals to P where the signals interfere.

Fig. 1.17 shows an example of occurrence of interference effects. The loudspeakers are driven from the same source. At points equidistant from the speakers (on the line *AB*, which represents the entire midplace) there is complete *constructive interference* if the speakers are driven *in phase*. There are also other points *P* where the waves arrive in phase and interfere constructively. These other points of constructive interference are located wherever the difference in distance to the speakers is a whole number of wavelengths.

$$|x_1 - x_2| = \lambda, 2\lambda, 3\lambda..... \qquad ...(1.9)$$

At other points *P*, the differing distances x_1 and x_2 result in the waves possibly arriving at *P out of phase*, even if they started out in phase at the speakers. The listening environment might therefore have *dead spots* where partial or complete destructive interference occurs for a particular wavelength λ. Maximum *destructive interference* occurs at points where

$$|x_1 - x_2| = \frac{\lambda}{2}, \frac{3\lambda}{2}, \frac{5\lambda}{2} ..... \qquad ....(1.10)$$

If the speakers emit a mixture of many different wavelengths some points *P* might show destructive interference of *one* wavelength and constructive interference of *another* wavelength. The *critical factor* in determining the maxima and minima of sound intensity is the *path difference*. $|x_1 - x_2|$.

Interference is characteristic of wave motion and its appearance in any phenomenon testifies to the *wave nature* of that phenomenon.

In any interference pattern the amplitude varies in a regular manner. Since the intensity of the wave is proportional to the square of its amplitude we see that the rate of flow of energy is not distributed uniformly as in a single wave. *In the interference pattern energy is diverted from the regions of destructive interference and appears at the regions of constructive interference.*

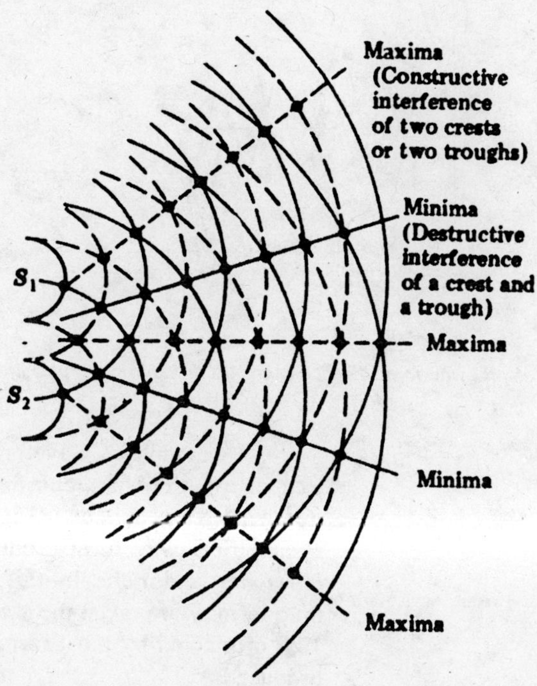

Fig. 1.18. Interference patterns from two sources showing points of maxima and minima. Dotted lines represent troughs and solid lines crests of the waves.

1.13. SUPERPOSITION OF WAVES

When two or more waves *exist simultaneously* in the same medium, each wave travels through the medium *as if the other were not present.* In the sense of propagation through the medium neither wave affects the other. However, at a point where two waves of the same kind reach simultaneously, the medium will have a displacement that is the *sum* of the displacements of the individual waves. Here *displacement* refers to the departure from normal to that property of the medium which varies as the wave passes through the medium. If the property is a complex quantity, the sum is a *vector sum.* Let ϕ_1 and ϕ_2 represent the displacements of the individual waves. Then at every point in the medium and at each instant of time the resultant displacement is

$$\phi_\gamma = \phi_1 + \phi_2 \qquad ...(1.11)$$

1.14 COMPLEX WAVES

The waveforms which represent speech and telegraph signals are of complex shape. Nevertheless if they *repeat the same pattern*, complex waveforms can be analysed into two or more components each of which is itself of *pure sine - wave form.* It is for this reason that *a.c.* theory and transmission line theory have been developed on the assumption of a pure sine-wave form.

In any complex waveform the lowest frequency component is known as the *fundamental frequency :* other sinusoidal components whose frequencies are *exact multiples* of the fundamental frequency are called *harmonics.* The first harmonic coincides with the fundamental frequency.

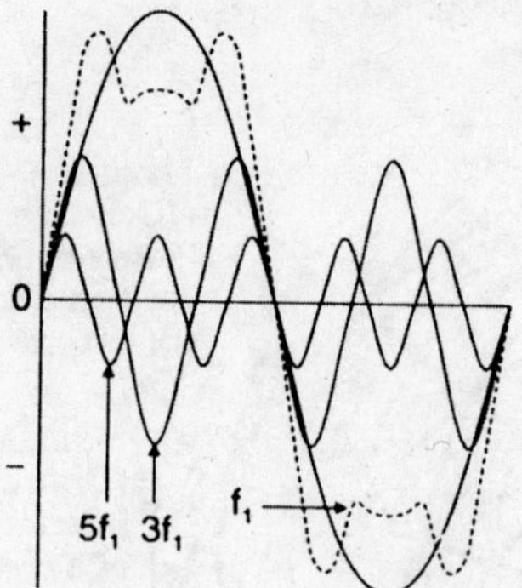

Fig. 1.19. Complex waveform: Fundamental plus third harmonic plus fifth harmonic.

A complex waveform containing fundamental frequency f_1, third harmonic frequency $3f_1$ and fifth harmonic frequency $5f_1$ is built up in Fig. 1.19. This shows the general symmetrical waveform produced by the presence of *odd harmonics* and in the limit, if all odd harmonic frequencies upto infinity are taken the *square topped waveform of* Fig. 1.20 originated by most forms of telegraph transmitters is produced.

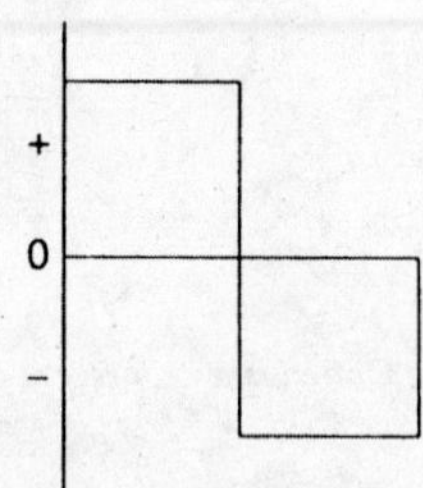

Fig. 1.20. Square topped waveform.

Complex waves which include *even harmonics* only are shown in Fig. 1.21 for the fourth harmonic, second harmonic and fundamental.

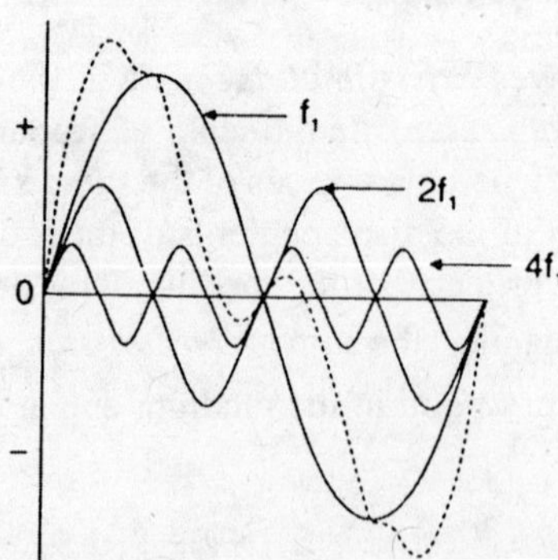

Fig. 1.21. Complex waveform : Fundamental plus second harmonic plus fourth harmonic.

Harmonic components of a complex wave may differ from the fundamental in three ways : (*i*) in relative peak values, (*ii*) in frequency and (*iii*) in phase displacement. It will be noted that the complete complex wave or ''group'' *repeats* its instantaneous values at the *fundamental frequency.*

The complete analysis of a complex wave is stated by Fourier's Theorem: *any complex wave which is a repeating function can be split into components each of which is a pure sine wave whose frequency is some whole-number multiple of the fundamental frequency.*

1.15 BEATS

The interference of two sound waves is also apparent when the frequency of one wave is *near to* that of the other. The *interaction* of the waves causes a periodic change in intensity (loudness) of the sound. This can be explained by adding graphically the amplitudes of the two waves at every instant of time, (Fig. 1.22). The resultant amplitude is a maximum when the waves are *in step* and a minimum when they are *out of step*. If these frequencies differ by (say) five vibrations per second then the waves get in and out of step five times per second and the number of pulses or beats heard per second is five. In general

$$\text{Beat frequency} = f_1 - f_2 \qquad \ldots(1.12)$$

Up to a beat frequency of 16 Hz, we hear a note of frequency *intermediate* between that of the two sources, waxing and waning in loudness. At beat frequencies above this we hear the two notes *separately* but they produce a disagreeable sound, which gets more dissonant as the beat frequency rises. *Eventually the beat frequency itself is heard as a separate note.* This plays an important part in harmony.

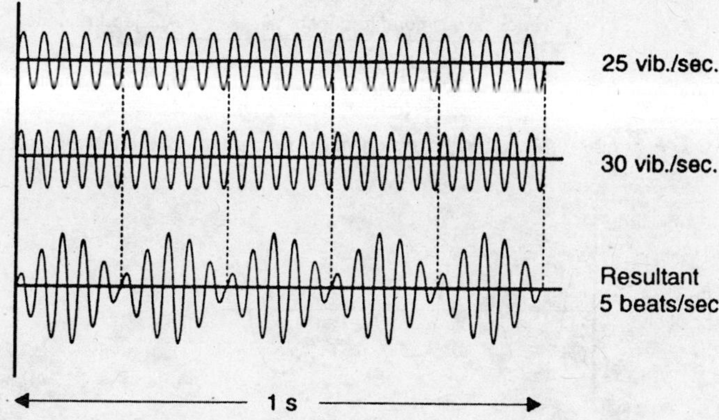

Fig. 1.22. The beat frequency is equal to the difference between the two frequencies sounding.

The effect is used to *tune* a piano string to the same frequency as the tuning fork, and to tune one string of a guitar with another. *The tension of the string is varied until the beat frequency reduces to zero.*

1.16 FORCED VIBRATIONS

When a vibrating tuning fork is held close to the ear the sound is heard quite loudly but the loudness falls of rapidly as the fork is moved away. Since the prongs are of very small area, *we may regard the fork as an approximate point source of energy,* and allowing for attenuation the loudness of the sound is inversely proportional to the square of the distance from the source. In this respect a point source of sound may be compared with a point source of electromagnetic waves. Both obey an *inverse square law.*

If, however the stem of the fork is pressed against a table top, the sound can be heard clearly all over the room. Under these conditions the table is set into *forced vibrations* and then acts as a large *extended source.* It may then by considered as equivalent to a large number of point sources, all of which contribute to the loudness of the room at any point in the room.

Although the sound is louder, it does not last long. The amplitude of the table top is very much less than that of the tuning fork, but its large areas transfer energy to the air at a much greater rate than the small area of the prongs.

1.17 RESONANCE

The best way to set a child's swing in motion is to give it small pushes *in time with* the natural period of the swing. This is an example of a general principal in physics called resonance.

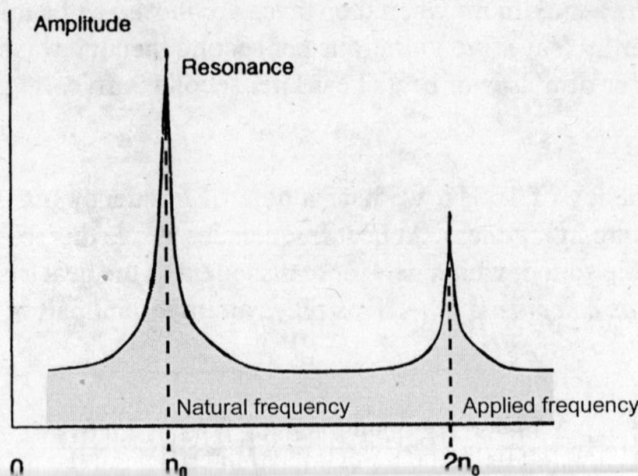

Fig. 1.23. Effect of the frequency of the forced vibration on amplitude.

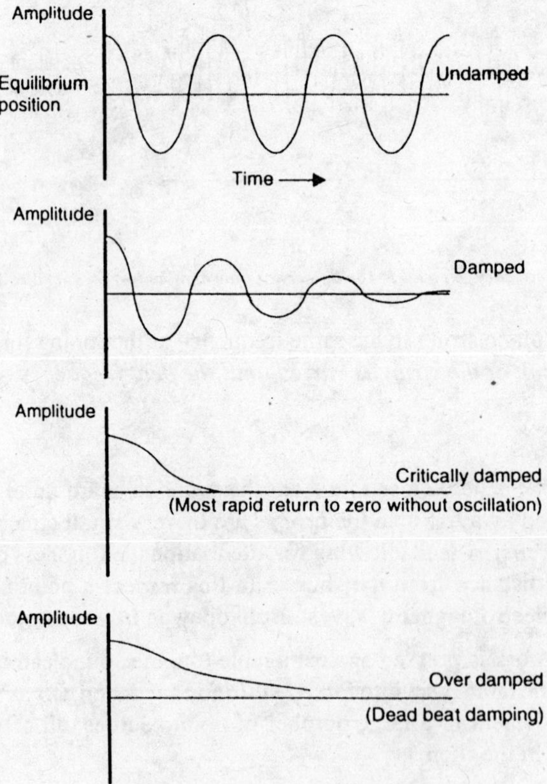

Fig 1.24. The graph shows the behaviour of a particle which is displaced from its equilibrium position and released under varying degrees of damping.

Resonance is said to occur whenever a particular body or system is set into oscillation at its own natural frequency as a result of impulses received from some other system which is vibrating with the same frequency.

Under these conditions it is possible for a very large amplitude of vibrations to be set up.

We find applications of the same principle in other branches of physics. The action of tuning a radio set is to adjust the value of capacitance in a circuit until it has the same natural period of oscillation for electricity as that of incoming signal. The small e.m.f. set up in the aerial is then able to build up a similar e.m.f. of large amplitude in the *tuned circuit.*

The tendency to resonate can be reduced by producing a means of *absorbing* the undesirable energy of vibration.

Instruments with moving pointers, such as balances and electrical meters, are *damped* to curtial oscillation about the final rest position. The damping may be provided by viscous, frictional, electrical or magnetic forces.

It is important to provide just the right degree of damping. This is usually the damping which returns the system to equilibrium position in the shortest possible time without oscillation. The condition is called *critical damping.* Fig 1.24 illustrates the effect of varying degrees of damping.

1.18 ECHOES

Echoes are produced by the *reflection* of sound from a hard surface such as a wall or cliff.

Let us suppose that a person claps his hands when standing some *distance* from a high wall and listens for the echo. The time which elapses before the echo arrives will depend on the distance away from the wall. Now in order that the echo may be heard *separately* from the original clap it must arrive at least 0.1 second later. Since sound travels at about 330 m/s, the reflected wave must have travelled a total distance of at least 33 m, and consequently the minimum distance of the wall must be about 17m.

When the reflecting surface is less than this distance the echo follows so closely upon *direct sound wave* that they cannot be distinguished as separate sounds. One merely receives the impression that the original sound has been prolonged. This effect is called *reverberation.*

OBJECTIVE TYPE OUESTIONS (A)

1. Which of the following is transmitted by a wave
   (*a*) amplitude (*b*) velocity
   (*c*) energy (*d*) momentum.
2. It is possible to distinguish between transverse and longitudinal waves by studying the property of
   (*a*) interference (*b*) diffraction
   (*c*) reflection (*d*) polarization
3. The relation between frequency n wavelength λ and velocity of propagation v of the wave is
   (*a*) $v = n\lambda$ (*b*) $n = v\lambda$
   (*c*) $n = v/\lambda$ (*d*) $n = 1\lambda$
4. The wavelength of sound is 10cm. Its frequency is
   (*a*) 3.3 KHz (*b*) 3.3 MHz
   (*c*) 330 Hz (*d*) 33 Hz
5. Sound waves having the following frequencies are audible to human beings
   (*a*) 5 Hz (*b*) 27,000 Hz
   (*c*) 5,000 Hz (*d*) 50,000 Hz

6. The velocity of sound is largest in
   (*a*) air (*b*) water
   (*c*) steel (*d*) vacuum

FILL IN THE BLANKS (B)

1. A wave is a ________ that travels through a medium.
2. ____________ may be transmitted by waves.
3. Where the waves arrive out of phase the interference is called ________.
4. Wavelength λ is the distance between two adjacent particles that are in the __________.
5. For resonance there is a ___________ transfer of energy and a resultant __________ sound.
6. In air at ordinary temperature the speed of sound is approximately ______.
7. The speed of sound in any medium depends on the _______ of the medium and upon its ________.
8. A sound wave may be _____________ if the speed of the wave is not the same in all parts of the medium.
9. An _______ occurs when reflected sound waves return to the observer in 0.1s or more after the original ___________ reaches him.
10. The intensity level is defined by an arbitrary scale that corresponds roughly to the _________ of ______.

CONVENTIONAL QUESTIONS (C)

1. Describe the primary difference between longitudinal and transverse waves.
2. What possible changes could occur to the velocity, frequency and wavelength of a wave as it goes from one medium to another?
3. What is the difference between progressive and standing waves?
4. What decides whether a medium will transmit longitudinal or transverse waves?
5. When transverse waves travel along a rope, each particle moves only at right angles to the rope. What is it that moves along the rope?
6. Why are waves of various sorts so useful in conveying information from one place to another?
7. Suggest practical uses that can be made of the fact that sound waves travel through solids with greater ease than through liquids or gases.
8. A sound produced by a tuning fork can be greatly amplified if the base of the vibrating fork is held against the top of a table. Explain why this happens.
9. What fundamental condition must exit for an object to resonate when a sound wave falls on it?
10. Explain why sound will not travel through vacuum.
11. Give some examples of the role that resonance plays in musical instruments. Under what circumstances should resonance be avoided in these instruments.
12. How are beats useful in tuning a musical instrument?
13. Justify the statements that the speed of sound in air increases by about 0.61m/s for each centigrade degree rise in temperature from 0°C.
14. Explain the laws of reflection of sound and explain how an echo may be produced.

15 Explain the difference between longitudinal and transverse waves. State a practical examples of each type.
Account for the fact that a distant lightning flash is seen before the thunder is heard.

16. State two essential differences between the modes of propagation of sound and light waves.

NUMERICAL QUESTIONS (D)

1. What is the wavelength of a radio station broadcasting at 1,000 Hz frequency ? Speed of radio waves is equal to 3×10^8 m/s. **(300 m)**
2. Assume that Young's modulus for silver is 7.75×10^{10} N/m^2 at 20° C. If it has a density of 1.05×10^3 kg /m^3, how fast does sound travel through silver ? (2.67×10^3 m/s.)
3. Copper has a density of 9 g/cm^3 and a bulk modulus of 1.2×10^{11} N/m^2. What is the speed of a wave through it ? (3.6×10^3 m/s.)
4. A person standing 99 m from the foot of a tall cliff claps his hands and hears an echo 0.6 *s* later. Calculate the velocity of sound in air. (330 ms)
5. What are the wave lengths in air of the piano's lowest note (frequency 27. 5 Hz) and its highest (frequency 4,186 Hz) ? Speed of sound is equal to 330 m/s. (12 m; 7.9 cm)
6. A car, sounding its horn at a frequency of 256 Hz, approaches a wall and the echo from the wall has a frequency of 288 Hz. How fast is the car moving? (Speed of sound = 330 m/s). (19.4 ms)
7. Two closed organ pipes are 785 and 80 cm long. What is the frequency of the beats produced between their fundamental tones ? (Speed of sound = 330 m/s). (2 beats /s)
8. Two waves of the same frequency but of amplitude in the ratio 1 : 3 are superposed. What is the ratio of maximum to minimum intensity ? (4:1)
9. Two tuning forks of frequencies 512 Hz and 520 Hz are sounded together. What is the time interval between the maximum sound and the minimum sound ? (1/8 s)

2

VIBRATIONS OF STRINGS AND COLUMNS

Most waves do reach a terminal point or boundary, and changes occur there. In particular, there is almost always reflection of the wave at the boundary of the medium, and hence a wave travels back into the first medium in a manner that is largely determined by the boundary conditions. When waves are reflected at a boundary, they travel back into the initial medium and there the two waves are superposed. Such superposition must always be considered when waves travel in a bounded medium.

2.1 STATIONARY WAVES

A sound wave which is *reflected at right angles* to a boundary travels back along its own path and produces an *interference effect*, which we call a *stationary or standing wave.* This wave has the same frequency and wavelength as the reflected progressive wave, but the particles in its path move in a different way.

The condition imposed on the wave is that *the reflected and incident waves must sum to zero at the rigid boundary.* At the instant $T = 0$ in Fig. 2.1 the incident wave has zero amplitude at the boundary and, therefore, so has the reflected wave. At this instant, they seem to produce an increased amplitude in some positions. We can predict their subsequent positions by moving the waves equal distances in their respective directions of travel as shown as Fig. 2.1.

There are several important features of stationary waves.

(*i*) There are particles in certain positions which we call *nodes* which do not themselves move (hence the name stationary wave) but which are at the centre of the compressions and rarefactions.

(*ii*) The other particles appear to *bounce back and forth* between the nodes.

(*iii*) The particles between any two nodes are *in phase.*

(*iv*) The amplitude of the particles between successive nodes varies from zero at the nodes to a maximum at points midway between the nodes called *antinodes.*

Stationary waves are formed when waves confined to one direction are *efficiently reflected* as in a pipe or a rod or string. The formation of a standing wave is closely connected with resonance of vibrating objects. *An object resonates at those frequencies for which the standing waves fit exactly into the dimensions of the object.*

2.2 VIBRATIONS IN STRINGS

Strings vibrating transversely often act as a source of sound. The velocity of a progressive wave in a string is the same for all wavelengths and it varies with the *tension* and the *mass per unit length* as follows :

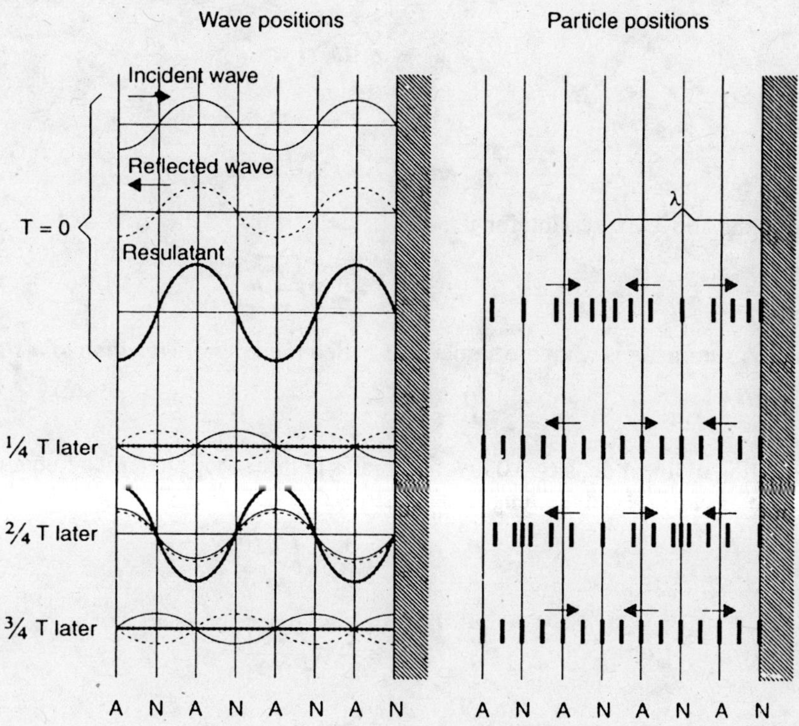

Fig. 2.1. The formation of a stationary wave by the interference of the incident and reflected waves.

$$\text{Velocity, } v = \sqrt{\frac{\text{Tension}}{\text{mass/unit length}}} = \sqrt{\frac{T}{m}} \quad \ldots(2.1)$$

A stationary wave is formed when the progressive wave is reflected at the fixed ends of the string, which act as nodes. The *simplest mode of vibration*, and the one which accounts for the greatest part of the energy in a string, has a single antinode between two nodes.

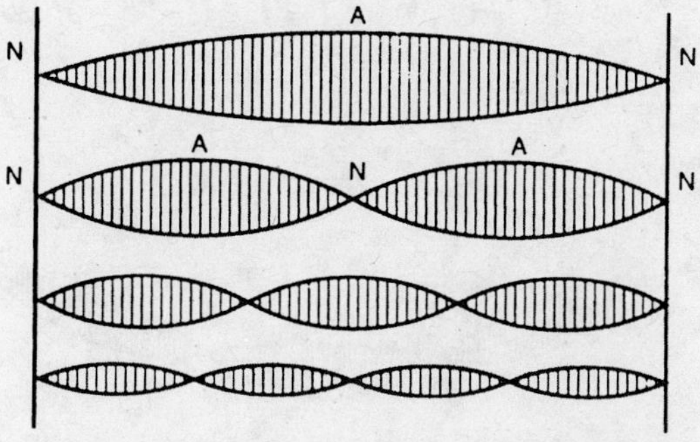

Fig. 2.2. Modes of vibration of a string under tension

In this case, the wave vibrates at its fundamental frequency f_1 and the length of the string is equal to half the fundamental wavelength λ_1 of the stationary wave, Fig. 2.2. A complete wavelength contains two loops.

$$l = \lambda_1/2$$

$$\therefore \lambda_1 = 2l$$

$$\text{and } f_1 = v/\lambda_1 = v/2l$$

Substituting the above expression for v

$$f_1 = \frac{1}{2l}\sqrt{\frac{T}{m}} \qquad \text{...(2.2)}$$

This equation summarizes what are sometimes called *the laws of vibration of a stretched string, i.e.*

(*a*) $f_1 \propto 1/l$ (*b*) $f_1 \propto T$ (*c*) $f_1 \propto 1/m$

Example 2.1 : A string of linear density 0.03 g/cm vibrates in three loops to emit a frequency of 600 Hz while it is under a tension of 480 *N*. Find the length of the string.

Solution.

$$f_3 = \frac{3}{2l}\sqrt{\frac{T}{m}}$$

$$600 = \frac{3}{2l}\sqrt{\frac{480}{0.03}}$$

$$l = 1\text{m}$$

Example 2.2 : A stretched wire 75 cm long emits a fundamental frequency of 300 Hz. By how much should the length be changed so that it emits a fundamental note of 450 Hz ?

Solution : Using the formula $f_1 l_1 = f_2 l_2$

$$l_2 = \frac{f_1 l_1}{f_2}$$

$$= \frac{300 \times 75}{400} = \frac{225}{4} = 56.25 \text{ cm}$$

$$\Delta l = l_1 - l_2 = 75 - 56.25 = \mathbf{18.75} \text{ cm}$$

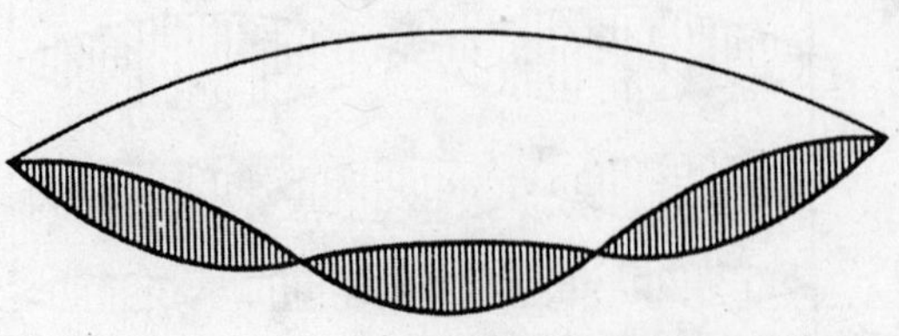

Fig. 2.3. Several waves may be present in the string at the same time

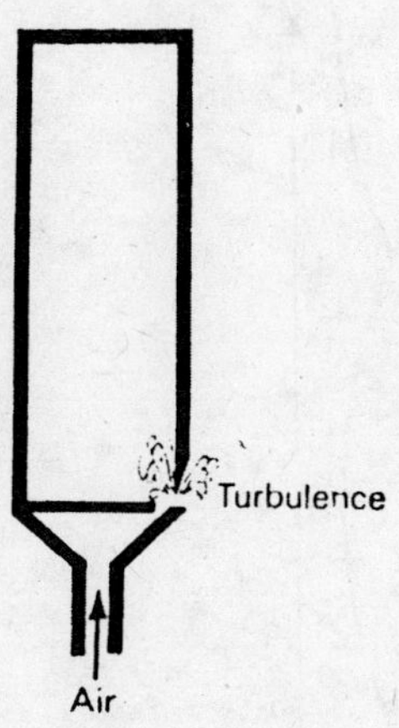

Fig. 2.4. The air in a pipe is set into vibration by the turbulence produced as a jet of air meets a sharp edge.

Other patterns or *modes of vibration* which produce nodes at the supports are possible, Fig. 2.2. These frequencies are present when the string is vibrating, but their amplitudes are much smaller than that of the fundamental frequency, Fig. 2.3 *These secondary frequencies are called harmonics, and they are classified according to their ratio to the fundamental frequency.* Thus frequencies of $2f_1$, $3f_1$ and $4f_1$ are the 2nd, 3rd and 4th harmonics respectively of frequency f_1, which is itself the first harmonic.

Plucking the string in the *middle* tends to favour the *odd harmonics*, which have an antinode in the middle of the string. A higher proportion of the energy is contained in the higher harmonics when the string is plucked nearer to the support. The difference in the sound of the string plucked in these two positions can be clearly heard. It is possible to *emphasize one particular harmonic* by touching the string at a point between the supports and thereby imposing a node at the right place for the selected frequency.

2.3 STATIONARY WAVES IN AIR COLUMNS

A column of air may be made to vibrate by means of a jet of air blowing against a sharp edge at one end of the pipe, Fig. 2.4. The disturbance acts as an antinode,sending waves up the column, where they are reflected, producing a stationary wave with a node at the rigid end. *Only those waves that fit into the tube with an antinode at one end and a node at the other resonate and produce audible sound.* The fundamental frequency and other frequencies emitted by the pipe are shown in Fig. 2.5.

When the end of the pipe is open, sound waves are still reflected. A compression reaching the open end is suddenly released producing a large particle amplitude or antinode at the end. The frequencies which resonate in an *open-ended tube* are shown in Fig. 2.6.

2.4 LONGITUDINAL AND TRANSVERSE VIBRATIONS IN RODS

When a rod is struck or rubbed, stationary waves are formed by reflection at its ends. A rod will resonate to longitudinal vibrations which have a node where the rod is supported and an antinode at any free end, Fig. 2.7. The length of the rod and the way it is supported determine the wavelength and therefore the frequency of the sound emitted.

If the rod is flexible, it can be made to vibrate transversely in which case it becomes *a reed.*

2.5 MUSICAL INSTRUMENTS

Wind instruments, such as organs, depend on the resonance of air columns. *The pitch of a note emitted by a column is generally controlled by changing the length of the resonating column. In wood wind instruments, the physical length is fixed, but their effective length is varied by opening and closing holes and valves.* The organ has a large number of pipes, which can be separately selected. *In brass instruments, we increase their length by opening valves which introduce new sections of tube.* The trombone has *a telescopic U-shaped tube which can slide in and out.*

The resonance in some, such as clarinets and saxophones, is controlled by *reeds*, while in others such as the trumpet,the *vibration of the player's lips* controls the resonance.

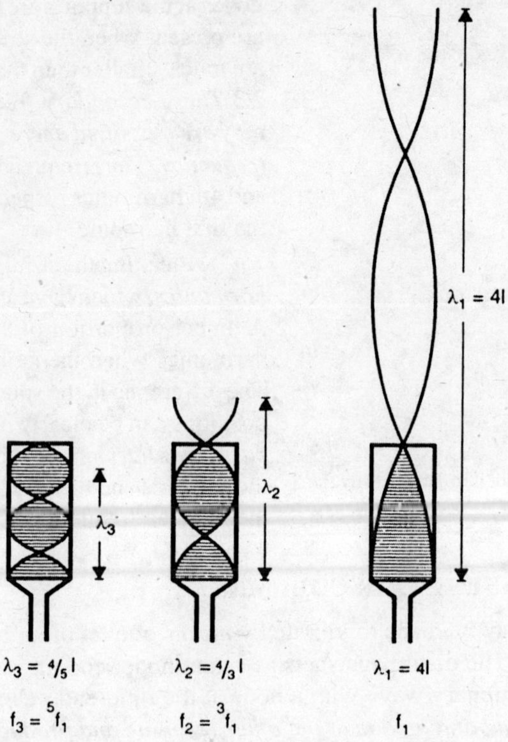

Fig. 2.5. Normal modes of vibration of air in a closed pipe.

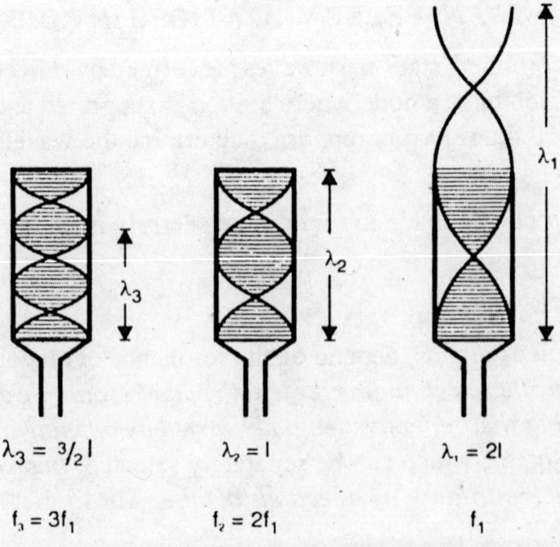

Fig. 2.6. Normal modes of vibration of air in an open-ended pipe.

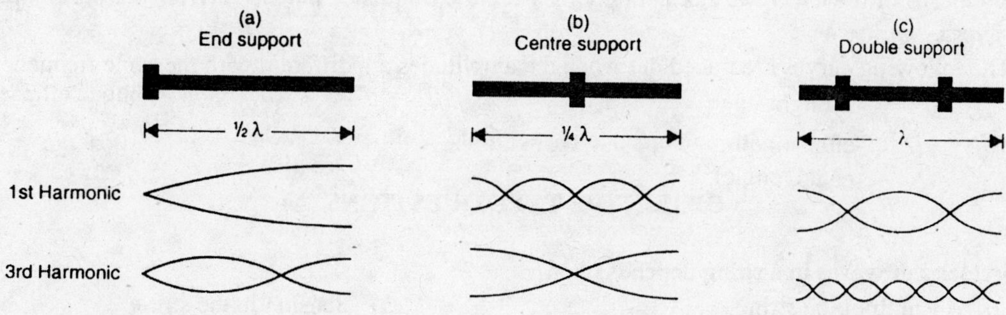

Fig. 2.7. Modes of vibration of a rod showing the absence of even harmonics.

2.6. LISSAJOUS FIGURES

A method of *frequency measurement* employs the cathode ray oscilloscope (CRO). A microphone converts the sound to an alternating potential, which is connected across the *Y* plates of the CRO. This controls the *vertical deflection* of the cathode ray beam. A *horizontal deflection* is imposed on the beam by a sweeping potential within the instrument and applied to the *X* plates. The frequency of the source can be deduced from the *resulting pattern* on the cathode ray tube, Fig. 2.8.

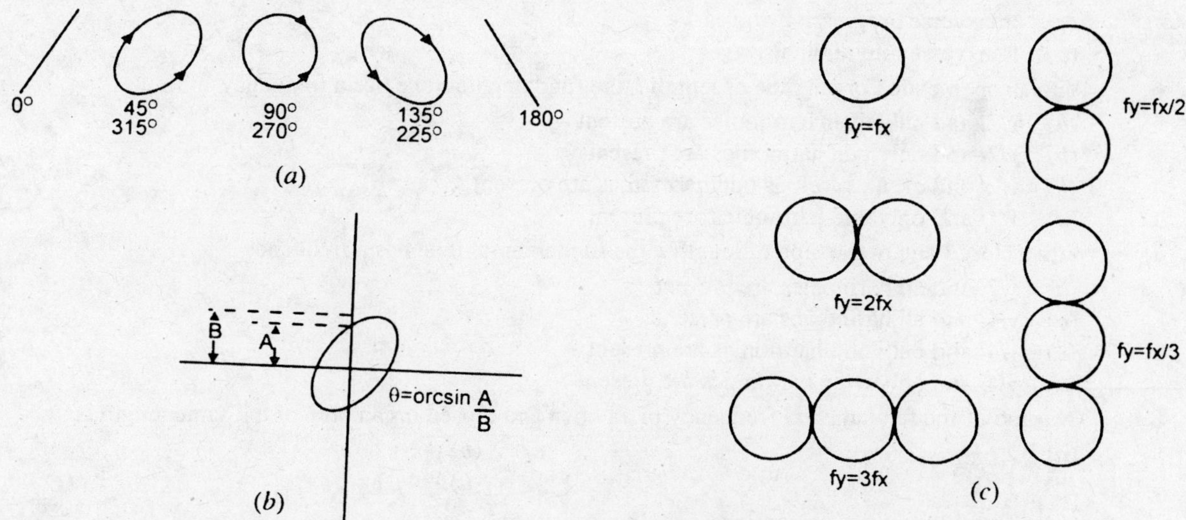

Fig. 2.8. Cathode ray tube patterns (a) Measurement of phase angle between two sine wave voltages at 45 degree intervals. (b) General phase angle measurement (c) Lissajous figures.

Using the *internal time base* the horizontal deflection of the cathode ray beam increases gradually from left to right and then returns instantaneously to zero. If a sinusoidal alternating potential is applied to the *X* plates, the resulting cathode ray patterns are called *Lissajous figures*. The ratio of the frequencies applied to *X* and *Y* plates can be deduced from these figures.

By means of the CRO, we can not only measure the frequency but also *investigate the complexity of the waveform.*

The following curves are traced out when the amplitudes are different with the same frequency.

(*i*) 0 - straight line

(*ii*) $\pi/4$ - oblique ellipse

(*iii*) $\pi/2$ -ellipse with coincidence (If amplitudes are equal, circle)

OBJECTIVE TYPE QUESTIONS (A)

1. Velocity of waves in a string depends upon
   (*a*) length of the string (*b*) tension in the string
   (*c*) density of the surrounding medium (*d*) temperature of atmosphere
2. If the frequency of vibration of a string is increased by a factor two, then tension in the string will be
   (*a*) half (*b*) double
   (*c*) one fourth (*d*) four times
3. The length of a stretched string is doubled and the tension is four times. The frequency will
   (*a*) become twice (*b*) be reduced to half
   (*c*) remain constant (*d*) becomes one-fourth
4. The frequencies of the harmonics of a string are
   (*a*) unrelated (*b*) of the same pitch
   (*c*) in the ratio 1 : 2 : 3 (*d*) in the ratio 1 : 3 : 5
5. A resonating column of air contains
   (*a*) stationary longitudinal waves
   (*b*) stationary transverse waves
   (*c*) transverse progressive waves
   (*d*) transverse longitudinal waves
6. With an open ended organ pipe of length *l*, the fundamental tone has a frequency
   (*a*) $v/2l$ and only even harmonics are present
   (*b*) $v/2l$ and only odd harmonics are present
   (*c*) $v/2l$ and even as well as odd harmonics are present
   (*d*) $v/4l$ and only odd harmonics are present
7. With a closed end organ pipe of length *l*, the fundamental tone has a frequency
   (*a*) $v/2l$ and all harmonics are present
   (*b*) $v/4l$ and all harmonics are present
   (*c*) $v/4l$ and only odd harmonics are present
   (*d*) $v/4l$ and only even harmonics are present
8. The ratio of the fundamental frequency of an open and closed organ pipe of the same length is
   (*a*) 2 : 1 (*b*) 1 : 2
   (*c*) 1 : 1 (*d*) 4 : 1

FILL IN THE BLANKS (B)

1. In case of a closed organ pipe __________ harmonics are present.
2. The frequencies of fundamental mode of vibration in open and closed pipes of equal lengths are __________.
3. A pipe open at both ends will produce __________.
4. Vibrations produced in a sitar wire are __________.

5. For an open organ pipe of length l the wavelength of the fundamental note is equal to ___________
6. In the case of standing waves ___________ can be transferred past a node.

CONVENTIONAL QUESTIONS (C)

1. How are standing waves produced ?
2. Is there a transfer of energy through the medium when a stationary wave is produced in it? Explain.
3. Sketch the fundamental wave and the second and third harmonics for (*a*) a vibrating string and (*b*) a vibrating rod clamped at one end.
4. Can more than one set of standing waves be present simultaneously within a single medium.
5. How might one locate the positions of nodes and antinodes in a vibrating string ?
6. How might fine dust on the body of a violin be used to show how the violin is vibrating ?
7. Why does an empty vessel produce more sound than a filled one ?
8. Explain why sound produced by a vibrating tuning fork becomes louder when its stem is placed in contact with a table.

NUMERICAL QUESTIONS (D)

1. The mass of a 4.0 metre wire is 0.01 kg. and it is stretched by a force of 400 newton. What is the speed of transverse waves in the wire ?

   (400 m/s)
2. Two perfectly identical wires are in unison. When the tension in one wire is increased by 1%, then on sounding them together, 3 beats are heard in 2 seconds. Calculate the initial frequency of each wire.

   (300 / s)
3. Find the frequencies of the fundamental note and second harmonic of a 2 m organ pipe with its end open and closed (Neglect end -correction Velocity of sound = 33 m/s)

   (open 82.5 ; 165 Hz closed 41.3, 123.9 Hz)
4. A steel wire of length 23.9 cm and diameter 0.8 mm vibrates with a frequency of 256 Hz. What is the tension in the wire ? (Density of steel = 7800 kg /m^3)

   (58.8 N)

3

HEARING

Hearing is the natural method for the reception of speech, music and noise and in particular is the ultimate significant termination of most acoustic systems involving the artificial reproduction of sound. Therefore,in order to properly design acoustical systems for the reproduction of sound, one must know something about the basic characteristics of hearing. The treatment in this chapter is limited to the capabilities of the human ear for receiving these sounds as well as other wanted sounds such as musical sounds,and interference produced in this reception by unwanted sounds or noise. The ultimate interpretative characteristics of the human hearing mechanism are not expressible in terms of objectively determined physical parameters. Instead they must be determined by operational experiments leading to statistical predictions as to the capabilities of an average ear under assumed or known conditions.

3.1 THE EAR

The ear has three sections--the outer, the middle and inner ear, Fig 3.1. The *outer ear* funnels the sound waves to a membrane called the eardrum, which is made to vibrate. In the *middle ear*, the vibrations of the ear drum are transmitted to a smaller membrane at the entrance to the inner ear. At the same time the pressure variation is increased about sixty times. This amplification is achieved partly by means of three bones in the middle ear, which act as levers, and partly because of the difference in area between the eardrum and the inner membrane.

The *inner ear* converts the mechanical vibrations to nerve impulses. The *conversion* takes place in a tapering spiral resonator,filled with a liquid and called the *cochlea*. The *cochlea is divided longitudinally by the basilar membrane*. This membrane is covered with nerve endings which *transmit* responses to the brain, the place where *perception* occurs.

The ear is situated near an organ which,by being sensitive to fluid pressure, gives animals a sense of *balance*. The organ consists of three semi-circular canals, and probably developed with the ear from a common, more primitive organ.

3.2. SOUND ENERGY

The energy in the sound wave, Fig 3.2,is the basic *stimulus* for the *sensation* of loudness. The rate at which *sound energy* is given out by a source is measured in joules per second,i.e; *watts*. A small radio receiver emits sound at an average power of about 10^{-7} W or 0.1 μW;while an amplifier suitable for a small auditorium will deliver an average of about 1 μW.

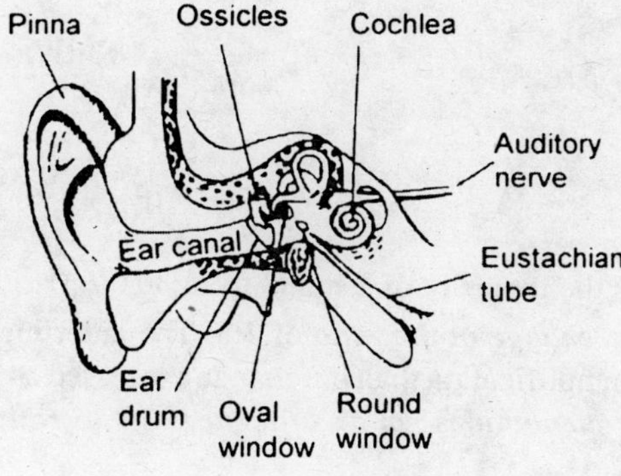

Fig 3.1. Structure of the ear.

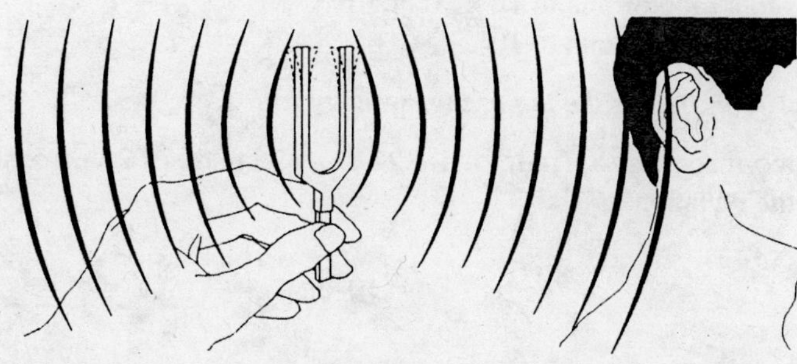

Fig 3.2. The energy in the sound wave is the basic stimulus for the sensation of loudness.

The *intensity* of a sound at a certain place in the path of a sound wave is ***the number of watts passing at right angles through a unit area of intercepted surface.*** As the sound energy travels away from the source, it spreads out and is *attenuated,* i.e; its intensity reduces. If there is no frictional loss of energy, and no obstacles or reflecting surfaces, the sound intensity for a point source of sound is attenuated according to an *inverse square law.* In this case the intensity I at a radius r from the source is given by:

Intensity $$I = \frac{\text{power of source}}{\text{area of sphere}}$$

$$I = \frac{P}{4\pi r^2} \qquad \ldots(3.1)$$

In practice,these conditions are rarely satisfied, but the law serves as a rough guide.

Example 3.1 : What are the intensities at distances of 5 and 10 m from a loudspeaker emitting sound at a power of 3μW uniformly in all directions:

Solution: Intensity
$$I = \frac{\text{Power}}{4\pi r^2}$$

At 5m,
$$I_1 = \frac{3}{4\pi 5^2} \approx 10^{-2}\mu W/m^2$$

At 10m,
$$I_2 = \frac{3}{4\pi \cdot 10^2} \approx \frac{1}{4}10^{-2}\mu W/m^2$$

If an amplifier increases the intensity of a sound from $4\mu W/m^2$ to $5\mu W/m^2$, we can express this either by a *change* of $1\mu W/m^2$ or as a *change in the ratio* of 4:5. The latter method of expressing the change as a ratio tells us far more about the amplification than the change expressed as a difference. Therefore, *in acoustics we express amplifications and attenuations not as differences but as ratios.*

Intensity ratio
$$= \frac{I_1}{I_2} \qquad \ldots(3.2)$$

A unit of ratio or *relative intensity* is called *Bel* and is equal to a ten-fold increase in intensity.

1 Bel corresponds to a ratio of one to ten or 10^1

2 Bel corresponds to a ratio of one to 10×10 or 10^2

n Bel corresponds to a ratio of one to 10^n

The relative intensity in Bels is the log of the intensity ratios.

One decibel (*dB*) corresponds to a ratio of 1:1.26 or an increase of 26 percent. To calculate a relative intensity (*RI*) in *dB* use the equation

$$RI = 10\log_{10}\frac{I_1}{I_2} \qquad \ldots(3.3)$$

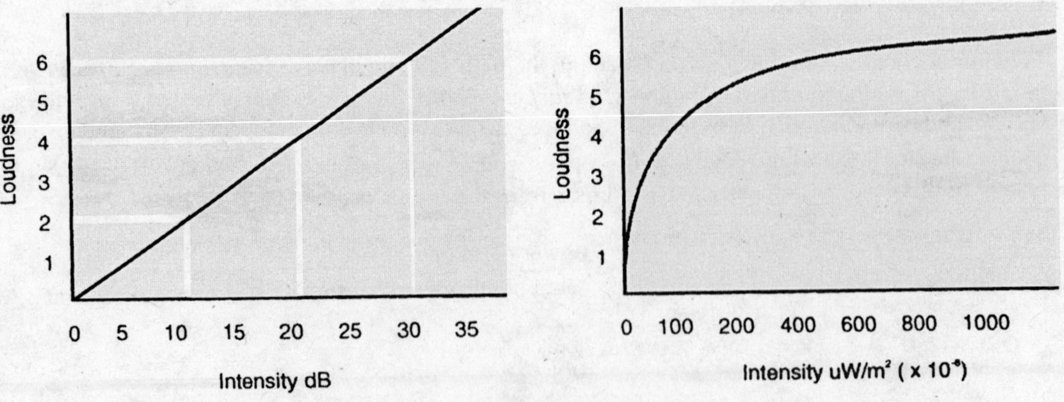

Fig. 3.3. A comparison of the loudness response of the ear and the intensity measured (a) in dB, (b) in μW/cm².

This may be regarded as the definition of the *decibel. Negative relative intensities denote attenuation.*

The human *sense of hearing* grades loudness in a way which broadly approximates to the decibel scale (Fig. 3.3). Also, the decibel happens to be a convenient size, because it is about *the smallest change in intensity which the ear can detect.* A further advantage is that *relative intensities are directly additive and subtractive.*

When two amplifiers are cascaded their gains multiple to give the overall gain of the combination. It is thus often convenient to have a *logarithmic* measure of gain, which will simply add.

Thus, when an amplifier contains three stages of amplification, that is, stages producing amplifications A_1, A_2 and A_3 dB, the total amplification A_T is given by

$$A_T = A_1 + A_2 + A_3 \qquad ...(3.4)$$

Because of the convenience of the dB system of expressing relative intensities, the system has been adopted so that it can indicate absolute values in intensity. This is done by defining a standard intensity and then expressing any other intensity in decibels above this *standard level.* The intensity chosen is that of 10^{-6} W/m^2, *a sound which is very near to the lowest intensity of sound audible to the human ear.*

Example 3.2 : The intensities of a sound are amplified from 10μW/m^2 to 150 μW/m^2. Calculate the amplification produced in dB.

Solution: Relative intensity $= 10 \log I_1/I_2 = 10 \log 150/10$

$= \mathbf{17.6\ dB}$

Example 3.3 : An acoustic panel reduces the intensity of a sound from 60μ/m^2 to 8.0 μW/m^2. Express the attenuation in dB.

Solution: Relative intensity $= 10 \log \dfrac{I_1}{I_2} = 10 \log \dfrac{8}{60}$

$$= 10(\log 8 - \log 60) = 10(1.249) = -\mathbf{8.8\ dB}$$

Example 3.4 : Express the noise of intensity 2×10^{-6} W/m^2 in decibels above the standard (10^{-12} W/m^2).

Solution : Relative intensity $= 10 \log \dfrac{2 \times 10^{-6}}{10^{-12}}$

$$= 10 \log 2 \times 10^6 = \mathbf{63.0\ dB}$$

3.3 LOUDNESS

For a given frequency, the more intense the note, the greater the stimulus to cochlea and the louder the note is. In order to appreciate the conception of loudness, we must face three problems:

(*i*) Loudness is a matter of *personal judgement,* which may vary from one person to another.

(*ii*) The human response to sound energy is *not directly proportional* to the energy content.

(*iii*) The loudness of a sound is dependent on its *frequency* as well as its *intensity*

Although the ear is not able to judge the *loudness ratio* of two sounds objectively, it is able to judge the *equality of loudness.* Using this ability, we can solve the first problem by having large number of people make a variety of sound comparisons and can thus, arrive at an *average response* to intensity. We are then able to define the response of an average ear, in the same way as illumination engineers define the response to colour of an average eye. Such statistical methods can act as very useful aids to greater precision of measurement.

When a sound is doubled in *intensity*, an observer will by no means judge it to have doubled in *loudness*. In fact, about a tenfold increase in intensity is needed to double the loudness. *The response of the average ear is much nearer to the decibel system of intensity measurement than it is to the absolute intensity.* For this reason we use the decibel of relative intensity as a measure of loudness. The zero on this scale is the *threshold of audibility* for a note of 1000 Hz ($10^{-6}\mu W.m^2$). The loudness of some common sounds varying from the audibility threshold to the *pain threshold*, at which a sound imparts physical pain, is given in Table 1.2.

There remains the problem that equally intense sounds of different frequencies are not judged equally loud by an observer. The relative intensity of a note in dB is a good measure of loudness for frequencies near to 1000 Hz, but *the ear is less sensitive to sounds of high and low frequency.* The extremes of frequency are not so loud as their relative intensities would indicate. The curves shown in Fig 3.4 are drawn through *those intensities which sound equally loud.* Note that *the curves are approximately horizontal at very high intensities.*

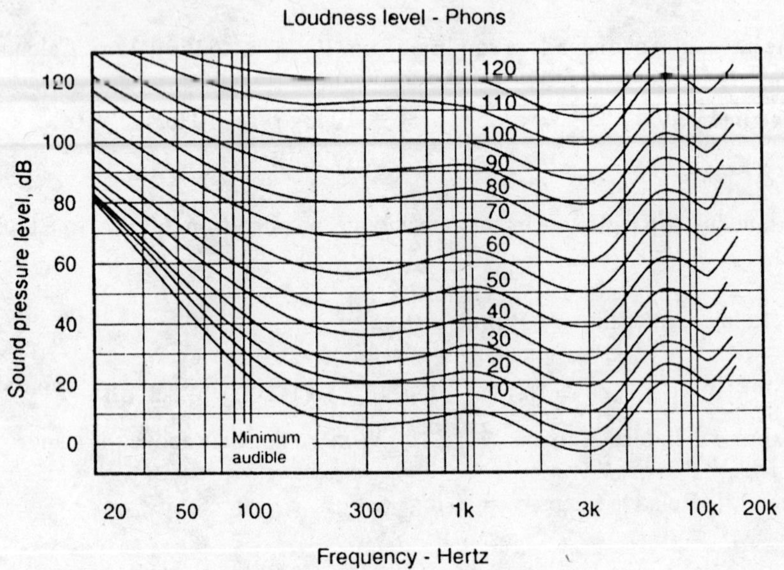

Fig 3.4. Equal loudness contours. These contours reveal the relative lack of sensitivity of the ear to bass tones, especially lower levels. Inverting these curves gives the frequency response of the ear in terms of loudness level.

We can express the loudness of a sound in a unit, the phon, which is independent of the frequency. *The loudness of a sound in phons is equal to the intensity in decibel above threshold of the 1000 Hz note which sounds equally loud.* The loudness of a note of a single frequency may be obtained by plotting its positions on the graph of Fig 3.4 and thus determining which curves of equal loudness it lies on.

A sound pressure level of 30 dB yields a loudness level of 30 phons at 1000 Hz, but it requires sound pressure level of 58 dB more to sound equally loud at 20 Hz. The 90 phon curve rises only 32 dB between 1000 Hz and 20 Hz. This is shown in Fig 3.5.

Many experiments conducted with hundreds of subjects and many types of sound have yielded that *for a 10 dB increase in sound pressure level the average person reports that loudness is doubled.* For a 10 dB

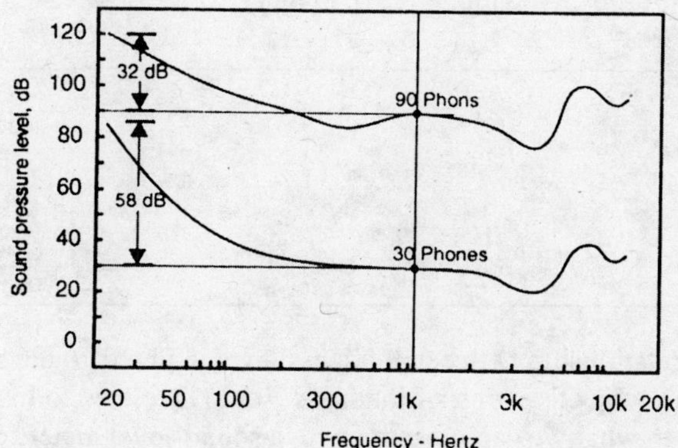

Fig. 3.5. A comparison of the ear's response at 20 Hz compared to that at 1000 Hz. At a loudness level of 30 phons, the sound pressure level of a 20-Hz tone must be 58 dB higher than that at 1000 Hz to have the same loudness. At 90 phons loudness level, an increase of only 32 dB is required. The ear's response is somewhat flatter at high loudness levels.

decrease in sound level, subjective loudness would be half. A unit of subjective loudness has been adopted called the sone. *One sone is defined as the loudness expressed by a person listening to a tone of 40-phon loudness level.* A sound of *2 sone* is twice as loud, and 0.5 sone half as loud.

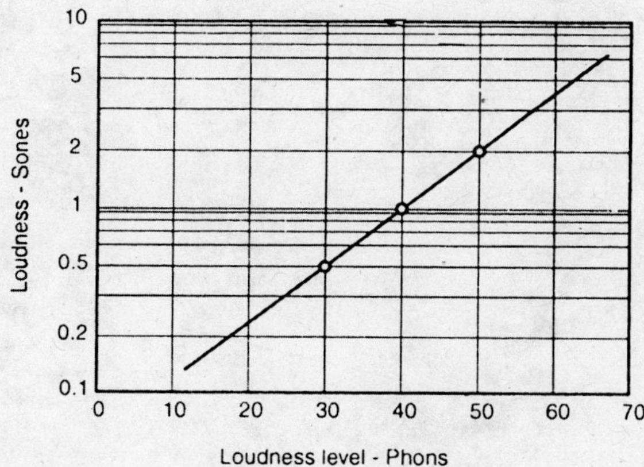

Fig. 3.6. The relationship between the physical loudness level in phons to the subjective loudness in sones.

Fig 3.6 shows a graph for translating sound pressure levels to loudness in sones. One point on the graph is the very definition of the sone the loudness experienced by a person hearing a 1000-Hz tone at 40-dB sound pressure level or 40 phons. A loudness of 2 sones is then 10 dB *higher*, a loudness of 0.5 sone 10 dB *lower*. A straight line is then drawn through these three points, which can be extrapolated for sounds of higher and lower loudness.

Table 3.1 Loudness Level in Phons vs Loudness in Sones

Loudness Level Phons	*Subjective Loudness Sones*	*Typical Examples*
100	64	Heavy truck passing
80	16	Talking loudly
60	4	Talking softly
40	1	Quiet room
20	0.25	Very quiet studio

Table 3.1 shows the relationship between loudness level in phons to the subjective loudness in sones. It is well to realise that a true subjective unit of loudness (sone) is related to loudness level (phon), which is in turn related by definition to what we can measure with a sound-level meter.

3.4 PITCH

The human ear is far more sensitive to *differences in frequency* than it is to differences in intensity of source of sound.

In some frequency ranges, a trained ear can detect differences in frequency of less than 1 percent. The *lowest frequency note* which produces a perception of pitch is about 20 Hz. Below this frequency, we are only aware of individual pulses. The human ear loses its sense of pitch, and indeed its sensitivity to any sound above a frequency of about 20,000 Hz.

The sensitivity of the ear of an adult to *high frequencies* reduces regularly with advancing years. Dogs are sensitive to sounds at frequencies as high as 40,000 Hz, while the bat, which relies on sound much in the same way as we rely on vision, can detect vibrations of around 50,000 Hz.

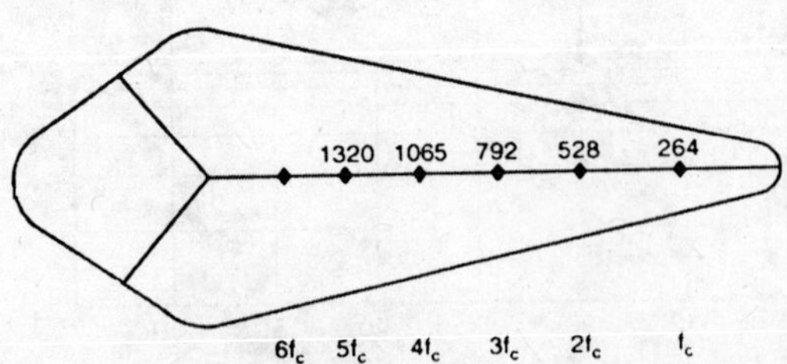

Fig. 3.7. Schematic impression of a pattern of cochlea response for the single frequency 264.

Each individual frequency of sound *stimulates* the nerves in a number of places along the cochlea, the main centres of stimulation being nearer to the open end for the higher frequencies (Fig 3.7). An *increase* in sound intensity causes both an increase in response and a spreading of the area of the nerves affected. This may render the ear less sensitive to pitch or even cause an apparent change in the pitch of a source of constant frequency. This serves to emphasize the subjective nature of pitch.

Pitch, a *subjective term*, is chiefly a function of frequency, but is not *linearly* related to it. Because pitch is somewhat different from frequency, it requires a *subjective unit, the mel.* Frequency is a *physical term* measured in *hertz. A reference pitch of 1000 mels has been defined as the pitch of a 1000 Hz tone with a sound pressure level of 60 dB.*

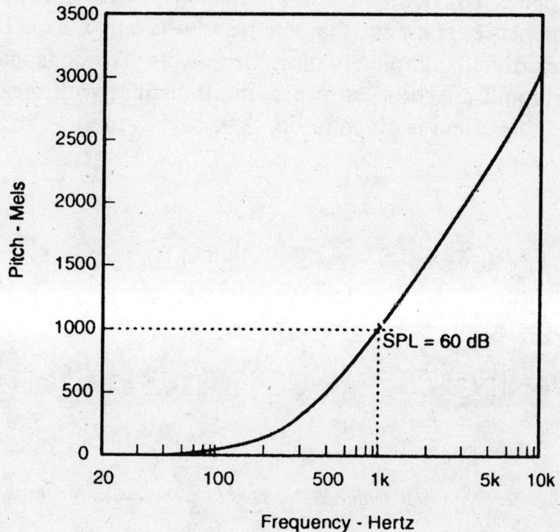

Fig. 3.8 Pitch (in mels, a subjective unit) is related to frequency (in Hz, a physical unit) according to the solid curve obtained by juries of listeners.

The relation between pitch and frequency, determined by experiments with juries of listeners is shown in Fig 3.8. Note that on the experimental curve *1000 mels coincides with 1000 Hz*, which tells us that the sound pressure level for this curve is 60 dB.

The subjective pitch of a *complex tone* containing overtones of appreciable magnitude depends not only on the frequency of its lowest component but also *on* the frequencies and amplitudes of its overtones.

3.5. TIMBRE

The *timbre* of a musical sound is the subjective characteristic that makes it possible for us *to distinguish between two tones having the same intensity level and fundamental frequency but different waveforms.* In other words, it expresses our ability to recognize the sound of a violin as different from that of a trumpet, even when the two instruments are sounding the same note with equal loudness.

Like loudness and pitch, *timbre is a complex characteristic* for although it is *primarily* dependent on the waveform of the tone being heard, it is also a function of its intensity and frequency.

Various *adjectives,* such as mellow, brilliant, reedy, brassy etc. have been used by musicians to describe different types of tones, but no serious attempt has been made to establish a set of subjective units of timbre, similar to those established for loudness and pitch.

3.6 HEARING LOSS WITH AGE

An infection or accident which damages the middle ear can affect the sound transmission mechanism and result in *deafness*. The sufferer from this defect will hear some sound by *bone conduction* to the cochlea through the head. If, on the other hand, the nerves in the cochlea are damaged and not functioning, no sound will be heard at all, and deafness will be *complete*.

As we get older, the sensitivity of our ears to high frequency sound reduces. This can cause *partial* deafness, which is first apparent as a difficulty in hearing speech clearly. Our identification of words depends far more on the *consonants* such as *b, p, t, f* etc. than on the *vowels a, e, i, o, u*. This is also true of the written word, and some systems of shorthand completely omit the vowels. The consonants are formed by sounds of the higher audible frequencies and are the first to become indistinct with the reduction in high frequency sensitivity. An impression of this effect is given in Fig. 3.9.

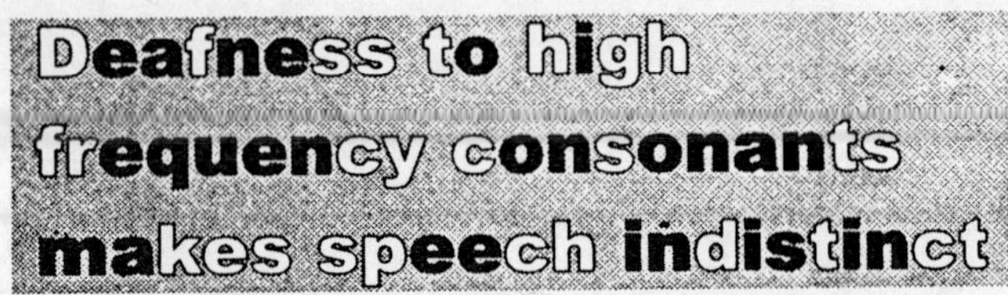

Fig. 3.9 An impression of the effect of partial deafness. Deafness to high frequency consonants makes speech indistinct.

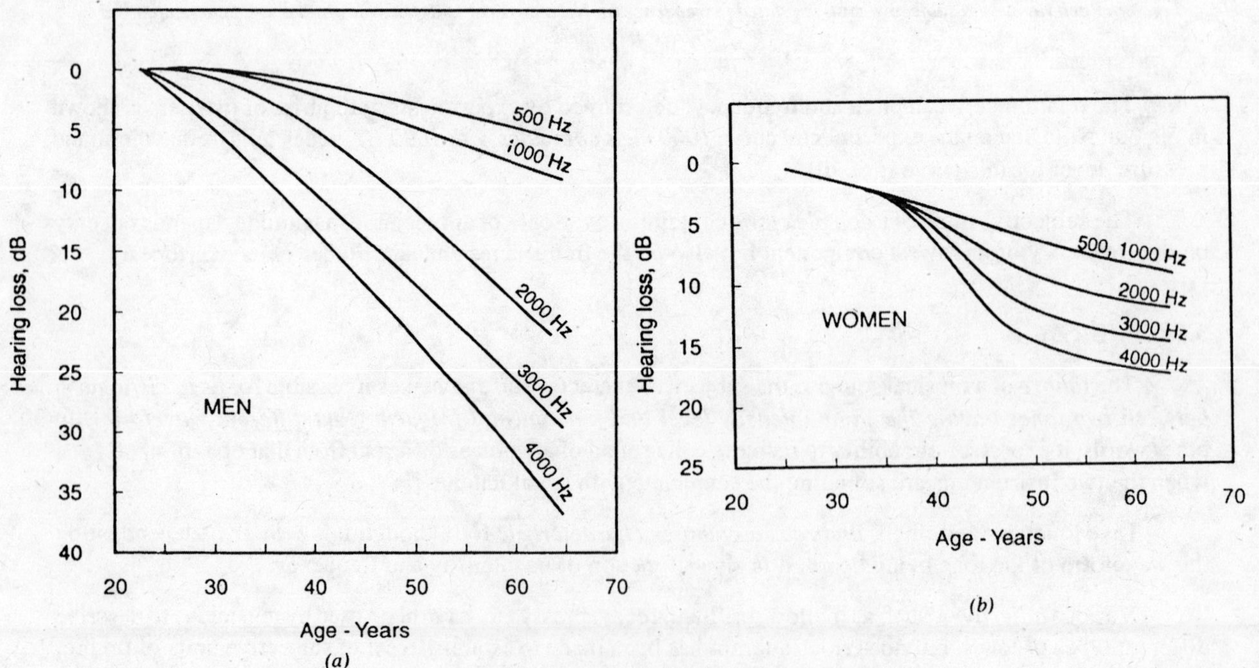

Fig. 3.10 Age-related hearing loss for (a) men and (b) women.

This type of deafness can be corrected by amplifying the higher frequency sounds by means of a *hearing aid.* With only one ear aided, however, the ear loses the sense of *direction of sound* which two ears can convey. The use of a hearing aid in one ear also reduces the ability of an observer to *listen selectively* to one sound against a background of noise.

The results of three separate studies on the average deterioration of hearing activity with age are summarised in Fig. 3.10. There seems to be a pronounced difference in this type of loss between men and women. *The greater loss observed in large populations of men can be attributed, partially at least, to the noisier places in which men often work.*

3.7 OCCUPATIONAL AND RECREATIONAL DEAFNESS

The hearing of workers in industry is now protected by law. The higher the environmental noise, the less the exposure allowed. (Table 3.2) Researchers are feverishly trying to determine what *noise* exposure workers are subjected to in a given plant. This is not easy as noise levels fluctuate and workers move about, but wearable *dosimeters* are often used to integrate the exposure over the work day. Industries are hard pressed to keep up with changes in regulations, let alone the installation of *noise shields* around offending equipment and keep *earplugs* in or *ear muffs* on the workers. *Nerve deafness* resulting from occupational noise is recognised as a distinct health hazard.

Table 3.2 Permissible Noise Exposure Times

Sound Pressure Level, dB. A-Weighting, Slow Response	*Maximum Daily Exposure Hours*
85	10
90	8
92	6
95	4
97	3
100	2
102	1.5
105	1
110	0.5
115	0.25 or less

Because there is no replacement known for damaged sensory cells in the inner ear, it would seem wise to protect the ones we have

OBJECTIVE TYPE QUESTIONS (A)

1. The unit of relative intensity is called
   (*a*) bel (*b*) phon (*c*) sone (*d*) mel
2. The loudness of a note depends on
   (*a*) frequency (*b*) intensity (*c*) (*a*) and (*b*) (*d*) none of these
3. Timbre of a note is a function of
   (*a*) intensity (*b*) frequency (*c*) (*a*) and (*b*) (*d*) none of these

4. Damage to the middle ear results in
   (*a*) partial deafness (*b*) complete deafness
5. If the nerves in the cochlea are damaged deafness will be
   (*a*) partial (*b*) complete
6. As we get older the sensitivity of the ear to high frequency sounds
   (*a*) increases (*b*) decreases

FILL IN THE BLANKS (B)

1. The outer ear ______ the sound waves to a membrance called the eardrum, which is made to __________.
2. The inner ear converts the mechanical vibrations to __________.
3. The energy in the sound wave is the __________ for the sensation of _____.
4. The sound intensity for a point source is attenuated according to an ________.
5. The human sense of hearing grades loudness in a way which broadly approximates to the ________
6. Loudness is a matter of ________, which may ________ from one person to another.
7. The ear is less sensitive to sounds of ______ and ________ frequency.
8. The human ear is far more sensitive to ________ in frequency than it is to differences in ________ of a source of sound.

CONVENTIONAL QUESTIONS (C)

1. What is the difference between the loudness and intensity of a note ?
2. How it is possible for the ear to distinguish between the same note played on different musical instruments ?
3. Why it is necessary to use a logarithmic scale to measure differences in sound intensity
4. What is the difference between a decibel and a phon?
5. Discuss the factors which determine (*a*) pitch (*b*) timbre and (*c*) loudness of a note produced by an instrument and heard by the ear.

NUMERICAL QUESTIONS (D)

1. A sound pressure level is 78 dB. What is the sound pressure ? (20 μ Pa)
2. The acoustic power output of a loudspeaker is measured and found to be 0.62 W. What is the power level?

   (108 dB)
3. A Shure model 578 omni directional dynamic microphone open-circuit voltage is specified as - 180 dB for the 150 ohm case. They also specify that OdB = 1 volt per μ bar. What would the open-circuit voltage be ?

   (0.1 millivolt)

4. A constant directivity horn is rated at a sound pressure level of 115 dB on axis at 1 metre with 1W into 8 ohms. If the input were decreased from 1 W to 0.22 W, what would be the sound pressure level at 1 metre distance?

(108.4 dB)

5. An input of 1 W produces a sound pressure level of 115 dB at 1 metre. What is the sound pressure level at 6.1 metres ?

(99.3 dB)

6. A line amplifier (600 ohms in, 600 ohms out) has a gain of 37 dB. With an input of 0.2 volts, what is the output voltage ?

(14.16 dB)

7. The amplifier in Example 6 has a bridging input of 10,000 ohms impedance and an output impedance of 600 ohms. With a 50 mV input, an output of 1.5 *V* is observed. What is the gain of the amplifier?

(29.5 dB)

4

SPEECH, MUSIC AND NOISE

Speech, music and noise are the three categories into which practically all important natural sounds may be grouped. Therefore, in order to properly design acoustical systems for the reproduction of sound, one must know something about the basic characteristics of speech. The treatment in this chapter is limited to a consideration of the characteristics of sounds produced by the human voice. The acoustical characteristics of speech sounds as well as those of music and noise may be measured with considerable precision by the use of standard acoustical instruments including microphones, frequency analysers, oscilloscopes, etc. The results of such measurements may be expressed in terms of precise physical parameters including frequency, power, pressure level, time duration etc.

4.1 THE VOICE MECHANISM

The *acoustic energy* associated with speech originates in the chest muscles which, by contrast, force air from the lungs up through the various components of voice mechanism. This steady stream of air may be

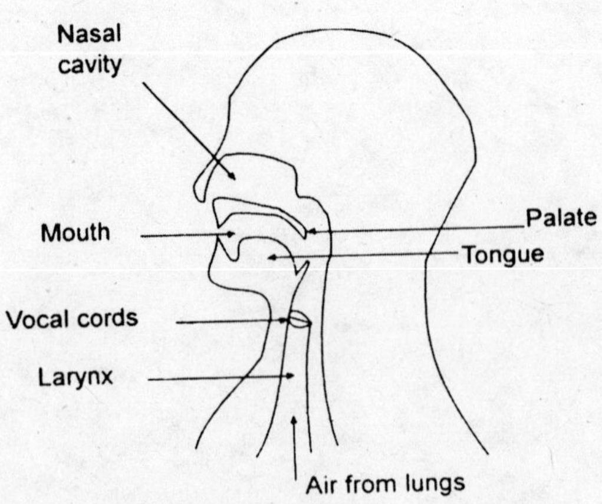

Fig. 4.1 Important elements of the voice mechanism.

looked upon as a *carrier of energy* which must be *modulated* in its velocity and correspondingly in pressure in order to produce sounds. The requisite modulation is accomplished in two basic ways leading respectively to *voiced* and *unvoiced* sounds.

Voiced sounds include the vowels of ordinary speech as well as the tonal characteristic of the singing voice. The primary modulating agent for voiced sounds is the larynx across which are stretched the *vocal cords*. The vocal cords are two membrane like bands forming a diaphragm with a slit like opening which modulates the air stream as it vibrates open and shut. The length of this opening (about 2.5 cm in men and 1.5 cm in women) and the tension to which the vocal cords are stretched determines the fundamental frequency of this modulation. The action of the vocal cords produces a *sawtooth* type of variation in velocity and correspondingly of pressure in the modulated stream of air, which, upon analysis into a Fourier series of sinusoidal waves is found to contain a large number of harmonically related frequencies.

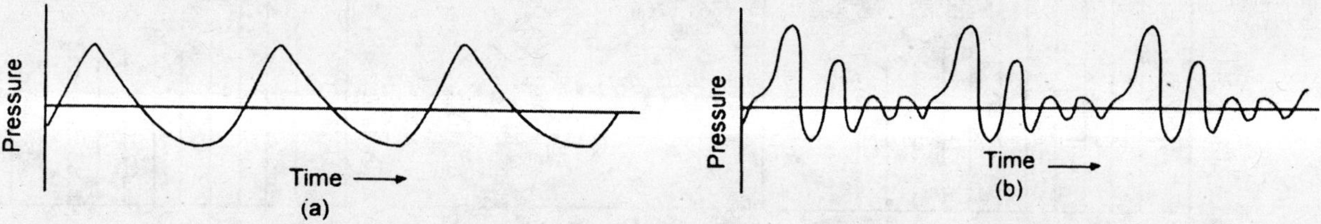

Fig. 4.2 (a) Acoustic pressure output at vocal cords. (b) Acoustic pressure output for vowel a as in father.

The numerous resonating cavities and orifices of the nose, mouth and throat form an *acoustic network* which further modulates the pressure wave of Fig. 4.2 (*a*). Many of these parameters are *controllable at will, i.e.* by changing the position of the tongue or altering the configuration of the lips, a wide variety of voiced sounds may be produced.

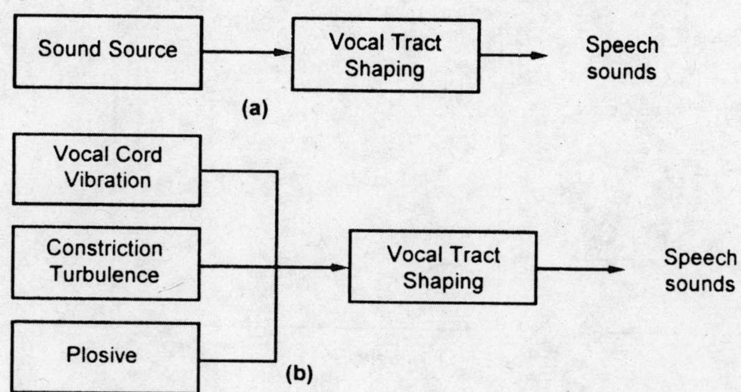

Fig. 4.3. (a) The human voice is produced through the interaction of two essentially independent functions and a time -varying -filter action of vocal tract. (b) The sound source can be broken down into vocal - cord vibration for voiced sounds, the fricative sounds resulting from air turbulence, and the plosive sounds.

It is also possible for the voice mechanism to produce sounds without the use of vocal cords. Such sounds are called *breath sounds.* For instance, a steady forcible exhaling of the breath will produce a hissing sound like that of escaping steam. This is apparently caused by turbulences set up in the flow of air through the numerous irregularities along the vocal tract. Additionally, they include such *unvoiced fricative consonants* as *f* and *s* as well as the *unvoiced stop consonants* (plosives) *p, t,* and *k.* Here the basic vibrations are produced by modulating the air stream by means of the tongue, teeth and lips. An analysis of the unvoiced types of

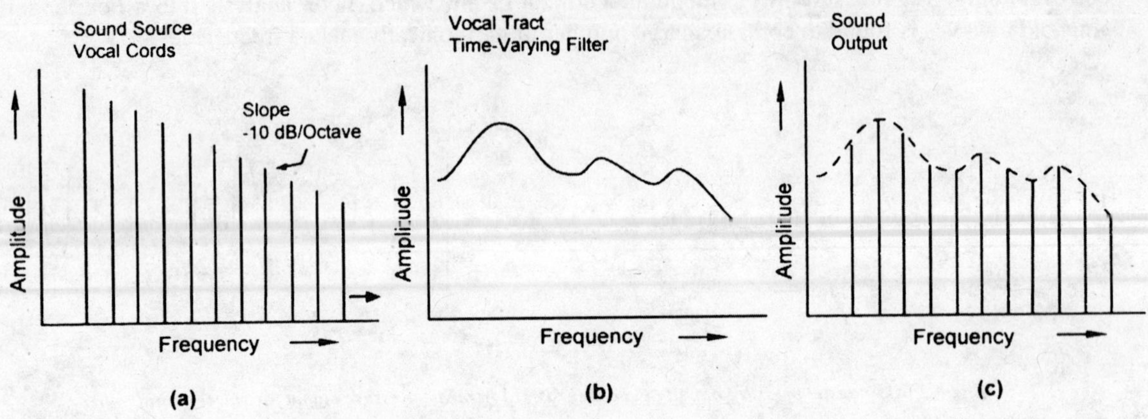

Fig. 4.4. Diagram showing the production of voiced sounds such as vowels. (a) The vocal cord sound is primarily a line spectrum with a - 10 dB/octave slope. (b) The time - varying filter action of the vocal tract shows resonances at odd quarter wave lengths of the tract acting as an acoustical pipe. (c) The filter action of (b) shapes the vocal cord sound of (a) to give the output sound.

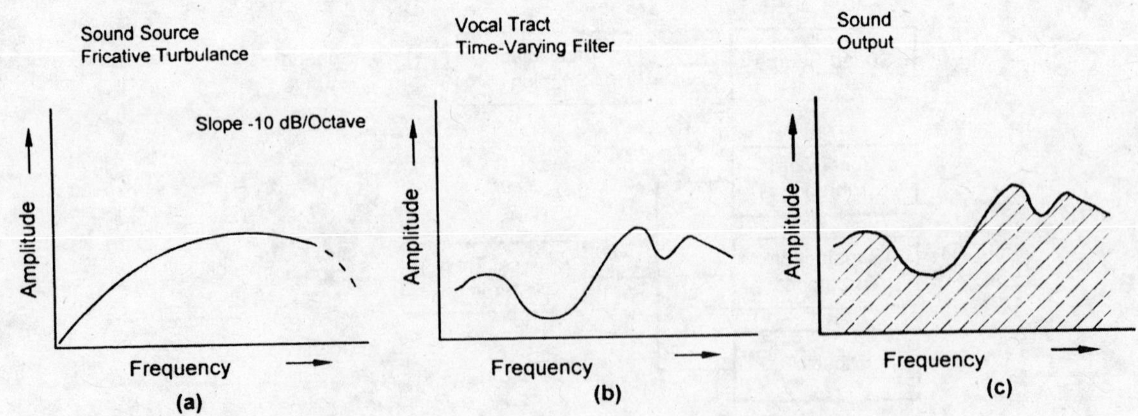

Fig. 4.5. Diagram of the production of unvoiced fricative sounds such as f, s, v and z. (a) The distributed spectrum of noise due to air turbulence resulting from constrictions in the vocal tract. (b) The time - varying filter action of the vocal tract. (c) The output sound resulting from the filter action of the distributed sound of (a).

sounds reveals a band of practically continuous frequency coverage largely confined to the upper portion of the audible frequency range. All of the three sources of speech (voiced sounds, fricative sounds and the plosive sounds) and the time-varying-filter action of vocal tract are illustrated in Fig. 4.3.

One of the many amazing things about the human body is the high degree of efficiency associated with the *multiple use of organic systems.* The functions of *eating, breathing* and *speaking* all take place in relative simultaneous harmony. We can eat, breathe and talk practically at the same time through the *inter-working* of muscle action and valves without food going down the wrong path. If we sometimes try to do too many things at once, the system is momentarily thwarted, and we agonise as a bit of food is retrieved from the wrong pipe.

4.2 FREQUENCY RANGE REQUIRED FOR SPEECH

Two factors determine the frequency range required for satisfactory speech transmission : the intelligibility / frequency distribution and the energy / frequency distribution. Special listening tests called *articulation tests* have been devised to measure *intelligibility.* In one form of articulation test, a group of listeners is asked to record syllables as they hear them announced over a loudspeaker as the frequency range of transmission is reduced in steps. The number of correctly recorded syllables as a percentage of the total number transmitted at each step is termed the articulation efficiency. The syllables are unrelated to one another (*i.e.,* they are nonsense syllables). In this way, results due to good guesswork are reduced. It is found in practice that *an articulation efficiency of 80% corresponds to about 100% intelligibility for normal speech.*

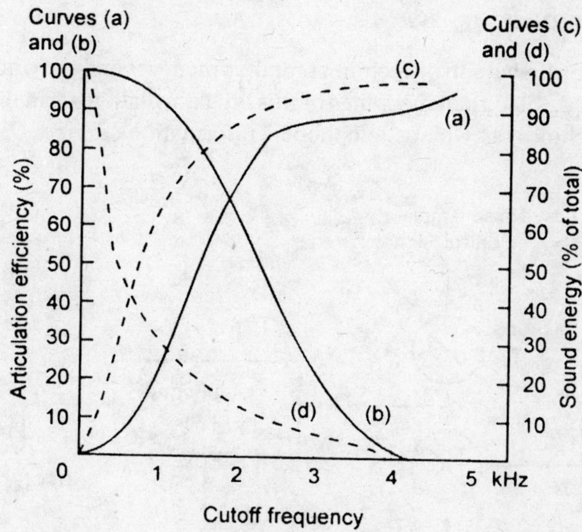

Fig. 4.6. (a) Articulation efficiency as a function of upper cutoff frequency; (b) articulation efficiency as a function of lower cutoff frequency; (c) sound energy as a function of upper cutoff frequency; (d) sound energy as a function of lower cutoff frequency.

Curves of articulation efficiency against frequency are shown in Fig. 4.6 (*a*) and (*b*). It can be seen that to maintain a high articulation efficiency, frequencies up to at least 2.5 kHz must not be suppressed (Fig. 4.6 *a*). It can also be seen that all frequencies below 1.5 kHz can be suppressed without drastically reducing the intelligibility (Fig. 4.6 *b*). The conclusion is, therefore, that *the intelligibility is contained in the high frequency components (1.5 to 2.5 kHz) of speech.*

As much as 80% of total energy is transmitted to listeners, even though all frequencies above 1 kHz are suppressed (Fig. 4.6 *c*). Suppressing all frequencies below 1 kHz reduces the energy transmitted to 15% (Fig. 4.6 *d*). The conclusion is, therefore, that *the energy in speech sounds is contained in the low frequencies.*

The *frequency limits* adopted internationally on the basis of such tests are *300 to 3400 Hz*. This range passes about 60% of the total energy and allows about 85% articulation efficiency.

Since the type of tests involved cannot fix the frequency limits precisely, these may be different for specific applications. For example speech transmission for *entertainment purposes* may require *80 to 8000 Hz* in order to capture the naturalness of the speaker's voice.

4.3 SYNTHESIZED SPEECH

Mechanical *speaking machines* date back to 1779, when Kratzenstein of St. Petersburg constructed a set of *acoustical resonators* to emulate the human mouth. These were activated with *reeds* such as those of a mouth organ. He was able to produce reasonably recognisable vowel sounds with a contraption. Wolfgang von Kempelen of Vienna did a much better job in 1791, which Wheatstone later improved upon. This machine used *bellows* to supply air to a leather tube that was manipulated by hand to simulate mouth action and included an 'S' whistle, a 'SH' whistle, and a nostril cutoff valve. Alexander Graham Bell patented a procedure for producing speech in 1876. One important precursor of the modern digital devices for synthesizing speech was the *analog Voder* from Bell Laboratories that was demonstrated at the World Fairs in New York (1939) and in San Francisco (1940). It took a year to train operators to play the machine to produce simple but recognizable speech.

4.4 DIGITAL SPEECH SYNTHESIS

Techniques for *storing* human speech in computer memory and *playing* it back under specified, fixed conditions are widely used. Electrical machines of this type now talk to us in the form of language translators, calculators, spelling machines, as well as telephone - information services. We shall be seeing a stream of

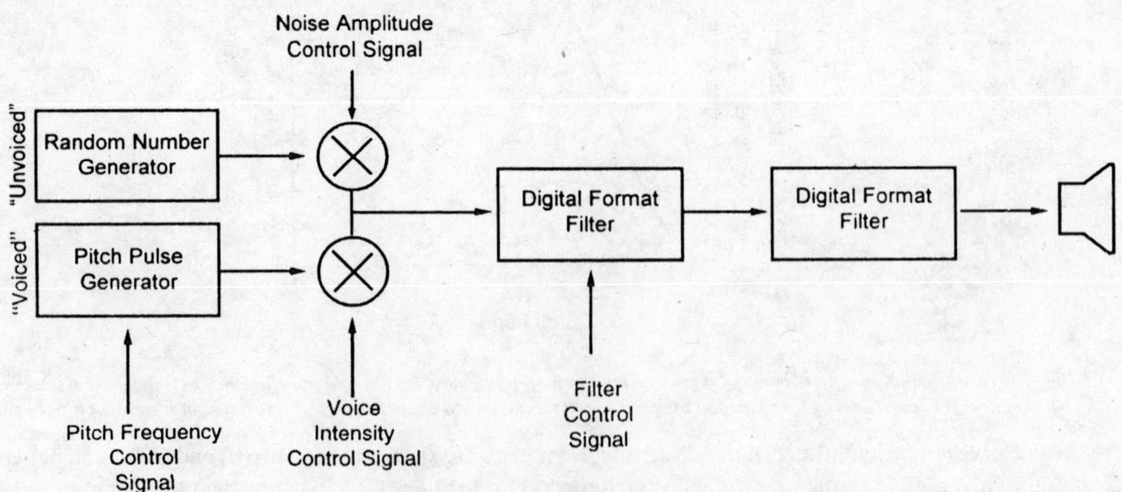

Fig. 4.7 A digital system for synthesizing speech. Note the similarity to the models of Figs. 4.3, 4.4 and 4.5.

other *answering - back applications* in the days ahead. This does not apply to speech synthesis, only storage and recall from memory on demand.

It is interesting to note that to program a computer to talk, a model of speech production is necessary and that the models of Fig. 4.3, 4.4 and 4.5 have been applied in just this way. Fig. 4.7 shows a diagram of a *digital synthesis system.*

A random-number generator produces the digital equivalent of the 's-like' sounds for the *unvoiced components.* A counter produces pulses simulating the pulses of sound of the vocal cords for the *voiced components.* These are *shaped* by time-varying digital filters simulating the ever-changing resonances of the vocal tract. Special signals *control* each of these to form digitised speech, which is then *changed* to analog form in the digital-to-analog converter.

4.5 DIRECTIONALITY OF SPEECH

Speech sounds do not have the same strength in all directions. This is due primarily to the *sound shadow* cast by the head. A question arises as to just how such directionality can be measured. Should a sound source be placed in the mouth? Well, the mouth itself is a continuous source of speech sounds, so why not use these speech sounds for the measurement of directionality ? That is exactly what Heinrich Kutruff has done with the results shown in Fig. 4.8. *Because speech sounds are highly variable and extremely complex, careful* averaging is necessary to give an accurate measure of directional effects.

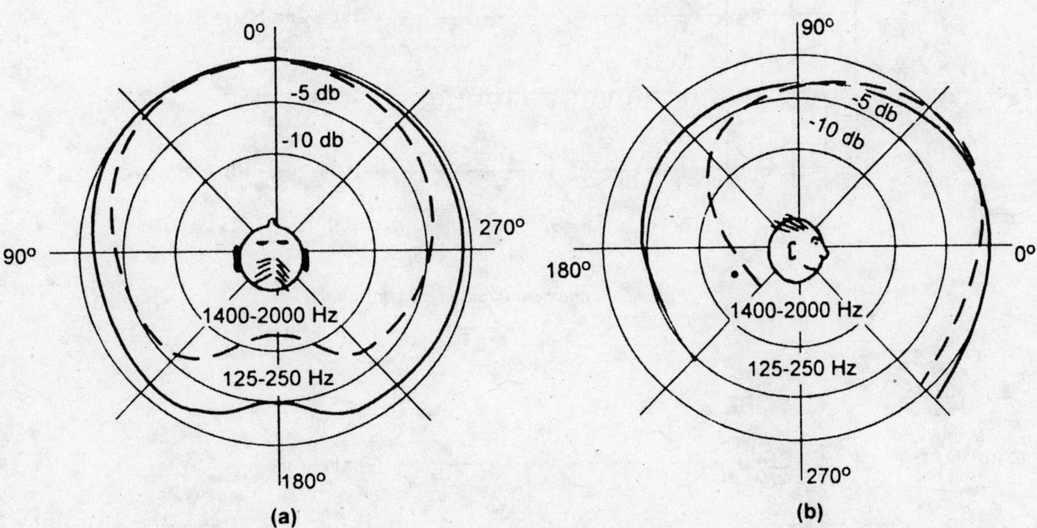

Fig. 4.8. Human voice directionality measured using the voice as a sound source. (a) Front-to-back directional effects of about 12 dB are found for the important speech frequencies. (b) In the vertical plane, the front-to-back directional effects for the 1400 to 2000 Hz band are about the same as for the horizontal plane.

The *horizontal directional effects,* as shown in Fig. 4.8 (*a*), show only a *modest* directional effect of about 5 dB in the 125 to 250 Hz band. This is to be expected because the head is *small* compared to wavelengths of 4.5 to 9 feet associated with this frequency band. There are significant directional effects however, for the 1400 to 2000 - Hz band. For this band which contains important speech frequencies, the front-to-back difference is about 12 dB.

In the *vertical plane* Fig. 4.8 (*b*), the 125 - to 250-Hz band shows 5 dB front-to-back difference again. For the 1400-to 2000 Hz band, the front-to-back difference is also about the same as in the horizontal plane, except for the torso effect. The discrimination against high speech frequencies picked up on a lapel microphone becomes obvious in Fig. 4.8 (*b*), although the measurements were not carried to angles closer to 270 degrees.

4.6 MUSIC AND NOISE

A sound of *regular frequency* is called a tone or *musical note,* and music is a combination of such sounds. Certain combinations of notes do, however, produce an effect of emotional tension, or *dissonance,* and this has played an important part in the vocabulary of music, throughout the ages, in varying degrees.

Noise in relation to music is not easy to define, but fairly general agreement may be expected with the definition that *noise is a sound or a combination of sounds of constantly varying pitch or irregular frequency.*

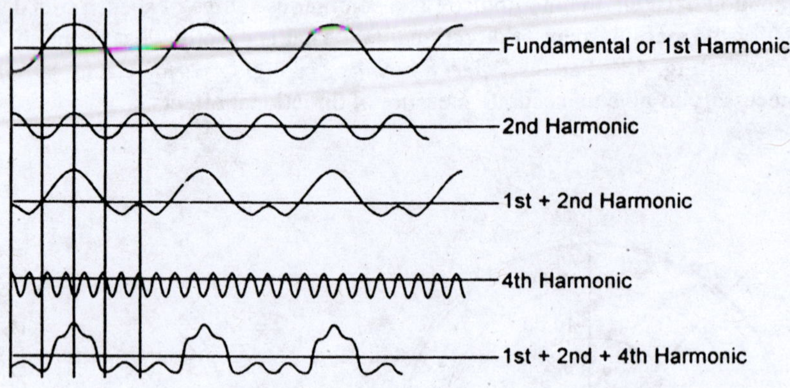

Fig. 4.9 Fundamentai and higher harmonics.

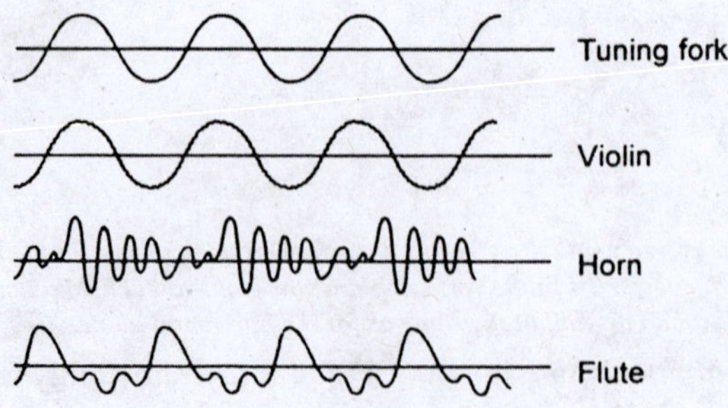

Fig. 4.10 The ear can distinguish between different instruments playing the same note from their waveform.

The ear is sensitive to the intensity and frequency of a sound wave, but perhaps more amazing is its *ability to detect differences in the complexity of a sound vibration.* The various kinds of instruments in an orchestra can be identified even when they are playing simultaneously, and instruments of different kinds playing the same note can be distinguished easily. *The sound energy emitted by every instrument is divided differently between its fundamental frequency and the higher harmonics,* Fig. 4.9. Each harmonic produces subtle differences in the pattern of responses in the cochlea and the brain interprets these as differences in tonal quality, Fig. 4.10.

4.7 MUSICAL SCALES

The ratio of the frequency of two notes is called the *musical interval* between the notes

$$\text{Musical interval} = \frac{f_1}{f_2} \qquad \ldots(4.1)$$

The basic unit of musical interval is the octave which denotes a frequency ratio of 2:1. Thus, there is an octave interval between the notes of 264 Hz and 528 Hz, and between the notes of 220 Hz and 440 Hz. The octave may be sub-divided into twelve equal ratios r such that

$$r^{12} = 2 \qquad \ldots(4.2)$$

i.e.

$$r = 1.06 \qquad \ldots(4.3)$$

In terms of octaves a frequency ratio f_1 / f_2 may be expressed as

$$f_1/f_2 = 2^{\text{octaves}} \qquad \ldots(4.4)$$

Hence, octave

$$= \log_2 (f_1/f_2) = \log_{10}\left(\frac{f_1}{f_2}\right) / \log_{10}^2 \qquad \ldots(4.5)$$

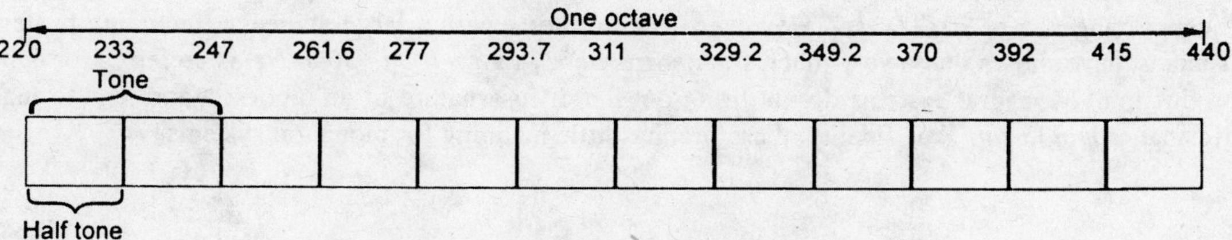

Fig. 4.11 Frequencies on the scale of equal temperament.

Thus, we can construct a scale, the *scale of equal temperament* composed of particular frequencies separated by this interval, Fig. 4.11. Each of these intervals is called a half tone or semitone, two half tones making a *tone*, and there being six complete tones to an octave. We letter the seven notes *A, B, C, D, E, F,* and *G,* using the same letters for notes an octave apart. The intermediate frequencies between the notes *A, B, C,* etc; are referred to either as *sharps* (#) of the lower note or as *flats* (Δ) of the upper note.

Example 4.1. A musical scale that is not based on any one key note but *has the same character independent of the starting note* is called an equally tempered scale. In this scale an octave is divided into 12 *half tones.* What is the interval between adjacent notes of equally tempered scale ?

Solution : Let α be the interval between the notes of an equally tempered scale. Then

$$256\ (\alpha \times \alpha \times \alpha \ldots\ldots 12 \text{ times}) = 512$$

$$\text{or } \alpha^{12} = \frac{512}{256} \qquad \alpha = (2)^{1/12}$$

$$\alpha = 1.0595$$

Several notes played together form a *chord.* When the frequencies of the notes in these chords are in one of the simple ratios 1:2, 2:3, 3:4, 4:5, 5:6, they produce a high degree of *consonance,* the actual degree we recognise being a matter of convention and, ultimately, of personal taste.

A series of notes arranged such that their fundamental frequencies have a definite ratio is called a musical scale.

In 1558, Zarlino constructed a musical scale by introducing *eight notes* between an octave. These eight notes constitute *major diatonic scale.* The first note (lowest frequency) is called *key note.* The following gives the names of eight different notes.

Table 4.1 Major Diatonic Scale

Notation	C Sa	D Re	E Ga	F Ma	G Pa	A Dha	B Ni	C′ Sa
Frequency key note 256	256	288	320	341 1/3	384	426 2/3	480	512
Interval key note	1	9/8	5/4	4/3	3/2	5/3	15 / 8	2

4.8 FREQUENCY RANGE REQUIRED FOR MUSIC

Direct *frequency analysis* of the output of a large orchestra shows that the sound energy covers the frequency range 15 to 20,000 Hz. However, listening tests with a large audience, including professional musicians, have shown that a very much smaller frequency range *160 to 3000 Hz,* is preferred for *monaural* reproduction. Monaural systems do not bring out the diffuse nature of an orchestra and tend to make the performance *unnatural.* True fidelity, therefore, has little meaning for monaural systems.

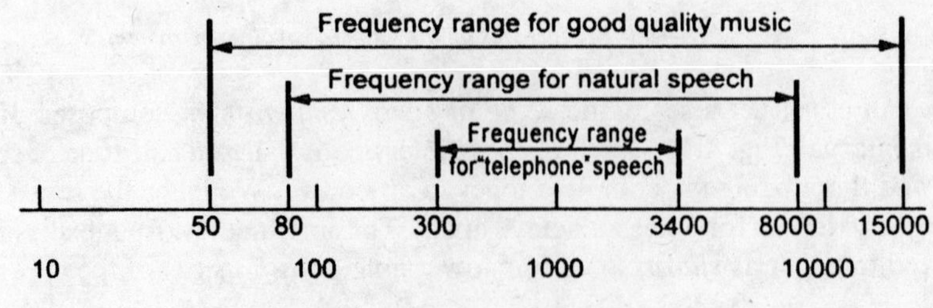

Fig. 4.12 Frequency range requirements for various types of audio signals.

Any system may be called a *high-fidelity system* provided it meets certain minimum requirements, such as minimum bandwidth, minimum amplitude versus frequency distortion, minimum harmonic distortion, and minimum dynamic range, while still not producing true fidelity.

Separate tests in which the sound passed directly from an orchestra to an audience through an *acoustical frequency filter* showed that the audience had a distinct preference for the full frequency range. In view of the conflicting evidence, and with the introduction of *stereophonic systems,* a frequency range of 50 to 15000 Hz is accepted as satisfactory for reproducing music. The various frequency-range requirements are summarized in Fig. 4.12.

Digital audio has brought some major revisions of our thinking in regard to dynamic range. The dynamic range in a digital system is directly related to the range of binary digits (bits) as shown in Table 4.2.

Table 4.2 Theoretical Dynamic Range of Digital Systems.

Number of Binary Digits	Dynamic Range-dB
4	24
8	48
12	72
16	96

4.9 NOISE

Noise is the name given to *unwanted sound.* Noise is truly something in the ear of the ''behearer.'' One person's scintillating music is another person's noise, especially at 2 A.M. It is a two-way street and fortunately, a good wall that protects a studio area from external noise also protects neighbors from what goes on inside. The psychological aspect of noise is very important - *acceptable if considered a part of a situation - disturbing if considered extraneous.*

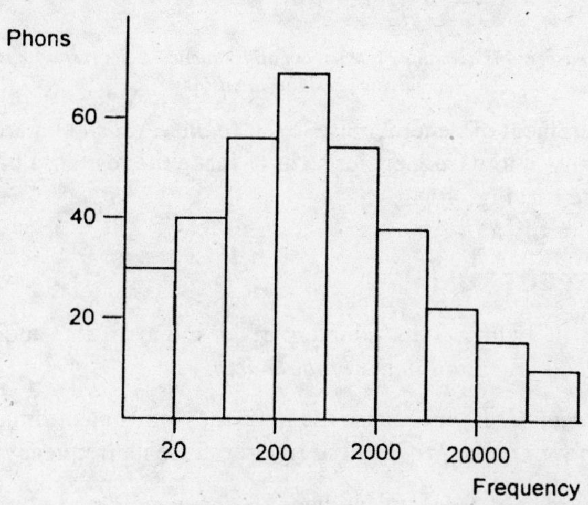

Fig. 4.13. A noise profile is obtained by measuring the loudness of each frequency band.

The level of noise is an important feature of human environment. It can cause effects ranging from minor irritation to permanent deafness, sometimes without the knowledge of people subject to it. Thousands of people in this country willingly work, and are allowed to work, under noisy conditions which specialists agree will cause some degree of permanent deafness. It costs money, however, to suppress noise and in the case of machines, it may result in a loss of power. Inevitably, *there is a conflict between economic necessity and the need to limit noise.*

If standards are to be setup and become accepted, it is necessary that *convenient methods* of measurement are developed. A *noise meter* is basically an amplifier whose input is a non-directional microphone and whose output is recorded on a meter which has been standardised and calibrated in decibels of intensity. The best instruments incorporate a series of *filters,* which allows the loudness of the different frequency bands present to be measured separately. These are then converted to phons of loudness with the help of equal loudness curves and plotted as a *noise profile.* For safety, the loudness of all the frequency ranges should be below the level at which aural damage is possible.

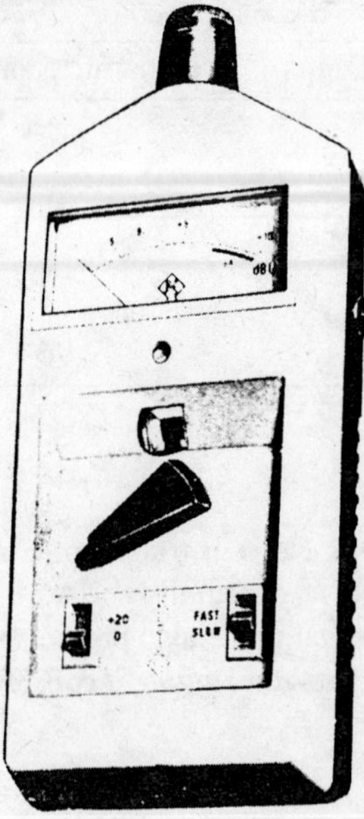

Fig. 4.14 The sound level meter measures the combined intensity of all frequencies. Its response is made nearer to that of the ear for medium and low intensities.

For a less exact measurement of general noise level, *sound level meters* are available which, by the use of filters, attenuate the intensity at low frequencies so as to make the response of the instrument approximate to the response of the average ear. Fig. 4.14.

4.10 THE CONCEPT OF SPECTRUM

The visible spectrum of light has its counterpart in sound in the audible spectrum, *the range of frequencies that fall within the perceptual limits of the human ear.*

We cannot see *far-ultraviolet light* because the frequency of its electromagnetic energy is too high for the eye to perceive. Nor can we see the *far-infrared light* because its frequency is too low.

There are likewise sounds of too low frequency - *infrasound* and sounds of too high frequency - *ultrasound* - for the ear to hear.

4.11 WHITE AND PINK NOISE

In the audio literature references to *white* and *pink* noise dominate. These two terms arose because there are two types of spectrum analysers in common use. One of these is the *constant bandwidth analyser.* No matter whether the analyser is scanning the low frequencies or the high frequencies the *passband remains a fixed number of hertz wide.* When random noise is scanned in frequency by such an analyser, its uniform distribution of energy yields a flat curve. Random noise, we have discussed so far, is white because *its counterpart in light is white light.* The energy of *white light* is uniformly distributed throughout the *visible spectrum.* The energy of *white noise* is uniformly distributed throughout the *audible spectrum.*

The second type of spectrum analyser has a *bandwidth that increases with frequency.* Examples of this are 1/3 octave, 1 octave, and other analysers with passband widths a constant percentage of the band centre frequency. One octave centred at 125 Hz (89 to 177 Hz) spans 88 Hz and has a bandwidth of 70.7 percent $\left(\frac{88}{125} \times 100\right)$. An octave centered at 8 kHz (5,656 to 11,312 Hz) spans 5656 Hz and also has a bandwidth of $\left(\frac{5650}{8000} \times 100\right)$. If random noise signal is analysed with a filter of this type, the spectrum graph would tilt upwards

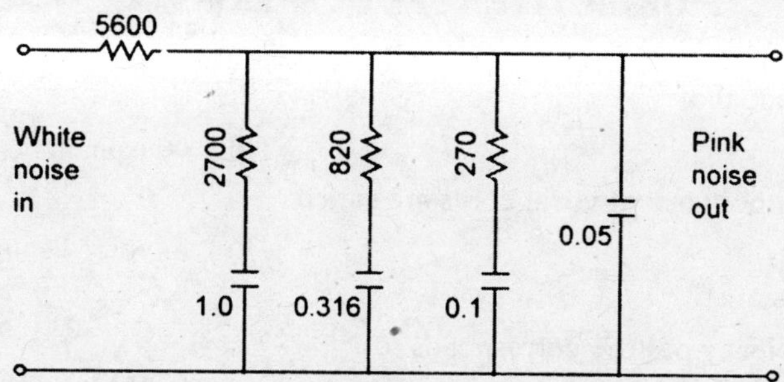

Fig. 4.15. A simple filter for changing white noise to pink noise. It changes random noise of constant energy per Hz to pink noise of constant energy per octave. Pink noise is useful in acoustical measurements utilizing analyzers having passbands with bandwidth of a constant percentage of the centre frequency.

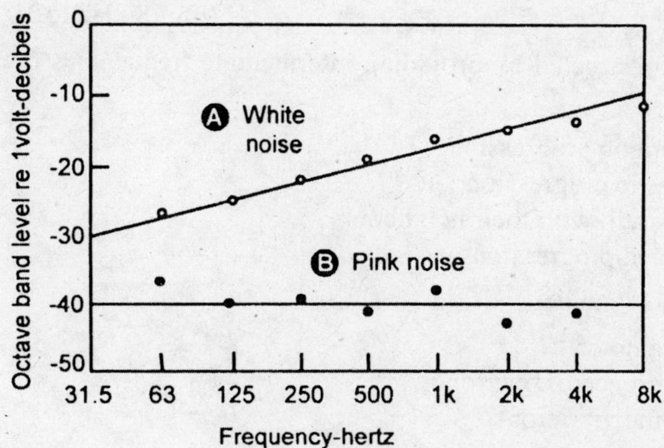

Fig. 4.16 White noise, having uniform distribution of energy with frequency, shows the spectrum of (A) when analysed with octave filters. The spectrum rises at 3 dB/ octave because the octave filters have widths with ever increasing number of Hz. Pink noise shows a flat spectrum when analysed with octave or fractional octave filters. This flatness makes pink-noise convenient for certain acoustical measurements.

with a slope of 3 dB/octave. Because noise signals are so desirable in acoustical measurements, random noise is commonly filtered to give a 3 dB/octave downward slope. Such a random noise signal with a 3 dB / octave downward slope is called *pink noise* because the low frequency end (the red end of the visible spectrum) is emphasized.

Converting white noise to pink noise is accomplished by sending it through a so-called *pink-noise filter*. Such a filter is shown in Fig. 4.15 and is nothing more than four resistors and four capacitors connected as shown

Fig. 4.16 shows some rough measurements made in analysing white noise (*A*) with an octave analyzer. There is constant energy per hertz. In going from one octave to the next higher octave, the number of hertz embraced is *doubled* and the octave energy for white noise is therefore doubled. The increase is then 10 log2 = 3 dB or *3 dB per octave slope.*

Fig. 4.16 (*b*) repeats the octave analysis of white noise as it is passed through filter of Fig. 4.15. The filter characteristic falls off with frequency at 3 dB / octave, compensating for the rising white noise spectrum of Fig. 4.16 (*a*) and resulting in the flat, pink noise characteristic of Fig. 4.16 (*b*).

OBJECTIVE TYPE QUESTIONS (A)

1. Voiced sounds include the
   (*a*) vowels (*b*) consonants
2. Sounds produced without use of vocal cords are called
   (*a*) voiced sounds (*b*) unvoiced sounds
   (*c*) breath sounds
3. An articulation efficiency of 80% corresponds to
   (*a*) 50% intelligibility (*c*) 75% intelligibility
   (*c*) 90% intelligibility (*d*) 100% intelligibility
4. Intelligibility is contained in
   (*a*) 20 Hz to 1 kHz (*b*) 1.5 kHz to 2.5 kHz
   (*c*) 4 kHz to 6 kHz (*d*) 8 kHz to 10 kHz
5. A musical scale is constructed by providing intermediate frequencies between a note and its octave which
   (*a*) form an arithmetic progression
   (*b*) form a geometric progression
   (*c*) bear a simple ratio with their neighbours
   (*d*) form a harmonic progression
6. Quality of a musical sound depends on
   (*a*) frequency of a note
   (*b*) presence or absence of overtones
   (*c*) amplitude of the vibrations
7. Persons with sweeter sounds have
   (*a*) greater number of overtones
   (*b*) lesser number of overtones
   (*c*) no overtones
   (*d*) neither more nor less overtones

8. In the musical sa, re, ga the frequency of the notes is
   (*a*) sa > re > ga (*b*) sa < ga
   (*c*) the same (*d*) sa < re < ga < ma

9. An air box attached to a musical instrument increases the
   (*a*) pitch of the sound (*b*) intensity of the sound
   (*c*) quality of the sound (*d*) shrillness of the sound

10. A set of tones whose frequencies are integral multiples of the fundamental frequency are called
    (*a*) overtones (*b*) harmonics
    (*c*) beats (*d*) Doppler frequencies

11. The tones that are separated by three octaves have a frequency ratio of
    (*a*) 3 (*b*) 6
    (*c*) 8 (*d*) 16

12. The same notes being played on the Sitar and Veena differ in
    (*a*) pitch (*b*) quality
    (*c*) both pitch and quality (*d*) none of these

FILL IN THE BLANKS (B)

1. The energy of sound is contained in ____________________.
2. The frequency limits adopted internationally for satisfactory transmission of speech are ____________________.
3. A sound of ____________________ is called a tone or musical note.
4. Noise is a sound or combination of sounds of constantly varying ____________ or irregular ______________.
5. The ratio of the frequency of two notes is called the ____________ between the two notes.
6. The basic unit of music interval is the ____________________
7. An octave denotes a frequency ratio of ____________________
8. Several notes played together form a ________________.
9. The dynamic range in a digital system is directly related to the __________.
10. Noise is the name given to ___________ sound.
11. The visible spectrum of light has its counterpart in sound in the _________ spectrum.
12. The energy of white noise is ____________ distributed throughout the _________ spectrum.

CONVENTIONAL QUESTIONS (C)

1. What is the difference between noise and musical sound ?
2. Define a bel.
3. Why the roar of a lion is different from the sound of a mosquito ?
4. Two musical sounds of the same loudness and pitch may produce different impressions on the human ear. What is the cause of difference ?

5. Explain the voice mechanism. How different types of sound are produced?
6. Discuss the factors which determine the frequency range of speech required for satisfactory sound transmission.
7. Explain the following:
   (*a*) articulation (*b*) acoustic network (*c*) musical scales
8. Explain with the help of a diagram a digital system for synthesizing speech.
9. How are musical scales formed ?
10. Explain the significance of an equally tempered scale. What is the relation between frequencies on the scale of equal temperament ?
11. Draw the diagram of a filter for converting white noise to pink noise. Mark the value of each component.
12. Briefly explain the different kinds of spectrum analysers.
13. Distinguish between white and pink noise.

5

MICROPHONES

In practically all modern acoustical work, the oscillations in sound pressure are picked up by some form of receiving electroacoustic transducer, such as a microphone, which converts them into similar electrical current or voltage oscillations. The latter either may be electrically amplified for immediate reconversion into sound energy or may be stored in some form, such as on magnetic tape, for later analysis of playback. When operating in air such transducers are known as microphones, and when operating in water as hydrophones. Microphones serve two principal purposes. First, they are used for converting music or speech into electrical signals which are transmitted or processed in some manner and then reproduced. Second they serve as measuring instruments converting acoustical energy into electrical currents which actuate indicating meters.

5.1 MICROPHONE

All *sound recording* starts with the use of microphones, often in very large numbers. Professional recording engineers, for examples, may use 32 or more microphones to record a large orchestra, with the output from each microphone being *separately* recorded on a wide strip of recording tape. The *placing* of each microphone, the *amplitude* of recording from each microphone, and the subsequent *mixing* of the sound from each track to form two tracks of a *stereo recording* or four tracks of a discrete *quad recording* are operations requiring great skill and experience.

A microphone is a device of the class called *transducers* which converts sound waves in air into electrical waves of the same frequency and wave shape. To do so, the microphone must make use of either the *pressure* of the air waves, or the *velocity* at which the air moves. Because of this, we have two types of microphones, the pressure - operated types, and the velocity-operated types.

5.2. CHARACTERISTICS OF MICROPHONES

There are many types of microphones available. Each has certain advantages and disadvantages. The type of material to be reproduced, the placement of the microphone, whether it is to be used indoors or outdoors, the frequency response desired, and a number of other factors, affect the choice of a microphone.

The basic types of microphones, grouped according to their *principle of operation* are : carbon, crystal, dynamic, ribbon and capacitor.

Each of these has its own characteristics with respect to : (1) output level, (2) frequency range (3) output impedance and (4) directivity. These characteristics ultimately determine whether or not a particular microphone is suitable for a given application.

1. *Output Level* : The output level of a microphone governs the amount of amplification that must be available for use with the microphone. The output level of microphones is usually given in dB preceded by a minus sign. *The minus sign means that the output level is so many dB below the reference level of 1 milliwatt for a specified sound pressure.*

The unit of sound pressure used for rating microphones is referred to as a *bar.* A bar is equal to a sound pressure of 1 dyne per square centimetre. *Speech* provides sound pressures between 0.4 and 15 bars. For *music* the pressure ranges from 0.5 bars to 1250 bars.

Microphones are rated in a number of different ways, and this often causes confusion. If ratings are given in any manner other than in bars, it is a good idea to *convert* their output level rating to dB below 1 milliwatt for a sound pressure of 1 bar (see Table 5.1).

Table 5.1. Comparison of Microphone Ratings

Rating Given	Correction Factor
dB below 1 mW / 1 bar	0 dB
dB below 1 mW / 10 bars	– 20 dB
dB below 1 volt / 1 bar	2 dB
dB below 1 volt / 10 bars	– 18 dB

A microphone with a *low output level* necessitates the use of an *amplifier with greater gain,* which, in turn increases the possibility of noise and hum.

2. *Frequency Response* : The frequency response of a microphone is a rating of the fidelity of relative output voltage which results from sound waves of different frequencies. The simplest way to find a complete picture of the frequency response characteristics of a microphone is to plot a curve of its output voltage vs input frequency. *Since good modern microphones are relatively flat over their range, it is often considered sufficient to specify the range over which their output does not vary more than plus or minus 1 or 2 dB.*

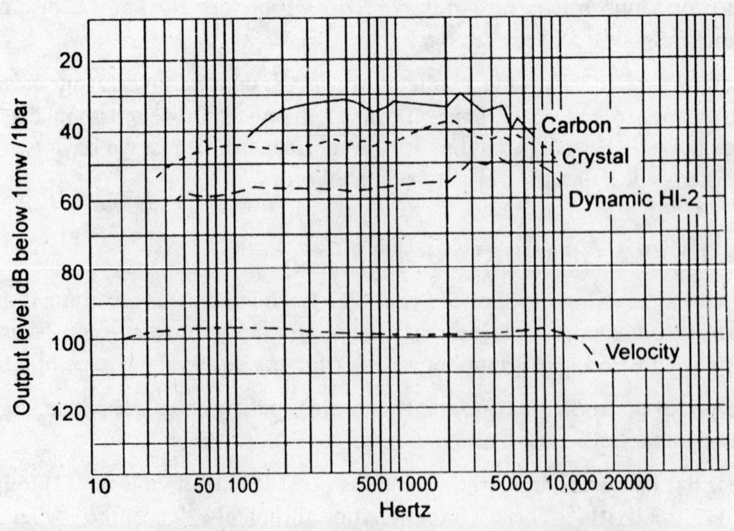

Fig. 5.1 Frequency response curves for typical microphones.

For ordinary home high-fidelity use, a microphone *frequency-response curve* should be reasonably flat between 40 and 10,000 Hz. With systems designed specifically for speech reinforcement, a lower limit of 150 Hz and an upper limit of 5,000 Hz is entirely satisfactory. Where it is desired to reproduce music with the highest possible fidelity, the frequency response should be flat (within 2 dB) from about 40 to 15,000 Hz Fig. 5.1 shows the response of several types of microphones.

3. *Output Impedance* : A microphone, like any other component with electrical inputs or outputs, has a value of impedance. When a microphone is connected to an amplifier, a *complete circuit* is made and electric current flows whenever a sound causes the microphone to generate an electrical voltage.

For most high quality microphones impedance is *low,* a few ohms ranging up to a hundred ohms or so, but as little as a fraction of an ohm in a ribbon microphone. Only capacitor and ceramic crystal microphones have *high* impedances.

The importance of microphone impedance to us is not a matter of the precise value but of *the ability of the microphone and the recorder to be matched together.* High impedance microphones must be connected into a recorder with high impedance input, otherwise both the signal amplitude and the frequency range will be adversely affected.

In general, when a ceramic or capacitor microphone is connected to a low impedance amplifier input, the output from the amplifier is small and the lower frequencies are reduced disproportionately, producing an unpleasant shrill sound on playback. Cassette recorders of good quality are not as a rule designed to take *high impedance microphones,* except when such microphones are supplied with the recorder, and if, for some reason, such microphones are used, a *high impedance preamplifier* must be used between the microphones and the recorder.

Low impedance microphones and inputs are much more common, one good reason is the *hum problem.* The frequency of the supply voltage, 50 Hz, is, like any other A.C. signal, radiated from any wire on which it is present, and can also be picked up on any wire. You can prove this for yourself by touching the input of an amplifier with the volume control turned half-way up. The resulting amplified hum sounds very loud in the loudspeakers, showing that the *hum signal* must be at least as strong as the signals normally fed into the amplifier. *Radiation* through the air at this low frequency is not an efficient process, however, as the house wiring is not long enough to act as an efficient antenna (aerial), and the hum signal picked upon on a piece of wire behaves as if it had come from a high impedance source. If the input of the cassette recorder amplifier is at high impedance, then the whole hum signal can be *transferred* to it unless every piece of wire and every connector is efficiently *shielded* from the hum by *earthed metal casings.* If, on the other hand, the input to the cassette recorder amplifier is at low impedance, the hum signal, if picked up, is *greatly reduced in amplitude* even if no screening is used.

4. *Directivity:* Microphones do not respond equally to sound reaching them from all directions. Their *frequency response characteristics* also vary, depending on the *angle* at which the sound reaches them. A microphone may respond *equally* to all frequencies between 40 and 10,000 Hz when the sound is originating *directly* in front of it, while the high-frequency response falls off rapidly as the sound originates further to *either side.* Where it is necessory to pick up sound from all directions, the directional characteristics of all microphones are not suitable.

The way in which a microphone responds to sounds coming from different directions is plotted on a circular graph which is known as a *polar diagram.* The centre of the circle is the zero point and concentric circles indicate successively higher levels of response as they move outward. The top of the circle is the *front* and the bottom the *back* of the microphones, and the straight lines radiating from the centre denote the *corresponding angles.*

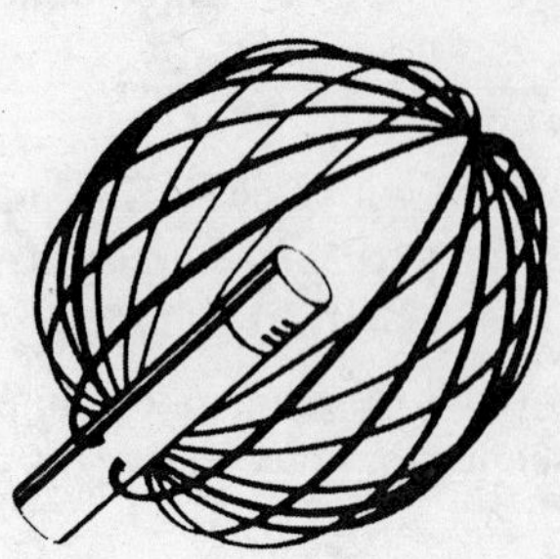

Fig. 5.2. Omnidirectional pickup pattern

Fig. 5.3 Cardioid pickup pattern.

A true *omnidirectional* microphone would have a plot of a perfect circle, Fig. 5.2, hence there would be little point in making a diagram. Few manufacturers in fact publish polar diagrams of such instruments. Actually, the pickup is rather less at the *rear* than at the *front*, so the plot would have a smaller radius at the bottom than at the front. *An omnidirectional microphone is most commonly used for general recording, and is the best for recording sound sources that move about a lot.*

When the directional response of a *pressure gradient* microphone is plotted, the shape of the figure is something like a heart, with the *maximum* response in the front curving around the sides to a *minimum* near the rear. It, therefore, is termed *cardioid* (Greek, kardia = heart), Fig. 5.3.

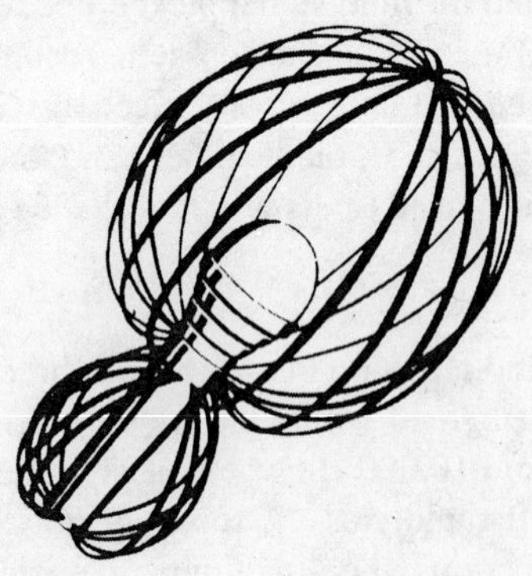

Fig. 5.4 Hyper or Super-cardioid pickup pattern.

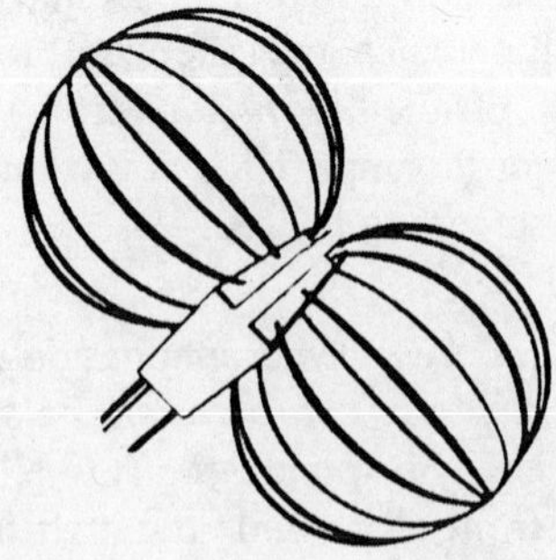

Fig. 5.5 Figure of - Eight pickup pattern.

To achieve a greater rejection of sounds coming from the sides, the *vents* in some models are designed to admit a higher sound pressure to give more cancellation at the rear of the cone or diaphragm. The result can be seen in the polar diagram, Fig 5.4, as a narrower forward lobe, and the response is known as *hyper* or *supercardioid.* One effect of this is that rear sounds can actually exert more pressure on the back of the cone than on the front, so movement is produced, although it is negative, that is a high pressure region produces forward instead of backward movement of the cone.

This appears on the diagram as a small lobe sometimes marked with a *negative sign* at the rear or below the centre position. It means that there is some pickup at the rear although not so much as at the front, but *the point of minimum pickup is nearly always about 120° from the front.* The rear lobe varies considerably with frequency and is sometimes found with cardioid as well as the hyper-cardioid.

If both sides of the ribbon of a velocity microphone are open to free air it follows that sounds from front or back will have exactly the same effect. In fact, really *there is no front or back.* As the ribbon is actuated by particle velocity instead of pressure, it responds only when *facing* the sound source. Sounds coming from the sides exert equal force on both sides of the ribbon and so have no effect. It is therefore *bidirectional,* and its diagram consists of two near circular lobes looking like a figure 8, hence the designation of a *figure of - eight response,* Fig. 5.5.

Example 5.1. The voltage gain of an amplifier when it feeds a resistive load of 1.0 kΩ is 40 dB. Determine the magnitude of the output signal voltage and the signal power in the load when the input signal is 10 mV.

Solution.

$$A_P = 20 \log \frac{V_2}{V_1} = 40$$

$$\log \frac{V_2}{V_1} = 2.0$$

$$\frac{V_2}{V_1} = 100$$

$$V_2 = 100 \times 10 = 1000 \text{ mV} = \mathbf{1.0\ V}$$

$$P_2 = \frac{V_2^2}{R_L} = \frac{1.0^2}{1000} = 0.001 \text{ W} = \mathbf{1\ mW}$$

Example 5.2. Express the power dissipated in a 15-Ω resistor in decibels relative to 1 mW when the voltage across the resistor is 1.5 V r.m.s.

Solution.

$$P_2 = \frac{V^2}{R} = \frac{1.5^2}{15} = 0.15 \text{ W} = 150 \text{ mW}$$

$$A_P = \frac{P_2}{P_1} = \frac{150}{1} = 150$$

and

$$A_P = 10 \log 150 = 10 \times 1.276$$

$$= \mathbf{21.76\ dB}$$

Example 5.3. An audio amplifier produces 20 watts output across an 8-ohm resistance when a 5-millivolt signal is applied to its input across a 1-megohm resistor. Determine the decibel gain.

Solution. The output power, $P_1 = 20$ watts.

Input power, $$P_2 = \frac{E_2^2}{R_2} = \frac{(0.005 \text{ volts})^2}{10^6 \text{ ohms}} = \frac{25 \times 10^{-6}}{10^6} = 2.5 \times 10^{-11} \text{ watts}$$

Hence, $$\text{dB} = 10 \log \frac{P_1}{P_2} = 10 \log \frac{20 \text{ watts}}{2.5 \times 10^{-11} \text{ watt}}$$

$$= 10 \log 8 \times 10^{11} = 10 \times 11.903 = \mathbf{119.03 \text{ dB}}$$

Alternatively, the output voltage

$$E_1 = \sqrt{P_1 \times R_1} = \sqrt{20 \text{ watts} \times 8 \text{ ohms}}$$

$$= \sqrt{160} \text{ volts} = 12.65 \text{ volts}$$

$$\text{dB gain} = 20 \log \frac{E_1}{E_2} + 10 \log \frac{R_2}{R_1}$$

$$= 20 \log \frac{12.65 \text{ volts}}{0.005 \text{ volts}} + 10 \log \frac{10^6 \text{ ohms}}{8 \text{ ohms}}$$

$$= 20 \log 2530 + 10 \log 1.25 \times 10^5$$

$$= 20 \times 3.403 + 10 \times 5.0969$$

$$= 68.06 + 50.97 = \mathbf{119.03 \text{ dB (as above)}}$$

If the resistance ratio (R_2/R_1) had been ignored, as *erroneous gain figure* of about 68 dB would have been obtained.

Example 5.4. A microphone has an output of –60 dB (with respect to a zero level of 6 milliwatts) and is connected to the 0.5-megohm input of a preamplifier. The preamplifier has a gain of + 40 dB. The signal

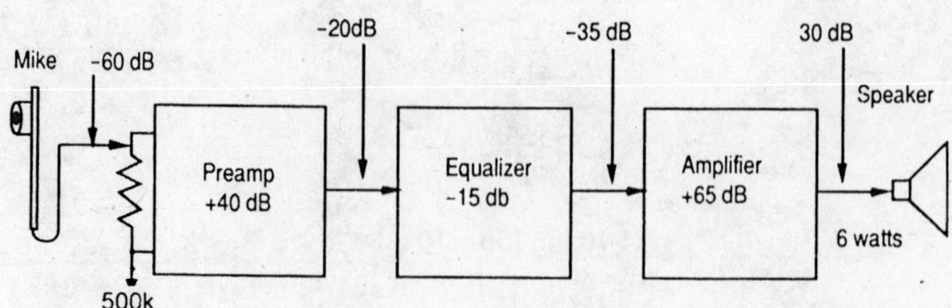

Illustration for Example 5.4.

then passes through an equalizer with an insertion loss of –15dB and through a main amplifier with a gain of +65 dB. If the output to the speaker is 6 watts, find the total power gain and the input voltage to the preamplifier.

Solution. Decibel gains and losses can be added algebraically. Hence,

$$\text{total dB gain} = +40 \text{ dB} - 15 \text{ dB} + 65 \text{ dB} = +90 \text{ dB}$$

The output to the speaker, therefore is

$$-60 \text{ dB (input)} + 90 \text{ dB (gain)} = +30 \text{ dB}$$

Since $$90 \text{ dB} = 10 \log \frac{P_1}{P_2},\ 9 = \log \frac{P_1}{P_2} \text{ and } \frac{P_1}{P_2} = \log^{-1} 9 = 10^9$$

Output power $$P_1 = 6 \text{ watts ; hence, } P_2 = \frac{P_1}{10^9} = \frac{6 \text{ watts}}{10^9}$$

$$= 6 \times 10^{-9} \text{ watts}$$

Input voltage $$E_2 = \sqrt{P_2 R_2} = \sqrt{(6 \times 10^{-9}) \times (0.5 \times 10^6)}$$

$$= \sqrt{0.003} \text{ volts}$$

$$= 0.0548 \text{ volts} = \mathbf{54.8\ mV}$$

5.3 CARBON MICROPHONES

For a *telephone system* important requirements are that: (*i*) the microphone shall be of convenient size; (*ii*) capable of mass production at low cost while possessing high sensitivity to operate from a simple battery; (*iii*) its performance must be stable and adequate to provide intelligible speech and articulation and (*iv*) it need not necessarily include the higher harmonic frequencies for reproducing.

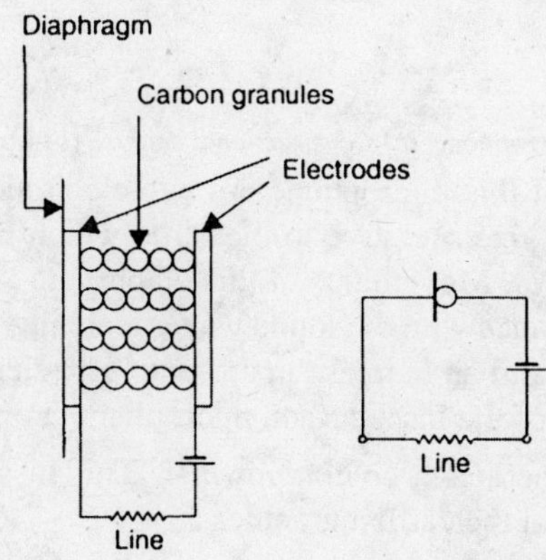

Fig. 5.6 Principle of carbon granule microphone.

On the other hand, microphones used for purposes such as *radio broadcasting* are relatively few in number and their cost is not a primary consideration: high fidelity reproduction up to about 10,000 Hz for natural speech and music transmission is essential.

An inset pattern of *carbon granule microphone* is in general use for telephone systems: the usual type is a *self-contained and sealed* microphone which can be readily and completely removed from the telephone instrument. Its operation depends upon the variation in *contact resistance* of the carbon granules when they are subjected to the pressure changes of sound waves. It follows that *this type of microphone does not produce an EMF but functions by modulating the current obtained from an external battery.*

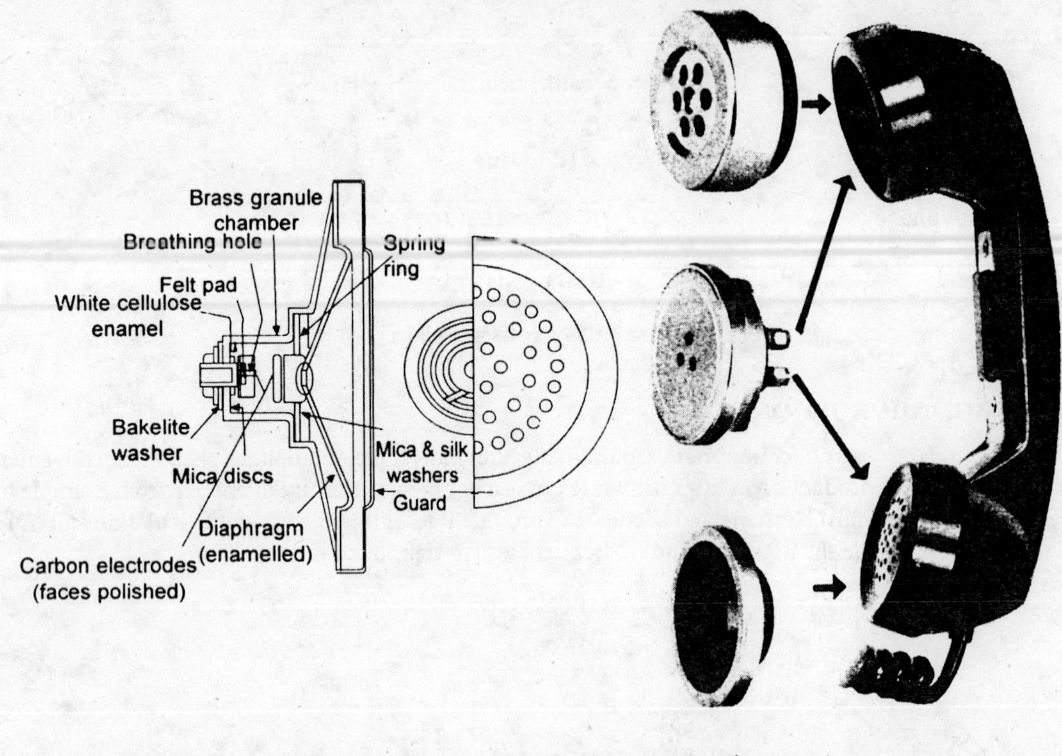

Fig. 5.7 Inset type carbon granule microphone. (a) Constructional details, (b) Carbon microphone in a telephone handset.

The essential components of this microphone are two electrodes between which are loosely packed carbon granules. One electrode is *fixed* relative to the other which carries a *diaphragm* to respond to the pressure changes of the sound waves. Movements of this diaphragm *vary the resistance* of the granules and so *control the line current* in accordance with the sound waves reaching the diaphragm. An increase in pressure produces a reduction in resistance and an increase in current. The elementary circuit arrangement is shown in Fig. 5.6. Constructional details of the inset carbon microphone are shown in Fig. 5.7.

The carbon microphone generates a *continuous hiss.* This hiss is due to small variations in contact resistance which take place between the carbon granules.

With carbon microphones *the electrical output is not directly proportional to the sound input level.* The practical effect of this non-linear distortion is to produce *harmonics* of the lower speech frequencies and these harmonics tend to *mask* higher frequencies normally present in the speech, resulting in *loss of clarity or articulation.*

The average *output level* of carbon microphone is of the order of - 30 dB (see Fig. 5.1). The best carbon microphones have a *frequency response* of approximately 60 to 7,000 Hz. They are substantially non directional, although their high frequency response above 300 Hz usually falls off at angles exceeding 40 degrees from the front of the microphone.

When the maximum output level is required from a microphone, the carbon microphone is often used. The frequency response characteristics of the carbon microphone are poor and cannot be used for high-fidelity work.

5.4 CRYSTAL MICROPHONES

Certain crystals, such as rochelle salt and quartz; possess the property of *generating small EMFs when subject to stress or strain.* This effect is utilised in what is known as the *crystal microphone.*

The construction of a crystal microphone is shown in Fig. 5.8. A thin finger shaped slice of crystal is secured at one end by means of a compliant clamp, and the apex of a cone is made to bear against the other. A sound pressure wave causes the cone to alternately press against and bend the crystal slice and release it. Thus, corresponding voltages are generated across the slice. A pair of contacts is fixed to opposite surfaces to take off the signal.

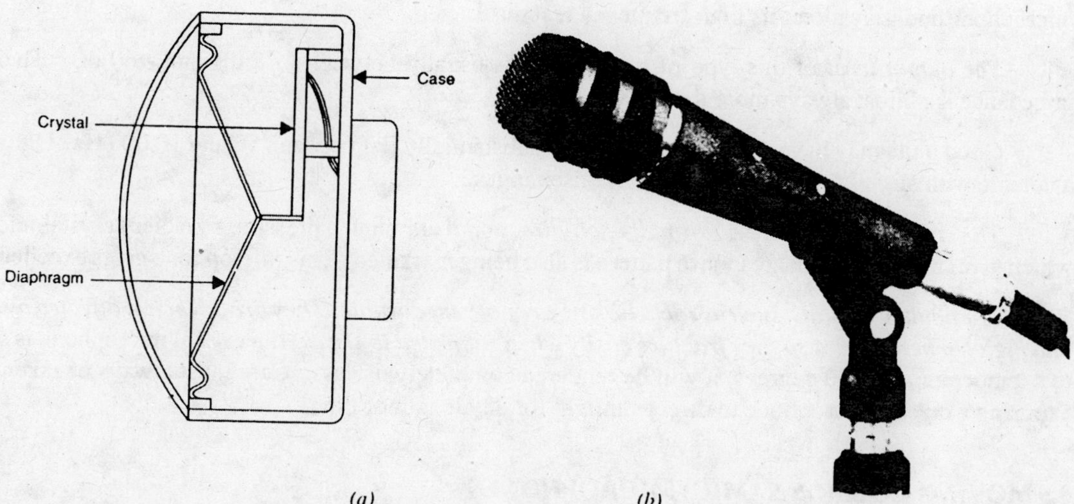

Fig. 5.8 Simple crystal transducer. Crystal slice is secured at one end with compliant clamps and bending action caused by pressure from the cone generates an EMF across the slice.

An improvement is obtained if the single slice of crystal is replaced by *two slices cemented together.* Then, when pressure is exerted, one slice is compressed while the other is stretched. Thus *equal and opposite* voltages are produced which, being *in series* like the cells of a car battery, give *double* the output. Any non-linearity which may arise due to the different mechanical strains between pressure and release is also thereby compensated. The double crystal unit is termed a *bimorph.*

One of the snags with this type of transducer is *the mass which must be moved* by the sound pressure acting on the cone. This consists of the mass of the cone plus that of the crystal or that part of it which is moved. This restricts the frequency response to its upper end to around 10 kHz and also limits the transient response. In addition there are resonances due to the cone and the crystal.

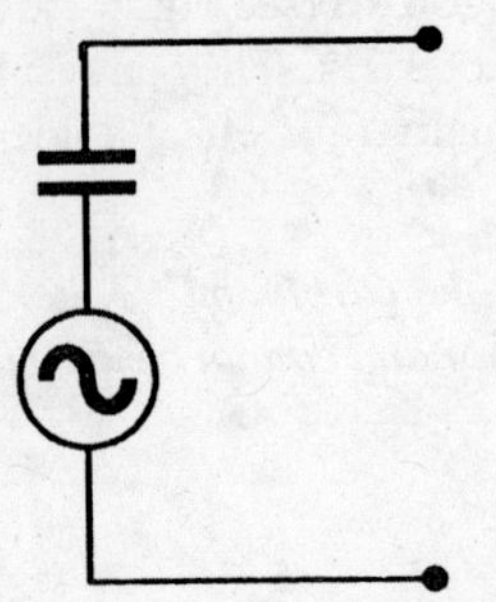

Fig. 5.9 Equivalent circuit of a crystal microphone consisting of an AC source in series with a capacitor.

With some of the better microphones the cone does not actuate the crystal directly but through a *cantilever.* According to the dimensions involved, the effect of the mass of the crystal and the mechanical resistance offered by its stiffness can be reduced, but at the same time so also is the amplitude of the transmitted vibrations, hence the signal output.

Another type of construction is the *sound cell* where several crystal elements are sealed together, this also being termed a *multimorph.* Here the cone is often dispensed with, the sound pressure waves acting *directly* on the crystal. Output is lower with this arrangement, but the frequency response is better and also the cone resonance is eliminated.

There is no DC path through a crystal microphone, the crystal being an insulator. Having the two electrical contacts on either side of the slice, the unit behaves as a capacitor. The *equivalent circuit* then, consists of a voltage source in series with a capacitor, Fig. 5.9. Capacitance values vary, but around 1,000 pF (0.001 μ F) is typical. This should be taken into account when considering *cable requirements.*

The crystal microphone is the type most widely used in lower cost installations. It has a relatively *high output level* and a *high impedance.* A long cable will reduce the output voltage available from a crystal microphone and may affect its high frequency response.

The *output level* of this type of microphone is usually between - 48 dB and - 60 dB. Their output impedance is almost always more than 100,000 ohms.

Good units may have a *frequency response* substantially flat between 50 and 10,000 Hz. Units are also available with slightly wider frequency response ranges.

The crystal microphone is normally non-directional although a pressure - gradient crystal microphone which gives a unidirectional response pattern is also being marketed. This microphone gives excellent results.

The natural crystals, such as rochelle salt, are not very durable. They are adversely affected by humidity and high temperature; also they fracture easily when subjected to shock. If a crystal microphone is subjected to a temperature of 130 degrees, it will be rendered completely useless. Care must always be taken to avoid exposing a crystal microphone to direct sunlight for any length of time.

5.5 MOVING COIL (DYNAMIC) MICROPHONES

We now come to one of the *most common* type of transducers, the principle being used both for microphones and loudspeakers.

This type of microphone (Fig. 5.10) works on the *generator principle.* The diaphragm carries a coil of wire placed in an intense magnetic field of *constant value* (B). Movements of the diaphragm consequent upon sound pressure changes result in an EMF ($E \propto BlV$) being generated in the conductor. This EMF is proportional to the velocity of motion (V) of the conductor in the air-gap: if the microphone is to have the same sensitivity at all frequencies it is necessary for the velocity of motion of the coil, due to a sound of given intensity to be independent of the pitch of the sound.

The ends of the coil are secured to two points on the cone and from these a pair of loose flexible wires is connected to terminal tags to which the microphone cable may be connected. So the electrical replica or *signal* as it is termed, can be conveyed to an external circuit.

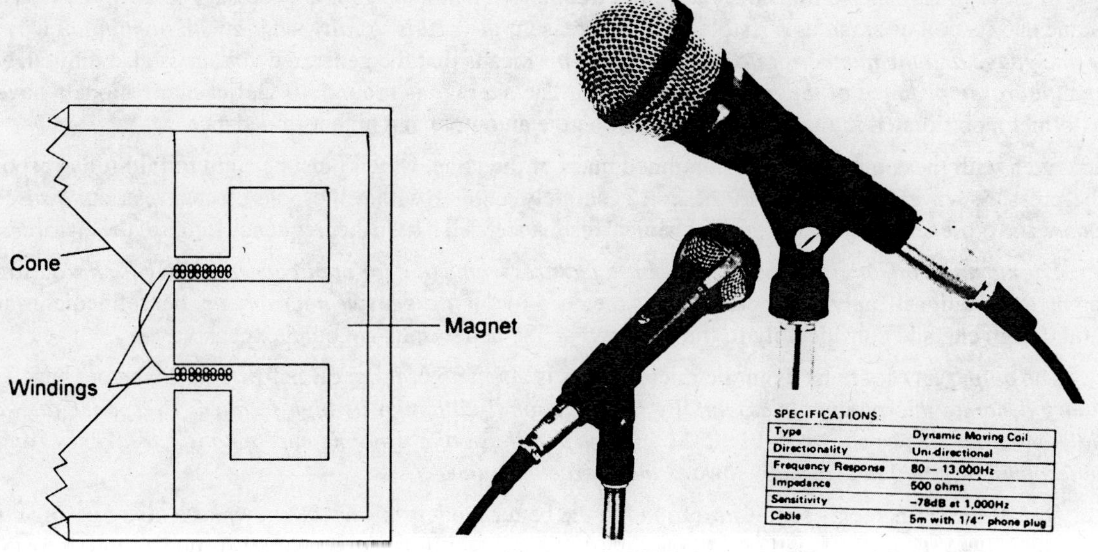

Type	Dynamic Moving Coil
Directionality	Uni-directional
Frequency Response	80 – 13,000Hz
Impedance	500 ohms
Sensitivity	–78dB at 1,000Hz
Cable	5m with 1/4" phone plug

Fig. 5.10 Moving coil microphone.

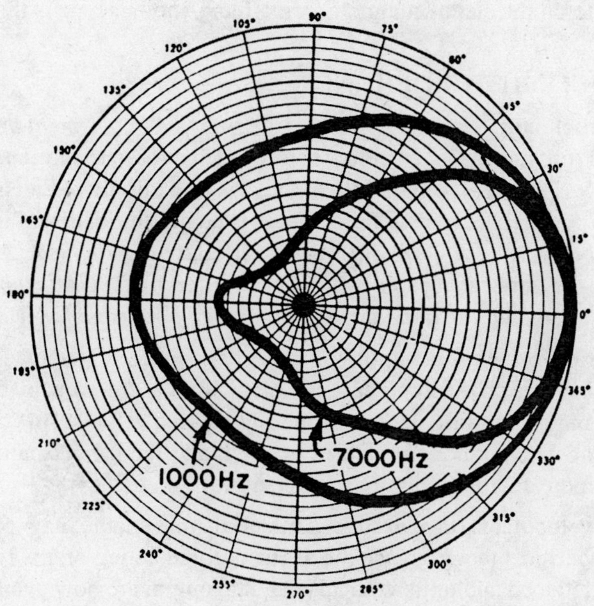

Fig. 5.11 Directivity of a dynamic microphone.

In order to respond to transients and high frequency sound waves it is necessary to keep the mass of the cone and its coil as small as possible. For this reason, *the coil is usually wound with aluminium wire of very fine gauge and the number of turns is limited,* This means that the generated voltage is also limited, and so *the natural impedance of the instrument is low.* The average is around 30 Ω, but many models have a transformer incorporated in the instrument itself to give an output at a higher impedance.

Even with these measures, the combined mass of the cone, which must be rigid to faithfully respond to the pressure waves and also keep the coil accurately centered within the pole-pieces, restricts *transient response* and more seriously places the mechanical resonance well within the frequency range of the instrument.

The moving-coil unit is reliable and robust, features which fit it for applications calling for hard usage. It can be made quite cheaply and therefore fills the need for an *inexpensive microphone*, but refinements and careful design can add considerably to the cost, and so the better units are quiet expensive.

The *output level* of most dynamic microphones is about 55 or more dB below 1 milliwatt per bar. *The ordinary dynamic microphone is essentially nondirectional, although its high-frequency response falls off rapidly on either side* as shown in Fig. 5.11. *To make full use of a dynamic microphone's frequency range, the microphone should face directly toward the source of sound.*

One of the refinements sometimes found in the better models is what is known as a *hum bucking coil.* AC operated equipment, especially that containing a mains transformer, is often surrounded by a magnetic field which oscillates at the mains supply frequency (50 Hz in India). If a moving coil microphone were used with such a field, a voltage of 50 Hz may be induced into the coil. Although this voltage would be quite small, so also is that produced by cone movement, hence it could be a significant level as compared to the wanted signal. After amplification the result would be an *audible hum.* The hum bucking coil is wound in the *opposite direction* to the moving coil and is positioned close to the moving coil capsule. Any hum field affects both the coils but as they are *connected in series and in opposite phase,* it is cancelled by the production of equal and opposite EMFs.

Moving-coil microphones are often described as *dynamic* but the term is sometimes applied to all transducers that operate on the electromagnetic principle. Another type is the ribbon unit.

5.6 RIBBON (VELOCITY) MICROPHONES

With these microphones, *the sound operates the transducing element directly without the use of a cone.* A ribbon of aluminium foil, *corrugated* to allow backward and forward motion, is mounted edge ways between two pole-pieces (Fig. 5.12). We have here, then, the classic example of electromagnetic generation, a single conductor moving in a powerful magnetic field. There is no multiplication of induced EMFs by the successive turns of a coil, *the ribbon can be regarded as a coil with but a single turn.* As a result, the output voltage is very low and so also is the impedance, something of the order of 0.1Ω. *All ribbon microphones therefore have a built in transformer to step up the impedance and voltage to a useable level.*

Even with the transformer, the output voltage for any given impedance value is still lower than that for a moving coil unit. It depends on the number of magnetic lines of force cut by the ribbon, which in turn depend on the strength of the magnet and the length of the ribbon. To achieve sufficient output, older microphones used large magnets and long ribbons which not only made them bulky and heavy but introduced ribbon resonances well within the audible range of frequencies.

More recent developments have produced materials that enable magnets to be made powerful yet light and also flared front apertures that increase the acoustic force acting on the ribbon. This has enabled ribbon and magnet size to be reduced and units with an adequate output are now available with ribbon sizes less than 1 inch in length and with a mass of less than half a milligram. *This has enabled an even smoother and extended frequency response to be obtained with freedom from resonance.*

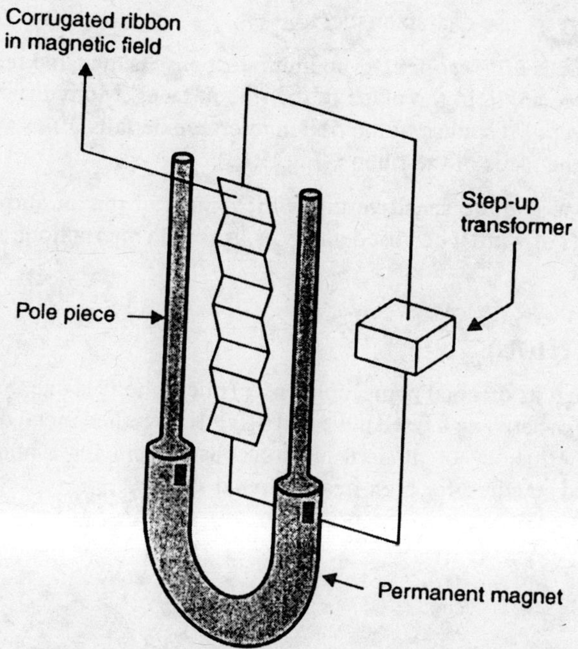

Fig. 5.12 Ribbon microphone consists of a single conductor (the ribbon) in a magnetic field and so always includes a transformer to step up voltage and impedance.

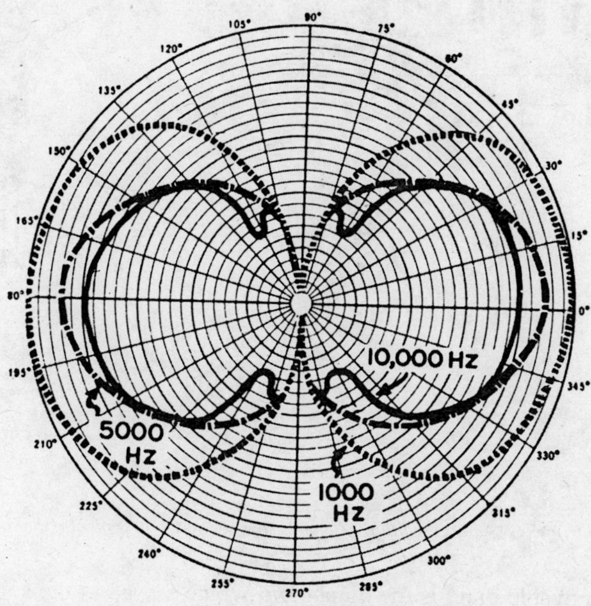

Fig. 5.13 Directivity of a ribbon microphone.

The *output level* of a velocity microphone is usually 60 dB below 1 milliwatt per bar. Generally ribbon microphones have excellent response characteristics.

The ribbon microphone is *bidirectional.* Maximum response is to sound reaching the front or back of the microphone at a 90-degree angle to the plane of the ribbon faces. More directional than the crystal and dynamic microphones the overall response of the ribbon microphone falls off as the angle of sound reaching it varies from 90 degrees to the faces of the ribbon (Fig. 5.13).

The ribbon microphone is quite sensitive to the movement of the air surrounding it, and it must be carefully protected from puffs of wind when used outdoors. A ribbon microphone should be at least 18 inches from the source of the sound.

5.7 CAPACITOR MICROPHONES

A capacitor (or to give it its original name, condenser) microphone is one that depends for its operation on the *variation of capacitance* between a fixed plate and a tightly stretched metal diaphragm. Its development represents a milestone in the history of modern electroacoustics and for a number of years this type of microphone was the accepted standard for high-quality sound systems.

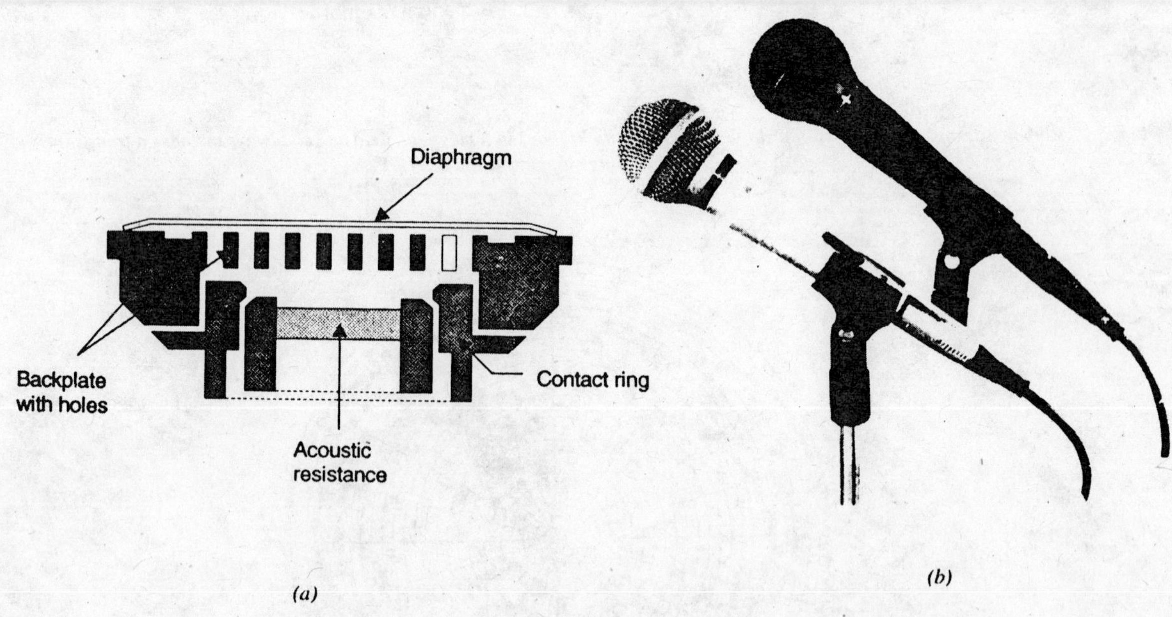

Fig. 5.14 (a) AKG Capacitor Microphone - cross section. (b) Typical capacitor microphones.

In this case the movable plate is the diaphragm which serves as the sound *pressure sensing element.* In order that it will respond well to transient and high frequency sound waves the diaphragm is made very *light* by forming it from a thin film of plastic material coated with a fine layer of metal. Aluminium is somtimes used for its lightness, but gold is also used for *anticorrosion properties* in the more expensive instruments. The diaphragm is sometimes embossed with a *pattern of shallow troughs,* which by varying the depth, width and number can be made to possess any degree of elasticity. Areas of stress are also thus reduced.

The fixed plate is mounted rigidly behind the diaphragm, and, in order to achieve as high a capacitance as possible very close to it. *Capacitance* can vary from 5-75 pF, but between 20-30 pF is most common.

There must be adequate passage of air behind the diaphragm, if the microphone is to work as a *pressure-gradient transducer,* so the whole area of the diaphragm cannot be mated with a corresponding area of backplate, while with others the plate is of lesser diameter than the diaphragm, thus giving an air gap around its edge. These passages open into the air chamber behind the fixed plate and from there through an acoustic-resistance element to the side vents.

For the capacitor microphone to function a voltage must be applied across it, and this also has the effect of pulling the diaphragm back towards the fixed plate by means of electrostatic attraction. Thus it is kept *taut* and *rigid*, a necessary characteristic if it is to respond faithfully to the incoming pressure waves, yet without any penalty in the form of increased mass.

The total capacitance of the unit is small, and therefore *the capacitance variations are minute.* In order to produce current flows of usable proportions, the applied voltage must be high. In some cases it can be over 100 V, but around 50 V is common. Some models are designed to work well down to 9 V, and thus receive power from a standard radio battery. These will work with even lower voltages but the sensitivity tends to fall off.

Different voltages will give rise to different diaphragm tensions and sensitivities. As the acoustic resistance in the rear access passage is fixed in value, it follows that any such diaphragm variation will produce a difference in the way it responds to a *pressure gradient* between front and back. Thus the polar diagram will be modified. *Some models make use of this principle to achieve a wide choice of polar diagrams with the same microphone, ranging from omnidirectional, through figure-of-eight and caridiod to hypercardioid merely by altering the polarising voltage.*

To conclude, the output level of the capacitor microphone is extremely low, and a high gain amplifier must be used with it. The amplifier should be mounted directly at the microphone, usually right in the microphone case. The capacitor microphone has very excellent frequency response and low distortion. *Because of the necessity of mounting an amplifier at or in the* microphone *case, the capacitor microphone is not recommended for ordinary high fidelity (hi-fi) work.*

5.8 ELECTRET MICROPHONES

In electret microphones *a permanent electrostatic charge is implanted into the metallised diaphragm, thereby eliminating the need of an external high voltage source.*

One way of doing this is to put the diaphragm sheet between the plates of an air-spaced capacitor which is then charged upto a high voltage. The sheet is then heated and allowed to cool with the charge still maintained across it. When it is removed, it has a permanent static charge which is the equivalent of an applied voltage of around 100 V. *This may be considered on a par with a permanent magnet which retains its magnetism after the energising field has been removed.*

Any electrostatic charge will *slowly leak away* due to the fact that there is no material yet known that is a perfect insulator. The charge in the diaphragm will also gradually diminish and so also the effectiveness of the microphone. Its life is measured like that of radioactive substances, as time taken to fall to half its previous level, the *half-life* as it is usually termed. *The maker's of electret microphones claim an expected half-life of between 100 to 1000 years, so users need have no worries on that score !* Conditions of *high humidity* though, may well accelerate the deterioration.

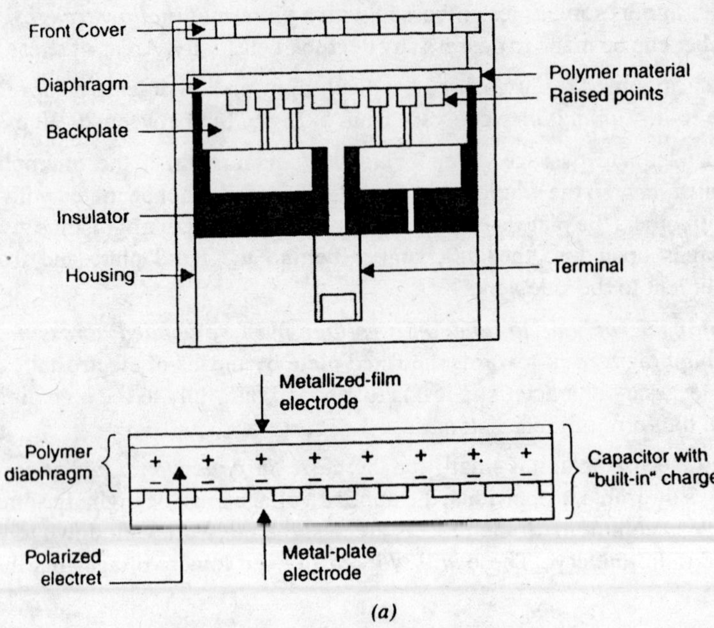

(a)

(b)

Fig. 5.15 (a) Construction details of electret microphone, (b) Electret units containing preamplifier and battery.

A *self-contained amplifier* is still necessary because the output impedance is of a similar order to that of the conventional capacitor microphones. This can be powered from a small battery cell. As the current taken by the amplifier is very low, just a fraction of a milliamp, it will have an *average life* of several thousand hours. This is in contrast to the battery life of capacitor microphones using DC converters to supply the polarising voltages which is in the lower hundreds.

Since their introduction, electret units have become very popular as they give reproduction approaching that of the conventional capacitor microphone but without the expense and complication of the external power unit. *Most of the built in microphones fitted to cassette recorders are of this type.*

5.9 GUN MICROPHONES

The *necessity* of having a microphone within a short distance of the sound source can in some cases be an inconvenience. In particular *with film and television work,* it is essential that the microphone is out of camera range, but there are many other applications where it is not possible to get very near to the source of sound.

An ordinary microphone will pick up sounds from quite a long distance, although the electrical output will be small because of the *square law reduction of sound intensity.* This in itself is no great drawback as extra gain can be provided in the amplifier. The real problem is that *unwanted sounds* which normally would be so weak as to pass unnoticed, now assume a significant proportion compared to the required sound. The extra amplifier gain used increases them too. When operating *outdoors* this means that distant traffic, aircraft, dogs barking, wind noise and other such sounds become obtrusive, while *indoors* reverberation from walls and ceiling becomes greater in proportion to the direct sound so giving an indistinct, distant effect.

For some while, professional engineers have been using the *gun* microphone, which is also known as the *rifle* and *barrel* microphone. *Sounds coming from all directions other than the one to which it is pointing are eliminated.*

Two types of construction can be used, and both make use of the *interference* principle.

Let us first of all imagine a single tube mounted in front of a pressure microphone but with a gap between the diaphragm and the end of the tube. Sound coming from the direction of the tube axis will pass down the tube and reach the diaphragm *simultaneously* with the pressure wave that arrives from outside the tube through the gap (Fig. 5.16 *a*). These will *reinforce* each other and so will actuate the diaphragm.

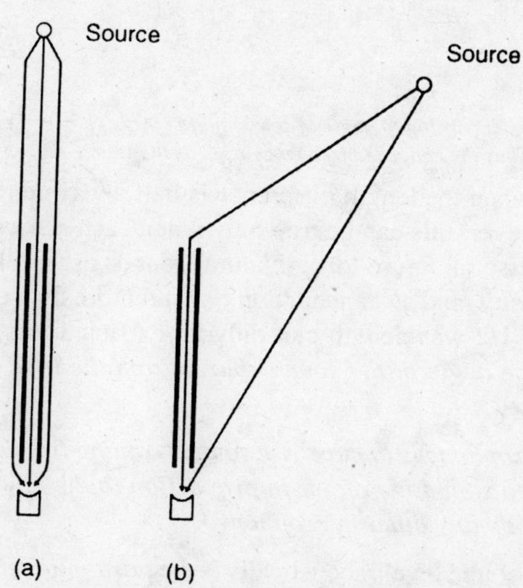

Fig. 5.16 (a) A single tube placed in front of a transducer, with a gap between them, directed toward a sound source. Sound travels down the tube to reach the diaphragm at the same time as that passing through the gap from outside. (b) When sound source is off axis, sound passing through the tube takes a longer path and so is delayed. When difference in path length equals a half wavelength of the sound there is cancellation.

Consider now a sound coming from a point well off axis (Fig. 5.16 *b*). It follows two paths, one *direct* through the gap and the other *indirect* through the tube. The latter is longer and so the sound wave following this path arrives later. The difference in length between the two paths is equal to the length of the tube in the case of a sound source at right angles to the tube axis, but is less for any angle smaller than a right angle (see Fig. 5.17).

When this difference is equal to half the wavelength of the sound, there will be *half-cycle delay* between the waves following the two paths; *the compression part of one will coincide with the rarefaction part of the other.* Thus they will be exactly out of phase and will cancel each other. The result is therefore zero pressure on the diaphragm.

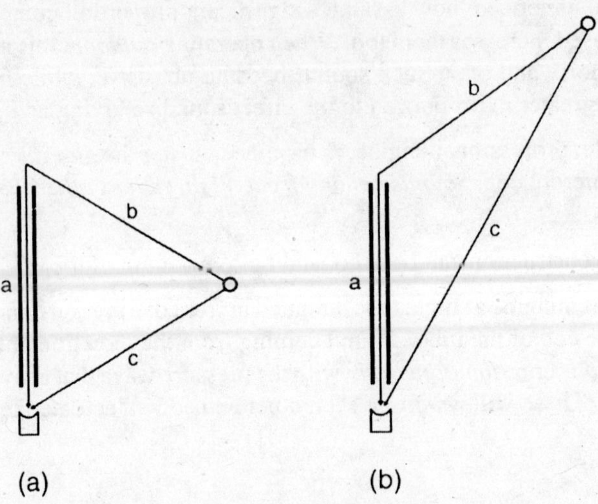

Fig. 5.17 Path-length difference varies according to angle of sound source. At (a) b = c, therefore a b - c = a; while at (b) b < c therefore a b - c < c. Thus the cancelled frequency rises as the source moves to a narrower angle.

This cancellation occurs when the length difference is half a wavelength and so also at 1 1/2, 2 1/2, 3 1/2 wavelengths and so on. However, this can be true only when each successive sound wave is identical to its predecessors, such as in the case of a pure tone. Natural sounds are rarely like this as there are constant changes of wave shape and intensity, and so cancellation beyond more than a cycle or two will not occur. In practice, the half and perhaps 1 1/2 wavelength can only be expected to suffer cancellation. *As the path difference varies according to the angle of the sound source from the tube axis, the exact frequencies also vary.*

The advantage of gun microphone is to cut down sounds coming from directions other than the desired one, which means that the wanted sound is greater in proportion to the unwanted one, but it is not actually greater than would be picked up by any other microphone.

A device which *magnifies* sound levels acoustically is the *parabolic reflector*. It works in a very similar manner to the way a concave mirror reflects light, but with additional complications. If a large concave surface is placed in the path of a sound pressure wave, the wave will be reflected and the direction of reflection will be governed by the angle of incidence between the wave and the surface.

In the case of a parabolic surface, the reflection from the whole surface area will converge on a single point which, as in the case of light, is termed the focal point. Thus the pressure is concentrated to a much higher intensity here. *A microphone is mounted facing inward to the reflector, with its cone or diaphragm at the focal point, so that most of the sound reaching it is reflected from the parabola.*

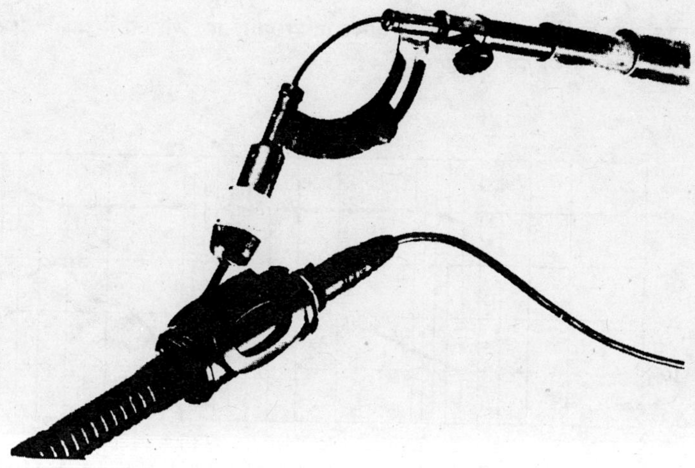

Fig. 5.18 The shotgun condenser microphone

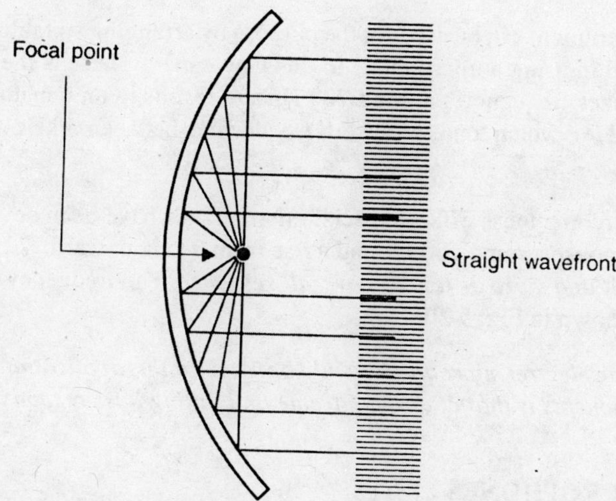

Fig. 5.19 Paths from a straight wavefront to a point of focus in front of a reflector must be of equal length so that sound arrives at the same time. This is ensured by using a parabola.

5.10 LAVALIER MICROPHONES

For many applications *mobility* is a primary requirement for a microphone. Gun microphones have some advantage over those permanently mounted on a stand as they can be used in the hand and carried around to the limit of the connecting cable length. As these can also be fitted to a stand with quick release clips, they have a versatility to which no doubt is due the popularity of this type of instrument.

A disadvantage is that *one hand of the user is always occupied and held up in an unnatural position.* This can impose a serious restriction on speakers or lecturers who may wish to demonstrate points with models or exhibits, use a blackboard or just be free to use gestures.

One answer to these problems is the lavalier microphone, which is *a small transducer suspended on the chest by means of a cord around the neck.*

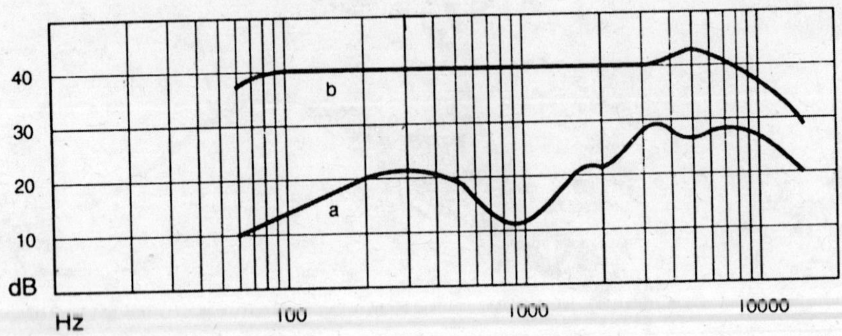

Fig. 5.20 Curve (a) shows the response of a typical lavalier microphone in free air but curve (b) gives the response when used as a lavalier. The trough at 700 Hz has equalised the chest resistance and the rising treble, the masking effect of the clothing.

Any type of instrument can be used for the purpose by arranging suitable fittings to take the cord. There are problems involved in using a microphone in this way. One of these is the *resonance of the human chest cavity* which emphasizes frequencies around 700 Hz and results in an unnatural boominess. Another is the *masking effect of clothing* which tends to absorb frequencies between 3 kHz and 10 kHz and give a muffled reproduction.

In order to overcome these effects special microphones have been developed for lavalier use. These have a dip in their response around 700 Hz and a rise from 3 kHz upward. *These characteristics balance the deficiencies and result in a more or less flat overall response.* The frequency curve in free air and that when used as a lavalier is shown in Fig. 5.20.

In this respect *an electret microphone would be ideal for this application as the frequency characteristic of most of these instruments is that of a rising treble and falling bass response.*

5.11 TIE-CLIP MICROPHONES

Yet another type of microphone with a similar purpose, but even more convenient, is now gaining popularity. This is the *tie-clip or lapel unit.* This is a tiny microphone supported by a small mount which can be clipped on to a tie, lapel, or other convenient part of clothing. One of the prime requirements for such an instrument is *small size and lightness.* This rather cuts out the moving-coil transducer as there is a practical limit to the minimum size of cone / coil assemblies, also the need for a powerful magnet makes it too heavy for this application.

Here again electret system comes to its own. These units can be made very small and are inherently light. Although they need an internal amplifier these can be formed on a tiny chip of silicon in the form of an *integrated circuit,* and so add hardly anything to the bulk or weight.

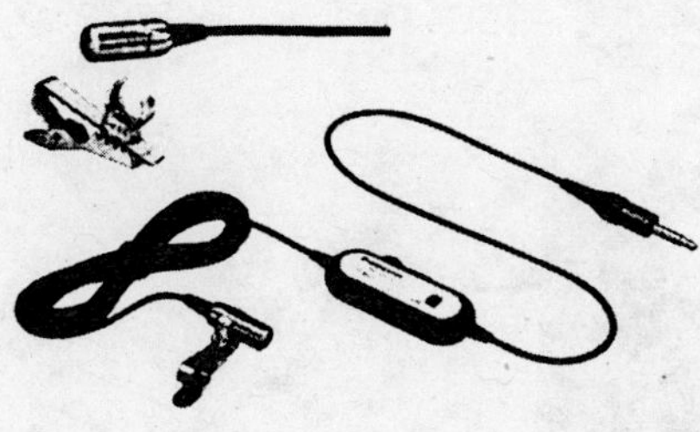

Fig. 5.21 Tie-Clip Microphone

A low voltage is needed to power the amplifier, but the current drain is very small, between 160-400 μA, so a small single cell can be used and still have an appreciable working life. Some microphones use a pen cell which is fitted in the microphone jack-plug, the current passing along the microphone cable, while others use a single mercury cell contained in the microphone itself. These are small disc-shaped cells, looking something like a *button*, and being only a few millimeters thick take up very little room. Battery life ranges from 5,000 to 10,000 hours or even more. In the case of capacitor transducers higher voltage and current is required, so this cannot be supplied from an internal battery but must be fed along the cable in some way.

As with the lavalier method of mounting, the frequency characteristic of the *electret* makes it very suitable for the purpose. *Chest cavity resonance* is less of a problem with the tie-clip or lapel unit because the microphone is not supported directly on the chest but stands off to some extent.

5.12 WIRELESS MICROPHONES

The ultimate in mobility is afforded by the wireless (radio microphone) because with this *there is no connecting cable and the user is free to move around over a distance of several hundred metres.* There are two basic types, one where the radio transmitter is contained within the casing of the actual microphone, and the other which takes the form of a slim pocket unit about the size of a wallet into which an ordinary microphone can be plugged.

The *integral microphone / transmitter unit,* Fig. 5.22, is rather larger than a normal gun microphone as batteries must be accommodated as well as the transducer and transmitter. In order to obtain sufficient power for the transmitter, the batteries are at least 9 V, but the size limits the capacity. Some 3 - 5 hours is the average life, but rechargeable batteries are often fitted to make the instrument more economic to run.

With the *separate pocket transmitter* a lavalier or tie-clip microphone can be used to give complete freedom to the user. The aerial takes the form of a short flexible lead which trails from the microphone. Usual length is a *quarter wavelength* at the permitted frequencies of the carrier wave.

The transmission is picked upon a special receiver which tunes to the frequency used and demodulates the signal delivering an audio frequency output which can then be applied to an amplifier or recorder in the normal way.

There are *fifteen frequencies* allocated for wireless microphones and all units work on any one of these. *Interference* is no problem because of the short range, it being unlikely that another user will be operating on the same frequency within about half a kilometre.

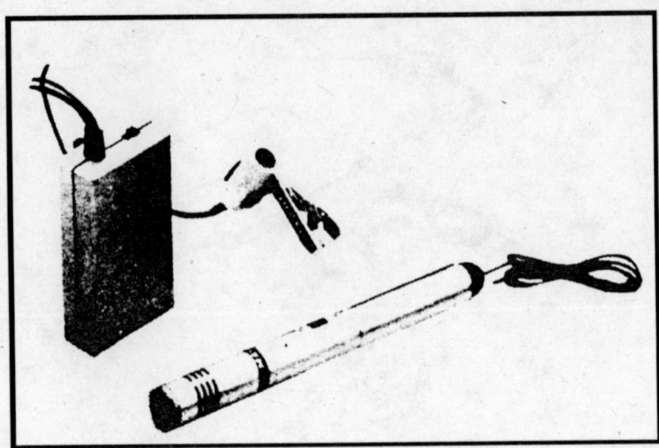

Fig. 5.22 Wireless microphone

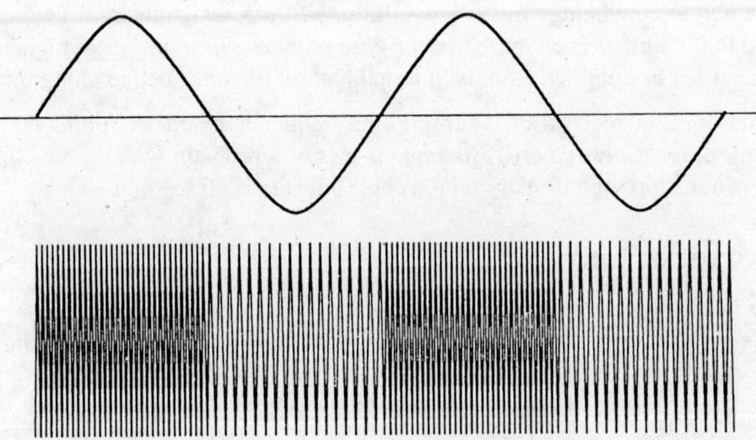

Fig. 5.23 Frequency modulation of a radio carrier (a) is the modulating signal and (b) the modulated carrier. Carrier frequency increases during one half-cycle of the modulating voltage and decreases during the opposite half-cycle.

The frequencies are in *four groups:* firstly a group with a wide bandwidth, 174.1, 174.5, 174.8, and 175.0 MHz. The second group is of narrow bandwidth, the frequencies being 174.6, 174.675, 174.77, 174.885, and 175.020 MHz. The third group is also of narrow bandwidth, being reserved for teaching deaf children in schools; these are 173.4, 173.465, 173.545 and 173.64 MHz. In addition, in certain circumstances, the frequencies of 174.65 and 174.95 MHz are allocated for communication on work sites. An ordinary F.M transmitter will not pick up wireless microphone transmissions.

The *narrow bandwidth* specification is for a deviation of ± 20 kHz, and is suitable for most speech applications. The *wide bandwidth* allocations allow a deviation of ± 75 kHz and give the better quality reproduction required by stage and cabaret artistes.

The transmitter *output power* must not exceed 50 mW in the case of narrow-band transmitters, and 10 mW with the wide band units.

Certain specifications also apply to the receiver. *Signal -to-noise ratio* must be better than 30 dB and *selectivity* such that a signal with a deviation of ± 10 kHz, 70 kHz away from the wanted signal in the case of narrow band receiver, and with a deviation of ± 2.5 kHz, at 200 kHz away from the wanted signal in the case of wide band receiver will not produce an increase of noise plus unwanted signal of more than 3 dB in the output.

An *interfering signal* of 3 mV should not give a signal in the output greater than 10 dB above noise level in the case of wideband receiver and 20 dB above with that of the narrow band unit.

It is possible for any receiver to generate and radiate a signal from the local oscillator which is part of the superheterodyne circuit universally used. The specification stipulates that *any such signal radiated from the receiver's aerial should not exceed 2.5μ W at any frequency.*

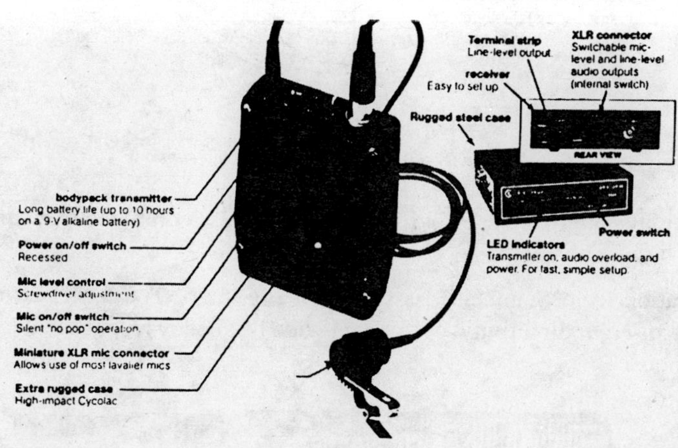

Fig. 5.24 Features of a wireless microphone.

These stringent requirements serve to protect the users from the effects of jamming and interference which undoubtedly would be common if such control were absent.

5.13 DUAL - UNIT MICROPHONES

Microphones are available which make use of two units to secure a particular directional pattern. A dynamic unit is often combined with a ribbon unit. Fig. 5.25 shows the directional pattern which results when a *bidirectional velocity unit* and a *nondirectional dynamic unit* are combined.

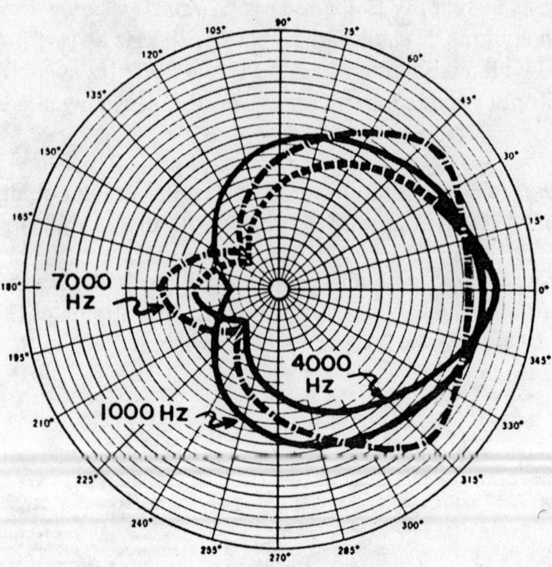

Fig. 5.25 Directivity pattern of a dual - unit microphone.

The resultant directivity pattern is a cardioid. Other units are also combined to secure specific directivity patterns.

A superb combination from Sennheiser is shown in Fig. 5.26. The MKH 30 is a pressure gradient microphone with a *figure-of-eight* directivity, optimising wide frequency response lateral sound rejection and extremely low-inherent noise.

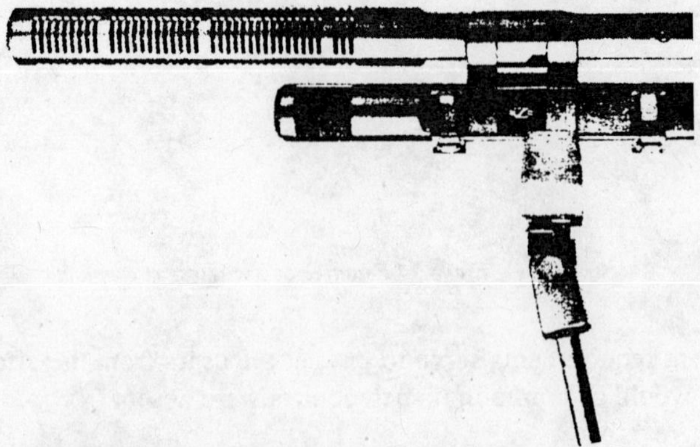

Fig. 5.26 A superb combination from Sennheiser.

The MKH30 is matched with the remarkable directivity of the MKH60, a *supercardioid* microphone, to enhance the stereo-image. Low frequency ambience and vibration pick-up is minimised by efficient roll-off-filters.

Each microphone can be used *separately* when stereo operation is not required.

Suggested applications of different types of microphones are given in Table 5.2.

Table 5.2. Microphones and their Applications.

Type	*Cost*	*Polar diagram*	*Operating principle*	*Quality*	*Suggested applications*	*Comments*
crystal	low	omnidirectional	pressure operated ; piezoelectric	low	nil	robust ; outdoor use ; high impedance
ribbon	medium/high	figure-of-eight	pressure gradient; ribbon diaphragm in magnetic field	excellent ; price a good guide	speech ; music, especially woodwind and strings	fragile ; not for outdoor use ; low impedance
moving coil	medium/high	omnidirectional or cardioid	pressure operated ; coil in magnetic field	can be very good	speech ; music, especially brass, bass drum, double bass	robust ; outdoor use; low or high impedance
capacitor	very high	omnidirectional, cardioid or figure-of-eight; can be switchable	pressure operated ; diaphragm varies capacitance ; polarising voltage	excellent	speech ; music, espcially brass and percussion	fairly robust ; outdoor use ; norm. low impedance

5.14 ELECTRICAL, MECHANICAL AND ACOUSTICAL ANALOGS

Any acoustical system can be represented in terms of an equivalent electrical or mechanical system. This is shown in Fig. 5.27.

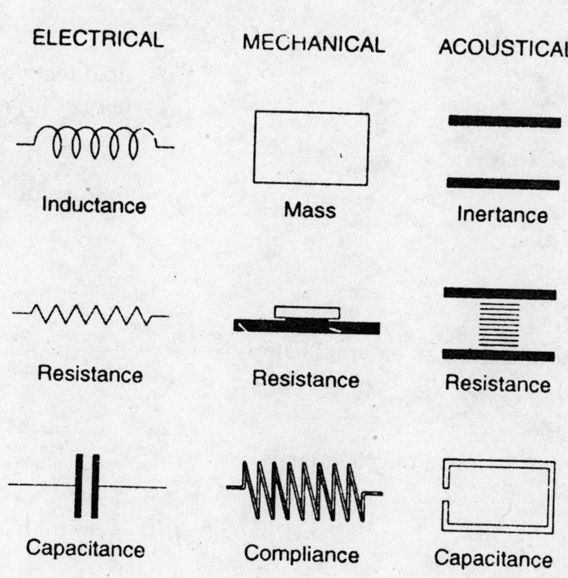

Fig. 5.27 The three basic elements of electrical system and their analogs in mechanical and acoustical systems.

The physicist freely uses these equivalents in setting up his mathematical approach in analysing a given system. For an example, the effect of a cabinet on the functioning of a loudspeaker is clarified by thinking of the air in the enclosed space as acting like a capacitor in an electrical circuit, absorbing and giving up the energy imparted by the cone movements.

Fig. 5.27 shows the graphical representation of the *three basic elements* in electrical, mechanical and acoustical systems. *Inductance* in an electrical circuit is equivalent to mass in a mechanical system and inertance in an accustical system. *Capacitance* in an electrical system in analogous to compliance in a mechanical system and *capacitance* in an acoustical system. *Resistance is resistance in all three systems,* whether it be frictional loss offered to air particle movement in glass fiber, frictional losses in a wheel bearing, or resistance to flow of current in an electrical circuit.

OBJECTIVE TYPE QUESTIONS (A)

1. Microphones are electroacoustic transducer's which convert
   (*a*) acoustic energy to electrical energy
   (*b*) electrical energy to acoustic energy
   (*c*) acoustic energy to mechanical energy
   (*d*) mechanical energy to acoustic energy

2. The unit of sound pressure used for rating microphones is
   (*a*) pascal (*b*) bar
   (*c*) watt (*d*) decibel

3. The polar diagram of a true omnidirectional microphone is
   (*a*) perfect circle (*b*) figure-of-eight
   (*c*) cardioid (*d*) none of the above.

4. Moving coil microphones are
   (*a*) active transducers (*b*) unidirectional
   (*c*) both (*a*) and (*b*) (*d*) neither (*a*) nor (*b*)

5. Moving coil microphones are
   (*a*) omnidirectional (*b*) unidirectional
   (*c*) neither (*a*) nor (*b*)

6. Carbon microphones
   (*a*) produce an EMF
   (*b*) modulate the current from an external battery
   (*c*) both (*a*) and (*b*)
   (*d*) neither (*a*) nor (*b*)

7. The natural impedance of a ribbon microphone is
   (*a*) very low (*b*) low
   (*c*) high (*d*) very high

8. The ordinary dynamic microphone is
   (*a*) omnidirectional (*b*) unidirectional
   (*c*) non directional (*d*) either (*a*) nor (*b*)

9. The polar diagram of a capacitor microphone can be modified by
   (*a*) varying the polarising voltage
   (*b*) varying the pressure gradient
   (*c*) both (*a*) and (*b*)
   (*d*) neither (*a*) nor (*b*).
10. The half-life of electret microphones varies from
   (*a*) 100 to 250 years
   (*b*) 250 to 500 years
   (*c*) 100 to 1000 years
   (*d*) 250 to 1000 years
11. The construction of gun microphones make use of the principle of
   (*a*) diffraction
   (*b*) interference
   (*c*) reflection
   (*d*) refraction
12. A device which magnifies sound levels acoustically is the
   (*a*) lavalier microphone
   (*b*) gun microphone
   (*c*) parabolic reflector
   (*d*) ribbon microphone

FILL IN THE BLANKS (B)

1. For most high quality microphones impedance is ______________.
2. Cassette recorders of good quality are not, as a rule, designed to take ______ impedance microphones.
3. Microphones do not respond equally to ______coming from all ________.
4. An ____________ microphone is most commonly used for general recording.
5. An inset pattern of carbon microphone is in general use for ________.
6. The carbon microphone generates a continuous ______________.
7. The double crystal unit is called a ____________.
8. The equivalent circuit of a crystal microphone consists of a ________ in series with a __________.
9. One of the refinements in better quality dynamic microphones is ________.
10. The output voltage of a ribbon microphone is __________ and so also is its __________.
11. In order to produce large capacitance variations the __________ must be high.
12. A self-contained amplifier is necessary with ________ microphones.

CONVENTIONAL QUESTIONS (C)

1. What are the characteristics of microphones ? Explain.
2. Explain the significance of a polar diagram.
3. With the help of a diagram explain the working of a dynamic microphone.
4. How does a ribbon microphone differ from a dynamic microphone ?
5. How does an electret microphone differ from a capacitor microphone ?
6. What are the limitations of crystal microphones ?
7. Explain the significance of a hum bucking coil.
8. In which microphones is a built-in transformer a necessity and why ?

6

HEADPHONES

Headphones are interesting devices. Headphone listening has many adherents. From a strictly utilitarian point of view there are many instances when Headphones are convenient, if not indispensible. Aesthetically, they can be regarded either as annoying tethers or as entrances to a world of private involvement with music more intimate than any loudspeaker can provide. Basically antisocial devices, they nonetheless bring peace of mind to many who otherwise find listening to music incompatible with household activities and noise. Headphones can make or break a multi-image presentation, business theater or multi-camera video production. When it operates as it should, it's largely a forgotten piece of equipment. But when it breaks down, especially during a program, it instantly causes the biggest headache any production person can ever have.

6.1. HEADPHONES AND HEADSETS

While a *headphone* Fig. 6.1 is defined as a listening device consisting of either one or two earphone receivers and a headband to hold them in place, a *headset,* Fig. 6.2, is essentially the same thing with a microphone attached. Headphones permit *one-way* communication. Headsets permit *two-way* communication.

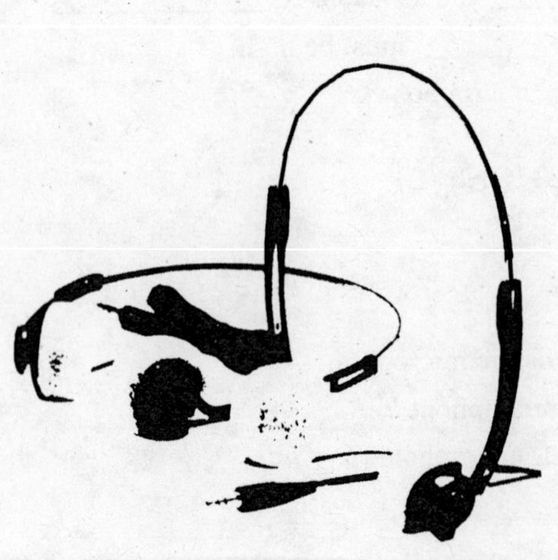

Fig. 6.1 Headphones permit one-way communication.

Fig. 6.2 Headsets permit two-way communication.

Particularly since the introduction of *stereo,* the use of headphones has become an important part of the high fidelity (hi-fi) scene, in some cases by choice and in others of necessity.

In the early days of wireless, headphones often described as *earphones* - provided the most practical way of listening to the then available signals. From 1923 onwards, with the emergence of public broadcasting and the availability of more powerful receivers, loudspeakers rapidly gained favour especially for family listening. Headphones, however, were still widely used with professional and communication equipment until well after World War II.

6.2. TYPES OF HEADPHONES

Headphones come in a vast assortment of shapes, sizes and types. Advertisements emphasize distinctive features even further, to the point where it is understandable that consumers get a little confused about the importance of features and forms.

In headphones, as in almost anything else, most of the differences exist merely as evidence of the manufacturer's struggle to be distinctive, to market unique products.

One can identify *three basic classes* of headphones: (1) supra-aural (2) circum-aural, and (3) intra-aural

(i) Supra-aural : Those that sit on the ears, usually pressing the external ear to the side of the head. Old versions of supra-aural headphones presented an uncomfortably hard bakelite or hard rubber surface to the ear. Nowadays *soft cushions are the rule*, some are foam or liquid filled and attempt an *airtight seal* to the irregularly shaped external ear. Others are fitted with large pads of acoustically transparent plastic foam and are designed in such a way that an airtight seal is not necessary.

(2) Circum-aural: Those with cushions that fit around the external ear, pressing directly against the head itself. In the best examples of circum-aural headphones the external ear is not deformed through contact with the unit.

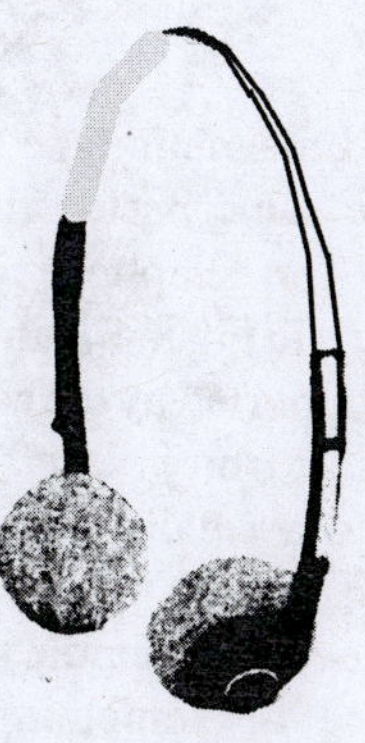

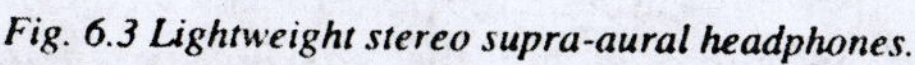

Fig. 6.3 Lightweight stereo supra-aural headphones.

Fig. 6.4 The HD 250 from Semheiser are of the circum-aural type.

(3) Intra-aural. These are in-the-ear or *insert earphones* of the type found in some aircraft, almost all hearing-aids and that are supplied for private listening to portable radios, tape recorders and TV's.

These classes describe headphones by their construction and how they are worn, but they tell us little about factors important to stereo listening.

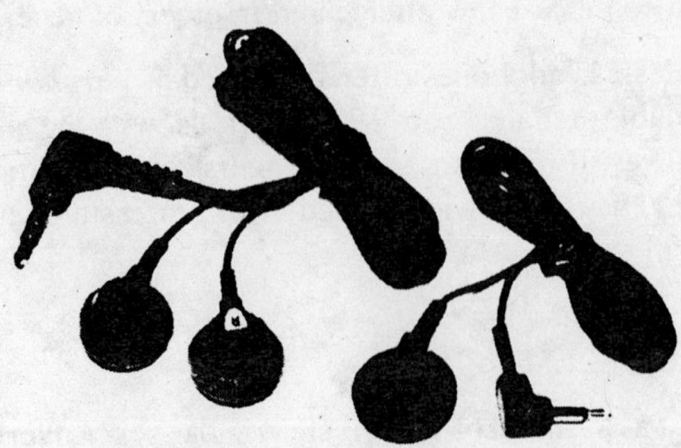

Fig. 6.5 Intra-aural (in-the-ear) earphones.

One important distinction among headphones can be made on the basis of whether they exclude external noise, and by the same token, whether they contain the musical sounds and prevent their leakage to the outside world. The former is important for satisfactory listening in a noisy environment; the latter might be important in hospitals, libraries, or perhaps at home, where even the low level leakage of music can be a distraction to someone else.

In order for a headphone to have *effective acoustic isolation* it must provide.

(*a*) an *airtight seal* with the external ear (supra-aural) or the head (circum-aural) and

(*b*) a *totally enclosed cup* over the ear, made of a rigid and relatively dense material.

These requirements eliminate headphones with electrostatic or electromagnetic membrane drivers (open backs) and porous cushion types (no airtight seal). Insert (intra-aural) earphones that fit the ear canal singly can be very effective at acoustic isolation.

6.3. MOVING IRON HEADPHONES

The so-called magnetic type (moving iron) headphones use the same basic principle as was originally adopted for telephone ear pieces or 'receivers.' As shown in Fig. 6.6, a thin flexible soft iron *diaphragm* is supported by the casing just clear of the tips of two soft iron *pole-pieces* which are attached, in turn, to a small *permanent magnet*. Magnetic attraction causes the diaphragm to be distended slightly towards the pole tips, while still leaving a clearance of something less than a half-millimetre.

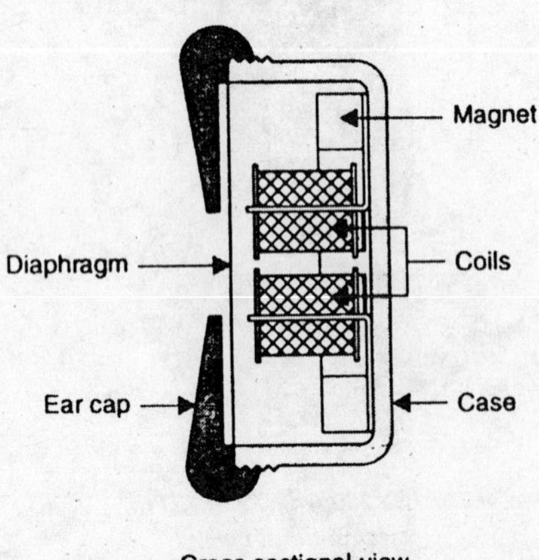

Fig. 6.6 While refined over the years, moving iron headphones still use the same principles as seen in the earliest telephone 'ear pieces'. They have no place in today's hi-fi scene.

The incoming audio current passes through coils wound around the pole-pieces. In the presence of an audio signal the *varying magnetic field* created by the current *interacts* with that from the *permanent magnet*, varying the *pull* on the diaphragm and causing it to *vibrate*, thereby *generating* corresponding sound waves.

Because of the inherent *mass and stiffness* of the metal diaphragm, headphones of this type suffer similar *limitations* to the old horn - type magnetic loudspeakers; a *natural resonance* at a few hundred Hertz, resulting in the so-called 'metallic quality.' They also exhibit quite *high harmonic distortion* due to the fact that the diaphragm drive is one-side or unbalanced.

6.4. CRYSTAL HEADPHONES

In the 1950s and '60s an assortment of crystal earphones appeared on the market, being essentially miniaturised counterparts of the crystal (or *piezoelectric*) loudspeakers.

Involving few metallic components, they could be made very small and light, as well as lending themselves to mass production. Many, infact, were designed to plug directly into the ear canal, being intended for use with portable radio sets and tape players. A few special quality models have been released over the years for monitoring situations but in general, *crystal headphones have slotted into a utility role rather than as a medium for hi-fi listening.*

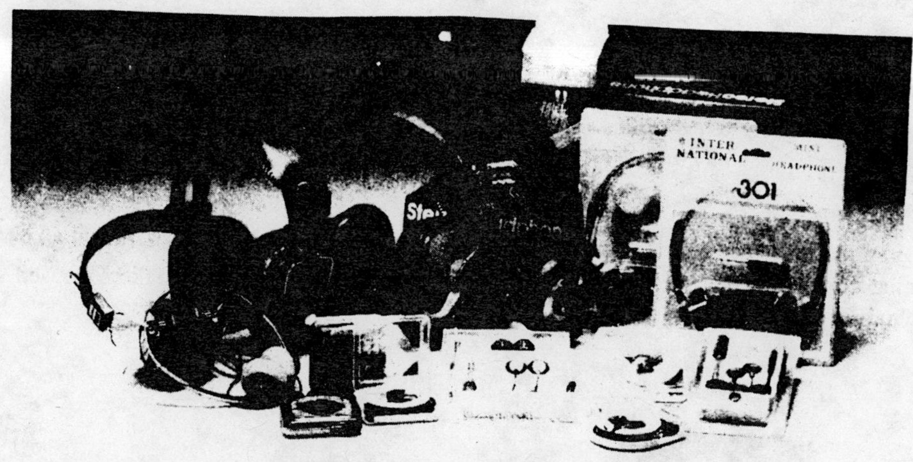

Fig. 6.7 Headphones and earphones in a wide variety.

As with crystal loudspeakers, *their impedance is predominantly capacitive,* measuring many thousands of ohms at low and mid frequencies. This may not present a problem where sufficient signal drive voltage is available, but *with ordinary low-impedance drive circuits, their effective sensitivity is likely to be poor compared with magnetic units.*

6.5 DYNAMIC HEADPHONES

The widespread adoption of dynamic or *moving coil* - loudspeakers in the 1930s raised the question as to whether it would be possible to use the same principle for headphones. If the *drive units* could be made small enough and light enough, they would presumably offer equivalent frequency response and much reduced distortion.

Moving coil earphones designed in 1936 and released in 1937 were discontinued in 1939 as too cumbersome and too expensive to justify continued production. But by 1967 twenty seven firms had entered the market and one-time descriptions like *heavy, cumbersome, and metallic* had also given way to *lightweight comfortable and high quality,* with headphone listening rapidly becoming an accepted part of the hi-fi stereo scene.

Fig. 6.8 Beyerdynamic headphones DT-990 with a very low mass of the diaphragm and moving coil assembly.

Because it was the most accessible approach, many of the dynamic headphones introduced to the post-war hi-fi market comprised, essentially, of two miniature (50-70 mm) loudspeakers housed in shells rather like the two halves of an electromagnetic unit egg.

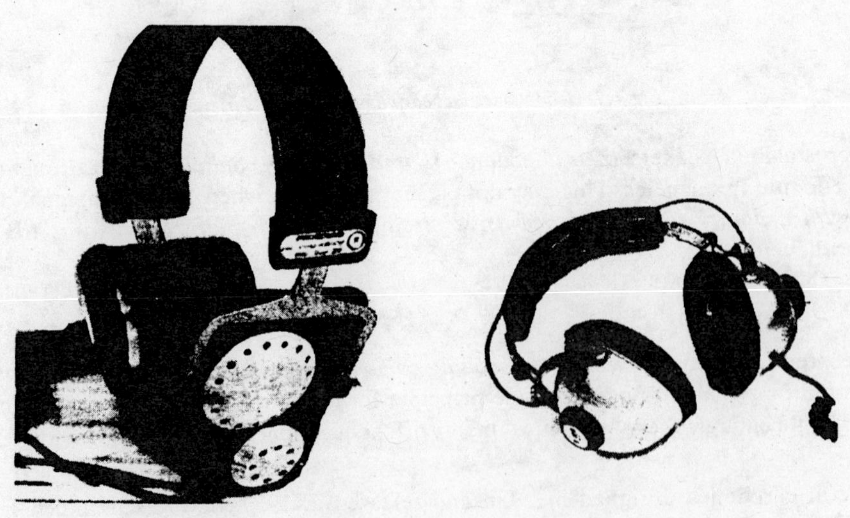

(a) (b)

Fig. 6.9 (a) Dynamic headphones with twin drive units in each ear piece (b) AKG Dynamic headphones.

Typically, the oval shells of a pair of middle - aged *Japanese-made Golding dynamic headphones* measure about 10 cm tall, 9 cm wide, and 5 cm deep. They are fronted by a gold anodised *grille* with two matching *pressure relief ports* at the rear. The *shells* are cushioned with a removeable pad, while the *headband* is adjusted and padded. While bulky and by no means light, the treatment gives them a soft and friendly feel.

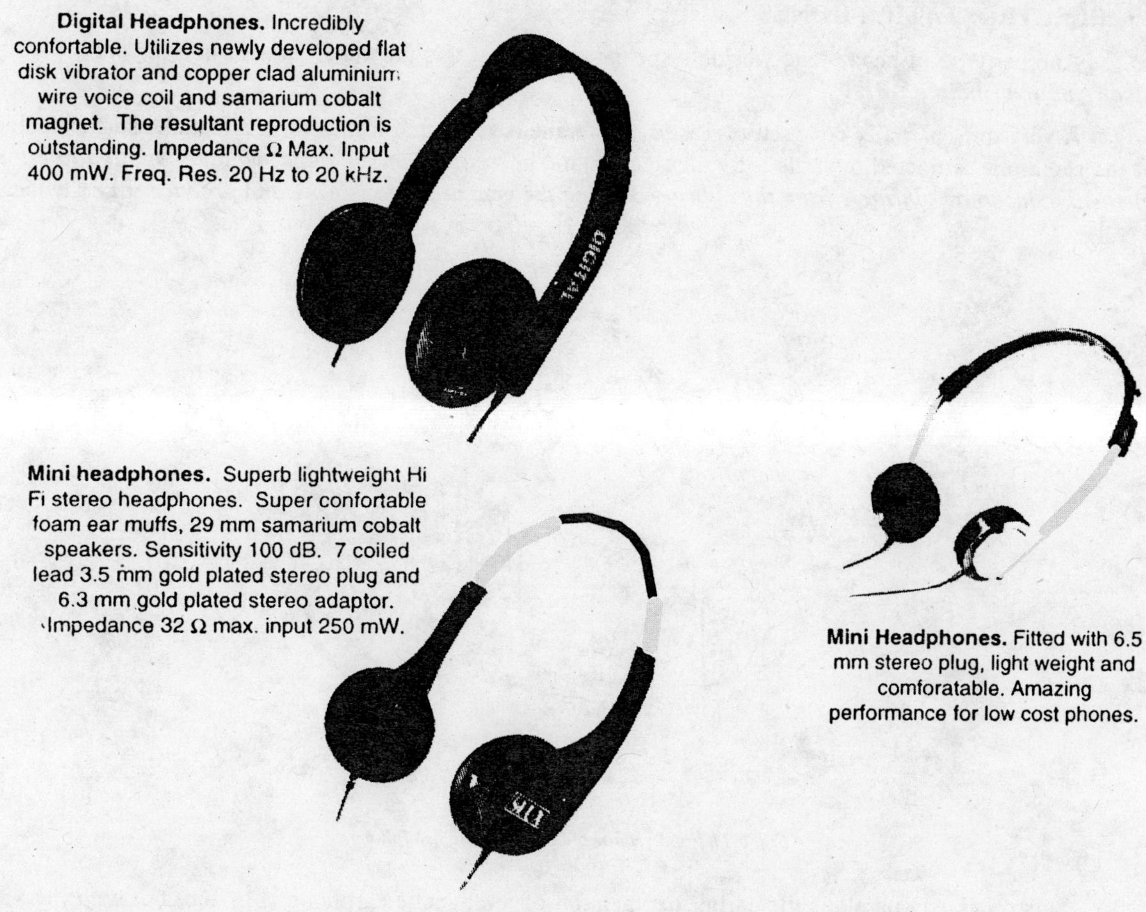

Fig. 6.10 Samarium cobalt headphones.

One can reasonably expect that equivalent models from reputable sources would incorporate drivers with magnets, cones and voice coils selected for the role and *with the shells physically styled and acoustically treated for the best available end result.* With outside noise largely excluded, and operating so close to the ears, the drivers can be expected to produce an ample *sound pressure level* (SPL) with only milliwatts of input thereby *minimising cone break-up and non-linear distortion.*

Being damped by both the front and rear air loading, cone resonance effects are held in check, flattening frequency response. Subjectively, *high frequencies are projected directly into the ear canal while the captive air inside the cushion surround also conveys low frequency pressure waves efficiently to the eardrums.* Most such headphones are capable of sound quality far better than is available from traditional magnetic earphones.

While any number of such headphones are still in use, many people find them too cumbersome for prolonged listening and in a warm environment, too hot around the ears. *Now-a-days most stereo headphones are much smaller and lighter, using specially developed uitra-compact drivers. Some retain paper based cones; others retain a dome type diaphragm similar in configuration to the squawker and tweeter loudspeakers. A few up market models include their own diminutive tweeter.*

6.6 ELECTROSTATIC PHONES

Another type of headphone warrants special mention at this juncture - those using the capacitive or *electrostatic principle.*

A very thin, partially conductive plastic membrane is supported just clear of a metallic surface being, at the sametime, attracted towards it by an electrostatic charge. *An audio signal voltage, applied to either element, adds to or subtracts from the charge, causing the membrane to vibrate and produce sound waves.*

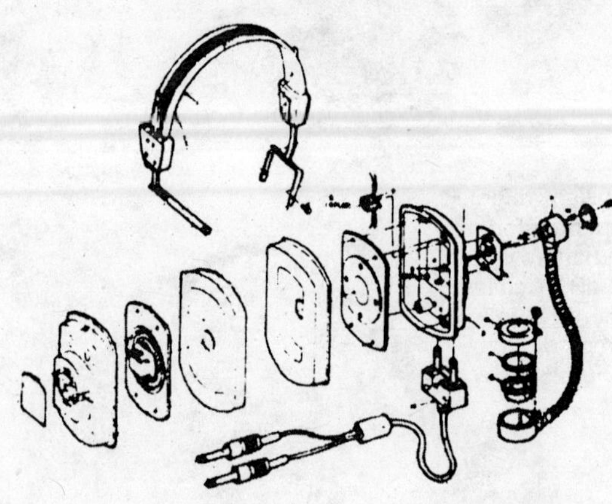

Fig. 6.11 Field repairable electrostatic headphones.

This obviously parallels the earlier explanation of a magnetic earphone (Fig. 6.o) but with the vital difference that *a plastic membrane can have much less mass and stiffness than a metal diaphragm, thereby minimising problems with system resonance and uneven response over the working range.* It also lends itself to a balanced configuration minimising harmonic distortion.

In early model electrostatic headphones, as with their headphone counterparts, a *high DC polarising voltage* had to be fed to the phones along with a *fairly high audio drive voltage.* This meant, in practice, that electrostatic headphones could be simply plugged into a socket provided for dynamic headphones. An *external adopter* was required, containing a mains powered high voltage supply and a step-up transformer for the audio drive signal.

6.7 ELECTRET ELECTROSTATIC HEADPHONES

In the mid 1970s *permanently polarised* or electret elestrostatic headphones made their appearance, with *Sennheiser* Models 2000 and 2002. While the need for a mains powered HT supply had been eliminated, they still need an external black box containing *twin step-up transformers,* to permit operation from the amplifier voice coil output terminals.

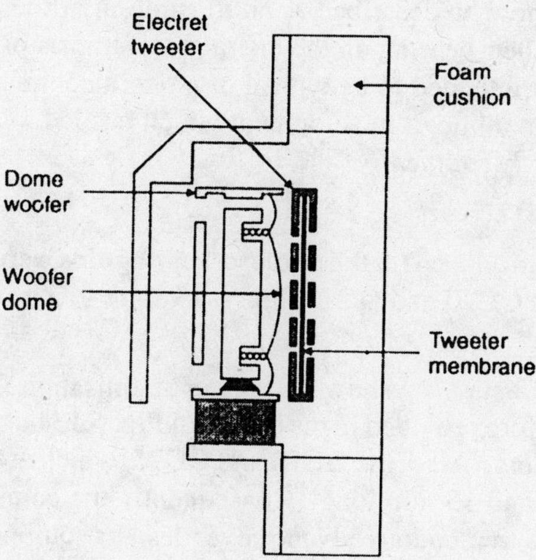

Fig. 6.12 An AKG - 4 K headphone combining a dome type dynamic woofer with an electret electrostatic tweeter, fed by a miniature step-up transformer. It can operate directly from a normal dynamic headphone feed socket.

In fact, Sennheiser made a virtue out of necessity by arranging for the adapter to feed two separate pairs of electrostatic headphones as well as providing *loudspeaker/headphone switching*. The company also included *protection circuitry* to limit the sound pressure level from the phone to 117 dB, alongwith *LED indicators* to show when the circuitry was being activated.

About the same time, *AKG* brought out its two-way K4 headphones which, despite their modest dimensions were a *two way design combining a dynamic driver and electret tweeter crossing over at 4.5 kHz.*

As indicated in Fig. 6.12, the *woofer* is deep inside the casing. Using a *highly efficient rare earth / cobalt magnet* and a *moulded polycarbonate diaphragm,* it can operate directly from a normal amplifier headphone socket. Its rated *low end* response extends to 16 Hz but, at the *top end* the body of the tweeter serves to block physically its residual output above the *nominal crossover frequency.*

The *tweeter* uses the electret principle, with a 12.5 micron chromium - sputtered foil optimised for the frequency range above 4.5 kHz. The higher drive voltage required is provided by *a* 1:5 step-up transformer, which can be made extremely small by virtue of the fact that it needs to handle only frequencies above 4.5 kHz, falling off steeply below that.

For the design, AKG claimed the best of both worlds-*the ability to operate from a normal headphone drive circuit and the reputed sonic transparency of an electroacoustic transducer for the top end.*

6.8. HEARING AIDS

The OC57, OC58, OC59 and OC60 form a range of transistors specially developed for hearing aids. *These transistors are cylindrical in shape and are only 4 mm long and 3 mm in diameter.* They are already available to the industry and will be generally available in due course. These types are in keeping with the general trend towards miniaturisation of the components of the hearing aid.

Using these transistors, a complete hearing aid can be mounted in spectacle frames, or as a *clip-on unit* concealed behind the ear or in the hair or clothing.

Three hearing-aid circuits will now be described as an illustration of the general design principles of audio amplifiers, and in particular for their bearing on the operation of transistors *at low levels* and *from low supply voltages.* These circuits are not intended to be typical of modern commercial practice. The first two circuits are *RC coupled* ; they represent a line of development which has led to extremely compact designs, such as the thrid circuit, which is *directly coupled.*

1. Four-Transistor RC-coupled circuit

Before special transistors were introduced for the purpose, hearing aids were designed using the standard small-signal a.f. transistors, that is, the OC70 and OC71. Fig. 6.13 shows a circuit of this type, in which the supply voltage is 2.4 V.

The usual considerations of D.C. stability show that better stabilisation is required at lower collector supply voltages. Good stability is therefore provided in the circuit of Fig. 6.13 but this necessarily means using a large number of resistors and capacitors. Also, the OC70 and OC71, which are approximately cylinderical in shape, are about 5 mm in diameter and 15 mm long. Consequently the complete hearing aid is *bulky by modern standards,* although of course a tremendous advance over designs equipped with subminiature valves, as regards both size and battery consumption.

Later on, some reduction in size was made possible by substituting, in this circuit, the OC65 for the OC70 and the OC66 for the OC71. The OC65 and OC66 were in a metal construction, which was about 3mm × 4 mm in cross-section and 7mm long. These two transistors were the forerunners of the OC57 series.

Demoded though this circuit is, it illustrates many of the earlier techniques and therefore will be discussed more fully than its present commercial position would warrant.

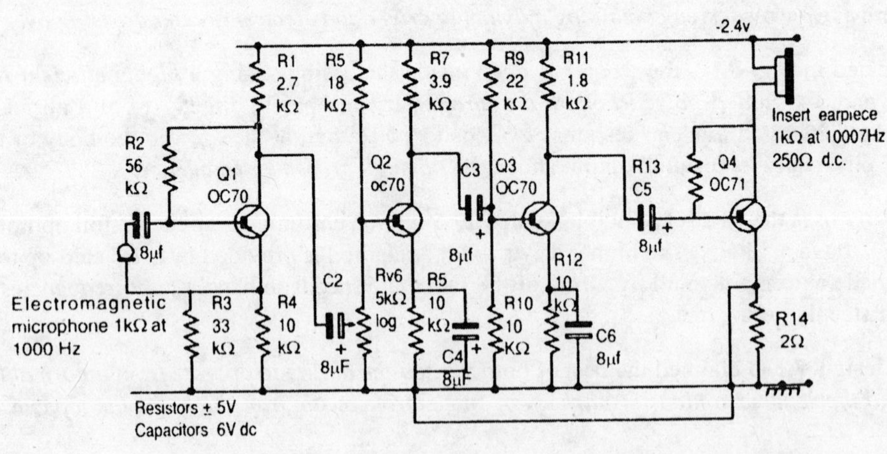

Fig. 6.13. Four-transistor RC-coupled hearing aid.

Noise.

The first two stages are operated at currents of only 0.3 mA to minimise noise. The gain control RV6 is sited between the first and second stages to ensure a *low noise level,* while keeping this control clear of the feedback loop.

Input Stage.

The input impedance of the amplifier is about 1kΩ. This matches the impedance of the electromagnetic microphones, which is specified as 1kΩ and 1000 Hz. R_2 is returned to the collector instead of the supply and provides A.C. and D.C. feedback. The unbypassed emitter resistor R4 does not contribute A.C. feedback, since the input is applied between base and emitter.

Driver Stage.

The collector current in the driver transistor Tr3 in 0.5 mA. A low collector load resistance is required in this stage to ensure *adequate swing* up to full drive without clipping.

Output Stage.

The output stage is designed round an OC71 to provide an *output* of nearly 2mW to the earpiece. Since in class A the maximum theoretical efficiency is 50% a *collector dissipation* of about 4mW is required. This is provided by a collector current of 2mA and a collector-emitter voltage of 2V, the load impedance being 2V/2mA = 1 kΩ (at 1000 Hz). The D.C. resistance of the earpiece is 250Ω, which at 2mA drops 2 × 250 = 0.5 V, and leaves nearly 2V across the transistor.

The collector current is set to 2mA by the choice of the feedback resistor R13, the value of which should be α′ times the 1kΩ load impedance. Thus for a transistor having an α′ of 47, R13 should be 47 kΩ.

R13, in addition to biasing the transistor, provides A.C. and D.C. feedback, the *D.C. stability* being statisfactory with this method of stabilisation because of the higher collector current of 2mA, and because R13 is chosen to match individual transistors.

Feedback Loop.

In addition to the A.C. feedback provided by R2 in the input stage and R13 in the output stage, the gain is *stabilised* by 12 dB of negative feedback taken over the last three stages. A feedback voltage proportional to the output current is taken from across R14 in the emitter of the output stage, and injected into the second stage in series with the bypassed emitter resistor R8. The 2Ω resistance for R14 can be made from approximately 7 in. of Eureka wire 0.0092 in. in diameter and having resistance of 10.6 Ω/yard (1 in. = 2.54 cm, 1 yd. = 36 in.).

Performance.

The *power gains* in the fours stages are 17, 21, 18 and 30 dB, giving a basic total of 86 dB. As there is 12 dB of loop feedback, the overall power gain reduces to 74 dB.

The *frequency response* of the amplifier itself is 2 dB down at 150 Hz relative to the response at middle frequencies, the electrical response being virtually flat from 1000 Hz to 7 kHz. *The overall acoustical response of a complete hearing aid would depend primarily on the microphone and earphone.*

The performance is satisfactory up to an ambient temperature of 40°C (104°F).

Total-harmonic distortion, measured at a test frequency of 400 Hz is 5% at full output.

Current drain is 3.5 mA with the correctly matched value for R13.

6.9. THREE-TRANSISTOR RC-COUPLED CIRCUIT

The use of the latest hearing-aid transistors in an RC-coupled circuit is illustrated by Fig. 6.14.

In this circuit the electrical *power gain* is about 85 dB using only three transistors. The *earpiece* should have an impedance of 650Ω and a D.C. resistance of 200 Ω, and the *microphone* should be a magnetic type of 2kΩ impedance. The loss in the microphone and earpiece is about 35 dB, so that the acoustical or *air-to-air gain* is 85 – 35 = 50 dB.

In the OC58 output stage, the collector current is about 2mA, and at fully supply voltage the *output power* is then about 0.5 mW, which is sufficient for this application. R7 does not have to be set for individual transistors.

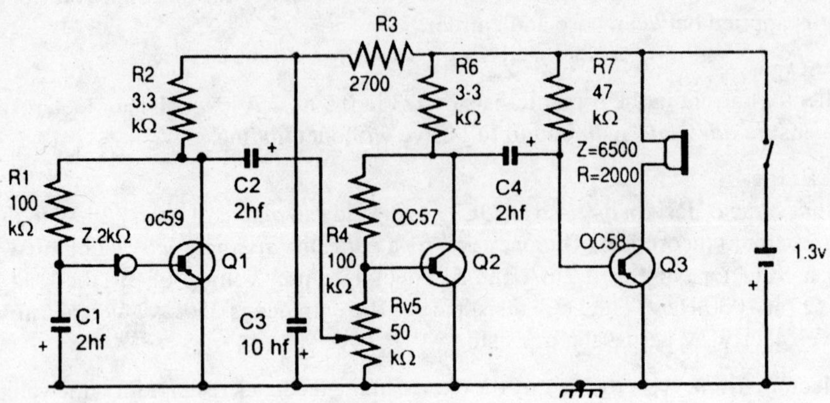

Fig. 6.14. Three-transistor RC-coupled hearing aid.

In the OC57 driver stage and OC59 input stage the collector current is 0.25mA. R1 and R4 provide sufficiently good *D.C. stability* to permit a wide range of ambient temperature. R4 introduces some A.C. feedback, but there is little loss in gain, because the input impedance falls and reduces the loss in the coupling network. C1 prevents A.C. feedback in the input stage.

By siting the volume control between the input and driver stages, the amplification of contact noise is reduced.

With a mercury cell the decoupling components for the first stage, R3 and C3, become unnecessary.

Average current drain is 2.7mA, and a life of about 100 hours can be obtained from a Mallory cell RM625. *Total-harmonic distortion* is 5% at 0.4 mW, and the electrical *frequency response* is flat to within 0.25 dB from 100 Hz to 4 kHz.

6.10. THREE-TRANSISTOR DIRECTLY COUPLED CIRCUIT

The previous circuit, if it is used with a mercury cell so that R3 and C3 may be omitted, still contains five fixed resistors and three capacitors. The need for a *subminiature hearing aid* which can be concealed in a pair of spectacles or a hair slide has led to the consideration of *direct coupling* for the amplifier. Such a circuit is shown in Fig. 6.15, and apart from the usual microphone, earpiece, battery and volume control, the only components required are the three transistors, three resistors and one capacitor.

The circuit is experimental rather than a production prototype. A pre-requisite of any large-scale production of a hearing aid utilising this circuit would be an examination of the effects of the production spreads of the transistors.

Choice of Circuit.

The only directly coupled circuit which will give a useful power gain with three transistors consists of three grounded-emitter stages in cascade. Stabilisation is provided by a D.C. negative-feedback loop. This feedback loop governs the choice of circuit. Overall feedback is applied from output to input, so that the D.C. gain of the whole amplifier is used to compensate the temperature-dependence of the transistor characteristics.

The volume control is sited so as not appreciably to affect the D.C. working conditions.

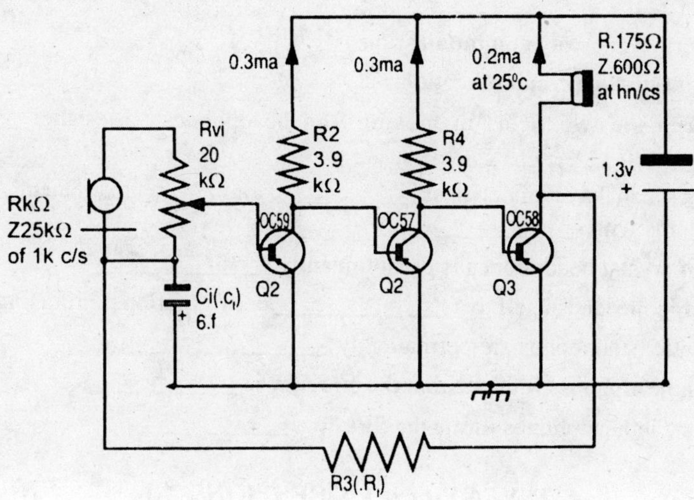

Fig. 6.15. Three-transistor directly coupled hearing aid.

OBJECTIVE TYPE QUESTIONS (A)

1. Moving iron headphones work on the principle of

   (*a*) magnetic attraction (*b*) magnetic repulsion
   (*c*) both (*a*) and (*b*)

2. Moving iron headphones suffer from

   (*a*) natural resonance (*b*) high harmonic distortion
   (*c*) non-linear distortion (*d*) none of these.

3. The drivers of dynamic headphones can be expected to produce

   (*a*) ample sound pressure level
   (*b*) insufficient sound pressure level

4. Electrostatic headphones

   (*a*) suffer from system resonance
   (*b*) are free from system resonance

5. Which one of the following suffers from high harmonic distortion

   (*a*) moving iron headphones
   (*b*) moving coil headphones
   (*c*) electrostatic headphones
   (*d*) crystal headphone

6. Which one of the following headphones need polarising voltage

   (*a*) moving coil (*b*) electrostatic
   (*c*) moving iron (*d*) crystal

FILL IN THE BLANKS (B)

1. ____________ permit one-way communication.
2. __________ permit two-way communic^tion.
3. Interaction between the two fields in moving iron headphones varies the ____________ on the diaphragm causing it to ___________.
4. Plastic membranes can have match less __________ and _____ than metal diaphragms thereby minimising problems with ___________.
5. The impedance of crystal headphones is predominantly _____________.
6. Electrostatic headphone lends itself to a _____________ configuration minimising _______________
7. Electret electrostatic headphones are permanently _______________.
8. LED indicators in headphones indicate that the circuitry is ____________.
9. Protection circuitry in headphones limits the SPL to a ____________.

CONVENTIONAL QUESTIONS (C)

1. Differentiate between headphones and headsets.
2. What are the different types of headphones based on their construction ?
3. Explain the working of a dynamic headphone.
4. Discuss the working of electrostatic headphones. How do they differ from other headphones?
5. What are the peculiarities of electret electrostatic headphones?
6. Discuss the limitations of various types of headphones ?
7. What are the requirements of a hearing aid?
8. Explain the working of a directly-coupled three-transistor hearing aid circuit.

7

LOUDSPEAKERS

We have progressed a long way since early designers fixed a crude horn on to a telephone earpiece to make the first loudspeaking telephone. Since then, of course, there has been the invention of the moving coil loudspeaker and many other drive principles have been adopted to further the quest for more perfect reproduction. The science of loudspeakers has been refined to the point where the response of a drive unit, at least at low frequencies, can be predicted with great accuracy. Just as the choice of a microphone dictates what goes into the recording, the loudspeaker has the last word over the quality of what comes out.

7.1 IDEAL LOUDSPEAKER

There are a number of *interrelated factors* that must be considered in the design of an effective transducer for converting *electrical energy* into airborne *acoustic energy*. These include electroacoustic efficiency, uniformity of frequency response, linearity of amplitude response, transient response, power handling capacity, size, durability and cost. An *ideal loudspeaker*:

(*i*) Would have an *electroacoustic efficiency* approaching 100 percent.

(*ii*) Would have an *acoustic output response* that is independent of frequency over the entire audible range.

(*iii*) Would introduce neither harmonic nor intermodulation *distortion* into its output.

(*iv*) Would *faithfully reproduce* transients as well as steady input signals.

(*v*) Would be capable of producing a *nondirectional radiation pattern.*

(*vi*) Would be of *as small a size as is possible* considering the required acoustic output.

No single transducer has been designed that is capable of satisfying all the above requirements. Of the many devices developed for the radiation of acoustic energy into air, the two most widely used are the *direct-radiator* or *dynamic loudspeaker* and the *horn loudspeaker.* Both of these loudspeakers utilise the *electrodynamic coupling* that exists between the motion of a vibrating surface, called the *cone* or *diaphragm* and the current in a so called *voice-coil*. Additional types of *electromechanical coupling* that are used for this purpose include *electrostatic coupling* in electrostatic loudspeakers and *electromagnetic coupling* in telephone receivers.

The speaker system itself can be divided into three *functional parts:*.

(*a*) *The electromagnetic part,* consisting of the voice coil and the field magnet. Audio frequency electric current in the coil causes mechanical motion of the cone or diaphragm on which it is mounted. This part is often referred to as the *driver* or *motor* of the system.

(*b*) *The mechanical part,* on which the driving coil is usually mounted and which is set into mechanical motion by the audio frequency electric current in the *driving coil.*

(*c*) *The acoustic part,* which transmits the sound energy developed by the mechanical part of the area served by the system in the most efficient and faithful manner possible. This takes the form of a *baffle* or *enclosure* with a horn being a form of enclosure.

A complete understanding of the operation of the speaker systems requires a sufficient view of the flow of acoustic energy from the output amplifier stage to the listener.

7.2 BASIC LOUDSPEAKER TYPES

When broadcasting first began, the only available devices for converting sound energy into electrical energy, and vice versa, were those used in the ordinary telephone. Microphones were usually of the carbon types. The hearing device, or telephone receiver, was a simple electromagnet energised by electrical signals. These signals caused a thin sheet of magnetic material (the telephone diaphragm) to vibrate *in synchronism.* Wearing headphones, however, was not exactly conducive to listening pleasure, besides the reproduction wasn't of the best quality. So, engineers started to search for *some sort of hearing device that could be separated from the head of the listener and still supply enough volume to make listening easy.* Hi-fi wasn't really thought of at that time.

Two problems faced the early engineers; *amplifying* the electrical signal enough to drive the loudspeaker, and designing an *off-the-head speaker* that could reproduce the electrical signals in *the form of sound.*

Fig. 7.1 Full range of speakers.

One of the first speakers was simply a large *horn* attached to an ordinary telephone earphone mechanism. This speaker had fairly good volume, but frequency response was poor.

In another early type of speaker, a *vibrating reed* was used instead of the thin disk of magnetic material (the diaphragm of the earphone). The free end of the reed was attached to a speaker cone, and the reed's vibration caused the whole cone to vibrate.

A third type of early speaker had a *balanced pivoted armature* located between a pair of magnets. Electrical impulses through the magnet coils caused up-and-down vibration. The armature movement was mechanically coupled to the speaker cone with a stylus, thus producing the sound.

There are three types of speakers in modern use: crystal, electrostatic (condenser) and electrodynamic (or simply dynamic) speakers.

7.3 CRYSTAL LOUDSPEAKERS

Rochelle - salt crystals have the property of becoming physically distorted when a voltage is applied across two of their surfaces. This property is the basis of the crystal type of speaker driver, illustrated in Fig. 7.2.

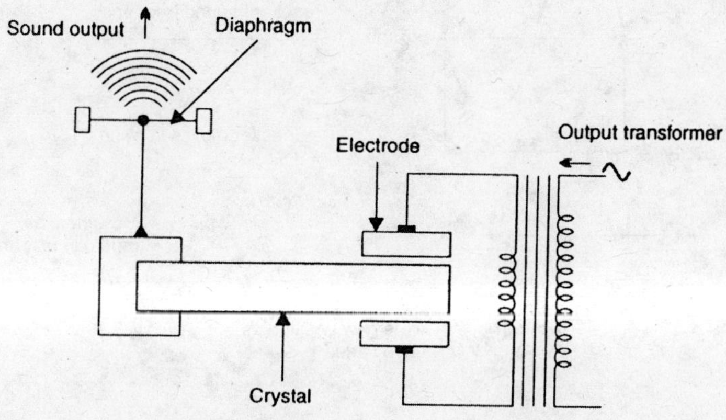

Fig. 7.2 Schematic representation for a crystal type speaker.

The crystal is clamped between two electrodes across which the audio frequency output voltage is applied. The crystal is also mechanically connected to a diaphragm. The deformations of the crystal caused by the audio frequency signal across the electrodes cause the diaphragm to vibrate and thus to produce sound output.

Crystal speakers have been impractical for reproduction of the full audio-frequency range because *the input impedance is almost completely capacitive.* Thus it is difficult to couple power into them. At high audio frequencies, the reactance becomes lower ($X_c = 1/2\pi fC$) and the relative amount of power smaller. In the bass range, stresses on the crystals are very great, and crystals have been known to crack under stresses.

Consequently, crystal units have found some use in *tweeters* (the high-frequency portion of dual-speaker units) and rarely even in this application because their response is not linear.

7.4 ELECTROSTATIC (CONDENSER/CAPACITOR) LOUDSPEAKERS

This type of speaker operates on the principle that DC voltage between two parallel metal plates causes these plates to attract or repel each other. The amount of attraction or repulsion depends on the applied voltage. If one of the plates is a flexible metal, it will bend. But *the amount of attraction and repulsion is not directly proportional to the voltage applied.*

For example, consider the movable and fixed plates of Fig. 7.3 with *no voltage* applied. Now suppose we apply a *slowly varying AC voltage* to both plates. As the voltage increases from zero, the potential difference between the two plates also increases. This, in turn, produces an increasing force of attraction between the plates, so that the movable plate bends towards he fixed plate. As the AC voltage decreases once more to zero, the attractive force decreases, and the movable plate moves back to its original position. But, now we have the second half of the AC cycle, in the *negative direction*. All that this means to the metal plates is that the positive and negative voltages have switched plates. The attractive force is still there, and it is still the same. So, we get another bend in the movable plate on the negative half of the AC cycle. Thus, for one full cycle of AC we have two bends in the movable plate in effect, a *frequency doubling*. A 2 kHz signal would give us a 4 kHz note.

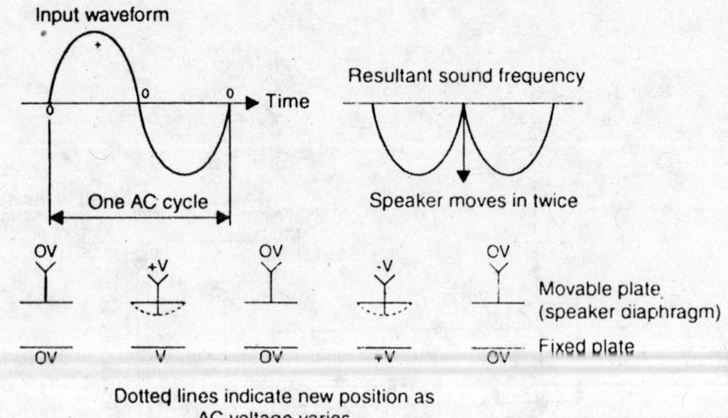

Fig. 7.3 Frequency doubling in an unpolarised loudspeaker.

To overcome frequency doubling, we polarise the speaker, that is, we apply a high voltage (7,000 volts or so) as a sort of *DC bias.* (Fig. 7.4). The voltage exerts a steady attraction between the two plates, so that now - with no signal - *the movable plate is bent slightly toward the fixed plate.*

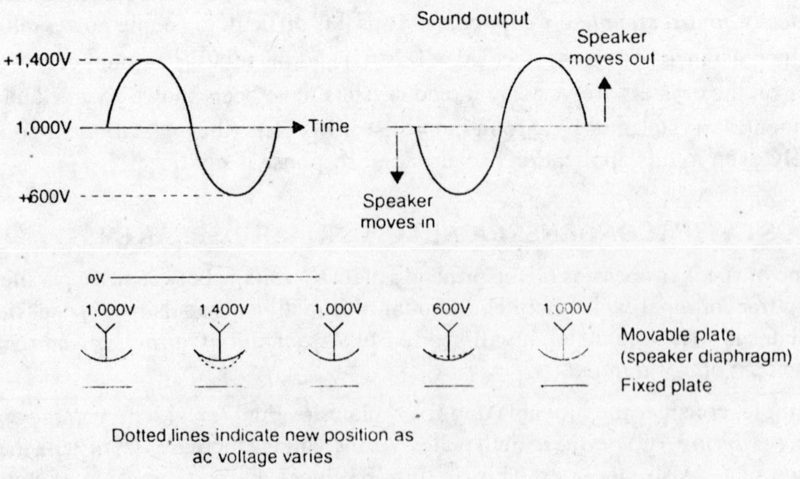

Fig. 7.4 Frequency doubling eliminated by DC polarisation.

Now suppose we apply a 400 V AC audio signal to the speaker. As the positive half cycle of the signal increases from zero the voltage between the plates *rises* from 1,000 V toward 1,400 V and the movable plate bends from its original position toward the fixed plate. As the AC passes its peak and returns to zero, the voltage between the plates *drops* from 1,000 V to 600 V. Instead of moving again toward the fixed plate, the movable plate moves farther away. *So we have a situation in which the bending of the movable plate is identical to the AC swing and there is no frequency doubling.*

A detailed view of a modern electrostatic speaker is shown in Fig. 7.5. The practical speaker of today uses push-pull, with a built-in-step-up transformer to work from the ordinary 8 ohm amplifier output tap. The polarising voltage is applied to the centre or movable plate through a resistor that keeps the voltage *stable* during variations in the signal voltage. The signal voltage is applied to the two outside plates. Because the

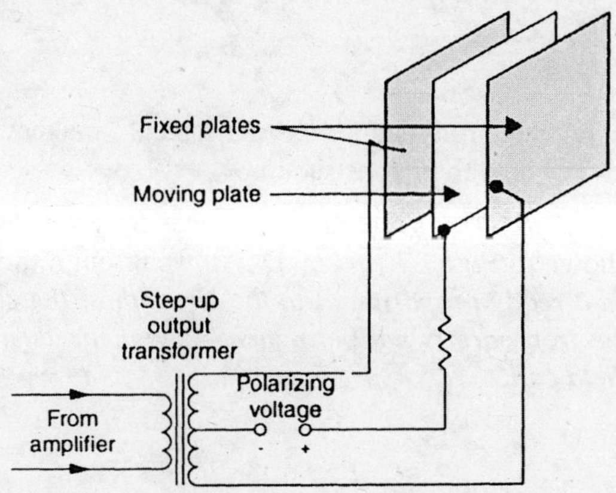

Fig. 7.5 Working principle of a typical electrostatic loudspeaker

Fig. 7.6 A two way electrostatic utilising a separate woofer and tweeter.

diaphragm is *centered* between the two plates that attract it equally, there is no bending when there is no signal. Also, *because of the push-pull action the diaphragm can move twice as far in response to signal voltages for the same amount of compression of the dielectric material.*

The major weakness of the electrostatic speaker *requires that the DC bias be much larger than the applied audio signal.* In practical speakers, 1,000 to 1,200 volts may be used. Further, when we get into the bass frequency ranges, a great deal of power would be required to get enough output. To produce such power, the speaker area would have to be very large. So, even though full range electrostatic speakers have been constructed *in practical use electrostatic speakers have been mostly confined to frequencies above 1,000 Hz.*

The step-up transformer and the high voltage polarising supply is usually built right into the modern electrostatic. Often the electrostatic unit and its matching woofer are sold together as a complete system.

Some high class systems use electrostatics to reproduce the high frequencies. Koss uses electrostatics on some of their stereo headphones.

7.5 DYNAMIC LOUDSPEAKERS

There are two varieties of dynamic loudspeakers : electrodynamic and permanent magnet (PM) speakers. Both work in exactly the same way, the difference is in their construction.

The *electrodynamic* speakers has a *soft iron* magnetic circuit, non retentive of magnetism, around whose centre leg, a large, multilayer field coil is wound, as shown in Fig. 7.7. When DC flows through this field coil, it magnetises the iron core. *A magnetic flux field directly proportional to the strength of the current through the coil is thus set up across the airgap. The iron core is not permanently magnetised, it stays magnetised only as long as current flows through the field coil.*

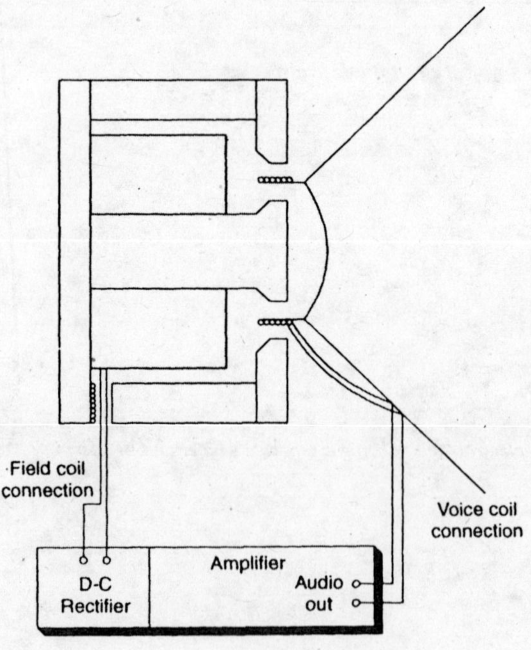

Fig. 7.7. Electrodynamic Speaker.

Improvements in permanent magnet materials have made the electrodynamic speaker practically obsolete, but some still-exist in vintage radios. Note that these use the field coil as part of a choke filter in the power supply, a good example of killing two birds with one stone. The electrodynamic speaker has disappeared completely so far as hi-fi is concerned, the permanent magnet speaker reigns supreme.

7.6 PERMANENT MAGNET LOUDSPEAKERS

The *most popular* type of loudspeaker today is the permanent magnet dynamic type. Because of its comparative simplicity of construction and design, the precision that may be built into it, the ease with which it is interfaced with other equipment, its easy adaptability to many different applications, and its comparative freedom from electrical trouble, the *dynamic loudspeaker* has found acceptance in all kinds of reproducing systems. It is found in the smallest pocket radios and is a major component of the most elaborate theater systems.

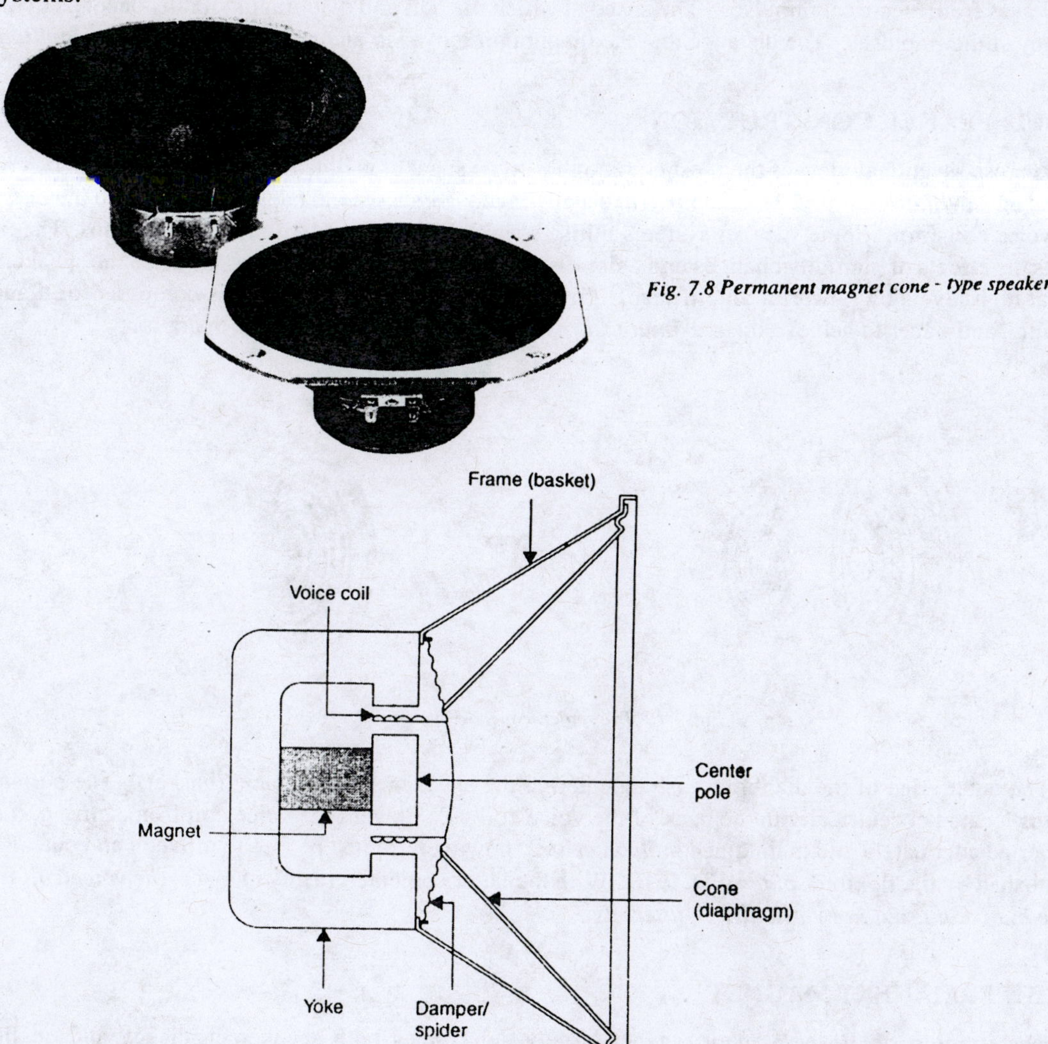

Fig. 7.8 Permanent magnet cone - type speakers.

Fig. 7.9 Cutaway views of typical PM cone-type speaker.

Just about all hi-fi woofers are of the permanent magnet (PM) type, Fig. 7.8. Cutaway view of the PM cone type speaker is given in Fig. 7.9. The *cone* (diaphragm) is energised by a *moving coil.* The woofer's magnetic field is supplied by a *permanently magnetised and highly magnetic alloy* instead of the iron-cored coil used in electrodynamic speakers.

The PM speaker contains a very light coil of wire affixed to the diaphragm and located concentrically around, within, or in front of the centre of the permanent magnet. The coil (voice coil) is free to move in the field of the magnet. Electrical impulses, varying at an audio rate, are applied to the *voice coil* by the amplifier. Because these impluses are constantly changing in amplitude and direction, a changing magnetic field is set up in the voice coil. This field *reacts* with the constant field of the permanent magnet. The result is that the voice coil moves further into the gap when the fields are opposite and attract, and farther out of the gap when they are alike and repel. This causes an *in-and-out movement of* the diaphragm; consequently, we obtain sound waves from electrical impulses. The speed at which the coil and diaphragm vibrate depends upon the *frequency* of the impulses. The distance that the diaphragm moves in and out depends on their *amplitude.*

7.7 LOUDSPEAKER CONSTRUCTION

A cross--sectional view of the most common PM type speaker is illustrated in Fig. 7.9. The *voice-coil* is wound on a *cylindrical form.* The *diaphragm* usually made from a special paper, is attached to the *outerrim* of the voice coil-form. Some speaker systems utilise speakers that have aluminium diaphragms. The metal prevents the effects of humidity changes and helps dissipate the thermal energy present when the speakers are driven at high levels by powerful amplifiers. High quality speakers use diaphragms composed of titanium, aluminium, and paper to achieve the *maximum stiffness* required for their mode of operation.

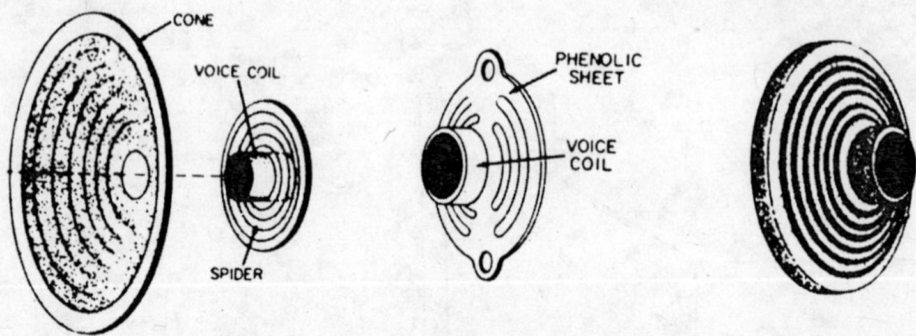

Fig. 7.10 The spider, cone, and voice coil.

The outer edge of the diaphragm is cemented to the speaker's metal frame (basket). The permanent magnet is located cocentrically in the back of the voice coil. To prevent the voice coil from shifting, it must be supported adequately and maintained at *dead centre;* however, it must be free to move in and out. This is accomplished by the flexible *spider,* Fig. 7.10. With the latter cemented firmly in place, *the voice coil is free to move in and out, but not vertically or laterally.*

7.8 THE PERMANENT MAGNET

The strength of the permanent magnet largely determines both speaker efficiency and quality of reproduction. The magnet must furnish the most powerful magnetic field in order to maximise the voice coil movement for any given signal. In general, *the stronger the magnet, the better the power handling capacity of the speaker, and the better the reproduction.*

•Enclosure: Air-tight •Speaker: 31 cm woofer, 10 cm mid-bass, 6 cm mid-high, 2.2 cm tweeter •Impedance: 6Ω •Max. input: 150W •Freq. range: 25 ~ 40 kHz •Sound pressure level: 90 dB •Dimensions: 395 × 740 × 380 mm •Weight: 40 kg

Fig. 7.11. A typical speaker system.

Woofers are speakers designed to reproduce the bass, or *low frequency*, portion of the audio-frequency spectrum. For woofers, the designer does not necessarily choose the strongest available magnet. At a given point, depending on a number of other design factors, adding more magnetic flux will reduce the *bass performance* of a woofer because the moving system becomes *overdamped*. The designer, therefore, chooses the magnetic flux level that provides the best compromise between *efficiency* and *bass response*. For a hi-fi speaker system this is very seldom the level at which the magnetic flux is as strong as possible.

Mid-range speakers are designed to reproduce the mid frequency portion of the audio frequency spectrum. *Tweeters* are speakers designed to reproduce the treble, or *high frequency*, portion of the audio frequency spectrum. As far as the magnetic flux in mid ranges and tweeters is concerned, the designer usually does try for the strongest flux field possible, of course, within the cost restrictions. A typical speaker system is shown in Fig. 7.11.

The mid-range speaker in a three-way system is often referred to as a *squawker*.

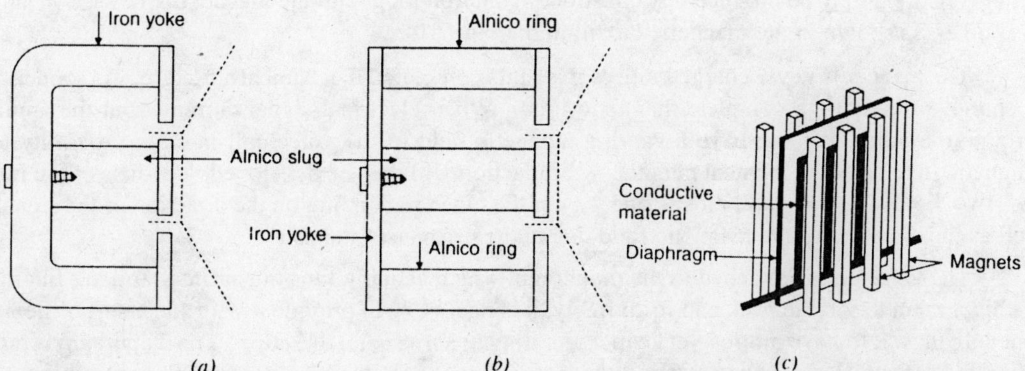

Fig. 7.12. Shapes of magnets in PM speakers : (a) slug- type, (b) ring-type, and (c) spaced - bar type.

Permanent magnets in speakers come in various *shapes*. One example is the magnet in the form of a *solid cylindrical slug*, as shown in Fig. 7.12 (*a*). A U-shaped iron yoke completes the magnetic circuit, except for the voice coil gap. Another type, similar in outward appearance, uses a *hollow cylindrical slug* with an E-shaped yoke, with almost the entire magnetic circuit composed of Alnico (Fig. 7.12 (*b*)). A third type, Fig. 7.12 (*c*), uses *spaced permanent magnets*. Here, instead of a voice coil, the diaphragm itself (which consists of Mylar stretched taut over a frame that has copper wire glued in a square - wave pattern to its surface) is placed in the magnetic field.

Ceramic magnets have several advantages over those made of metal. They are lighter, stronger, and less expensive to produce than metal magnets of comparable size. Their development has resulted from the need for *lighter* and *smaller* magnets, but *ones that have strong fields*. Their principle of construction is simple. During the period when the ceramic is fired, that is, molded to its final shape while eliminating the impurities, the ceramic is subjected to a a very strong magnetic field. Once, the ceramic is cooled, this magnetic field is retained, and we have a very strong permanent magnet.

7.9. THE VOICE COIL

The voice coil is the only thing within the speaker that carries any electrical current or signal. It is energised directly from the amplifier. The *voice coil, as its name implies, is the part of the speaker that does the talking.*

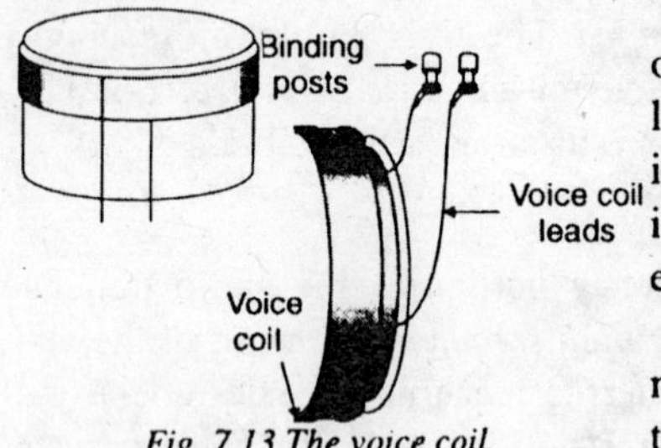

Fig. 7.13 The voice coil.

The voice coil, Fig. 7.13, consists of several turns of wire wound on a supporting bobbin. Depending on the functional design of the loudspeaker, the *coil* itself may be copper or aluminium wire, although insulated aluminium ribbon is also used. In the case of the latter, the ribbon is wound on edge with the flat surfaces of neighboring turns adjacent to each other and all the turns held together by a binding cement.

The bobbin, or the *voice coil former* upon which the wire is wound, may be made of a strong grade of thin paper, wound around on itself several times to provide a rigid cylinder. Sometimes the voice coil is wound on aluminium or duraluminium formers, and in some designs the voice coil formers are made of rigid paper, reinforced by an aluminium ring around the outer edge.

Should the voice coil warp or become twisted, it might contact the core of the magnet. When this occurs, a rasping noise develops, rendering the speaker useless. To offset this possibility, many speakers now contain *warp-resistant, aluminium base voice coils.*

The voice coil of a dynamic loudspeaker can speak only when it is immersed in a magnetic field. Such a magnetic field may be produced by constructing an iron loop with the magnet in one section and an air-gap in another, as shown in the magnetic circuit in the Fig. 7.14.

The magnet is never charged unless it is in its completed mechanical structure. As soon as the magnet is charged, it sets up a complete magnetic field. When electrical signal current from the amplifier flows through the voice coil, we have a varying magnetic field of the voice coil in close proximity to the fixed magnetic field of the permanent magnet. Motor action will thus be developed. Because of the interaction of the two fields the voice coil moves one way or the other, depending on the direction of the signal current as well as the direction of the magnetic field due to the permanent magnet.

The voice coil is attached to the diaphragm which actually fans the air into motion. In order to allow the diaphragm to vibrate back and forth freely, it is necessary to provide it with some sort of flexible support that will allow it to have motion yet keep it vibrating in a true axial direction. The diaphragm is provided with a flexible area at its outer edge *sufficiently compliant* to allow the diaphragm to flex in and out.

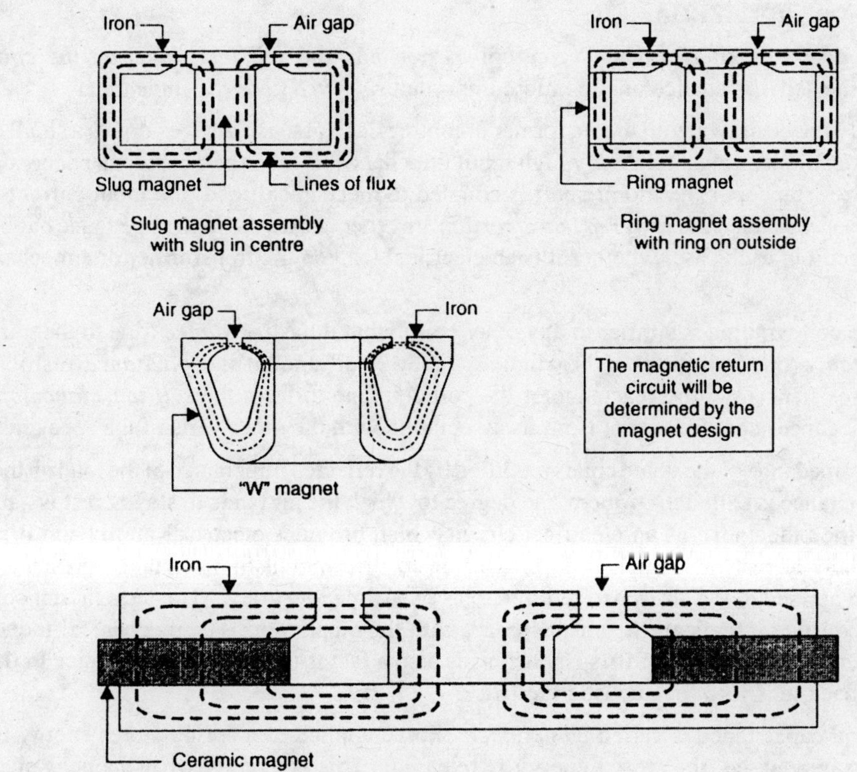

Fig. 7.14 The magnetic circuit.

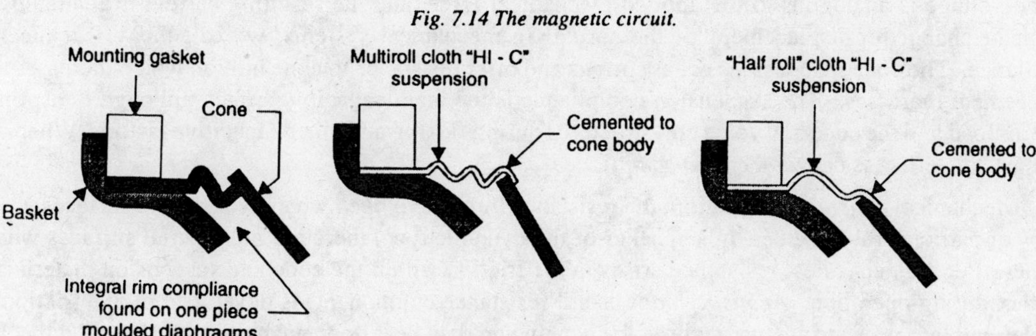

Fig. 7.15 Various types of diaphragm suspensions.

These compliances, Fig. 7.15, may be either half-roll or multi-roll and are cemented to the main body portion of the cone. They are referred to as *high compliance type* (HI-C) and, because of their looseness, permit the cone to move over abnormally large excursions.

Sometimes the *rim compliance* is accomplished by providing an annular ring of soft chamois leather, which is cemented both to the basket edge and to the paper diaphragm. A general term used for this edge compliance is the *surround* because it literally surrounds the speakers.

7.10 LOUDSPEAKER IMPEDANCE

The impedance looking into the voice coil is not only the *self impedance of the coil* itself but a combination of the self-impedance and the more important *reflected acoustic impedance.*

A parallel may be drawn with a transformer or motor. Each of these devices draws a small current when operating *unloaded*, indicating a relatively high input impedance. When the transformer secondary is *loaded* by an electrical resistance, or the motor shaft is coupled to mechanical load, the input current rises and *the input impedance of each device is lowered in proportion.* In other words, the load impedance has been *reflected* into the input circuit in each case, whether it is an electrical load on the transformer or a mechanical load on the motor.

The voice coil winding is similar to any other coil in that it has *resistance* (due to the wire used in the winding), *inductance* (due to the turns of the windings), and a small amount of *capacitance* (distributed between the turns). The resistance and the reactance of the coil combine to form the *self-impedance of the winding* without any impedances coupled into it from its association with the other parts of the speaker.

The self-impedance of the voice coil is modified by the reflected impedance of the load on the diaphragm. Mechanical inductance is called *inertance.* The degree to which the air tends to stay at rest is a measure of its inertance. It is the inductance of an electrical circuit which provides electrical inertia and it is the current which tends to stay at rest or in motion in proportion to the amount of inductance present. Mechanical inductance when applied to the air is also referred to as *acoustic inductance.* The term inertance is especially applied to the acoustic system and the air in contact with the diaphragm. The mechanical inductance of the cone and voice-coil structure and of its suspensions is also a factor in the input impedance to the voice coil, and is reflected back to it with the acoustic inertance.

Mechanical capacitance is called *compliance.* Force applied to a spring stores energy in the spring. When the spring is released, the stored energy is released. This is exactly what happens electrically in a capacitor in which energy is stored by the flow of current into the capacitor by application of a voltage. The applied voltage is analogous to the applied mechanical force, and the resulting current is analogous to the motion or change or displacement of the spring. In mechanical systems, we call this effect mechanical compliance. The cone suspensions act as springs and offer resistance to cone motion which increases as cone displacement increases. The suspension compliance is the main capacitive effect, although the springiness of the air load and the cone and voice coil structures during flexing add other capacitive factors. When applied to the air, this effect is called *acoustic compliance.*

Mechanical *resistance* is friction; the resistance force developed when two or more surfaces, layers, or groups of particles rub together. In a speaker of the dynamic type, there are no material surfaces which rub together. Purely mechanical resistance arises in the friction within the cone and suspension materials when they flex during operation. *Acoustic load* (useful resistance component) is developed by the friction of the particles and layers of the air surrounding the cone when they bear upon each other or along the mechanical surfaces of the speaker assembly *when motion is imparted to the air* in the form of acoustic vibrations.

Mechanical components of *impedance* have the same relationships among themselves as exist among their counterparts in the electrical circuit. *Power is dissipated only by the resistive component. The inertance and compliance produce mechanical reactance which varies in the same way as electrical reactance varies.*

7.11 ACOUSTIC IMPEDANCE AND RESONANCE

The resistances and reactances of the system (including acoustic, mechanical and electrical effects) combine in the *effective impedance* looking back into the voice coil. This combination is best visualised by an *equivalent circuit* illustrated in Fig. 7.16. The *efficiency of power transfer* can be seen to be dependent on the proportion of the impedance represented by R_A, which represents actual acoustic power dissipated in overcoming air friction and in radiating the acoustic power.

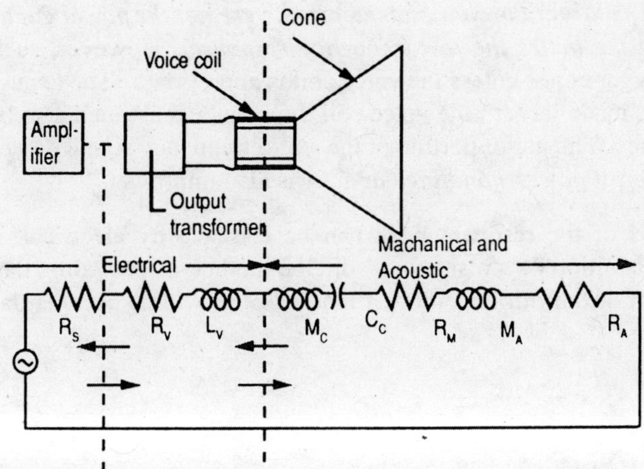

Fig. 7.16 Equivalent circuit of a loudspeaker

As in a purely electrical system, the capacitive reactance C_c resonates with the combined inductive effects L_V, M_c and M_A at some frequency called the *resonant frequency of the speaker*. At the resonant frequency, all reactance is cancelled out of the system and output and efficiency are far greater than what they are for other frequencies.

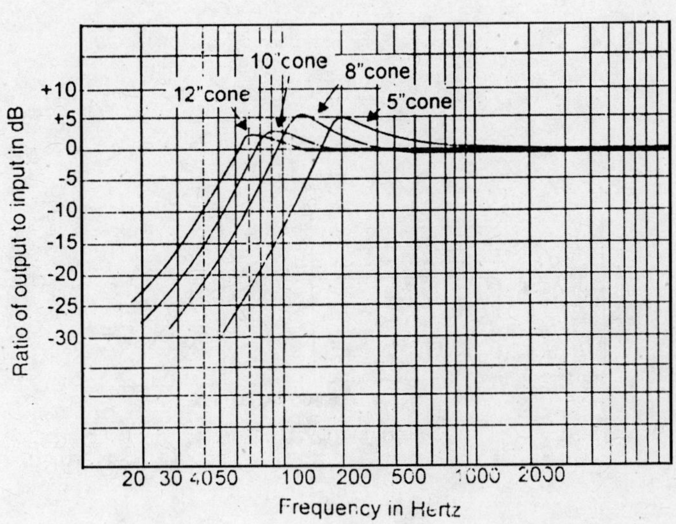

Fig. 7.17 Effect of cone size and resonance on low-frequency response.

The curves of Fig. 7.17 show the effect of speaker resonance on response. The response rises to a peak at the resonant frequency, then it falls rapidly at lower frequencies. The shape of the response curve shows how the resonant frequency can be used as an indication of the *limit of low-frequency response.*

The size of the cone is important because it influences both the low frequency response and power-handling capacity. *The larger the cone diameter is, the greater the power capacity for all low frequency components combined and the better the low frequency response.* However, such improvements are not necessarily derived from larger cones unless the voice coil is appropriate. The acoustic impedance offered to the cone rises as the cone is made larger ; the voice coil impedance must then also be made larger for proper energy transfer and efficiency. The major portion of the audio frequency signal power is in the low-frequency components, so that the *overall power-handling capacity* is also improved.

Undesired sharpness of the resonant peak can be lessened by electrical, mechanical, or acoustic *damping.* Damping is the addition of a resistive load. One method of providing damping is through proper design of the *output stage* of the amplifier. Another method of providing speaker damping is through proper design of *speaker enclosure.*

7.12 WOOFERS

There are two types of low-frequency speaker, the commonly known *woofer,* and the more recent addition the *subwoofer.* The latter is used for the reproduction of frequencies *below* those produced by the woofer, and it is generally purchased as an *add-on* to an existing system.

The low frequency speaker provides the *bass* of any hi-fi system. Its sole purpose is to reproduce the low frequency notes of the program source. The prime requisite for low-frequency reproduction is a *large diaphragm,* the larger the better. *The smallest diaphragm for any halfway decent woofer is 8 inches ; for a subwoofer it is 12 inches.* In addition to large size, the diaphragm must be of *fairly heavy construction.* Light diaphragms just can't hold up under the vibrations encountered under the lower audio ranges.

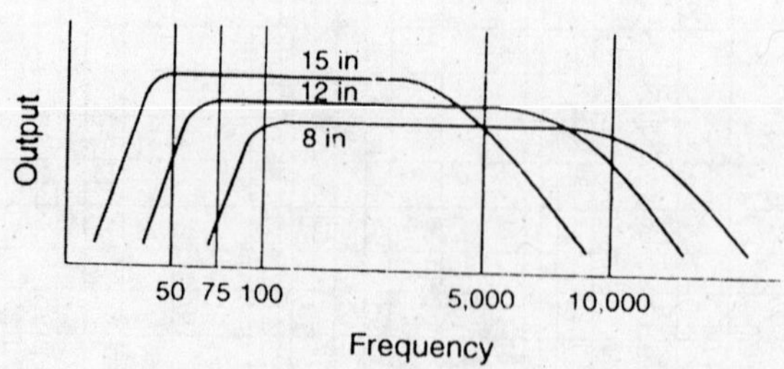

Fig. 7.18 The popular sizes of speakers have overlapping characteristics of response. The large speakers have more extended lows, the smaller ones more extended highs.

A woofer must be able to vibrate back and forth very easily, *i.e.* have *high compliance.* One way to accomplish this is to have the diaphragm loosely connected to the frame. The gasketing that holds the periphery of diaphragm to the frame/basket is fastened so that it barely keeps the diaphragm from slipping loose, but no more as shown in Fig. 7.19. With this construction it takes less force (acoustical power in this case) to move the diaphragm any particular distance (less power required for a given excursion).

Rather than the loose suspension system, the cone is supported by a very flexible material so that it can be moved very easily by the voice coil. *The suspension is tight but the sine wave at the diaphragm edge is made very flexible.*

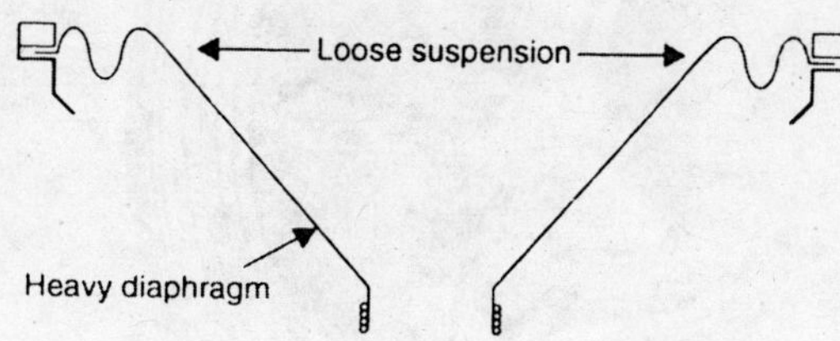

Fig. 7.19. Suspension of woofer diaphragm. For good reproduction, a woofer should have a large diameter, stiff diaphragm, high compliance and a large voice coil.

A woofer must also have a *a large voice coil* to handle considerable heat. The larger the voice coil, the more the current produced by the amplifier output circuit and, therefore, the more the power the woofer can handle.

Finally a *strong magnet* can be of great help to move the heavy voice coil and cone assembly too well. The better the woofer, the heavier the magnet assembly (unless it's ceramic).

To sum up, *a good woofer must have a large, heavy diaphragm, a strong magnet, high compliance, and a large voice coil.*

7.13 MID RANGE AND EXTENDED RANGE SPEAKERS

The *mid-frequency loudspeaker* is supposed to have a good response at the frequencies located between those covered by the woofer (low-frequency speaker) and the tweeter (high-frequency speaker). However, some *extended range speakers* can be used as general purpose units. An adequate general purpose speaker, Fig. 7.20 must have good response at both the low and the high ends of the audio range, not to mention in the middle.

Tests by various manufacturers reveal the following data: A *15 inch* diameter loudspeaker will give excellent low frequency response but will be sadly deficient in the highs. A *12 inch* diameter unit will still give good low frequency response, halfway decent middle frequency response, but not too good treble response. *The 8 inch unit, one of the most popular today, is the smallest that will permit fairly good response on low frequencies, and good response at the middle and high frequencies.* This is shown in Fig. 7.18.

Conventional extended range speakers are now available in frequency ranges claimed to extend as far as 45 to 15,000 Hz, although *they need a good enclosure to cover the range.* To obtain this type of response, the loudspeaker must be designed so that the diaphragm will vibrate well at the low and middle frequencies, yet have some sort of *compensation* to permit good response at high frequencies.

The top illustration in Fig. 7.22 shows the converse type of diaphragm. Here the curvature of the diaphragm allows the apex to *isolate* itself without the need for special *decoupling devices.* The centre diaphragm shows the *apex decoupler* built right into the apex area. The lower diaphragm illustrates the joining of two discrete diaphragms to each other by means of a *flexible connection*. Thus one diaphragm can vibrate while the other one doesn't.

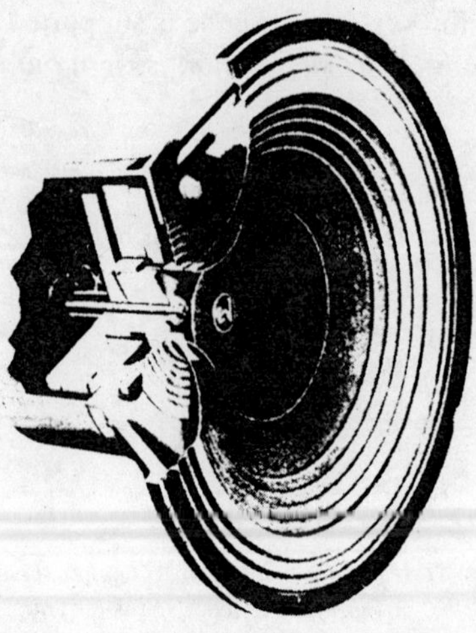

Fig. 7.20 Cut - away view of a general - purpose speaker.

Fig. 7.21 A three-way speaker system.

In the *duplex* type or *coaxial speaker*, we have a high frequency horn centered within the framework of the low-frequency speaker, as shown in Fig. 7.23. However, *there is only one voice coil;* thus it more properly falls into a single-unit extended - range category. The switch from high - to low frequency response is made by a *mechanical cross over* built right into the speaker framework.

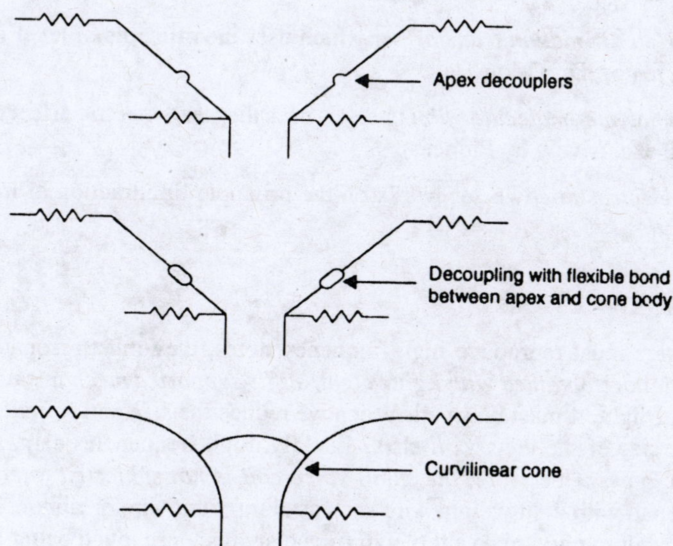

Fig. 7.22. Various diaphragm apex, decompling methods. By using mechanical dividing methods, the apex of the cone can be separated to handle high frequencies while the cone vibrates as a single unit for the lower ones.

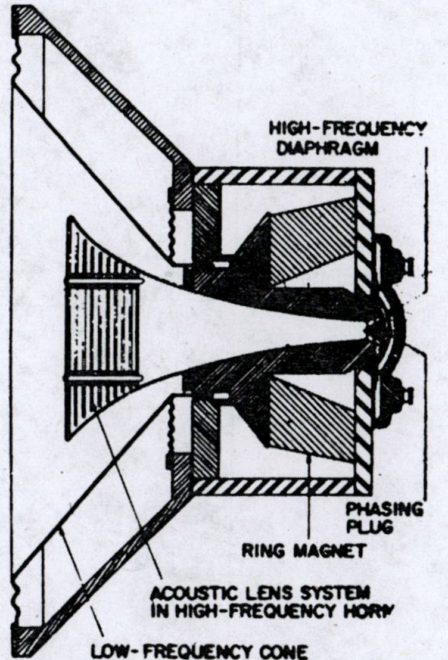

Fig. 7.23 A dual concentric speaker.

7.14 HIGH FREQUENCY LOUDSPEAKERS

There are two main types of high frequency speakers; the well-known *tweeter* and the more recent *supertweeter.* Supertweeters can be add-ons or they can be *integral with the system.* Six basic high frequency speakers (tweeters) exist.

(*i*) The *cone* is a physically disincentive version of the woofer.

(*ii*) The *dome,* so called because of its dome-shaped diaphragm.

(*iii*) The *horn,* so named because it is a horn.

(*iv*) *The Heil air-motion transformer* which uses the principle of lever in its operation, named after its inventor, Dr. Oskar Heit.

(*v*) *High polymer molecular film* tweeter, uses thepiezo-electric effect for its principle of operation (used exclusively by Pioneer).

(*vi*) The *electrostatic* tweeter works on the principle of attraction or repulsion between two metal plates.

7.15 CONE TYPE TWEETERS

Since *tweeters* must reproduce high-frequency notes, they must resonate at high frequencies. *High rcsonant frequencies are obtained with lightweight, stiffly supported mechanisms.* To make the diaphragm of a cone-type tweeter light, it must be small. When we reduce the size and weight of the diaphragm we must, in turn reduce the size of the *voice coil* also. Luckily, high frequencies carry only a comparatively small amount of electric power, therefore, *the small voice coil is not subjected to electrical overload.* Without exception, it is wound with light weight wire such as aluminium wire or ribbon. The lightness of the moving system provided by aluminium makes the high frequency response much better than if copper were used.

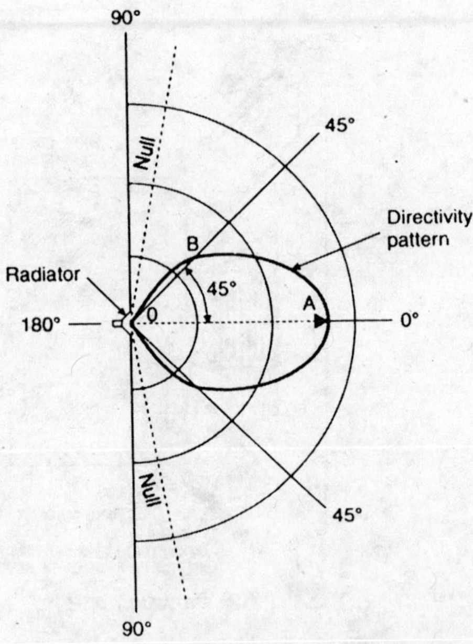

Fig. 7.24 Radiation pattern for a typical cone at one frequency.

Cone type radiators tend to concentrate radiation of the high frequency components of sound in a narrow cone about the axis of the radiator. The degree of directivity of a speaker is indicated by a directivity pattern Fig. 7.24. The axis of the radiator is considered the reference line with an angle of zero degrees. *Directivity patterns* are normally shown as a top view in the horizontal plane through the radiator axis. A cone in free space should have the same pattern in a vertical plane.

The line *OA*, Fig. 7.24, indicates by its length that the sound radiated along it is a *maximum* compared with that in any other diversion. At a 45° angle, the line *OB* is a measure of the relative sound intensity in that direction. Since *OB* is only half as long as *OA*, a listener along *OB* would listen only about half the volume compared to what a person along *OA*, at the same distance from *O*, would hear. At angles near 90°, the pattern indicates *minimum* (zero) radiation. In any practical setup, such a *zero area* would not exist because sound waves reach there by *reflection.*

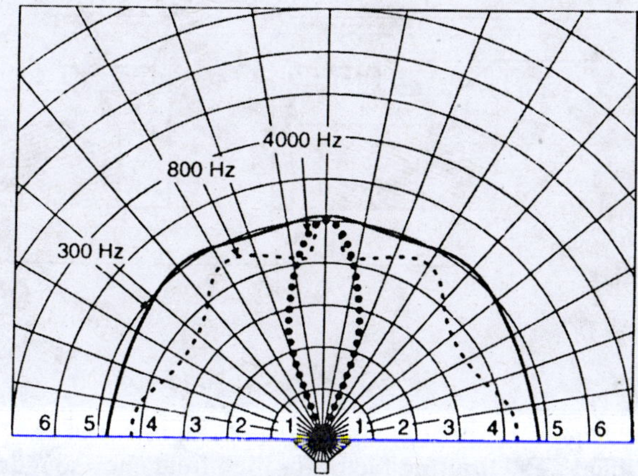

Fig. 7.25. Variation of directivity with frequency for a typical cone radiation.

Because, *directivity normally varies considerably with frequency*, a complete diagram (Fig. 7.25) must show separate patterns for each of at least several frequencies. Fig. 7.25 depicts variation of directivity with frequency for a 12-inch cone, assuming that the speaker is mounted in an infinite baffle. *Notice how much narrower the radiation pattern is at highs than at lows.*

A single cone-type tweeter distributes high frequency sounds unevenly. It *lobes* the higher frequencies directly out in front and tends to cause a drop off at the sides. This effect can be overcome by arranging two or more cone tweeters as shown in Fig. 7.26. In this way, *overlapping individual lobes from separate speakers cover the listening area.*

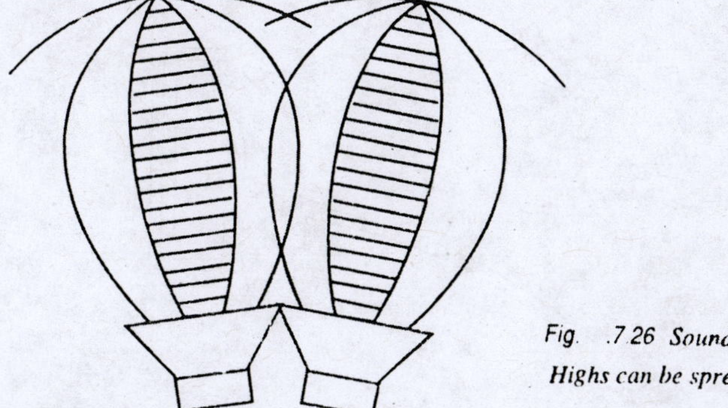

Fig. 7.26 Sound distribution of 2-tweeter array. Highs can be spread out by angling the speakers.

7.16 DOME TYPE TWEETERS

Uniformly dispersed flat energy response begins with a speaker systems ability to radiate sound at all frequencies evenly in all directions. *Even dispersion* of sound energy means that the sound emanating from the program source will sound the same for listeners in all parts of the room.

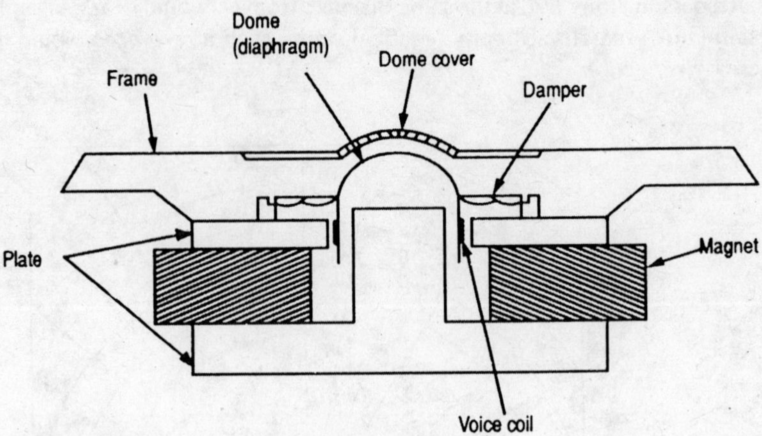

Fig. 7.27 Dome-type tweeter.

For *low frequency sounds* this problem of dispersion is not of practical consequence, since they are very nearly omnidirectional. The limiting factor for high frequency sounds is that a speaker will begin to be directional when its circumference equals the wave-length of the frequency being reproduced. *Directionality increases as the wavelength decreases with respect to the speaker's dimensions.* The laws of physics dictate the most direct approach to the problem of even dispersion of high frequency energy; *the drivers used must be as small as possible.* Dome tweeters, Fig. 7.27, are designed according to this principle in order to use these physical laws to the listener's advantage.

7.17 HORN TYPE TWEETERS

To obtain resonable output from a loudspeaker, we must vibrate large amounts of air. For this, we must usually have fairly large vibrating surfaces, such as the cones in woofers. The larger the cone surface, the

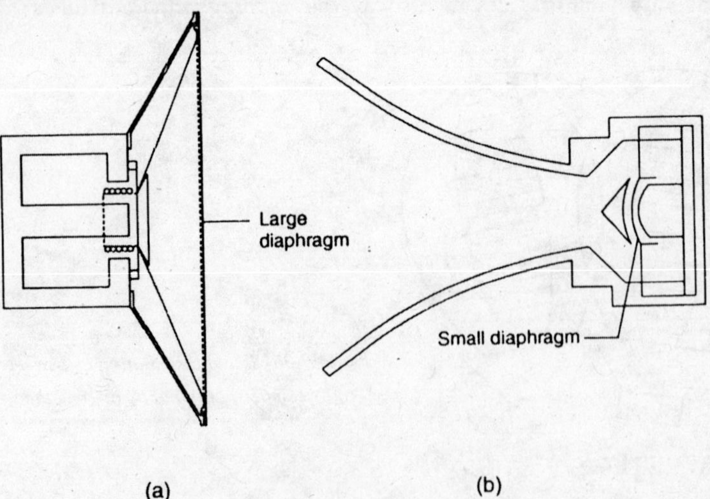

Fig. 7.28. Cone and horn-loaded tweeters (a), when used as a direct radiator will produce approximately the same sound output as the horn radiator (b) with approximately only 20 percent of the direct radiator power because of the small light diaphragm of the latter's compression driver.

greater the output. But the tweeters cone (diaphragm) must be small to attain its high-frequency response. Thus only a small amount of air can be moved, reducing the output power.

We can increase the acoustic output from any type of diaphragm if we couple directly to a horn, converting the system to a horn loaded one. Fig. 7.28 shows the relative difference in size between the diaphragms of a cone-type tweeter and a horn-loaded one. The driving force of the voice coil of the latter is distributed between the small mass of the diaphragm and the mass of air in the horn. Since air weighs much less than paper or metal the overall load on the voice coil, for the same acoustic output as that of the cone type tweeter, can be greatly reduced. Also, *for the same electrical input, the output of the horn loaded system is greater.*

A *horn* is a tube so flared (tapered) that the diameter increases from a small value at one end called the *throat* to a large value at the other end called the *mouth.* Horns, Fig. 7.28 (*b*), have been used for centuries for increasing the radiation of the human voice and musical instruments.

The horn does acoustically what the cone does mechanically. It couples the small voice coil area to a large area of air. In this way, the horn acts as an *acoustic transformer* and converts the relatively high impedance at the throat and driver.

The horn is a fixed physical boundary for its enclosed column of air and does not vibrate itself. Acoustic energy fed to its throat must therefore be obtained from a vibrating diaphragm which converts mechanical motion from the driver voice coil to acoustic energy. Although *the cone-type radiator acts as both diaphragm and radiator and transducer from mechanical to acoustic energy, the horn acts only as a radiator, with both input and output energy being acoustic.*

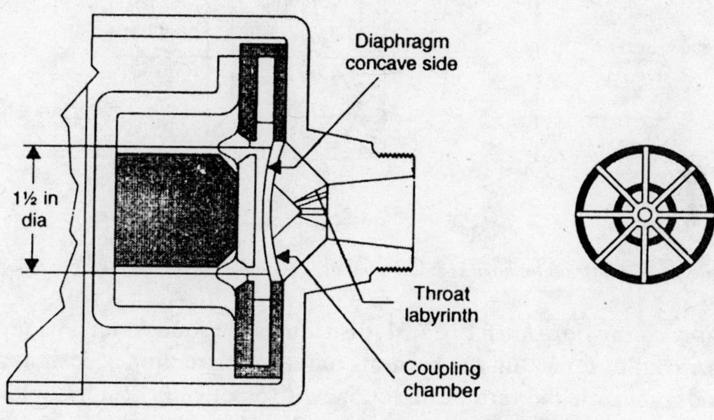

Fig. 7.29 Cross section of a compression horn driver unit.

For maximum efficiency, horn driver units, such as the one shown in Fig. 7.29, are of the *compression type.* The back of the driver is completely sealed to provide a stiff air cushion behind the diaphragm. The sealed-in stiff air cushion is the underlying principle of acoustic suspension, or air-suspension speaker systems.

Any device that radiates energy into three-dimensional space has certain very specific *directional properties.* For example, an unshaded light bulb is pretty much omni-directional, casting illumination in most directions. Like lamps, speakers have definite directional characteristics. At very low frequencies, any speaker is virtually omnidirectional. *The directional characteristics of a speaker change with the frequency of the sound being generated, particularly with how the wavelength of the sound relates to the physical size of the speaker's* diaphragm. For a flat, circular cone, dispersion will be virtually omnidirectional for frequencies with wavelengths that are more than 4 times the diameter of the cone. Dispersion narrows to approximately 60° when the wavelength equals the diameter, and to 30° when the wavelength is half the diameter. A speaker with a 2 - inch radiating surface will be practically omnidirectional at 1,000 Hz, but will radiate a 10,000 Hz tone in a beam not much more than 30° wide.

Horns can be, and are, designed for controlled wide - angle distribution. Such horns are square or rectangular. These horns achieve similar results in different manners.

The *diffraction horn*, Fig. 7.30, operates on the principle that sound coming from a narrow - slit, which is small compared to the wavelength, emerges in a cylindrical wavefront pattern from the slit. That is, *the wavefront diffracts out of the narrow slit*. This type of horn has smooth angular response with no lobes or valleys in the sound pattern.

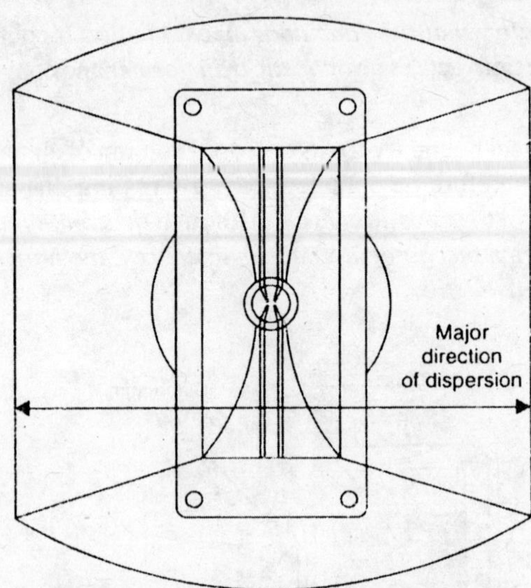

Fig 7.30 The mouth of a diffraction horn is small compared with wavelength, making it a horizontal - dispersion slit source.

The *reciprocating flare horn* Fig 7.31, distributes the sound over a wide angle by reversing the direction in which pressure builds up within the horn. Its mouth construction is designed for horizontal dispersion. The horn first expands rapidly in the vertical direction, with practically no horizontal expansion. The sound pressure travelling along such a channel finds it relatively easy to expand vertically. But in trying to expand horizontally, the sound pressure builds up along the side walls of the horn. The result is that *when the sound finally comes out of the horizontally shaped mouth (i.e., the reciprocating flare), it does so with exceptionally wide horizontal dispersion.*

The *sectoral/multicellular horn*, Fig. 7.32 is similar to the reciprocating flare horn except that the mouth of the horn is compartmentalized to diffuse the sound over a wider area. What we have here is a sort of *shower head* that spreads the sound much like a shower head spreads the water coming from it.

7.18 HIGH FIDELITY

There are several schools of thought concerning high fidelity - *'play it exactly as performed school'*, *'play it so it sounds real school'*, and *'play it the way I like to hear it'* school. There is logic in all three approaches. We must accept that there are as many types of high fidelity as there are listeners. In order to satisfy these concepts, it is necessary to provide a sufficient variety of component parts to make possible many different kinds of system. We have discussed speakers from the point of view of their general application to the hi - fi field.

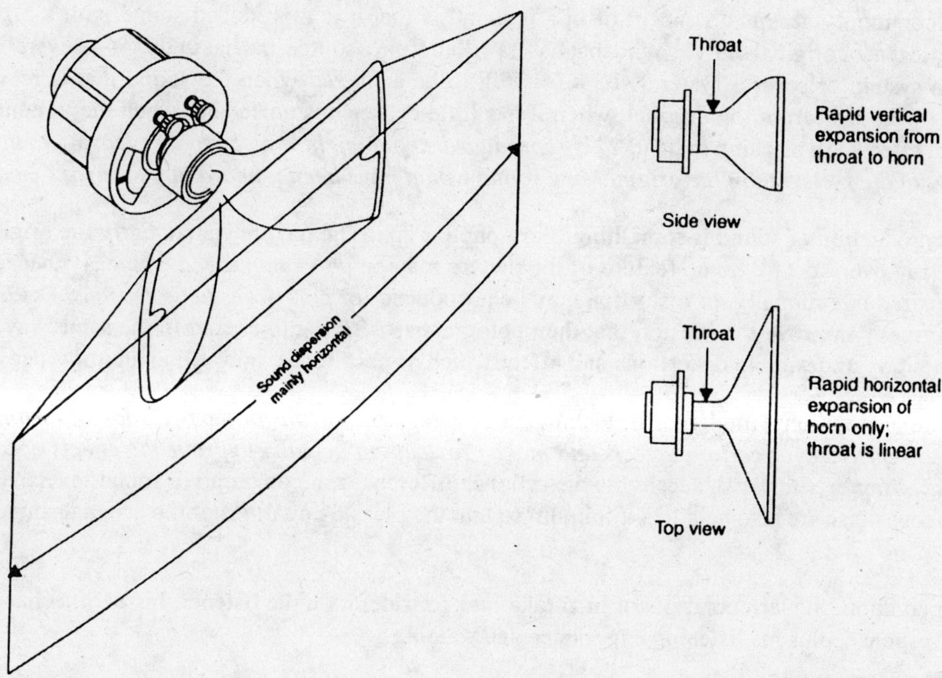

Fig 7.31 The reciprocating flare horn looks like a diffraction type but distributes sound in an opposite manner.

Fig 7.32 The multicellular horn

There is as yet no exact scientific operational definition of a hi - fi system. Standards and specified measurements of performance of a system have not been possible to establish because of the *limitations* of the human ear and because of *variations* in human taste, room acoustics, system distortions, noise and comparative volume levels.

A commonly accepted concept of hi - fi sound is that it is reproduced sound with a high degree of similarity to that of original or live sound that has travelled from a source and has undergone several conversions through a system or several systems. Hi - fi is felt to be achieved when the sound that is reproduced has negligible distortion from the original, when it has little extraneous noise, and when the volume levels and acoustical effects are pleasing to hear. *This reproduced sound might even be more pleasing to the listener at the output of the system than the original live sound would have been if heard at its source.*

Reproduction of sound is something like a photograph. The picture cannot carry the original scene to the viewer in every detail. Some feature of the picture may be *de - emphasized* whereas other features may be *emphasized* intentionally, or distortion may be introduced for *purely aesthetic reasons.* Distortion of this type can greatly improve the illusion that the photographer is trying to create. In the same way, the picture can be spoilt by undesirable distortions and effects, such as poor focus, poor film and improper lighting.

Like photography, modern hi-fi techniques encompass *controls for modification of the original (live) sound to compensate for certain defects and make provision to actually improve the effects according to an individual- listener's tastes.* Undesirable distortions, differences in comparative sound levels, and injection of extraneous noise are also held to a minimum so that the pleasing qualities of the original sound will not be reduced.

In addition, modern concepts of hi-fi take into consideration the listener, his ear mechanism, and his nervous response, plus his listening experience and training.

The word presence is used to describe the degree of realism of the reproduced sound. This term suggests that the reproduction is so real that the listener can feel the presence of the source that is causing live sound, even though the source may be extinct.

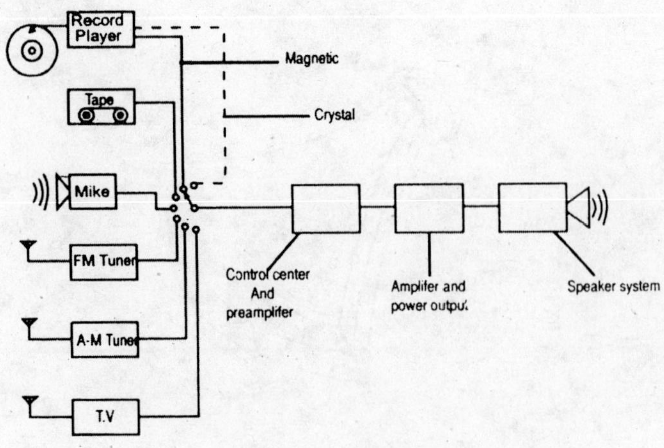

Fig. 7.33 Block diagram of a high-fidelity system.

A complete *hi-fi system* may be divided into functional sections as illustrated in Fig. 7.33. The way in which the output differs from an input or a desired ideal output is called *distortion.* Distortion may be created in anyone or more of these sections. If more than one section is causing distortion, the final output sound may reflect the *sum of the distortions from all distorting sections.* A section may be purposely designed to *introduce* distortion of a type which *compensates* for inherent distortion in another section. For example, bass and treble boost circuits can be used to offset (at least partially) the falling off of the response of a speaker at the highest and lowest frequency portions of the response range.

The hi-fi system is somewhat like a chain, which is likely to be limited in overall performance by its weakest section; but the chain analogy breaks down in the foregoing case in which *the distortion introduced by one section may be used to compensate for distortion in another section.*

The *speaker system* is the weakest link in the hi-fi equipment chain because two conversions of energy must take place-electrical-to-mechanical-to acoustic. Such energy conversion is known as transduction, and the devices which effect it are known as *transducers.*

Input devices such as phono pickups and microphones are also transducers, and have many of the same weaknesses as speakers, though to a lesser degree, because of the relatively *low power levels* at which they operate. Input devices provide transduction between sound input (or physical motion of a phono-pickup needle) and electrical output, just the reverse of the action in speakers.

The *amplifier system* can also contribute distortion. The *voltage amplifier* stages are the least troublesome. The *power amplifier* stage is an important contributor to the overall distortion in the system.

If we have performance goals to aim at we will best know what to look for in *practical but less than perfect equipment.* Keeping in mind the types of imperfections and distortions they can introduce, we may now summarize the features of a *theoretically ideal system.*

1. Interpret, amplify, compensate and reproduce sound components of any and all frequencies in the audible range with good efficiency.
2. Add negligible frequency components not in the original sound.
3. Distribute the sound in such a way that its sources would appear to be located nearly the same as they were in the original and so that the quality of the sound would be independent of the location of the listener with respect to the speaker system.
4. Allow negligible unnatural delay of some frequency components relative to others.
5. Reproduce without resonance effects or hangover, the sudden large changes in sound volume level.

7.19 MULTI SPEAKER SYSTEMS

We will now concentrate on the more specific and specialised units that become component parts in the more expanded hi - fi reproducer systems.

Multi speaker systems have much to offer for good hi - fi reproduction in the way of characteristics that are virtually impossible to obtain from a single wide range speaker *Advantages* derived from multispeaker systems are due to the fact that with two, three, or four speakers in the reproducing system, *we have better control of the overall performance characteristic of the system through control of the individual component speakers.*

The situation may be likened to the difference between having only one ceiling lighting fixture in a room to provide overall illumination and having several lamps in corners and on tables, to provide more adequate light coverage in *specific areas* where light is most needed and in *degrees* best suited to those areas.

Since multispeaker systems are composed of combinations of special purpose speakers, the multispeaker system as a whole is a very efficient system. Multispeaker systems have the following advantages:

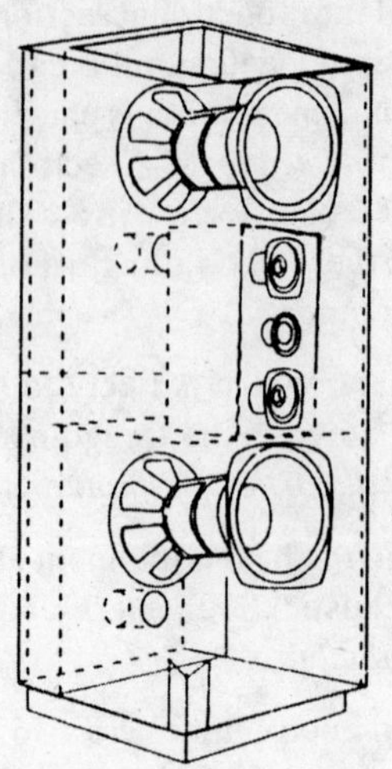

Fig. 7.34 Each cabinet of the multispeaker system is divided into five separate chambers, of which two are of the band-pass bass reflex type.

Fig. 7.35 A typical multispeaker system.

(*i*) Multispeaker systems have *reserve power handling capacity* necessary for high program bursts.

(*ii*) Multispeaker systems *reduce intermodulation distortion.*

(*iii*) Multispeaker systems *may be balanced one against the other by means of volume controls to* give that particular feeling of concert hall reality that most pleases the listener.

(*iv*) A *smooth over all response* may be obtained from multispeaker systems.

(*v*) Multispeaker systems have *compatible 'roll off' ; or 'cut off' characteristics* ; speaker ranges overlap.

(*vi*) Multispeaker systems provide *flexibility of performance.*

7.20 CROSSOVER NETWORKS

In multispeaker systems, in which specialised speakers are used for different frequency bands, *it is necessary and desirable to segregate different bands of frequencies into the respective speakers designed to handle them.* This segregation of the various bands of acoustic energy ensures *optimum utilisation of audio power*, resulting in better overall performance of the system.

The simplest type of network consists merely of a single *capacitor*, as illustrated in Fig. 7.36 (*a*). The fact that the reactance of a capacitor is inversely proportional to frequency is employed to distribute the audio signal. The tweeter and woofer voice coils are connected in *series* and a capacitor is connected across the woofer. The value of capacitance is made such that at frequencies above the range of the woofer the reactance of C becomes so low that it shunts the woofer, which acts as a bypass capacitor.

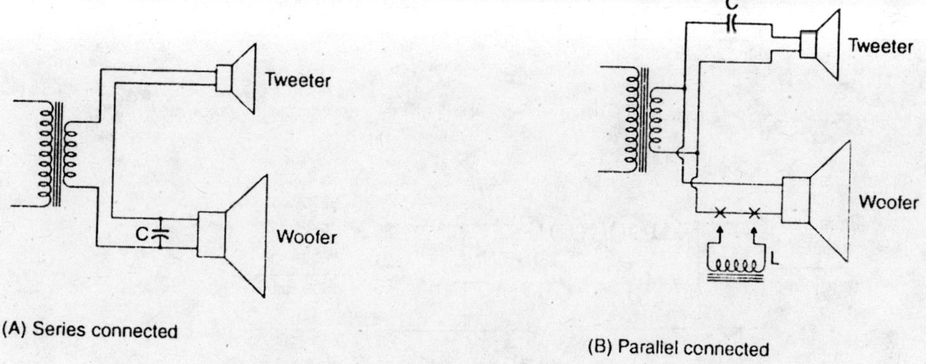

Fig. 7.36 Simple divider circuits employing single reactances.

Low frequency components can be kept out of the tweeter, if a *parallel* connection of voice coils is used with capacitor in series with the tweeter circuit, as shown in Fig. 7.36 (*b*).

Fig. 7.37 Typical crossover networks.

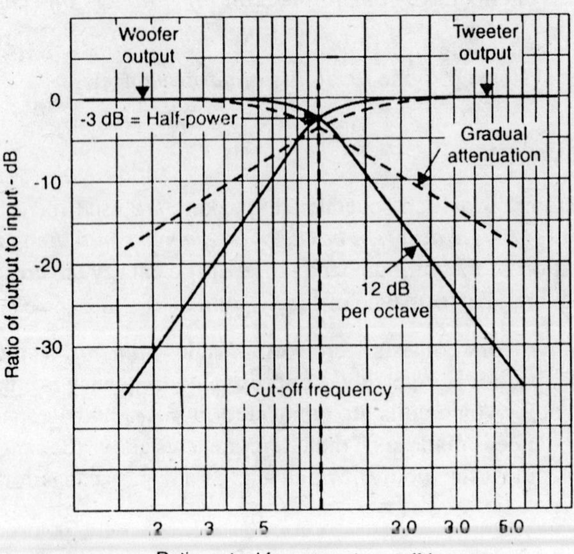

Fig. 7.38 Crossover network response curves.

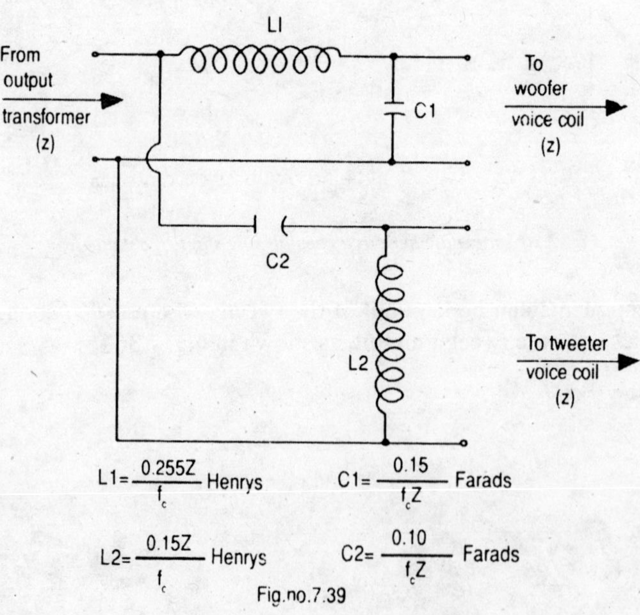

Fig. 7.39 A filter type crossover network.

Inductances can be used along with capacitors to make the crossover network more complete. For example, in Fig. 7.36 (*b*) the inductance (*L*) can be connected in series with the woofer leads as shown. The inductance, the reactance of which increases with frequency, *chokes* the high frequency components out of the woofer, and the capacitor (C) *blocks* low frequency components out of the tweeter. The values of C and L must be such that the reactance in each case is a little lower than the voice-coil impedance in the frequency range to be attenuated.

Although the crossover range should not be too narrow, simple reactance circuits are ordinarily too broad in the changeover region. A combination of *low pass filter* (for the woofer) and a *high-pass filter* (for the tweeter) is usually employed. *With this type of circuit, much more rapid attenuation can be made near the crossover frequency than is possible with simple capacitor and inductor arrangements.*

Attenuation of 12 dB per octave is considered proper in most applications. Gradual crossover arrangements attenuate at about 6 dB per octave. Filters with sharper cut off than this can be constructed by the use of additional components, but power losses in the filter become excessive and the additional sharpness is not necessary anyway.

A typical crossover network response graph is shown in Fig. 7.38. The curve of woofer output crosses the curve of tweeter output response at the crossover frequency. This intersection is at the 3 dB or *half-power level,* at the crossover frequency, half the output power is being fed to each unit. The individual response characteristics of the woofer and tweeter must *overlap* substantially.

One of the most popular types of crossover networks is shown in Fig. 7.39 alongwith the formulas for calculating the values required for any crossover frequency f_c and speaker impedance Z.

Example 7.1. Find the component values for $f_c = 500$ Hz and $Z = 8$ ohms.

Solution:

$$L_1 = \frac{0.255 \times 8}{500} = \mathbf{4.08\ mH}$$

$$L_2 = \frac{0.15 \times 8}{500} = \mathbf{2.4\ mH}$$

$$C_1 = \frac{0.15}{500 \times 8} = \mathbf{37.5\ \mu\ Fd.}$$

$$C_2 = \frac{0.10}{500 \times 8} = \mathbf{25.0\ \mu\ Fd.}$$

7.21. IMPEDANCE MATCHING

It will be useful to give a simple recapitulation of transformer theory. A transformer represents a straight forward application of the phenomenon known as *electromagnetic induction,* whereby any change in the magnetic flux linking an electric circuit is accompanied by an induced electromotive force in that circuit. The *value* of this induced voltage will, at any instant, be proportional to the rate of change of the flux, while its *direction* will always be such as to oppose the change producing it. The change of linkages may be produced by movement of the electric circuit with respect to a magnetic field which is constant, as in a dynamo, or by varying the current in a stationary electric circuit. When the induced e.m.f. produced in one circuit is due to a current varying in an adjacent circuit, as in a transformer, it is said to be due to the *mutual inductance* between the two circuits. The e.m.f. induced in a circuit by a varying current flowing in the circuit itself is said to be due to the *self-inductance* of the circuit. The inductance of a conductor can be *increased* by winding it in a coil and *further increased* by winding the coil on a core of ferromagnetic material.

In general a transformer consists of two insulated coils of wire so associated that the magnetic flux due to a current in one coil is effectively linked with the turns of the other coil. The coils are normally wound on a core of ferromagnetic material which greatly increases the magnetic effect of the current. Usually one of the windings, the one to which the input is connected, is called the *primary* and the other is called the *secondary*. In some instances additional secondary windings may be provided to give alternative outputs.

An elementary transformer corresponding to the form used by Faraday in his original experiments is shown diagrammatically in Fig. 7.40. An alternating current, I_p, flowing in the primary winding will cause an alternating magnetic flux, indicated by the dotted lines, to circulate in the ring core. In so doing, this alternating magnetic flux will link with the turns of both windings and induce, assuming an ideal transformer, an e.m.f. E, in each turn, proportional from instant to instant to the rate of change of current in the primary. This induced voltage will act in such a direction so as to tend to oppose the change of fluse and in the primary winding it will take the form of a back e.m.f. If N_p is the number of turns in the primary winding the total back e.m.f. will be equal to N_pE_1 and neglecting any losses in the trnasformer this will be equal to the impressed e.m.f. E_p. In other words, stability is established when the primary current has a value equal to that required to induce a back e.m.f. and opposite to the applied e.m.f.

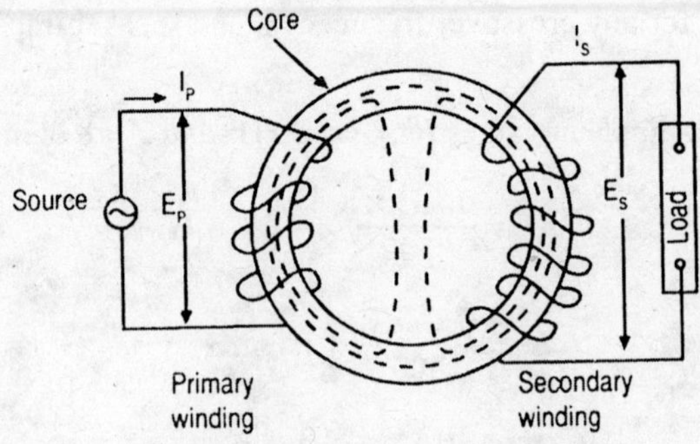

Fig. 7.40. Elementary Transformer

Similarly, if N_s is the number of turns in the secondary winding, the total e.m.f., E_s, induced in the secondary winding is equal to N_sE. We thus have

$$E_p = N_pE \quad \text{...(7.1)}$$

and

$$E_s = N_sE \quad \text{...(7.2)}$$

and therefore

$$\frac{E_p}{E_s} = \frac{N_p}{N_s} \quad \text{...(7.3)}$$

Again assuming an ideal transformer having no losses, the power in the secondary circuit will be the same as that at the input to the primary winding ; thus we have

$$I_pE_p = I_sE_s \quad \text{...(7.4)}$$

so that

$$\frac{E_p}{E_s} = \frac{I_s}{I_p} \quad \text{...(7.5)}$$

We therefore have the following relationships :

$$\frac{N_p}{N_s} = \frac{E_p}{E_s} = \frac{I_s}{I_p} \qquad ...(7.6)$$

Thus the ratio of the primary and secondary voltages is equal to the turns ratio, while the ratio of the primary and secondary currents is the inverse of the turns ratio. The transformer therefore provides a ready means of converting an alternating supply voltage to another value, *either higher or lower* depending on the requirements, and this constitutes one of its most common uses. Where a.c. mains are used to provide power supplies to telecommunication equipment, transformers also serve the purpose of *isolating* the equipment from direct connexion to the mains, both on grounds of safety and to prevent unwanted earth potentials reaching the equipment.

Apart from their uses in connection with power supplies, transformers find very many uses in telecommunication circuits. In signalling and switching circuits they may be used for isolating tow parts of a connection for direct current while permitting the transmission of speech currents between the two parts. They may also be used to connect an earthed circuit to one which has both sides *balanced* with respect to earth.

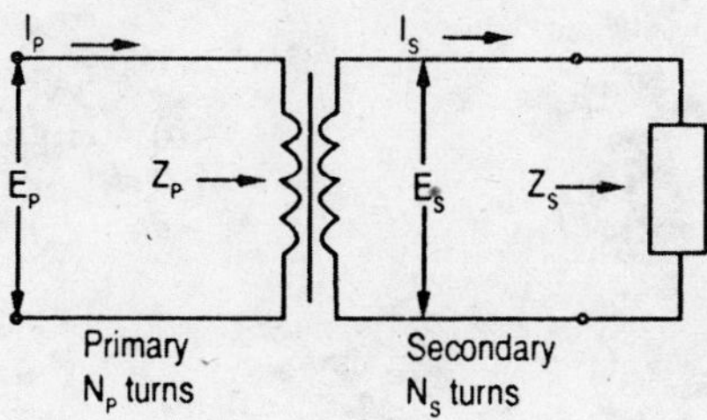

Fig. 7.41. Impedance transformation

A further use of a transformer is for *impedance matching.* This is to allow a low-impedance circuit to be connected to a high impedance circuit or vice versa without loss of power. In this case *the turns ratio required is equal to the square root of the ratio of the impedances to be matched* as can be seen from the following :

Let Z_p = Impedance as measured at terminals of primary winding (Fig. 7.41.)
Z_s = Impedance of circuit connected to secondary winding

Then $$\frac{E_s}{I_s} = Z_s \text{ and } \frac{E_p}{I_p} = Z_p$$

Therefore $$\frac{Z_s}{Z_p} = \frac{E_s}{I_s} \times \frac{I_p}{E_p} = \frac{E_s}{E_p} \times \frac{I_p}{I_s} \qquad ...(7.7)$$

But $$\frac{E_s}{E_p} = \frac{I_p}{I_s} = \frac{N_s}{N_p}$$

Therefore $$\frac{Z_s}{Z_p} = \left(\frac{N_s}{N_p}\right)^2 \quad \ldots(7.8)$$

So that $$\frac{N_s}{N_p} = \sqrt{\frac{Z_s}{Z_p}} \quad \ldots(7.9)$$

OBJECTIVE TYPE QUESTIONS (A)

1. A loudspeaker converts

(*a*) Electrical energy to mechanical energy
(*b*) Acoustic energy to electrical energy
(*c*) Mechanical energy to electrical energy
(*d*) Electrical energy to acoustic energy.

2. The input impedance of a crystal loudspeaker is

(*a*) Resistive (*b*) Inductive
(*c*) Capacitive
(*d*) A combination of (*a*) and (*b*).

3. Crystal speakers are used in

(*a*) Low frequencies (*b*) High frequencies
(*c*) Mid frequencies (*d*) Both (*c*) and (*d*).

4. Electrostatic speakers need a

(*a*) D.C. polarising voltage (*b*) A.C. polarising voltage
(*c*) Either (*a*) or (*b*) (*d*) Neither (*a*) nor (*b*).

5. The d.c. bias (polarising voltage) in electrostatic speakers is of the order of

(*a*) 500 to 1000 V (*b*) 1100 to 1500 V
(*c*) 1000 to 1200 V (*d*) 500 to 1200 V.

6. High class sound systems use electrostatics to reproduce

(*a*) Low frequencies (*b*) Mid frequencies
(*c*) High frequencies (*d*) Both (*a*) and (*b*).

7. High quality dynamic speakers use diaphragms composed of

(*a*) Paper (*b*) Titanium
(*c*) Aluminium (*d*) None of the above.

8. Woofers are speakers designed to reproduce

(*a*) Bass (*b*) Treble
(*c*) Neither, (*a*) nor (*b*) (*d*) Either (*a*) or (*b*).

9. Tweeters are speakers designed to reproduce

(*a*) Low frequencies (*b*) High frequencies
(*c*) Mid frequencies (*d*) Both (*a*) and (*c*).

10. The smallest diaphragm for any half-way decent woofer is
    (*a*) 4″ (*b*) 6″
    (*c*) 8″ (*d*) 12″.
11. High resonant frequencies are obtained with
    (*a*) Light weight (*b*) Stiff suspension
    (*c*) Loose suspension (*d*) Heavy weight.
12. A horn-loaded unit
    (*a*) Increases the acoustic output
    (*b*) Decreases the acoustic output
    (*c*) Multiplies the acoustic output
    (*d*) None of these.

FILL IN THE BLANKS (B)

1. The principle of operation of a loudspeaker is the ___ of that of a microphone.
2. Crystals have been known to __________ under stresses.
3. To overcome frequency doubling we __________ the speakers.
4. The voice coil is free to move ______ but not vertically or laterally.
5. The stronger the magnet, the ______ the speaker.
6. ________ magnets have several advantages over those made of metal.
7. The _________ is that part of the speaker which does the talking.
8. Mechanical inductance is called ____________.
9. Mechanical ____________ is called compliance.
10. Undesired sharpness of the resonant peak can be lessened by __________.
11. Directivity normally varies considerably with _________.
12. The ___________ does acoustically what the cone does mechanically.

CONVENTIONAL QUESTIONS (C)

1. Explain the difference between a microphone and a loudspeaker.
2. What are the requirements of an ideal loudspeaker ?
3. What are the different types of loudspeaker ? Explain the working of any one of them with the help of a neat sketch.
4. Discuss the limitations of ceramic and electrostatic speakers. What are their fields of application ?
5. Draw the equivalent circuit of a dynamic loudspeaker and explain the significance of each component.
6. Explain the following terms :
    (*a*) Compliance (*b*) Inertance
    (*c*) Reflected impedance
7. What type of voice coil and voice coil suspension will you use in
    (*a*) Woofers (*b*) Tweeters

8. Explain in detail the working of a permanent magnet loudspeaker.
9. What is the difference between a cone-type tweeter and a dome-type tweeter ?
10. Explain why the efficiency of an indirect radiating loudspeaker is greater than that of a direct radiating loudspeaker ?
11. Discuss the relative merits and demerits of horn type and cone type speakers.
12. Explain the working of either horn type or cone type loudspeaker with the help of a neat sketch.
13. Why multispeaker systems are required ?
14. Explain the difference between a subwoofer, woofer, mid-range speaker, super tweeter and tweeter.
15. What are the advantages of multispeaker systems ?
16. Explain the working of a simple crossover network.
17. With the help of a block diagram explain a hi-fi system.
18. Explain the following briefly.
    (*a*) Presence (*b*) Distortion
    (*c*) Input devices (*d*) Amplifier system
19. What is significance of a transformer in impedance matching ?
20. How will you match a low impedance to a high impedance ? Give a practical example.

8

SPEAKER BAFFLES AND ENCLOSURES

An essential part of any loudspeaker system is the enclosure in which the loudspeaker is to be used. An enclosure for a loudspeaker is often called a baffle, and perhaps technically speaking, baffle is a more descriptive word than enclosure. The prime purpose of any enclosure is to provide the proper acoustic circuit for the loudspeaker to work with, so that maximum efficiency and best performance can be obtained from the combination. In the effort to provide this acoustic circuit, the sound coming from the loudspeaker is routed into certain paths and prevented from going into other paths by blank walls put in its way. Thus the term baffle as used in a technical sense connotes a means of routing the sound energy. Since most of today's baffles are built into more or less complex box like structures the words baffles and enclosures are used interchangeably.

8.1. ACOUSTIC DOUBLETS

The cone of a loudspeaker moves as a *unit.* When the cone moves *forward,* the air in front of it is *compressed* and the air in the back of the cone is *rarefied* The sound released from the rear of the cone is of opposite phase to that released from the front of the cone. If the sound from the rear is allowed to flow so that it meets the sound from the front *cancellation* takes place and the response of the speaker drops off sharply.

Such an arrangement, Fig. 8.1, is called an *acoustic doublet* in which the two sides of the diaphragm are represented by a negative and a positive pole radiating respectively a negative and a positive wave (see Fig. 8.1 (*b*)). If these doublet sources are close together, it is obvious that the negative wave from one will be directly cancelled by the positive wave from the other at almost all points in space. This is true, however, if the poles of the doublet are so close that their spacing is only a small fraction of the length of the radiated wave. This condition is readily recognised in any speaker when tested on a workbench *without benefit of any baffle at all.* Under this condition, the *doublet mode of operation* is most effectively working, with the result that *there is a complete absence of low frequency response.* To prevent low frequency cancellation resulting from doublet operation, we should place some *barrier* between the front and the back.

The wavelength of sound becomes longer as the frequency decreases. Front-to-back interference is worst at the lowest frequencies and ordinarily marks the *cut off frequency* of the speaker mounting. At *higher frequencies,* the compressions and rarefactions are so closely spaced that there is no definite general cancellation action as at the *low frequencies.*

8.2 BAFFLES

By extending the edges of the cone with some rigid flat material, we make it necessary for sound waves from the rear to travel out to the *edges of the material* before they can meet the sound waves from the front and interfere with them. The added material is called, a *baffle,* and its principle is illustrated in Fig. 8.2.

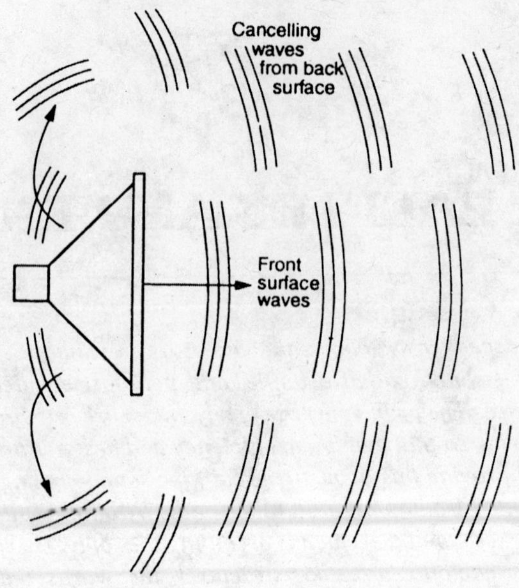

(a)

Diaphragm moving in this direction creates a pressure pulse in front and rarefied section in back

Rarefaction

Pressure

One wavelength

One wavelength

At any point the positive and negative intensities practically cancel each other resulting in no sound pressure

Theoretical diaphragm with positive pulse from one side, and negative pulse from other side

When wavelength is large compared to pole spacing, radiation is effectively "dipole" type and cancellation takes place

(b)

Fig. 8.1 Doublet operation of unbaffled diaphragm.

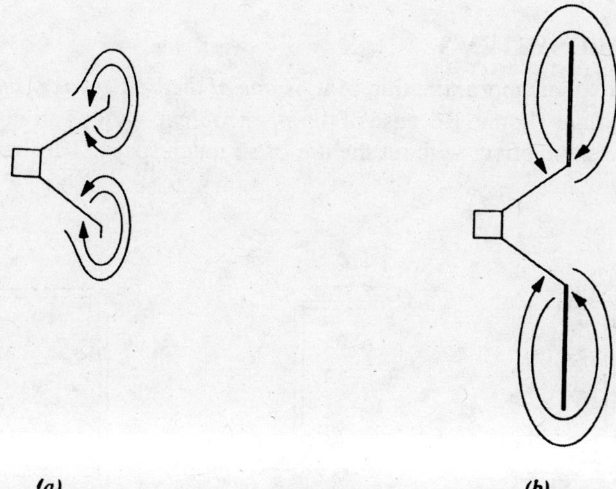

Fig. 8.2 Baffle eliminates doublet action by increasing back-to-front wave path length by one-half wavelength.

With the speaker alone Fig 8.2 (*a*), the front and rear waves must travel only along one side of the cone to meet at the edge. This path is so short that a speaker alone without baffle will usually not reproduce much below about 350 Hz. Fig 8.2 (b) shows the situation with a baffle added. The length of the interference path is increased by the width of the baffle material on each side of the cone.

Baffles should be made of good sound insulating material and should be soft enough to prevent rattle. Soft woods are satisfactory but a material like Celotex is more appropriate. The speaker must be securely fastened to the baffle, and the baffle must be rigidly mounted to prevent rattle.

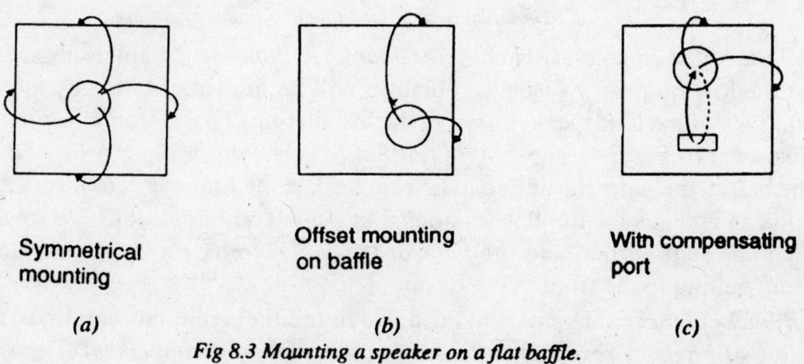

Fig 8.3 Mounting a speaker on a flat baffle.

If a baffle is to be of limited size, the speaker should not be mounted in the centre. The centre is a bad position because the path lengths to all four edges are the same and the frequency components, on one side, at which the path length is one wavelength are severely attenuated, Fig 8.3 (*a*). If the speaker is moved towards one *corner*, Fig 8.3 (*b*), the uniformity of response is much better because the path length to each edge is different and the interference attenuation is distributed. *The interference does not have to take place around the edge of the baffle, but sometimes is deliberately made to take place through a port,* as illustrated in Fig 8.3 (*c*). The design principle of such an arrangement is to *equalise* the response peak due to speaker resonance by the sharp attenuation around that frequency by sparing the port so that the sound travels a half wavelength.

8.3. INFINITE BAFFLE SYSTEMS

An infinite baffle or an approximation of it, is *one of the best speaker mountings.* However, its size is a disadvantage in an ordinary home. Because of the *size problem,* various arrangements have been developed in an attempt to get the same effect without the use of so much space. This has led to the design of *speaker enclosures.*

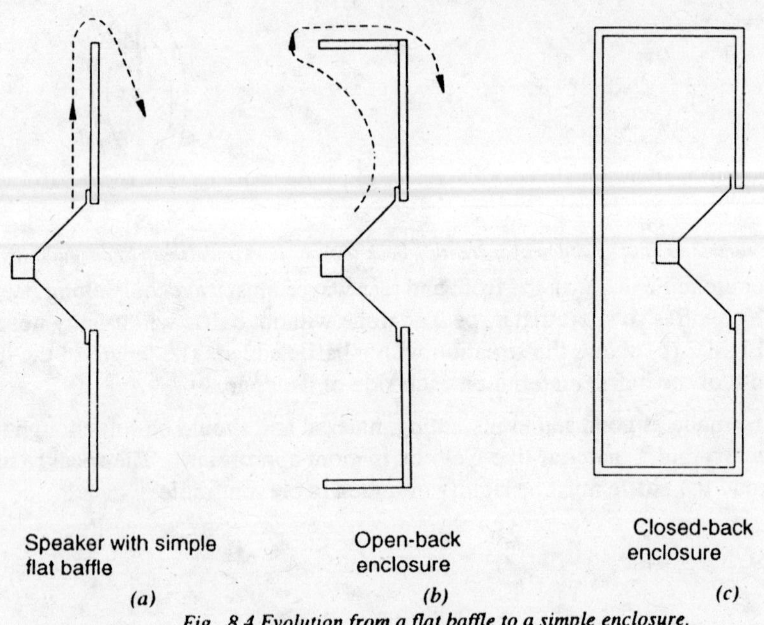

Fig. 8.4 Evolution from a flat baffle to a simple enclosure.

A true infinite baffle completely eleminates doublet action. The complete absence of doublet operation well ensure that the low frequency response obtained will be limited not only by the baffle, but by the low frequency capabilities of the loudspeaker used in the installation. The evolution from a flat baffle to a simple enclosure is illustrated in Fig 8.4. The *simple flat baffle* is shown in Fig 8.4 (*a*). To reduce the maximum dimension of the baffle, the outer portions can be bent back at the four edges to form an *open box,* Fig 8.4 (*b*). The path length is as great as for the flat versions, but the lateral dimensions are smaller. In Fig 8.4 (*c*) the process is carried one step further, and *the back of the enclosure is enclosed,* the back prevents any sound from the rear from getting to the front.

The *open-back cabinet* arrangement of Fig 8.4 (*b*) is the one commonly employed for radio and television receivers. *The low - frequency response is lacking in this type of arrangement.* The small speaker used and the cabinet are limiting factors. Large console models employ larger speakers and larger cabinets, but the path length is seldom sufficient to allow reproduction below 150 Hz, unless some special cabinet design is used. The open - back box speaker has yet another disadvantage. *Any sudden sound peaks of any frequency or sounds of low frequency can shock - excite the box into oscillation at the resonant frequency.* This accounts for the fact that many console radio receivers and some of the earlier juke boxes emitted a *constant booming sound* during reproduction of music.

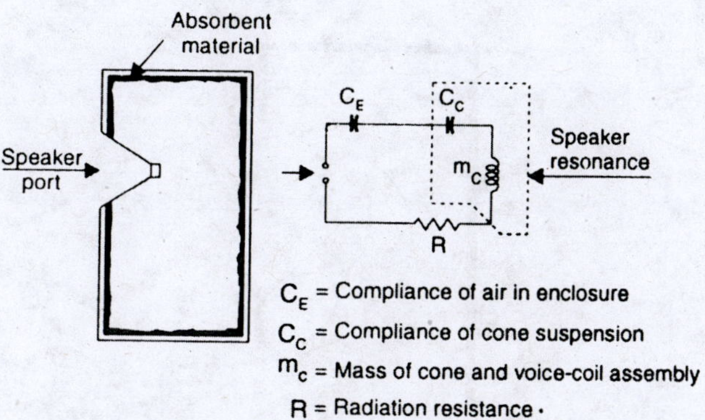

Fig. 8.5 Equivalent circuits of closed - box enclosure.

As soon as the box is *closed* as in Fig 8.4 (*c*), the air in it is no longer free to move in the open as in the open -back cabinet. Instead, the action of the cone causes pressure changes in the cabinet rather than combination of pressure and velocity. This means that the compliance of air is an important factor. Compliance is *acoustic capacitance* and combines with the compliance of the cone suspension in such a way as to raise the resonant frequency of the system to a value higher than that of the speaker alone.

The mass (inductive effect) of the cone, the compliance (capacitive effect) of the cone suspension, and the compliance (capacitive effect) of the air in the box are all effectively in series with each other, Fig 8.5. *An important feature of the closed box is the fact that, because there is no motion of air in and out of it, there is no inertance or inductive effect. There is only compliance, or capacitive effect.* Accordingly, the box does not resonate itself as do some other types. All it does is enter the resonant circuit of the speaker unit and raise the resonant frequency of the system above that of the speaker. *Generally, a closed box of a given volume will raise the resonant frequency of the system by a given percentage above the resonant frequency of the speaker.*

Because the closed box keeps back radiation from getting around to the front and interfering, it is frequently referred to as the *infinite baffle.* The closed box enclosure is not at all equivalent to a true infinite baffle unless it is so large that its effect on the resonant frequency is negligible.

In the infinite - baffle enclosure, all radiation from the rear of the speaker is eliminated. This minimises interference but also makes systems of this type *less efficient* than others which utilise the sound from the rear.

One way to overcome this loss of efficiency is to use two or more identical speakers in the enclosure. *When two or more speakers operate on the same audio frequency and are mounted close to each other, they interact.* Sound energy from one speaker reinforces the vibration of the cone of the other speaker and vice versa, in what is known as mutual coupling. The action is similar to that in a multielement antenna array. The result is that *the efficiency of the combination is much better than the efficiency of either speaker alone.*

Even speakers of the same model are seldom exactly alike, because of tolerances in manufacture. These tolerances are actually an advantage because the difference in the resonant frequencies of the respective speakers tends to distribute the speaker resonance effect and make low - frequency response more uniform.

8.4 BASS - REFLEX SYSTEMS

In another, more popular arrangement, *radiation from the rear of the cone is used to reinforce the front radiation rather than cancel it.* This arrangement is shown in Fig. 8.6, and is known as the bass - reflex enclosure. It is the same as the closed box except that an opening is cut below the speaker. This opening, or *port,* allows air to flow in and out of the box as the speaker cone moves back and forth.

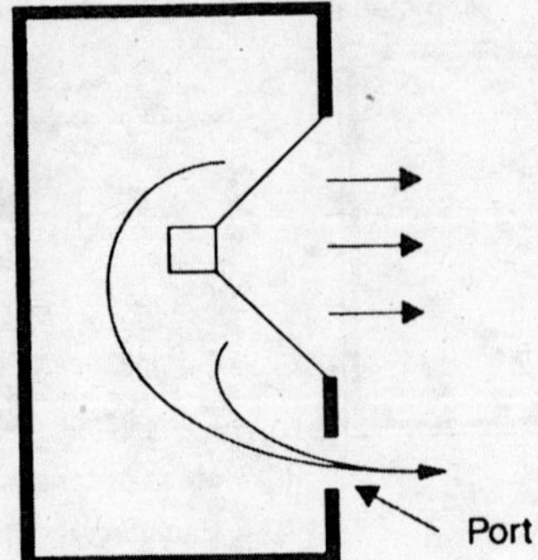

Fig. 8.6 Port in bass reflex cabinet is cut so that back waves from speaker reinforce front waves at low frequencie

Addition of the port adds acoustic inductance (inertance) to the load offered to the speaker by the cabinet, because of the motion of air through the port. The flow of air through the port is also accompanied by friction, which is an acoustic resistance effect. *The port inertance, friction, and enclosure compliance form a parallel-resonant circuit,* Fig 8.7. This circuit is connected effectively in series with the series - resonant circuit of the suspension and mass of the cone of the speaker.

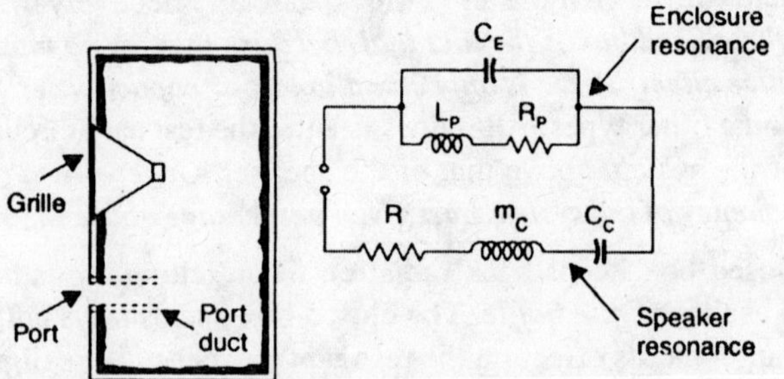

L_p = Port inertance (acoustic inductance)
R_p = Acoustic resistance of port

Fig. 8.7 The bass - reflex enclosure and its equivalent circuit.

The effective length of the path of the sound from the rear of the cone is primarily dependent on the nature of the impedance offered by the enclosure itself, as expressed in the *equivalent circuit.* The design is simply a matter of keeping the resonant frequency of the enclosure in proper relation to the resonant frequency of the speaker. *Because the enclosure offers parallel resonance, and the speaker series resonance, one of these resonances can be made to offset the other.*

Because the bass - reflex enclosure reverses the phase of the back radiation and makes it additive to the front radiation at low frequencies, it is often referred to as an *acoustic phase inverter.*

Some designers prefer to make the bass - reflex type of enclosure with *distributed ports.* Instead of one port of calculated size, a number of smaller ports are used. *The areas of distributed ports add up to the proper calculated port area.*

8.5 ACOUSTIC LABYRINTH SYSTEMS

The labyrinth is *a tuned pipe with one open end.* When the wave from the speaker reaches the end of the pipe, it spreads out into the listening environment causing a sudden pressure drop which reflects back through the pipe to the speaker as a rarefaction. At the *quarter wave frequency* (λ/4) of the sound, the air in the mouth of the pipe is at *minimum velocity* but *maximum pressure* (Bernoulli's theorem). This condition produces an accompanying maximum rarefaction reflected back to the speaker cone. Here, the antiresonant action of the pipe offers maximum damping to the speaker.

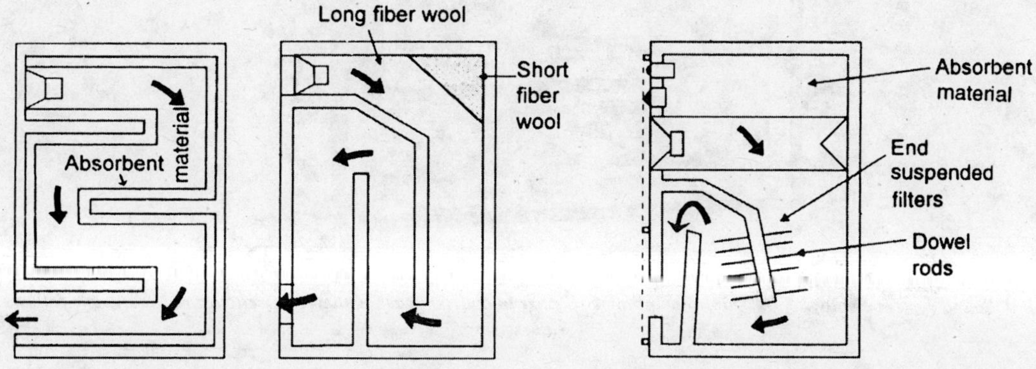

Fig 8.8 Drawings show, interior details of original labyrinth (1936), transmission line (1965) and current design of acoustic labyrinth speaker. In all cases, the small arrows indicate the direction of the rear radiation from the speaker.

At the frequency at which the length of the pipe is a *half - wavelength* (λ/2) of the sound, the air in the mouth is at low pressure but high velocity. Because there is no sudden change in pressure as the wave moves out of the pipe, there is no antiresonant action by the pipe, and the speaker cone is able to move freely. And, because of the λ/2 phase shift (180°) within the pipe, the emerging wave is *in - phase* with that coming from the front of the cone, adding to the speaker's output.

The Olney labyrinth (1936) utilised a pipe cut to λ/4 at the speaker's resonant frequency. The λ/4 action of the pipe damped the speaker's resonance, reducing boom; the λ/2 action added to the speaker's output in the octave above resonance. Olney chose the λ/2 frequency to be about 75 Hz. His 8"speaker had a 50 Hz free-air resonance reduced to 40 Hz by the labyrinth. Olney noted the presence of objectionable resonances, which were produced as harmonics of the fundamental resonance of the enclosure. To counter this he lined the walls with sound absorbing material which combined with bends in the pipe, eliminated the output from the labyrinth at frequencies greater than 150 Hz.

Bailey developed *stuffed transmission line* (1965). The theory behind it is that an infinitely long acoustic line, or pipe, will absorb the back wave from the speaker without troublesome reflections that can produce standing waves behind the cone. A line of infinite length, of course, would be impractical; so, Bailey substituted *stuffing* for length. His long - fibre wood stuffing attenuated the back wave so effectively that the sudden pressure changes at the mouth, with their attendant reflections, were avoided. Extremely low frequencies were not attenuated and escaped to increase bass response. Bailey claimed, that his transmission line, with its less abrupt low-end cutoff, produced less ringing than conventional ported enclosures.

The current design uses a tapered pipe with variable density damping. This transmission line damps three resonances-air in the top chamber resonating with the mass and compliance of the driver tube resonance, and free air resonance of the driver - thus broadening and extending the bandpass.

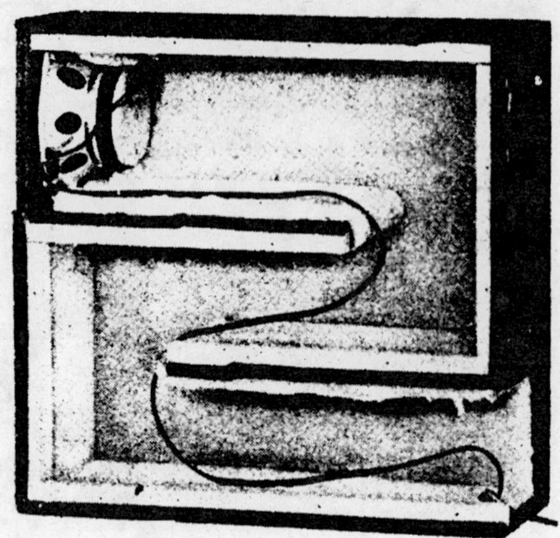

Fig. 8.9 Sound system employs 5 1/4" high-compliance speaker in quarter-wave tunnel reflex enclosure. Rear wave follows path of the wiring.

8.6 FOLDED - HORN SYSTEMS

Outdoor speaker systems and a great many movie theater sound systems employ speaker coupling systems known as *exponential horns.* Such horns with their constant rate of flare provide an *acoustical match* between the cone of the loudspeaker and the air to which sound energy must be coupled (see Fig. 8.10).

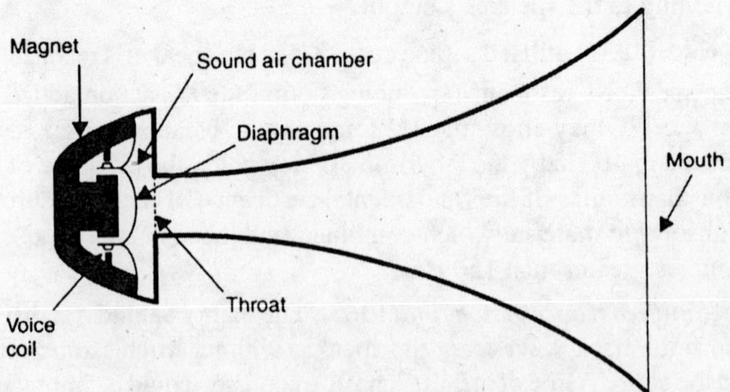

Fig. 8.10 Cross-sectional view of a horn-type reproducer.

To be effective horns must be designed to a length that is determined by the lowest cutoff frequency desired. For example to be able to use a horn design down to 32 Hz the horn would have to be nearly 9 ft. long. Even the average listening room would not accommodate one of these. Nevertheless, *horn designs provide by far the greatest speaker efficiency attainable among the systems discussed.* Fundamental horns are often used in tweeter or high frequency arrays in multi-element speaker system designs.

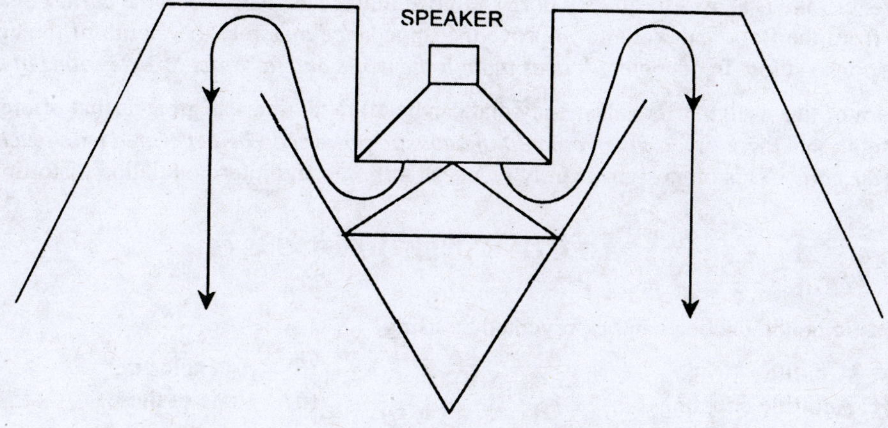

Fig. 8.11 Front-loaded folded horn design gives small cabinet dimensions.

For use at lower frequencies, many manufacturers have evolved variations of the horn principle called the *folded horn* design. A cross - sectional view of such a design is shown in Fig. 8.11. *In this folded configuration, the design lends itself to more acceptable furniture concepts in speaker enclosures.* Well-designed systems based upon the folded - horn principle are quite popular because of the vast amounts of sound they can produce for a given amount of amplifier driving power.

8.7 CORNER FOLDED HORN SYSTEMS

Shown in Fig. 8.12 are top and side cross-sectional views of *the most enduring horn design* - the Klipschorn - invented and patented by Paul W. Klipsch.

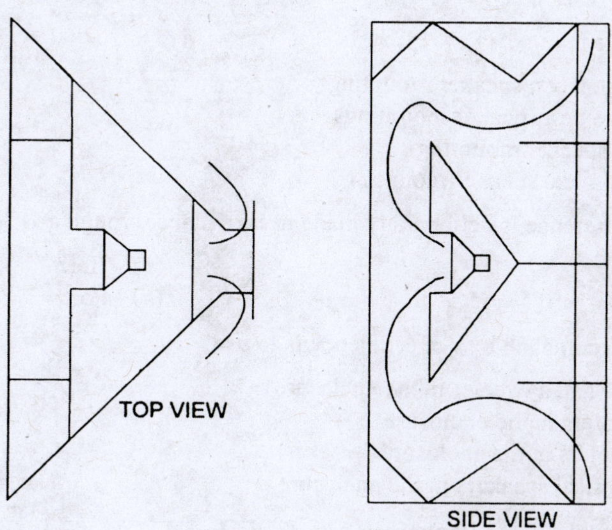

Fig. 8.12 Top and side views of Klipschorn design. The walls and the floor of the room extend the horn principle.

The enclosure is a low - frequency horn so folded that it can be placed in the *corner* of a room to utilise reflections from the floor and walls to improve the impedance match at the mouth of the horn and thereby increase response at low frequencies. *It is as though the walls and floor act as extensions of the horn itself.*

A horn of this design offers efficiency that can be 10 to 50 times as great as that offered by an infinite - baffle design ; and therefore, *a given amount of acoustic power can be generated with much less excursion of the speaker cone.* This often significantly reduces harmonic and intermodulation distortion.

OBJECTIVE TYPE QUESTIONS (A)

1. Acoustic doublet action can be prevented by using

(*a*) A baffle (*b*) An enclosure
(*c*) Acoustic coupling (*d*) None of these.

2. If a baffle is to be of limited size, the speaker should not be mounted in the

(*a*) Corners (*b*) Centre
(*c*) Either (*a*) or (*b*) (*d*) Neither (*a*) nor (*b*).

3. A true infinite baffle completely eliminates

(*a*) Speaker resonance (*b*) Doublet action
(*c*) Acoustic coupling (*d*) None of these.

4. Many console radio receivers and some of the early juke boxes emitted a constant.

(*a*) Booming sound (*b*) Metallic sound
(*c*) Wavering sound (*d*) None of these.

5. Compliance is

(*a*) Acoustic capacitance (*b*) Acoustic inductance
(*c*) Acoustic resistance (*d*) None of these.

6. An infinite baffle is

(*a*) One of the worst speaker mountings
(*b*) One of the best speaker mountings
(*c*) An ideal speaker mounting
(*d*) An economical speaker mounting.

7. Sometimes interference is deliberately made to take place through a

(*a*) Cone (*b*) Iorn
(*c*) Baffle (*d*) Port.

8. One way to overcome the loss of efficiency is to use

(*a*) A woofer and a tweeter in the enclosure
(*b*) Two woofers in the enclosure
(*c*) Two tweeters in the enclosure
(*d*) Two identical speakers in the enclosure

9. A bass reflex enclosure is often referred to

(*a*) An acoustic lens (*b*) A series tuned circuit
(*c*) An acoustic phase inverter (*d*) A parallel tuned circuit.

10. The current design of labyrinth systems uses a
   (*a*) Tapered pipe with variable density damping
   (*b*) A cylindrical pipe with variable density damping.
   (*c*) A tapered pipe with constant density
   (*d*) A cylindrical pipe with constant density.

FILL IN THE BLANKS (B)

1. In an acoustic doublet mode of operation, there is a complete absence of ________________ response.
2. In a closed box enclosure there is only ______________ effect.
3. The low-frequency response is lacking in _____________ cabinet arrangements
4. In a bass reflex enclosure radiation from the rear of the cone is used to _______________ the front radiation.
5. Horn designs provide the ______________ speaker efficiency.
6. In a corner folded horn system the walls and the floor act as _____________ of the horn itself.

CONVENTIONAL QUESTIONS (C)

1. What is the need for a baffle ?
2. Explain briefly the different types of baffles.
3. How infinite baffles were evolved ?
4. Differentiate between an open box and a closed box enclosure.
5. What is an acoustic doublet ? Explain the doublet mode of operation. How can doublet action be eliminated ?
6. Discuss the action of an infinite baffle. How can an enclosure act as infinite baffle ?
7. Explain the working of an acoustic labyrinth system with the help of a neat sketch.
8. Compare a folded horn system with a corner-folded horn system.
9. Draw and explain the equivalent circuit of a closed - box enclosure.
10. Explain the following :
   (*a*) Acoustic phase inverter (*b*) Distributed ports.

9

MONO, STEREO AND QUAD

High fidelity means the reproduction of sound with a high degree of faithfulness to the original. For decades, this re-creation of live concert performance in your home has been the goal of audio industry. The whole chain of hardware - from the microphones which recorded the sound to the loudspeakers that reproduced the sound in your living room - has been improved to the point where there is practically no difference in the live and recorded sound. The audio industry had actually arrived at this point, using a single channel of sound (monophone) back in the late 1950s.

9.1 MONOPHONY

The word *monophony* is derived from two Greek roots : *mono* meaning one and *phone* meaning sound. Thus the combined roots mean one sound or, as we use it in hi-fi parlance, *one-source sound.* To be more specific, monophony (mono) signifies, *one-source sound, one amplifier channel, one speaker system.*

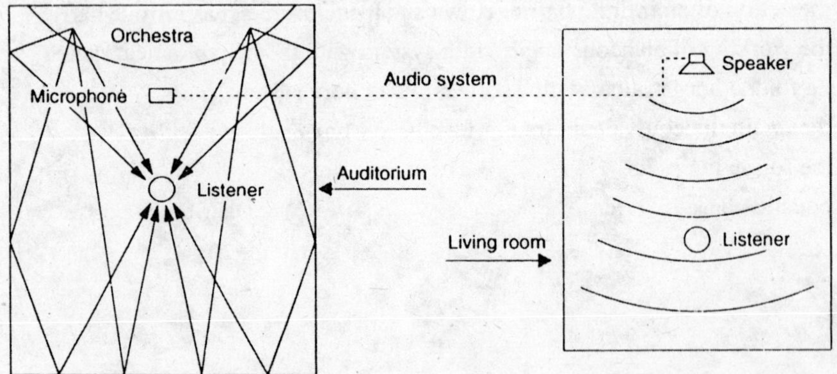

Fig 9.1 Basic operation of single channel (monophonic) recording and playback system.

Yet another term used in describing monophonic sound is *monaural.* Since mono means one and auris means, ear, the root of monaural is *one-ear.* The term monaural was used for some time to describe what we now know as monophonic. However, due to limited scope of the word monaural the word monophonic, meaning *one channel from start to finish,* has superseded it.

Fig 9.1 is a simplified illustration of how an orchestra was recorded, back in the late 1950's. Sometimes multiple microphones were used in a session, but they were mixed down to a single signal which eventually made its way to the listening room environment of your home. But *there was still something missing from the concert-hall environment.*

9.2 STEREOPHONY

The word *stereophony* is derived from two Greek roots, *stereo* meaning solid and *phone* meaning sound. Thus stereophony denotes solid or three dimensional sound - *sound coming from different sources, at different locations, with different volume levels* (two or more separate channels and speaker systems).

The other term often associated with hi-fi systems is *binaural.* This word is derived from two roots: *bi* meaning two and *auris* meaning ear. Thus binaural may be literally translated as *two ear sound.*

In a *monophonic sound system,* the sound usually emanates from only one location when being reproduced. Dual speakers, large horns, and location of the speaker in a corner of the room, can be used to *spread* the sound so that it is difficult to place the sound source at one point. However, *there is no stereo effect until two separate channels utilising several microphones, amplifiers and speaker systems are used.* While monophonic sound may be very pleasing to listen to, stereophonic sound has advantages for the music lover that cannot be equalled by monophonic systems.

The association of binaural sound with stereo-sound arises from the fact that the binaural system was an early method of obtaining stereo effect. *The basic idea behind binaural techniques is the fact that we, as human beings, have a sensation of direction in the sound we hear, because our two ears work separately.* The sound sensation to the brain from one ear is kept separate from the sound sensation from the other ear, and both are transmitted to the brain through separate auditory nerves. The brain compares the two auditory signals received by the ears and, from the differences between them, determines the *direction* from which the sound came.

The theory behind *stereophonic systems* (binaural systems) is that if the two sound signals, which would be heard by a pair of human ears at the source, are transmitted, reproduced in the same relationship, and applied to the corresponding ears at a remote location, all the *directional effects* of direct line listening will be preserved.

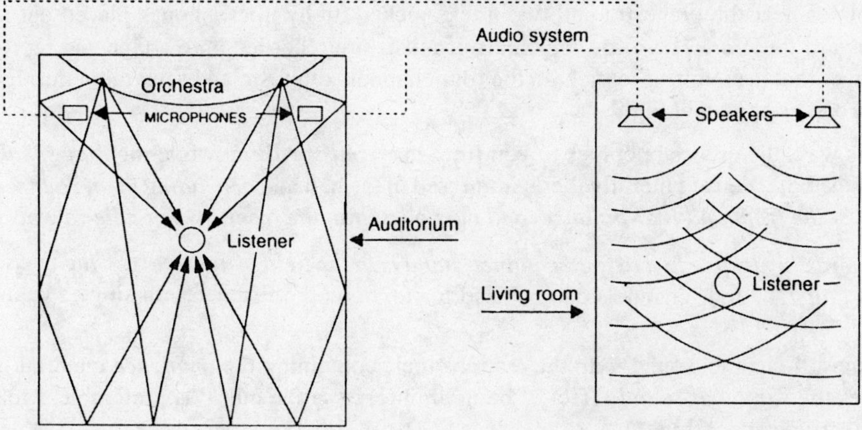

Fig 9.2 Two-channel (stereophonic) recording and playback is shown in this diagram.

Stereo was developed in the late fifties and increased the *concert-hall feeling* considerably. As Fig 9.2 shows, an orchestra was no longer beamed to the listener from a single point but was dispersed across a line stretching *between* two speakers in a virtual curtain of sound. The orchestra now had *breadth.* The recorded orchestra was now beginning to sound like its *live counterpart.*

Modern stereophonic sound, with its directivity and depth properties, adds the *third dimension* to the sound. It makes clear distinction between foreground, middle and background, as well as between right, middle and left sound sources. *Stereo has therefore been able to produce a greater amount of clarity and instrumental sound colour than monophonic sound.* The reception of complex sound sources is also made possible. The directional effects and small time delays from echoing and reverberation of the elemental sounds of music are separately channeled from source to ear, thus *providing high-quality simulation of live music.*

However, one still did not have the illusion of *being there.* The problem pertains to how our ears and brain interpret sound - psycho-acoustics. In a concert hall, we are immersed in the sound coming from all directions: the *direct sound* from the orchestra on stage, the *reflected sound* bouncing off the side walls, the ceiling, and the rear wall; and the *sounds of audience* clapping, talking, coughing, or moving in their seats. All of these sounds are present in the hall during a live concert.

Acoustic engineers have always been concerned with the *liveness* or *ambience* of a particular hall. *Four-channel sound was conceived as a means of fooling your brain into thinking you are at a live performance. It is an illusion of being there.*

The entire electronics industry is a dynamic one because of change; and *each change has improved the overall enjoyment of the consumer.* From radio to television, from black-and-white TV to colour TV, from tubes to transistors, from transistors to integrated circuits, from wire to magnetic tape - each step has caused some problems, but they were easily overcome. The advent of 4-channel does not mean that 2-channel sound is obsolete any more than colour television killed black-and-white television.

9.3 QUADRAPHONY

The word *quadraphony* is derived from two roots, *quad* meaning four and *phone* meaning sound. Thus quadraphony denotes *four-channel sound;* see Fig 9.3.

Let us now look at the various ways of achieving 4-channel sound. *The most straight forward method is called discrete.* This is a copy of the master tape which consists of the two tracks of music that was fed to the microphones near the orchestra and two tracks picked up by microphones placed out in the hall itself - generally toward the *rear.* These four channels of music are recorded onto a tape and reproduced by means of a tape player that is equipped to pick up the four channels of music and send them through four amplifiers to four speakers, as shown in Fig 9.4.

In this way, the orchestra comes to you from the *front speakers,* while the *rear speakers* re-create the ambience of the hall. You are literally there, immersed in the hall and *depending upon your seating preference, you can adjust the front and rear balance, and put yourself in the front row, middle row or away in the back.*

An interesting by product of four-channel sound is in the field of pop music which was never performed in a concert hall. The four channels can be used to *surround* the listener with singers or instrumentalists in the group.

At times, 4-channel sound with the rear channels containing the *ambience* material is pleasing, while at other times the *surround-sound* effect of being immersed in the music is preferable. Musical enjoyment , like art, is a very personal thing.

The *matrix method* of achieving 4-channel sound consists of encoding four channels of information into two channels by *mixing* them together in a complex phase and amplitude relationship; see Fig 9.5

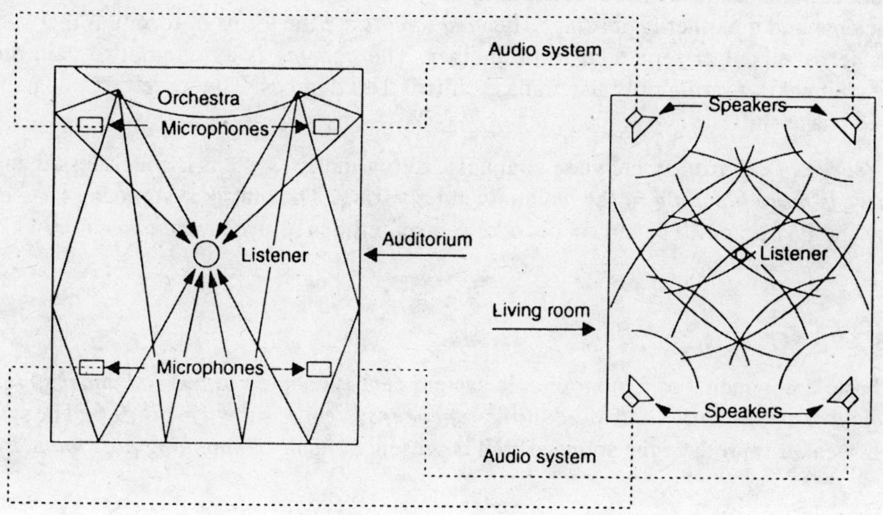

Fig 9.3 Four-channel sound reproduction gives an illusion of concert-hall ambience.

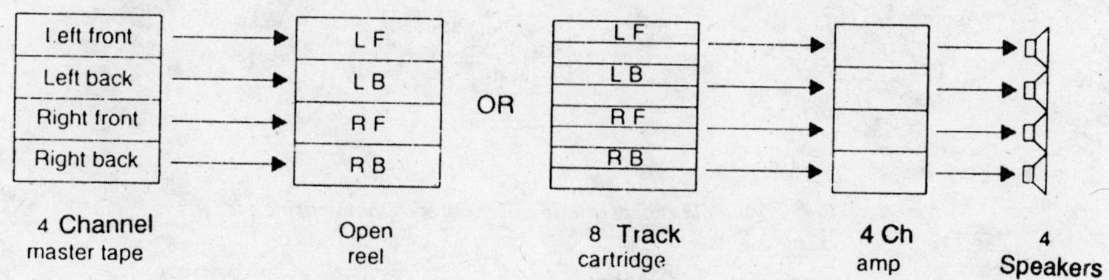

Fig 9.4 The discrete four-channel systems keep the individual channels separate.

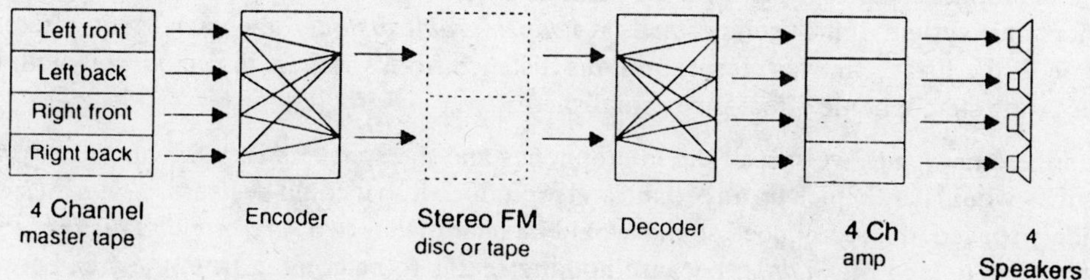

Fig 9.5 In a matrix system, the four channels are combined into two as illustrated here.

These two channels of information can then be pressed into a normal stereo record, broadcast over an FM-stereo station, or recorded onto a 2-channel stereo tape. When you play these two channels of music through the proper equipment, including a *matrix decoder,* the two channels will be restored somewhat to the original four channels.

Various companies have introduced matrix decoders. There are two main factors to be considered - one is the coefficients and the other is phasing. The *coefficients* are the terms in formulas that specify how much of each channel is mixed or separated from another. The *phasing* is an attempt to gain more distinctness between the channels. A commonly used phase shift is 180 degrees. The more complicated matrix circuits use 90 degree phase shifts.

The *degree of separation* between channels from a matrix system is much less than with a discrete system. *There is more blending of the channels in matrixing.* The matrix system can be used for 2-channel records, tape and FM stereo. The matrix decoder can sometimes be used to *enhance* normal 2-channel stereo music.

9.4 STEREO PRACTICE

The basic components of a monophonic system can be adapted to *two - channel stereo* by adding yet another *channel of amplification* and an additional *speaker system* as shown in Fig. 9.6. The system is designed to have each speaker reproduce the sound which is present at its *corresponding position* at the source.

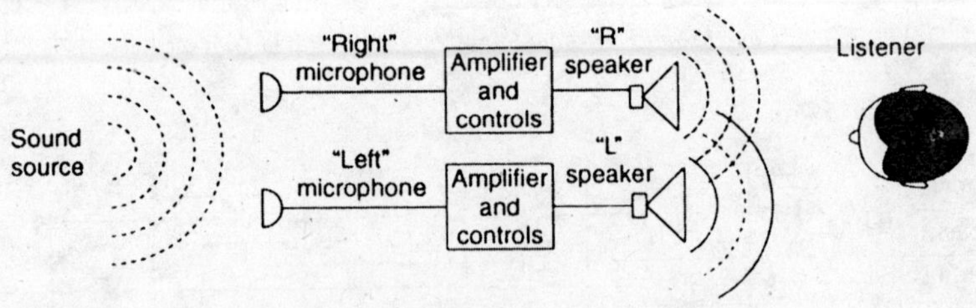

Fig. 9.6 Simple representation of the stereophonic principle.

In stereo practice, proper *positioning* of the equipment and audience is an important consideration. Another important consideration is speaker and microphone *placement* - how far apart the microphones should be from each other when *recording* and how far the speakers should be from each other when *reproducing* the recording.

If the microphones and speakers are *too close* to each other, they will appear as one, and we will then have a monophonic system. If the listener stands at a *considerable distance*, they would still effectively appear as one source. If the listener moves closer, until his distance from speakers is *comparable* with the distance between them, the stereophonic effect is re-established.

If, on the other hand, we spread the microphones and the speakers *very far apart,* they would be so distant that they would hardly pick up any sound and reproduce it so it could be heard. *Slightly closer together,* but still widely spaced, the reproduced sound would be heard from two *separate and distinct* sources. This effect is known as *hole in the middle,* because nothing seems to be coming from the area between the two speakers, whereas the centre of the orchestra would be at this point at the source.

To overcome the hole-in-the-middle effect, stereo may employ any number of channels, Fig. 9.7. An *elaborate arrangement* may provide for pickup of an orchestra with six to twelve microphones proportionately spaced around the orchestra, with a corresponding number of amplifiers to reproduce the program. The amplifiers feed the speakers which are placed at the points where microphones were located in the original recording.

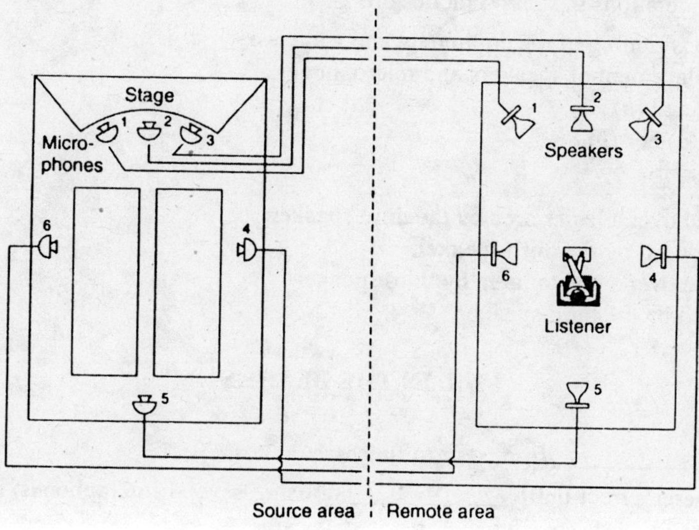

Fig. 9.7. An elaborate stereo system.

To provide signals for the *third speaker*, three microphones, *properly spaced* from left to right, may be employed at the sound source. The signal from the *middle microphone* is mixed with the signals from the left and right microphones and the resultant signals are fed to the left and right channels respectively. Thus *the left and right channels contain a certain proportion of the effects of the centre microphone output.*

The signals at the output of the pre-amplifiers may be divided so that the majority of the output from either channel is fed to its respective left or right speaker, but less than one-third of the output of each channel is fed to the center speaker, which operates at a lower input level than the other two.

The stereo effect is produced over a greater area by the third speaker. The listener can move towards or away from the left or right speakers without losing the stereo effect. The illusion of a *curtain of sound* spreading across the three speakers is developed.

OBJECTIVE TYPE QUESTIONS (A)

1. Mono signifies

   (*a*) Two sound sources, two channels, and two speaker systems.
   (*b*) One sound source, one channel, and one speaker system.
   (*c*) Either (*a*) or (*b*)
   (*d*) Neither (*a*) nor (*b*).

2. Stereo

   (a) Increases the concert hall feeling considerably.
   (b) Decreases the concert hall feeling considerably.
   (c) Enhances the concert hall feeling.
   (d) None of the above.

3. Important considerations in stereo practice are
   (*a*) Proper positioning of the equipment and audience
   (*b*) Proper placement of speakers and microphones
   (*c*) Both (*a*) and (*b*)
   (*d*) Neither (*a*) nor (*b*).

4. The stereo effect is
   (*a*) Produced over a lesser area by the third speaker.
   (*b*) Not produced by the third speaker.
   (*c*) Produced over a greater area by third speaker.
   (*d*) Enhanced by the third speaker.

FILL IN THE BLANKS (B)

1. Mono means ____________ from start to finish.
2. There is no stereo effect until ____________ utilising several microphones, amplifiers and speaker systems are used.
3. Quadraphony is an extension of ____________.
4. The most straight forward method in quad is called ____________.
5. An interesting by product of four-channel sound is in the field of ________.
6. The degree of separation between the channels from a matrix system is ________ than that with a discrete system.

CONVENTIONAL QUESTIONS (C)

1. Explain the difference between a mono and a stereo system.
2. How four - channel sound has evolved ?
3. What are the different methods of achieving four channel sound ?
4. What are the factors to be considered in matrix decoders ?
5. Discuss the significance of third speaker in eliminating hole-in-the-middle.
6. With the help of a simplified diagram, explain the working of an elaborate stereo system.

10

ACOUSTICS

We usually pay too little attention to the part that the home listening room plays in the general problem of obtaining hi-fi. Probably the main reason for this attitude is that there doesn't seem to be much we can do unless we are willing to rebuild or redecorate the room. If, however, we can gain a basic understanding of what it takes to make a good listening room we will find that there is much we can do to ensure that the money invested in hi-fi equipment doesn't go down the drain. A tremendous amount of knowledge has been accumulated regarding the acoustic problems of concert halls which can in turn, be applied to our own listening areas.

10.1 REFLECTION OF SOUND

When a free sound wave travelling through air strikes a uniform, large, solid, plane surface it is reflected in the *same manner* as a light ray. The following facts in connection with reflection of sound are to be noted.

1. The analysis of reflection from *curved surfaces* may be derived by applying the laws of reflection of light.
2. A concave surface leads to the concentration of reflected waves at certain points. Hence *concave surfaces may be provided to work as reflectors.*
3. A convex surface tends to spread the reflected waves. Hence, convex surfaces may be used with advantage *to spread the sound* throughout the room.
4. The study of laws of reflection of sound helps in *selecting the proper shape of the room with regard to distribution of sound in that room.*
5. The assumption that sound is reflected in the same manner as light is true only for *limited circumstances* and hence great caution should be exercised while applying the laws of reflection of sound.

10.2 REVERBERATION

A sound, once started in a room, will *persist* by repeated reflections from the walls until its intensity is reduced to the point where it is no longer audible. If the walls are good reflectors of sound waves —for example, hard plaster or marble — the sound may continue to be audible for an appreciable time even after the original sound stops. The repeated reflection that results in this *persistence* of sound is called *reverberation.*

In an auditorium or classroom, excessive reverberation may be highly undesirable, for a given speech sound or musical tone will continue to be heard by reverberation while the next sound is being sent forth. The practical remedy is to cover the walls with some *sound absorbent material*, usually a porous substance like felt compressed fiber board, rough plaster or draperies. The *regular motions* of air molecules, which constitute the sound waves, are converted into *irregular motions* (heat) in the pores of such materials, and consequently less sound energy is reflected.

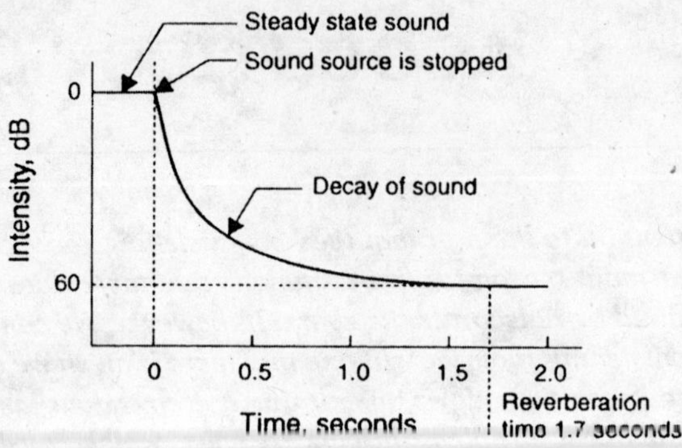

Fig 10.1 Growth and decay of sound in a room.

Suppose that a sound whose intensity is one million (10^6) times that of the faintest audible sound (threshold of hearing) is produced in a given room. The time it takes this sound to die away to inaudability is called the reverberation time.

$$m = 10 \log_{10} \frac{I_1}{I_2} \qquad \text{...(10.1)}$$

It is found that for an average speaker the ratio $I_1 / I_2 = 10^6$, so that $m = 60$ dB. *The time for reverberation is thus defined as the time required by the reverberant sound to decay one- millionth of its initial value. This naturally corresponds to a drop of 60 dB.*

Some reverberation is desirable, especially in concert halls; otherwise the room sounds too *dead*. For a moderate sized auditorium the reverberation time should of the order of 1 to 2 *s*. For a workroom or factory, it should, of course, be kept to much smaller values, as sound deadening in such cases results in greater efficiency on the part of workers with much less attendant nervous strain.

The selection of the correct time of reverberation is called the *optimum reverberation time* and it will depend on the acoustics of the room. Table 10.1 gives the relation between reverberation time and acoustics of the room. *If the reverberation time is too long it results in overlapping of speech, if it is too short, it produces the effect of deadness.*

Table 10.1 Reverberation Time and Acoustics

Reverberation time in seconds	Acoustics
0.50 to 1.50	Excellent
1.50 to 2.00	Good
2.00 to 3.00	Fairly good
3.00 to 5.00	Bad
Above 5.00	Very bad

Table 10.2 gives the recommended values of optimum reverberation time for halls designed for various purposes after considering the indicated audience strength.

Table 10.2 Optimum Reverberation Time for Halls

Activity in hall	Optimum reverberation time in seconds	Audience factor
Churches	1.8 to 3.0	Two-thirds
Cinema theaters	1.3	Two-thirds
Large halls	2.0 to 3.0	Full
Law courts; Committee rooms; Conference halls	1.0 to 1.5	One third
Music concert halls	1.6 to 2.0	Full
Parliament house; Assembly halls; Council chambers	1.0 to 1.5	Quorum
Public lecture halls	1.5 to 2.0	One-third

10.3 ABSORPTION OF SOUND

When a sound wave strikes a surface, part of its energy is *absorbed* by friction, part of its energy is *transmitted* and the remaining part of its energy is *reflected.* As reverberation directly depends on the loss of energy of sound waves due to friction, it is of great importance. This property of a surface by which sound energy is converted to some other form of energy is known as absorption and the *absorption coefficient* of a surface indicates the degree to which this surface affects the absorption of sound. It is thus *the ratio of the energy absorbed by the area to the energy striking the area.*

The value of the coefficient of absorption will depend on the frequency of the sound. Table 10.3 gives the value of coefficient of absorption for some of the common surfaces. These values correspond to the normal

Table 10.3 Absorption Coefficients

Material	Absorption coefficient per m^2
Open window	1.00
Ventilators	0.10 to 0.50
Brick wall 40 cm thick	0.03
Plaster on wall surface	0.02
Glass against solid surface	0.03
Marble	0.01
Stage curtain	0.20
Linoleum or concrete floor	0.03
Solid wooden floor	0.09
Framed wooden floor	0.13
Plywood on battens	0.17 to 0.26
Window glazed	0.18
Curtains in heavy folds	0.40 to 0.75
Metal	
Individual Objects	
Audience, adults in wooden seat	0.46
Plain seat	0.02
Covered seat	0.10 to 0.20

frequency of 500 Hz. It may be noted that *the coefficient of absorption for an open window is taken as unity.* This is very easy to understand as sound waves approaching an open window must completely pass through it.

10.4 SABIN'S EQUATION

Prof W.C. Sabin of Harvard University, U.S.A. carried out a number of experiments in rooms of different sizes and was able to establish the following formula which is known as *Sabin's equation.*

$$\text{In MKS units } t = \frac{0.161\, V}{a} \qquad \ldots(10.2)$$

where t = reverberation time in seconds

v = volume in m^3

and a = total absorbing power in m^2 – sabins

$= (a_1 S_1 + a_2 S_2 + a_3 S_3 + \ldots)$ absorption units of individual objects.

S_1, S_2, S_3, etc; being coefficients of absorption of respective surfaces a_1, a_2, a_3 etc;

$$\text{In FPS units: } t = \frac{0.049V}{a} \qquad \ldots(10.3)$$

The unit of sound absorption is m^2 - Sabin and it is equal to sound absorption of one square metre area of a fully open window. Since V and a can be worked out to the plans and specifications of the structure, it is possible to design a structure for any desired value of reverberation time.

10.5 LISTENING ROOM CHARACTERISTICS

A room can be compared to a loudspeaker enclosure. Both the room and the enclosure are surrounded by six surfaces. The differences between the two lie in their *size* and *composition.* Yet the similarity in construction is great enough to warrant a comparison.

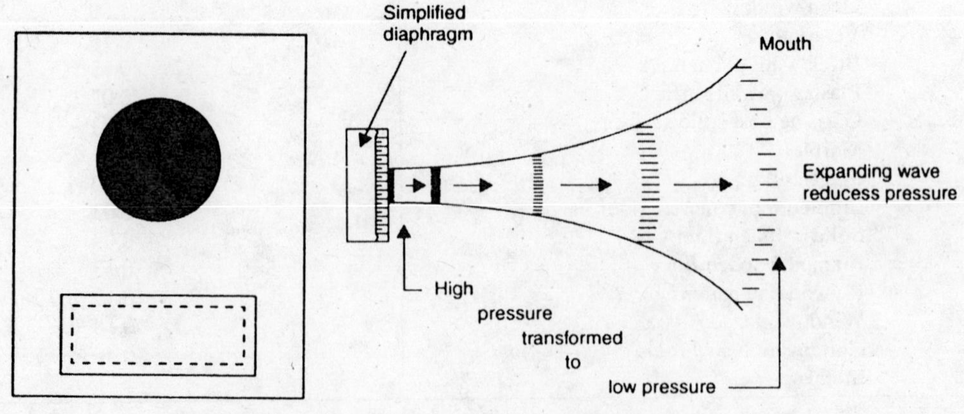

Fig. 10.2 Two general categories of loudspeaker.
(a) The resonant type (b) The non-resonant type.

Loudspeakers fall into two general categories. There is first the *resonant* enclosure, such as the bass-reflex type, with the characteristic resonance determined by the enclosures size and the port in the baffle. Then there is the *non-resonant* type, which employs the largest possible horn mouth to minimise any possible resonant condition.

Similarly, rooms may be thought of as resonant or non-resonant, with the *degree* of resonance or non resonance dependent upon their shape, size and the materials contained in them.

Unlike the speakers with a single major resonance, *rooms and enclosures may, in general, have several resonances or modes of vibration.* Rooms that are roughly rectangular in shape may have many natural resonances, depending on the room's length and other dimensions. These other dimensions can also develop their own resonance modes. Thus we can expect that our listening room will have several natural resonances that are completely a function of its *size and relative dimensions* and partly a function of its *liveness.*

10.6 LIVE ROOMS

Live rooms are those with a great deal of resonance. The best example of a live room is a tile-lined bathroom. The hard smooth surfaces of the tiles reflect sound almost as well as mirrors reflect light. The *constant and repeated reflection of sound* from one hard wall to the other, with little absorption, builds up the sound around the ear. This quality of room liveness is a function of the room's normal mode of vibration. Such a room may be *detrimental to smooth audio reproduction.*

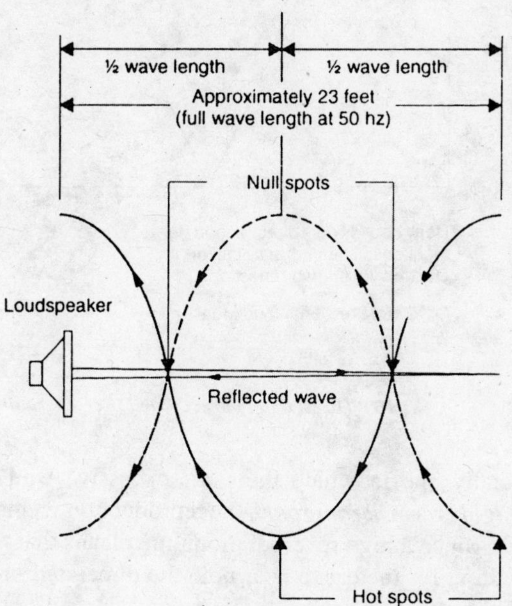

Fig 10.3 Low frequency hot and null spots.

The reflection of sound from one wall to the other and the room's normal mode of vibration set up *hot spots* and *null spots* of sound energy for different frequencies and for different areas of the room. *Rooms vibrate and respond most readily to low frequencies.* Thus, as shown in Fig 10.3, a standing wave may be produced for a particular frequency, resulting in peaks at some points and complete cancellation at others.

This condition often causes disagreement as to what two people hear in the same room. For instance, if one were at a spot where there was *reinforcement* due to the standing wave, he might hear *mostly bass notes.* If some other person were standing at a point of *cancellation,* he would hear *far less bass notes.*

To make a *truly objective listening test,* therefore, a hi-fi system must be heard from many different positions in the room. Your *location* in the listening room is of considerable importance. The *loudspeaker's position in the room will also affect the final results. A live room may produce many response irregularities that materially affect the system's smoothness.*

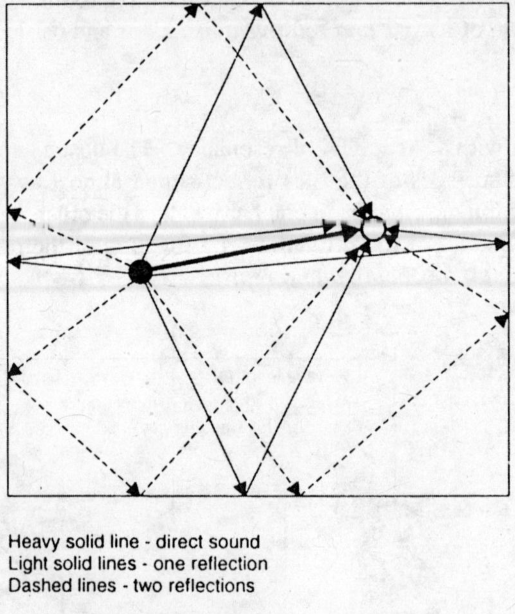

The ear receives sound from an omnidirectional speaker through many paths, direct and reflected, as shown here

Fig 10.4 Direct and indirect sound (Reverberation).

A *live room*, especially one in which there are excessive wall reflections (Fig 10.4), has a long reverberation time. In the *highly reverberant room,* the reproduction seems louder than normal for an amplifier with moderate audio power. Since a high reverberation time means that *the sound is not being absorbed,* the reflected sound is there waiting for the ear to hear both the direct and indirect sound. This makes the sound louder to the ear (Fig 10.5).

Another characteristic of live rooms is that the directivity characteristic of the reproducing unit is considerably changed by the room's reflective properties. With highly reverberant walls, successive reflections greatly diffuse the several radiating beams from the loudspeaker. *This effectively increases the uniformity of reproduction of the higher frequencies throughout the room.*

In a too-live room, then, we may say that the articulation will deteriorate, that irregularities of low-frequency response will occur, that the system will sound louder, and that the sound will be more diffused.

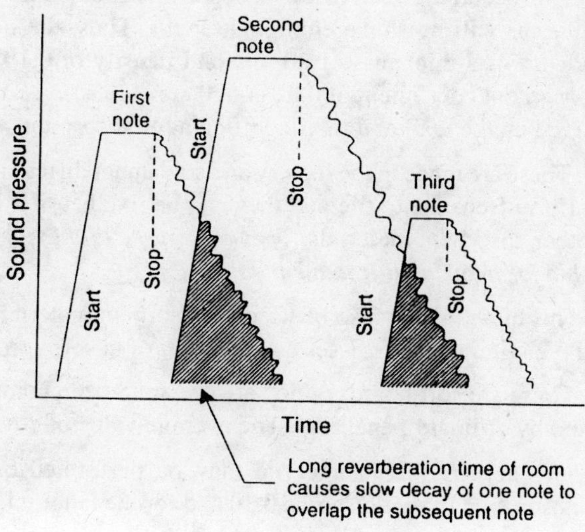

Fig 10.5 Long decay time.

10.7 DEAD ROOMS

The dead room is characterised by the high degree of sound it absorbs. If the walls were totally absorbing, all the sound would be lost since no waves would be reflected. Thus, *a totally absorbent room would not only attenuate the wall-to-wall reflections but would also reduce the room's normal mode of resonance.* An overdamped (dead) room wastes any acoustic power produced in it.

In a *dead room*, the system's high-frequency distribution depends solely on the loudspeaker and baffle combination. If the speaker high-frequency beam is sharp, the listener's position in the room will determine *how many* highs and *how much* of them will be heard. For the low frequencies, the dead room exhibits fairly uniform distribution characteristics with no standing waves. Therefore, few low-frequency peaks or null points appear throughout the room.

Room size is closely related to the amount of reverberation desired. *Generally, small rooms should have less reverberation time than large rooms.* If we are in a small room, we are closer to both the loudspeaker and the sound-reflecting surfaces. Reflections will therefore reach us sooner than those from the walls of a larger room. Consequently, *if we want a clear separation between notes, the time for the reflected sound to reach us in the small room must be reduced.*

10.8 ABSORBENT MATERIALS

Various types of absorbent materials are available in the market under different trade names. The value of coefficient of absorption is supplied by the manufacturer. Following are some of the most common types of absorbent materials.

1. *Hair felt* : This material was used by Prof. Sabin in his experimental works. The average value of coefficient of absorption of 25 mm thick hair felt is 0.60.

2. *Acoustic plaster* : Also known as *fibrous plaster*, it includes granulated insulation material mixed with cement. If the quantity of cement is *more* than that required, the plaster will not have

sufficient pores to become effective for acoustics. If the quantity of cement is *less* than that required, the plaster will not have enough strength. Thus, the quantity of cement should be carefully decided. For a thickness of 20 mm and density of 0.10 g / cm^3, the acoustic plaster possesses an absorbent coefficient of 0.30 at 500 Hz. *Acoustic plaster boards* are also available. They can be fixed on the wall and their coefficient of absorption varies from 0.15 to 0.30.

3. *Acoustic tiles* : These are made in the factory and sold under different trade names. The absorption of sound is uniform from tile to tile and they can be fixed easily. However, acoustical tiles are costlier than other absorbent materials. *They are most suitable for rooms in which only a small area is available for acoustical treatment.*

4. *Strawboard* : This material can also be used as absorbent material. With a thickness of 13 mm and density of 0.24 g/cm^3, it possesses a coefficient of absorption of 0.30 at 500 Hz.

5. *Pulp boards* : These are soft boards which are prepared from compressed pulp. They are cheap and can be fixed by ordinary panelling. The average value of coefficient of absorption is 0.17.

6. *Compressed fibre boards* : This material may be perforated or unperforated. The average coefficient of absorption of the former is 0.30 and for the latter it is 0.52. It has a density of 0.30 g/ cm^3.

7. *Compressed wood particle board* : This material is provided with *perforations* and it can be painted also. With a thickness of about 13 mm, the average coefficient of absorption is 0.40.

8. *Perforated plywood* : This material can be used by forming composite panels with mineral wool and cement asbestos, or mineral wool and hardboard. *It is generally suspended from trusses.* The average value of coefficient of absorption for the former is as high as 0.95 and for the latter it is about 0.20.

9. *Wood wool board* : This material is generally used with a thickness of 25 mm and has density of 0.40 g / cm^3. The average value of coefficient of absorption is 0.20.

10. *Quilts and mats* : These are prepared from mineral wool or glass wool and are *fixed in the form of acoustic blankets.* The coefficients of absorption of such quilts and mats depend on the thickness, density, perforations, mode of fixing nature of backing and frequency of sound.

10.9 ACOUSTIC DESIGN OF AUDITORIUMS

The following *conditions* should prevail in an auditorium possessing good acoustical properties.

- The initial sound should be of good intensity or loudness.
- The sound which is produced should be evenly spread over the whole area covered by the audience.
- The initial sound should be clear and distinct.
- The initial sound should reach the audience with the same frequency and intensity.
- All undesired sound should be reduced to such an extent that it will not interfere with the normal hearing of speech and music.

The following *factors* play an important role in the acoustic design of an auditorium.

- *Volume* : The best guide for deciding the volume of a proposed auditorium will be the *detailed study of an existing auditorium used for similar purposes.* However, it is found from various considerations that the volume of auditorium *per seat* should be very near to 4.2 m^3 and it should not usually exceed 5.7 m^3 per seat.

- *Shape* : It involves *geometrical aspects* of the hall. The paths followed by reflected sounds are traced and concentration of sound waves, if any, are noted. *Concave walls* are not good for acoustic purposes as they tend to concentrate sound waves. *Plain walls* are better. But *convex walls* are excellent and are used to reduce the possibilities of echoes to the minimum possible extent.
- *Sound absorption* : Adequate absorbing surface should be provided in the hall *to control the reverberation.* The acoustical design of an auditorium will be incomplete without the provision of such a surface. *Careful study of the hall should be made before recommending the type of absorbent material and its location in the hall.*

10.10 ACOUSTICS OF STUDIOS

The term studio is used to indicate *a room where sound is picked up by a microphone* and it includes radio-broadcasting station, television station and sound-recording studio. The acoustic design of a studio requires the following *special precautions* to be taken for its efficient working.

- The *noise level* in the studio should be brought down to 20 to 30 dB.
- *Echoes* should be completely eliminated.
- The finish on the *outside surfaces* should be of a sound reflecting material.
- *Partition walls and exterior walls* of the studio should be sufficiently rigid to resist vibrations and to prevent resonance.
- The finish on the *interior surfaces* of walls, ceilings and floors should be of absorbent materials. This will help in reducing time of reverberation.
- Provision of windows in the studio should be minimum so as to *prevent transfer of external noise* into the studio.
- Suitable sound insulation should be provided to air-conditioning machinery and such other equipments, so that *equipment vibration* is not carried upto the working chambers of the studio.
- *In a multi-studio building, it is preferable to locate all the studios on the same floor.* In no case. studios should be located one above the other. *There should be a minimum space of one floor between two consecutive studios in elevation.*
- Heavy curtains and draperies may be used with advantage to *control the time of reverberation* in the studio.
- In order to *control optimum reverberation time* (acoustical conditions) of the studio, arrangement of hinged panels (Fig 10.6) in walls or rotable cylinders (Fig 10.7) should be made.

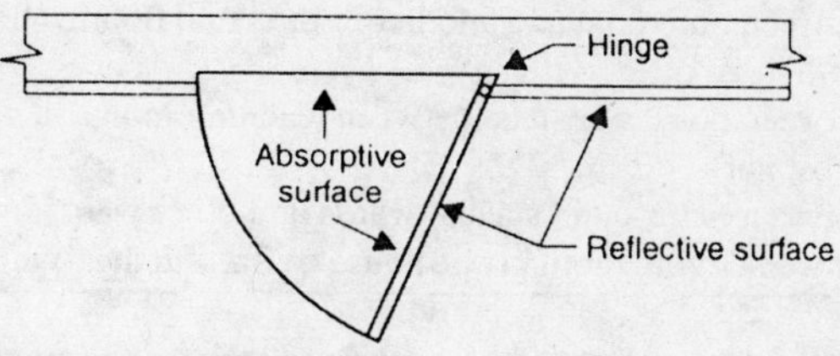

Fig. 10.6 Hinged panels

In the case of a *hinged panel* arrangement, a hinge is provided and one surface of the hinged panel is *absorptive* while the other is *reflective.* Hinged panel construction is cheap and easy to maintain. In place of hinged panel, *revolving panels* with a hinge at the center may also be provided and two different absorbing materials may be provided on both faces of the revolving panel. In the case of *rotable cylinders,* drums of required diameter are suspended from the ceiling. The three surfaces of the drum, each subtending an angle of 120° at the centre are provided with different absorptive materials. The cylinders can be rotated by rack and pinion arrangement and thus a desired combination of two surfaces can be exposed to the room.

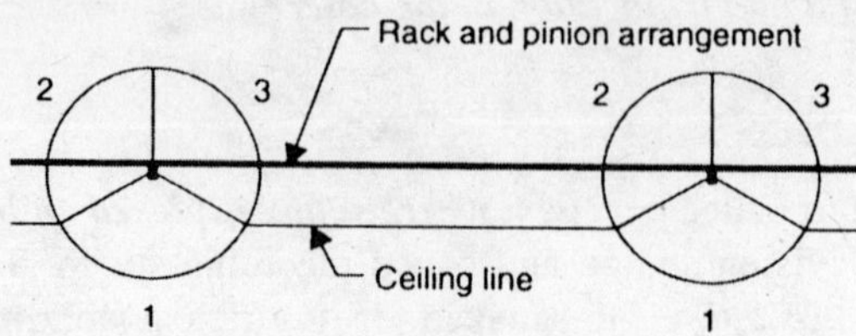

Fig 10.7 Rotable cylinders.

10.11 SOUND INSULATION

The term sound insulation or *sound proofing* is used to indicate the reduction in the level of sound when it passes through a building component.

Sound insulation and sound absorption are quite different terms. The purpose of a sound-insulated construction is to *prevent* the sound from passing through it. *Hard materials* in general are poor sound absorbers but they are good sound insulators.

The purpose of a sound absorbent material is to *reduce* the sound reflected from a surface. Hence, *porous materials* in general are good sound absorbers, but they are poor sound insulators.

Insulation of sound is measured in an adjoining room while *absorption of sound* is measured in the same room.

Table 10.4 shows the *desirable levels* of sound between individual rooms for air-borne sound.

Table 10.4 Sound Insulation between Individual Rooms

Overall sound insulation in *dB*	Situation of Rooms
30	Between one room and another in the same floor or flat; between one room and another room in an office.
40	Between houses or flats; between teaching rooms in a school; between normal wards in a hospital
45	Between extra-quiet special wards and other wards in a hospital.
50	Between living room in one house or flat and the living rooms and bed rooms in another.

As the airborne sound passes through any structure, loss of sound intensity takes place. This is expressed in terms of *sound reduction index* (SRI) or transmission loss (TL). Sound reduction index of some materials is given in Table 10.5

Table 10.5 Sound Reduction Index

Material	SRI
6 mm Plywood	20 dB for 5 kg/m^2 mass
18 mm Chipboard	25 dB for 10 kg/m^2 mass
6 mm Glass	30 dB for 20 kg/m^2 mass
11 mm Plywood (lead bonded)	35 dB for 25 kg/m^2 mass
50 mm Wood wool with 2 × 12 mm plaster	27 dB for 80 kg/m^2 mass
112 mm Brick with 12 mm plaster	45 dB for 200 kg/m^2 mass
225 mm Brick with 2 × 12 mm plaster	55 dB for 500 kg/m^2 mass

10.12 NOISE

There are two distinct types of noise; indoor noise and outdoor noise.

Indoor noises are created in the adjacent room or in the same room where noise is noticed. The sources of indoor noises are moving of people, crying of babies, playing of radios, banging of doors, traffic in staircase, movement of furniture, conversation of the occupants, operation of cisterns and water closets, noise of typewriters, working of machines nearby, etc.

Outdoor noises are created from nearby streets and the largest source of outdoor noise is generally the automobile traffic on the road. The other sources of outdoor noises are railways, aeroplanes, loudspeakers, and machines in nearby factories etc.

Noise is transmitted to the room through walls, ceilings or conduits. The origin of transmitted noise may be air borne or due to impact.

The maximum level of noise which neither annoys the occupants nor damages the acoustics of the building is termed the *acceptable noise level* in the building. It depends on

(*a*) Nature of the noise

(*b*) Type and use of the building

(*c*) Time of fluctuation of the noise

(*d*) Background noise

Table 10.6 gives the *generally acceptable noise levels* for different structures from the point of view of economy, comfort, and practical considerations of the conditions prevailing in our country.

Table 10.6 Acceptable Noise Levels

Type of building	Acceptable noise level in dB
Radio and TV studios	25 to 30
Auditoriums and music rooms	35 to 40
Small office, courtroom, libraries	40 to 45
Schools	45 to 50
Residential buildings and restaurants	45 to 55
Large public offices, banks and stores	50 to 60
Factories	60 to 65

OBJECTIVE TYPE QUESTIONS (A)

1. Concave surfaces may be provided to act as
   (*a*) Absorbers (*b*) Reflectors
   (*c*) Insulators (*d*) None of the above.
2. Convex surfaces may be used to
   (a) Absorb the sound (*b*) Spread the sound
   (c) Concentrate the sound (*d*) None of the above.
3. The coefficient of absorption of an open window is taken as
   (*a*) 1 (*b*) 2
   (*c*) 1/2 (*d*) 1 1/2.
4. Rooms and enclosures may have
   (*a*) A single resonance (*b*) Several resonances
   (*c*) Two resonances (*d*) None of the above.
5. Rooms vibrate and respond most readily to
   (*a*) Low frequencies (*b*) High frequencies
   (*c*) Mid frequencies (*d*) None of the above.
6. To make a truly objective listening test, a hi-fi system should be heard from
   (*a*) A single listening position
   (*b*) Different listening positions
   (*c*) The centre of the listening room
   (*d*) A corner of the listening room.

FILL IN THE BLANKS (B)

1. -------- of sound is called reverberation.
2. Too long a reverberation time results in ------------ of sound.
3. Too short a reverberation time produces the effect of --------- .
4. The ratio of the energy absorbed by the area to the energy striking the area is called -------------
5. The coefficient of absorption depends on the ------------ of sound.
6. The unit of sound absorption is -------------- .
7 Liveness of room may be ---------- to smooth audio reproduction.
8. A live room may produce many response ----------- .
9. In a too live room articulation will ------------ .
10. Small rooms should have -------- reverberation time than large rooms.
11. The noise level in a studio should be brought down to -------- .
12. Heavy curtains and draperies may be used to control the ---------- .
13. Porous materials are good sound ---------- .
14. Hard materials are good sound ----------- .
15. Insulation of sound is measured in an ---------------- .

16. Absorption of sound is measured in the -----------

CONVENTIONAL QUESTIONS (C)

1. Disuss the significance of reverberation.
2. What will happen if the reverberation time is

   (*a*) Too long (*b*) Too short.
3. Explain the difference between an acoustically dead room and an acoustically live room.
4. Define

   (*i*) Sabine (*ii*) Absorption coefficient

   (*iii*) Optimum reverberation time (*iv*) Sound reduction index.
5. What are the characteristics of a listening room ?
6. Differentiate between a sound absorber and a sound insulator.
7. What are the factors that you will consider while designing an auditorium ?
8. Give three examples each of use of sound absorbers and sound insulators
9. What are the types of noise encountered in architectural acoustics ?

11

DISC RECORDING AND REPRODUCTION

When Edison first developed phonograph recording, he used vertical motions of the recording stylus to record sound vibrations. This came to be known as the 'hill and dale' method of recording. However, in such a system, the 'hills' in the record groove had to lift the playback arm and cartridge. Since early arms and cartridges were relatively heavy, this motion caused excessive wear. Because of this, the record industry adopted the lateral recording method, in which the needle goes from side to side in accordance with the sound vibrations. All commercial monophonic records for home use are laterally recorded, and monophonic pickups are designed accordingly. The year 1946 marks the start of the disc hi-fi. It wasn't until 1952 that a binaural record and a special tone arm with which to play it appeared. The first stereo-disc was demonstrated in 1957. The same year saw the invention of the moving-magnet stereo pickup.

11.1 MAKING THE TAPE

The live performance is recorded on tape, Fig. 11.1. Often, each instrument and/or singer is given a *separate microphone* and a *separate channel*. Professional tape recordings may have as many as 32 channels so that the sound of each instrument can be *individually monitored* - balanced, adjusted, and enhanced. Different channels can even be recorded at different times.

Microphone placement is important. Experience has shown where the optimum positions are.

The multiple channels recoraed in the studio are mixed down to stereo before the record is cut. At this, the mixing stage, record producers make many decisions that determine the end sound quality of a performance.

In the *standard recording procedure Fig. 11.2*, the microphones feed the signals to preamplifiers From there the signals go to amplifiers, thence to a tape recorder. In recent innovations with *noise reduction systems* the process is much more complex but offers greater reduction of noise.

The tapes run at *speeds* of 15 or 30 inches per second (ips) to ensure high quality and accurate response.

11.2. MAKING THE RECORD

When the *master tape* is played, its magnetic patterns are converted into electrical signals. These are fed through amplifiers to a *cutter head* on a lathe. The cutter head *converts* these signals into mechanical motion and causes a *cutting stylus* to move as dictated by the signals on the tape. Thus, an *incredibly complex groove* is inscribed into a *master disc*. Moulds are then made from which the records are stamped.

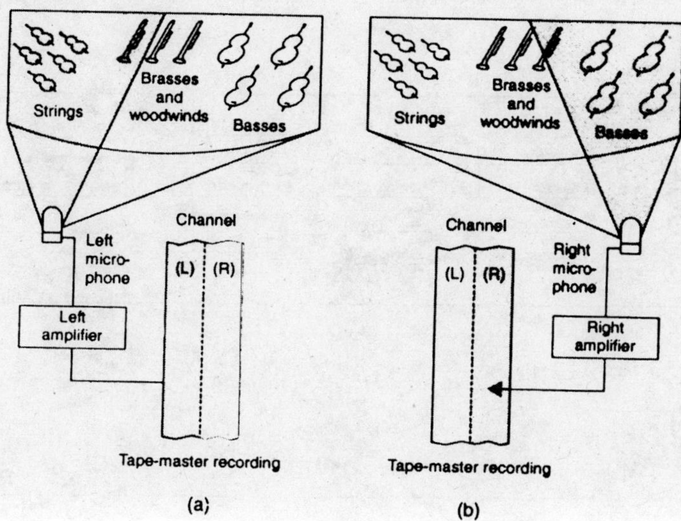

Fig. 11.1. Tape-master recording

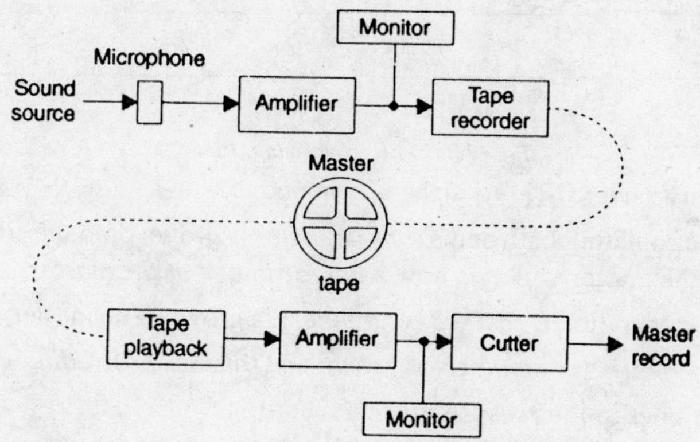

Fig. 11.2. Standard recording procedure.

11.3. PROFESSIONAL RECORD MANUFACTURE

1. The *recording session* may be in a studio or an external location A simple two-microphone balance may be used, or a multi microphone set-up.

2. Multi-mic as well as *two mix sessions* can be recorded directly onto twin-track 1/4 in tapes.

2a. *Multi-mic sessions* are more commonly recorded on multi-track machines (1 or 2 in tape : 8, 16, 24 or 32 tracks).

3 & 3d. The best takes are *edited* together to produce a complete performance.

4. An edited twin-track tape is called a *master* and will usually be copied for safety.

4a. A multi-track tape is copied *via* a mixing desk by a process known as *reduction,* to make ...

5. ... a *twin-track production master.*

6. The twin-track tape goes to the cutting room for *transfer to lacquer*.

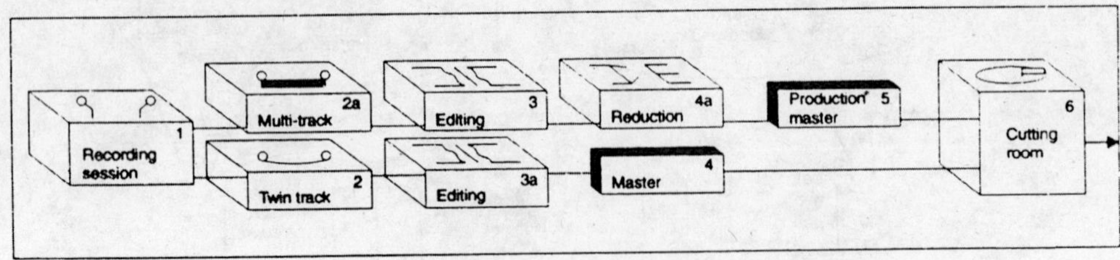

Fig. 11.3 (a) Studio

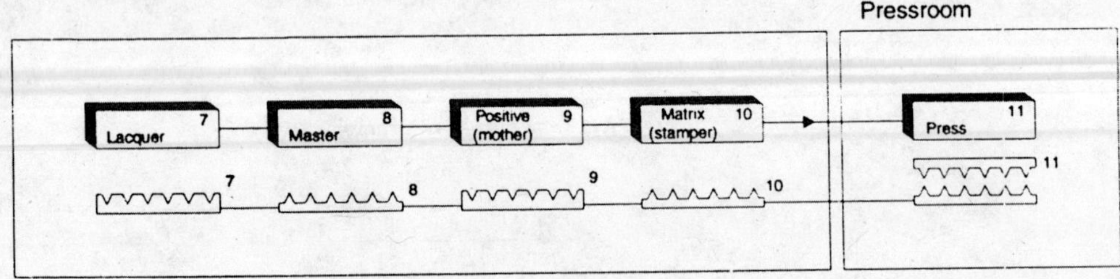

Fig 11.3 (b) Electroforming (Plating).

7. The lacquer is *sprayed with silver* to make it conductive.
8. It is put into an electroplating bath and a *nickel master* is grown onto the silver coating. The lacquer is *discarded* leaving the master which is now a *mirror image of the master.*
9. By a further plating operation a *positive* (or *mother)* is grown. The master can be re-used.
10. From the positive a *matrix* (or *stamper)* is grown and this has *protruding grooves* like the master.
11. A *pair of stampers* goes to the press to make a record.

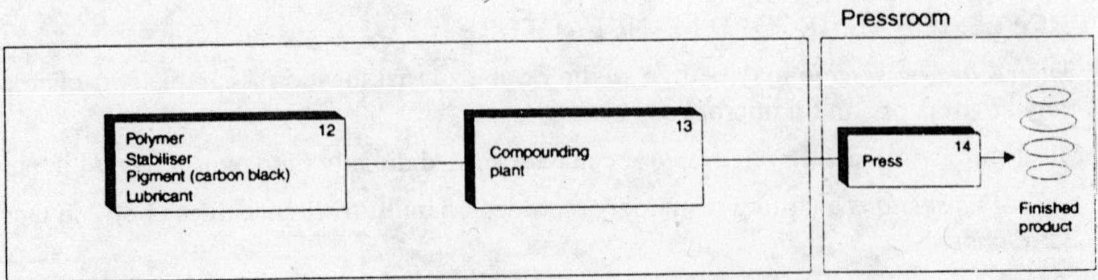

Fig 11.3 (c) PVC manufacture.

12. The main *constituents* of the vinyl are a vinyl copolymer, a stabiliser, a lubricant and a pigment. The latter is usually carbon black, to produce the *traditional record coloring*.

13. The materials are *compounded* and fed to the press by extrusion or as cakes of small dice-like particles; from these the *final product* is pressed.

14. Manual, semi automatic and fully automatic *presses* are all used in record manufacture today.

11.4. STEREO PICK-UP TECHNIQUES

The most general type of stereo pickup technique is called *time-intensity pickup*. This means that the signals in the two stereo channels will vary in both intensity and time according to the *difference in direction and in distance* of the sound source from the two microphones.

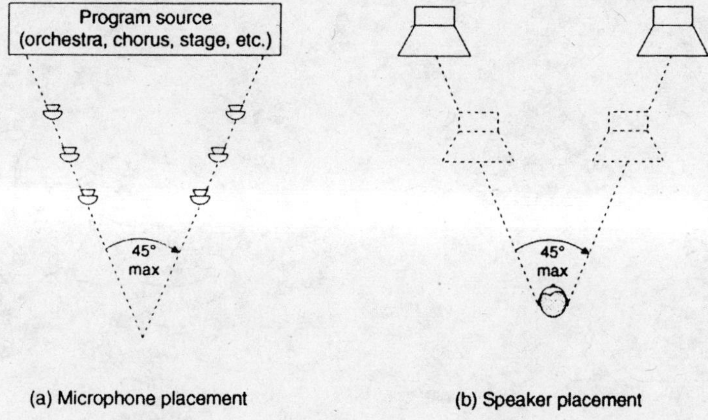

Fig. 11.4 Microphone and speaker placement.

The *stereo effect* can be obtained by picking up the sound with two microphones spaced a certain distance apart and located in front of the sound source. If they are too close the stereo effect is lost. Experiments have led to the conclusion that *there is a definite optimum relation between the microphones separation and their distance from the source. A similar relation exists between the listeners' speakers and their distance*

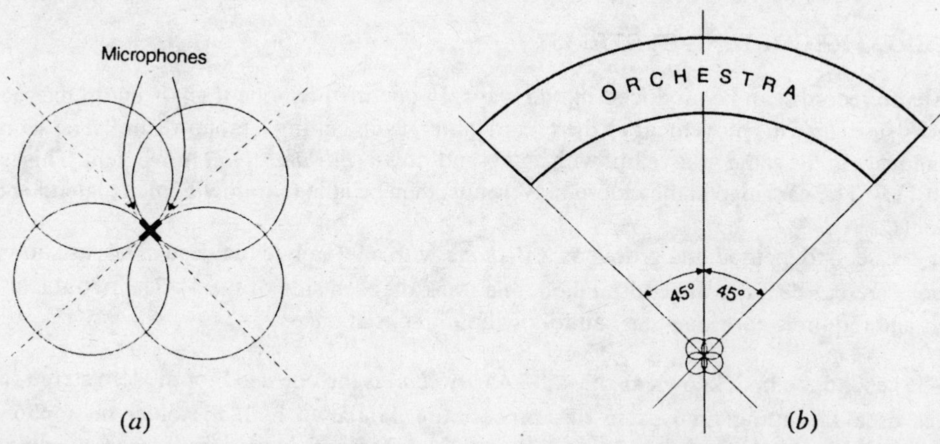

Fig. 11.5. (a) Polar characteristics of coincident microphones. (b) Coincident microphones used to record an orchestra

from him. This is illustrated in Fig. 11.4. In practice the *angle* at the listener should not be less than 30 degrees nor exceed 45 degrees. This relation keeps the distance of the listener *from* the speakers approximately the same as the distance *between* the speakers and ensures sufficient *time intensity variation* to provide stereo effect.

A good rule to follow is to *spread* the microphones as far apart as you can and still pickup an appreciable amount of sound from the *middle portion* of the source. Usually the pattern will be such that they can be placed along the sides of a triangle (see Fig 11.4).

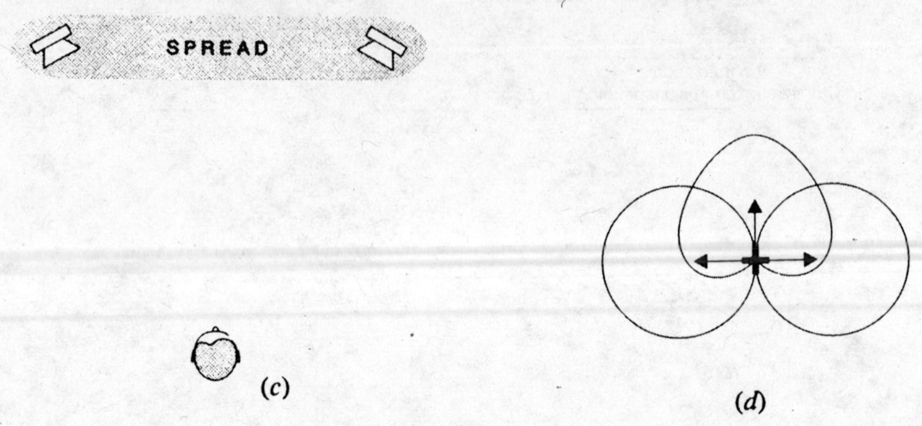

Fig. 11.5. (c) The technique shown results in a spread of sound during reproduction between speakers. (d) Cardiod and bi-directional microphones are combined in the middle-side recording system.

Yet another method, referred to as the *intensity-difference system,* is sometimes used in stereo recording. In this system, the microphones are not spaced; however, they have a directional characteristic as shown in Fig. 11.5. *The stereo effect is obtained by proper orientation of the microphones.* One method is to mount the microphones at 90 degrees to each other and 45 degrees to the center of the source.

11.5 STEREO RECORDING SYSTEMS

Stereo records can be produced in two ways. In one method, one track is cut in the record groove by the side-to-side (*lateral*) movement of the record cutter, as in cutting a mono-record. The second track is cut simultaneously in the same groove but with an up-and-down *(hill-and-dale)* movement. This is shown in Fig. 11.6 and 11.7. The cartridge in the record player must then be able to follow *both* the lateral and hill-and-dale movements.

In the second method, the groove is a *V-shape* with a 90° angle between sides, as shown in Fig. 11.8. Both tracks are cut be a quasilateral method, one track on each side of the V. The two tracks are, therefore, 90° apart and require a cartridge that can follow both tracks at once.

The second method, known as the 45° – 45° *system*, is the one used for modern stereo recordings. The principles used in cutting grooves in this process are similar to both the older monophonic lateral and hill-and-dale methods. However, there is one important difference. *In modern stereo recording a single stylus (cutter) simultaneously cuts an independent information channel into each wall of the record groove.*

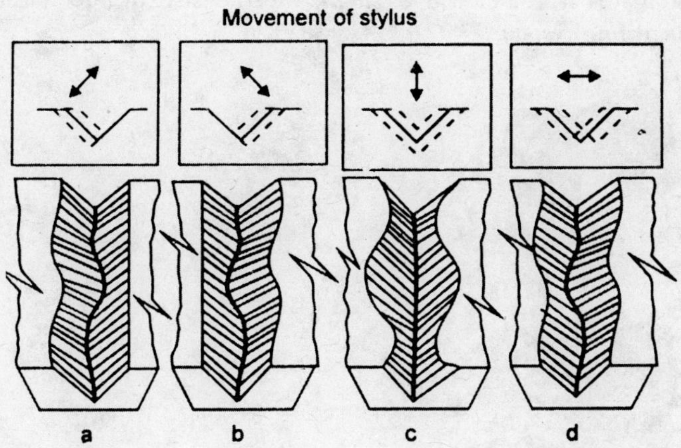

Fig. 11.6. Stylus movement in stereo-record grooves.

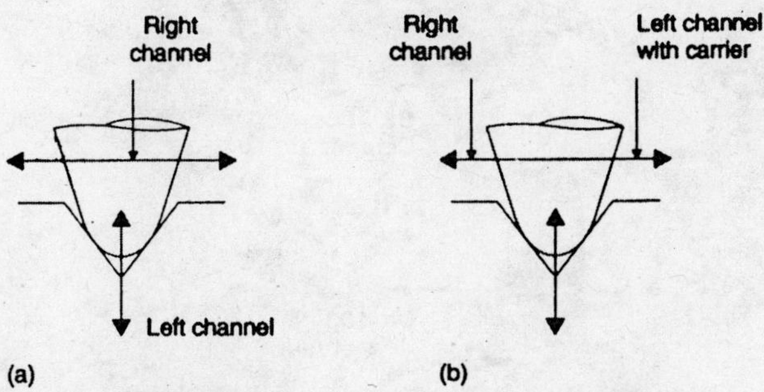

Fig. 11.7. Other systems of stereo recording. (a) Vertical/lateral system. (b) Frequency modulated lateral system.

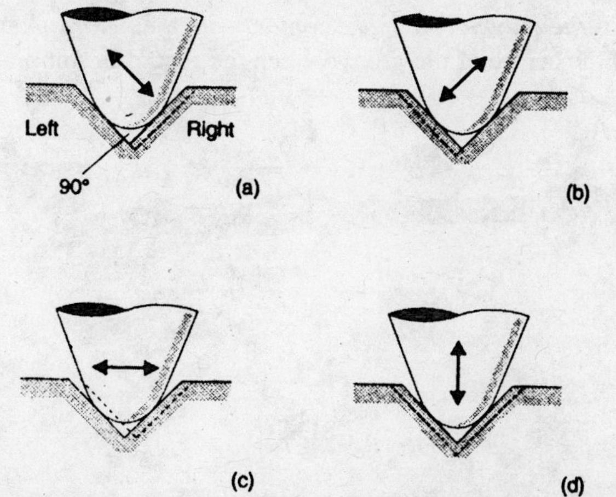

Fig. 11.8 Movement of the stylus in a stereo groove. (a) Right-hand channel modulated. (b) Left-hand channel modulated. (c) Both channels modulated equally and in phase. (d) Both channels modulated equally but in opposite phase.

11.6 EXTRACTING THE MUSIC

There is only one way to extract the music and that is with a cartridge. A *cartridge* consists of three basic elements : the *stylus*, the *cantilever*, and the *generating system.*

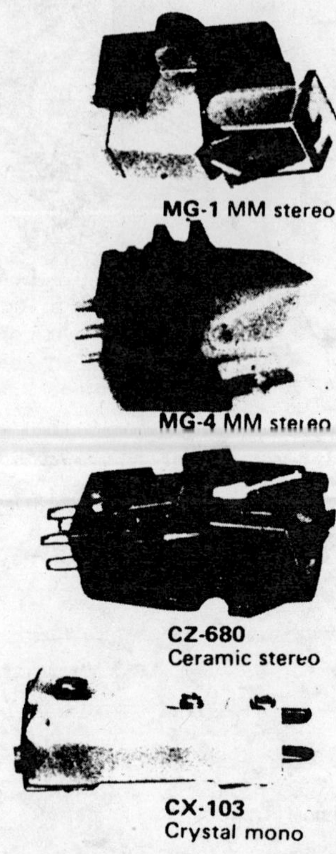

Fig. 11.9. Mono and stereo cartridges.

1. *The Stylus* : *The stylus is the only part to make contact with the record.* A force must be exerted by the tone arm to keep the stylus in the centre of the groove when the record is spinning (see Fig. 11.9).

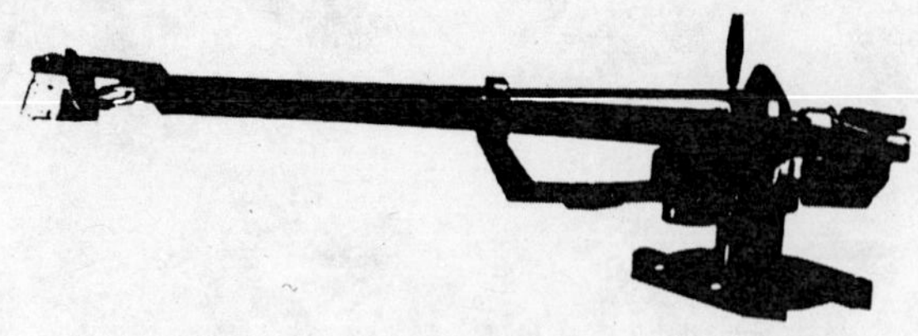

Fig. 11.10 The Tonearm.

This *tracking weight* is usually low-between one and two grams - but because the stylus makes contact with less than one millionth of a square inch of the record surface *it exerts tremendous pressure* : 6000 pounds per square inch ! With such enormous pressure any *roughness* or *irregularity* in the stylus will cause *record damage.* The stylus must also be hard or it will wear out quickly. For these reasons, high-quality styluses are made from the *hardest material* known to man - *pure diamond.*

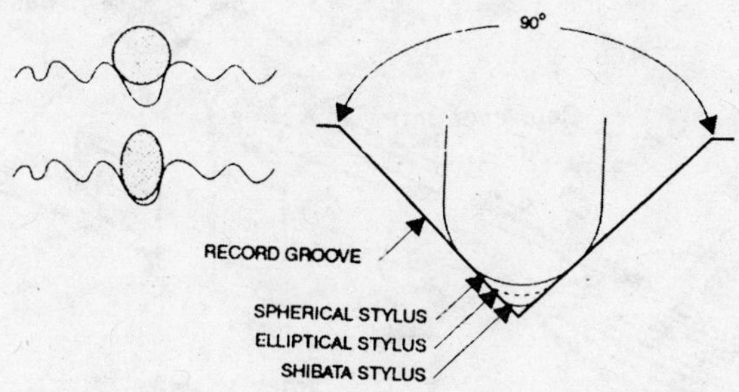

Fig. 11.11. In theory, an elliptical point reads the recorded waveform more accurately than does a conical point.

There are three common shapes of styluses, Fig. 11.11 spherical stylus; elliptical stylus, and fine-line or line-cut stylus.

(*i*) *Spherical Stylus:* The *spherical* (also called *conical*) shape is the *least expensive* to manufacture and is used in low cost cartridges. Unfortunately, *the stylus cannot accurately trace the highest musical tones.*

(*ii*) *Elliptical Stylus :* The more expensive *elliptical* stylus has a *narrow profile* enabling it to follow the petite undulations of the groove with greater precision. *It is used in the majority of high quality cartridges.*

(*iii*) *Fine-Line Stylus :* A third stylus shape, which goes under various names such as *fine-line* and *line cut,* provides more faithful tracking of the groove. It has an *even narrower profile than the elliptical shape* and makes contact with a wider area of the groove, thus reducing the tremendous pressure the stylus exerts on the delicate record surface. This shape also reduces record wear.

2. The Cantilever : The precious diamond moves at tremendous speed through the complex groove of a spinning record. A delicate armature, called *the cantilever, transmits its motions to the generating system.* (See Fig. 11.12).

In order for the cantilever to relay the minute movements of the stylus, it should be *light weight* so it can respond rapidly to the twists and turns of the record groove. And it must be *rigid* because any bending or flexing adds distortion and sound colouration. Obviously, light weight and high strength require great ingenuity in design and materials.

3. The Generating System : High fidelity cartridges are based on the law of physics which states : when a material capable of conducting electricity is *set into motion* in a magnetic field-or when a magnet is near a conductive material (such as copper coils) _ electrical signals are generated. *The direction and speed of the movement determines the ultimate strength of the generated signal.*

Thus *the generating system of a cartridge converts the movement of the stylus and cantilever into what is termed the output signal.* There are three ways of doing this : moving the magnet, moving the iron and moving the coil.

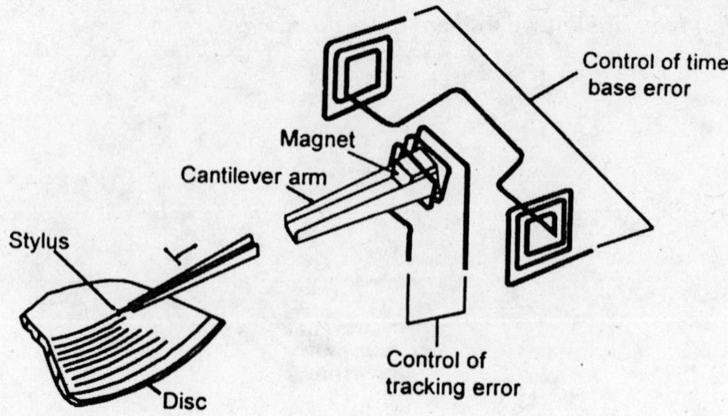

Fig. 11.12. Cantilever arm.

11.7 TYPES OF MICROGROOVES

A *mono-microgroove* is formed on the disc—a lacquer blank, the first stage in the manufacturing process by a cutter which moves laterally (side-to-side) only. This is shown in Fig 11.13 through 11.16.

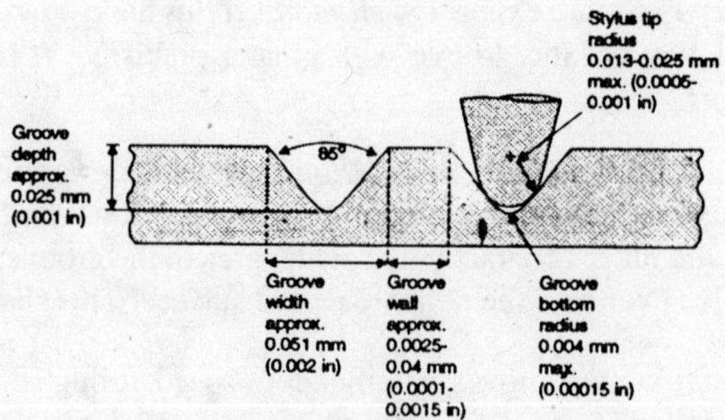

Fig 11.13. The dimensions of the groove in a long-play (LP) record.

The reproducing stylus will move from side-to-side when tracing the modulated groove, as indicated in Fig. 11.14, and it is easy to visualise that this is the only movement required for the reproduction of a *single channel of information.*

The *depth* of the groove does not change. It is of course, usual to speak of grooves, although there is only one continuous spiral. *On micro groove discs there are 150-300 grooves to the inch.*

In *stereo recording,* an unmodulated groove would be essentially the same as that on a mono disc. It is, however, required to accommodate *two separate channels of information* and to arrange for these to be traced by one stylus. Therefore, a signal is cut on each groove wall: one signal is geometrically at 90 degrees to the other but at 45 degrees to the surface of the disc. In a recording of any *complex sound,* the stylus follows a variety of slanting, vertical and lateral movements which depend on the *relative amplitudes and phase of the signals.*

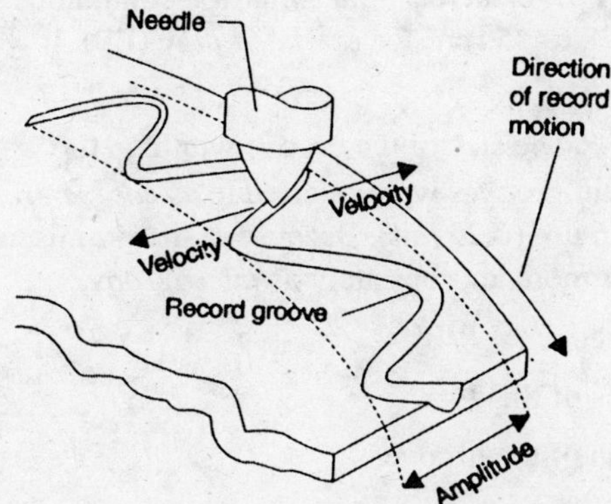

Fig. 11.14 The reproducing stylus is moved laterally by the modulated groove.

Fig. 11.15 The groove of a gramophone record enlarged about 100 times.

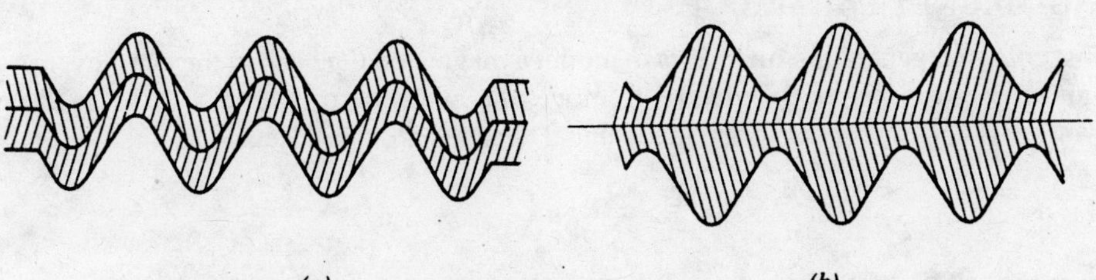

(a) *(b)*

Fig. 11.16 Mono grooves with (a) lateral modulation and (b) vertical modulation.

Where as a stylus of 0.001 in tip radius is used to trace mono discs, a smaller tip is used for the more complex stereo groove. The correct size for most high fidelity applications is 0.0005 in. and it is advisable to use this size for all pickups provided that the playing weight is suitably low. Since the tip radius is half that for mono, the tip area is also smaller by a factor of two. This means that the pressure on the stereo groove is about twice that on the mono groove, and *stereo pickups must therefore track at a smaller weight.*

Primitive 12 inch discs with 78 rpm speed could give a playing time of 3 minutes only. This was too short a playing time. It was increased to 6 minutes by introducing equalisation. This was still considered to be short. Two factors combine to increase the playing time necessary for continuous playback : the speed of rotation (rpm) and size of the groove.

1. *Speed of rotation* : To increase the time for continuous playback the *speed of rotation was decreased* from 78 to 45 rpm for *extended play* (EP). It was further reduced from 45 to 33 1/3 rpm for *long play* (LP).
2. *Size of the groove* : The size of the groove width on top was reduced from 150 μm (6 mil) to 50 μm (2 mil). 150 μm grooves were referred to as *coarse grooves:* The size of stereo grooves was brought down to 6 μm (0.25 mil). Because of their miniscule size stereo grooves are referred to as *fine grooves* or more appropriately as *microgrooves.*

Maximum number of grooves = R/W ...(11.1)

where R = radius of the disc

and W = width of the groove

assuming that the grooves on adjacent spirals just touch each other.

Problem 11.1 : Find the number of grooves per cm. on a disc of 300 mm. if the groove width is 50 μm.

Solution : Radians (R) of the disc $= 150 \text{ mm.} = 150 \times 10^{-3}$ m

Groove width $(W) = 50\ \mu\, m = 50 \times 10^{-6}$ m

Maximim number of grooves $= \dfrac{150 \times 10^{-3}}{50 \times 10^{-6}} = 3000$

Number of grooves per cm $= \dfrac{3000}{30} = 100$

The actual number of grooves will depend on the *spacing* between adjacent grooves.

11.8. MOVING-MAGNET CARTRIDGES

The most common generating method is to mount a magnet on the back of the cantilever. As the motion of the stylus and cantilever causes the magnet to move, the magnetic properties change in the nearby coils and signals (voltages) corresponding to the mechanical changes are generated.

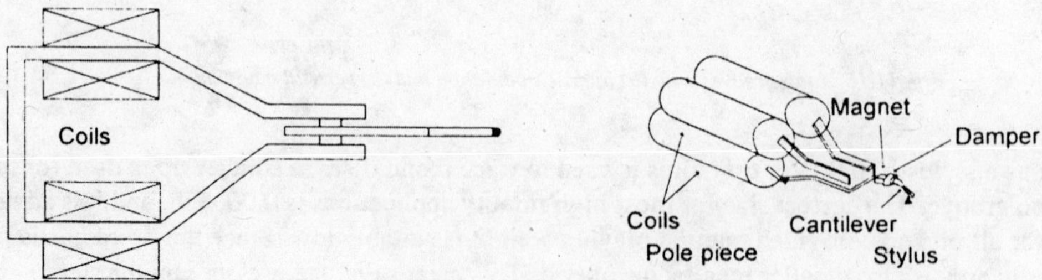

Fig. 11.17. Moving-magnet cartridge.

Moving-magnet cartridges are *less costly to manufacture* since the coils can be machine-wound. They have high output and *when the stylus wears out simply slide the old stylus assembly out of the body of the cartridge and slide a new one in,*

The magnetic pickup is a *current-operated device.* The construction of magnetic pickups varies greatly. The movable magnet is *damped.* This is accomplished in a number of ways, depending on the construction of the particular cartridge.

The coil is connected directly to the input of the preamplifier. The *frequency range* of a magnetic pickup is greater than that of the crystal type. The finest magnetic pickups have a frequency range of from 50 to over 15,000 Hz. The *distortion* content of a fine magnetic pickup may be as low as 0.1 percent at 400 Hz, and 1 percent to 4 percent at high frequencies. A typical moving-magnet cartridge is shown in Fig. 11.17

11.9 MOVING-IRON CARTRIDGES

To improve tracking in the groove and achieve low distortion the mass of the cartridge parts that move the moving mass—must be lowered.

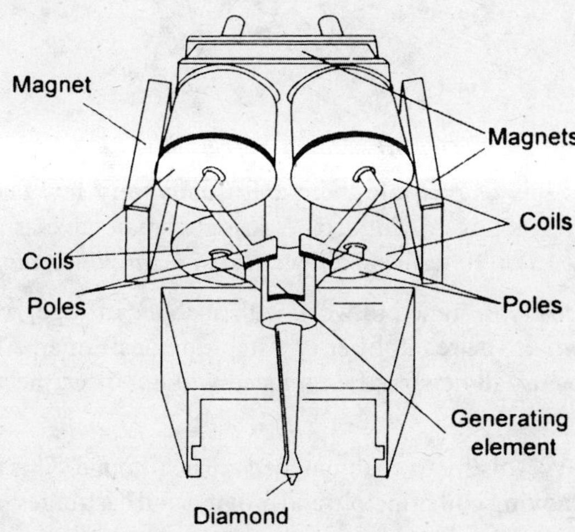

Fig. 11.18. Variable reluctance cartridge.

This can be done by removing the magnet from the rear of the cantilever (as in the moving magnet types) and substituting a tiny, very light, hollow piece of iron. The iron piece moves in the field of a stationary magnet, changing that field and generating a signal in the coil, Fig. 11.18.

As the armature moves toward or away from a stationary magnet pole, the *reluctance* (magnetic resistance) of the magnetic path changes. These cartridges are also referred to as *variable reluctance cartridges.* The voltage generated is proportional to the rate of movement in the groove.

The *advantages* gained by lowering the moving mass are better tracking, lower distortion with all the advantages of the moving magnet types; moderate cost, high output and removable stylus assembly. This type of cartridge also requires the use of a preamplifier.

11.10. MOVING COIL CARTRIDGES

The most sophisticated and most expensive way to generate a signal is the moving coil system. In this system carefully hand-woundcoils are mounted on the cantilever and move in the field of a stationary magnet. The signal is generated directly, with a minimum of distortion because there are fewer losses in the transfer from mechanical to electrical signals.

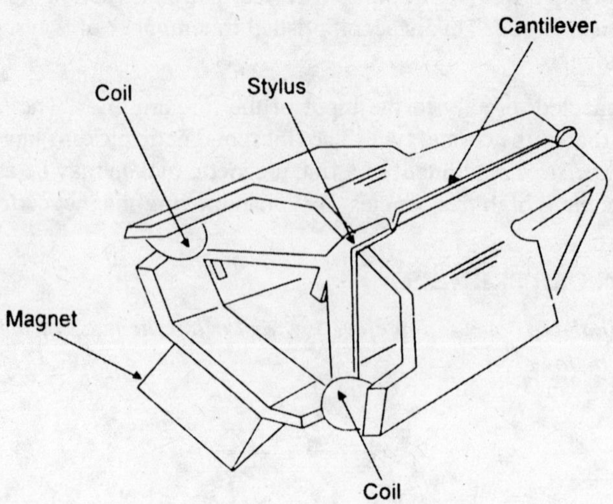

Fig. 11.19. Moving coil cartridge

To keep moving mass as low as possible there are usually very few turns of the coil. For example, stationary coils (as in moving magnet or moving iron cartridges) may have as many as 3,000 turns of wire. The moving coil may have only 15 or 20 turns, and *fewer turns mean lower output voltage.*

Unless an amplifier is already designed to work with such a cartridge, *moving coil (dynamic) models require an accessory to boost output* - a preamplifier or a step-up transformer. Also, because the coils are part of the moving system, the stylus usually cannot be replaced without disconnecting the tiny wires inside the cartridge.

Moving coil cartridges have a distinctive, unmatched musical sound. This may be because the recording cutter head also operates on the moving coil principle and moving coil cartridges provide an exact complement.

A dynamic cartridge is shown in Fig. 11.19. The response of this cartridge extends beyond 20,000 Hz, and it gives almost as much output as a good reluctance pickup.

11.11 CERAMIC CARTRIDGES

The ceramic cartridge, shown in Fig. 11.20, utilises the so-called *piezo-electric effect.* Movement of the cantilever, which is fixed to the stylus, applies varying forces - twists and pressures - to a special crystalline material which in turn generates the electric voltage and subsequent current. The rigid armatures required to exert this twisting force naturally increase the moving mass, and the ceramic elements are sometimes affected by heat and moisture. Although commonly used in inexpensive record players, they do not fulfill the discriminating demands of real hi-fi.

As a circuit component, a crystal cartridge acts in the same way as does a capacitor and can be considered as such. Extremes of temperature drastically affect the operation of a crystal. *The maximum sensitivity of a crystal is usually at about 75 degrees.* As the temperature rises above or falls below 75 degrees, the sensitivity of the crystal falls off slowly. Cartridges using barium titanate have much improved temperature characteristics. *Ceramic pickups* have overcome this difficulty and are now commonly used in place of crystals.

11.12 FEATURES OF STEREO CARTRIDGES

In stereophonic applications, one of the most critical components is the pickup cartridge.

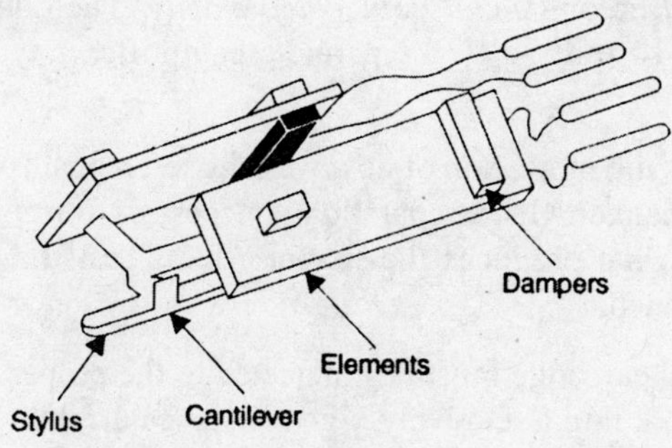

Fig. 11.20 Ceramic Cartridge

The *weight* of a cartridge and pickup arm assembly is transmitted to the record as *needle force.* The maximum allowable needle force without excessive record wear is related to the *size of the needle tip.* Wear is dependent on pressure, which is force per unit area. The smaller the needle tip, the less the needle force must be to prevent wear. Too small a needle-tip size would result in *loss of tracking;* that is, *the needle would skate over the record instead of staying in the groove.* Thus the minimum needle tip size is kept to 0.5 mill, and the range of needle force is roughly from 1/2 to 6 grams.

Frequency response is usually stated in terms of frequency range and the deviation, in dB, of the response over that range. The high limit of the range should extend to near 20,000 Hz or beyond to take full advantage of the better recordings.

The *output voltage* of a cartridge is important, not only because it is used to calculate how much amplification is required, but also because the higher the output the more chance there is of having a good *signal-to-noise ratio.* Output voltage does not establish the criteria unless the level at which the needle is driven is also specified. Standard records are used to provide the drive for output voltage tests. Manufacturers

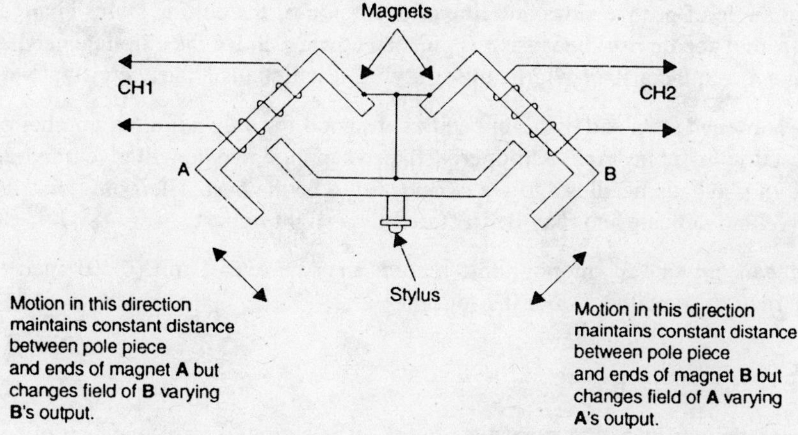

Fig. 11.21. Basic stereo-cartridge operation.

state output for either 5 cm/sec or 10 cm/sec *needle velocity.* The output should be higher for the greater velocity. For outputs below 10 millivolts, a separate preamplifier may be necessary, depending on the gain of the system used.

Channel separation is the indication of how well the left signal is kept out of the right channel, and the right signal out of the left channel. Good separation is necessary for good stereo effect, since the difference between the two signals is what produces the spatial effect. Tests have shown that a minimum of 15 dB separation should be maintained.

To operate properly, a cartridge must be connected to the proper amplifier *input impedance.* Ceramic and crystal pickups must work into a relatively high resistance (15 kil ohms to several megohms) compared with magnetic cartridges (5 to 100 kilohms). Each cartridge manufacturer specifies the specified load characteristics for his products.

Compliance is a measure of how easy it is to move the needle in the direction it must be driven during playing. It is measured in the distance, in millionths of a centimeter, that the needle can be pushed by a force of 1 dyne in that direction. Stereo cartridges, Fig. 11.21, should have a high compliance in all directions of motion. That is why stereo cartridge manufacturers specify both *lateral and vertical compliance.* It is because of the low vertical compliance of most of the monophonic cartridges that *stereo records should never be played with monophonic pickups.*

Channel balance is given as the difference in dB between the output of the two channels for the same drive. The optimum arrangement is a system that is naturally balanced throughout and balance deviation of 1 dB or less should be considered good.

11.13. PLAYBACK NEEDLES

To minimise wear and secure the maximum useful life from a recording, a *properly shaped needle,* Fig. 11.22, must be used. A properly shaped needle will also minimise background noise.

The difference in shape between the cutting stylus and playback needle causes trackability distortion and pinch distortion. *Trackability distortion* is overcome by the use of an elliptically shaped needle. When an elliptically shaped needle tracks the groove, its points of contact *nearly coincide* with the original points of cutting stylus contact at any given position. *Pinch-effect distortion* is caused by the changing width of the groove as the angle of groove sides and line of excursion of the cutting stylus change. As the groove width narrows, the round needle rises because it is pinched upward. Since the elliptical needle rides almost the same points of contact as the constant-width cutting stylus, it maintains a fairly constant vertical position.

Plain, hardened steel and osmium needles are good for only a limited number of plays and, therefore, are seldom used in hi-fi playback. A properly shaped sapphire needle will give hundreds of plays. Diamonds are also used in playback needles and are considered to be the best. Their only disadvantage is that they are very expensive and delicate and may be fractured by a slight impact.

The *needle-tip size* for monophonic records may be over 1 mil (0.001 inch), while needle tips for stereophonic records are either 0.5 or 0.7 mil.

11.14 TONE (PICKUP) ARMS

A cartridge can deliver its full performance only if it is mounted on a tone arm of equally high quality, see Figs. 11.10 and 11.23. *The tone arm's primary function is to hold the stylus in its path while it travels across the record groove and to apply the necessary tracking force while compensating for other unwanted forces.*

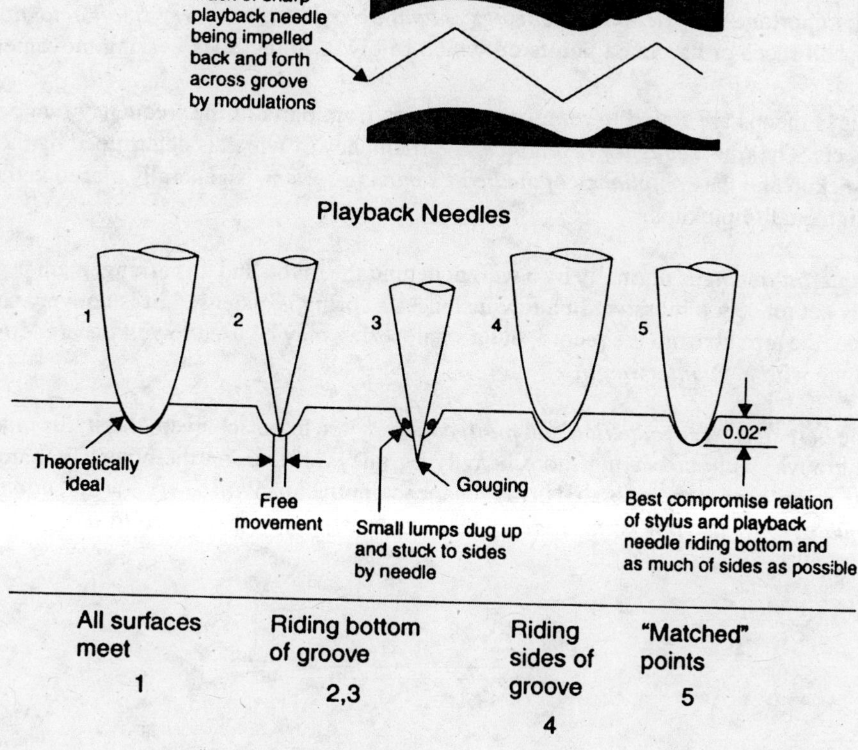

Fig 11.22. Properly and improperly shaped playback needles.

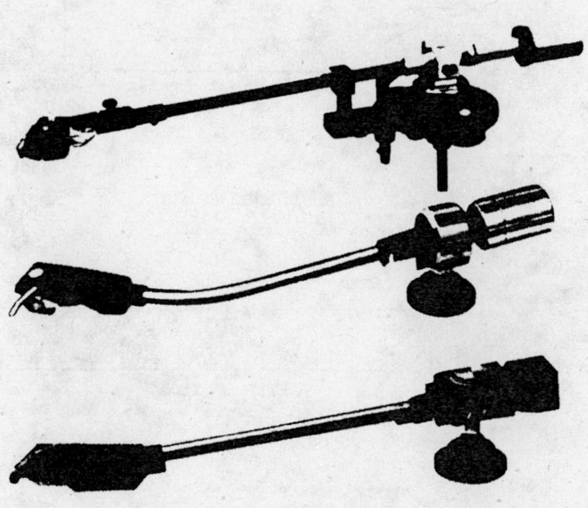

Fig. 11.23. Modern tone arms for use with light and small heads or cartridges.

Modern tone arms, for use with light and small heads or cartridges, Fig. 11.23, are themselves of *small mass,* and have the characteristics which will permit low playing weights to be used without mishap. It is particularly important that *friction of bearings should be reduced to a very small amount* and the best arms have either ball races or hardened points on which to pivot. Sometimes bearing movement is *damped* by a viscous fluid.

Various means are found to remove resonances from parts of the frequency range where they would have ill effects. The *low frequency resonance,* the frequency of which is determined by the effective mass of the whole pickup and the *compliance of the head's moving system,* is generally placed in the 10-20 Hz region in recent high quality pickups.

Counterbalancing is normally by a weight behind the pivot, and this arrangement is generally adequate if the arm is not made too massive. In a few instances a spring is also used. It is not very satisfactory to use a spring to pull the arm up from the record, but a small spring may be used to pull the arm down after balancing has been done with a *counterweight*

If the *best stereo reproduction* and *minimum wear* are to be achieved, the stylus must ride correctly in the record groove, without bearing more heavily on one wall than on the other. Techniques to ensure this degree of accuracy are not incorporated in the cheaper components. *With a very few exceptions, record changers are inadequate from this view point; indeed, they are rarely suitable for serious listening to stereo.*

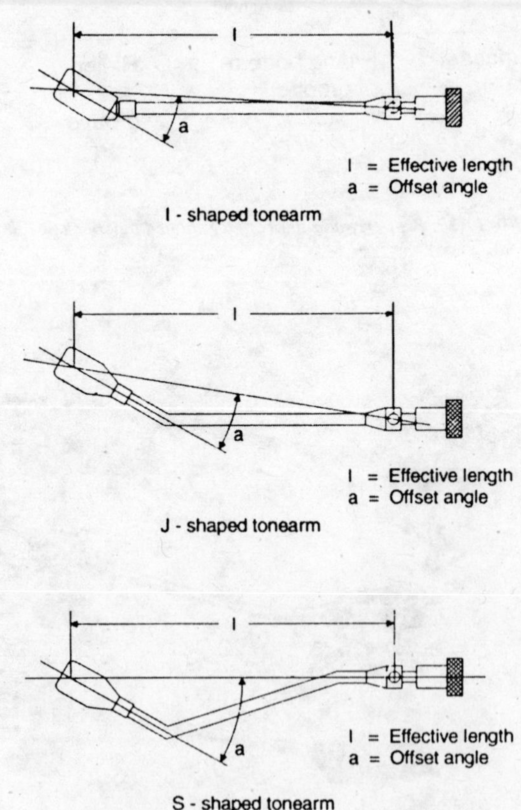

Fig. 11.24 Offset angle of conventional tone arms.

Accurate *tracking* will not be obtained if the arm is subject to the *side-thrust* which is commonly encountered in disc reproduction. Conventional arms are offset and arranged so that the stylus will fall in front of the record centre. This is done to minimise *tracking error.* The result is normally a maximum tracking error of 3 degrees with minimum error at the centre. Unfortunately, this arrangement, allied to the friction of the stylus in the groove, causes the arm to *swing inwards*. This aspect of arm design has received considerable attention in the last few years and in some modern tone arms a simple device is incorporated to correct side thrust. For instance, a small weight can be arranged in such a way that it will just correct the side thrust by pulling on the arm.

The cartridge assembly is at an angle (α) to the arm's axis, the line connecting the stylus tip to the arm's pivot. This is called the *offset angle* and its purpose is to minimise the tone arm's tracking error. *Tracking error* is the angle that the cartridge axis deviates from a tangent line to the record at any given point on the record. The smaller the tracking error, the better.

Some form of mechanical aid for *lifting and lowering* is essential if a very small playing weight is used. A number of arms have in-built devices, but separate accessories can be obtained for arms which lack this facility.

11.15 TURNTABLE

The term turntable refers to a composite of parts; the *platter* on which the record rests, the *motor and drive system*, the *tone arm*, the *cartridge* and the *housing* for the entire assembly.

The platter and drive assembly revolve the record at a specific speed. The tone arm holds the cartridge and suspends it over the record. The stylus in the cartridge traces the record groove. The housing contains the various components and stops vibrations from adversely affecting performance.

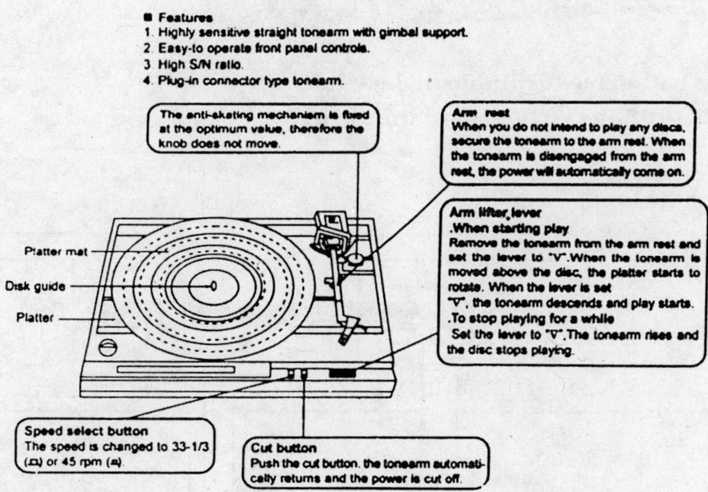

Fig. 11.25 Hitachi Model HT-12 Turntable

The *basic turntable* has fewer inbuilt mechanical devices to degrade performance _ which means *sound quality* will be improved. *Fewer parts also means there is less likelihood of break down.*

A basic turntable plays only long play (LP) records, has one speed (33.33 rpm), is manual, and usually devoid of cosmetic frills. In other words, it's a *single play unit.*

Just because a turntable is basic doesn't necessarily mean it's cheap. Most high performance, high cost models are of the basic type and are quite spartan in appearance.

All turntables revolve at 33.33 rpm, the speed required for LP record reproduction. For playing *singles* –170 mm and 300 mm diameter-a unit with 45 rpm is also required. Record collectors with archives of old discs may also need 78 rpm.

The primary consideration with turntable speed is that it is consistent. The motor and drive system must revolve the platter (and the record) at the exact required speed. Any deviation in speed is known as speed error or *wow and flutter*.

All turntables exhibit some form of *speed error* but most slowing or speeding effects are so miniscule they are inaudible (if they are not you can hear them quite clearly).

Turntables with a feature called *quartz lock* incorporate electronic speed correction, where the speed of the platter is continually referenced against the unwavering oscillation of a quartz crystal. When a variation in speed is sensed, the circuitry compensates for it.

The belt-drive turntable in Jer-Mate's HS-101 has an S-shaped tone arm

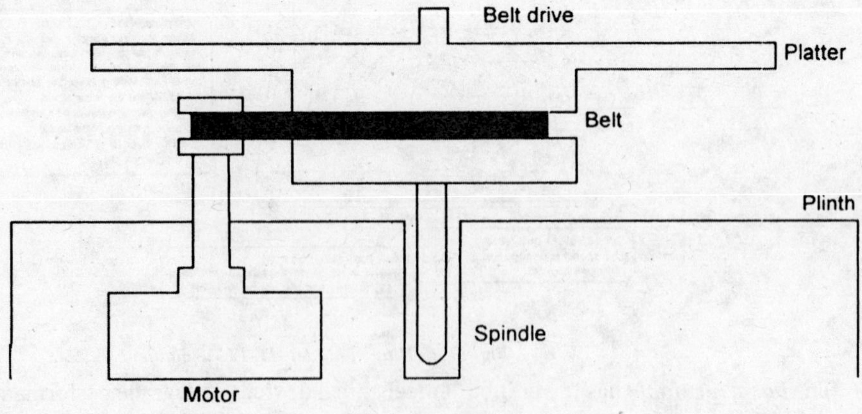

Fig. 11.26 Belt drive turntable

The way speed is changed is also a factor. *Electronic switching is silent and doesn't cause as many undesirable vibrations as mechanical switching.*

Belt drive produces better performance while direct drive permits a greater degree of sophistication. Both drive systems are comprised of a number of parts-the motor, its connection to the platter and the spindle on which the platter sits.

In a *belt drive turntable* the motor turns a spindle with a pulley attached to it. A belt wrapped around the pulley and the platter rotates the latter at the desired speed (see Fig. 11.26.)

The *motor* in any turntable is a notorious source of noise and vibration which can degrade performance. The *neoprene belt* in a belt drive system helps to *decouple* the motor from the platter and thereby reduce the transmission of these unwanted energies. The belt also helps *isolate* the stylus from outside influences.

In a *direct drive turntable*, Fig. 11.27 and 11.28, the platter rests on a spindle which is connected to the motor. There are no belts or pulleys and the platter is driven directly, hence its name. The platter's connection to the motor means *vibrations and noise from the turntable and outside can be more easily transmitted.*

Performance from belt drive and direct drive turntables is related to price. It is reasonable to expect an expensive direct drive model would outperform a cheap belt drive unit and *vice-versa.*

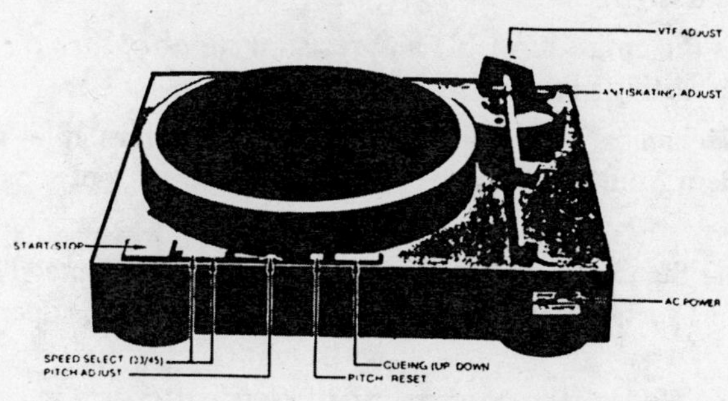

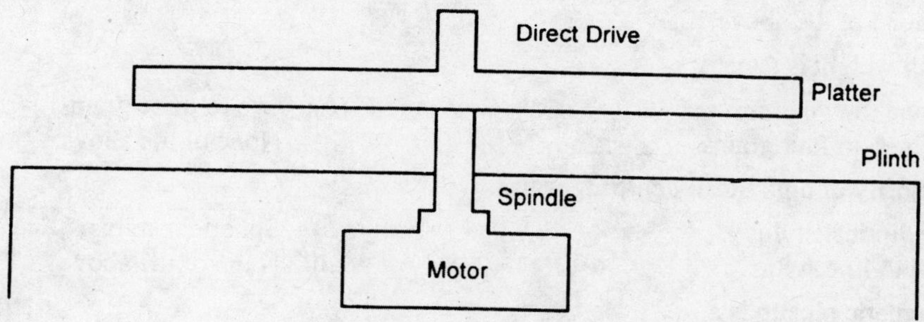

Fig. 11.27. Direct drive turntable.

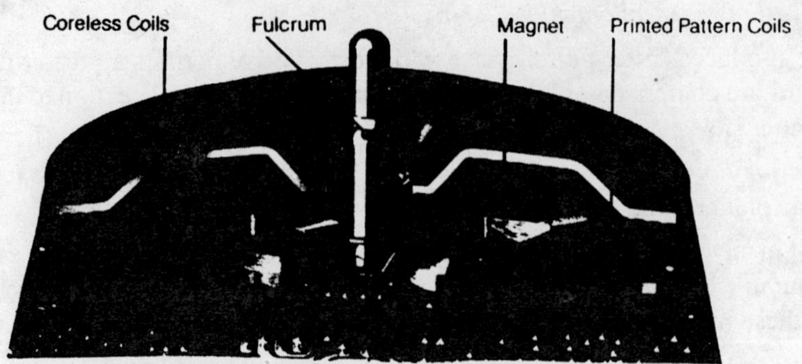

Fig. 11.28 Construction of Coreless Direct Drive Motor

Every turntable needs a *housing* to contain its components and like the other sections of the turntable it is comprised of a number of parts (see Fig 11.25); the baseboard, plinth, feet and cover. Manufacturers frequently use materials which tend to block or restrict the transmission of vibrations. In addition to its engineering significance, the housing dictates the *overall aesthetics* of the turntable.

The *baseboard* supports the tone arm, platter housing, function controls and sometimes the motor. It must be rigid and solid but more importantly it must be *isolated* from the main plinth. Some manufacturers use springs, rubber or other materials to *restrict* transmission of vibrations, but the effectiveness of the design is more important than the design itself.

The *plinth* supports the entire turntable and rests on *shock-absorbing* feet. These feet form a barrier where the turntable and its resting place meet.

Footsteps set up mechanical vibrations which travel to the turntable. The housing's feet must absorb these vibrations to stop them from reaching the stylus. It's handy if footsteps are *adjustable* so the turntable can be accurately levelled.

The *cover* does more than keep out the dust. In most designs it is made from heavy duty perspex which prevents airborne vibrations such as sound waves interfering with stylus operation

OBJECTIVE TYPE QUESTIONS (A)

1. Stereo records are produced by
- (*a*) lateral movement of the record cutter
- (*b*) vertical movement of the record cutter
- (*c*) lateral and vertical movement of the record cutter
- (*d*) none of the above.

2. Tracking weight is usually
- (*a*) one to two grams
- (*b*) two to three grams
- (*c*) three to four grams
- (*d*) None of the above

3. The majority of high quality cartridges use
- (*a*) elliptical stylus
- (*b*) Spherical stylus
- (*c*) fine-line stylus
- (*d*) None of the above

4. The magnetic pickup is a
- (*a*) current-operated device
- (*b*) voltage-operated device

5. To improve tracking the moving mass of the cartridge must be
   (*a*) lowered (*b*) raised
6. The most sophisticated and expensive way to generate a signal is
   (*a*) the moving magnet system (*b*) the moving iron system
   (*c*) the moving coil system (*d*) none of the above.
7. Ceramic cartridges utilise
   (*a*) electromagnetic induction (*b*) piezo-electric effect
   (*c*) either (*a*) or (*b*) (*d*) neither (*a*) nor (*b*).
8. The weight of a cartridge and pickup arm assembly is transmitted to the record as
   (*a*) tracking force (*b*) needle force
   (*c*) both (*a*) and (*b*) (*d*) neither (*a*) nor (*b*)
9. The output voltage of a cartridge is important because
   (*a*) it is used to calculate how much amplification is required
   (*b*) it forms the basis for signal-to-noise ratio
   (*c*) both (*a*) and (*b*)
   (*d*) neither (*a*) nor (*b*).
10. Stereo cartridge manufacturers specify
    (*a*) lateral compliance (*b*) vertical compliance
    (*c*) both (*a*) and (*b*) (*d*) neither (*a*) nor (*b*).
11. The difference in dB between the output of the two channels for the same drive is called
    (*a*) tracking error (*b*) tuning error
    (*c*) channel balance (*d*) none of the above.
12. Trackability distortion is overcome by the use of
    (*a*) an elliptically shaped needle. (*b*) a spherical shaped needle
    (*c*) either (*a*) or (*b*) (*d*) neither (*a*) nor (*b*).
13. Tracking will not be obtained if the arm is subject to
    (*a*) vertical thrust (*b*) lateral thrust
    (*c*) neither (*a*) nor (*b*) (*d*) either (*a*) or (*b*).
14. In-built mechanical devices
    (*a*) upgrade performance (*b*) degrade performance.
15. Turntables with quartz lock incorporate
    (*a*) electronic speed correction (*b*) mechanical speed correction.

FILL IN THE BLANKS (B)

1. The multiple channels recorded in the studio are ________ before a record is cut
2. In the standard recording process microphones feed the signals to ______.
3. When a master tape is played its magnetic patterns are converted to ______.
4. The stereo effect is obtained by proper ________ of the microphones.
5. The stylus is the only part to make ________ with the record.
6. The cantilever transmits its motions to the ________ system.

7. The actual number of grooves will depend on the ________ between adjacent grooves.
8. Moving coil cartridges require an accessory to ________ output.
9. The maximum sensitivity of a crystal is at about ________ degrees.
10. To operate properly a cartridge must be connected to the proper amplifier ________.
11. Stereo records should never be played with ________ pickups.
12. Most tone arms have small ________.
13. Any deviation in speed is known as ________ and ________.
14. Footsteps of a turntable must be ________.

CONVENTIONAL QUESTIONS (C)

1. Describe the various steps involved in making records.
2. Briefly explain stereo pickup techniques.
3. What are the different systems of stereo recording ?
4. Explain the following :
   (*a*) Tracking weight (*b*) Tracking error (*c*) Cantilever
   (*d*) Channel balance (*e*) Channel separation. (*f*) Micro groove
5. With the help of a suitable sketch explain the working of a moving coil cartridge.
6. What are the features of a stereo cartridge ?
7. Briefly explain the parts of a turntable.
8. What is the difference between a belt drive turntable and a direct drive turntable. Which drive is preferred and why ?

12

SOURCES OF DISTORTION IN DISC EQUIPMENT

Some distortion is always present in the sound reproduced from a gramophone record or magnetic tape. It is inevitable that part of this distortion will be distortion recorded on the disc or tape itself, but the loudspeaker, pick-up and recording head also make substantial contributions. A carefully constructed amplifier is far from being the weakest link in the chain: in fact : it is improbable that the limit to the quality of reproduction which can be obtained will be set by the performance of the amplifier. High quality reproduction will only be achieved if careful attention is also paid to the sound input source to the amplifier.

Even if perfect associated equipment could be obtained, it would not be possible to eliminate distortion entirely. There would still be some distortion - a comparatively small amount it is true- which is bound to occur when discs or tapes are used.

12.1 LOUDSPEAKERS

The choice of a *suitable loudspeaker* is very important for both disc and tape reproduction. While it would be wrong to assume that price is the only guide available, normally the higher grades of loudspeaker will justify their cost by giving correspondingly better reproduction. The choice should never be made hastily. It is not sufficient to listen to one or two records because it is easy to construct a system which will sound impressive at first hearing. *Extended periods of listening to music are desirable to discover if mental fatigue is likely to be caused by the system. Furthermore, because it is easier to judge speaker quality by listening to comparatively simple sounds rather than to full orchestral recordings, the system should also be tested by assessing the fidelity with which good - quality broadcast speech is reproduced.*

A simple *direct - radiator loudspeaker* - that is, a single cone unit handling the entire range of frequencies can give excellent results when mounted in a *suitable enclosure*. It is important to remember that *this enclosure should always be that specified by the speaker manufacturer* since it is not possible to design a suitable cabinet without knowing accurately the basic performance of the loudspeaker unit. *One of the main disadvantages of all types of simple direct radiators is that the high - frequency response is very directional.* This fact, however, can to some extent be overcome by using the loudspeaker in conjunction with some form of reflector. The reflector can be built into the cabinet, or a corner of the room can be used for the purpose. *The effect of the reflector is to distribute the high-frequency components in a more random manner so that with the adequate treble boost provided by the amplifier, good results will be obtained.*

There is no doubt, however, that even the simplest of *dual systems* is much better than a single unit. Such a dual system may consist of two direct radiators - for example, a 12 in. unit for the *bass - end of the spectrum* and a 5 in unit for the *middle and treble components* - used in conjunction with a suitable *cross over network.* While the effective frequency response may not be very much increased by this type of dual system, *various forms of distortion will be reduced considerably, added to which there will be a better distribution of high frequencies from the smaller source.*

(a) (b)

Fig. 12.1 (a) A dual system is much better than a single unit (separate units). (b) Two- way crossover network.

Dual systems using twin direct radiators are available in two forms ; either with two completely *separate* units, Fig. 12.1 each mounted in its own enclosure or, alternatively, with the high frequency unit mounted on some form of modified frame, making it an *integral* part of the low frequency unit, as shown in Fig. 12.2.

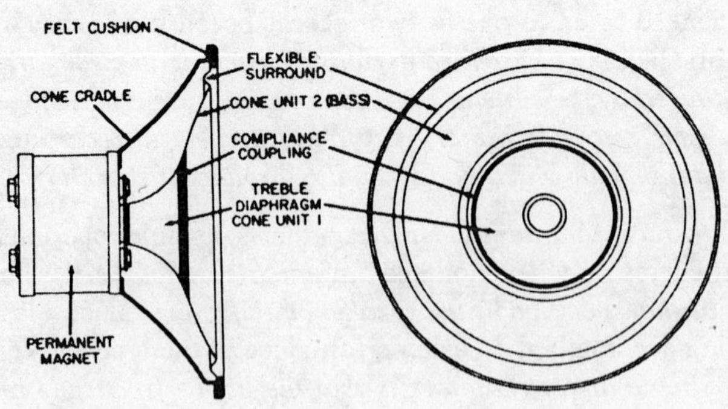

Fig. 12.2 An integral dual system.

Greater improvement may be obtained by using *three loudspeaker units* together with a *three-way crossover network,* Fig. 12.3. This extends the frequency range and reduces distortion still further, as well as providing a greater range of distribution patterns.

Fig. 12.3 31 cm, 3-way Speaker System with 3-way crossover network

The most complicated form of dual loudspeaker system is that which is *horn loaded over its entire range*. This may be achieved by using separate horn assemblies, each associated with a loudspeaker unit covering part of the frequency spectrum. Alternatively loudspeakers of the *concentric type* can be used. Such

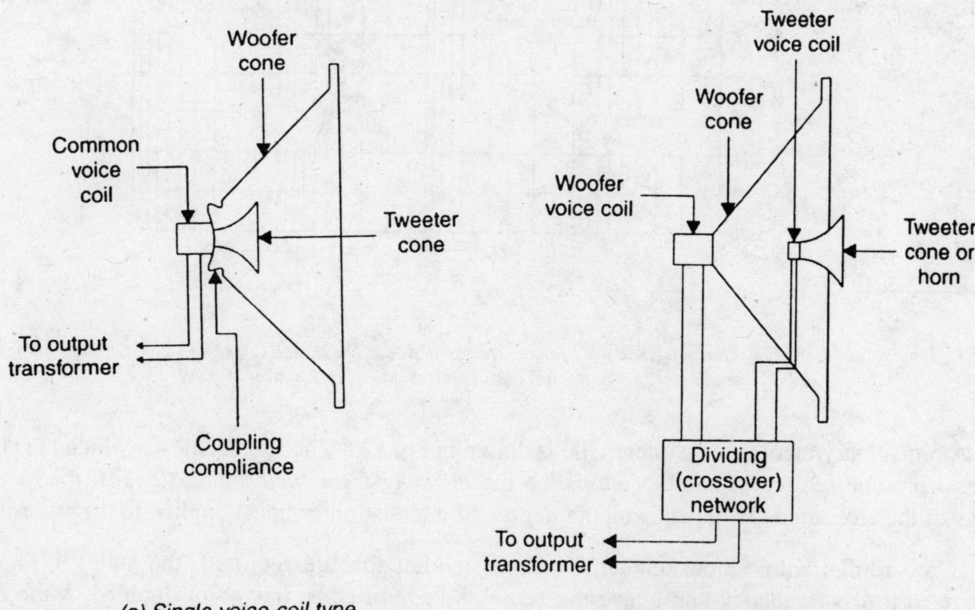

Fig. 12.4 Basic arrangements employed in coaxial speakers.

systems are always expensive but they have certain advantages; the *electroacoustic conversion efficiency* is high, the *acoustic damping* for the entire audio - range is much improved over that of any simple direct-radiator system and the larger size of the bass horn gives better *realism*. This is shown in Fig. 12.4.

12.2 DISC EQUIPMENT - RECORDING CHARACTERISTICS

Magnetic recording heads are normally used for cutting discs ; the stylus is attached either to the moving iron armature or to the moving coil. The *velocity* at which the stylus vibrates (measured as the stylus passes its equilibrium position) is *directly proportional* to the recording-signal strength, so that. for a constant voltage, the velocity is constant, and the amplitude or *width* of the groove is *inversely proportional* to the signal frequency. This is known as *constant - velocity recording*.

If the amplitude at high frequencies is chosen to give an acceptable signal-to-noise ratio, the amplitude at low frequencies will be excessive. Distortion will be high and too much space will be needed between adjacent rings of the groove to ensure that break-through from one ring to another does not occur. Thus, *some restriction of the amplitude is desirable at these low frequencies and, to achieve this, bass signal voltages from the recording amplifier are attenuated before being recorded.* Recordings in which the maximum width of the groove is restricted to some limit are known as *constant amplitude recordings*.

Because of the difficulty in maintaining a satisfactory signal-to-noise ratio if a constant velocity characteristic is used at high frequencies, *it is normal to boost treble signals from the recording amplifier before recording them.* Consequently, a recording characteristic indicating the variation of recording voltage with frequency can be divided into three sections: the bass section showing *voltage attenuation*; the middle section showing-*constant velocity recording* ; and the treble section showing *voltage boost*.

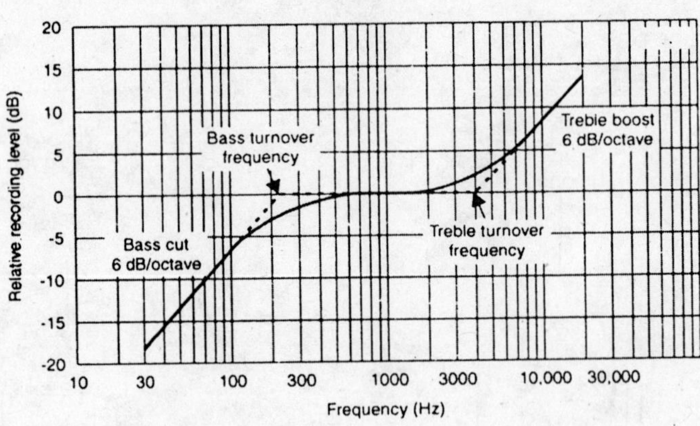

Fig 12.5 Hypothetical frequency characteristic used for recording on discs. The dotted lines show constant amplitude and constant velocity regions ; the full lines show practical realisation of these.

An imaginary recording characteristic is drawn in Fig 12.5 The frequencies at which the characteristic changes from one section to another are called the *turnover frequencies*. The *slope* of the bass and treble sections of the characteristic depends on the *degree* of attenuation or boost applied to the recording signal.

If recordings with a constant amplitude of modulation are required, the stylus velocity must be proportional to the frequency and hence the signal voltage must be halved or doubled as the frequency is halved or doubled and this gives a characteristic slope of *6dB per octave*.

Actual recording characteristics differ considerably from the imaginary ones of Fig. 12.5. The straight lines of the characteristics cannot be achieved in practice, and *actual characteristics are more like the continuous curve* shown in Fig. 12.5.

The bass and treble turnover frequencies in actual curves are defined as those frequencies at which the ratios of the recorded voltages to the true signal voltage are $1 : \sqrt{2}$ and $\sqrt{2} : 1$ (that is, - 3 dB and + 3 dB) respectively. *The true signal voltage is assumed, for reference purposes, to be the level of voltage recorded at 1 kHz.*

In most practical characteristics, the middle horizontal section is discarded, and only a point of inflection at 1 kHz is retained. The bass and treble sections of practical curves are not produced indefinitely with a constant slope, but are flattened at the limits of the range of audible frequencies.

12.3. DISC EQUIPMENT-PLAYBACK CHARACTERISTICS

If a *magnetic pick-up head* is used to reproduce the sound recorded on discs, the output voltage from the head will be proportional to the velocity of vibration of the needle. Consequently, the output from a magnetic head used with recordings made to characteristics of the type shown in Fig. 12.5 will increase with frequency in the bass and treble regions. (This will not be so with *crystal pick-up heads* as the output of these is proportional to the amplitude of modulation). Thus, *correction or equalisation will be required during playback amplification to restore the true level of the signal voltage, and this equalisation should be the converse of the attenuation and boost applied while recording.*

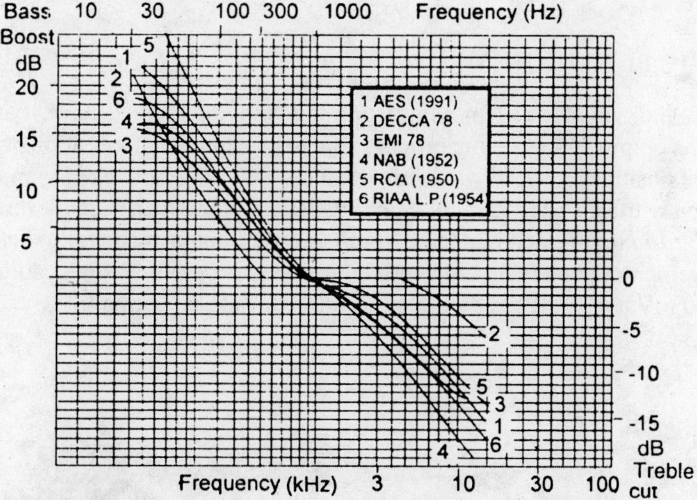

Fig. 12.6 Commercial playback characteristics for disc equipment.

The recording characteristics used by the different recording companies before 1955 followed, to a greater or lesser extent, the curve shown in Fig. 12.5, but the *differences* were large enough to make equalisation a matter depending greatly on the records to be played. Many companies issued their own equalisation characteristics, and *corrective networks* had to be designed with these in mind. Examples of the playback characteristics of the major recording companies are shown in Fig. 12.6.

In 1955, however, the majority of these companies agreed to adopt the characteristics of the *Recording Industries Association of America* (RIAA) which are drawn in Fig. 12.7.

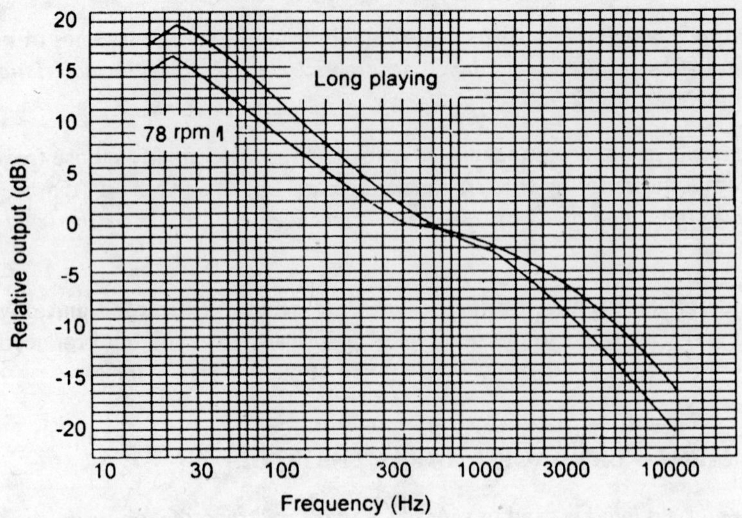

Fig. 12.7 RIAA playback characteristics for standard and microgroove discs.

Consequently, there is no need with up-to-date recordings, to adopt the pick-up head and record to each other by using separate equalising networks for each brand of record. Instead, one network (usually in the form of a *frequency-selective feedback network)* can be incorporated in the amplifier, of course, with the two characteristics for standard and microgroove records, separate networks will still be required for these.

12.4 MAGNETIC PICK-UP HEADS

Magnetic pick-up heads (moving-iron and moving- coil) have basically *constant - velocity characteristic,* the output voltage being proportional to the velocity of vibration of the pick-up stylus (measured at its equilibrium position). *Output voltages* which are quoted for *comparing* magnetic pick-up heads should therefore be given with reference to *stylus velocity.* This reference can be made directly by quoting the velocity; alternatively the *ratio* (expressed in decibels) of the actual output voltage to that which would be obtained at some standard velocity (1 cm/s) should be given. Hence, an output voltage could be given as, say, 100 mV at 3.16 cm/s or 100 mVat + 10 dB, and both expressions would be equivalent.

Fig. 12.8 The Zenn MC 2.5 Boron Moving-Coil Cartridge

Moving-iron magnetic pick-up heads can be divided into three classes, governed by their outputs (measured at a stylus velocity of 3.16 cm/s):

1.	Low-output heads	Voltages below 20 mV
2.	Medium output heads	Voltages between 20 and 100 mV
3.	High-output heads	Voltages above 100 mV.

The basic output of a *moving-coil pick-up head* is of the order of millivolts. The lower level of output results from the desirability of lightness of the pick-up unit and hence the restriction on the number of turns of the coil.

Both types of head are suitable for high-quality equipment, although the *intrinsically low output* of the moving-coil type necessitates the use of a *high-ratio pick-up transformer* between it and the amplifier. The *impedance* of magnetic heads is of the order of kilohms at a signal frequency of 1 kHz, and the *required loading* (that is, the input impedance of the amplifier) for the rated output of the heads is of the order of 100 kΩ. *Their loading should be independent of frequency,* otherwise the input signal to the amplifier will also vary with frequency.

12.5 CRYSTAL PICK-UP HEADS

In a *crystal pick-up head* a strain is produced in the piezo-electric crystal by the vibration of the stylus as it follows the modulation of the recorded groove. The *amplitude* of this modulation governs the output voltage of the head.

The rated *output voltage* of a crystal pick-up head is always quoted at a particular frequency. This is essential because, for comparison purposes, a recording made to a constant velocity is used and the amplitude of modulation - and, therefore, the output voltage - of such a recording is inversely proportional to the frequency.

A typical output voltage from a high-output crystal head would be 1 V at 1 kHz. The output at a frequency of 300 Hz would therefore be about 3 V. It is thus evident that *attenuation is desirable at the lower frequencies to provide a balanced output and to prevent the input stage from being over-driven.* The highest - quality crystal heads have outputs considerably lower than this - a typical value being 0.5 V at 1 kHz.

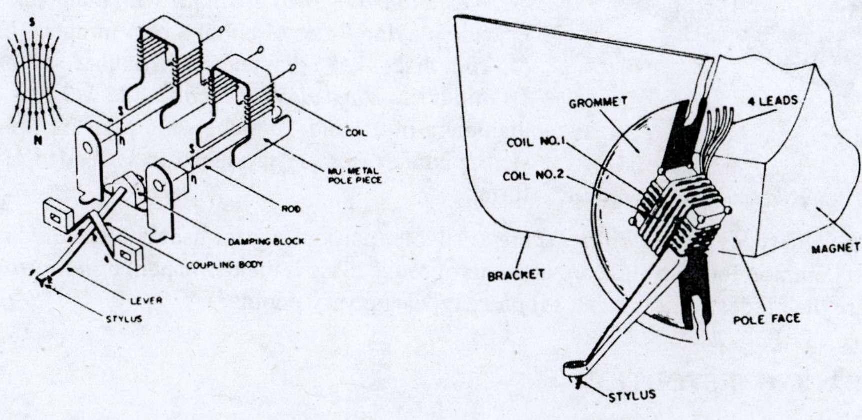

Fig. 12.9 Construction of stereo pickup cartridges: (a) moving - magnet type (b) moving-coil type.

The *impedance* of a crystal pick-up head is equivalent to a *capacitance*, the value of which is of the order of 1000 pF. The optimum load resistance of a crystal head will depend on the *method of loading*. If it is loaded so that its characteristic is similar to that of a magnetic pick-up head the load resistance is of the order of 100 kΩ. If the characteristic is not modified, then an input impedance of the order of megohms is desirable.

12.6 STEREO PICK-UP HEADS

Pick-up heads for use with stereophonic equipment can be either the magnetic or the crystal type. The *construction* of these pick-up heads, however, will differ considerably from that of monaural heads. They must be capable of responding to the two-dimensional modulation of the recording groove and of translating the response into *two separate signals* where as monaural heads have only to follow a lateral displacement and convert it into a single signal. Consequently, there must be two signal generators in each stereophonic head and these must have very *similar characteristics*. They will, therefore, both be of the *same type* - both crystal or both magnetic. Also, the pick-up stylus will require a *vertical compliance* comparable with its *lateral compliance* so that the tip of the stylus can follow the contours of the recording groove faithfully and provide *balanced stimuli* for the twin generators.

The *output voltages* obtained from these generators are very much lower than the voltages derived from monaural heads of the same type. *Moving-iron* stereophonic heads will produce voltages of the order of millivolts for each signal and *crystal heads* will give signals of the order of 100 mV. The *impedance* of each section of a stereophonic head will be of the same order as that of the same type of monaural head - about 1 kilohm for magnetic heads and the equivalent of about 1000 pF for crystal heads.

12.7 DISCS

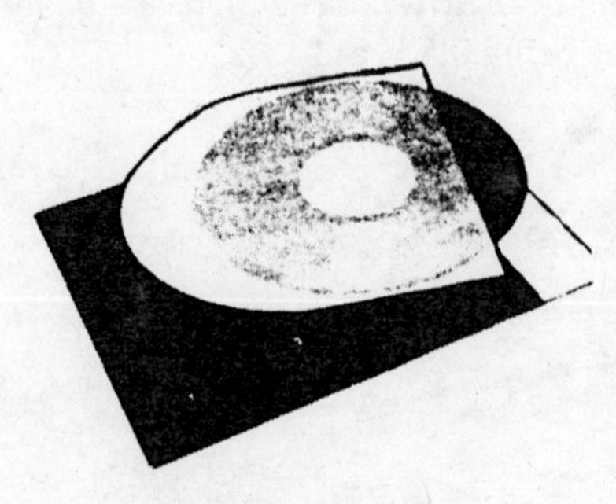

Fig. 12.10 Special anti-static inner sleeve.

Records are normally made for replaying at three different *speeds*: 78, 45 or 33 1/3 rpm. The first or *standard* type of disc is pressed in a hard non-pliable material called *shellac*. Record *wear* with this material is relatively high even under good playing conditions. Wear with *vinyl*, the pliable material used for 45 and 33 1/3 rpm, *microgroove* type of disc, is very slight under comparable conditions. The main trouble with vinyl records is the tendency to collect dust through electrostatic attraction.

Standard records are made with diameter of 10 or 12 inches and provide playing times of about 4 or 5 minutes. *Extended-play records* (45 rpm discs) have diameters of 7 inches and playing times of about 10 minutes. *Long-playing records* (33 1/3 rpm discs) are made with diameters of 10 and 12 inches and play for up to about 30 minutes; stereophonic discs are produced in Extended-play and Long playing forms.

Several types of *pick up stylus* are available (particularly for use with standard monaural records) but the most common for both standard and microgroove discs is the permanent or *jewel-tipped* stylus. *Diamond* stylus are the hardest wearing, but *sapphire* types are very popular.

12.8 TRACING DISTORTION

One form of distortion which occurs with *monaural disc recordings* is that resulting from the difference in the shape of the recording groove and the rounded playback stylus. The way this distortion arises can best be seen by considering the *width* of the recording groove.

When no sound is recorded, the groove is unmodulated and the cutting face of the stylus is at right angles to the *length* of the groove. The *width* of the groove is uniform and is the full width of the cutting edge.

With a signal however, the stylus will be displaced, from the unmodulated position and the cutting face is at an acute angle to the direction of the groove. Thus the *width of the groove is not uniform.* This is illustrated in Fig. 12.11, which shows a sine-wave groove, very much magnified.

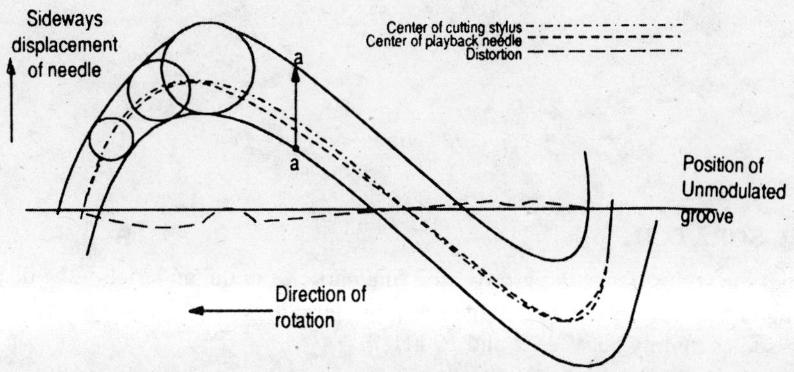

Fig. 12.11 Illustration of tracing distortion (sine - wave modulation of recording groove is greatly magnified).

The width of the groove in the direction a-a always corresponds to the full width of the cutting edge of the stylus. The *true width* of the groove - that is, the width at right-angles to its length - therefore depends on the *angle* between the modulated groove and the direction of the unmodulated groove, and will be smaller for greater values of this angle. The true width of the groove is the same as the width of the cutting edge only at the *peaks* of the sine wave.

If distortion is to be avoided, the sideways movement of the playback stylus should reproduce exactly the movement of this cutting stylus. The circles in Fig. 12.11 represent the point of the pick up stylus as it rests in the groove during playback, and the centres of these circles should therefore lie on the line traced by the centre of the cutting stylus. The figure shows that this condition is not fulfilled: the dotted line traced by the centre of the pick-up stylus is not the same as the chain line traced by the centre of the cutting edge. The amount of distortion resulting from this *lack of coincidence* depends on the difference between these two lines, measured parallel to a-a. *There is no distortion at the peaks of the sine wave, nor in the unmodulated position.*

Tracing distortion consists of *odd harmonics.* The distortion increases as the modulation becomes *more spiky.* Thus, it increases with the amplitude of modulation - one reason why constant amplitude recording is necessary at low frequencies. It also increases with *frequency,* which is the reason for restricting the amount of treble boost used when recording. The distortion is also greater if the *modulation is compressed,* as it will be near the centre of the record.

12.9 PINCH EFFECT

The sine-wave groove of Fig. 12.11 also illustrates the so-called *pinch effect* which results from the *varying width* of a modulated groove on a monaural recording. Because of this varying width, the pick-up stylus will have to ride up and down as it traverses the groove, Fig. 12.12.

At the end of a microgroove recording, the maximum vertical movement may be as much as 11% of the maximum lateral displacement. The output of some pick-up heads varies slightly with this vertical movement so that pinch effect can be an additional source of distortion. As the needle has to move up and down twice during each cycle, *the distortion is of even harmonic order.*

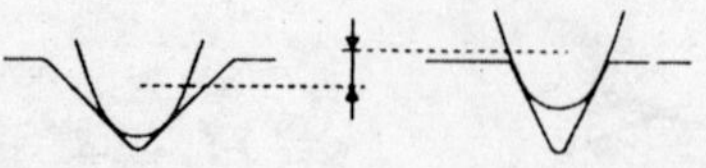

Fig. 12.12 Illustration of pinch effect.

12.10 NEEDLE-SCRATCH

As the pick-up stylus traces the groove, the fine particles in the material of the disc cause the stylus to make *small irregular movements.* The voltage setup by these movements is amplified and is heard as needle scratch. The noise lies mainly between 2 and 10 kHz.

Fig. 12.13 Needle wear.

Fig. 12.14 Dual Automatic Turntable features disc diameter selector and damped up / down cueing levers (far left foreground) and antiskate and tone arm leveler controls (near left foreground). Cast platter is a full 12 in. in diameter and comes with neoprene ribbed disc mat.

Needle scratch will be particularly bad when the diameter of the needle is *too small* for the groove. For good reproduction the stylus should rest on the two sides of the groove. If it rests right *in* the trough, it may ride up the walls, thus giving excessive scratch and additional distortion. If it rests *on top* of the groove only, it cannot follow the groove correctly, and there will then be a tendency for the stylus to *skate* across the record.

12.11 TRACKING ERROR

Another important difference between the making and replaying of discs is that the pick-up stylus as it traverses the record does not follow the same path as the cutting stylus. The cutting stylus is set to move inwards in a straight line along a radius of the disc, whereas the pickup stylus must necessarily be mounted on a pivoted tone arm, and must cross the record in an arc of a circle. The playback stylus is thus not always held at right angles to the groove and the output will vary slightly. *This effect is greater for large records, and can be reduced by using longer pick-up arm.*

Fig. 12.15 Tracking distortion can be reduced by using a longer pick-up arm.

12.12 TURNTABLE DRIVING SYSTEMS

It is essential that the mechanical driving systems of disc-playing equipment should be accurately made, especially with stereophonic equipment. Rotation of the turntable must be extremely steady. These requirements must necessarily be very severe if the mechanical system is to match the performance expected from the acoustical and electronic equipment.

In a high-quality installation, in which the loudspeaker is separate from the main chassis, the system can be mounted solidly to the motor board, and the motor board itself floated on rubber to ensure that no mechanical vibration is transmitted to the cabinet.

An unsuitable system can spoil the loudspeaker output in three objectionable ways : wow, flutter and rumble.

Wow and flutter are produced by variations in the speed of the turntable. Those variations occurring at low frequencies are termed *wow : flutter* results from faster fluctuations. Slow variations can be caused by spindles and other rotating parts loosely mounted or off-centre, so that the wow resulting is heard at regular intervals. Unevenness of the drive, changes of friction (catching or slipping) and record slip can give rise to *intermittent wow.* Small irregularities are normally experienced when driving from a motor. Sudden changes in speed as little as 0.5% will be clearly noticeable as flutter when a long note is being played. A sufficiently steady speed is usually achieved by providing a *heavy turntable,* or some other moving part to act as a *flywheel.* Care should be taken to see that the *speed - changing mechanism* is not subject to undue wear. This can lead to wow or flutter after much use.

Rumble results because the *mechanical parts* run noisily. These parts should run sufficiently quietly for no objectionable rumbling to be heard at full gain when there is no other input signal.

OBJECTIVE TYPE QUESTIONS (A)

1. Dual systems using twin direct radiators are available as
   (*a*) Two completely separate units (*b*) an integral unit
   (*c*) either (*a*) or (*b*) (*d*) both (*a*) and (*b*).
2. Crossover networks
   (*a*) extend the frequency (*b*) compress the frequency range
   (*c*) reduce distortion (*d*) increase distortion.
3. The velocity at which the stylus of magnetic - recording heads vibrates
   (*a*) is directly proportional to the recording signal strength
   (*b*) is exponentially related to the recording signal strength
   (*c*) is inversely proportional to the recording signal strength
   (*d*) is not related to the recording signal strength.
4. Before recording, bass signal voltages from the recording amplifier are
   (*a*) attenuated (*b*) boosted
   (*c*) neither (*a*) nor (*b*)
5. Before recording, treble signals from the recording head are
   (*a*) attenuated (*b*) boosted
   (*c*) neither (*a*) nor (*b*)
6. The slope of the bass and treble sections of the recording characteristic depends on the degree of
   (*a*) attenuation applied to the recording signal (*b*) boost applied to the recording signal
   (*c*) (*a*) or (*b*). (*d*) (*a*) and (*b*).
7. Corrective networks are
   (*a*) frequency selective (*b*) amplitude selective
   (*c*) (*a*) or (*b*) (*d*) (*a*) and (*b*).
8. Tracking error can be reduced by using a
   (*a*) medium sized pick up arm (*b*) Smaller pick-up arm
   (*c*) longer pick-up arm.

FILL IN THE BLANKS (B)

1. Tracking error is greater for __________ records.
2. Wow and flutter are produced by variations in __________ of the turnable.
3. Rumble results because mechanical parts __________ noisily.
4. Needle scratch will be particularly ________ if the diameter of the needle is ________ for the groove.
5. Distortion due to pinch effect is of ________ harmonic order.
6. Tracing distortion consists of ___________ harmonics.

CONVENTIONAL QUESTIONS (C)

1. Discuss the significance of a good speaker system for disc and tape reproduction.
2. What are the dissimilarities between recording and playback characteristics ?
3. Explain the difference between standard, EP, and LP records.
4. Briefly explain the following
   (*a*) tracing distortion
   (*b*) pinch effect
   (*c*) tracking error
   (*d*) wow, flutter and rumble.

13

MAGNETIC RECORDING AND REPRODUCTION

Cassette recording was originally devised as a cheap, simple-to-operate system for tape recorders designed to appeal to a mass market. At a time when the sales of reel-to-reel recorders were encountering resistance due to the cumbersome threading procedure, the size of the reels and the weight of the machines, cassette recording, with the simple snap in loading mechanism, rapidly gained in popularity with the result that cassette recorders now account for practically all home recorder sales outside the semi-professional and hi fi markets

13.1 MAGNETIC RECORDING

All of the work in the recording studio makes use of the magnetic tape recorder. The *ideal recorder* would be a memory bank of perfect accuracy with an unlimited capacity for storing separate groups of information in a constant time relationship. The ideal recorder must be able to store the sounds heard in the monitor speakers so that they can be played back at a later date, exactly *re-creating* the sounds heard from the mikes. *Magnetic tape is at present the accepted and the most practical means to approach this ideal.* It has advantages over direct recording into disc in that multitrack formats, editing and over dubbing are possible.

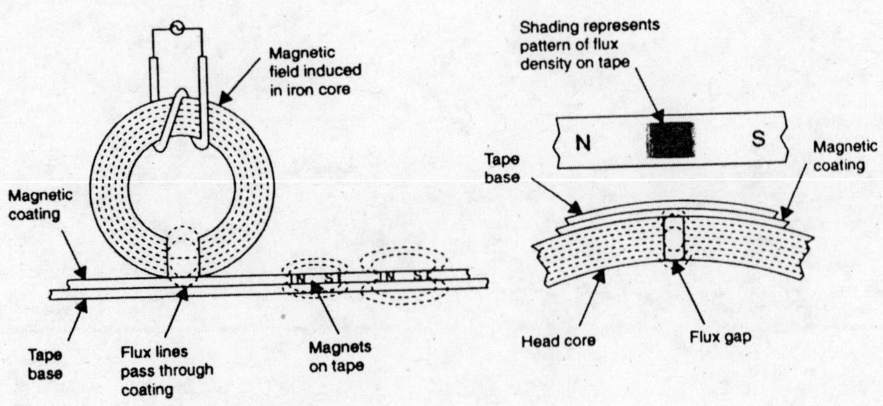

Fig. 13.1 Magnetic recording process.

The theory of tape recording is based on relating physical lengths of tape to periods of time. By playing back each section of a length of tape at the *same speed* at which it was recorded, the original rthythm and duration of each sound and the spaces between sound are preserved. *The best way of ensuring that the time spectrum remains unchanged whenever the tape is played is to record and play it back at a constant speed.*

Sound vibrations can be recorded as a pattern of varying degrees of magnetisation on a plastic ribbon *impregnated* with a plastic material. The tape moves at a *constant speed* across the poles of an electromagnet, which are less than 0.1 mm apart. The electrical output from a microphone is supplied to the electromagnet and *controls* the magnetic field between the poles. Thus, the section of tape *bridging* the poles at any instant is magnetised to a degree controlled by the sound vibration. (Fig. 13.1)

13.2 MAGNETIC PLAYBACK

All methods of recording and reproduction rely on a *sequence of reversible processes* with sound vibrations at one end of the sequence and a lasting change produced in some other property at the other end.

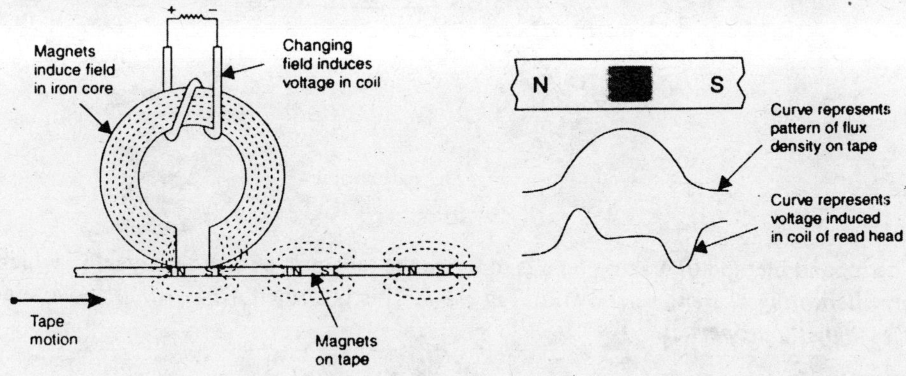

Fig. 13.2 Magnetic playback process.

During playback the magnetised tape recording is moved at the *same speed* past the poles of the magnet and induces a field in the magnet. This, in turn, induces a current in the windings, which can be amplified to *reproduce* the original sounds, as shown in Fig. 13.2. In practice, all these changes are *superimposed* on a very high frequency current of about 80 kHz, which acts as a *carrier wave* for the low frequency vibrations.

The *amplitude* of the voltage across the head coil depends on the *rate of change of flux* in the core assuming *constant speed.* This is, in turn, dependent on both the *strength* of the recorded magnet and the *sharpness* of the recorded magnet and the *sharpness* of definition at its ends.

The process of magnetic playback is thus the *reverse* of the process of magnetic recording.

13.3 MAGNETIC ERASING

The erasing of magnetic tape is done in one of three ways. The first, and simplest, method is to place an entire reel of the tape in a *strong alternating magnetic field* as shown in Fig. 13.3. This leaves the tape in a somewhat *unpredictable magnetic state.* It is termed *bulk erasing* and is more useful for removing recorded information from tape.

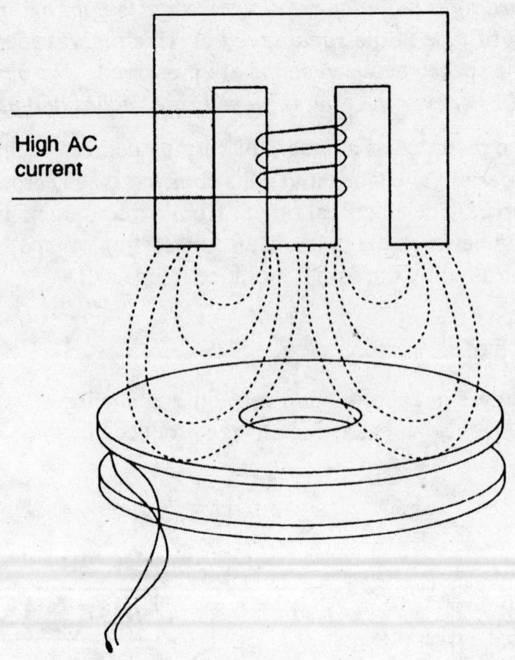

Fig. 13.3 Bulk erasing.

The second method of erasing tape is to move the tape past a recording head in which there is a *high - frequency alternating current*, as shown in Fig. 13.4. This is termed *AC erasing*, and essentially *neutralises the tape's magnetic properties.*

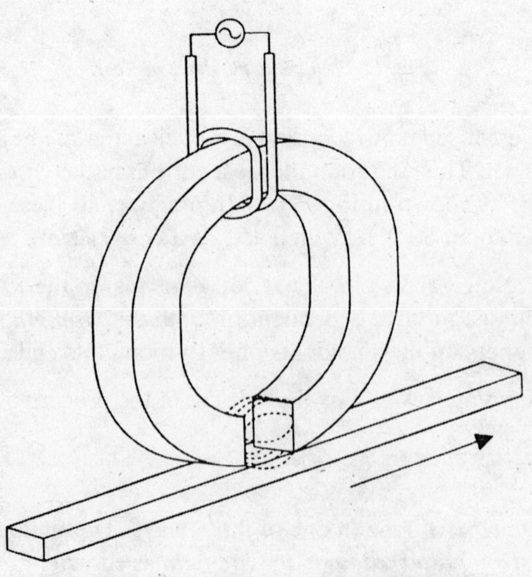

Fig. 13.4 AC erasing.

The third method is to move the tape past a recording head in which there is *direct current* as shown in Fig. 13.5. This is termed *DC erasing*, and essentially *magnetises the tape uniformly along its entire length*, thus turning it into one long thin magnet.

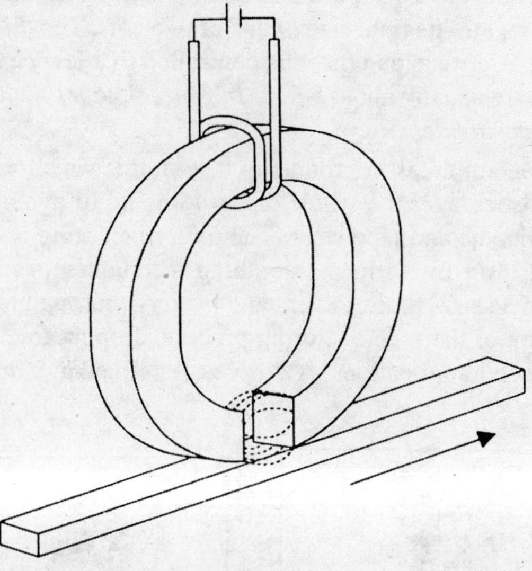

Fig. 13.5 DC erasing.

13.4 THE RECORDING MEDIUM

There are two types of recording medium in general use. The first uses a thin coating of *magnetic oxide* and the second a thin layer of *metallic magnetic film.*

The magnetic oxides which are most often used are those of iron, usually *gamma ferric oxide* (Fe_2O_3) but sometimes *ferrous oxide* (Fe_3O_4). More recently *chromium dioxide* has been employed to achieve better packing density and greater output signals. All these oxides are produced in the form of fine *acicular particles* with a length to diameter ratio of at least ten to one. Typical lengths of these particles range from a few microns to a few tens of microns.

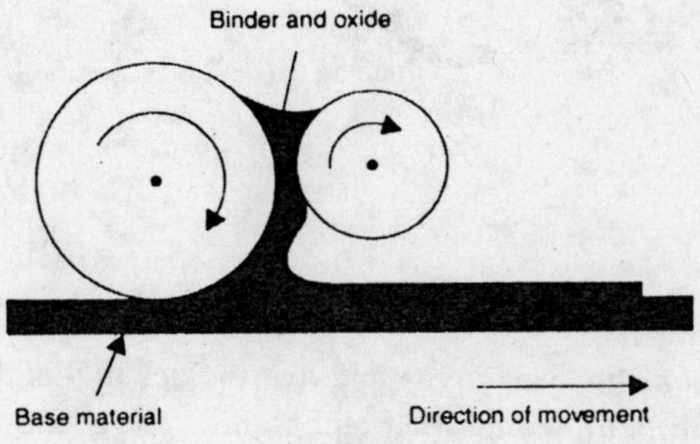

Fig 13.6 A method of coating tape.

The process of *coating* magnetic tape is illustrated in Fig 13.6. The oxide is mixed with a fluid containing a resin type *binder* and applied in a thin layer to the tape base material. The tape then passes through a drier to evaporate the solvents. This is often done in the presence of a magnetic field to *orient* the oxide particles for better performance. After drying the tape is passed into a machine which *slits* it from its original width, which can be as much as 1.5m, into its final size of one or two cms. The *thickness* of the coating produced for memory system applications is critical and must be controlled to a few percent. Typical coatings range from 1 to 25 μm in thickness and are usually about 5 μm. *In general, tapes with the thinnest coatings are capable of the highest information packing density.*

Some of the main problem areas in producing tape in this way are those of obtaining a coating with uniform magnetic characteristics over the whole tape width and of avoiding clusters of magnetic particles which give rise to noise, dropouts and tape wear. The *uniform coating* is obtained by careful control of the original layer of oxide and resin by various smoothing techniques to remove irregularities. *Clusters* are minimized by choice of particle size, basic solvent and by very thorough mixing in a ball mill before use. By careful control and inspection of the manufacturing process, dropout free lengths of greater than 500 m are regularly produced for computer applications. A *dropout* is defined as a decrease in signal of 50%

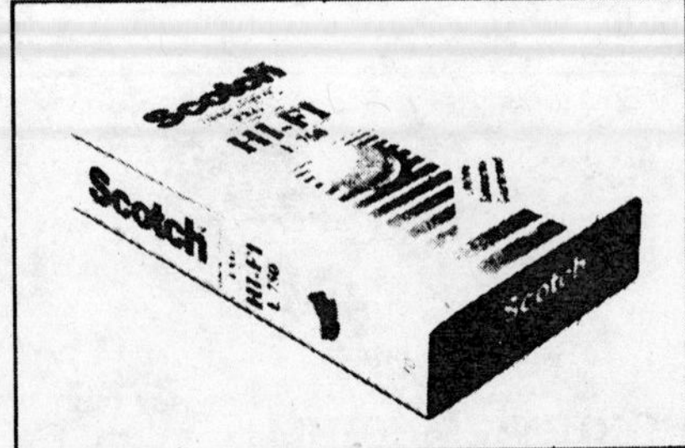

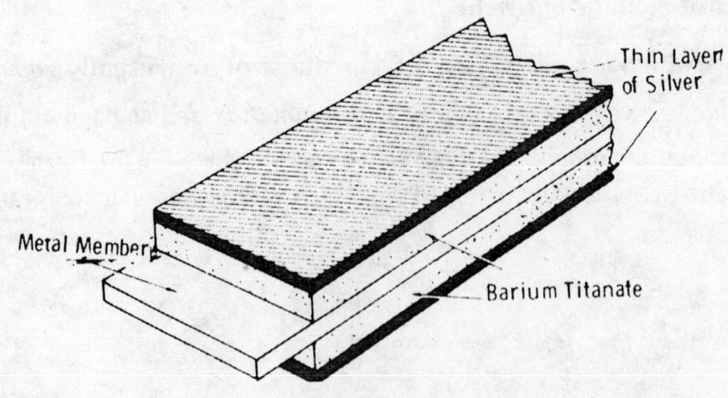

Fig. 13.7 Magnetic tape

Metallic thin magnetic films have an advantage over oxides in that they can be *controllably produced in much thinner layers* by electroplating and other techniques. They tend, however, to be more sensitive to mechanical forces and therefore are usually deposited on to *rigid bases*. The most frequently used films are Nickel-Cobalt alloys and these are produced in thickness ranging from 0.2 to 2μm. *A wide variety of magentic characteristics can also be produced by varying the composition of the alloy and by adding very small quantities of other elements.* Great care must be taken to prepare a smooth base for the deposition of the thin film in a *stress-free state.* This, of course, is necessary with oxide coatings. The *smooth surface* which results from the use of metallic films enables recording heads to run closer to the magnetic surface and thus achieve better performance. The main manufacturing problems are again those of achieving a uniform magnetic coating over a large surface and of preparing this surface to receive the film.

Plastic tape is made in various types to satisfy various *performance and price requirements.* For example, a relatively thick tape (one-and-one-half mil) is used when *strength and dimensional stability* of the tape are primary considerations. When optimum tape-to-head contact is required (for optimum data resolution) a thinner tape (1 mil or less) having greater flexibility is used. When *optimum reliability* is required, a tape that is fully certified and heavy duty is used. A *fully certified tape* is one that has been factory tested to ensure that the entire area of the tape used for recording has minimum number of "bad spots" that cannot be recorded correctly. A *heavy-duty tape* is one that uses a "hard" binder (in the oxide coating) that is highly resistant to wear. Typical heavy-duty tape is capable of being passed over the head more than 500 times before the data must be re-recorded, and more than 20,000 times before showing signs of wear. Even when sign of wear is evident, the tape can be *cleaned and reconditioned to* enable continued usage.

13.5 MAGNETIC HEADS

Most of the developments which have taken place to improve the quality and usefulness of magnetic tape recording have centred on *recording and replay heads*, the successors to the old horse shoes. The shape of the tape head is roughly that of a horse-shoe magnet wound with a coil of wire, *but with the gap between the pole pieces reduced to an incredibly small distance.* Tape head gaps are now measured in thousandths of a millimeter (μ *m*) or millionths of an inch, and *the size of the gap, particularly for replay, must be as small as can be kept consistent in manufacture, remembe*ı*ing that the gap dimension must be the same for each part of the tape covered by the gap and must not vary greatly from one sample of head to another.*

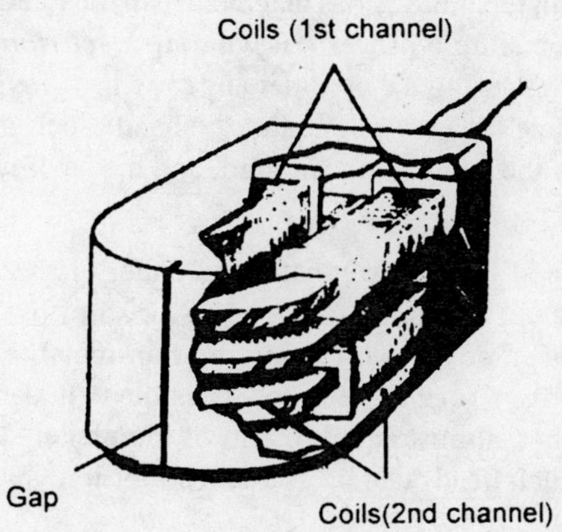

Fig 13.8 A modern tape head is fundamentally an electromagnet with a small gap across which the flux is very intense during recording. The small head can be used for replay.

The reason for keeping the gap so small is that *the ability to record and replay high-frequency sounds depends critically on this gap size and on the speed at which the tape is taken past the gap.*

Imagine, for example, that the gap is one thousandth of an inch (0.025 mm). Each magnetized portion of the tape must therefore be at least 0.001 in (0.025 mm) in length, so that *a complete wave* would need a strip of tape at least 0.002 in (0.050 mm) long to record. Now if we are trying to *record* a frequency of 5000 Hz (5kHz) meaning 5000 complete waves per second, the *length of tape* which must pass the head per second is 5000 × 0.002 in which is 10 in/s (25.4 cm/s). For a frequency range of up to 10kHz, which is only starting to be tolerable by modern standards, the *tape speed* for the 0.001 in (0.025 mm) gap would have to be at least 20 in/s (50.8 cm/s), and this might prove only *barely adequate.* Such high tape speeds were used in early recorders, but they require enormous lengths of tape on large reels if recordings of any useful length are to be made. *The use of modern micro-gap heads has now made possible the use of tape speeds as low as 1 7/8 in/s (4.76 cm/s) as used in cassette recorders.*

Fig 13.9 Teac reel-to-reel recorder.

A corresponding improvement in technology has taken place in the materials used for tape heads. Early tape recorders used heads made by the simple process of winding *steel ribbon* round a cylinder to several layers of thickness, welding the ends, removing the cylinder and sawing a *gap* in the layers. A coil of insulated wire was then wound round to complete the construction of the head. Such *primitive methods* are no longer used, because it is impossible to attain the small gaps required, and the steel wears rapidly as the tape, which is quite abrasive, passes over it.

Modern tape heads, Figs. 13.8 and 13.11, are constructed either of *special quality nickel-steels,* which have excellent performance, but wear out in, perhaps, five years or so of use, or from *modern non-metallic materials.* Modern techniques have been devised for using the non-metallic but magnetic materials called *ferrites,* which are then finished to a *glassy surface.* It is now possible to make such materials into heads with *remarkably small gaps* and performance almost equal to that of the nickel-steels but highly resistant to the rubbing of the tape, so much so that such heads can be guaranteed for ten years or more.

Though some quality *reel-to-reel machines,* Fig 13.9, use separate record and replay heads, usually with different gap dimensions, *it is usual on cassette recorders to have only one head for both recording and for replay.* This considerably limits the *creative possibilities* of the cassette recorder, since it is not possible to transfer recordings from one track to another, nor to record on one track while another is playing. Such requirements, however, belong more properly to the world of professional or semi-professional recording, whereas cassette recorders, Fig 13.10, are exclusively instruments of amateurs interested more in reproducing music than in creative work.

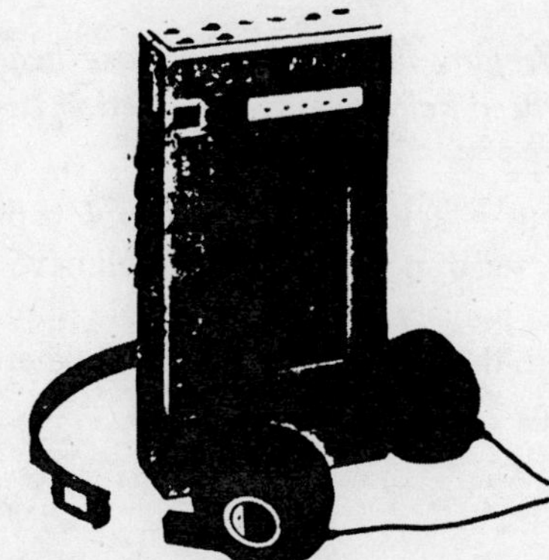

SOUNDMATE T-811 AM/FM STEREO AND CASSETTE PLAYER

Range	AM 530 – 1,600 kHz FM 88 – 106 MHz
Functions	Play, Stop, Rew, F.F., Eject, Talk line
Track type	4-track 2-channel stereo
Frequency response	40 – 12,000 Hz
Wow & flutter	0.15% (WRMS)
Tape radio selector	By switch
Tape speed	4.75 cm/s
Fast winding time	approx. 130 sec. with C-60 cassette
Power output	40 mW × 2 max
Power source	4.5V DC (3 pcs. AA batteries)
Dimensions	31 × 97 × 134 mm

Fig 13.10 Cassette recorder

1 Newly-Developed Rec/Play Head
2 Compact(10mm)design with Fewer Parts
3 High Performance and Low Cost
4 Newly-Developed Compact Auto-Reverse Head
5 Sophisticated Design Features Reduced Dimensions
6 Program Search Compatability & Low Inductive Noise
7 Ultra-Compact(9mm)Auto-Reverse Play Head
8 Metal-Capable Hi-B Hard Permalloy

New Design Features High Efficiency with Fewer Parts
Newly-Developed Microcassette Magnetic Erase Head
Metal-Capable

Fig 13.11 Magnetic heads

Yet another requirement of a recorder is *erasure*, since one of the attractions of magnetic tape recording is that unwanted recordings can be erased and replaced by new recordings. Erasure in any kind of magnetic recording is carried out by another head using a *much wider gap* and wound with a coil to which is supplied a current signal of much higher frequency than any recorded sound signal. Frequencies ranging from 75 to 120 kHz are now common. Such a signal applied to a comparatively wide-gap head cannot be recorded on the tape, and has the effect simply of *jumbling* the magnetism of the tape particles so that the information is erased by removing the alternate changes in direction of magnetisation of the particles in adjacent pieces of tape. In normal use, and inevitably in cassette recorders, *the tape passes over the erase head before passing over the record/replay head, and the erase head is activated only during recording, so that the tape is automatically erased before being recorded.*

13.6 CASSETTES

The cassette is actually quite a *complex* piece of machinery. For example take the *shell.* In order to operate properly, the dimensions of the shell have to be extremely accurate. A good shell is precision molded from high impact polystyrene. Many people think that all cassettes are alike. It is not so. The quality of the cassette used with the cassette recorder will determine whether or not the customer gets his money's worth out of the system.

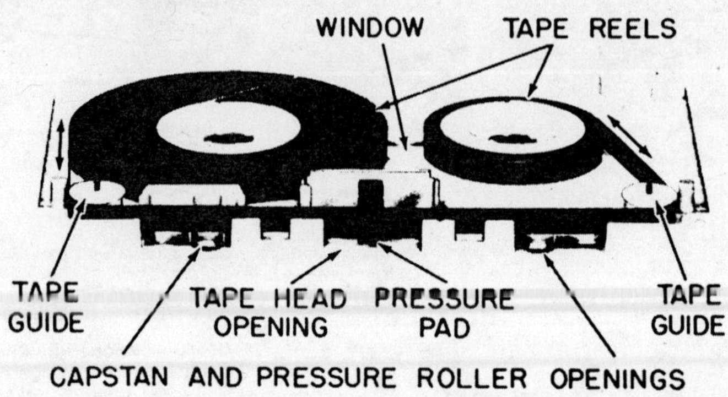

Fig. 13.12 Mechanical features of a cassette.

A cassette deck supplies only motion and electronics to the tape- Every thing else (Fig. 13.12), the tape, guidance system, tensioning devices, lubrication for the tape, and head-to-tape contact arrangement is *built into* the cassette itself. The cassette deck, Fig. 13.13, is no better than the cassette put into it.

Each of the component parts, shown in Fig 13.14, has *a very specific function*, and each must perform that function *with complete reliability and great precision.* The shell, for example, must be molded to the exact dimensions called for in ISI specifications. If the tolerances are *too wide,* you won't be able to get the cassette into the deck. If they are *too narrow,* you won't be able to get proper alignment with the heads once the cassette is in the deck. The *guide rollers* are crowned and flanged for proper alignment with the heads and spin on stainless steel bearings. The *pressure pad* supplies the right amount of pressure for a good head-to-tape contact. The pressure pad assembly is felt backed with a *plated phosphor bronze spring*. You can see how much tape is left through the *window* at the top. The *slip-sheet* (liner) lubricates the roll of tape as it moves back and forth in the cassette shell. The *leader* is pinned into the hub. The tighter you pull the leader the tighter it grasps. The *tape* is held on to the *supply and take up hubs.* The *two halves* of the shell join smoothly and precisely in the centre of the *working face.* The cassette is *screw assembled.* Exploded view of a cassette is given in Fig. 13.14.

Cassettes are marked in terms of the *duration of recorded programme.* A two track cassette marked C60 will have a program of 30 minutes on *each track.*

The total thickness of the tape and its coating determines what length of tape can be wound onto a cassette. If only a *short playing time* is needed as in a C60 cassette (30 minutes each half) then a comparatively thick tape and coating can be used. The thicker tape makes for longer tape life with less risk of breakage or stretching. To obtain *longer playing times,* thinner tapes and coatings must be used, so that the risk of breakage and stretching is greater. One should not make valuable recordings on C 120 tape because their thin material is much more liable to stretch, deform, or break during rewind or play. Cassettes of different playing times-from the same manufacturer will give different recording qualities, with different frequency ranges, noise levels and distortion levels for the same reading of VU meters during recording.

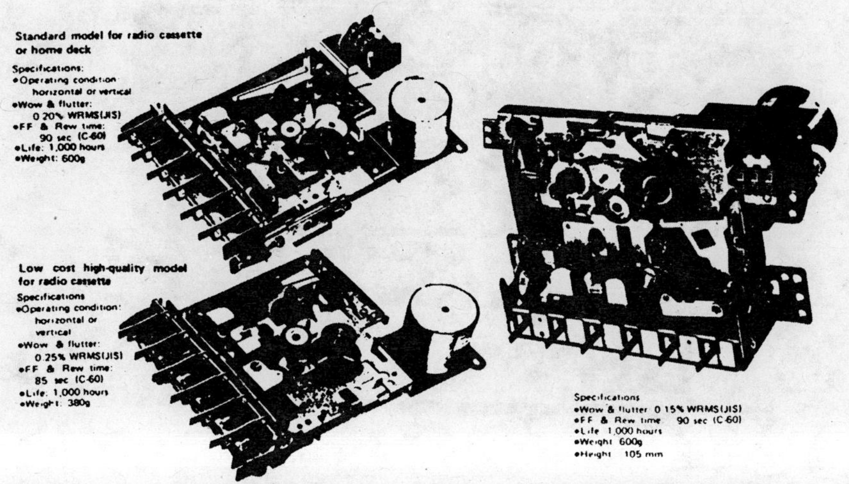

Fig 13.13. Top loading and front loading cassette decks.

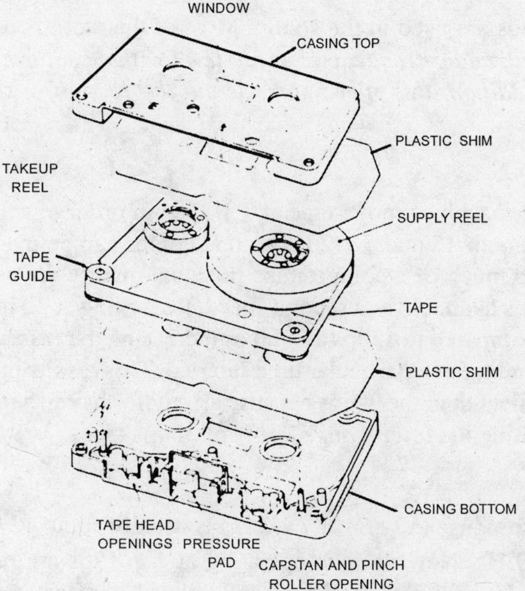

Fig. 13.14 Exploded view of typical cassette.

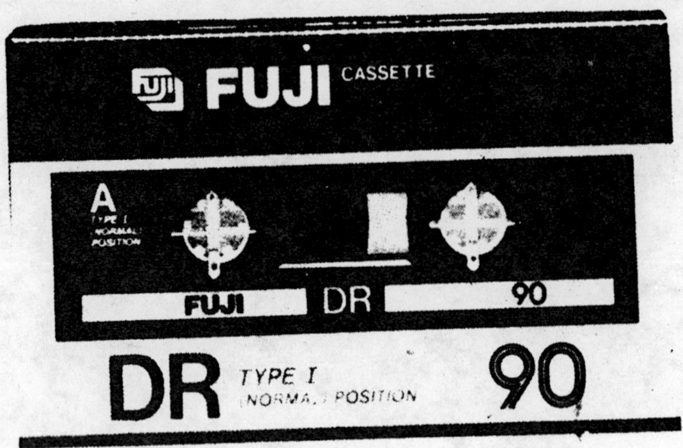

Fig 13.15 Fuji C90 Cassette

13.7 TAPE SPEEDS

Several different tape speeds are used in the studio. Most of the work is done at 15 ips (inches per second) because *this speed allows all audio-frequencies at full level without saturating the tape and produces a good signal-to-noise ratio. In addition, this speed spreads out the recording signals far enough apart for easy editing.*

Consumer hi-fi tape machines move tape at 7 1/2 ips in order to save tape. As a result most 1/4-inch studio recorders operate at both 15 and 7 1/2 ips so that copies can be made for clients at the proper speed. The 7 1/2 ips produces too much of a compromise in signal quality for use in recording multitrack master tapes, so 30 ips is sometimes available as a second speed. Recording at 30 ips produces a signal-to-noise ratio increase of about 3 dB as compared to 15 ips as well as increasing the *resolution* of the recording by doubling the length of tape corresponding to a particular time interval. This results in better *transient response.* The 15 ips recording is more prevalent than the 30 ips because all studios have 15 ips capability, while not all of them have the 30 ips speed, limiting the *interchangeability* of 30 ips tapes.

The upper limit of response (±2 *dB*) at 7 1/2 ips is usually about 12,000 to 18,000 Hz and at 3 3/4 ips between 7,500 and 12,000 Hz. New designs operating at 1 7/8 ips are providing comparatively excellent frequency response. Don't let a few hundred cycles of upper-limit frequency response be a final criterion in choice of a machine, because manufacturers differ in the rigidity and manner of making their tests. *Listening is the best test.*

Table 13.1 Recording Time-Hours

Tape Length (Feet)	1 7/8 ips (Reel & Cassette)	3 3/4 ips (Reel & Cartridge)	7 1/2 ips (Reel)
Recording one direction (Single-track mono, dual-track stereo, or 4-track quad)			
150	0.25 (Cassette)	0.125	-
300	0.5 (Cassette)	0.25	-
600	1	0.5	0.25
900	1.5	0.75	0.375
1,200	2	1	0.5
1,800	3	1.5	0.75
2,400	4	2	1
3,600	6	3	1.5
Recording both directions (Dual-track mono or 4-track stereo*)			
150	0.5 (Cassette)	0.25	-
300	1 (Cassette)	0.5	-
600	2	1.	0.5
900	3	1.5	0.75
1,200	4	2	1
1,800	6	3	1.5
2,400	8	4	2
3,600	12	6	3

* Double these times for 4-track mono

The speeds which have been standardised for tape are :

15 ips	38 cm/s
7 1/2 ips	19 cm/s
3 3/4 ips	9.5 cm/s
1 7/8 ips	4.75 cm/s

Further reduction in speed is not possible due to problems of *poor resolution* associated with *weak output.*

Lower speeds, 9.5 and 4.75 cm/s, are used for long play in domestic cassette recorders. *Higher speeds* are used in good quality studio tape recorders.

13.8 MAXIMUM USABLE FREQUENCY

One cycle of variation covers a distance of one wavelength (λ) on the tape. If T is the time period in seconds then

Distance covered = $(\lambda)/T$ (speed)

If S denotes the speed of the tape, we have

$$S = \lambda/T \qquad ...(13.1)$$

$$= fT \qquad \text{(since } T = 1/f\text{)}$$

For *optimum output,* the gap-width should be half a wavelength of the maximum usable frequency (f_m).

∴ $$f_m = S/2G \qquad ...(13.2)$$

Equation 13.2 gives the maximum usable frequency (f_m) in terms of tape speed S and gap width G. Thus, *increasing the tape speed increases the maximum usable frequency corresponding to reasonably good output.*

Problem 13.1 : If the gap width is 0.0254 mm and the tape speed is 25.4 cm/s find the maximum usable frequency.

Solution : Gap width $G = 0.0254 \text{ mm} = 254 \times 10^{-7}\text{m}$

Tape speed, $S = 25.4 \text{ cm/s} = 0.254 \text{ m/s}$

Maximum usuable freuency,

$$f_m = \frac{0.254}{254 \times 10^{-7} \times 2}$$

$$f_m = \frac{254 \times 10^{-3}}{2 \times 254 \times 10^{-7}}$$

$$= \mathbf{5{,}000 \ Hz}$$

Problem 13.2 : With a gap-width of 6 microns and a tape speed of 19 cm/s. Calculate the maximum usable frequency.

Solution : Gap width $G = 6 \text{ microns} = 6 \times 10^{-6}\text{m}$

Tape speed, $S = 19 \text{ cm/s} = 0.19 \text{ m/s}$

Maximum usable frequency,

$$f_m = \frac{0.19}{2 \times 6 \times 10^{-6}}$$

$$f_m = \frac{19 \times 10^{-2} \times 10^{6}}{12} = 1.5833 \times 10^{4} \text{ Hz}$$

$$= \mathbf{15833 \ Hz}$$

13.9 TRACK CONFIGURATIONS

The term *single track recording* indicates that only one track is being recorded on the tape. We might have a dual-track tape, but if only one track were utilised, it would be a single-track recording.

The width of standard reel-tape is 1/4 inch (250 mils). Usually single-track recordings do not cover the entire width of the tape. If they did, we might not be able to erase the tape completely. Thus single-track recording widths usually vary anywhere from 0.118 to 0.245 in (118 to 245 mils), whereas the erase track is always about 0.006 in (6 mils) wider. If both the record and erase tracks were exactly alike in width, any misalignment would prevent complete erasure. Also *by making the erase track wider we make sure that any leakage flux extending beyond the actual recording flux is also erased.*

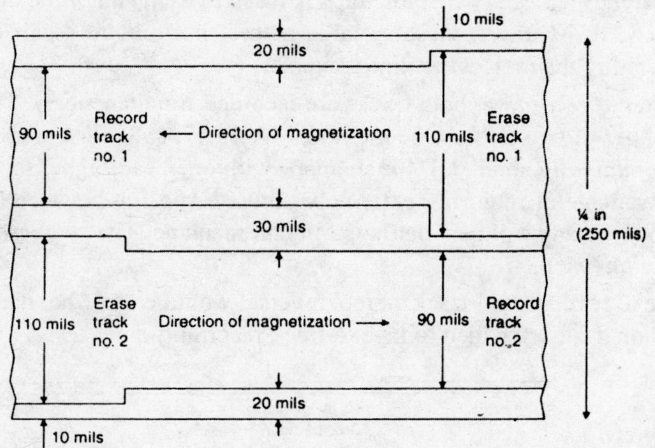

Fig. 13.16 Dual-track 1/4 in tape showing record and erase track widths and placement (monophonic recording).

Dual-track recordings can be either monophonic or stereo. Basically, they call for recording two individual tracks on the 1/4 inch tape as shown in fig. 13.16 In the *monophonic mode*, two methods of dual-track recording are possible. On older machines, one track is recorded, and when the end of the tape is reached, the supply and take up reels are *interchanged manually,* so that one becomes the other. Thus, if we recorded on the top portion of the tape on the first go-round, when we switched reels, we would again record on the top.

Another method is used on more recent machines. The machine *automatically reverses* itself at the end of the first recording. Then the tape winds itself back in the opposite direction. The *reversing mechanism* can be actuated mechanically or electrically. In *mechanical methods* when the tape is fully unwound after the first recording, the tension on the tape actuates a switch. This kicks the motor system into reverse, automatically changing the functions of the supply and take up reel. At the same time, the direction changing mechanism shifts the recording/playback head up or down to place it in contact with the track being recorded or played back. This is true only if a *single head* is used. If *dual-track heads* are stacked on top of one another, up and down shifting is eliminated.

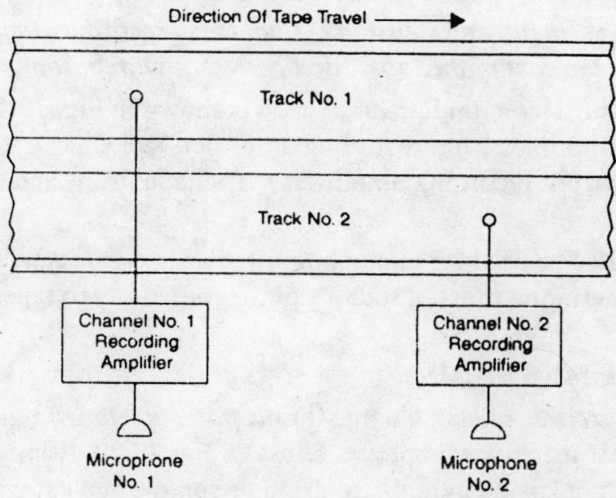

Fig. 13.17 Dual-track stereo recording.

In *electrical methods* a piece of metallic tape is fused to each end of the uncoated side of tape. When the tape reaches the end of its travel, these metallic pieces short circuit a pair of contacts thus closing an electrical circuit and setting the motor drive into reverse.

In *dual-track stereo recording,* both tracks are recorded simultaneously. The recording procedure can be accomplished in either of two ways. One method, Fig. 13.17, calls for the use of two *discrete* (separate) heads, *staggered longitudinally* about 1 1/4 inch apart, with one head higher than the other. This, the older system, has the disadvantage of requiring extremely accurate *spacing* between the respective head gaps to ensure that any stereo tape can be played on the particular machine. The *staggered method* has given way to the stacked head, or *in line method.*

It is impossible to reverse dual-track stereo, reversal would erase what had been recorded previously. Hence, a new dimension has been added to tape, 4-track recording.

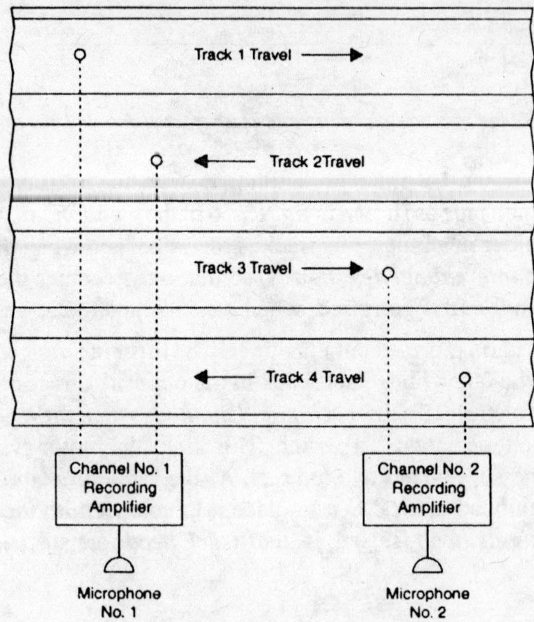

Fig. 13.18 Four-track stereo recording.

Four-track recording is simply an extension of dual track recording to permit reversal of the tape, add to record-playback time, cut the cost of the tapes, and permit quad recording.

The method of recording stereo on four-track tape is shown in Fig. 13.18. Note that tracks 1 and 3 are recorded in *one direction* (solid lines from recording amplifiers) whereas tracks 2 and 4 are recorded in the *other direction* (dotted lines from recording amplifiers). The tape itself, and the track spacing are shown in Fig. 13.19.

Four-track tapes can also be used for monophonic recording. One can record upto eight hours of music on a single 7 inch reel. Furthermore, most of today's pre-recorded stereo tapes are four-tracks types.

13.10 TAPE TRANSPORT MECHANISM

The term *tape transport* is a collective name for the *parts of a tape recorder that control the movement and handling of the tape.* All machines employ a *capstan,* Fig 13.20, to move the tape past the heads at a precise speed. A capstan is a precision cylindrical drum or spindle against which the tape is passed in order to maintain a *constant tape speed.* A pressure roller holds the tape, by spring pressure, against the capstan.

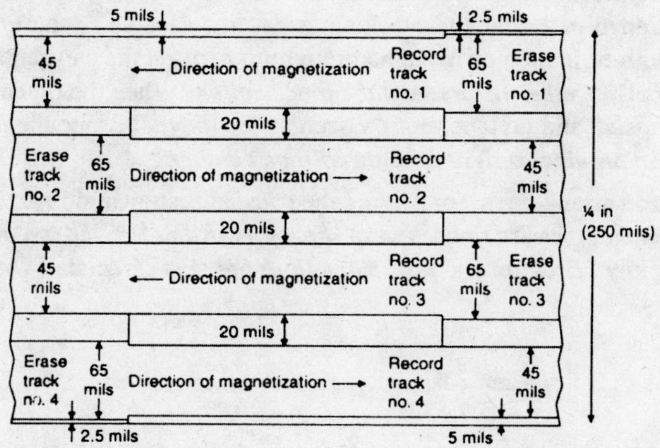

Fig. 13.19 Four-track 1/4 inch tape showing stereo record and erase track widths and placement

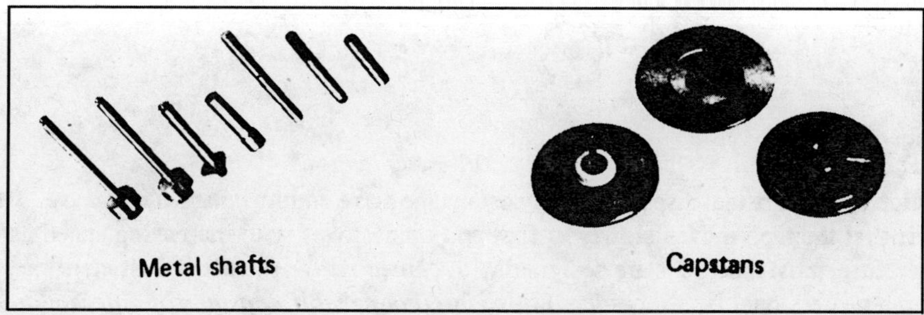

Fig. 13.20 Capstans and shafts

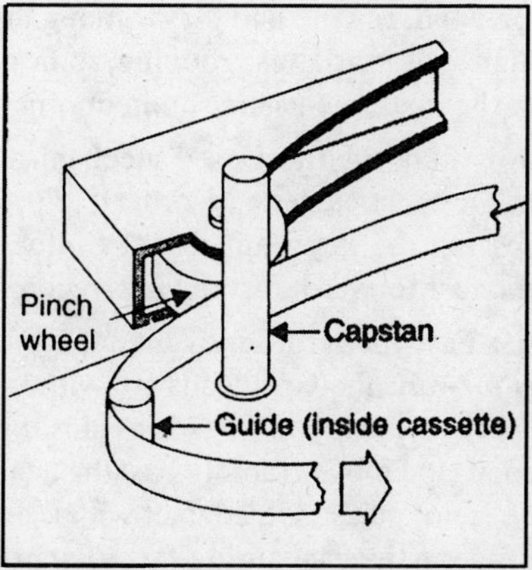

Fig. 13.21 Pinch roller mechanism operation.

The capstan in a typical *pinch roller* tape drive mechanism rotates constantly under drive motor power. A pinch roller is positioned in a movable bracket close to the capstan so that it can be *withdrawn from the capstan* a few thousandth of in inch or it can be brought to *bear on* the capstan as shown in Fig. 13.21. *The magnetic tape passes between the capstan and the pinch roller.* When the pinch roller is withdrawn the tape does not contact the capstan and no tape motion occurs. To move the tape the pinch roller is forced against the tape and *the capstan provides a tight capstan-to-tape contact.*

The most common arrangement for driving the capstan is shown in Fig. 13.22. The tape transport also provides for fast movement, forward or backward (*Fast forward* or *Rewind*) so that any desired portion of the tape can be located rapidly. *Braking* and *pause functions* are also associated with the tape transport.

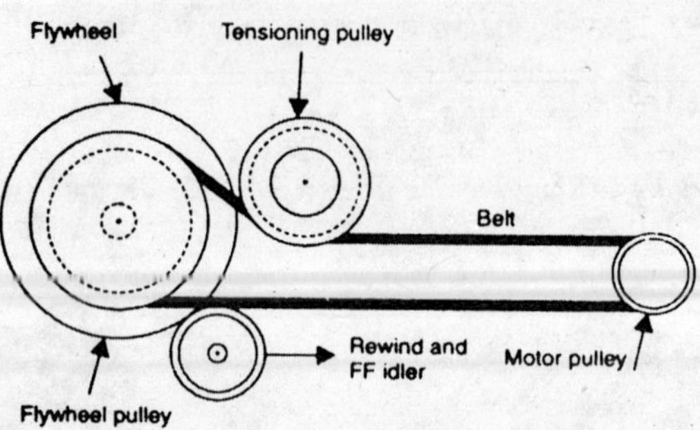

Fig. 13.22 Driving the capstan.

Low priced recorders that dispense with capstan tape drive employ *take-up reel drive.* The disadvantage of reel drive is that the tape moves slowly at first and then moves with increasing speed as the supply reel unwinds. Therefore, most machines are designed with *capstan drive.* Capstan diameters vary from 3/32 inch to 1 inch. Some professional tape recorders utilise *interchangeable capstans of different diameters* in order to change the speed of tape travel. The *pressure* exerted on the tape between capstan and roller must be sufficient to prevent slippage. However, excessive pressure accelerates wear of surfaces and bearing.

13.11 MECHANICAL CONTROLS

Two types of controls are fitted to cassette recorders, those which are mainly *mechanical* in action (though they may be electrically operated), and those which are *purely electronic* and act on the signal. Among the first group are the controls of tape wind, rewind and play; among the second group are the recording level controls and the record on/off switch. In addition to this grouping, some controls are used in everyday operation, some seldom, others (usually within the instrument) are set up during initial testing or in later servicing.

Taking the mechanical controls first, the most used mechanical controls are those which control the winding of the tape on the cassettes. On most cassette recorders, at present manufactured, these controls are operated by piano-type keys; depressing the key until it locks into position switches the control on. The mechanical keys of this group are the fast forward, fast rewind, record/play, pause, stop and cassette eject.

Fast Forward and Rewind : Fast forward winds the tape on so rapidly so that *a place later in a recording can be found.* During *fast forward* the tape heads are withdrawn from the tape and the amplifier is switched out of action so that *no replay can take place.* This avoids the twittering and clucking sound which would otherwise be heard as a recorded tape moves rapidly past the heads. With the heads withdrawn the *idler wheel* is also withdrawn and the tape is not pressed in contact with the capstan spindle. This avoids having to spin the capstan at high speed (causing rapid wear) during fast forward. The tape drive during fast forward is carried by applying drive to the pick-up reel on to which the tape is being wound within the cassette. This is done through the splined shaft over which the centre of the reel fits when a cassette is placed in the recorder. This is shown in Fig. 13.23.

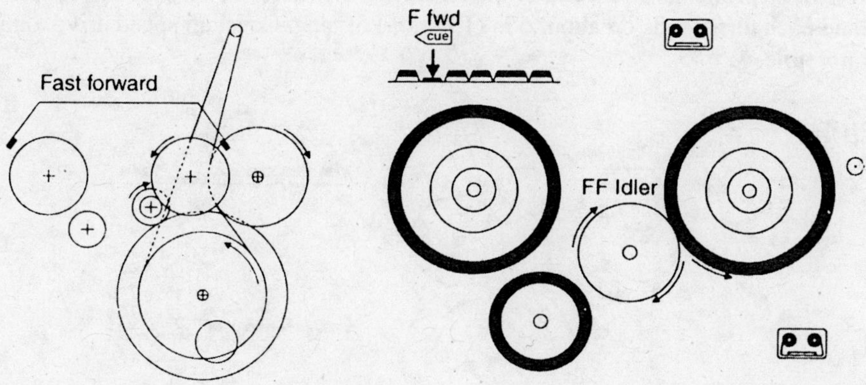

Fig. 13.23 Fast Forward winds the tape on rapidly so that a place later in recording can be found.

The arrangements for *rewind,* Fig. 13.24, are similar, though it is a common practice to make rewind *slower* than the fast forward. Rewinding of the tape is done so that *a recording that has just been made can be heard all the sooner.* In general, it is used more than the fast forward.

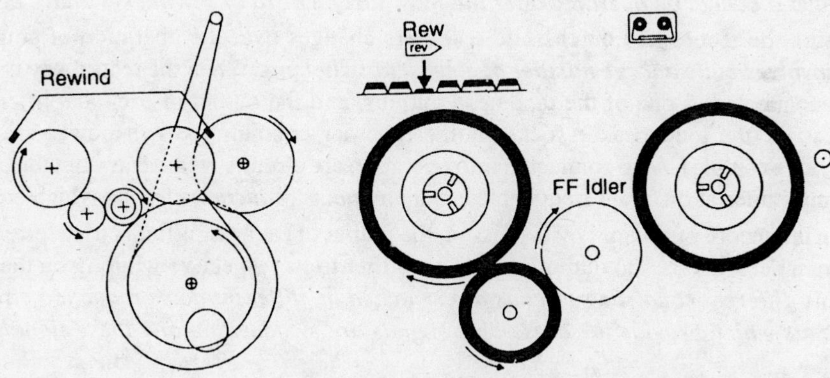

Fig. 13.24 When a recording has just been made the cassette can be rapidly Rewound so that the replay can be heard all the sooner.

Play and Record : The *play* key carries out a large number of electrical and mechanical operations. During its operation *power* is applied to the capstan spindle motor and to the amplifier, the idler wheel moves so as to *pinch* the tape against the capstan spindle and *the heads move forward to contact the tape,* pressing it against the pressure pad housed inside the cassette. At the same time some *drive* must be supplied to the forward wind spindle engaged in the cassette pickup reel, so that the tape which is moved on by the capstan is taken up and evenly wound onto the pickup reel. Where *only one motor* is used, this usually means a *friction drive* is applied by a pulley held between the motor spindle and the drive spindle by a light spring. *Some slip* *must be allowed for, as the takeup spindle can't be turned at a constant rate.* When the takeup reel is empty, its radius is that of the wheel, about 1/2 inch (12 mm), so that one turn of the reel winds on the tape by about 3 in (75 mm). If the takeup speed is arranged so as just to keep the tape winding on at the beginning of the reel then there will be problems at the end of the reel when the radius of the take up reel plus that of the tape is about 1 in and each turn winds on about 6 in (157 mm) of tape. Constant speed drive to the take up reel is therefore not possible.

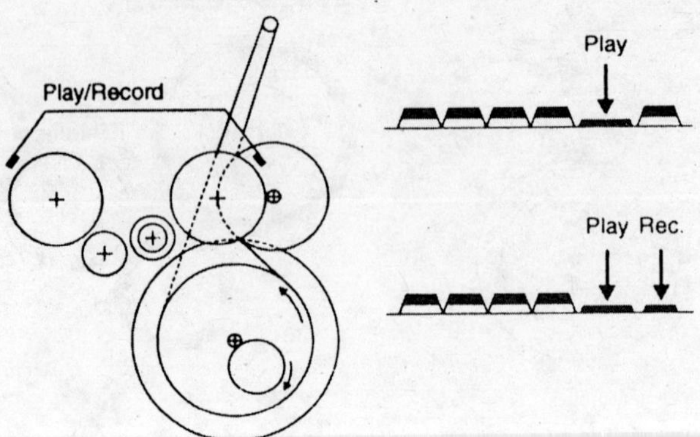

Fig. 13.25 The play key carries out a large number of electrical and mechanical operations.

When a *separate motor* is used for rewind, these awkward mechanical operations do not arise, and a light tension can be applied to the tape by connecting the motor to a voltage rather lower than is normally used for driving the take up reel.

The *record* key or button is an electronic control which is *mechanically interlocked* with the other controls so that *it cannot be operated after the play, fast forward or rewind keys have been operated.*

Pressing the record key on cassette recorders changes over the operation of amplifiers from play to record and *involves quite a large number of electrical switching.* When the record key is *up* the input of each amplifier is connected to one of the tape head outputs, and the output of each amplifier is connected to the output socket, or the loudspeaker socket if the recorder contains its own power amplifier stages (as in amplidecks, for example). Also connected into operation are circuits which carry out the process of *equalising* the signal amplitudes at different frequencies, and any *noise reduction circuits* which are switched on.

When the record button or key is pressed, the output of each amplifier is transferred to the microphone, radio and gram deck inputs and output of each amplifier to its respective winding on the recording head. At the same time *different equalising circuits are switched in, different noise reduction circuit paths (if fitted), and the circuits which provide the erase-bias signals are switched on and these signals are coupled to the tape heads.*

Pause, Stop and Eject : The *pause control,* used on most cassette recorders, *enables a short pause to be made during recording without using the stop control,* which would disengage the heads and disengage the drive. Where no pause control is fitted, short stops are inconveneninent as they require the manipulation of the play and record keys together to start recording again. Using a pause control eliminates this two finger action, with the advantage that the volume level of recording signal occurring during the pause can be checked. The action of the pause control may be purely mechanical or purely electrical. In the *purely mechanical type,* the idler wheel is pulled away very slightly from capstan spindle, and the heads are left in contact with the tape, with the capstan spindle still revolving. This ensures a *rapid restart* when the pause is operated again to press the idler wheel against the capstan again, though the acceleration of the tape as the drive is taken up may cause a squeal to be recorded. Some recorders which use capstan motors controlled by low voltage electronic circuits have *purely electrical type* pause controls which *interrupt the supply* to the capstan motor and switch off the tape amplifier momentarily as the motor is restarted. *Care has to be taken with, the electrical type, so that it is not held for long periods.* The risk is that the idler wheel pressed in one stationary position against the capstan spindle will develop a flat portion which will from then on cause wow on both recording and playback. There is no risk of this happening with the mechanical method.

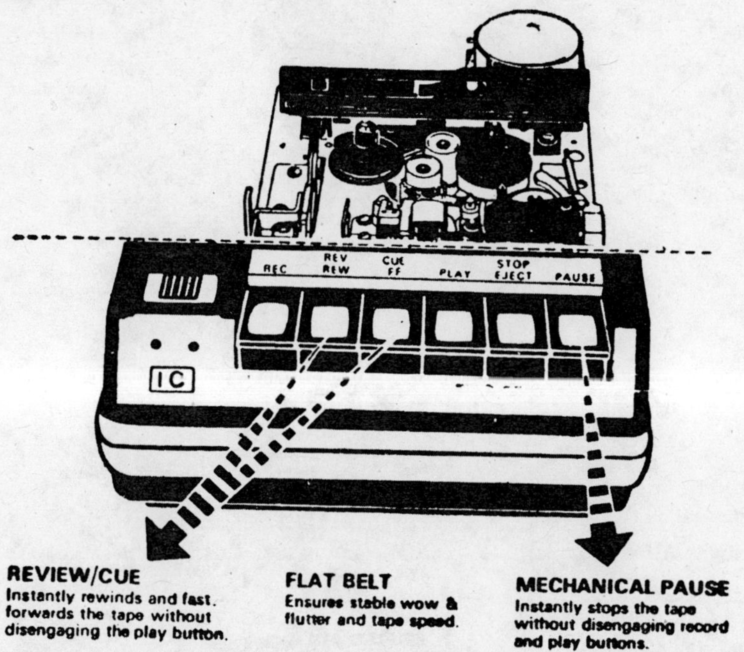

Fig. 13.26 The pause control enables a short pause to be made during recording without using the stop control, which would disengage the heads and disconnect the drive.

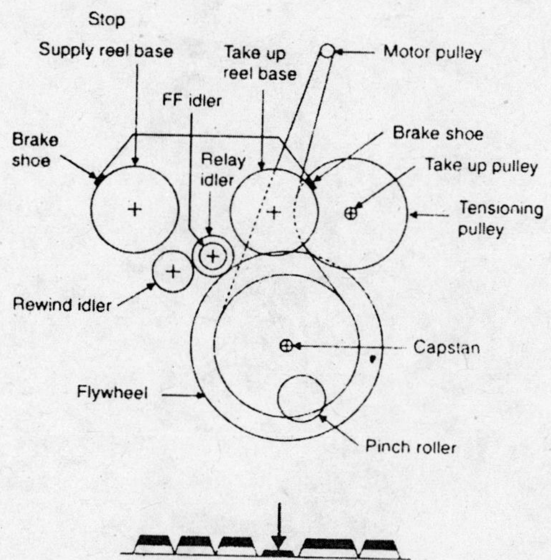

Fig. 13.27 The stop control operates on fast forward, rewind, normal play or record and its action is to return all functions to the stop position, reversing the sequence of operations which took place when any of the keys had been pressed.

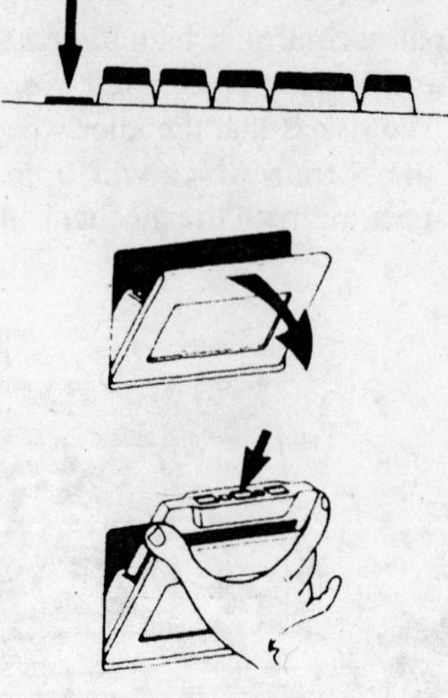

Fig. 13.28. Inserting the cassette.

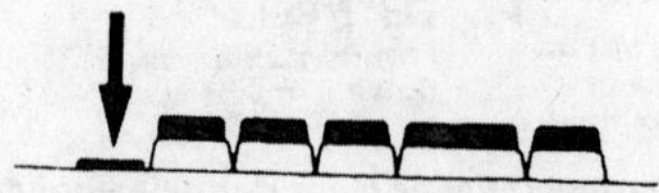

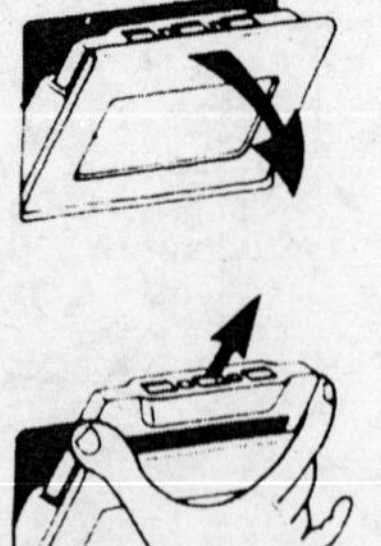

Fig. 13.29 Removing the cassette.

The *stop key* operates on fast forward, normal play, record or rewind, and *its action is to return all functions to the stop position, reversing the sequence of operations which takes place when anyone of the previous keys has been pressed.* The piano key controls are made *interlocking and spring loaded* so that when any key is pressed down, it has to be moved *against* the force of a spring and once in the operating mode, is *locked* in position by a ratchet action. Further more once a control, such as fast forward for example, is operated other controls which carry out different operations *cannot be depressed,* so making it impossible to operate the keys normally. This stop key does not lock but *releases* all the ratchets holding any other keys so that when the stop key is operated any other key which has been locked down will spring back. In addition, *using the stop key always cuts off the amplifier.*

The *eject key* or catch releases the spring lid which covers or holds the cassette in place. In some models *the lid opens and the cassette is sprung away from the driving shafts.* In others the cassette is held in grooves in the lid, and *pressing the eject key will lift the lid and the cassette with it* making cassette removal even simpler.

The eject key may be *separate* or it may be *combined* with the stop key, so that a *gentle push* on the stop key stops operation and a *further movement* against a stronger spring causes ejection.

13.12 MISCELLANEOUS FEATURES

Brakes are usually applied to quickly stop the reels. When the recorder is stopped the reels should stop immediately, otherwise the tape may *spill over.* In cassette recorders small plastic or metallic bars attached to an arm are used as brakes. When the stop key is operated, the brakes press against both the reels and stop their movement. The brakes are released when any of the keys, other than the stop key, is pressed. One such braking arrangement is shown in Fig. 13.30.

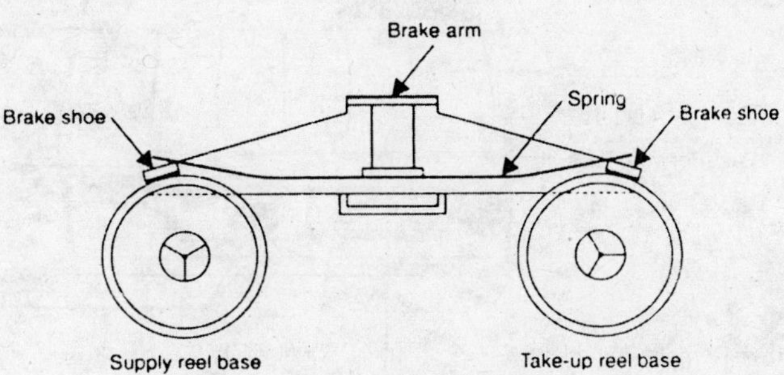

Fig. 13.30 Braking arrangement.

All stereo cassette recorders incorporate some sort of *auto stop* or warning signal to operate at the end of the tape, or when jamming occurs. They come into action when the take-up spool ceases to turn with the motor still activated. Autostops perform for whole of the usual stop operation, withdrawing the heads and idler and switching the amplifier off.

A few recorders incorporate *auto reverse* switch. When the end of a cassette tape is reached, the signal generated from the detector at the take-up spool causes the tape to reverse and *a second set of heads,* acting on the other half of the tape width comes into action, so playing the *other half of the tape.*

Cheap portable cassette units often use no precise method of *locating the position of recording* on the cassette, and rely on the crude *scaler* of tape-reel radius which is engraved on the cassette itself. Such a method is quite unsuitable for serious work and a *digital counter* of some sort is fitted to all stereo cassette units in the medium to high quality class.

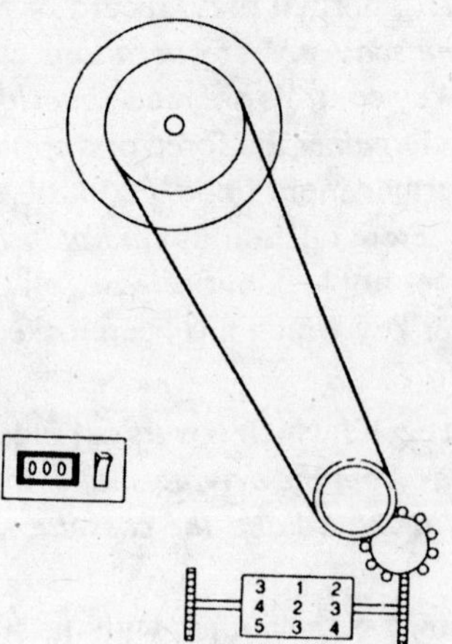

Fig. 13.31 Locating the position of recordings on the cassette.

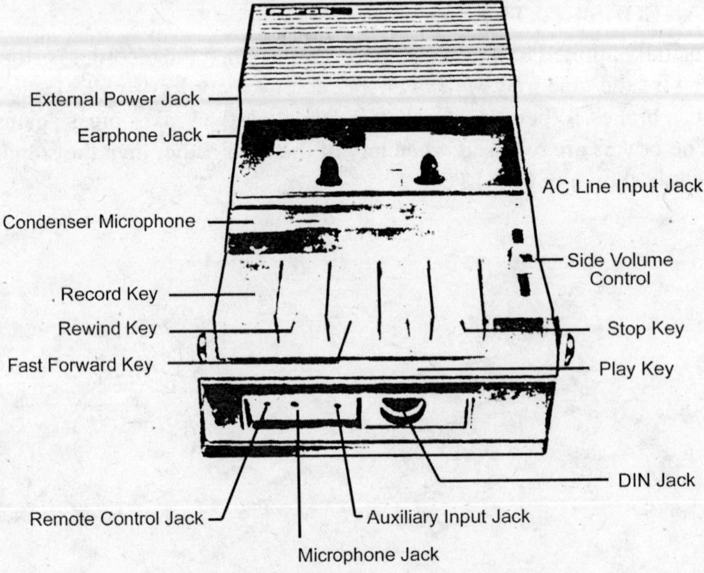

Fig. 13.32 Controls and facilities on paros - 101C portable cassette recorder.

The usual method of operating the counter is by a belt from the take-up shaft, as shown in Fig. 13.31, so that the *count rate* is rapid at the start of recording a cassette and slow at the end. Because of *belt slip* the count is not exact, but the position of a recording can usually be found with very little error, often only a second or two out of place. It is useful to know what the *maximum count* is for a cassette side, a typical value being 680 for a C90 cassette side.

The *controls and facilities* on 'paros 101C' portable cassette recorder are shown in Fig. 13.32.

13.13 ELECTRONIC CONTROLS

Of the purely electronic controls, the *recording volume controls* are of greatest importance in use. They may take the form of a *single* volume control together with a balance control, or, more commonly, of *separate* volume control for two stereo-channels. Some recorders use *rotary controls,* others use the currently fashionable *slider controls,* but the aim in each case is to control the amplification of the input signal being recorded, so that the best *recording level* is achieved. Because of the inevitable differences between the two amplifiers of the stereo pair and differences at the signal source the volume control settings for the two channels may not be identical, but should be close to each other in setting.

One reason for the popularity of slider controls is that they can be set close together so that both L (left) and R (right) volume controls can be *altered together,* a procedure which is very difficult when separate L and R rotary volume controls are used. When rotary controls are used, a single volume control operating on both channels, together with a *balance control* to alter the amplification of one channel relative to the other, is preferable.

Many recorders provide separate pre-amplifiers for the microphone inputs, and in such cases usually provide separate recording volume controls. Where such separate controls are provided, *signal mixing* of microphones and other inputs is possible.

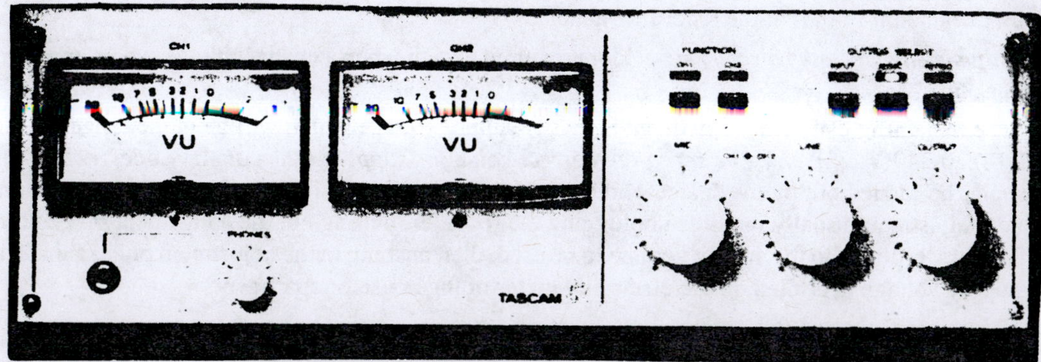

Fig. 13.33 V.U. meter scalers. These meters are practically always scaled in decibels, so that the spacing between numbers varies along the scale.

The amount of signal reaching the tape heads is measured by the *recording level meters,* often called VU (volume unit) meters. The precise quantity which these meters measure depends very much on the design of the recorder, but the principle is always the same : the recording level is adjusted by using the recording volume controls until the maximum level reached by the needles of the meters is to a *red line* or a '0dB' mark on the meter scale. In practice, the best way of interpreting the meter readings can only be found by experience.

Stereo cassette recorders intended for serious use will inevitably incorporate *active noise reduction circuits.* As far as the user is concerned, the noise reduction switch will be on for all recording from disc or microphone, and all radio recording *except where noise reduction has already been carried out on the radio signal.* The noise reduction switch will similarly be on for the replay of any recordings made with the same machine, but may be off for other replayed cassettes. When the cassette being replayed has not been recorded using the *Dolby process* (the only noise reduction process used for pre-recorded cassettes) the noise reduction switch should be off during replay, otherwise there will be a serious loss of high frequencies. Cassettes which have been recorded using the Dolby process should be played back with the Dolby noise reduction switched on.

Some recorders have been manufactured using noise reduction circuits on *replay only.* Where these are of the DNL (*Dynamic Noise Limiter*) type they may with some advantage be *left switched on at all times* (they are automatically taken out of circuit on recording). A few low and medium quality cassette recorders incorporate switches labelled *noise reduction* which merely switch in a *treble filter,* cutting high frequency response. These are not worth using except for recordings of speech, and *should normally be switched off.*

Input and output adjustment controls are, however, very useful if a cassette deck (not having loudspeaker outputs of its own) is to be matched to any possible amplifier unit. At the input it is usual to have several different inputs of different sensitivity levels. Nevertheless, the tape output from some amplifiers may deliver a signal which is too large for the cassette recorder input, since the volume controls of the cassette recorder are located after the first stage of amplification, and an overload of this first stage cannot be prevented by an adjustment of the recording volume controls. An *input adjustment control* enables large signals to be reduced in amplitude before reaching the first stage of the recording amplifier.

An *output adjustment control* can also be useful where the cassette recorder is being used with an amplifier of different design. By adjusting the output control, the replay of the cassette recorder can produce the same volume at the loudspeakers with the same amplifier settings as are used for disc or radio, so avoiding having to make large changes to the setting when switching from one signal source to another.

Most cassette recorders will have a *bias amplitude control* so that the value of bias can be adjusted for a particular tape. *This is not a user control, and is part of the circuitry within the case.* No attempt must be made to adjust this control except by an experienced electronic engineer suitably equipped with instruments to check the amount of bias at the recording head.

Adjustments of Dolby circuits are seldom required except when a change of tape-type is made for some reason or other.

The remaining electrical adjustment is that of supply voltage which will be in the range 110-115-120 V or 220-230-240 V. *This must be set to the correct value for supply in the country concerned.* The setting will usually be carried out by the dealer who sells the recorders and will be correct for the country in which it is sold. Adjustment usually requires unplugging the recorder, pulling out the adjustment plug and rotating it until the *arrow* points to the supply voltage to be used, then pushing in the adjustment plug again. The value of the supply voltage is printed on the electricity meter of the house or apartment.

13.14 ERASURE GUARD

Most recorders now incorporate an *erase prevention device* which automatically ensures that pre-recorded cassettes cannot be accidentally erased by recording over them. The system uses a *mechanical interlock* with the record key or button, and is operated by a spring-loaded plunger at the back (opposite the heads) of the cassette holder.

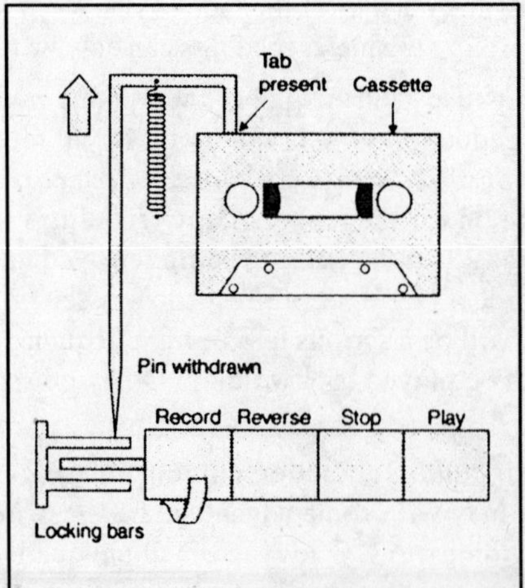

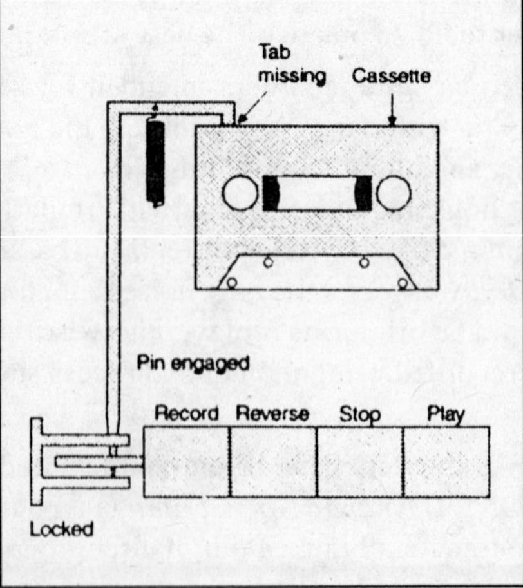

Fig. 13.34 Erase prevention mechanism.

With no cassette in place, the plunger is extended, and the lever connected to it internally locks the record control. When an *un-recorded cassette* is placed in the holder the back of the cassette presses the plunger in, unlocking the record control. *Pre-recorded cassettes* have a slot cut in this position at each side (the rear left-hand side, looking at the cassette facing its pressure pad, with either half playing) so that the plunger is not pressed in, and *the record control remains locked in the off position.*

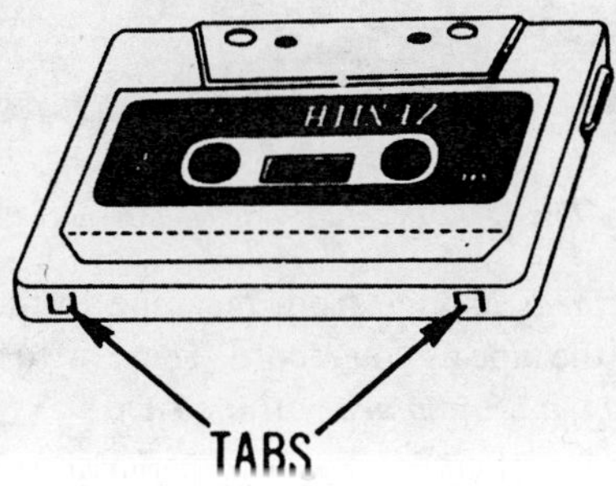

Fig. 13.35 Erasure guard.

Unrecorded cassettes have a *breakable tab* at this position, Fig. 13.35, so that when a valuable recording is made it can be protected against accidental erasure by breaking off the tab on the back left-hand corner. If a change of plan requires that this recording is to be erased after all, *a piece of sticky tape can be placed over the slot, so that the plunger is pressed in.*

13.15 TAPE RECORDER TYPES

A vast variety of tape recording devices can be found in today's market. There are two basic machines; the complete recorder and the deck.

The *complete recorders* are totally self-contained units. They contain the recording, playback and erase mechanisms, the necessary playback amplifiers and loudspeakers. All that is required is a reel, cartridge, or cassette of blank or pre-recorded tape. But you won't find many *amplidecks* (as they are commonly known as). High fidelity speakers won't fit into such small cases.

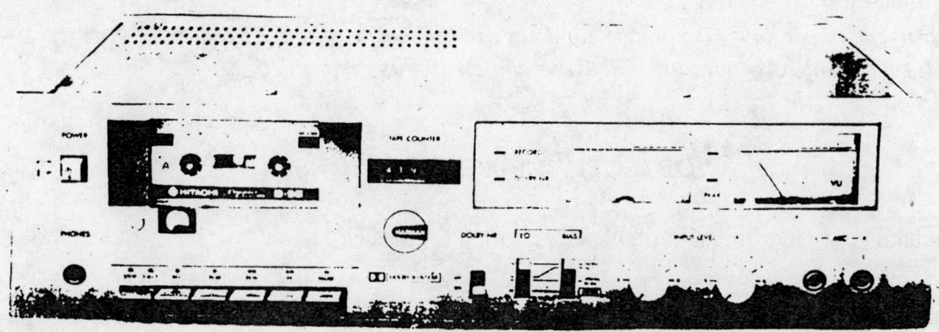

Fig. 13.36 Front loading stereo cassette tape deck - HARMAN.

The second type, the deck, can be subdivided into two classes. The first is the better known one, the *record playback type* shown in Fig. 13.37. Except for the speakers, these units come equipped with everything yo need to record and playback. The other subdivision is the *playback only deck* or tape player. This type is used only with pre-recorded tapes because it contains no recording mechanism.

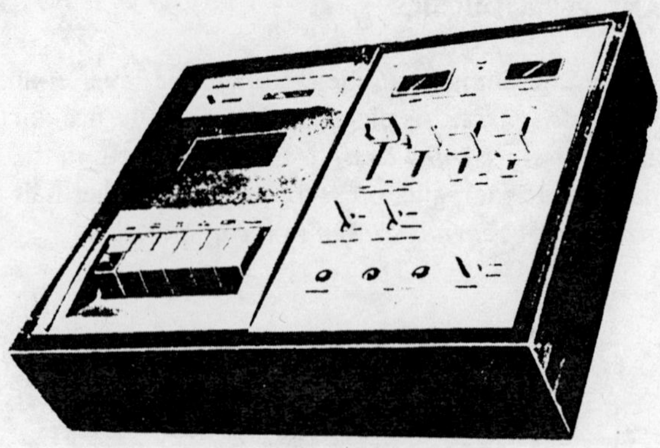

Fig. 13.37 Record/Playback-deck-paros.

The number of *operating features* change from machine to machine. Some offer *sound-on-sound* operation. Almost all let you monitor the tape as you record. There are other features too numerous to mention. Just remember, *the more options you want the more will it cost you.*

13.16 TAPE VS DISC

A spool of tape provides an unvarying standard of performance throughout its length. A disc's performance, in respect of frequency response and distortion falls off to some extent towards the inner grooves. A low overall distortion can be achieved during tape reproduction, and a superior crosstalk performance is possible in both the recording and reproduction of tape.

Normally tape has a long life and is not easily damaged. There need be no difference between disc and tape in respect to noise levels; this applies when both are new, but tape may well become noisier than disc after prolonged use.

The *handling* of tape, a disadvantage for many users, can be avoided by enclosing the spools in cassettes. These are especially useful for tapes which are to remain undisturbed, but the method is of little interest to hobbyists who require easy access to the tape.

Tape is more expensive than disc if high quality music recording is the aim. For other purposes the cost depends on such factors as tape speed, number of tracks and skill at editing tape. The expense is more apparent if tape reproduction alone, using commercial tape records, is the requirement. Stereo is likely to be more costly than mono.

As far as the music-lover is concerned, discs are the most attractive because of the extensive recorded repertoire. It seems unlikely that tape will ever compete in this respect.

OBJECTIVE TYPE QUESTIONS (A)

1. All methods of recording and playback rely on a sequence of

(*a*) reversible processes (*b*) irreversible processes

2. Tapes with thinner coatings are capable of the

(*a*) lowest information packing density (*b*) highest information packing density

3. Metallic thin films

(*a*) can be controllably produced in much thinner layers

(*b*) cannot be controllably produced in much thinner layers

4. The size of the gap, particularly for replay
   (*a*) must not be small (*b*) must be small.
5. The use of modern micro-gap heads has now made possible the use of tape speeds as low as
   (*a*) 3 3/4 ips (*b*) 1 7/8 ips
   (*c*) 7 1/2 ips
6. It is usual for cassette recorders to have
   (*a*) separate heads for recording and replay
   (*b*) a single head for recording and replay
7. Erasure is carried out with a head with a much
   (*a*) smaller gap (*b*) wider gap
8. A cassette deck supplies
   (*a*) only motion to the tape
   (*b*) only electronics to the tape
   (*c*) only motion and electronics to the tape
9. High speeds are used in
   (*a*) domestic cassette recorders
   (*b*) good quality studio tape recorders
10. Increasing the tape speed
   (*a*) increases the maximum usable frequency
   (*b*) decreases the maximum usable frequency
11. The width of standard reel tape is
   (*a*) 1 in (*b*) 1/2 in
   (*c*) 1/4 in
12. The reversing mechanism can be actuated
   (*a*) mechanically (*b*) electrically
   (*c*) (*a*) or (*b*) (*d*) (*a*) and (*b*)
13. Tape transport controls the
   (*a*) movement of the tape
   (*b*) handling of the tape
   (*c*) movement and handling of the tape
   (*d*) none of the above
14. The maximum count for a C90 cassette side is
   (*a*) 480 (*b*) 570
   (*c*) 680 (*d*) 760

FILL IN THE BLANKS (B)

1. All methods of recording and reproduction rely on a sequence of ________.
2. The amplitude of voltage across the head coil depends on the ________ in the core.
3. The magnetic oxides which are most often used are those of ___________.
4. All these oxides are produced in the form of fine ___________.
5. A wide variety of magnetic characteristics can be produced by varying the _________ of the _________.

6. The tape passes over the __________ before passing over the record/replay head.
7. The tighter you pull the leader the tighter it __________.
8. To obtain longer playing times ________ tape must be used.
9. Lower speeds are used for long play in ________ cassette recorders.
10. Higher speeds are used in good quality ________ tape recorders.
11. In dual-track stereo recording both tracks are __________.
12. Four track recording is simply an extension of __________ recording.
13. The record key is an electronic control which is mechanically ________ with the other controls.
14. The pause controls enable a __________ to be made during recording without using the stop control.
15. A ________ counter of some sort is fitted to all stereo cassette units in the middle to high quality class.
16. The amount of signal reaching the tape heads is measured by ________ meters.

CONVENTIONAL QUESTIONS (C)

1. Explain the principle of magnetic recording and reproduction.
2. What are the different methods of erasing the tape ?
3. Explain the method of manufacture of anyone of the recording mediums.
4. What is the difference between a record head, a playback head, and an erase head ?
5. List the parts of a cassette with the help of a neat sketch.
6. Derive the relation between maximum usable frequency, tape speed and gap width for optimum output.
7. What are the different track configurations for mono and stereo.
8. Explain the mechanical controls of a cassette recorder.
9. Explain the electrical controls of a cassette recorder.
10. How does the erase prevention device operate ? Why is it required ?
11. What are the different types of tape recorders.
12. What are the advantages and disadvantages of tape over disc ?

14

SOURCES OF DISTORTION IN TAPE EQUIPMENT

A stereo cassette recorder does not justify its place in a hi-fi system on grounds of convenience. As with any other part of the hi-fi system, it must conform to hi-fi standards of frequency range, distortion levels, and signal-to-noise (S/N) ratio. It is because of the deficiencies of cassette recorders in these vital standards that acceptance of cassette recording in hi-fi has been slow to come; the rate of improvement in quality has now made these standards attainable. High quality cassette recorders are now generally accepted as a part of a complete hi-fi system, used to record material and compile programs not available on disc, and to replay such material with a quality of reproduction which compares well with other units of the system. Used in this way, stereo cassette recording brings new and fascinating possibilities to the enjoyment of music which after all, is the whole aim of hi-fi.

14.1 RECORDING AND PLAYBACK HEADS

Generally speaking, tape decks (with the exception of some of those used with the so called professional equipment) *combine the duties of recording and reproduction in a common head.* The requirements of the recording and the playback heads are fundamentally the same and, *as they are not required simultaneously economy is served by combining them.*

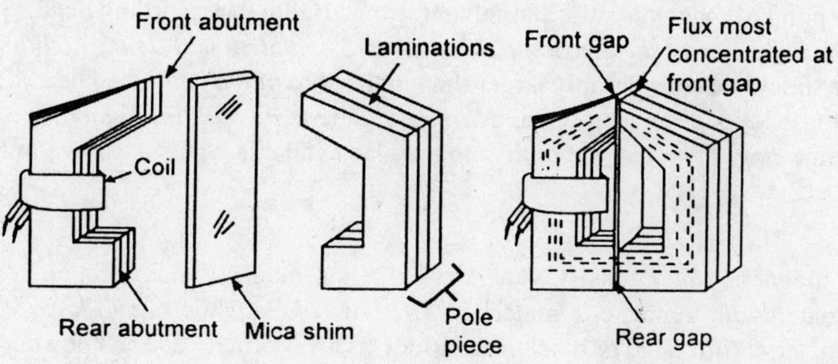

Fig. 14.1 Construction of a single-channel magnetic head.

Modern heads are usually constructed of *two semicircular stacks* of high-permeability laminations of about half to one inch in diameter. The stacks are assembled symmetrically with a *head gap* (against which the tape passes) of about five ten thousandths of an inch or less, and an *auxiliary gap,* diametrically opposite the head gap of about *ten times* that width. Both gaps are filled with nonmagnetic metallic shims to prevent any accumulation in the gaps of magnetic material which may be rubbed off the tape by friction, and to maintain a *linear relationship* between the flux at the gaps and the current through the exciting coils of the head.

The *impedance* of these coils, which are placed symmetrically about each core, should, *for recording, be low compared with the source impedance.* This ensures, as far as possible a current which is independent of the frequency of the signal. For the *playback* process, however, *maximum output voltages are to be gained from high impedance windings. Any conflict between these requirements can be resolved by feeding the output from low impedance coils into a step-up transformer.*

Stereophonic heads consist fundamentally of two monaural heads mounted one on top of the other. *Each head forms part of a complete playback system, the input signal for which is derived from one of the tracks of specially recorded stereophonic tapes.* Facilities for domestic stereophonic recording are seldom provided because of the difficulties to be encountered and the advanced technique required for such an operation.

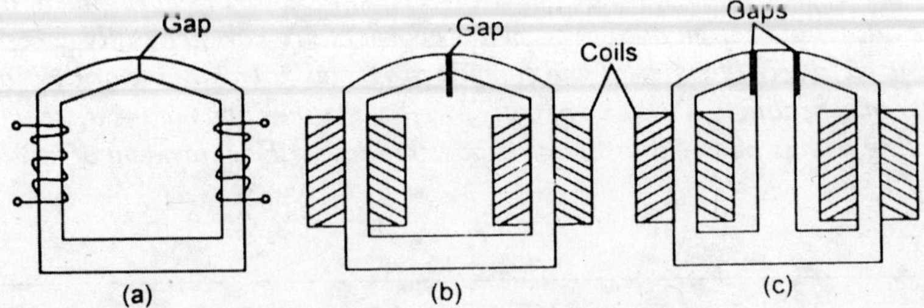

Fig. 14.2 Comparison of record and erase-heads. (a) Record - head cross section ; (b) single-gap erase-head cross section ; (c) double-gap erase-head cross section.

14.2 ERASE HEADS

Although similarities exist between the erase head and the record head they differ in construction in two principal respects. Firstly, *appreciable power is required (about 2 to 4 W) for efficient erasure,* while that necessary for the recording head is about 1 mW. Hence the core of the erase head must be of a material having a higher point of magnetic saturation than the material in the recording head. Secondly, *the longer the tape is under the influence of the erasing field, the more complete will be the cleaning.* Thus the gap in the erase-head core should be considerably larger than in the core of the combined head. The erase gap is usually about fifteen thousandths of an inch wide. *Normally there is no auxiliary gap in the core of the erase head.* Erase heads with a *single gap* and a *double gap* are also available. This is shown in Fig. 14.2.

14.3 TAPES

Modern magnetic tapes consist, generally, of a non-magnetic base (paper, cellulose acetate or p.v.c. for example) coated with a magnetic material. *The base obviously needs to have strength and suppleness.* The magnetic material (often red or black iron oxides) is a very finely divided powder mixed in some binding substance (lacquer, for instance), and *the coating applied to the base material has to be very smooth.*

Unevenness in the surface of the tape will tend to cause broken contact with the heads and consequently an *undulating level of recording.* The magnetic material needs a *high coercivity* to prevent, in particular, very large demagnetization losses at treble frequencies, and a *high remanence* to give a good level of recording.

Tapes are usually supplied in 5 or 7 inch reels, giving *playing times* depending on the transport speed and the thickness of the tape, of between 10 and 90 minutes. The *standard width* of the tape is a quarter of an inch which normally allows for two (and sometimes four) adjacent recording paths. With stereophonic tapes *adjacent paths* are used for the twin signals comprising the program. The total playing time of such tapes is therefore reduced by half.

14.4 TAPE TRANSPORT SYSTEMS

The basic requirement of a tape transport system is that *a steady speed should be imparted to the tape.* Momentary changes in this speed will cause wow or flutter. *Wow results from slow variations; flutter results from fast variations; the dividing line is arbitrary.*

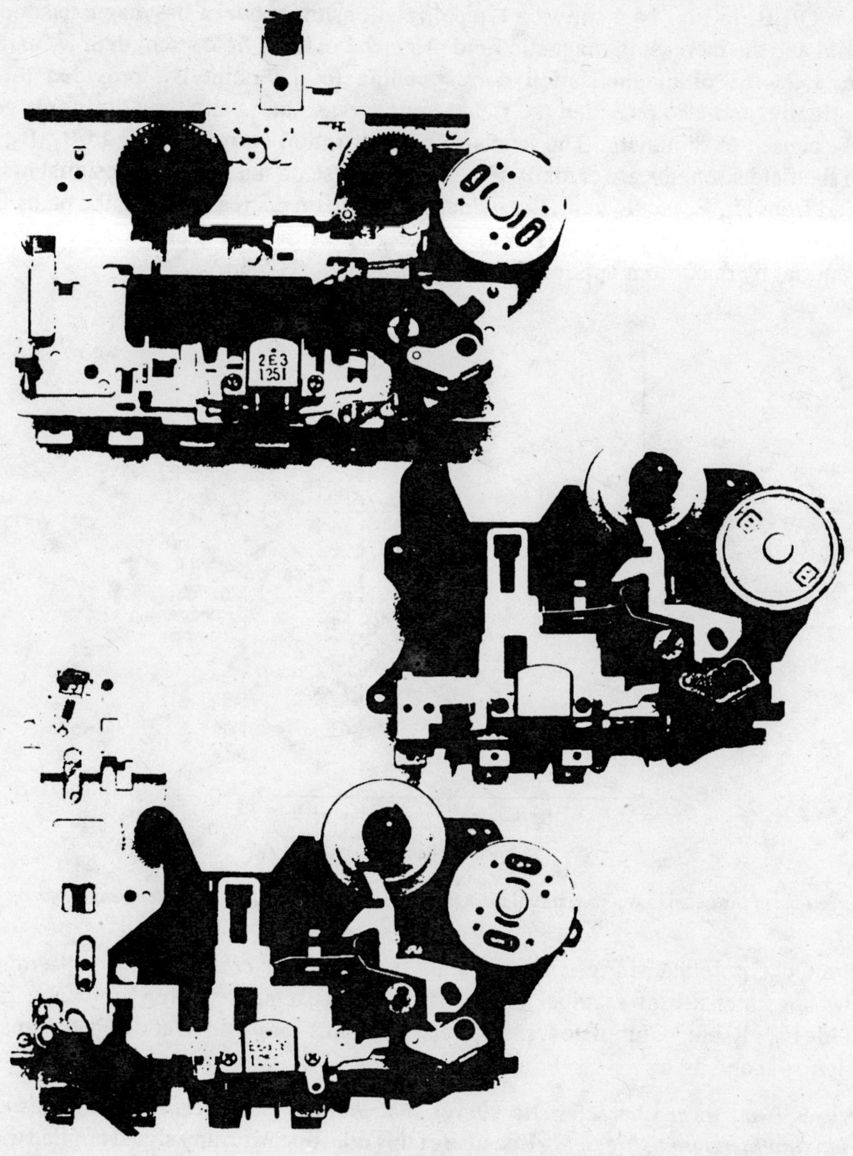

Fig. 14.3 Mini cassette mechanisms.

The actual speed of transport has a considerable bearing on the performance of the recording apparatus. *The higher the speed, the better the performance at the higher frequencies.* This is closely allied to the size of the gap in the recording head. But, of course, economically, higher speeds are a disadvantage ; the playing time of the tape is obviously reduced. Most transport systems are standardardised for playing speeds of either 1 7/8, 3 3/4, 7 1/2 or 15 inches of tape per second, with possibly, a choice of speeds.

14.5 AMPLITUDE DISTORTION

The *non linear relationship* between the magnetism residing on the tape and the magnetic field inducing it, will cause considerable amplitude distortion in the recording.

The curve OB_3B_5 in Fig. 14.4 shows a typical relationship between the magnetisation produced in a magnetic material and the increasing magnetic field produced in it. A field strength of H_2 units, for example, would produce a degree of magnetisation corresponding to the point B_2, provided the material was unmagnetised initially, and also provided the field strength was not, at any time during its general increase from zero to H_2, caused to diminish. The levels of magnetisation corresponding to B_1, B_2, etc. are those obtaining when the field strengths are actually H_1, H_2, etc. These do not represent residual magnetism. If the fields are reduced from H_1, H_2 etc; to zero, the induced magnetism decreases along the paths B_1R_1, B_2R_2, etc; respectively, and the points R_1, R_2, etc; denote the degrees of residual magnetism, or *remanence,* induced in the magnetic material by maximum applied field strengths of H_1, H_2, etc.

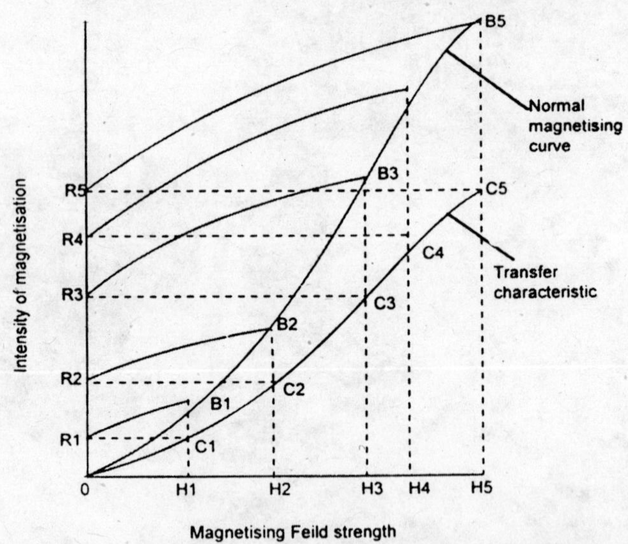

Fig. 14.4 Variations of instantaneous and residual intensities of induced magnetism with the strength of magnetising field.

The curve OC_3C_5 plotted in Fig. 14.4 gives a typical *transfer characteristic* for a magnetic material; that is, it shows the intensity of residual magnetism (R_1, R_2, etc.) resulting from any given maximum magnetising field (H_1, H_2, etc.) Complete transfer characteristics for positive and negative magnetising fields are shown in Figs 14.5 and 14.6.

It is obvious from these characteristic curves that *the degree of residual magnetism is not a linear function of the maximum magnetising field.* Because of this non-linearity, any signal applied to the microphone would, on recording suffer distortion. All signals would suffer bottom - bend distortion resulting in *even harmonics* of the fundamental appearing in the recording. Large signals would be further distorted by the magnetic saturation depicted by the flattening of upper sections of the characteristics, and the clipped reproduction would contain a high percentage of *odd harmonics* of the original.

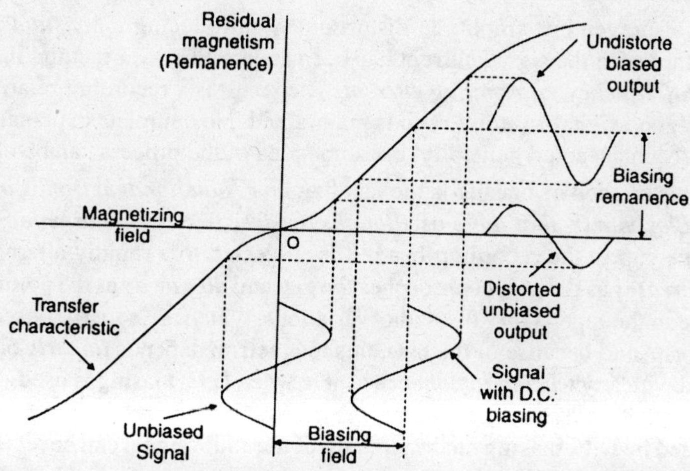

Fig. 14.5 Distortion occurring with no bias and its removal with D.C. biasing.

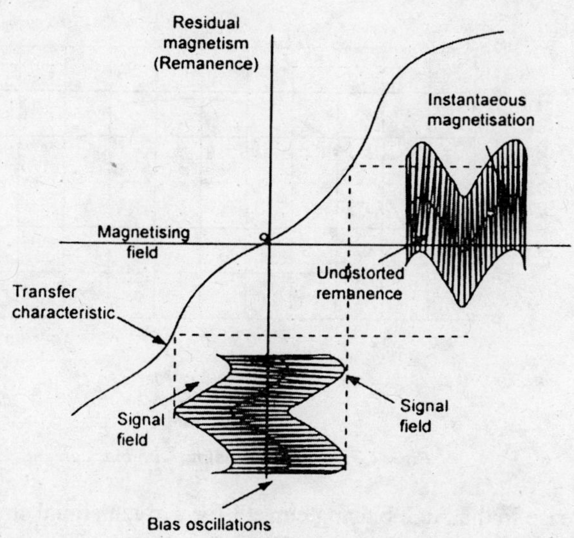

Fig. 14.6 Removal of distortion with A.C. biasing.

The *middle sections* of the arms of the transfer characteristics are, however, *approximately linear* and if the variations in the magnetic field strengths are *confined* to these portions, then the recording will be a relatively *undistorted replica* of the original sound. If a constant current is fed into the recording head together with the signal current, the effect is to *lift* the variations in the magnetising field *above the lower curvature* of the characteristic (Fig. 14.5). *The D.C. bias induces a constant degree of magnetisation in the tape on which the audio variations are superimposed.* Only the variations of the residual magnetism will appear on replaying the recording, so that the bias will not be translated into sound (in fact, *some noise does result from the biasing remanence).* But because of the limitation imposed by the saturation curvature of the characteristic, *the fact that the bias magnetism is added to the audio magnetism, means that the amplitude of the signal must be restricted quite considerably. The signal-to-noise ratio is thus lowered when D.C. bias is used.*

This method of preventing amplitude distortion by introducing a *constant biasing current* into the recording head together with the signal current has been superseded by a method in which the direct current is replaced by a *high-frequency, alternating current.* The result is a recording relatively free from harmonic distortion in which a good signal-to-noise ratio is maintained. No complete explanation of the mechanism of A.C. biasing has yet been accepted generally, but some idea of the process can be obtained from Fig. 14.6.

A feature of this method of biasing is the fact that *no residual magnetism is induced in the tape by the A.C. signal provided, of course, that these oscillations are free from even harmonic distortion.* As any point on the tape passes the gap in the recording head, it is subjected to a rapidly alternating magnetic field, the strength of which *increases* as the point approaches the gap and *dies away* as the point recedes. Such a process causes no remanence in the tape. Only the audio-variations, which are superimposed on the A.C. field cause any residual magnetism, and because of the bias, this magnetism depends *linearly* on the audio signal. Thus, the tape is not loaded with any *wasted remanence* as it is when D.C. biasing is used. In addition to this, it will be seen from Fig. 14.6 that *both linear sections of the transfer characteristic can be used.* It follows then that the limitations imposed by D.C. biasing on the strength of the audio signal can be relaxed considerably without increasing the danger of approaching the saturation level of the tape. *A.C. bias, therefore, reduces bottom bend distortion and, compared with D.C biasing gives an improved signal-to-noise ratio.*

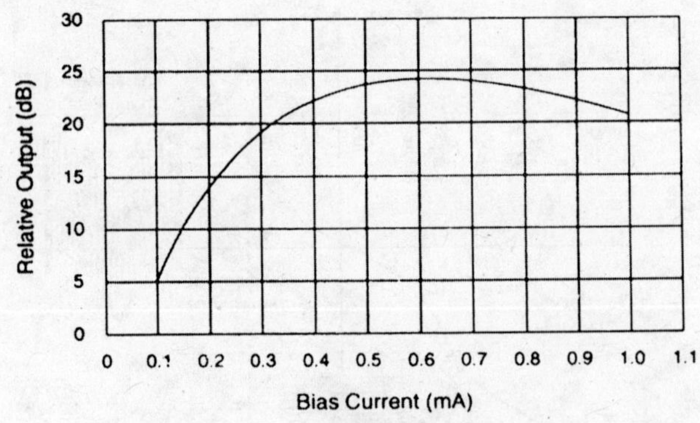

Fig. 14.7 Variation of output with bias current.

Initially an increase in the A.C. biasing current for a given signal strength lessens the distortion of the recorded signal and increases the level of recording. But *too much bias will cause top bend distortion and will also cause a lessening in the output, especially for treble signals.* The bias acts very much in the manner of the erase signal. A typical variation of the output with bias current is shown in Fig. 14.7.

14.6 ERASING

The field associated with A.C. biasing signal does not cause any residual magnetism in a tape passing through the field. If the tape is unmagnetised when the signal is applied, it will be unmagnetised when it has passed through the field.

If the amplitude of the A.C. signal is sufficient to saturate the tape, then even if the tape is not unmagnetised originally, it will be so when it has travelled through the field. This fact is used in the erase heads of most magnetic recording equipment. The tape is drawn through a strong local magnetic field arising from a high-frequency alternating current in the erase head. As each point of the tape travels into the field, it is saturated, and as each point recedes, it is subjected to a field strength which diminishes to zero some distance from the head. By this process it is *cleaned* of all previous magnetisation.

14.7 OSCILLATOR COIL

The bias oscillator should generate a sine wave with negligible harmonic distortion to keep noise to a minimum. *Symmetry in the waveform is essential.* An unsymmetrical sine wave has, in fact, a D.C. component made up of the difference between the positive and negative amplitudes of the sine wave. This will introduce noise in the recording and impart some wasted remanence to the tape.

It is usual for the oscillator transistor to provide both bias and erase currents. In most of the commercial tape decks, the *combined record-playback head* is usually of high impedance, while the impedance of the *erase head* may be high or low. To accommodate low-impedance erase heads, a separate oscillator output, in the form of a secondary winding to the oscillator coil, is required for *matching* purposes. Some degree of flexibility is possible if a *tapped secondary winding* is used, to suit all types of erase heads.

14.8 FREQUENCY RESPONSE

The degree of magnetisation of the tape which results from recording with a current through the head, which is constant at all frequencies of the form shown by the curve ABC of Fig. 14.8. The reduced intensity in the treble region is attributable in the main to *self-demagnetisation* of the tape.

To outline the *recording process* briefly - the tape is drawn steadily past the recording head, and the field produced by the signal current in the head induces some degree of residual magnetism in the section of the tape nearest to the head. The *amplitude* of the signal governs the amount of magnetism, and the *sense* (whether positive or negative) controls the direction of the magnetism in each section.

On the basis of molecular theory of magnetism, the molecular magnets in the tape are *aligned* by the magnetising field. The *strength* of the field governs how many of these magnets are brought into line. The *direction* of the field dictates whether the alignment is with north or south polar leading.

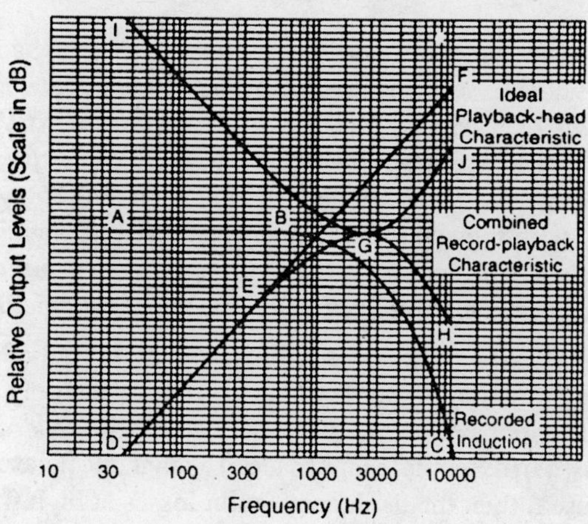

Fig. 14.8 Tape record-playback characteristics.

The field setup by a sinusoidal signal current in the head would change direction regularly, so that the *pattern* of magnetism induced in the tape would have adjacent sections of the tape with north and south poles leading alternately (see Fig. 14.9). *Thus, the poles of each section of the tape would be adjacent to the like poles of the neighbouring sections.*

The *length* of these sectional magnets in the tape depends on the *speed* at which the tape passes the head, and on the *frequency* at which the magnetising field changes direction. *For a constant transport speed the length varies inversely as the frequency of the signal.* For low-frequency signals the magnets are comparatively long, but for high-frequency signals they are short.

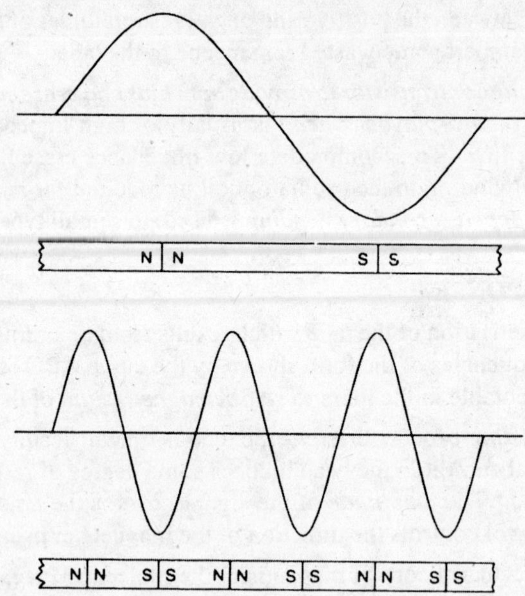

Fig. 14.9 Lengths of sectional magnets corresponding to low-and high-frequency signals.

For the *playback process,* the magnetised tape is drawn past the playback head at a *constant speed,* and the magnetic fields associated with the sections of the tape *move and cut* the coils of the head, thereby setting up voltages in the coils. *The rate of change of flux cutting the coils governs the voltage generated.* Thus, the rapidly changing fields set up by the sections of the tape on which high frequency sounds have been recorded must of necessity give rise to larger voltages than the sections containing low-frequency impressions. This is so, even if the impressions at both high and low frequencies have the same intensity of magnetisation. *The actual speed of the tape has no bearing on the relative levels of the output at the various frequencies. An increase in the transport speed simply multiplies all levels by the same amount.*

The rate at which the playback voltage rises with frequency is equivalent to *6 dB per octave.* In Fig. 14.8, the line DEF indicates the frequency response of the *playback head* only. (A tape which has been ideally magnetised is assumed for this response curve). The response for the *recording head* alone will depend on various practical considerations. For example, a tape made of a magnetic material of high coericivity is difficult to demagnetise, and if one is used, then the demagnetisation losses at high frequencies will be less than if a low-coercivity tape is used. *The use of a high-coercivity tape, of course, makes intentional erasing more difficult.*

14.9 EQUALISATION

The response for complete equipment, (that is, the response combining both *recording and playback* deviations) is indicated by the curve DEGH of Fig. 14.8. *To obtain an equalised output, from this response curve it is necessary to introduce boost at both ends of the frequency range.* For ideal equalisation the compensating curve must be a mirror image of the curve DEGH, so that the equalising response curve needs to be the curve IGJ.

There will be, in fact, further high-frequency losses encountered in tape recording. *The tendency of the A.C. bias signal to act as an erasing signal during the recording process is more pronounced at high frequencies because the remanence is less deeply seated at these frequencies. Also, the physical dimensions of the gap in the playback head cause losses when the size of the sectional magnets and the gap are commensurate.* (This is usually referred to as *gap effect*). These combine to produce a more accentuated drop in the response curve in the treble region.

Unfortunately, *equalisation* is not simply a matter of compensating for bass and treble deviations. Several performance requirements are in conflict, and a compromise is the best that can be achieved. A wide frequency response, low distortion or a high signal-to-noise ratio can each be obtained at the expense of the others. Over a restricted range, an increase in bias current, for example, lessens distortion, but causes treble attenuation and tends to lessen the signal-to-noise ratio. Furthermore *equalisation is a function of the physical properties of the tape used.* The equalisation provided with any combination of tape deck and amplifier may, for example provide a level response over a wide frequency range with one brand of tape, but may give rise to a large treble peak with another brand.

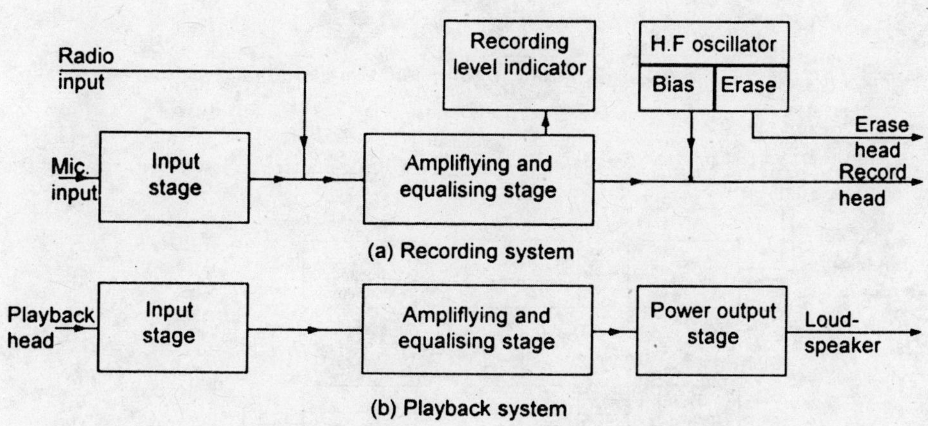

Fig. 14.10 Block schematic diagram of tape recording and playback systems.

14.10. BLOCK DIAGRAM

A tape amplifier can be represented essentially by the block schematic diagram of Fig. 14.10. Normally, the design of both (recording and playback) units will follow general A.F. practice, although attention must be paid to certain requirements of the equipment. *Precautions will be required to minimise hum-pick-up, and the overall signal-to-noise ratio should be at least 40 dB.*

The attenuation of the bass response in a tape recorder takes place mainly in the playback head. The treble response suffers its greatest losses in the recording process. Thus, in order to load the tape more or less evenly, *it is best to provide treble boost in the recording amplifier and to apply bass boost in the playback amplifier.* Some treble accentuation may also be required in the playback amplifier, but this has to be limited in order to avoid excessive amplification of the background noise.

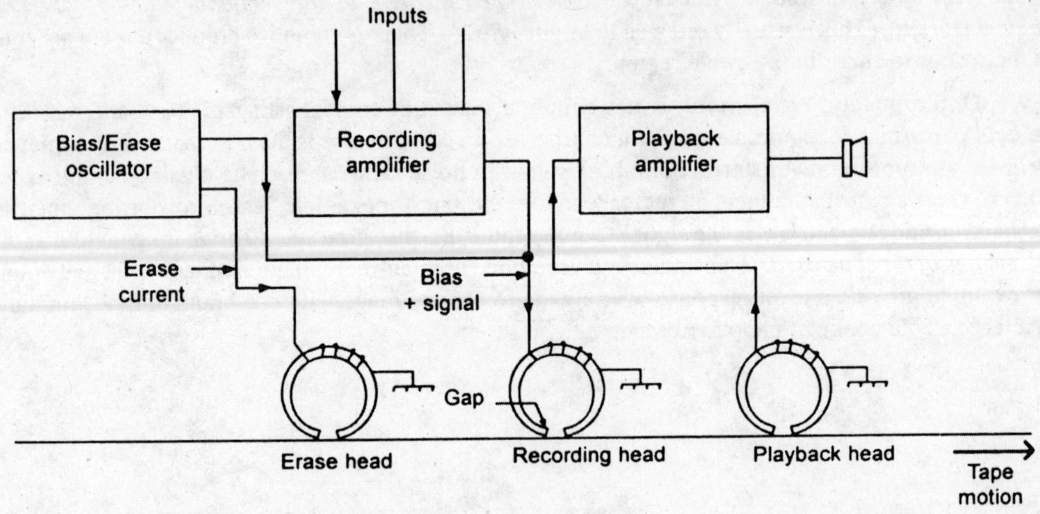

Fig. 14.11 Basic elements of a tape recording system with separate recording and playback heads.

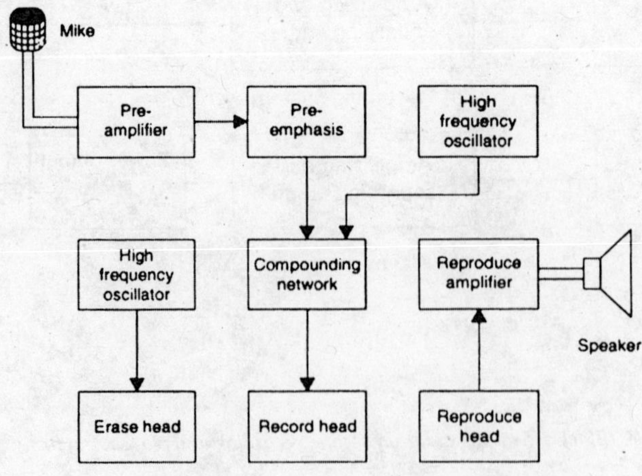

Fig. 14.12 Basic elements of a tape recording system with a combined recording and playback head.

Basic elements of a tape recording system with separate recording and playback heads is given in Fig. 14.11 and that of a tape recording system with a combined recording and playback head is given in Fig. 14.12. Both the systems are self-explanatory.

OBJECTIVE TYPE QUESTIONS (A)

1. Tape decks combine the functions of
(*a*) recording (*b*) reproduction
(*c*) (*a*) and (*b*) (*d*) none of the above

2. Modern heads are usually constructed of two semi-circular stacks of high-permeability laminations assembled
(*a*) symmetrically (*b*) asymmetrically

3. Each stereophonic head forms part of a complete
(*a*) recording system (*b*) playback system
(*c*) (*a*) or (*b*) (*d*) (*a*) and (*b*)

4. Power required for efficient erasure is approximately
(*a*) 1 to 2 W (*b*) 2 to 6 W
(*c*) 2 to 4 W (*d*) 4 to 8 W

5. Erase heads are available with a
(*a*) single gap (*b*) double gap
(*c*) (*a*) and (*b*) (*d*) none of the above

6. Unevenness in the surface of the tape will cause
(*a*) increased level of recording
(*b*) decreased level of recording
(*c*) undulating level of recording
(*d*) none of the above

7. The higher the speed the better the performance at
(*a*) all frequencies (*b*) low frequencies
(*c*) medium frequencies (*d*) high frequencies

8. The non-linear relationship between the magnetism residing on the tape and the magnetic field inducing it will cause considerable
(*a*) phase distortion (*b*) harmonic distortion
(*c*) amplitude distortion (*d*) intermodulation distortion

9. Bottom-bend distortion results in
(*a*) even harmonics of the fundamental
(*b*) odd harmonics of the fundamental

10. With D.C. biasing signal-to-noise ratio is
(*a*) unaffected (*b*) increased
(*c*) lowered

11. With A.C. biasing
(*a*) no residual magnetism is induced in the tape
(*b*) residual magnetism is induced in the tape

12. Too much bias will cause
    (*a*) top bend distortion (*b*) bottom bend distortion
13. Too much bias causes lessening of the output, especially for
    (*a*) bass signals (*b*) treble signals
14. Symmetry of the waveform of the oscillator coil is
    (*a*) essential (*b*) not essential
15. It is usual for the oscillator transistor to provide
    (*a*) bias current (*b*) erase current
    (*c*) bias and erase currents
16. For a constant transport speed, the length of the magnet varies
    (*a*) directly as the frequency (*b*) inversely as the frequency
17. To obtain an equalised output, it is necessary to introduce boost at
    (*a*) lower end of the frequency range
    (*b*) top end of the frequency range
    (*c*) both ends of the frequency range
    (*d*) none of the above.
18. Equalisation is a function of the
    (*a*) physical properties of the tape used
    (*b*) chemical properties of the tape used
    (*c*) composition of the tape used
    (*d*) dimensions of the tape used
19. The attenuation of bass response in a tape recorder takes place in the
    (*a*) recording head (*b*) playback head
20. The treble response suffers its greatest losses in the
    (*a*) playback process (*b*) recording process

FILL IN THE BLANKS (B)

1. Both gaps are filled with non magnetic shim to maintain a __________ relationship between the flux at the gaps and the current through the exciting coils of the head.
2. Appreciable power is required for __________ erasure.
3. The longer the tape is under the influence of the erasing field the more __________ will be the erasure.
4. The base of the magnetic tape needs to have ______ and suppleness.
5. Tapes are usually supplied in ______ or _____ inch reels.
6. For stereophonic tapes ______ paths are used for twin signals of the program.
7. The basic requirement of a tape transport system is to impart a ________ to the tape.
8. Too much bias will cause _______ distortion.
9. Some degree of flexibility is possible if a ___________ secondary is used to suit all types of erase heads.
10. The use of a high-coercivity tape makes __________ erasing more difficult.
11. Physical dimensions of the gap in the playback head cause losses when the size of the sectional magnets and the gap are _________.
12. It is best to apply ____ boost in the recording amplifier and to apply ____ boost in the playback amplifier.

CONVENTIONAL QUESTIONS (C)

1. Explain the construction of a magnetic head with the help of a suitable diagram.
2. Differentiate between recording, playback and erase heads.
3. What are the factors which influence the recording and playback characteristics of a tape recorder ?
4. How are the deviations in the recording and playback characteristics of a tape recorder compensated for?
5. What are the requirements of a tape transport system ?
6. Explain the functioning of a tape system with the help of a block diagram.

15

AMPLIFYING SYSTEMS

Remarkable progress has been made in the field of sound recording and reproduction. The introduction of high fidelity disc and tape recordings and the improvement in the design of playback equipment have set new standards for the discerning listener and have consequently created the need for systems providing amplification of the highest quality. Because of the variety of signal sources in existence, the range of application of these systems has had to be wide. Therefore pre-amplifying circuits which cater for pick-up and playback heads, F.M. tuner units and microphones have assumed great importance. Furthermore, the emergence of stereophonic recordings and pick-up heads has created a demand for dual channel amplification and this has been fulfilled in equipment consisting of either self-contained stereophonic amplifiers or specially designed stereophonic pre-amplifiers coupled to pairs of conventional power amplifiers.

15.1 MONAURAL SYSTEMS

The choice of line-up is dictated by the *strength of the signal* to be reproduced. For example, a 10 W circuit incorporating volume and tone controls can be used satisfactorily *without a pre-amplifier* if input signals greater than 500 mV are available. Such signals can be obtained from high-output crystal pick-up heads or F.M. radio tuner units. A 3W circuit with volume and tone controls can be used *directly* with sources which give signal voltages greater than 100 mV.

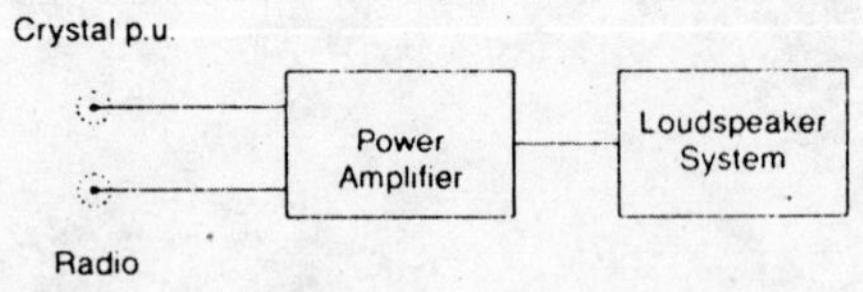

Fig. 15.1 Direct use of power amplifiers with high input signal sources.

However, signal voltages from low output crystal pick-ups, magnetic pick-ups and tape-recording heads (which can have values of upto 500 mV but are usually between 2 and 100 mV) are normally too low for direct use with the power amplifiers, and a *pre-amplifier* is necessary to increase the sensitivity of the system. In addition, discs and tapes are deliberatly recorderd in such a manner that the strength of signals derived from them will be a *function of frequency*. Consequently, the strength of these input signals will usually vary in some way not related to the original sound patterns and, to obtain realism from such recordings, some degree of compensation is required in reproducing these sounds. *This equalisation is best incorporated in a pre-amplifier, so that the scope of application of power amplifiers can be increased without any resultant fall in the standard of performance.*

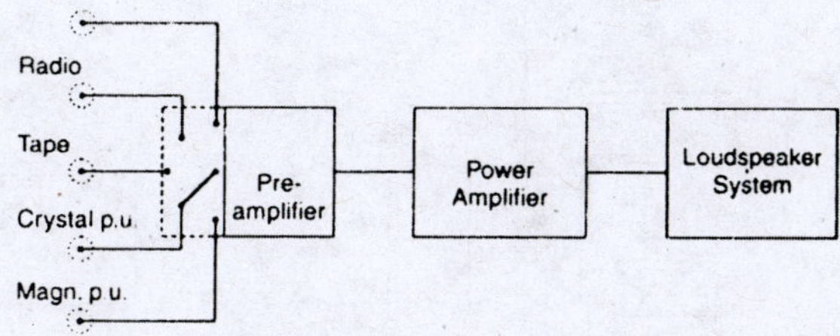

Fig. 15.2. Use of pre-amplifiers and power amplifiers for low-level signals.

The simple arrangement for the 10 W and 3 W amplifiers, cited as examples, connected directly to the input source is shown in Fig. 15.1. The use of pre-amplifiers is indicated in Fig. 15.2. The 3 W tape circuit comprises a *self-contained record-playback system* for microphone and radio input sources. The circuit is also suitable for replaying pre-recorded tapes. A *program-monitoring signal* is available when recording and, during playback, this source will provide a *low-level signal* for driving a power amplifier if an output power higher than 3W is required.

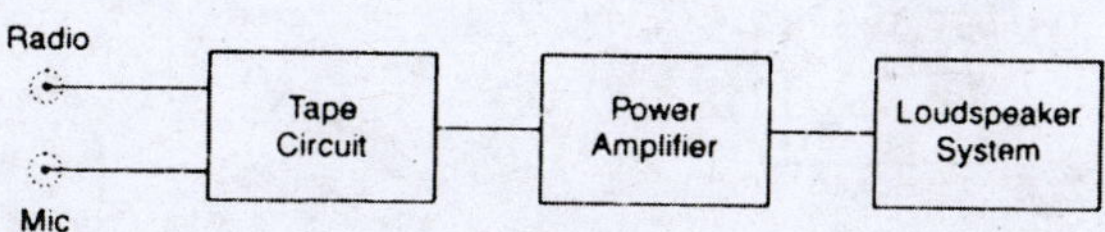

Fig. 15.3 Tape circuits and power amplifiers arranged for use with radio or microphone signals.

The *tape pre-amplifier* provides recording facilities comparable with the self-contained unit but, on playback, the circuit only provides an *equalised signal* for driving a power amplifier. A simple arrangement for the combination of tape pre-amplifier and power amplifier is shown in Fig. 15.3. This line-up is suitable also for the 3W tape amplifier when the low-level playback output is used.

For a complete record / monitor - playback system using pick-up input sources, the tape pre-amplifier should be used in conjunction with line-up of Fig. 15.2. The new arrangement is shown in Fig. 15.4. It is important that the tape deck in both arrangements should always be switched to the *playback position* before the connection from the tape output to the pre-amplifier input is made. If the tape recorder is in the *record position* when the connection is made *oscillation* will occur.

When *recording*, the input signal is fed to the appropriate input channel of the pre-amplifier and then from the program-recording output of the pre-amplifier to the input of the tape circuit. The program can be *monitored* by using the normal output of the pre-amplifier to drive the power amplifier. For *playback*, the equalised output from the tape pre-amplifier is fed via the *Auxiliary input channel* to the main amplifying chain.

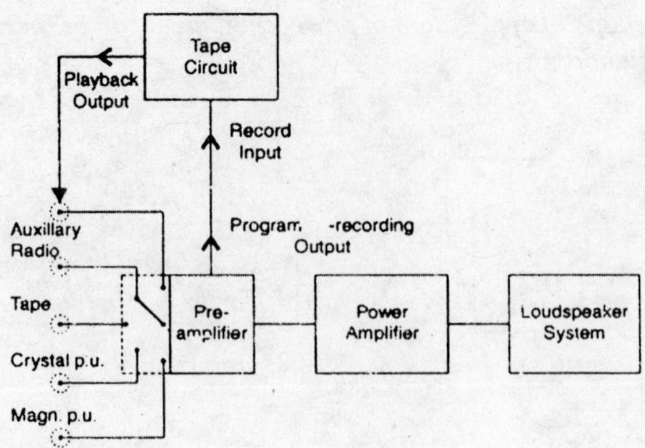

Fig. 15.4 Arrangement of pre-amplifiers and power amplifiers will complete tape record / monitor playback facilities.

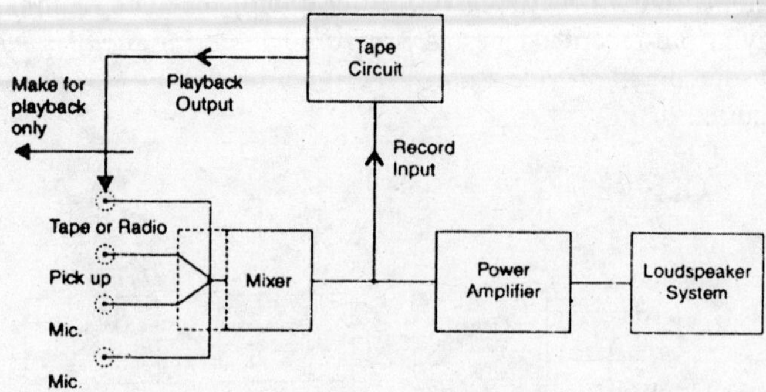

Fig. 15.5 System for program recording, monitoring and replaying with provision for mixing input signals.

To feed *several inputs* into an amplifying system at the *same time,* the main pre-amplifier in Fig. 15.2 should be replaced by the input mixing pre-amplifier. Recording, monitoring and playback are possible with the new arrangement shown in Fig. 15.5. The output impedance of the mixer is low so that the inputs of the tape pre-amplifier and the power amplifier can be connected simultaneously to the mixer without affecting the quality of reproduction.

15.2 STEREOPHONIC SYSTEMS

For stereophonic reproduction, *two separate amplifying channels* are required, one for each of the twin signals obtained from the stereophonic signal source. In the simplest arrangement, a 2W or 7W stereophonic amplifier is connected to the source as shown in Fig. 15.6. If the stereophonic pre-amplifier is used; it should be coupled to two power amplifiers and loudspeaker systems as in Fig. 15.7.

Stereophonic systems should be suitable for reproduction from monaural sources and *compatibility switches* are usually incorporated for this purpose. For instance, the switching indicated in the pre-amplifier stage of Fig. 15.7 selects the input channels and also sets the equipment for correct or reversed stereophonic reproduction or for monaural reproduction using both power amplifier and loudspeaker systems. Facilities are also provided for *stereophonic radio signals* in anticipation of stereophonic transmissions.

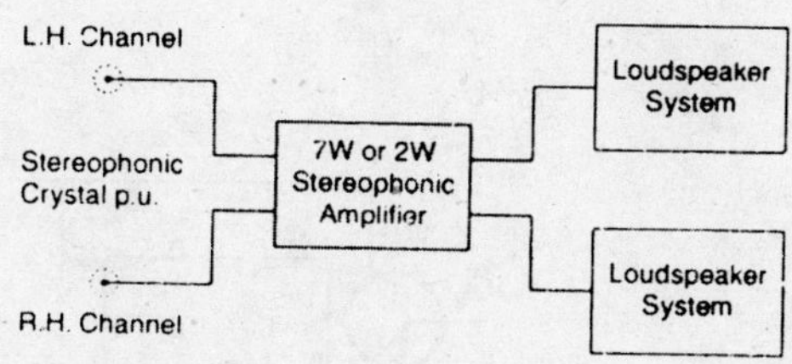

Fig. 15.6 Use of stereophonic amplifiers

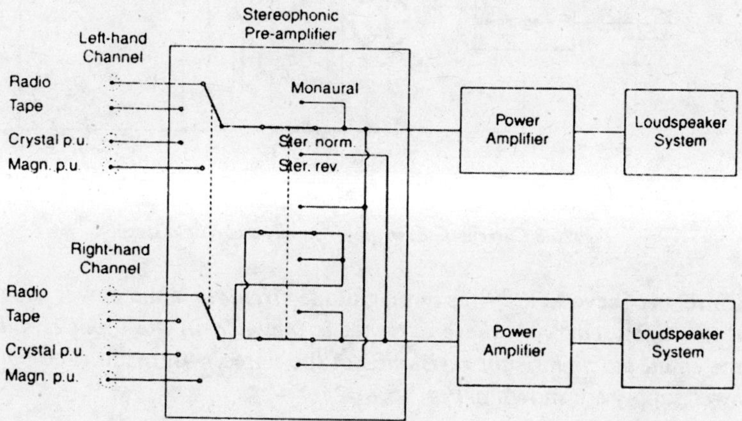

Fig. 15.7 Arrangement of stereophonic pre-amplifier with pairs of conventional power amplifiers.

15.3 POWER SUPPLIES

Facilities are provided in amplifier circuits for obtaining power supplies for a single-channel pre-amplifier or F.M. radio tuner unit. For stereophonic applications, the supply for each channel of the preamplifier can be taken from the power amplifier used with that channel.

A *separate* power stage can also be used for the pre-amplifier when it is used with any of the power amplifiers. Physically however, the extra unit may prove an encumberance and consequently *it may be preferable to use the supply stage of the main amplifier to provide the power requirements of the tape preamplifier.*

15.4. LAYOUT OF HIGH-QUALITY AUDIO CIRCUITRY

The circuits for high-quality reproduction function only as they should if they are carefully laid out and constructed. The *relative positions* of input, output and power supply, and the earthing and screening arrangements are of particular importance, and it is suggested that layouts should be designed with reference to *the code of good practice* outlined here.

General Layout : The *input sensitivity* of an amplifier is typically 3mV at 1kHz on the magnetic pick-up position, and the *output voltage* is of the order of 20V, implying a voltage gain approaching 10^4. *It is therefore essential to keep the output separated and screened from the input.* The magnetic field from the mains transformer may cause hum, so the mains transformer should be *as remote as possible* from the input.

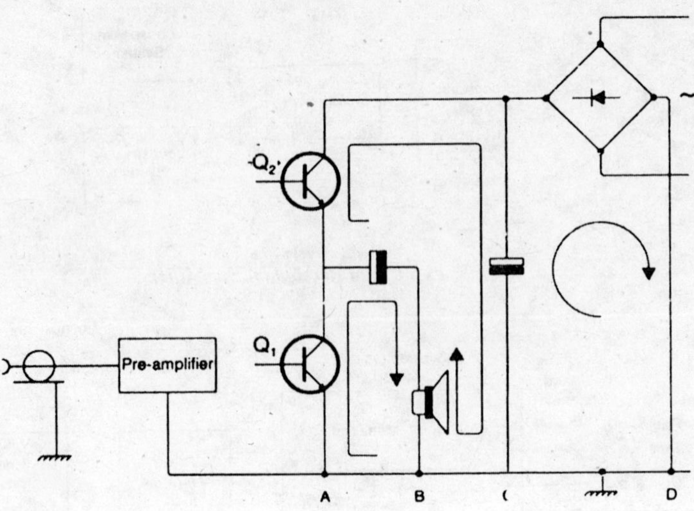

Fig. 15.8 Current paths in power supply and output stage.

Earthing : Currents of several amperes in magnitude *circulate* in the power supply and output stages. *It is important that no wiring carrying these currents is included in the input circuit,* otherwise hum or instability, due to the *small but significant resistance* of the wires, will result. The *paths* of currents in the output stage and power supply are shown in Fig. 15.8.

Between points A and B there is a voltage due to Q_1 , between B and C a voltage due to Q_2 ; and between C and D a voltage due to the power supply. The input can be earthed at A, but not at B, C or D. In practice it is usual to combine A, B, C and D in one *common earthing point.*

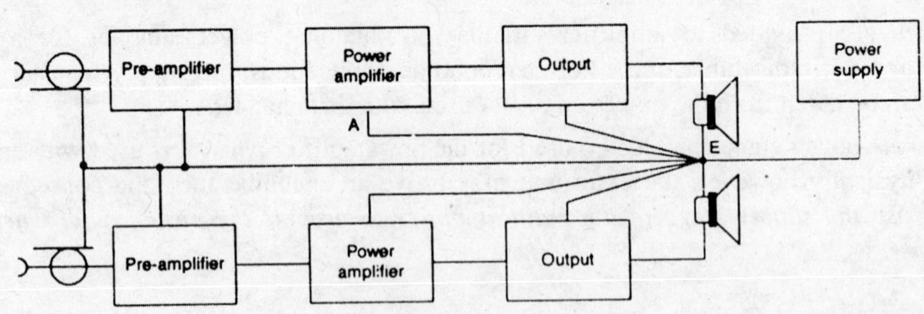

Fig. 15.9 Earthing arrangement for a stereo amplifier.

The earthing arrangement for stereo amplifiers is considerably more complex than for mono, since the single power supply and common earth for the two signal inputs make it more difficult to avoid *earth loops.* A typical arrangement is shown in Fig. 15.9.

Point E is the *common earthing point.* The voltage across the wire AE is effectively *in series* with the input to the power amplifier. This does to cause any trouble provided the wire is *short,* because the sensitivity at the input of the power amplifier is, say, 200 mV and the output-stage current is taken to the earth point *separately.*

Magnetic Fields : The principal source of magnetic fields in an amplifier is the *mains transformer.* Hum due to this cause will be at a frequency of 150 Hz because the *hysteresis* of the core results in the magnetic field being predominantly *third harmonic.* The magnetic field will induce a voltage in any circuit loop.

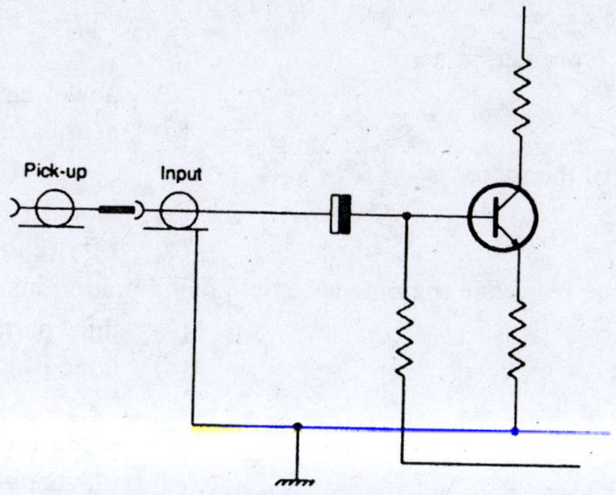

Fig. 15.10. Input loop of a pre-amplifier.

The *input loop* of a pre-amplifier is shown in Fig. 15.10. The area of this loop should be minimised, and it may also be necessary to *screen* the mains transformer by making the parts of the chassis surrounding it from mild steel rather than aluminium.

The magnetic field associated with the currents in the output stage may occasionally cause trouble. Again, *the remedy is to minimise the area of the circuit loops in the output stage* by, for example, running emitter and collector leads close together, in order to minimise radiation.

Hum : There are many possible causes of hum. A clue to the cause of hum may often be obtained by determining its frequency.

50 Hz hum may be due to stray pick-up caused by poor-screening earth loops or television break-through. The last is recognisable by the distinctive buzz of the field sync pulses, and also by the fact that it is accompanied by the sound bars.

100 Hz hum originates in the power supply and is often indicative of poor decoupling of the B+ supply or incorrect connection of the earth leads.

150 Hz hum comes from the magnetic field of the transformer.

Resistor Noise : Every resistor has inherent *thermal noise.* However, certain resistors, notably *carbon composition* types, give an excess noise which is proportional to the voltage across them. Although this excess noise does not usually contribute significantly to the noise in the transistor circuit, *it is advisable to use high-stability resistors for the collector load and base bias resistors of the input stage.*

OBJECTIVE TYPE QUESTIONS (A)

1. A pre-amplifier is necessary to
   (*a*) the frequency response of the system
   (*b*) the signal-to-noise ratio of the system
   (*c*) the sensitivity of the system
   (*d*) none of the above
2. Equalisation is best incorporated in a
   (*a*) pre-amplifier (*b*) power amplifier
   (*c*) either (*a*) or (*b*) (*d*) neither (*a*) nor (*b*)
3. The output impedance of the mixer is
   (*a*) low (*b*) moderate
   (*c*) high (*d*) very high
4. The hysteresis of the core results in the magnetic field being predominantly
   (*a*) second harmonic (*b*) third harmonic
   (*c*) fourth harmonic (*d*) none of the above
5. 100 Hz hum originates in the
   (*a*) mains transformer (*b*) recording head
   (*c*) power supply (*d*) playback head

FILL IN THE BLANKS (B)

1. On playback the pre-amplifier provides an __________ signal for driving a power amplifier.
2. The program can be monitored by using the ________ output of the pre-amplifier to drive the power amplifier.
3. For stereophonic reproduction two ________ channels are required.
4. Stereophonic systems should be suitable for reproduction from monaural systems and ______ switches are usually incorporated for this purpose.
5. The ________ positions of input, output and power supply, and the earthing and screening arrangements are of particular importance.
6. The area of the circuit loops in the output stage should be ________.
7. It is advisable to use _______ resistors for the collector load and base bias resistors of the input stage.

CONVENTIONAL QUESTIONS (C)

1. Explain the difference between monaural and stereophonic system.
2. What are the functions of a pre-amplifier? Why is it required and where in an audio system, is it connected?
3. Give a brief description of
   (*a*) input sensitivity (*b*) compatibility
   (*c*) earth loops (*d*) hum
   (*e*) resistor noise

16

NOISE REDUCTION SYSTEMS

Any sound which is an unwanted signal is noise. This noise can be kept to extremely low levels, but when sound is transformed into an electrical signal, noise can be added to it in ways over which we have very little control. The random movement of electrons in a wire or in the atmosphere will add noise to signals in wires or to signals being transmitted by radio. The pressing and replaying of clean discs add noise, and even the best amplifier will add some more noise. All of these sources can, however, be kept small compared to the noise of the recorded cassette tape.

If we play a blank tape we find that the noise level is very low. The noise originates mainly in the recording process. It is, in fact, the erase and bias signals which arranges the magnetism of the tape particles at random so as to produce the effect of noise on replay, and if a blank tape is passed through the recorder with the record button down, but no signal input, the increase of noise level when the recorded part is played can be heard. Part of this noise can be reduced by good design of the oscillator which produces the erase bias signal, and by good processing of materials used for the production of the tape.

16.1 NOISE REDUCTION

If we record on *wider sections of tape,* we find that the signal-to-noise ratio is much better, as when a cassette is recorded mono on the same equipment as is used to record stereo cassettes. The difference is caused by the random nature of the noise. If we compare two strips of tape, one recorded to *double* the width of the other, we would expect that because *twice* as many particles are magnetised by the signal in the wider strip, the replay signal would have *twice* the amplitude from the wider strip as from the narrow strip. But the noise from the wider strip is not twice as great, but nearer to 1.4 times as great (where 1.4 is *approximately* $\sqrt{2}$) This arises because noise is not a signal of definite shape, and particles in the tape may line up with their magnetism *opposing* each other just as readily as they line up with their magnetism *assisting*.

The problem of signal-to-noise ratio in tape recording is simply solved, therefore, by using wider sections of tape, so that a full width tape (1/4 in) should have a very good signal-to-noise ratio. Dividing a full width tape into two sections for a *half track stereo tape* reduces the signal-to-noise ratio in each channel, but the results are still *acceptable*. Further reducing the recorded width by half again to give *four-track stereo* (two-halves of stereo recording) degrades the signal-to-noise ratio still further, and another halving of width, needed to fit *two sets of stereo channels* on a cassette tape, has resulted in noise levels which are *unacceptable* when compared with the results obtainable from F.M. radio or from discs. It is worth noting that *all disc recordings start by being tape recording.*

Despite considerable improvements in tape materials and in the design of erase / bias oscillators, *the noise content of a cassette tape is too high unless some sort of noise reduction process is carried out on the signal at some stage.* Several methods of noise reduction have been devised. No high-quality cassette deck or recorder lacks a *noise reduction system.*

The type of noise found from different signal sources varies because of the *different causes* of the noise. Some receiver noise is *white* meaning, that it contains frequencies spread all over the audio range, *with no preferred frequencies.* The noise from *discs* is not white, but has a range of frequencies related to the speed of the disc past the recording cutter and the replay stylus. Noise of this sort can be shifted to a lower frequency by making the disc revolve at a low speed. Fortunately, *noise in lower frequency ranges is not particularly offensive to the ear,* so that change over from 78 rpm discs to 33 rpm discs caused a very great reduction in the apparent noise as heard by the listener. Reducing the speed of *tape* does not have effect, however, and the noise remains obstinately in the same frequency range which, unfortunately, happens to be *the range of frequencies to which the ear is very sensitive,* with the peak amount of noise at around 4.5 to 5.0 kHz.

In a few low priced cassette machines, a button marked noise reduction simply cuts the level of frequencies, above about 2.5 kHz, so reducing the noise and also removing much of the signal. The signal frequency presents a problem for *much of our useful musical information occurs in the same range as the noise and any attempt to remove these frequencies makes for a replayed signal of intolerably low quality.*

16.2 PRE-EMPHASIS AND DE-EMPHASIS

In *frequency modulation* the modulation or sideband frequencies often extend outward to as much as 15 kHz. This provides a much broader audio range than is usually found in conventional A.M. broadcasting. The higher audio frequencies produced by the program source are generally *weaker* than the lower audio frequencies, and thus produce a *smaller frequency deviation* in the carrier. In the normal transmitting process a certain amount of *interference* is encountered, such as atmospherics, which have little effect on the large -amplitude low frequency signals. However, with regard to the weak high frequency signals, the *signal-to-noise ratio* is low. To correct this condition, the Federal Communication Commission (FCC) standards call for a *pre-emphasis network.* This consists of a high-pass filter having a time constant of 75 micro-seconds. The pre-emphasis at the F.M. transmitter provides for an audio gain of 1 at 50 Hz, to a gain of 7 at 15 kHz. From 50 to 500 Hz, the rise is very slight, but from there it increases rapidly. Thus, *the net effect of pre-emphasis is to increase the signal-to-noise ratio at the higher audio frequencies.* Standard pre-emphasis and de-emphasis curves are illustrated in Fig. 16.1.

If the pre-emphasis were not *corrected* at the receiver, the sound signal would have a *heavy treble effect.* To compensate for this, the F.M. receiver employs a *de-emphasis network,* a low-pass filter. As in the case of the high-pass filter, the low-pass filter in the de-emphasis network also has a time constant of 75 micro-seconds.

The de-emphasis network can be placed in any one of the several places in the receiver, but practice finds it located at the output of the F.M. detector circuit. It is an RC filter that shunts the higher audio frequencies to ground while affecting the lower audio frequencies to a lesser extent. *The overall effect of the pre-emphasis and de-emphasis networks is to provide a noise free signal at the output of the receiver that is a true replica of the sound signal at the studio.*

This is the solution to the problem of noise in F.M. radio, adopted almost from the beginning of F.M. broadcasting. This type of noise reduction, so easily applied to broadcasting, requires that every receiver shall be fitted with the correct de-emphasis filter. *The system is a double-ended one, because both transmitted and received signals have to be processed by similar circuits.* Unfortunately the frequency range of the filters is not internationally standardised. In addition, the upper frequency limits of F.M. transmitters have been improved so much over the years, that the amount of treble boost now needed at the highest frequencies is *excessive,* and causes *distortion* at the transmitter.

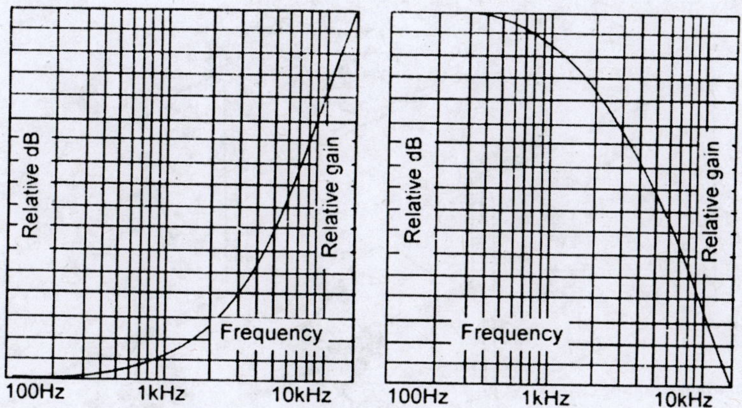

Fig 16.1 Pre-emphasis and de-emphasis

A similar problem of *excessive boost* makes the pre-emphasis and de-emphasis methods unsuitable for cassette recorders. Because of the low tape speed, and the limits of size of head gap, recording of the high frequencies is difficult enough, needing a considerable boost to the high frequencies in the recording amplifier just to obtain a *reasonable level response* at the output. The *extra boost* which would be needed for pre-emphasis makes the scheme impossible because the filter system used makes the boost greater at the higher frequencies, and the amplitude of the signal would lead to *over-recording and distortion.*

16.3 COMPANDERS

The professional solution to the problems of limited dynamic range and poor signal-to-noise ratio has been overcome by the use of *compression and expansion circuits* called companders, Fig. 16.2. The *compressor* circuit boosts weak signals relative to strong signals by a preset amount, so that the range of signal amplitude is less than it was in the original signal. The compressed signal is then recorded with the peak just reaching the *peak permitted recording level,* so that the weakest signals are at a level well above the noise level.

On replay, the signals are passed through the *expander* section, which exactly *reverses* the process, restoring the original range, and also reducing the noise caused by the recording process to very low levels.

If the compressing and expanding circuits exactly match each other, excellent results are obtained. In practice, all recorded or transmitted signals ar compressed to some extent, but the circuits used for compression are non-standardised so that we carry out the expansion correctly at the receiver or replay stages. In addition, the circuits used for compression and expansion in professional work are extremely complicated taking into account every requirement of recording.

Simple compression circuits may, far from reducing noise, even increase it. A good example is the use of *automatic level controls* in cheap cassette recorders, since these circuits are simple compressors. With no sound in, the gain of the recording amplifier is adjusted to maximum, so that the noise signal in the input circuits is amplified and recorded as if it were a signal. When a true signal appears, the gain is reduced so as to keep the signal near the peak recording level. The effect on replay is of music at an almost constant sound level with a *variable level of noise,* the noise suddenly becoming very loud during what should be the quiet passages of music. This is most undesirable, rather just the opposite of what is needed.

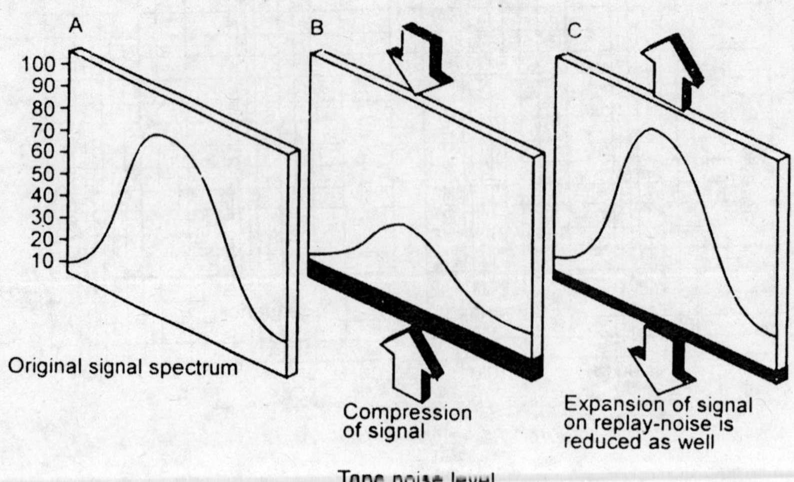

Fig. 16.2. Compander action.

16.4 NOISE REDUCTION SYSTEMS

The following noise reduction systems are widely used.

1. Dolby A system
2. Dolby B system
3. Philips DNL system
4. dbx system
5. Burwin system

Of all the noise reduction systems, the only two which have been applied widely to cassette recorders are the Dolby system and the Philips dynamic noise limiter (DNL) system. The only factor common to them is that *they seek to reduce noise particularly in quiet passages of signal where noise is most noticeable, and in the frequency ranges to which the ear is most sensitive.* Though the DNL system has some applications in playback of older recordings, the *Dolby process,* being a double-ended system, is more effective, much more widely used, and likely to be applied to F.M. stereo transmitters and to disc recordings also.

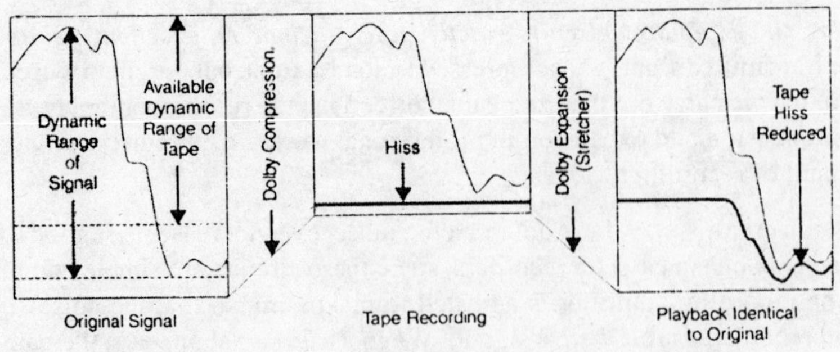

Fig. 16.3 The Dolby system reduces hiss by a compression and expansion process on low volume signals: loud signals remain unchanged.

Dynamic range in tape recording is limited at one end by *tape saturation* level and at the other end by *tape hiss.* Increasing the signal input above a pre-determined level will result in severe distortion without giving an increase of output level. Reducing the input level will not affect the hiss, which is of constant amplitude and independent of the incoming signal level.

Tape hiss is present on all recordings but it will not always be audible. An example is *pop music* which is compressed into a narrow dynamic range, has a high average level and is subjectively loud.

When the dynamic range is wide, as with the *solo piano,* if the recording level is set to give full modulation on peaks, the quietest passages will be *comparable* with the hiss level. If the hiss level becomes higher than the music, the music is *masked* by it; similarly when the music level exceeds that of the hiss, the hiss is masked.

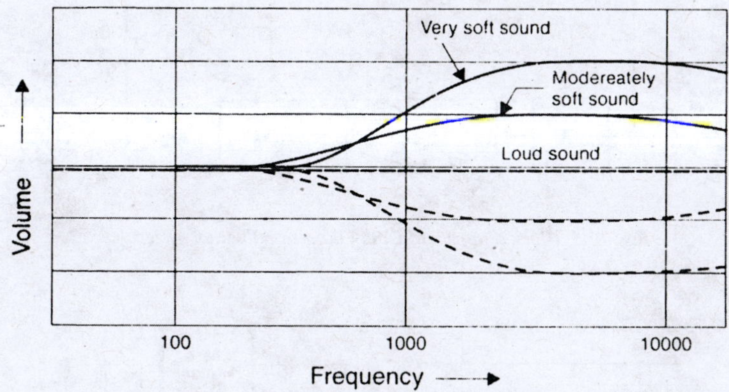

Fig. 16.4 The Dolby B's effect on frequency response during recording (solid lines) and playback (dashed). Mirror image playback action exactly cancels treble boost that was applied to low-level passages during recording, restoring the original flat response to sound.

To render the tape hiss unobtrusive it is necessary to reduce it in level compared with the quietest signals to be reproduced (most of which will be in the high and mid frequency range). One way of doing this is to apply a given amount of high frequency *pre-emphasis* during recording, similar to that used for F.M. transmission, and to use a corresponding *de-emphasis* on replay. The objections to this approach have already been discussed.

What is required is a processing circuit on record, the response characteristic of which will vary with both the frequency and the amplitude of the applied signal. As the signal level progressively falls below a given datum so the processor will selectively increase the gain of the system to compensate; *the amount of compensation will depend on the level and frequency of the incoming signal.* On replay an inversion of the process will restore the original dynamics and frequency range, and the previously emphasised low level signals will be attenuated together with the hiss.

The above principle, *with a number of elegant refinements,* is the basis of the noise reduction system devised by Dr. Ray Dolby. As the hiss problem is particularly acute in low speed machines using thin tapes, all top quality cassette recorders incorporate Dolby B circuitry. The original *Dolby A* system is intended for professional use and operates on the *entire frequency spectrum,* not just the highs. The *Dolby B* system is a consumer product system. In the Dolby B system, the high-frequency outputs of the pre-amplifier are amplified during recording. During playback, these frequencies are correspondingly attenuated. Since the hiss and general noise of recordings occur in the *high-frequency spectrum,* these too are attenuated, on the order of 10-50 dB.

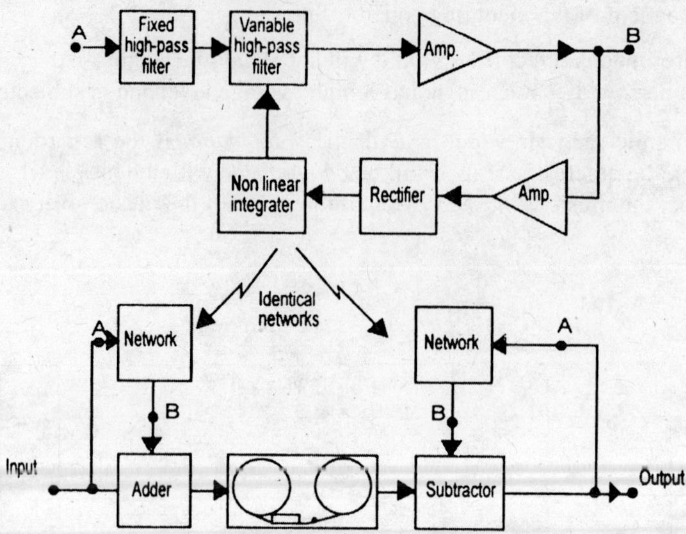

Fig. 16.5 Block diagram of Dolby B noise reduction system.

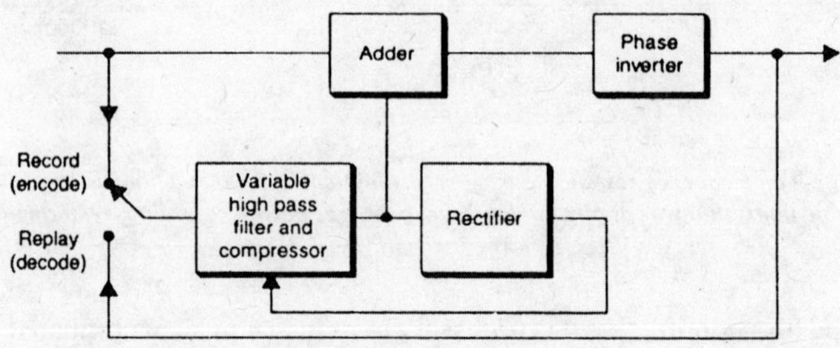

Fig. 16.6 Simplified block diagram of Dolby B system.

Fig 16.3 illustrates the reduction of tape hiss by Dolby B process. Fig 16.4 show the Dolby B's effect on frequency response during recording and playback. Fig 16.5 shows how the Dolby system works. Fig. 16.6 is a simplified block diagram which shows the Dolby principle more clearly. We can see that there is a *main program chain* and a side chain. The *side chain* has a compressor preceded by a high-pass filter. In the *encode* (record) mode signals below a given threshold are boosted by the compressor and added to the main chain. The degree of boost is dependent on the *control voltage* fed back from the rectifier to the variable filter. When the control voltage is zero, maximum boost is applied. *The control voltage increases with increasing signal level and when the control voltage is maximum, no boost is applied.* After processing the composite signal is recorded in the conventional manner

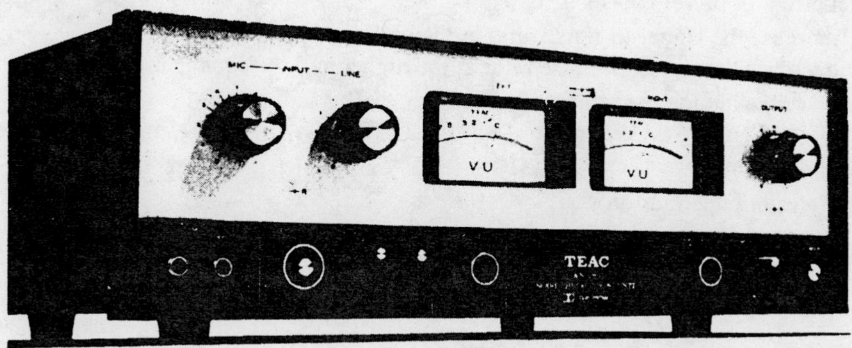

Fig. 16.7 Teac's AN-180 Dolby noise reduction unit. Dolby eliminates unwanted tape hiss and noise.

On *decode* (replay) the circuit operates with the processor in a feedback loop. This time a phase inverted signal is fed to the filter/compressor and the resultant is added to the main chain. Because of the *phase inversion* this addition is in fact a cancellation of the earlier processing and the original signal is retrieved with its dynamics and frequency range in tact, but with the hiss highly attenuated. It should be noted that *the hiss will only be reduced if it is a product of the record/replay process.* Any hiss on incoming program material will be interpreted as music and processed accordingly and faithfully reproduced.

The dbx noise reduction system uses dynamic range compression to encode the signal before recording, and a matched expander to restore original dynamics on playback. The *compression and expansion factor* is 2:1, and is linear in dB over a 110 dB range. A 20 dB change in program is recorded as a 10 dB change on the tape. Thus, a recording medium capable of only 55 dB signal-to-noise ratio-can record and reproduce the full dynamic range of live music, which often exceeds 100 dB. Noise reduction here is of the order of 30 dB.

The Burwen system works with original program material which has not been modified during recording. The fundamental principle is that noise which is quite audible during the absence of recorder material is effectively masked by high frequency material at moderate or loud volume; consequently any circuitry which can de-emphasize high frequencies, when there is no high frequency program material to mask the hiss, will also eliminate the important noise component. Because it is designed to work with material that has not been previously encoded, the Burwen system is of particular significance to owners of old 78 R.P.M. records. Noise reduction here is of the order of 6 to 8 dBm but that is without significant loss of high frequency program material.

OBJECTIVE TYPE QUESTIONS (A)

1. The problem of signal-to-noise ratio is simply solved by using.
- (*a*) wider sections of tape
- (*b*) narrower sections of tape

2. White noise contains frequencies spread
- (*a*) All over the audio range
- (*b*) over a restricted portion of the audio range

3. Noise in lower frequency ranges is
- (*a*) particularly offensive to the human ear
- (*b*) not particularly offensive to the human ear

4. The net effect of pre-emphasis is to
   (*a*) increase the signal-to-noise ratio at the higher audio frequencies
   (*b*) decrease the signal-to-noise ratio at the higher audio frequencies
5. Dynamic range in tape recording is limited by
   (*a*) tape hiss (*b*) tape saturation
   (*c*) (*a*) and (*b*) (*d*) neither (*a*) nor (*b*)
6. Dolby A system operates on
   (*a*) the entire frequency spectrum (*b*) just the highs

FILL IN THE BLANKS (B)

1. In a ________ system both the transmitted and received signals have to be processed by similar circuits.
2. Companders are ________ cum ________ circuits.
3. The expander circuit reverses the process of ________ circuit.
4. Noise reduction circuits seek to reduce noise, particularly in ________ passages of signal.
5. The compression and expansion factor in a dbx noise reduction system is ________ .
6. The Burwen system works with original program material which has not been ________ during recording.

CONVENTIONAL QUESTIONS (C)

1. What is the need for noise reduction systems ?
2. Explain the different sources of noise in tape recording.
3. Give a brief explanation of the different types of noise reduction systems.
4. What is the difference between Dolby *A* and Dolby *B* ?

17

RECORDER/AMPLIFIER CIRCUITS

A 4W tape recorder amplifier and a tape pre-amplifier are described, and the modifications necessary for stereo operation are considered. An automatic gain control circuit for use with tape recorders is also included.

A 6W universal amplifier is described. It incorporates an equalising pre-amplifier, an active tone control circuit and a BEL IC-CA810 in the power output stage. The performance characteristics of the amplifier are also given.

17.1 4W TAPE RECORDER

The 4*W* tape recorder, shown in Fig 17.1, uses silicon transistors in the first three stages to obtain *high sensitivity*. The output stage uses a *complementary matched pair*, transistor *AD* 161 (*n-p-n*) and *AD* 162 (*p-n-p*) which are also used for the *oscillator* during recording. *Equalisation* is provided for tape speeds of 7 1/2 ips (19.1 cm/s), 3 3/4 ips (9.5 cm/s), 1 7/8 ips (4.75 cm/s), and 15/16 ips (2.4 cm/s), which conforms to *CCIR. Standards* where appropriate.

Play back : The *BC* 149 or *BC* 109 has been chosen for the *input stage* because it is particularly suitable for use in audio input stages where *low noise* is essential. Its collector is directly connected to the base of Q_2; its emitter is connected via an *RC* network to the collector of Q_2. This *RC* network provides a path for negative feedback by means of which *equalisation* is effected. The frequency response curves on playback are shown in Fig. 17.2.

A D.C. feedback path is also provided from the emitter of Q_2 to the base of Q_1. The D.C. feedback along this path *stabilises* the working points of Q_1 and Q_2.

The other transistors in the tape recorder form a *conventional amplifier*. The complementary pair of Q_5 and Q_6, is driven by Q_4, which is itself driven by the *high-gain BC* 148 or *BC* 108 transistor. Potentiometer R_{23} is adjusted to give a *quiescent output stage current* of 5 *mA*.

Record : Different input circuits are needed to *match* the different signal sources. For example, the input circuit for a *crystal microphone* contains a shunt capacitor. C_1, and that for a *radio receiver* is a potential divider. A *moving coil micorphone* however can be connected directly because the amplifier has an input resistance of 200*K* Ω. During recording, Q_5 and Q_6 operate as an *oscillator* at 50*k Hz*, the *erase head being used as an oscillator coil.* The capacitive taps, C_{20}, C_{21}, and C_{22}, also provide the necessary conditions for sustained oscillation.

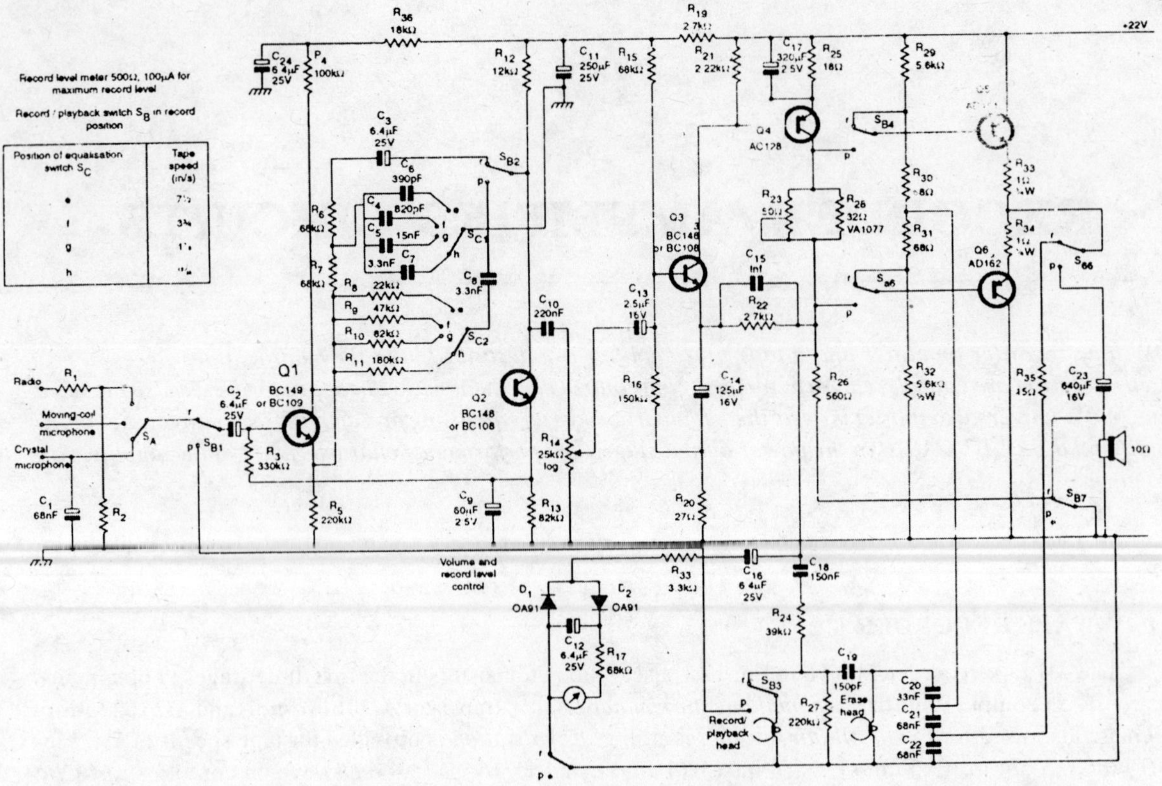

Record level meter 500Ω, 100μA for maximum record level

Record / playback switch S_B in record position

Position of equalisation switch S_C	Tape speed (in/s)
e	7½
f	3¾
g	1⅞
h	15/16

Fig. 17.1 4W Tape Recorder Circuit

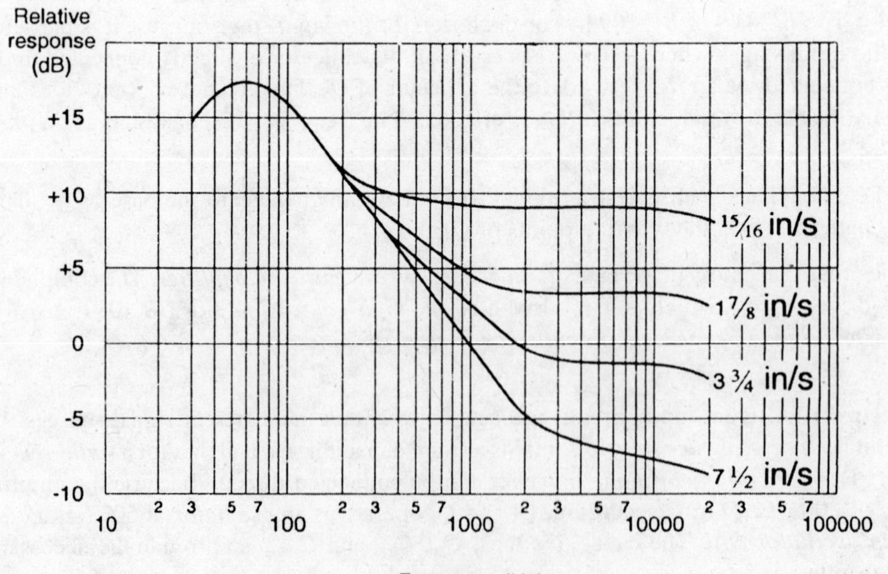

Fig. 17.2. Playback characteristic

The *recording level indicator* gives a reading proportional to the peak-to-peak voltage at the collector of Q_4. The full recording level with a sine wave is 4*V* r.m.s., which produces a current of 95 µ.A through the meter and 110 µA r.m.s. through the recording head.

During recording, the equalisation circuits connected to switch SC are not used; feedback is via the path C_3, R_6 and R_7, which has been designed to give *treble boost.* The frequency response curves on record are shown in Fig. 17.3.

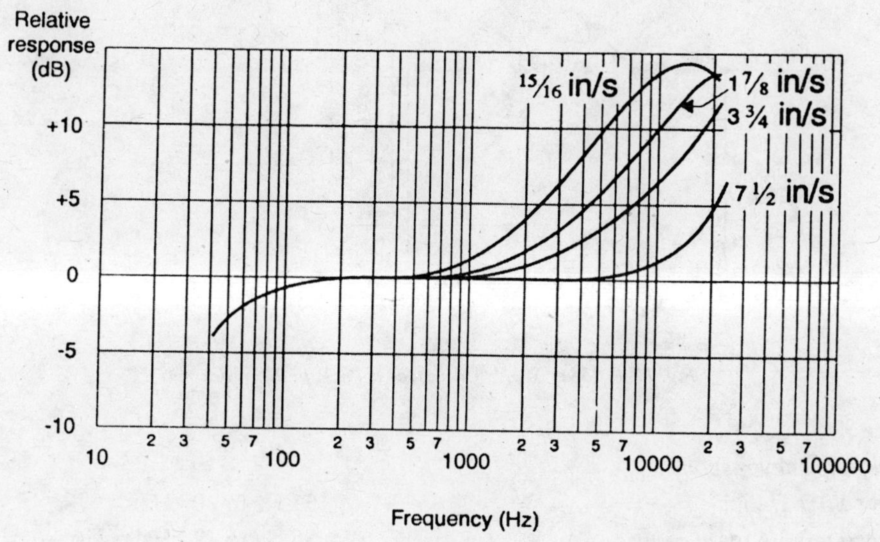

Fig. 17.3. Record Characteristic

The recording current is taken from the collector of Q_4 *Record/Playback heads* with a larger inductance are not suitable because they would require a larger bias voltage. The *recording head* has the following characteristics :

Quarter-track head

Gap	0.0001 in
Width	0.043 in
Spacing	0.136 in
Inductance at 1kHz	70 mH
Bias at 50 kHz	> 11 V
Recording current for 2 dB below 2% distortion	110 µA
Output at 1 kHz using E.M.I. test tape T.B.T.I	0.44 mV
Output at 2 kHz for 110 µA recording current	1.07 mV

Power Supply : The circuit for a simple power supply suitable for use with the above tape recorder is shown in Fig. 17.4. It contains a mains transformer with a secondary output of 17*V*. This is rectified by means of bridge rectifier type *BY* 164, and smoothed by means of a 640 μ*F* capacitor.

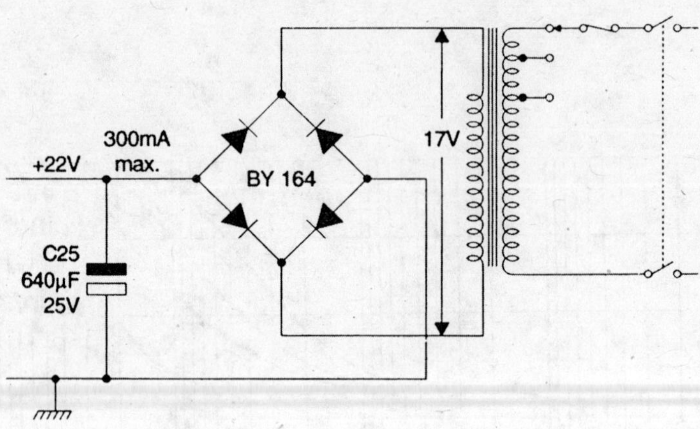

Fig. 17.4 Power Supply Circuit for a 4W Tape-Recorder.

• **Performance :**

Overall frequency response	
Tape speed = 7 1/2 ips	55 Hz to 20 kHz
Tape speed = 3 3/4 ips	55 Hz to 12 kHz
Tape speed = 1 7/8 ips	55 Hz to 6 kHz
Playback	
Output power	4 W R.M.S.
Sensitivity at base of Q_1, at 1 kHz and with tape speed of 7 1/2 ips	400 μV
Distortion for 4 W output	1.2%
Noise relative to 4 W, with volume control set so that a signal produced with 110 μA recording current gives 4 W output; 3 dB band width = 30 Hz to 15 kHz	
Tape speed = 7 1/2 ips	-60 dB
Tape speed = 3 3/4 ips	-57 dB
Tape speed = 1 7/8 ips	-55 dB
Tape speed = 15/16 ips	-50 dB
Record	
Sensitivity at base of Q_{11}, for 110 μA recording current, with volume control at maximum setting	100 μV
Distortion at full record level	1%
Recording current level for 4 W output, with volume control at maximum setting	40 μA

17.2 TAPE PRE-AMPLIFIER

Circuit Description : Fig. 17.5 shows a pre-amplifier circuit derived from the tape recorder circuitry already described. The oscillator stage in this circuit uses a complementary pair comprising one *AC* 176 and one *AC* 128 transistor.

Performance : The distortion measured at the collector of transistor Q_{4}, (*AC* 128) in the *playback position* is shown in Fig. 17.6. The measurements were made at a speed of 3 3/4 ips, although these characteristics vary little with speed. At a speed of 7 1/2 ips with the volume control at maximum, the output voltage is 6.55V for an input voltage of 400 μV. The total current drain on playback is 25 mA.

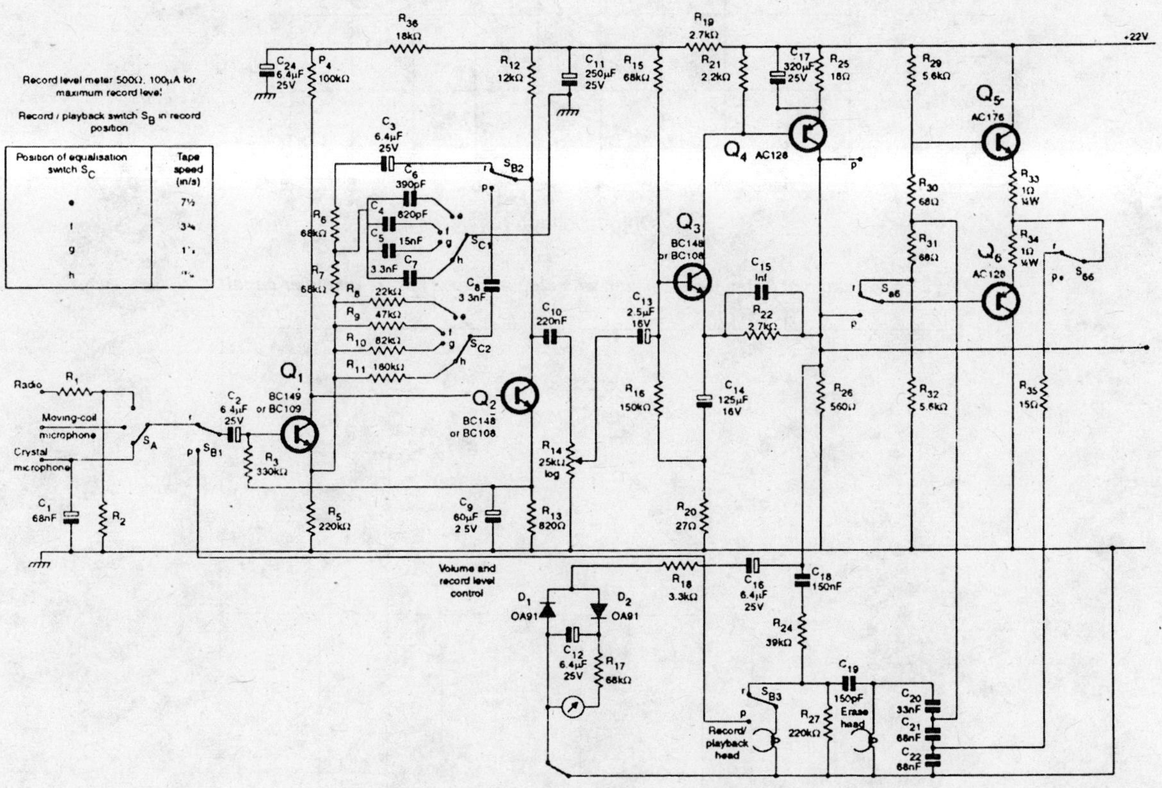

Fig. 17.5 Tape Pre-amplifier Circuit.

17.3 MODIFICATIONS FOR STEREO OPERATION

In a *stereo recording system,* the bias and erase frequencies of the two channels must be equal, if the production of *audio beat frequencies* is to be prevented. Therefore, the bias and erase power for the two channels must be derived from one common oscillator or from two *oscillators which have their frequencies locked together.* The arrangement shown is Fig. 17.7 can be used for the pre-amplifier of Fig. 17.5 only if the *AC* 176 / *AC* 128 pair is replaced by an *AD* 161 / *AD* 162 pair and the emitter resistors reduced accordingly.

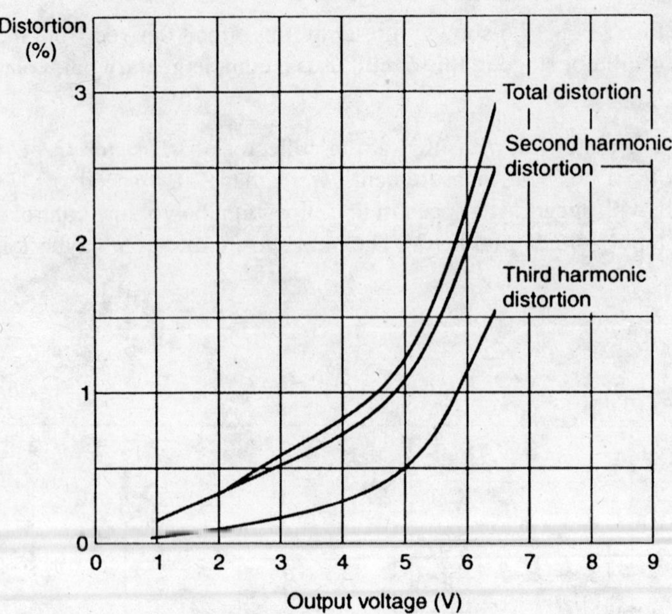

Fig 17.6 Variation of harmonic distortion with output voltage for pre-amplifier circuit.

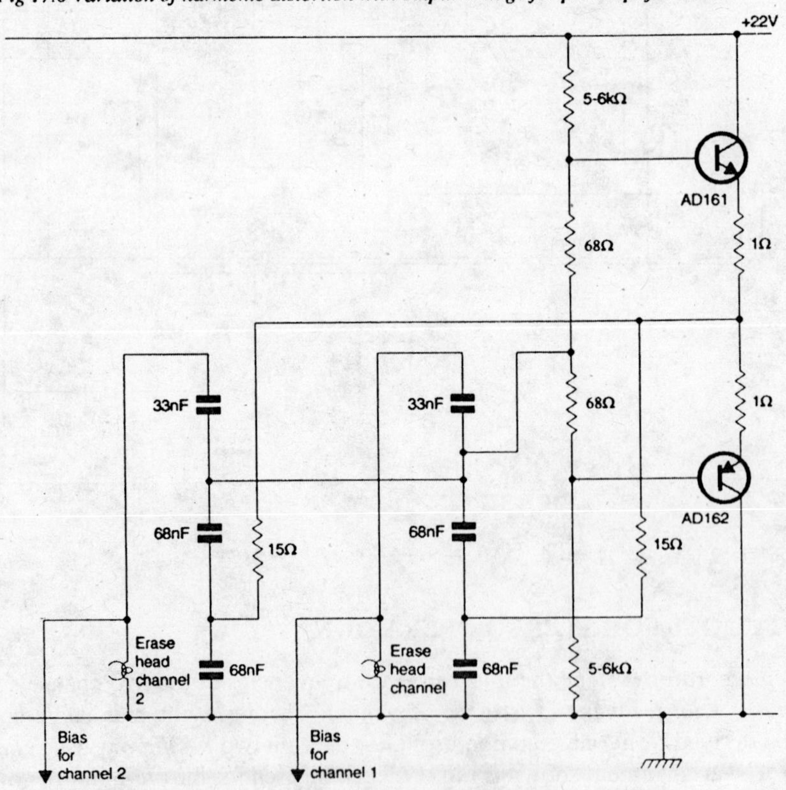

Fig 17.7 Bias and erase power from common oscillator for stereo operation.

In the alternate arrangement in Fig 17.8 two *AC* 176 / *AC* 128 pairs will operate satisfactorily.

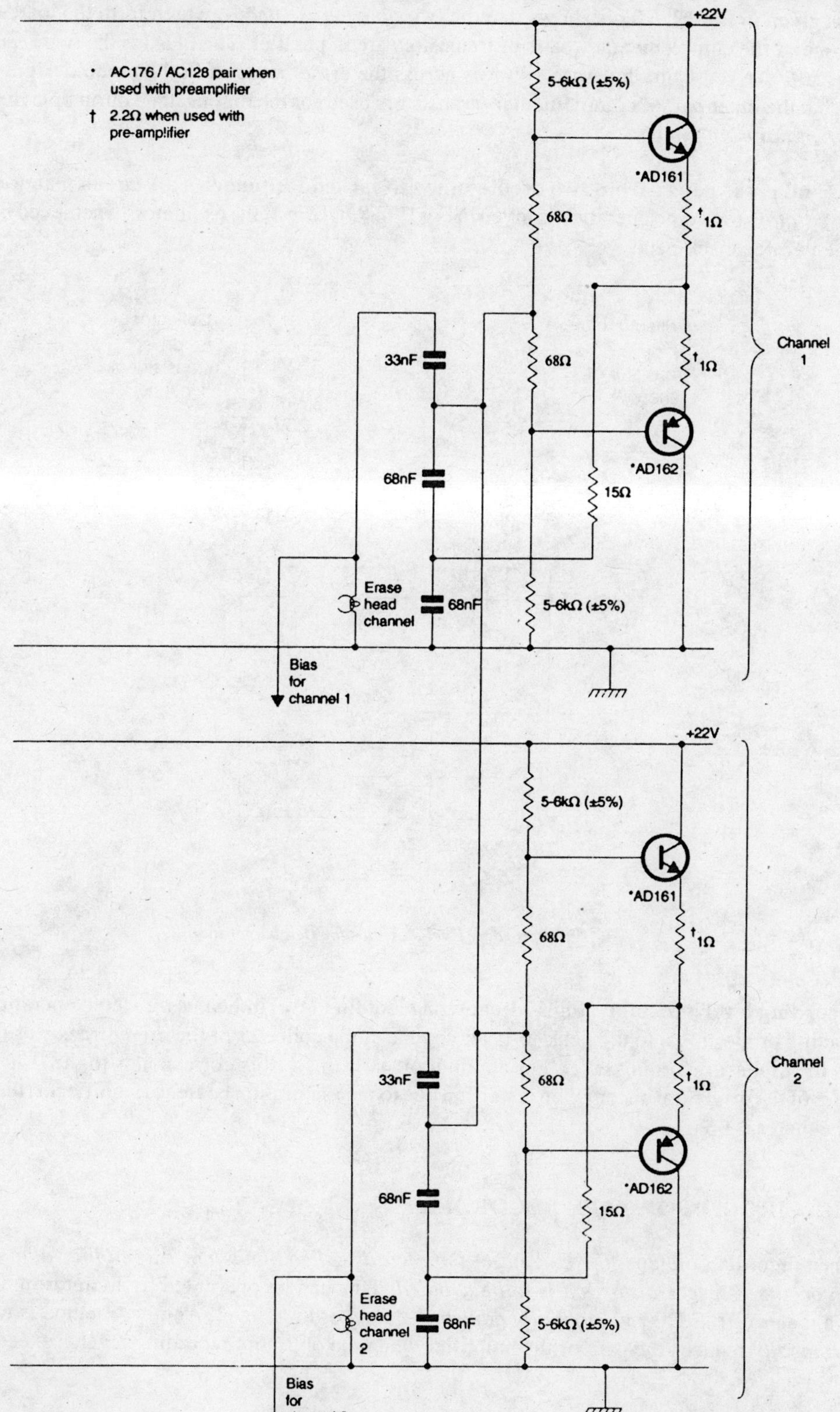

Fig 17.8 Bias and erase power from two oscillators with locked frequencies for stereo operation.

Bias and erase power from a common oscillator : A circuit that uses a *common oscillator* for stereo operation is given in Fig 17.7. This shows that the two *tuned erase heads,* which form the load and feedback circuits of one of the complementary pairs of transistors are in parallel. The bias for the two *recording heads* however, is still derived from the erase voltages across the erase heads as in the mono design. This has the advantage that the other pair of complementary transistors used for the output stage during playback may now be used as a *monitoring facility.*

Bias and erase power from two oscillators with locked frequencies : A circuit that uses *two locked identical oscillators* for stereo operation is given in Fig 17.8. *Locking* of the oscillators is achieved by connecting together the two feedback paths.

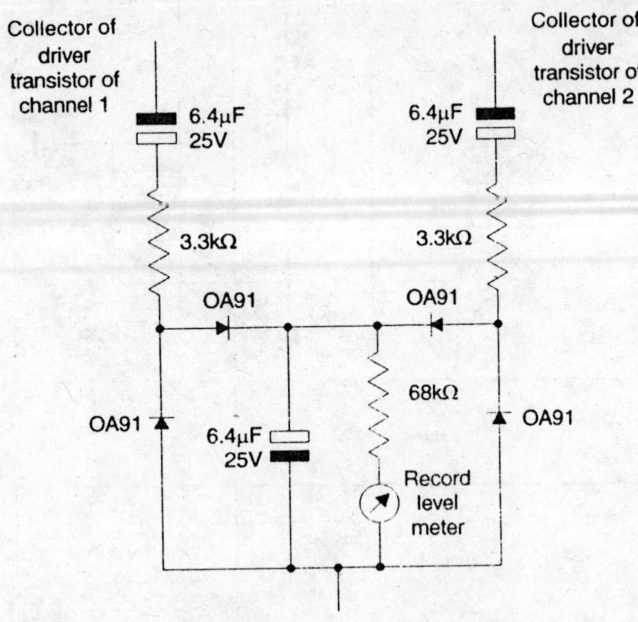

Fig 17.9 Recording level indicator for stereo operation.

Recording level indicator : Fig 17.9 shows a recording level indicator for stereo operation. The meter gives a reading proportional to the peak-to-peak voltage at the collector of the driver transistor in the channel which has the highest recording level. For an input of 4V r.m.s. (this corresponds to the r.m.s. voltage at the collector of the driver transistor when distortion due to tape saturation becomes significant) the D.C. current through the meter is 95 μA

17.4 A.G.C. CIRCUIT FOR TAPE RECORDERS

When recording on tape, *the current in the recording head must not exceed the value at which tape saturation occurs if severe distortion is to be avoided.* This can be prevented by monitoring the current by means of a meter or L.E.D., and *manually adjusting* the recording level. Another method is to use a circuit that *electronically* reduces the gain of the amplifier when the optimum recording level is exceeded.

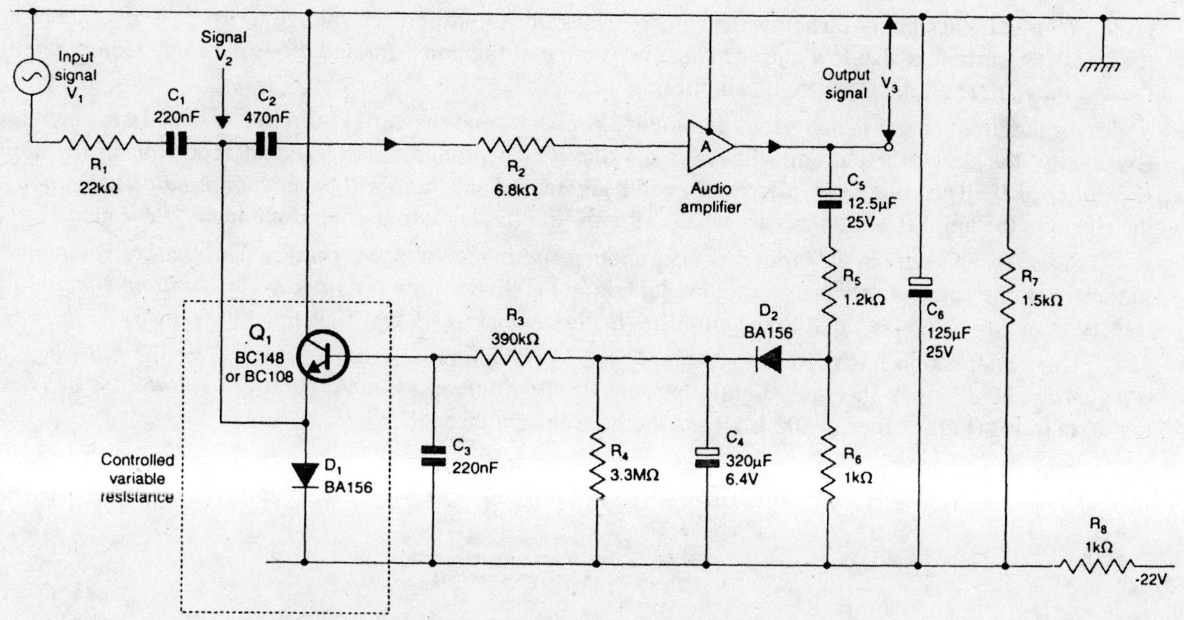

Fig 17.10 Automatic Gain Control Circuit

To ensure that the distortion is recorded for only a short time, the gain control must respond very rapidly to the overload signal. When the gain has been reduced, however, the *recovery time* must be long, for otherwise *volume compression* will result, and loud and soft passages of music will be recorded with equal loudness.

A gain control circuit that uses a minimum of components and can produce an attenuation of 40 dB within 150 milliseconds is shown in Fig 17.10. The recovery rate does not exceed 6 dB per minute. *With a faster recovery rate, the same circuit arrangement is suitable as a volume compressor in other applications.*

Circuit Description

Gain control is achieved by means of an attenuator containing a diode. *Controlling the direct current through the diode controls its slope resistance and consequently, the amount of attenuation.*

The severe harmonic distortion caused by the exponential characteristic of one diode can be reduced by using two diodes in a *reverse-parallel arrangement.* In the circuit in Fig 17.10, the base emitter junction of a transistor is used as one of the diodes. Because the transistor is needed as a D.C. amplifier, this arrangement saves the expense of one diode.

Part of the amplifier output is rectified and used to provide the *gain control signal.* The time constants of the rectifier circuit are such as to give the *high rate of attenuation* and the *long recovery time.*

In Fig 17.10 unit A is the *audio amplifier* and R_2 represents its input resistance, which should be about 7 KΩ.

The *variable attenuator* comprises R_1, C_1, D_1 and part of transistor Q_1. Because C_3 is a short circuit to A.C. signals, the base-emitter junction of the transistor is, in effect, in reverse-parallel with D_1 for all A.C. signals.

Resistors R_5 and R_6 form a *fixed attenuator* by means of which a constant fraction of the output is fed back to the *control circuit*. The feedback voltage is rectified by D_2, changes C_4, and is applied through R_3 to the base of Q_1. Hence, the voltage across C_4 controls the current through TR_1 and D_1.

With an input signal well below that which causes tape saturation, the voltage across C_4 is low. Therefore the transistor current is also low and the *slope resistance* of the transistor and diode are high. Consequently, the *shunting effect* on the input to the amplifier is negligible.

As the input signal is *increased* the voltage across C_4 and the current through Q_{11} will also increase. Eventually, the transistor current will reach a value which produces a significant reduction in the *slope resistance* of D_1. The voltage drop across R_1 will be increased and there will be an *appreciable attenuation* of the signal to the amplifier. This occurs when the amplifier has an output of approximately 2V r.m.s.

A *further increase* in the input produces another rise in the voltage across C_4. The consequent increase in transistor current now produces a rapid reduction in the diode *slope resistance*. The *shunting effect* of the transistor and diode on the input to the amplifier is greatly increased, and V_2 is much less than V_1.

Once attenuation has started, voltages V_2 and V_3 remain practically constant. Over the *full range of attenuation*, 40 dB, they increase slightly because of slight increases in V_{BE}, the forward voltage of D_1, at higher currents (of the order of 300 μA), and the base current through R_3.

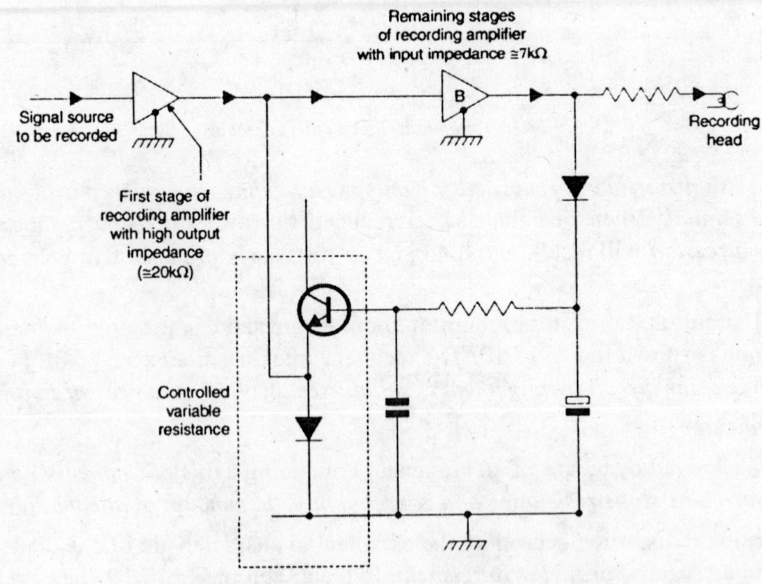

Fig. 17.11 Incorporation of gain control circuit in a tape amplifier.

The *attack time* is largely determined by the time constant of C_4 and the source impedance of the charging circuit, which is R_5 in parallel with R_6. For an attack time of about 200 ms, the source impedance should be 600 Ω.

The *recovery time* is determined by the *time constant* of C_4 and its discharge circuit - R_4 in parallel with R_3 and the *slope resistance* of TR_1 and D_1. Over a wide range of attenuation, R_3 is much larger than the slope resistances of the transistor and the diode. The *rate of discharge* of C_4 is then controlled by the value of R_3, and consequently, the *rate of change of attenuation* is constant. At very low transistor currents, the slope resistances of the transistor and the diode greatly exceed R_3, and the rate of discharge of C_4 via the transistor is very low. This tends to make the rate of change of attenuation also very low. Under these conditions, however, R_4 provides an *alternative discharge path.* Hence, the rate of change of attenuation has a low limit that is determined by the value of R_4. When R_3 is large with respect to the slope resistances of the transistor and the diode, the effect of R_4 is negligible.

The transistor is a *BC* 148 or *BC* 108 *n-p-n* planar type, which, with its *low leakage current* and high gain, makes this simple arrangement possible. The silicon diodes also are used because of their low leakage currents.

In Fig. 17.10, *the gain is controlled by attenuating the signal before it reaches the audio amplifier.* This has the disadvantage that, when the signal is attenuated, the noise originating in the first transistor of the amplifier is unaffected. *A better arrangement would be to have the attenuation after the first stage of the amplifier thus reducing noise and signal simultaneously. The arrangement is shown in Fig 17.11. A high voltage is required to operate the attenuating circuit; this will probably be available only at the output of the stage feeding the recording head.*

Use of Electrolytic Capacitors in Timing Circuits : The *maximum leakage currents* quoted in published data on electrolytic capacitors are too high for use in *long time constant* circuits. Experience suggests, that the leakage currents of capacitors *taken off the shelf* will be low enough or will soon fall to a suitably low value. Therefore, electrolytic capacitors of a suitable type are readily available. At the worst, a tape recorder which has been *stored for a very long time* will have an initial recovery time slightly shorter than normal, but this will rapidly approach the normal value.

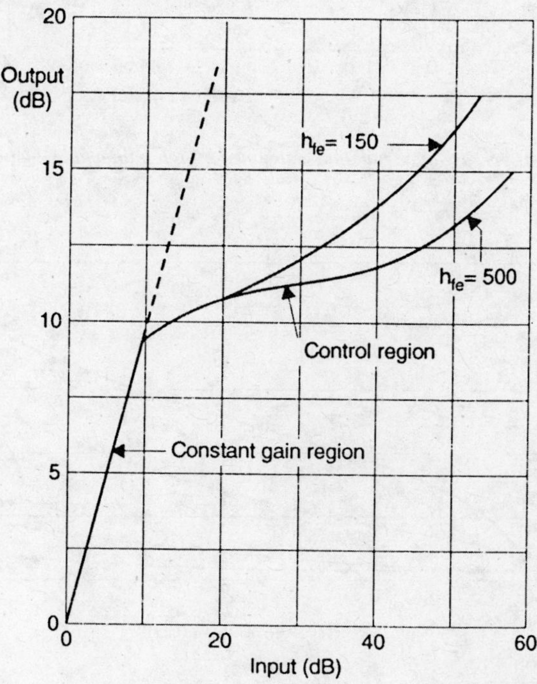

Fig. 17.12 Variation of output with input for high-gain and low-gain transistors.

Performance : When the circuit controls the gain, the signal V_2 is not maintained exactly constant nor completely independent of the amount of attenuation. This is shown in Fig. 17.12 by the input/output characteristics of the circuit with high-gain and low-gain transistors. When a *low-gain transistor* is used an attenuation of 40 dB is accompanied by an increase of 7 dB in the output with a *high-gain transistor,* the corresponding increase in output is 4-5 dB

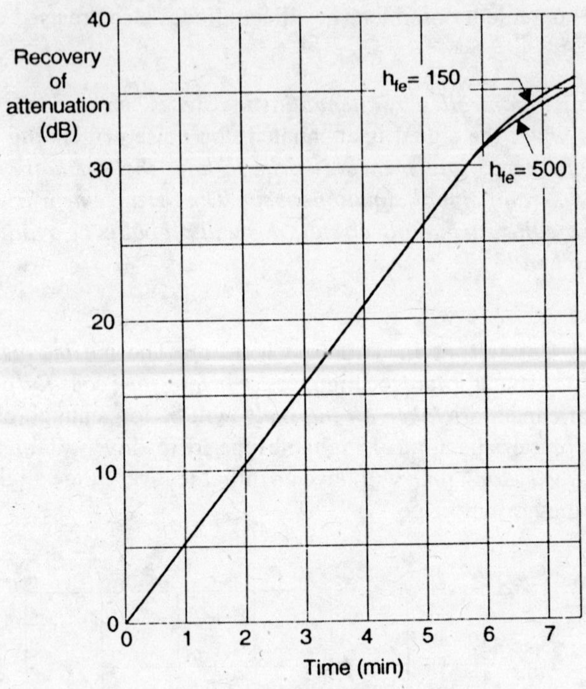

Fig 17.13 Decrease of attenuation with time for high-gain and low-gain transistors.

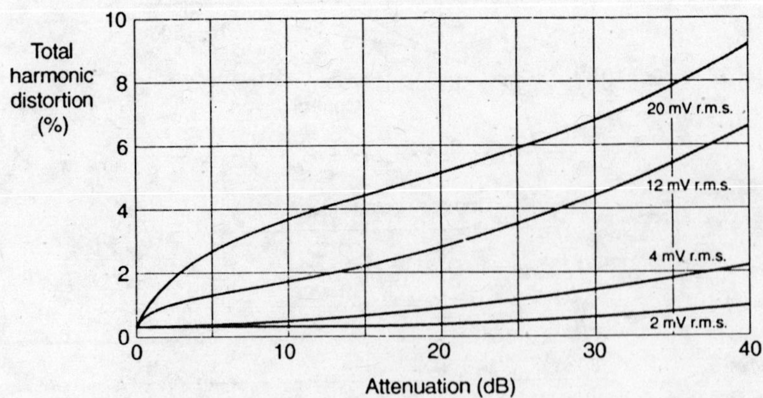

Fig 17.14 Variation of Total Harmonic Distortion of V_2 with amount of attenuation for various values of V_2

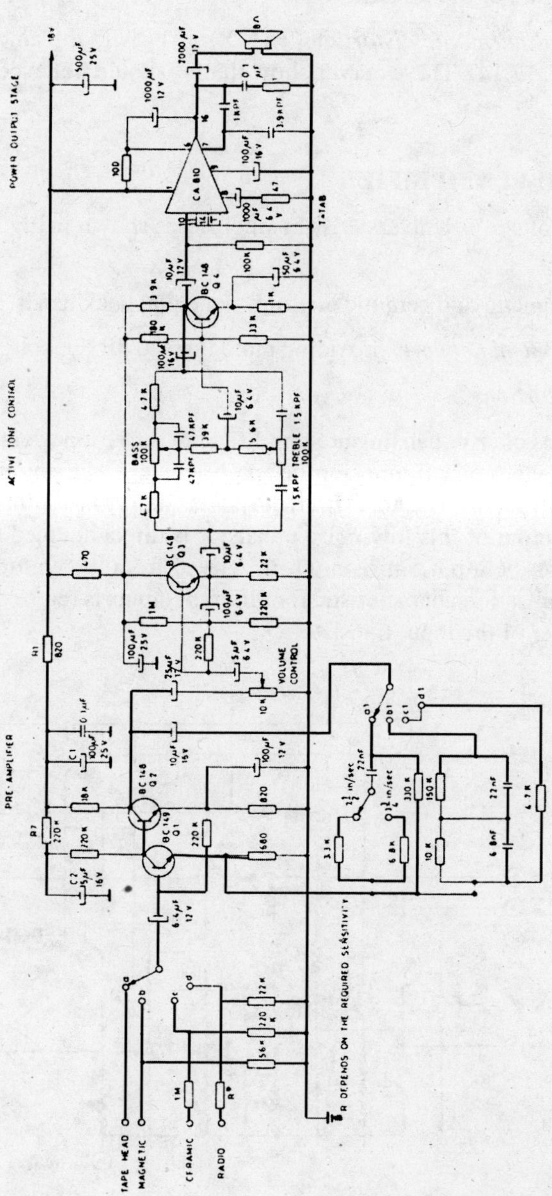

Fig. 17.15. 6 W Universal Hi-Fi Amplifier using BEL IC-CA 810.

The change of attenuation with time is shown in Fig 17.13. This diagram shows that the rate of change of attenuation is substantially constant over much of the range. It does not exceed 6 dB per minute and is almost independent of the gain of the transistor.

The variation of *total harmonic distortion* (T.H.D.) of V_2 with the amount of attenuation for various values of V_2 is given in Fig. 17.14. These curves show that V_2 should not exceed 4 mV r.m.s. if the T.H.D. is not to exceed 2%.

17.5 6 W UNIVERSAL HI-FI AMPLIFIER

The circuit diagram of a 6 W Universal Hi-Fi amplifier is shown in Fig. 17.15. The *salient features* of this circuit are :

1. *Inputs* for magnetic and ceramic pick-ups, tape playback heads and radio receivers.
2. *Active tone control network,* providing more than 20 dB dynamic range for both bass and treble.
3. Very low *distortion.*

The amplifier consists of an equalising pre-amplifier, an active tone control circuit and a CA 810 power output stage. To achieve better noise figures for the amplifier stage, the pre-amplifier uses a low-noise n.p.n transistor BCM 149(Q_1) at its input. The noise from the power supply is minimised by the decoupling network R_1 - C_1 and R_2 - C_2. The output of this low-noise transistor is direct coupled to another n.p.n. transistor BC 148 (Q_2). The base bias for the input transistor Q_1 is derived from the emitter of Q_2 to achieve better D.C. stability. The required equalisation characteristic for different inputs is realised by appropriate feedback from collector of Q_2 to the emitter of the input transistor Q_1.

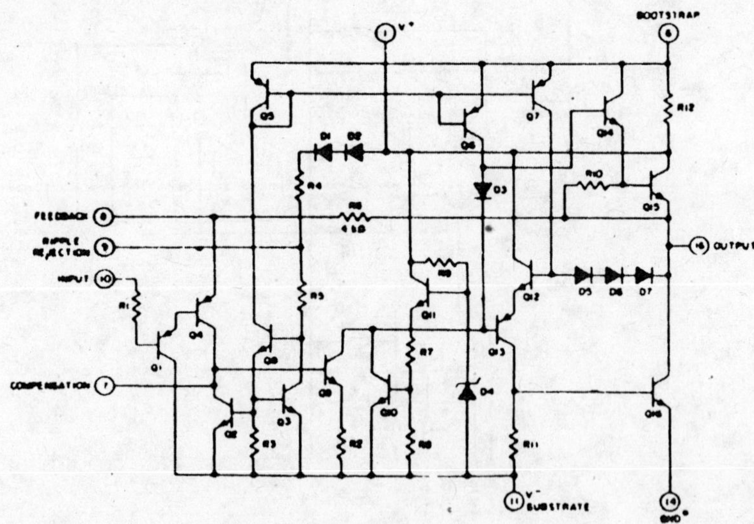

Fig. 17.16 Schematic diagram of IC-CA 810.

The output of the pre-amplifier is connected to the tone control circuit through a buffer transistor Q_3. The active tone control circuit uses an amplifier, Q_4, and a frequency dependent network to obtain the required tone control characteristics. The power output stage makes use of IC-CA 810 and delivers 6 W power into an 8 Ω load. Typical performance characteristics of this amplifier are given below.

Performance Characteristics

1.	Supply voltage	18 V
2.	Minimum output power	6 W
3.	Input sensitivity for $P_0 = 50$ mW	
	Radio	4 mV / 100 kΩ
	Ceramic pick-up	15.8 mV / 1 MΩ
	Magnetic pick-up	0.7 mV / 50 kΩ
	Tape input	0.8 mV / 20 kΩ
4.	Input sensitivity for maximum power output	
	Radio	35 mV
	Ceramic pick-up	185 mV
	Magnetic pick-up	6.1 mV
	Tape input	8.8 mV
5.	Zero signal current	12 mA
6.	Supply current at maximum output power	430 mA

18

ELECTRONIC MUSIC SYNTHESIZERS

While a great many people have recently become interested in electronic music, few of them have any knowledge of just what goes into generating the sounds of this new music. In view of the fact that the first Moog synthesizers were made available only in 1964, this is not surprising. While electronic musical instruments consist of basically simple circuits, or building blocks, the instrument itself is usually a complex, sophisticated device.

Electronic musical instruments differ markedly from conventional or traditional musical instruments. Instead of reeds over which streams of air must be blown, strings which must be bowed or strummed, and diaphragms that must be tapped or beaten to produce sounds, electronic instruments use electrons to generate the signals that make up the sounds and speakers for sound propagation. In appearance an electronic musical instrument resembles a computer more than a 'musical instrument'. They are often as complex as a computer to operate as well, demanding of the composer and or player an expertise beyond the knowledge of musicology.

Pioneers in the field of "new" music recognize that it has a deep spiritual basis. Electronic music is not simply another kind of music -like pop, country, etc; it is a different kind -emphasizing pure sound in addition to themes, randomness with imposed order and looseness as a complement to restricted thought. The fact that no one has yet devised a universal way to score this music is significant.

Even more significant is that electronic music encompasses many disciplines by combining different aspects of science, music ,and philosophy-even theater. People who look down on "music by machines" fail to realize that electronic music was not designed to replace humans; rather, it brings a powerful new tool to the hands of the musician.

18.1 TYPICAL GENERATORS

Periodic or *cyclic waveforms* -sine, square, sawtooth, and triangle waves-form the *backbone* of electronic music generators (see Fig. 18.1) Most instruments (synthesizers) contain as many as several dozen oscillators, each capable of simultaneously providing some, and frequently all, of the basic waveforms. Each waveform has a *distinctive sound quality* all its own, an important feature in electronic music. It is necessary to have at least a basic understanding of *overtones* (harmonics) to understand why the waveforms have different characteristic sounds.

The sine wave is harmonically the least complex of waveforms. It contains no overtones. This means that a 200 Hz sine wave concentrates all of its energy at 200Hz and no other frequency. A 200 Hz *triangle wave*, on the other hand, consists of the algebraic sum of many sine waves of different *numerically related frequencies and amplitudes*. The *major portion* of a triangle wave's energy is concentrated in the 200 Hz fundamental frequency, But since a triangle wave contains *a number of odd harmonics*, energy will also be channeled into the odd harmonic frequencies. (600, 1000, 1400Hz, etc.) *The higher the frequency, or harmonic, the smaller the amount of energy it receives.* The third harmonic receives 11 percent the energy contained in the fundamental, the fifth harmonic receives 4 percent, the seventh harmonic receives 2 percent, and the ninth harmonic receives only 1 percent.

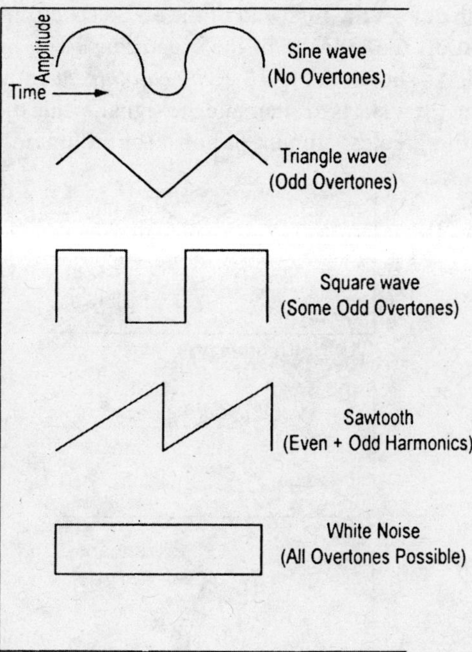

Fig. 18.1 These are the basic waveforms involved in the synthesis of music.

Square waves contain *some odd harmonics,* the content depending on whether the wave is perfectly square or rectangular. *Sawtooth waves* contain *strong even and odd harmonics.*

Waveforms with generous amounts of harmonics are easy to modify. By cutting out certain overtones, such as all harmonics above an arbitrary frequency, the essential nature and sound of the wave changes. Different waveforms can also be used to give *emotional shading* to a piece of music.

A sequence of notes consisting of sine waves will yield a different emotional response than would the same sequence of square wave notes.

The white-noise generator is a type of special-purpose audio generator used in electronic music. In theory, it generates a wave that contains all possible frequencies with all possible relationships at the same time. It sounds like falling rain or escaping steam.

18.2 BASIC MODIFIERS

Just as there are circuits that *generate* sounds, there are also circuits that *modify* or "process" those sounds. Two of the most basic processors in common use are the amplifier and the filter. The *amplifier* is extremely valuable in electronic music because it controls the dynamic aspect- crescendo and diminuendo-of a sound. The *filter* is a somewhat more complex circuit. It is an amplifier that controllably restricts the frequency response in one of three ways.

As a *low-pass* filter, it can cut off the high end of the audio spectrum. The *high-pass* configuration can be used when it is desired to cut off the low end of the spectrum. When both ends of the spectrum are to be cut off, a *bandpass* filter is used. As shown in Fig.18.2, *the cutoff frequency* of a low-pass filter or a high-pass filter is the frequency at which the filter starts to attenuate the signal, while the *resonant frequency* of a bandpass filter is the frequency at which the greatest amount of boost (or minimum amount of cut) is present.

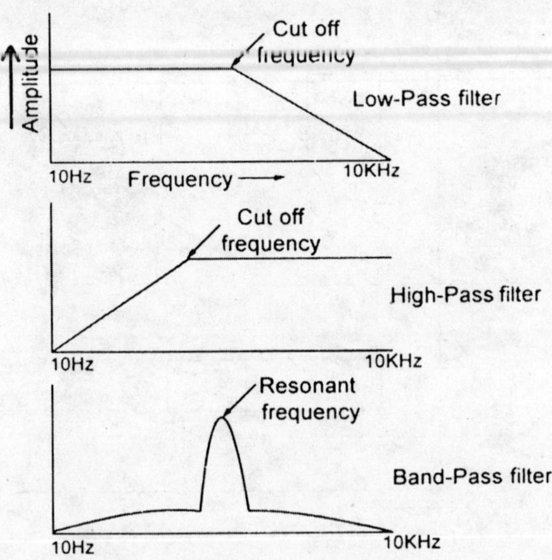

Fig. 18.2 Frequency characteristics of three basic types of filter circuits.

Because the timbre of a sound depends largely on its harmonic content and since the desired harmonic content lies above the fundamental frequency of a signal, *variable-cutoff low-pass filters that selectively attenuate certain overtones are used for modifying the timbre of various waveforms.*

18.3 VOLTAGE CONTROL

With the oscillator, amplifier, and filter, a sound can be *specified* according to waveform, pitch, timbre, and dynamic level. But without voltage control, a system consisting of these three building blocks has serious limitations, *By using voltage control, the major parameters of the building blocks are made to respond to a DC or an AC voltage.*

A voltage-controlled oscillator (VCO) has its pitch controlled not by turning a potentiometer, but by applying to it a control voltage. The cutoff frequency of a filter (VCF) or the gain of an amplifier (VCA) can be similarly controlled by a variable DC voltage.

For example, the amplifier shown in Fig.18.3 has a conventional audio input but instead of using a potentiometer level control, *a voltage is applied to the input to control the gain.* A 4-volt potential is sufficient to cut off Q1 and reduce the gain to zero. By slowly making the control voltage less negative, the gain will increase until the amplifier is operating at full gain at 0 volt. (In the oscillators used in Moog and ARP synthesizers, *a 1-volt change in control voltage causes a one-octave jump in frequency*; so, by applying stepped increments of 1/12 of a volt, the 12 tones of a conventional musical scale can be generated.)

Because it requires much technique to precisely specify frequency, amplifier gain, etc, simply by having the musician turn a potentiometer, most electronic music composers rely on voltage control to perform complex pitch, amplitude, and timbre changes. As a result, *specialized circuits designed solely to generate control voltages have come into being.* For example, a sub-audio triangle wave of, say, 0.5Hz makes an excellent control voltage. For one second, the signal rises to peak amplitude (2 volts in Fig. 18.4). Then for another second it tapers off to zero amplitude. Applied to a voltage-controlled amplifier, this signal would be heard at the output as a linear increase and decrease of equal duration in each direction.

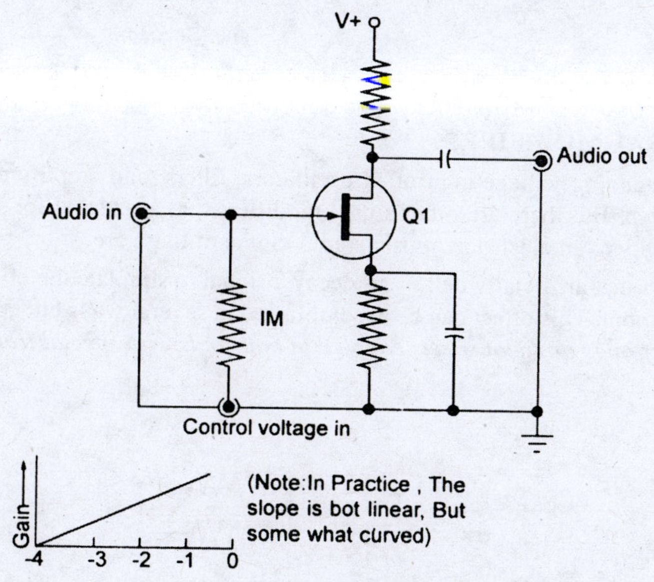

Fig. 18.3 Circuit of an amplifier with gain controlled by variable voltage.

The common *tremolo* circuit found in guitar amplifiers and organs is simply a sine wave-controlled VCA. *Vibrato* is generally a VCO that is controlled by a 6-Hz or so sine wave. Of course any waveform can be used to control a voltage-controlled system, but the problem with using only oscillators to generate the control voltages is that sometimes a *non-repeating waveform* is desired. This problem is overcome with the aid of an *envelope generator,* a device that can produce a control voltage with specified rise, on, and fall times.

18.4 ENVELOPE GENERATOR

The *timbre gate,* when broken down into a block diagram, can be illustrated as shown in Fig. 18.5. It consists of a voltage-controlled amplifier with an audio input and two audio outputs. An envelope generator with adjustable rise, on, and fall times generates the control voltage for the VCA.

The envelope generator can be triggered *manually* or *automatically* to generate an envelope. The automatic triggering unit furnishes pulses that initiate the envelope. (Sometimes this type of pulse control system is referred to as a clock or a timer circuit.) *Because of this timed feature, periodic control voltages of an unusual shape can be generated.*

Although there are many types of control voltage generators, one of the most popular is the *keyboard. It furnishes a string of control voltages to a control input with the various voltages selected by push button or key switches* (see Fig. 18.6). In most commercial synthesizer keyboards, a pulse output is also available whenever a key is closed.

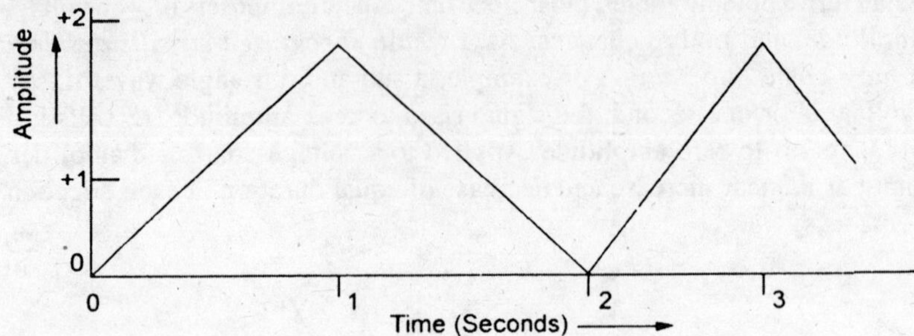

Fig. 18.4 Waveform of a triangle wave used to control output of amplifier

18.5 OTHER SIGNAL MODIFIERS

A system containing voltage-controlled oscillators, filters, and amplifiers represents *a sophisticated electronic music setup.* But there are other signal modifiers that are useful for specific functions and add a great deal of versatility to an electronic music studio. Some of these are:

Reverberators that artificially delay and decay a signal to simulate the effect of a sound bouncing off the walls of a large room. This effect can be accomplished in several ways but *generally it is done by feeding the signal through springs or metal sheets coupled at both ends with special transducers.*

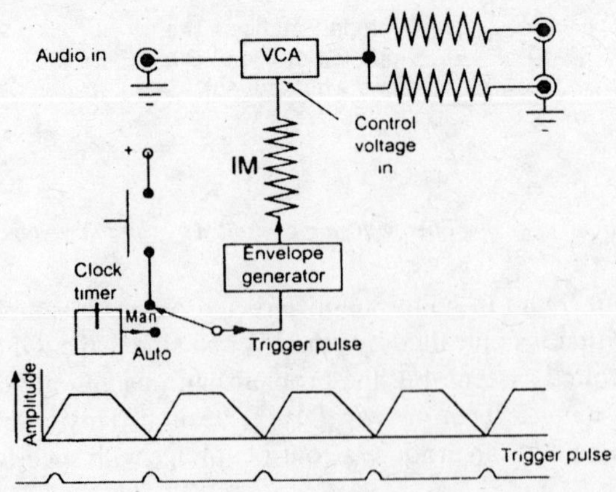

Fig. 18.5 Block diagram of a voltage controlled amplifier with two outputs.

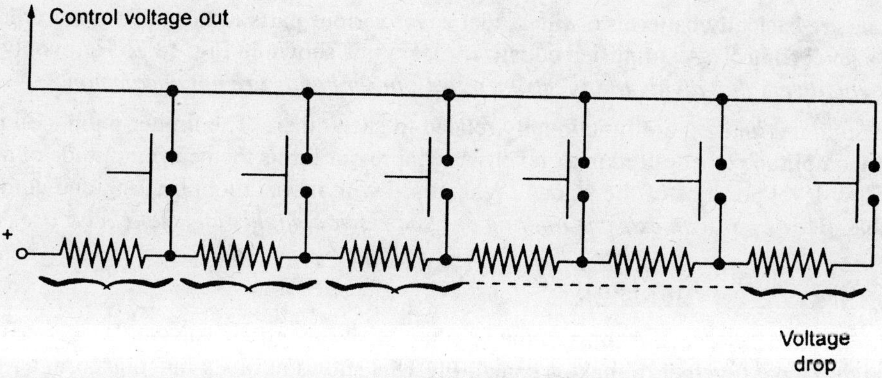

Fig. 18.6 Typical switch arrangement on a push button keyboard control.

Tape echo, a very precisely controlled delay system, *bestows a sense of spaciousness, delay, or repetition upon electronic music.* It also permits a musician to "accompany" himself.

Ring modulators have two inputs, one for an audio signal and another for a modulating (or carrier) signal. *At the output of the modulator appears the algebraic sum of the input frequencies, but both original signals are suppressed.*

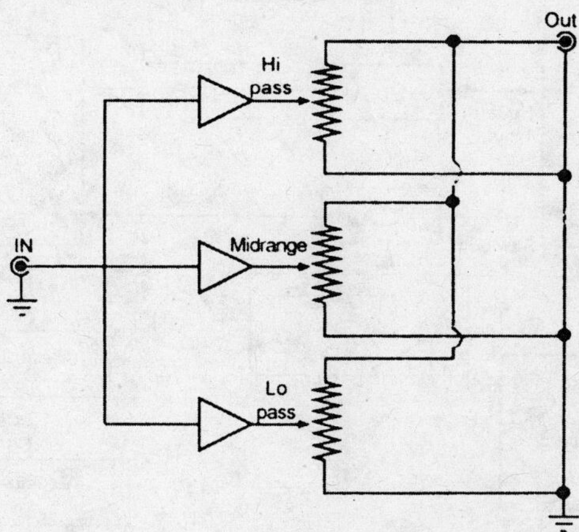

Fig. 18.7 Simplified equalizer diagram has 3 filters; 12 or 24 may be used.

Fuzztones and *waa-waa's* are also used as electronic modifiers. *The fuzztone generates harmonics from a signal with otherwise low harmonic content. The waa-waa is a variable-bandpass filter, its resonant frequency controlled by a footpedal.*

Mixers are not processors, but they are often used to adjust two or more signal levels to suitable proportions. They are also useful for adding together control voltages when control by more than one waveform is desired. *Mixers can be sophisticated multi-input devices containing echo buses, switchable reverberation, filtering arrangements, and other convenience features.*

Equalizers are basically batteries of filters that cover various parts of the audio spectrum, each output terminating in a level control. A simplified equalizer diagram is shown in Fig. 18.7. Here, only three filters are shown, *but equalizers that divide the spectrum into 12 or 24 bands are not uncommon.*

Compressors and *limiters* are functionally related to each other. The limiter's job is to maintain the maximum signal amplitude to a desired peak level. A compressor limits the peak amplitude of a signal but it also raises the low-level portions of the signal. When used with moderation, limiting and compression are hardly noticeable. However, *with extreme limiting or compression, highly unusual effects are obtained.*

18.6 ASSEMBLING A SYNTHESIZER

A modern synthesizer contains most of the building blocks mentioned above. To give an idea of how these blocks might be put together to make a sound, Fig 18.8 shows in block diagram form a representative method of interconnection (traditionally with the aid of *patchcords*). The heart of the system is the VCO, its frequency determined primarily by the control voltage from the keyboard. By properly tuning the keyboard, it is possible to generate a *chromatic scale.* In addition, a 6-Hz sub-audio sine-wave control voltage vibrato-modulates the primary frequency.

The master VCO has both square and sawtooth outputs. An envelope generator delivers a control voltage to a VCA and a VCF. This combination modifies the square wave. The sawtooth wave is processed by an envelope generator / VCA combination. Both waveforms, after being modified, are mixed down to one output, given some reverberation, and equalized to emphasize or cut any specific frequency components.

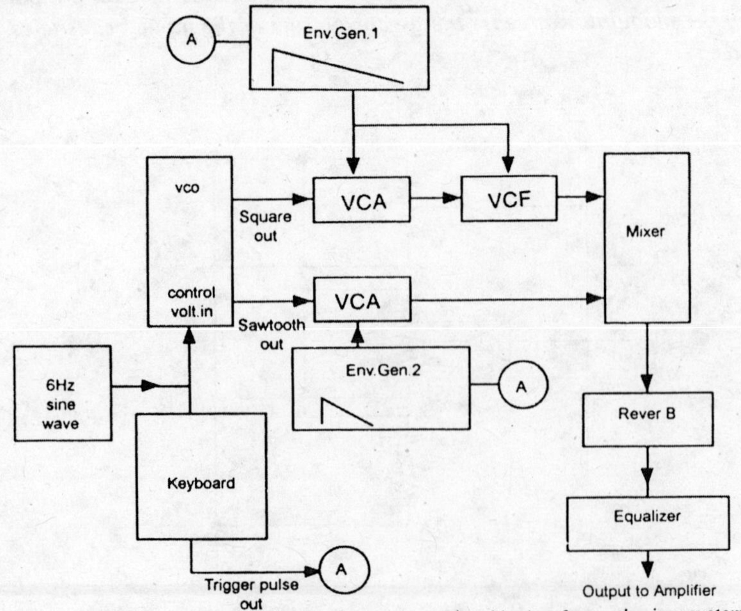

Fig. 18.8 One typical way of interconnecting blocks of a synthesizer system.

Striking a key initiates several events. First, a control voltage is applied to the VCO to generate a tone of specific pitch. Simultaneously, a pulse triggers the two envelope generators. Envelope generator 1 is set for a long decay, while generator 2 is set for a short decay. By controlling a VCF, envelope generator 1 also changes the harmonic content of the square wave so that, as the signal decays, overtones are removed. The overall effect is a note of a certain pitch, modulated by vibrato, whose attack is mainly a burst of sawtooth wave followed by a slowly decaying square wave, the timbre of which changes as it decays.

Changing any part of the system yields a radically different sound. Any subtle changes can give dramatically different results. This is perhaps why so many people refer to electronic music as a medium of infinite potential. *There are so many building blocks and their possible combinations so numerous that the problem for many composers is not what sounds to make, but what sounds not to make.*

OBJECTIVE TYPE QUESTIONS (A)

1. Sine wave is harmonically
   (*a*) the least complex of waveforms (*b*) the most complex of waveforms
2. A triangle wave contains a number of
   (*a*) even harmonics (*b*) odd harmonics
3. White noise generator generates a wave that
   (*a*) contains all possible frequencies
   (*b*) does not contain all possible frequencies
   (*c*) contains harmonically related frequencies
   (*d*) is a pure sine wave of continuously variable frequency
4. A low pass filter can cut
   (*a*) the lower end of the audio spectrum
   (*b*) the higher end of the audio spectrum
5. When both ends of the spectrum are to be cut
   (*a*) a low pass filter is used (*b*) a high pass filter is used
   (*c*) a band pass filter is used
6. In the oscillators used in Moog and ARP synthesizers a 1 volt change in control voltage causes
   (*a*) a one-octave jump in frequency
   (*b*) a two-octave jump in frequency
   (*c*) a three-octave jump in frequency

FILL IN THE BLANKS (B)

1. The higher the frequency, or harmonic, the ________ the amount of energy it receives.
2. In a voltage controlled amplifier a _______ is applied to the input to control the gain.
3. The common circuit found in guitar or amplifiers is simply a ______ controlled VCA.
4. Waveforms with generous amounts of harmonics are ______ to modify.
5. Variable cut off _________ are used to modify the timbre of various waveforms.
6. Changing any part of a synthesizer yields a ________ different sound.

CONVENTIONAL QUESTIONS (C)

1. What forms the basis of electronic music synthesis ?
2. What are the basic modifiers of sound ?
3. Explain the significance of voltage control.
4. Briefly explain the working of an envelope generator.
5. What are the different signal modifiers ?
6. How will you assemble a synthesizer ?

19

OPTICAL RECORDING AND REPRODUCTION

Compact Disc (CD) Players are a very specialised form of phonograph, record player, or turnbale. A CD player plays pre-recorded discs (carrying music, speech, etc.) through a conventional hi fi or stereo system (amplifier and loudspeakers). The disc is single-sided, 4.75 in (120 mm) in diameter, and can contain upto 60 min of hi fi stereo sound. The compact disc spins at a high rate of speed compared with a conventional audio record, and uses a light beam / optical pick-up instead of the standard stylus/needle pick-up arm. In addition to superior sound (to either analog LP or digital PCM recordings), CD players can provide immediate access to audio at any part of the disc. It is also possible to program CD players to play only selected portions of the audio material.

19.1 THE DISC

In the *Laser Vision System*, Fig. 19.1(*a*), which records *video* information, the signal is recorded on the disc in the form of a *spiral track* that consists of a succession of pits. The intervals between the pits are known as *lands.* The information is present in the track in *analog* form. Each *transition* from land to pit and vice versa marks a zero crossing of the modulated video signal. On the *Compact Disc,* Fig. 19.1(*b*), the signal is recorded in a similar manner, but the information is present in the track in *digital* form. Each pit and each land represents a series of bits called *channel bits.* After each land/pit or pit/land transition there is a 1, and all the channel bits in between are 0, (see Fig. 19.2).

The *density of the information* on the Compact Disc is very high; the smallest unit of audio information (the *audio bit)* covers an area of 1 / μm^2 on the disc, and the diameter of the *scanning light spot* is only 1 μm. The *pitch* of the track is 1.6 μm, the *width* 0.6 μm and the *depth* 0.12 μm. The *minimum length* of a pit or the land between two pits is 0.9 μm, the *maximum length* is 3.3 μm. The side of the transparent carrier material T in which the pits P are impressed the upper side during playback if the spindle is vertical is covered with a *reflecting layer R* and a *protective layer* P_r. *The track is optically scanned from below the disc at a constant velocity of 1.25 m/s.* The *speed of rotation* of the disc therefore varies from about 8 rev/s to about 3.5 rev/s (or 480 rpm to about 210 rpm).

19.2 PROCESSING OF THE AUDIO SIGNAL

For *converting* the analog signal from the microphone into a digital signal, pulse-code modulation (PCM) is used. In this system the signal is periodically sampled and *each sample is translated into a binary number.* From Nyquist's sampling theorem *the frequency of the sampling should be at least twice as high as the highest frequency to be accounted for in the analog signal.* The number of bits per sample determines the *signal-to-noise ratio* in the subsequent reproduction.

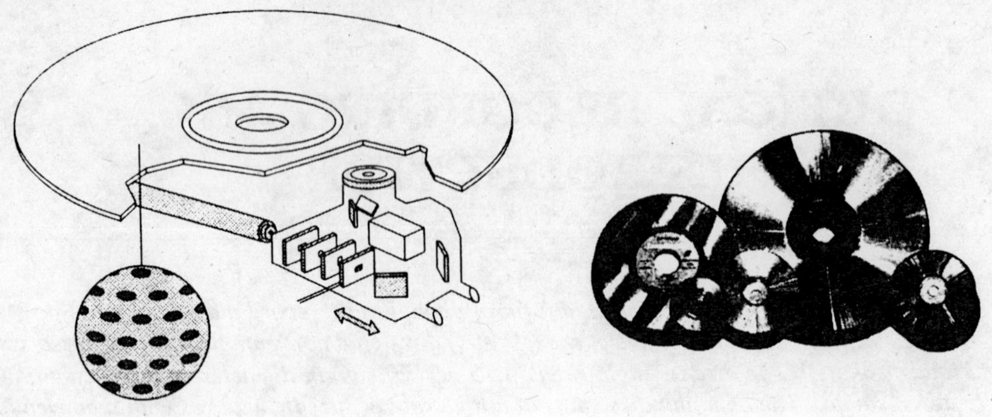

(a) *(b)*

Fig. 19.1 (a) Details of Philips Laser Vision system showing the optical pick-up and the disc microstructure and (b) Compact discs.

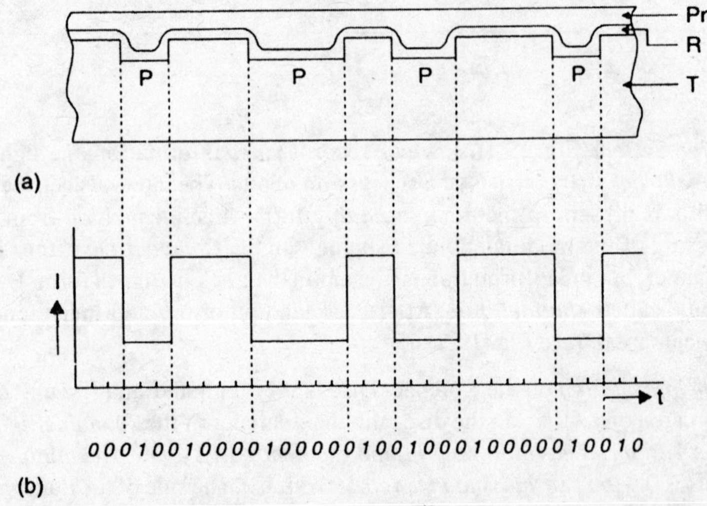

Fig. 19.2 (a) Cross section through a Compact Disc in the direction of the spiral track. T transparent substrate material, R reflecting layer, P, protective layer, P the pits that form the track. (b) I the intensity of the signal read by the optical pick-up (see Fig. 19.3), plotted as a function of time. The signal, shown in the form of rectangular pulses, is in reality rounded and has sloping sides. The digital signal derived from this waveform is indicated as a series of channel bits Ch.

In the *compact disc system* the analog system is sampled at a rate of 44.1 kHz, which is sufficient for the reproduction of the maximum frequency of 20 kHz. The signal is quantized by the method of uniform or linear quantization, the sampled amplitude is divided into equal parts. The number of bits per sample (these are called *audio bits)* is 32 ; i.e. 16 for the *left* and 16 for the *right* audio channel. This corresponds to a *signal-to-noise ratio* of more than 90 dB. The net *bit rate* is thus $44.1 \times 10^3 \times 32$ or 1.41×10^6 audio bits per second. The audio bits are grouped into *frames*, each containing six of the original samples.

Successive blocks of audio bits have blocks of *parity bits* added to them in accordance with a coding system called *Cross-Interieaved Reed-Solomon Code* (CIRC). This makes it possible to correct *errors* during the reproduction of the signal. The ratio of the number of bits before and after this operation is 3 : 4. Each frame then has *Control and Display* (C & D) bits added to it; one of the functions of C & D bits is providing the *information for the listener.* After this operation the bits are called *data bits.*

Next, the bit stream is modulated, that is to say the data bits ar translated into channel bits which are suitable for storage on the disc, (see Fig. 19.2 *b*). The *Eight-to-Fourteen Modulation* (EFM) is used for this purpose. In EFM code blocks of eights bits are translated into blocks of fourteen bits. The blocks of fourteen bits are linked by these *merging bits.* The ratio of the number of bits before and after modulation is thus 8 : 17.

Table 19.1 Names of the successive signals, the associated bit rates and operations during the processing of the audio signal.

Name	*Bit rate in 10^6 bits/s*	*Operation*
Audio Signal		PCM (44.1 kHz)
Audio bit stream	1.41	CIRC + (parity bits); Addition of C & D bits
Data bit stream	1.94	EFM; Addition of merging bits;
Channel bit stream	4.32	Addition of synchronization patterns.

For the synchronisation of the bit stream an *identical synchronisation pattern* consisting of 27 channel bits is added to each frame. The *total bit rate* after all these manipulations is 4.32×10^6 channel bits / s.

Table 19.1 gives a survey of the *successive operations* with the associated bit rates, with their names. From the magnitude of the channel bit rate and the scanning speed of 1.2 m/s it follows that the *length of a channel bit on the disc is approximately 0.3 μm*.

The signal produced in this way is used by the manufacturer to *switch On and OFF the laser beam* that illuminates the light sensitive layer on a rotating glass disc, called the *master.* A patterns of pits is produced on the disc by means of a photographic developing process. After the surface has been coated with a thin silver layer, an electroplating process is applied to produce a nickel impression called the *metal father.* From this father disc impressions called *mother discs* are produced in a similar manner. The impressions of the mother discs, called *sons or stampers,* are used as tools with which the pits *P* are *impressed* into the thermoplastic transparent carrier material T of the disc (see Fig. 19.2).

19.3 READOUT FROM THE DISC

The disc is *optically scanned* in the player. This is done by *AlGaAs semiconductor laser* Fig. 19.3 shows the optical part of the *pick-up.* The light from the *laser* La (wave-length 800 nm) is focused through the *lenses* L_2 and L_1 onto the *reflecting layer* of the disc. The diameter of the light spot S, Fig. 19.4, is about 1 *μm*. When the light falls on an *interval* between two pits, the light is almost totally reflected and reaches the four photodiodes D_1 to D_4 via the half -silvered mirror M. When the spot lands on a *pit* - the depth of a pit is about 1/4 of the wavelength in the transparent substrate material-interference causes less light to be reflected and an appreciably smaller amount reaches the photodiodes. When the output signals form the four photodiodes are added together the result is a *fairly rough approximation* to the rectangular pulse pattern present on the disc in the form of *pits and intervals.*

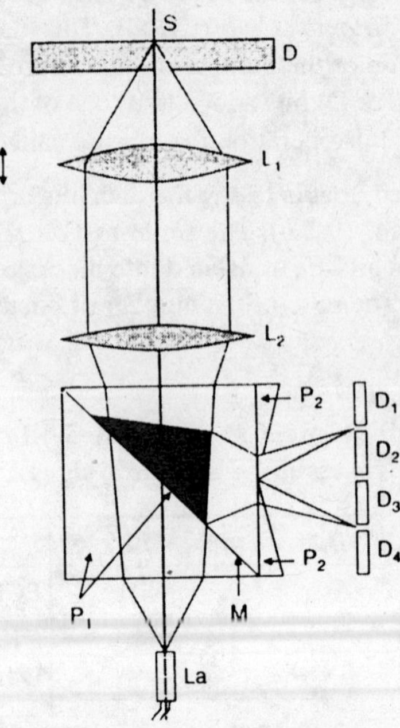

Fig. 19.3 Diagram of the optical pick-up. D, radial section through the disc. S laser spot, the image of the disc of light emitting part of the semiconductor laser La. L_1, objective lens, adjustable for focusing, L_2 lens for making the divergent laser beam parallel. M half-silvered mirror formed by a film evaporated on the dividing surface of the prism combination. P_1, P_2 beam splitter prisms, D_1 to D_4 photodiodes whose output currents can be combined in various ways to provide the output signal from the pick-up and also the tracking - error signal and the focusing error signal. (In practice the prism P_2 and the photodiodes D_1 to D_4 are rotated by 90° and the reflection of the mirror M does not take place in a radial plane but in a tangential plane).

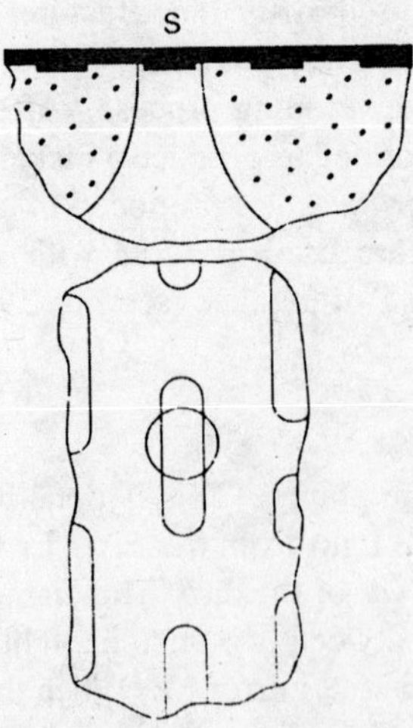

Fig. 19.4 A magnified view of the light spot S and its immediate surroundings, with a plan view. It can clearly be seen that the diameter of the spot (about 1 μm) is larger than that of the width of the pit (0.6 μm).

The optical pick-up shown in Fig. 19.2 is very small (about 45 × 12 mm) and is mounted in a *pivoting arm* that enables the pick-up to describe a radial arc across the disc, so that it can scan the complete spiral track. Around the pivotal point of arm is mounted a *linear motor* that consists of a combination of a coil and a permanent magnet. When the coil is energised the pick-up can be directed to any required part of the track, the *locational information* being provided by the C & D bits added to each frame on the disc. The pick-up is thus able to find *independently* any particular passage of music indicated by the listener. When it has been found the pick-up must then *follow the track accurately* -to within ±0.1μm without being affected by the next or previous track. Since the track on the disc may have some slight *eccentricity*, and since also the suspension of the turntable is not perfect, the track may have a maximum *side-to-side swing* of 300 µm. A *tracking servo system* is therefore necessary to ensure that the deviation between pick-up and track is smaller than the permitted value of ± 0.1μm and in addition, to absorb the consequences of small vibrations of the player.

The *tracking-error signal* is delivered by the four photodiodes D_1 to D_4. When the spot *S*, seen in the radial direction, is situated in the centre of the track, a *symmetrical* beam is reflected. If the spot lies slightly to one side of the track, however, interference effects cause *asymmetry* in the reflected beam. This asymmetry is detected by the prism P_2, which *splits* the beam into two components. Beyond the prism, one component has a higher mean intensity than the other. The signal obtained by coupling the photodiodes as $(D_1 + D_2) - (D_3 + D_4)$ can therefore be used as a tracking error signal.

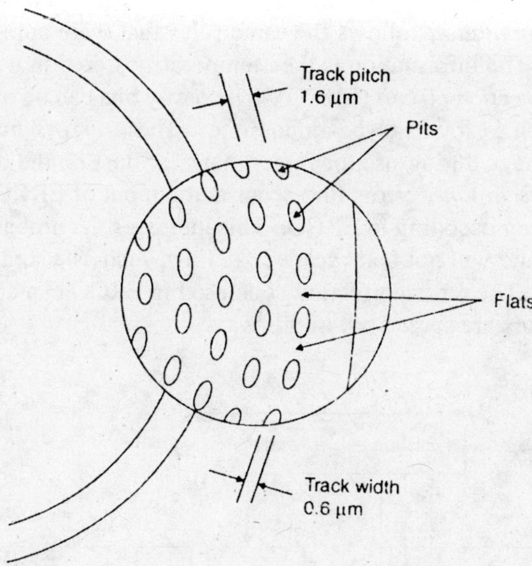

Fig. 19.5 The information on the compact disc is recorded in digital form as a spiral track consisting of a succession of pits. The pitch of the track is 1.6 µm, the width 0.6 µm, and the depth of the pit 0.12 µ m. The length of a pit or the land between two pits has a minimum value of 0.9 µm and a maximum value of 3.3 µ m.

As a result of the *aging or soiling of the optical system*, the reflected beam may acquire a slowly increasing, more or less constant asymmetry. Owing to a D.C. component in the tracking error signal, the spot will then always be *slightly off-centre* of the track. To compensate for this effect a *second tacking error signal* is generated. The coil that controls the pick-up arm is therefore supplied with an alternating voltage at 600 Hz, with an amplitude that corresponds to a *radial displacement* of the spot by ±0.05µ*m*. The output sum signal from the four photodiodes - which is at a *maximum* when the spot is at the centre of the track - is thus modulated by an alternating voltage of 600 Hz. The amplitude of this 600 Hz signal *increases* as the spot moves off-centre. In addition the *sign* of the 600 Hz error signal changes if the spot moves to the other side of the track. This second tracking - error signal is therefore used to correct the error signal mentioned earlier with a direct voltage. The output sum signal from the photodiodes, which is processed in the player to become the audio signal, is thus returned to its maximum value.

The *depth of focus* of the optical pick-up at the position S (see Fig. 19.3)is about 4μm. The *axial deviation* of the disc, owing to various mechanical effects, can be a maximum of 1 mm. It is evident that a servo system is also necessary to give correct focusing of the pick-up on the reflecting layer. The objective lens L_1 can therefore be displaced in the direction of its optical axis by a combination of a coil and a permanent magnet, in the same way as in a loudspeaker. The *focusing - error signal* is also provided by the row of photodiodes D_1 to D_4. If the spot is *sharply focused* on the disc, two sharp images are precisely located between D_1 and D_2 and between D_3 and D_4. If the spot is *not sharply focused* on the disc, the two images on the photodiodes are not sharp either and have also moved closer together or further apart. The signal obtained by connecting the photodiodes as $(D_1 + D_4) - (D_2 + D_3)$ can therefore be used for controlling the focusing servo system. The deviation in focusing then remains limited to $\pm 1\mu m$.

19.4 RECONSTITUTION OF THE AUDIO SIGNAL

The signal read from the disc by the optical pick-up has to be reconstituted to form the analog audio signal. Fig 19.6 shows the block diagram of the signal processing in the player.

In DEMOD the *demodulation* follows the same rules that were applied to the EFM modulation, but now in the *opposite sense.* The information is then temporarily stored in a *buffer memory* and then reaches the *Error-detection and correction circuit* (ERCO). The parity bits can be used here to correct errors, or just to detect errors if correction is found to be impossible. These *errors* may originate from defect in the manufacturing process, damage during use, or finger marks or dust on the disc. Since the information with the CIRC code is *interleaved in time,* errors that occur at the input of ERCO in one frame are spread over a large number of frames during decoding in ERCO. This increases the probability that the maximum number of correctable errors per frame will not be exceeded. A flaw, such as a scratch, can often produce a train of errors called an *error burst.* The error-correction code used in ERCO can correct a burst of about 4000 data bits, largely because the errors are spread out in this way.

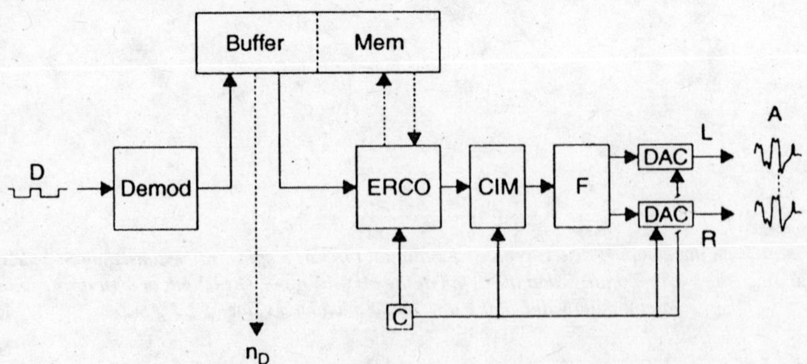

Fig. 19.6 Block diagram of the signal processing in the player. D, input signal read by the optical pick-up. A, the two output analog audio signals from the left (L) and the right (R) audio channels. DEMOD, demodulation circuit ERCO, error-correction circuit. BUFFER, buffer memory forming part of the main memory, MEM, associated with ERCO. CIM (Concealment, Interpolation and Muting) circuit, in which errors that are only detected since they cannot be corrected are masked or concealed. F. filters for interpolation. DAC, digital-to-analog conversion circuits. Each of the blocks mentioned here are fabricated in VLSI technology. C, clock generator controlled by a quartz crystal. The degree to which buffer memory capacity is filled serves as a criterion in controlling the speed of the disc.

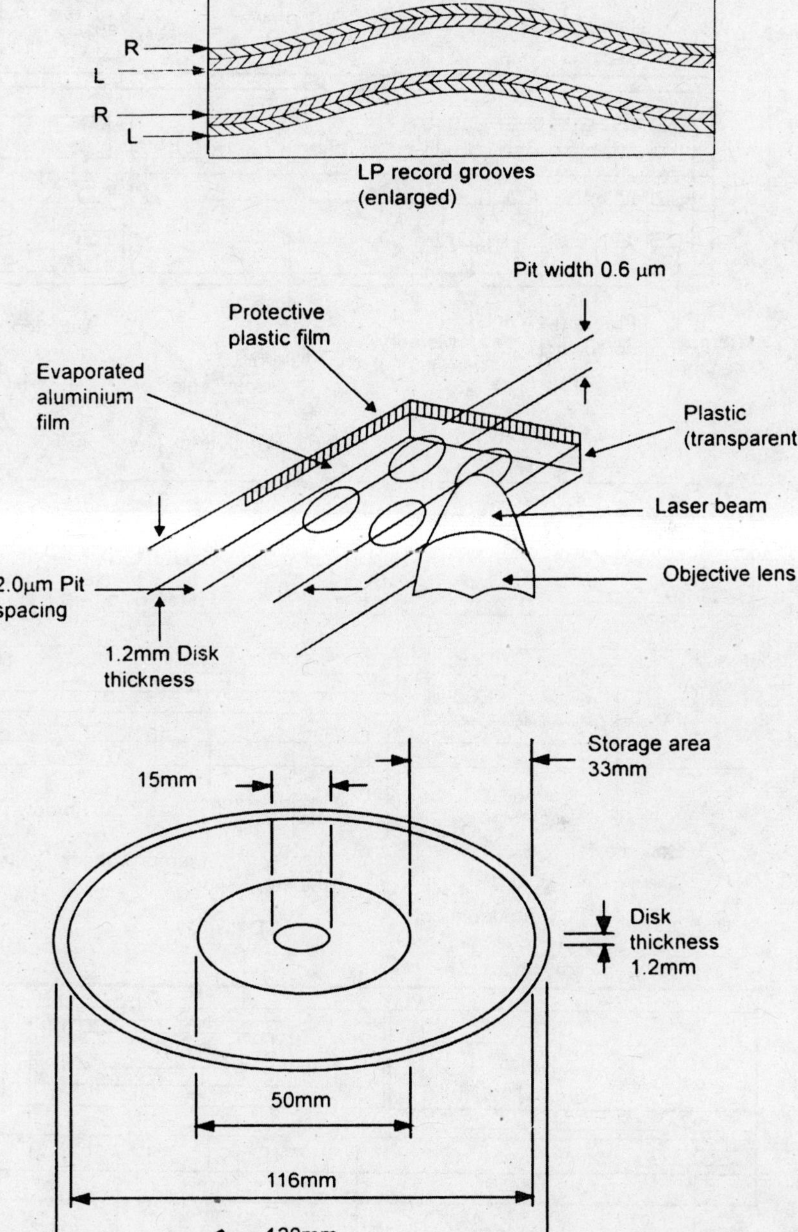

Fig. 19.7 Comparison of the grooves in a conventional LP record and the pits of a compact disc

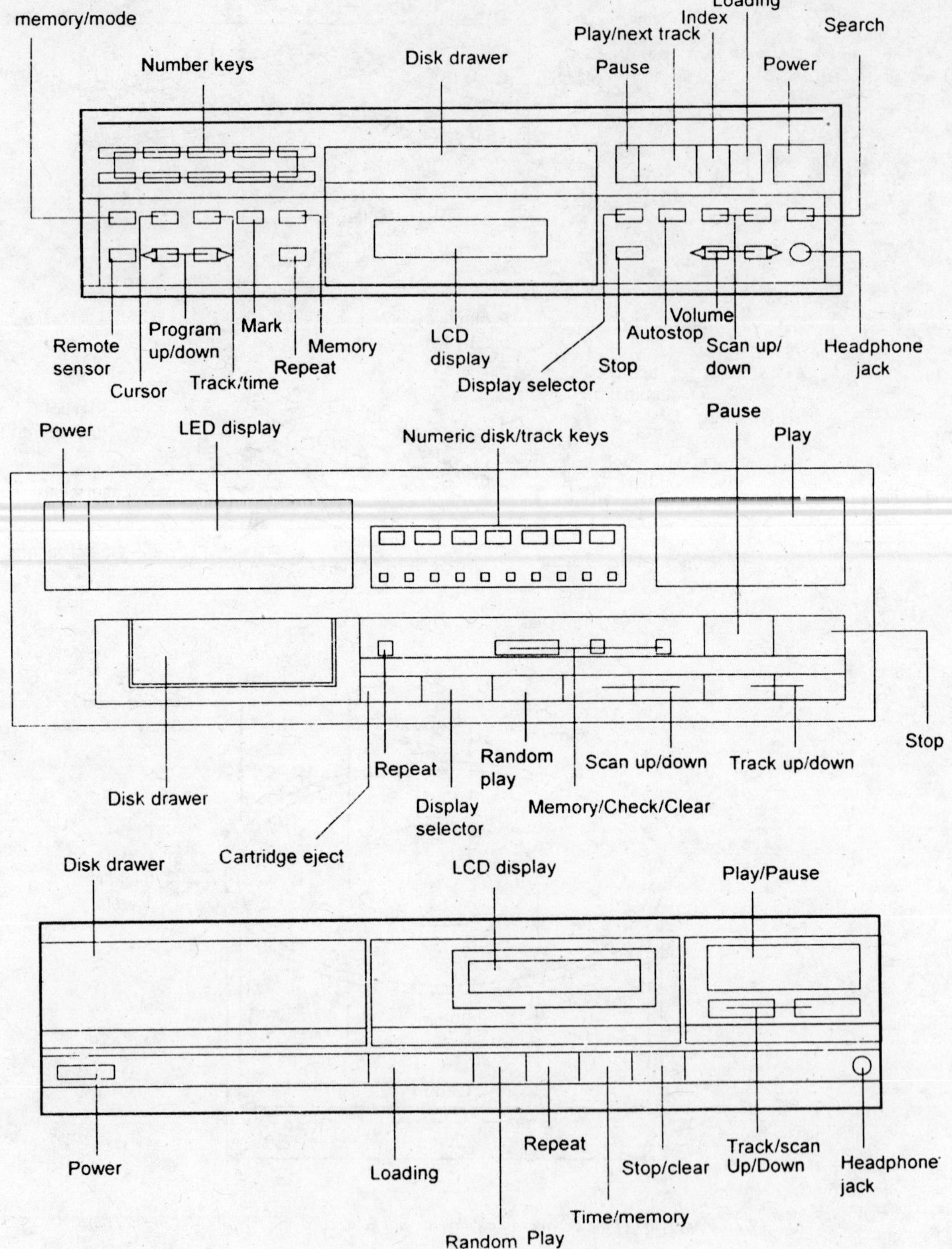

Fig. 19.8 Operating controls of typical CD Players.

If more errors than the permitted maximum occur, they can only be detected. In the *Concealment, Interpolation and Muting* (CIM) block, the errors detected are then masked. If the value of a sample indicates an error, a new value is found by *linear interpolation* between the preceding value and the next one. If two or more successive sample values indicate an error, they are *muted* (made equal to zero). At the same time a gradual transition is created to the values preceding and succeeding it by causing a number of values before the error and after it to decrease to zero in a particular pattern.

In the *digital-to-analog converter* (DAC) the 16 bit samples first pass through *interpolation filters F* and are then translated and recombined to re-create the original analog signal *A* from the two audio channels *L* and *R*. Since samples must be recombined at exactly the same rate as they are taken from the analog audio signal, the DACs and also CIM and ERCO are synchronised by a *clock generator, C,* controlled by a quartz crystal.

Fig. 19.6 also illustrates the *control of the disc speed* n_D. The bit stream leaves the buffer memory at a rate synchronised by the clock generator. The bit stream enters the buffer memory, however, at a rate that depends on the speed of revolution of the disc. The extent to which n_D and the sampling rate are matched determines the *filling degree* of the buffer memory. The control is so-arranged as to ensure that the buffer memory is at all times filled to 50% of its capacity. The analog signal from the player is thus completely free from *wow and flutter,* yet with only moderate requirements for the speed control of the disc.

A comparison of the grooves in a conventional LP record and the pits of a compact disc is given in Fig. 19.7.

The operating controls of typical Compact Disc (CD) players are illustrated in Fig. 19.8.

19.5 FILM SOUND TRACK

In the case of sound cinephotography; it is convenient to record the sound in the same way as the picture is recorded, that is, by the *photographic process.* The sound track of the film is a narrow strip down one side of the film, the capacity of which varies in accordance with the sound vibration.

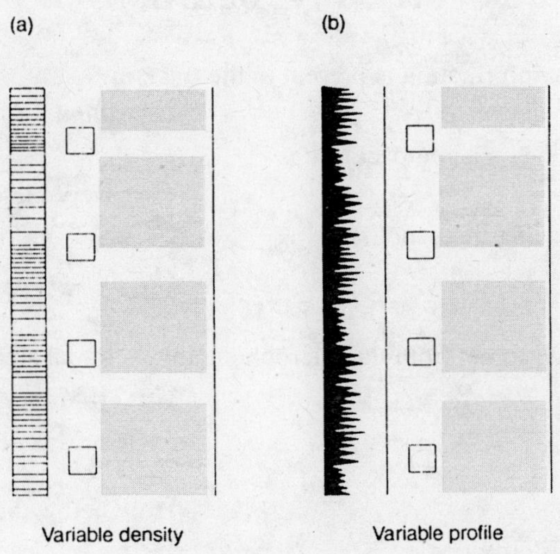

Fig. 19.9 Film sound track.

One way of producing this track is to use the output from the microphone to vary the light emitted from a gas discharge tube, which is focused on the unexposed film. Development of the film produces a track of *varying density,* Fig. 19.9 (*a*). The output of the microphone can vary the sound track exposure in other ways; for example, it can vary the width of an aperture restricting the light source. Alternatively the oscillating currents can be converted electromagnetically into the vibrations of a mirror, which directs a light beam to film the width of the track to varying degrees, Fig. 19.9 (*b*).

The method of regenerating the sound from this track is illustrated in Fig. 19.10. The density of the sound track at any point *controls* the amount of light shining on to a photocell. As the film moves, the current output from the photocell *varies* accordingly, and after amplification it is passed to the speaker.

If several speakers are to be controlled separately, as in the case of stereophonic sound systems, several tracks are needed. They may restrict the space available for picture projection, and so *magnetic sound tracks* which can be fitted into about 1/5 th of the space are fast replacing the optical ones.

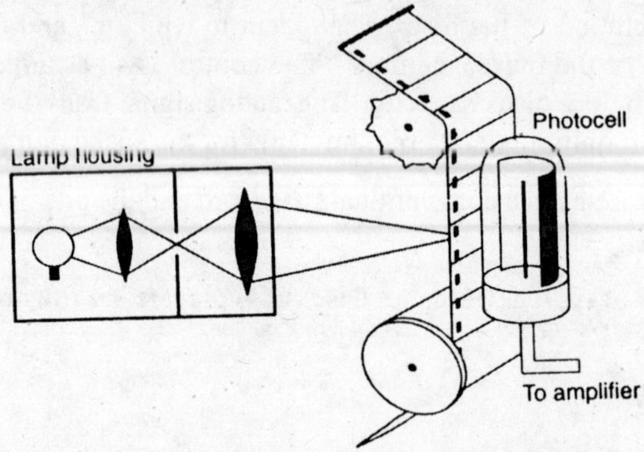

Fig. 19.10 Regenerating the sound from a sound track.

OBJECTIVE TYPE QUESTIONS (A)

1. On the compact disc, the information is present in the track in
(*a*) analog form (*b*) digital form

2. The density of information on a compact disc is
(*a*) very small (*b*) very high

3. The diameter of the scanning light spot is
(*a*) 10 μ*m* (*b*) 1 μ*m*
(*c*) 0.1 μ*m*

4. For converting the analog signal from the microphone into a digital signal.
(*a*) A.M. is used (*b*) F.M. is used
(*c*) PCM is used

5. The track is optically scanned from
(*a*) above the disc (*b*) below the disc

6. Locational information is provided by
(*a*) audio bits (*b*) channel bits
(*c*) parity bits (*d*) C & D bits

FILL IN THE BLANKS (B)

1. Each sample is converted into a ____________.
2. The frequency of sampling should be at least _______ as high as the highest frequency to be accommodated in the analog signal.
3. One of the functions of *C* & *D* bits is providing the __________ for the listener.
4. The optical pick-up is mounted in a ________.
5. The information is temporarily stored in a __________.
6. The extent to which n_D and the sampling rate are matched determines the __________ of the buffer memory.
7. The information with the CIRC code is ___________ in time.
8. _________ memory forms part of the main memory.
9. The analog signal is sampled at a rate of __________.
10. Parity bits are added in accordance with a coding system called ___.
11. The pick-up must follow the track accurately to within ________.
12. A flaw, such as a scratch, can often provide a train of errors called ________.

CONVENTIONAL QUESTIONS (C)

1. What is the difference between an LP record and a compact disc ?
2. Explain the following:
   (*a*) channel bits (*b*) audio bits
   (*c*) parity bits (*d*) data bits
   (*e*) merging bits
3. What is the function of the tracking-error signal ?
4. What is the function of the focusing-error signal ?
5. Explain the various steps involved in reconstitution of the audio signal.
6. Briefly explain the operating controls of a CD players.

20

COMMERCIAL SOUND

Now that fidelity has reached quite a high standard, most audio people prefer microphones that give the most transparent effect : make a recording where the illusion is so good that the original performers seem to ''come through completely''. Sound is commercial to the extent that a segment of the public has a preference for the same audio qualities, consciously or subconsciously.

But preferences change gradually, and usually in favour of the more realistic effect, because that, ultimately is the easiest to listen to. Artificial effects, ultimately become tiring. What may have started as a wonderful sound does not continue to satisfy because it cannot be identified as something real. This has been one of the problems that electronically - generated forms of music have had in gaining acceptance. And studios are becoming more and more involved in theses devices. We are living in days when purely electronic sounds, that are not intended to simulate some older sound form, are beginning to gain acceptance in their own right.

20.1 ELECTRIC GUITAR

The guitar, Fig. 20.1, was the first instrument to make popular use of electronics. Electric guitars use a suitable pick-up to transduce the motion of the guitar strings to electrical signals. These signals may be amplified and fed into one or more loudspeakers, and *usually provision is made to modify the character of sound in several ways.* The tone may be modified so that a guitar of a given form and size, with a fixed tone capability *without electronics*, can be made to sound like many instruments in the guitar family *through the use of electronics.* In addition, either vibrato or tremolo can be added to tones electronically, and a variety of *novelty effects,* such as ''fuzz tone'' can be produced.

A typical United States patent describes a system for electric guitars and the like as illustrated in Fig. 20.2 and as modified in Fig. 20.3. The *pick-up section* on the guitar may consist of one or more electromagnetic or piezoelectric transducers per string. The output of the pick-up section is in *electrical form* representing the tones, including all partials, generated by the vibrating strings of the instrument. The *design* of the pick-ups and their *locations* with respect to the strings influence the relative amplitudes of the partials of the tones, but the positions selected for the system of Fig. 18.2 provide a full complement of partials.

The electrical signals produced by the pick-up section in Fig 20.2 are amplified by an *isolating amplifier* and passed to a series of circuits which may be selected individually or in combination by means of switches. This series of circuits may include a circuit which passes all partials without coloration, a circuit which passes low frequencies with less attenuation one or more circuits that accentuate, one or more bands of frequencies (called *formant filters)*, and a circuit that favors the passage of high-frequency signals and attenuates low frequencies. This series of circuits enables the musician to select the sound of the acoustic guitar, rock-and-roll guitar, classic guitar, Hawaiian guitar, bass guitar, solo guitar, and other sounds. After the electrical signals have passed through one or more *coloration circuits*, they are again amplified and passed to a *modulator* circuit that makes use of a low-frequency (subsonic) oscillator to add a vibrato or tremolo effect or both *at a desired degree and rate*. The tone signals are again amplified and passed to one or more loudspeakers, with a suitable volume control provided.

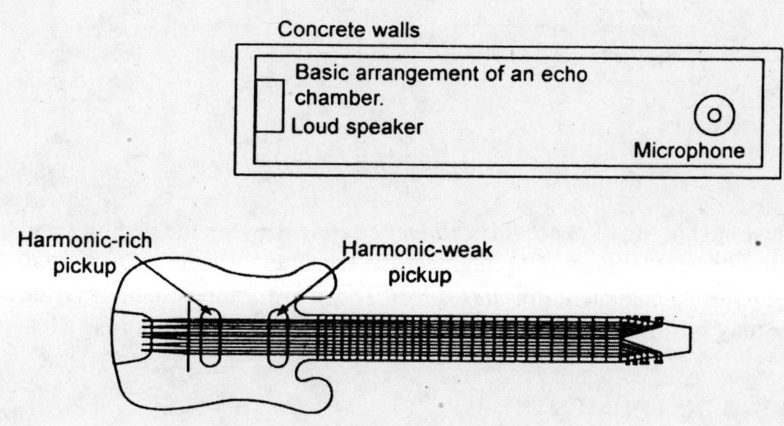

Fig. 20.1 Pick-ups on an electric guitar are frequently arranged as shown here to provide outputs of different timbre, from which the ultimate sound effect can be given quite a varied range.

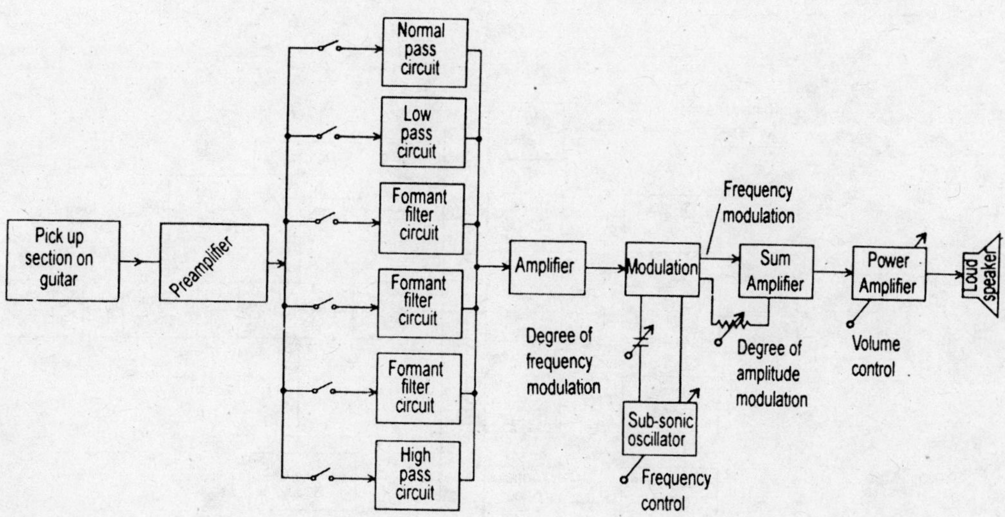

Fig. 20.2 Electronic system for an electronic guitar.

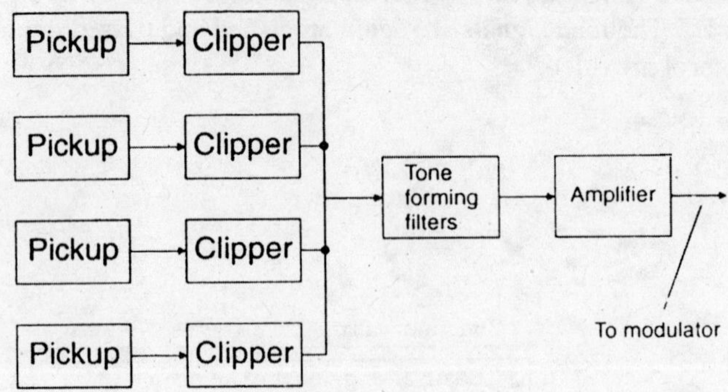

Fig. 20.3 Modified electronic system for an electric guitar.

The system illustrated in Fig 20.3 provides *clipper circuits* between the pick-ups and the tone-forming filters. The clipper circuits generally serve to augment the strength and number of partials so as to *provide greater possibilities for variety in the tone-forming filters.* The tone-forming circuits may be the same as those shown in Fig 20.2, or may be different. The resulting signals may be treated as in Fig 20.2.

20.2 ELECTRIC WIND INSTRUMENTS

The treatment discussed above can also be given to the tones of the other *stringed instruments* including pianos, and to wind instruments also.

In wind instruments only one pick-up is generally used, a sound-pressure microphone placed in the mouthpiece or in some other part of the instruments. The electrical signals from the pick-up microphone can be treated in ways similar to those illustrated in Fig 20.2 or 20.3. Other variations are common also.

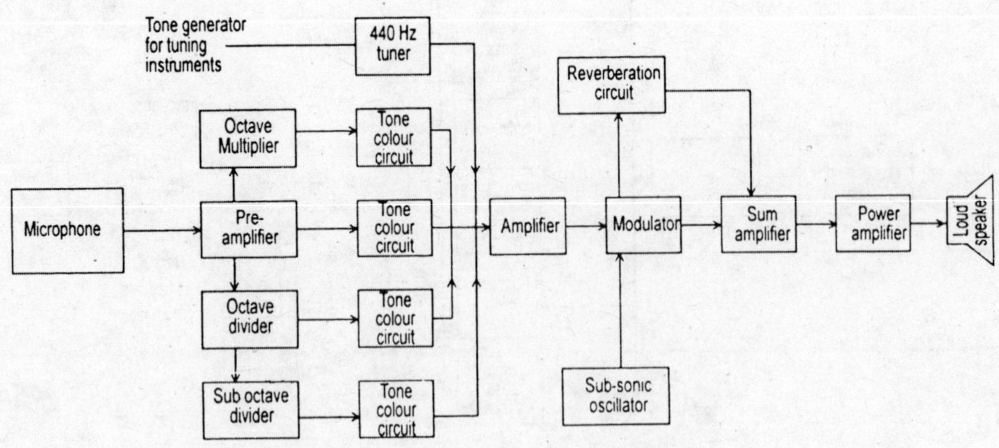

Fig. 20.4 Electronic modification to wind instruments.

The block diagram in Fig 20.4 indicates a simplified composite of functions offered on different commercial products to be used to either amplify or modify, or both amplify and modify the tones of individual wind instruments.

In addition to *tone modifications in timbre, tremolo or vibrato* embellishment and virtually unlimited sound power *it is possible for one instrument to sound as if it were as many as four octavely-related instruments playing simultaneously*. This is done by circuits that divide or multiply the fundamental frequency of the tone produced by the instrument played by the musician. Also, artificial reverberation or echo effects can be added to the tone.

The 440 Hz tuner shown in Fig 20.4 is an oscillator that can be used to produce standard note *A* for *tuning* the instrument being played.

Electronic systems to *amplify and modify* tone of individual musical instruments are being used in marching bands, concerts, motion picture sound tracks, recording, and many other ways.

20.3 RECORDING

In electronic music synthesizer systems, *a single series of tones* is generated at a time. As a result, all of the characteristics of a tone can be applied to each tone in a series of tones. This is, of course, impossible in the case of conventional musical instruments or in the case of all-electronic musical instruments because, in general, more than one tone is generated at a time. As a matter of fact, the entire composition is produced in one run. *The electronic music synthesizers provide the most powerful system for the generation of musical sound and the resultant production of recorded music yet invented.*

When a single series of tones is produced by a *manual* electronic music synthesizer, by a *programmed* electronic music synthesizer, or by a *computer*, the recording is essentially made with a synchronized multi-channel magnetic tape-recorder. In this way, the magnetic tape recorder can be synchronized with any picture or programme. *Some form of synchronized multi-channel recorder is employed in all types of electronic music synthesis.*

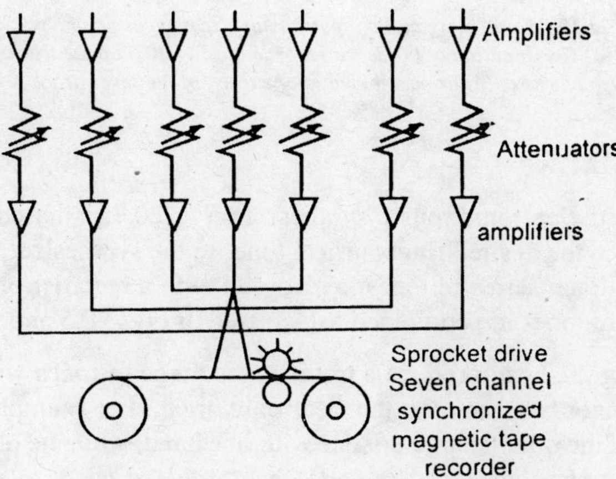

Fig. 20.5 Schematic diagram of a synchronized seven channel recorder for recording seven different and separate audio inputs or tone sequences on seven tracks of magnetic tape.

Seven different audio inputs representing seven series of tones are recorded on seven separate tracks, Fig 20.5. This magnetic tape is reproduced, and the audio signals in the seven channels are *mixed and recorded* on a single track of a seven track recorder, Fig 20.6 (*a*). When the seven tracks of this recorder are filled with seven different operations, the recorded tape is *reproduced and combined* to a single track representing 19 tone sequences, Fig 20.6 (*b*). If the recorder in Fig 20.6 (*a*) is a seven-channel recorder and all the tracks are filled, the total final result represents 343 tone sequences. These can be combined to a single track again, and so on.

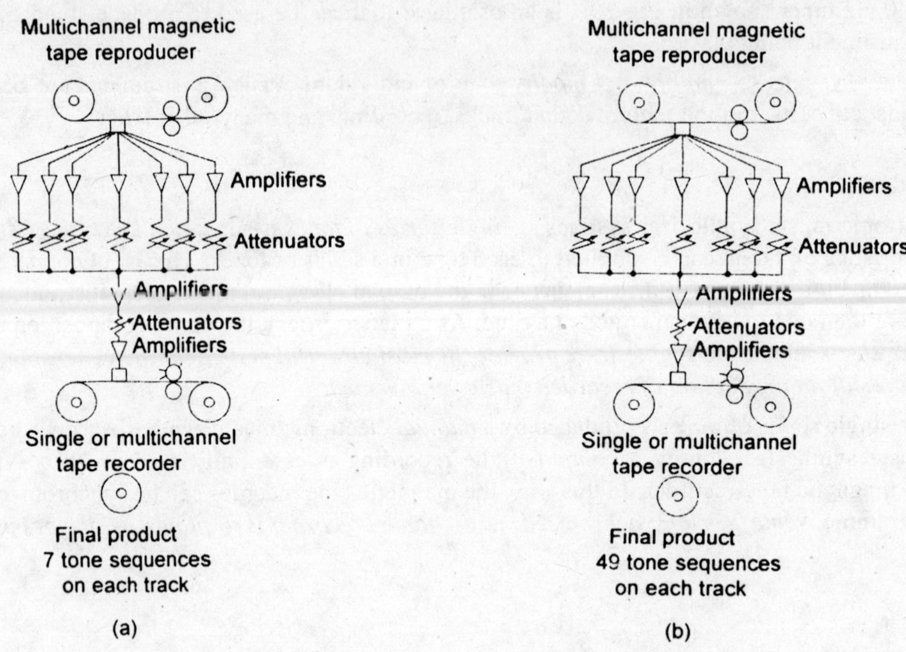

Fig 20.6 Combining audio inputs or tone sequences on one track.
(a) Combination of the seven recorded in Fig 20.5 on the track.
(b) Seven tracks of (a) are combined to one track of the seven-track recorder.

20.4 MANUAL SYNTHESIZER

The manually operated electronic music synthesizer, Fig 20.7, is based upon the generation of the properties of a tone to produce the desired final musical tone. In the synthesizer in Fig 20.7 *only one series of tones is produced at a time.* Each series of tones is recorded with a recorder of the type shown in Fig 20.5. The various different series of tones are combined as depicted in Figs 20.5 and 20.6 and as described above.

The synthesizer in Fig 20.7 can produce a tremendous range of tones with all manner of variation in each series of tones; this cannot be done with the electronic organ. For example, the performer changes the volume, timbre, and some of the other characteristics with one hand, while he plays on the keyboard with the other hand. Therefore, *great physical dexterity and considerable musical talents are required to produce acceptable music by means of the electronic music synthesizer.* The main reason for this is that each series of tones must possess *distinct musical qualities* in order to produce musical effects beyond the conventional. In addition, there must be a *perfect synchronism* of the different series of tones which is also carried out subjectively, that is by ear.

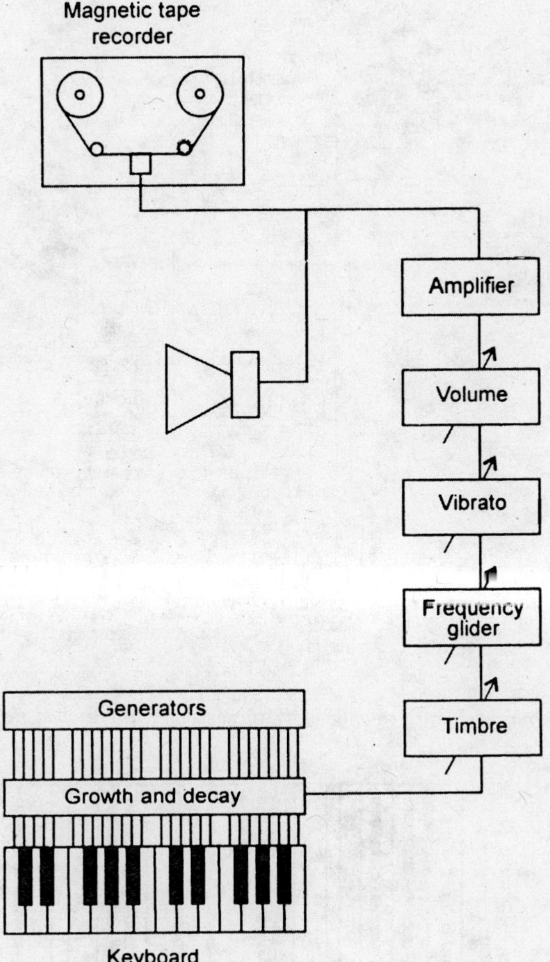

Fig 20.7 Diagrammatic representation of the elements of a manual electronic music synthesizer.

20.5 PROGRAMMED SYNTHESIZER

The use of a programmed electronic music synthesizer opens up an entirely new field for the production of recorded music, Fig 20.8. To synthesize or produce any musical tone whatsoever, the programmed electronic music synthesizer must provide the facilities for producing *all* the characteristics of a tone.

The program for each series of musical tones is recorded on the coded paper record, which controls all the functions of the electronic music synthesizer. *The information is recorded and stored in the paper record in the form of black ink lines.*

When the paper is run through the machine, the sensors scan the *paper record.* The paper record, Fig 20.9, shows that a *binary code system* is used. With the use of this system, 2 different *functions* can be obtained from 1 track, 4 from 2 tracks, 8 from 3 tracks and 16 from 4 tracks, and so forth.

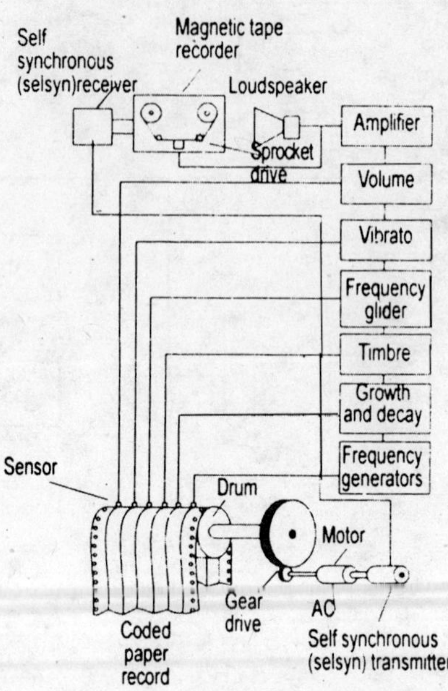

Fig. 20.8 Diagrammatic representation of the elements of a programmed electronic music synthesizer.

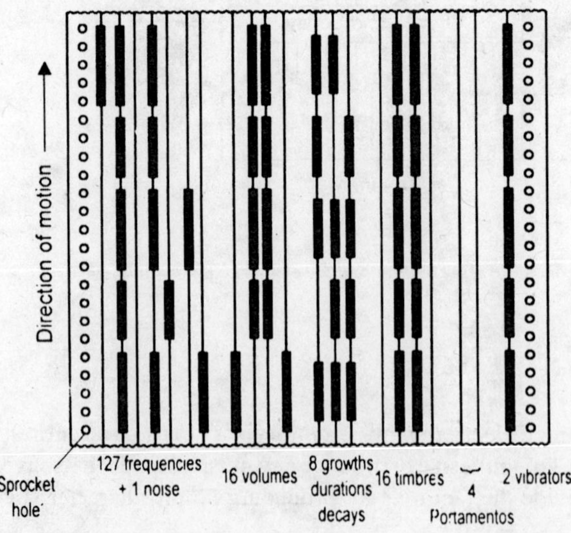

Fig. 20.9 Coded paper record for electronic music synthesizer.

The preparation of the record of Fig 20.9 is simple. The musician simply marks the paper with a wide ink pen to obtain the type of sound he desires. The marking of the paper can be done anywhere. After the record is completed, the musician runs the record through the synthesizer and listens to the series of tones by means of the loudspeaker. If he is not satisfied, he can make changes by adding new lines or by marking out existing lines with a white marker. In general, no changes are required for the frequencies because they are taken directly from the music. The *length* of the lines is determined from the note. Therefore, except for errors the frequency markings are straight forward. As mentioned, changes can be made in the other functions. After the record is satisfactory, the paper record is recorded on magnetic tape Fig. 20.5. This process is carried out for each series of tones until all of the series of tones of the composition have been completed. The method of combining the series of tones has been described earlier and shown in Figs 20.5 and 20.6.

Entirely new tone complexes and combinations which cannot be achieved with conventional instruments are possible with the programmed synthesizer. With conventional instruments the musician is *limited* to the use of lips, mouth, fingers, hands, and feet to perform the different functions. In the programmed electronic music synthesizer, there are no such limitations. In addition, in the case of manual operation, there is the problem of *synchronizing* the various series of tones. However, in the programmed electronic music synthesizer, as is evident from the direct drive between the paper record and the magnetic tape recorder, there is no problem of synchronizing the series of tones.

20.6 DIGITAL COMPUTER

Music may be synthesized by means of a digital computer. The key process involves a *digital-to-acoustic converter*. A schematic block diagram of the conversion process is shown in Fig 20.10. A magnetic tape is prepared by the computer, on which are recorded *successive digitized samples of the acoustic output.* The numbers are then converted to pulses, the amplitudes of which are proportional to the numbers. The pulses are smoothed by a low-pass filter to obtain the input for an audio loudspeaker. If the *sampling rate* is 20,000 per second, the top frequency from the loudspeaker will be 10,000 Hz. Each sample is produced from a four-decimal number. The signal-to-noise ratio will be of the order of 60 dB. *Within the limits of frequency range and signal-to-noise ratio, the converter can produce any sound, whatsoever, provided the appropriate sequence of digital samples can be generated.*

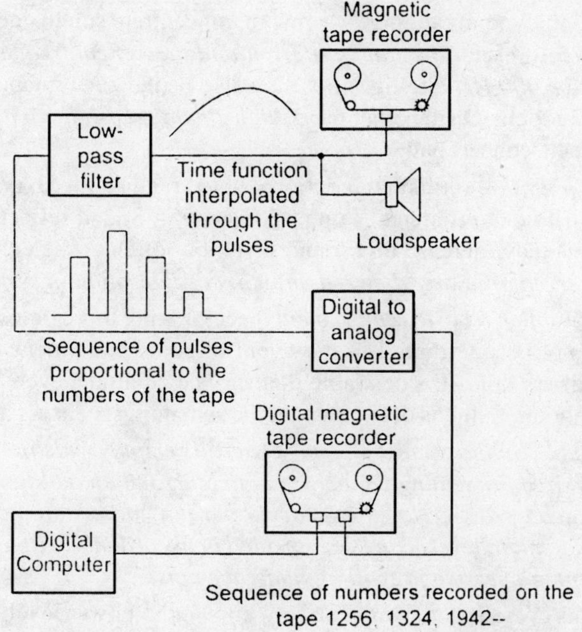

Fig. 20.10 Digital-to-acoustic converter

The basic form of the generating program is a scheme for producing a sequence of sounds representing individual instruments. The instruments are formed by combining a set of basic building blocks called *unit generators.* Appropriate combinations of these unit generators can produce sounds of almost any desired complexity.

The first step in the production of a musical selection is to punch a set of cards which *specify* the instruments of the orchestra. The instruments and the computing program are fed into the computer, which punches a card deck that is the *music generating program* or orchestra. Next a series of *note cards* or score must be prepared. These provide the properties of a tone.

As in the case of the programmed synthesizer, the process is simplified if the composer listens to one series of notes at a time, makes suitable modifications, and then *groups* the combination of tones after each one has been approved. Minor modifications can then be made in the group, if necessary. The series of tones are then recorded as depicted in Figs 20.5 and 20.6.

The exposition given above has described one process for the electronic synthesis of music employing a digital computer. As the programming for computers is simplified, the general procedure will also be simplified. However, the composer operating the synthesizer, regardless of the method employed for the electronic synthesis of music, is the final judge of the rendition. The analog and digital synthesizers can produce any tone, regardless of whether it has ever been produced before. Therefore, the method which will be selected is the one that provides the composer with the easiest means for translating what he has in mind into the final sound product.

20.7 PUBLIC-ADDRESS SYSTEMS

A public-address (PA) system consists basically of a microphone, amplifier and speaker(s) *to facilitate the communication of intelligible speech to groups of people.* Coverage of large groups may be the chief purpose of providing communication over *outdoor* areas or of providing sound reinforcement *indoors* with sufficient fidelity so that the sound system is unobtrusive.

Acoustic problems differ considerably for outdoor and for indoor installations. Directional horns, Fig 20.11, are often employed in outdoor systems to concentrate sound energy efficiently over the areas served. *Intelligibility is the usual performance criterion and the system frequency response is typically limited to a range of 300 Hz to 5,000 Hz.* Intelligibility may also be the performance criterion for an indoor system such as an airport waiting room. On the other hand, *high-fidelity reproduction is required for a sound-reinforcement system*, for instance a concert hall.

Outdoor installations utilising directional horns require approximately 2 watts of audio power per 1000 square feet and operate at efficiencies up to 50 per cent. Sound reflection from stadium walls, grandstands, fences or buildings may increase this sound level, but often at the expense of intelligibility. Therefore, *the acoustics of the served area are often a dominant consideration in system design.*

Reliability is often a basic requirement, necessitating more or less duplication of facilities, arranged so that service will not be interrupted in the event of some equipment failure. In other words, if one of the high-power amplifiers fails, it is desirable that service continue, even at a reduced sound level. Similarly, if a speaker or line should fail, it is desirable that the remaining speakers and lines continue to operate normally.

Indoor sound - reinforcement systems generally employ wide range speakers. A minimum audio power of 50 watts is required for seating capacities upto 500, and an additional 50 watts for capacities upto 1000, and so on. It is good practice for the control operator to be stationed in a room communicating with the served area, so that audio levels can be controlled in accordance with realistic requirements, and maintained at a level sufficient to override prevailing audience noise.

Installations operating at comparatively high audio power level, either outdoors or indoors, have been standardised in accordance with the *70.7 volt speaker matching system* established by the Electronic Industries Association (EIA). The system reduces losses in the wiring, and increases operating efficiency.

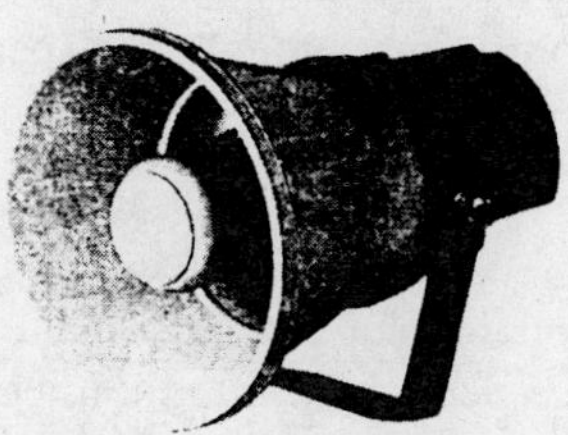

Fig. 20.11 Directional horns used in outdoor P.A. systems.

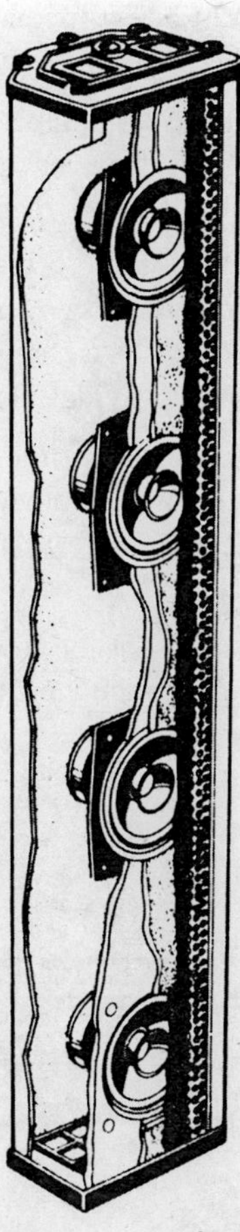

Fig. 20.12 A sound column comprising several woofers and tweeters.

20.8 PLANNING A PUBLIC ADDRESS SYSTEM

As a general rule, *planning* of a PA system starts with :

calculation of the area to be served,
estimation of the required audio power,
choice of the types of speakers to be utilised, and
estimating the approximate number of speakers.

As an illustration, most *office areas* can be served satisfactorily with 8-inch cone speakers rated at 1/2 watt each. Ceiling speakers are installed about 25 feet apart. *Ware houses or meeting halls* often require 5-watt speakers mounted from 25 to 50 feet apart. When noise is troublesome, 5 watt indoor horn speakers may be installed at necessary locations to maintain a satisfactory *audio-to-noise ratio.* Speakers serving *spectator groups in outdoor installations* should be placed near the audience. For distance coverage upto 1/2 mile, clustered projectors are employed. *School auditoriums* are best served by *sound columns,* comprised of several speakers mounted in the same vertical or horizontal housing, as shown in Fig. 20.12.

After the required audio power has been determined by adding the power ratings of speakers in the system, a suitable amplifier is chosen. *It is a good practice to operate an amplifier below its maximum - rated output level.* For example, if 6 horn - type speakers rated at 6 watts each and 5 cone type speakers at 3 watts each are used, the total audio power requirement is 51 watts. In turn, an 80 - watt amplifier would be a conservative choice, permitting the addition of a few special speakers *in the event they may be needed.*

Most PA amplifiers have provisions for one or more microphones and one or more other *sound sources such* as FM-AM radio, tuner outputs, TV sound output, tape-recorder and record-player outputs. *Level (fader) controls* may be provided to adjust the volume from each of several sources simultaneously. High-input impedances are generally provided and *matching transformers* can be employed to accommodate low-impedance sound sources. Fig. 20.13 illustrates a typical Audio Research Amplifier.

Most PA amplifiers have speaker outputs for a *constant voltage system* (either 25 or 70.7 volts) plus 4, 8 and 16 Ω outputs to match voice coils directly. A *25 volt system* is easy to instal because conduit is not required. On the other hand, a *70.7 volt system* provides comparatively high operation efficiency.

After the amplifier has been selected, suitable microphones should be considered. *The choice of microphones is chiefly determined by intended use.* In other words, no one microphone will perform equally well under all conditions. The basic considerations are :

(*a*) the pick-up element that is utilised
(*b*) the pick-up pattern that is optimum for the location, and
(*c*) the conditions under which the microphones have to operate.

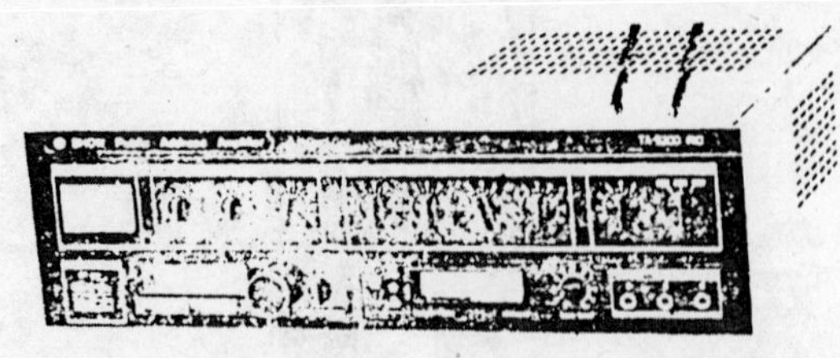

Fig. 20.13 A typical Audio Research Power Amplifier.

Ceramic microphones are comparatively inexpensive, have relatively high output and wide frequency response and resist climatic changes. *Dynamic microphones* resist shock and bad weather conditions, have superior frequency response and fidelity, and are well suited for orchestral or vocal reproduction. *Carbon microphones* are inexpensive and rugged, have high sensitivity and also have the disadvantages of inferior frequency response and comparatively high distortion. Typical microphones used in PA systems are shown in Fig. 20.14.

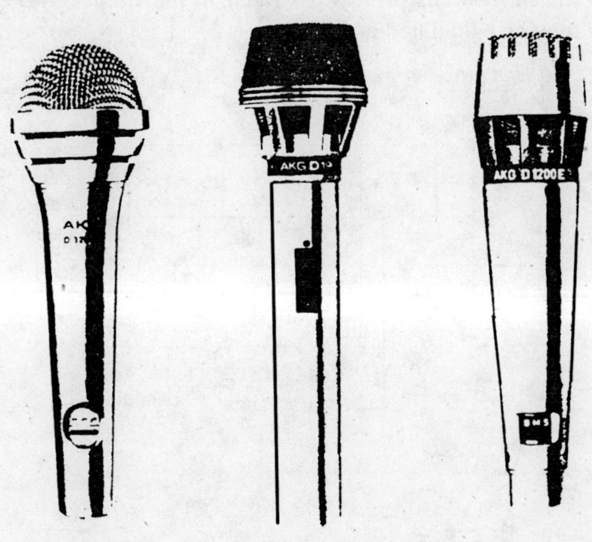

Fig. 20.14 Typical microphones used in PA systems.

A detailed description of different types of microphones is given in Chapter 5. Ceramic and dynamic microphones are generally recommended for the reproduction of speech and music. High-quality dynamic microphones, sometimes called *broadcast-quality microphones* have high -fidelity response and are commonly used with PA systems for auditoriums, stage groups, theaters, and concert halls. Carbon microphones are *utility-type transducers* and are used only in systems where economy is desired and high-quality reproduction is not necessary.

Microphones generally have substantially better characteristics than speakers. Microphones are also affected less by poor acoustics than are speakers. In many PA systems a wide choice of microphones is permissible whereas speaker type, placement and energisation may be critical and demanding.

20.9 SPEAKER MATCHING SYSTEMS

Speaker matching systems permit *higher operating efficiency* in audio power systems. *The 70.7 volt system is preferred for high-power PA networks.* Fig. 20.15 illustrates such a network for three speakers. A *matching transformer* is installed to match a speaker or a group of speakers to the 70.7 volt line. This is called a *constant - voltage system* because the line voltage is comparatively unaffected by switching speakers on or off. Network calculations are as follows :

1. Determine the *power rating* of each speaker.

2. Add the power ratings to find the *total power demand* and use a 70.7 volt amplifier with a rated power output at least equal to this demand.
3. Select a 70.7 volt *matching transformer* for each speaker, or for each group of speakers, Fig. 20.15, with appropriate primary wattage ratings.
4. Connect the *primary terminals* of each transformer across the 70.7 volt line from the amplifier output. Note that *a primary mismatch up to 25 percent is tolerable.*
5. Connect the *secondary terminals* of each transformer to its speaker, or group of speakers, observing the *matching ohms taps.*
6. In case the matching transformers are rated in *impedance values*, the primary wattage of a transformer may be calculated as follows.

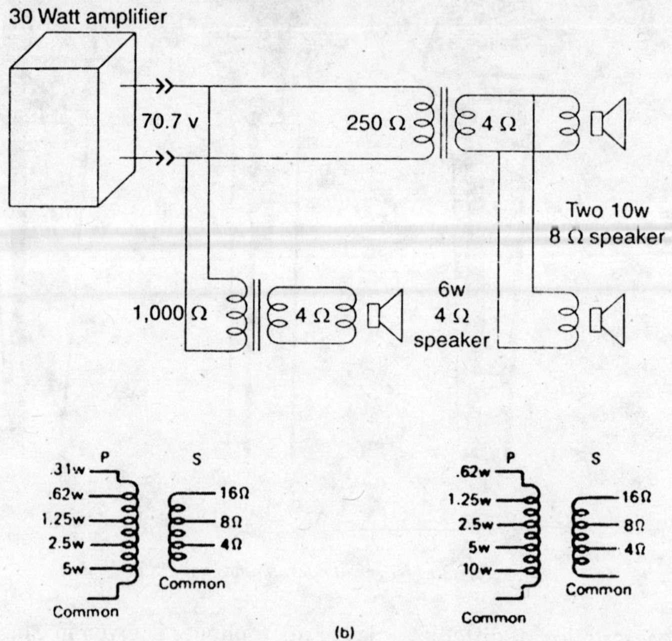

Fig. 20.15 A typical 70.7 volt speaker matching (a) configuration and (b) typical 70.7 volt transformer taps.

$$Z_P = \frac{70.7^2}{P} \qquad \text{...(20.1)}$$

where Z_P is the rated primary impedance and
P is the wattage rating of the speaker.

Problem 20.1 : Calculate the power rating of the amplifier in Fig. 20.15

Solution : The power demand of the speakers = 26 watts.

The amplifier rating = **30 watt (approx.)**

Problem 20.2 : In the above example, calculate the impedance of the primary windings.

Solution : The primary impedance for

$$\text{6-watt speaker} = \frac{70.7^2}{6} = \mathbf{833\Omega(approx.)}$$

A 1000 Ω impedance could be utilised.

The primary impedance for the 20-watt speaker combination $= \dfrac{70.7^2}{20}$

$= 250\ \Omega$**(approx.)**

20.10 PA SYSTEM CHARACTERISTICS

In *speech reinforcement systems* the listener hears sound from two sources. Part of the sound comes from the person making the speech; yet another part of the sounds comes from the PA system. The total sound that arrives is greater than that from either source alone, but the two portions do not arrive simultaneously. From a practical viewpoint, *the maximum tolerable time delay between the two arriving wavefronts is approximately 25 milliseconds.* A longer time delay is likely to result in a *confusing separation* of the two sound sources; the listener's attention jumps from one source to the other and becomes distracted. Since the human ear has a natural tolerance for echo effects, an echo tends to be disregarded under ordinary conditions. However, if the echo-arrives too long after the original wavefront, it is no longer disregarded and the listener becomes uncomfortably aware of the *split* sound sources. Therefore, it is good practice to plan a PA system so that no listener will be exposed to two arriving wavefronts that have a time delay greater than approximately 25 milliseconds. That is called the *Haas effect.*

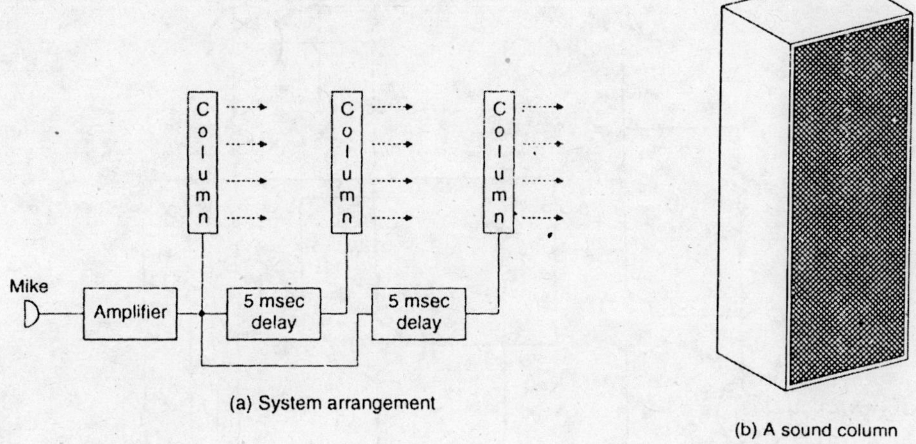

(a) System arrangement

(b) A sound column

Fig. 20.16 PA system with time-delay units.

A cluster of speakers, or a *speaker column,* will provide considerable sound coverage. However, in large buildings that have appreciable reverberation, a large speaker column may need to be *supplemented* by smaller columns placed farther back from the sound source. A supplementary speaker column, Fig. 20.16, can be installed with a *time-delay circuit* so that the sound from the first column arrives 5 millisecond before the sound from the second column is radiated. *The intensity of the delayed sound should not be more than 10 dB greater than that of the first - arriving sound, otherwise separation is likely to become evident.* Time delay units for PA systems generally employ a tape deck with spaced playback recording and playback heads, or a magnetic disc.

20.11 PA AMPLIFIERS

A PA amplifier may be designed as a unit, or it may consist of separate amplifiers. When more than one audio input source is utilised (two - microphones, for example) operating controls ar provided to *change*

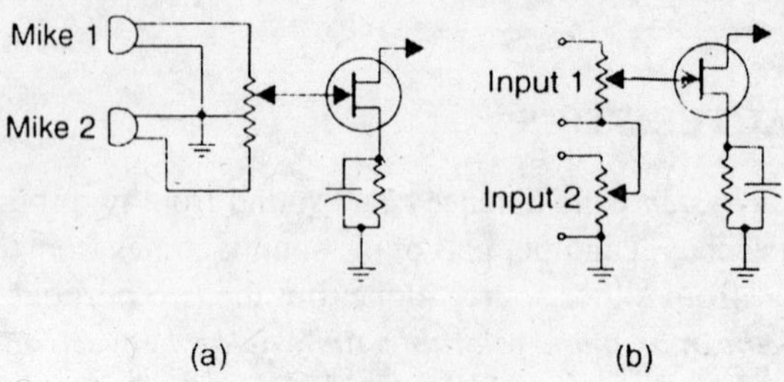

Fig. 20.17 Amplifier input arrangements; (a) a fader circuit that does not provide mixer action and (b) a mixer device with fader controls.

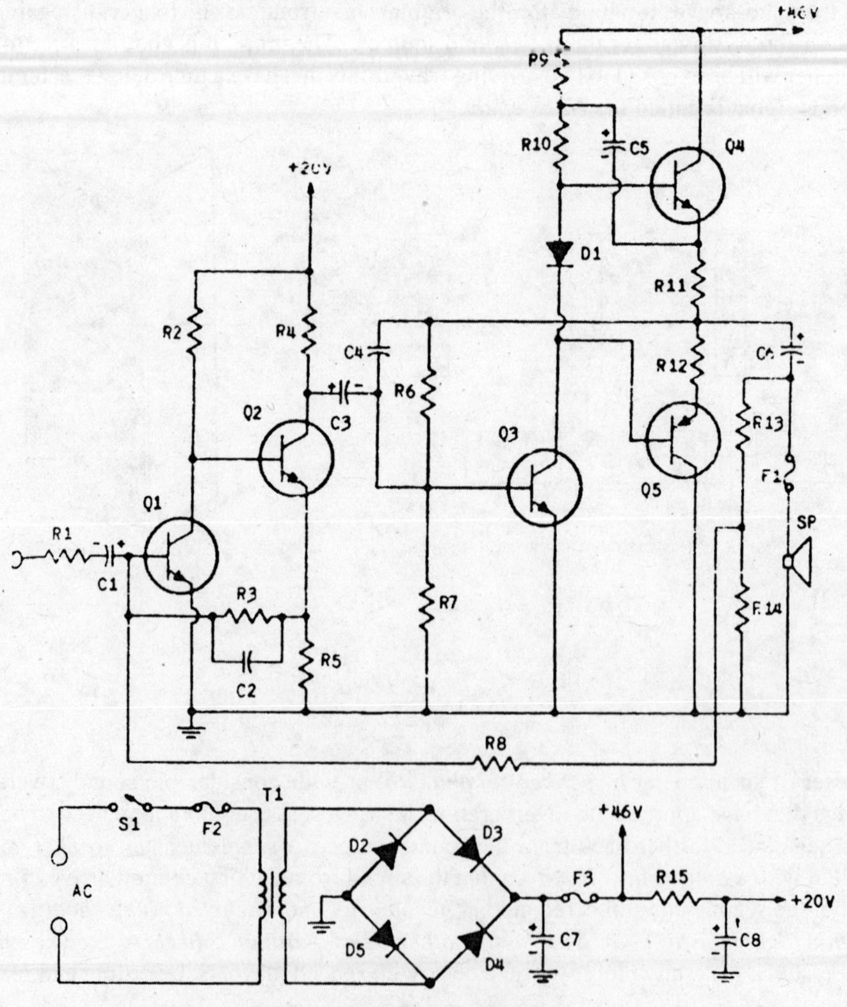

Fig. 20.18. A 25 Watt PA Amplifier.

from one source to the other, or to *mix* the outputs of the sources, The *mixer and fader* section is designed to provide *flexibility of operation* without sudden increases or decreases in volume. A *fader* serves to gradually reduce the volume of one source while gradually increasing the volume of another source. A *mixer* may or may not provide fader action.

A microphone amplifier provides preliminary step-up of the microphone output signal. The power amplifier is driven by an intermediate amplifier which is basically similar to a microphone amplifier except that it may employ high power transistors. If a power amplifier has sufficient gain it may be driven directly by the microphone amplifier.

The *complementary symmetry* public address audio amplifier, Fig. 20.18, is rated at 25 watts audio power. For stereo operation, the amplifier circuit is doubled, using the single power supply for both channels. Since this is a basic power amplifier, a suitable *pre-amplifier* is required. The balanced bridge power supply helps eliminate ripple, resulting in noise-free reproduction. A dual voltage output is obtained across dropping resistor R15.

This amplifier can be used in a number of applications, provided there is suitable drive from a pre-amplifier. Typical applications include high fidelity. It can also be used in an electronic organ, as a playback amplifier for a tape deck, and the 25 watts of audio power is sufficient for use as a musical instrument amplifier.

20.12 MEGAPHONES

Portable electronic megaphones, Fig. 20.19 are used by construction workers, law-enforcement officials, boat navigators and so on. The unit comprises a *reentrant horn speaker* with a microphone mounted on the back.

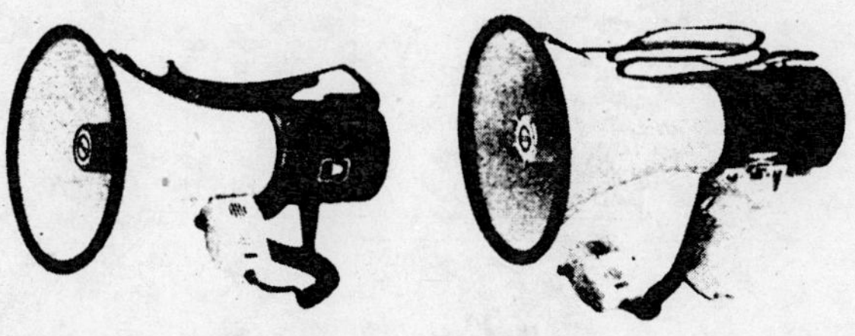

Fig. 20.19 Portable electronic megaphones.

A 5-transistor class B amplifier is typically employed to step up the microphone signal. The unit is powered with a 9-volt battery and has a maximum output of 8 watts with a sound range of 600 yards. A volume control is provided and a built-in siren signal source in available in the unit shown in Fig. 20.19. Yet another design is shown in Fig. 20.20. This is a 20-watt unit and provides more than twice as much as the portable unit. A switch is provided to change the circuit configuration into a sensitive directional listening arrangement. An input jack is available for feeding in a signal from a record player, tape player, or radio tuner.

20.13 INTERCOMMUNICATION EQUIPMENT

Intercommunication units are classified into master and remote types. A *master station* can call any one of the remote stations. On the other hand, a *remote station* can call the master station only. An intercommunication wiring system may employ master stations throughout. In this case, any station can call any other station. This type of intercom set has remote units with *beep buttons* that permit a caller to signal the master station even when the system is turned off.

Fig. 20.20 A 20-watt PA talk-listen portable unit.

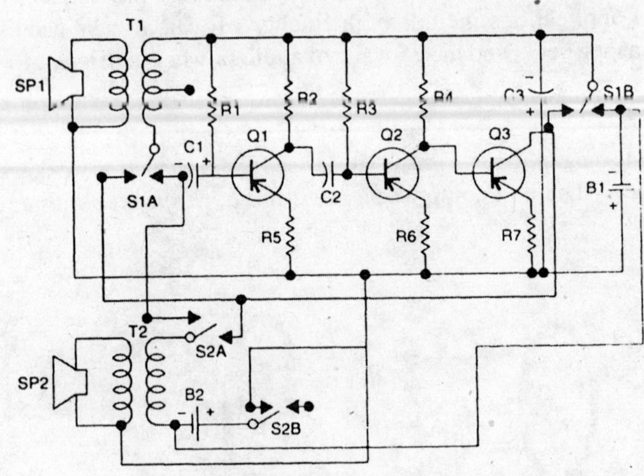

Fig. 20.21 Intercom

The intercom circuit shown in Fig. 20.21 is a 3-transistor amplifier with a slightly sophisticated switching circuit. *Each speaker serves the dual function of microphone and loudspeaker*, with the switches determining what that function is. When S1 is in *talk position.* T_1 feeds the base of Q_1 through C_1.SP_1 does indeed serve as a microphone. At the same time, S_{1B} is connected so that the output signal from Q_3 emitter circuit is connected to transformer T_2 feeding the signal to SP_2 which then functions as a loudspeaker. With the circuit operating in this fashion, whatever is spoken into SP_1 will be heard in SP_2. By reversing switch S_1, you *reverse* the process. S_2 also provides a *calling signal* by feeding the amplifier output signal back to the input, causing it to oscillate and howl.

The *remote unit* consists only of SP_2, S_2 and B_2. Additional remote units may be added, connecting them the same as the first one, using 4-conductor cable.

20.14 BACKGROUND MUSIC AND PAGING SYSTEMS

Background music may at first seem to be merely an amenity, but it has proven economic value. *Music-while-you-work* was introduced during World War II because it was found to boost production at a time when there was a manpower shortage. In stores, background music puts buyers in a happier mood and the sales go up. In airports and other public buildings, it may have no direct economic advantage, but as background music is now readily available, and a *paging system* is virtually a necessity, the surroundings can be rendered more pleasant by utilising the system for background music, while it is on stand by.

Background music is chosen to be *unobtrusive,* not requiring listener concentration for its enjoyment. And the reproduction is not so wide-range as either high fidelity in the house, or musical program broadcast. Reducing the response to *lower frequencies* enables smaller speakers (physical size) to be used and reducing the *upper frequencies* lends a pleasant roundness to the reproduction that helps ensure its unobtrusiveness.

Paging systems fit in quite naturally with background music, using microphones at certain control locations for feeding necessary paging announcements to the system. At an airport, for example, each airline will have a microphone to enable the counter to page customers who may have failed to pickup their tickets or other material or for whom telephone messages may be waiting.

Complex paging systems are built up of relatively simple units. Microphone inputs feed into immediate microphone amplifiers and power amplifiers feed sections of loudspeakers, and the outputs from groups of microphone amplifiers are fed to the appropriate power amplifiers so that correct groups of areas are covered with each group of announcements. One more thing in such a complex system : the *indication* whether the system, or any part of it, is free from a particular announcement. This is achieved by a *system of lights,* keyed by a microphone in use to indicate to all other microphone stations, that might want to use the same group or groups of speakers, that the system is in use.

A more sophisticated system might include an electrical (by means of *relays*) or electronic (by means of *signal-gating circuits) interlock* to ensure that accidental pressing of another key will not interfere with an announcement in progress.

OBJECTIVE TYPE QUESTIONS (A)

1. In electric guitars provision is made to

(*a*) amplify the character of sound
(*b*) modify the character of sound
(*c*) amplify or modify the character of sound
(*d*) amplify and modify the character of sound

2. Circuits that accentuate one or more bands of frequencies are called

(*a*) band pass filters (*b*) formant filters
(*c*) low pass filters (*d*) high pass filters

3. In electronic music synthesizer systems

(*a*) series of tones are generated at a time
(*b*) a single tone is generated at a time
(*c*) a single series of tones is generated at a time

4. Music synthesis by means of a digital computer involves a

(*a*) digital-to-acoustic converter

(*b*) acoustic-to-digital converter

(*c*) analog-to-acoustic converter

(*d*) acoustic-to-analog convertor

5. High-fidelity reproduction is required for a
   (*a*) public address system (*b*) intercom system
   (*c*) sound reinforcement system (*d*) paging system
6. School auditoriums are best served by
   (*a*) woofers (*b*) tweeters
   (*c*) sound columns (*d*) loudspeaker systems
7. The maximum tolerable time delay between two arriving wavefronts is
   (*a*) 5 ms (*b*) 15 ms
   (*c*) 25 ms (*d*) 35 ms
8. A sophisticated paging system may include
   (*a*) amplifiers (*b*) relays
   (*c*) oscillators (*d*) signal-gating circuits
9. A master station (intercom) can call
   (*a*) only one remote station
   (*b*) any one of the remote stations
   (*c*) only two remote stations
   (*d*) any two of the remote stations

FILL IN THE BLANKS (B)

1. In an electric guitar provision is made to alter the _______- of sound in many ways.
2. The pick-up section on the guitar may consist of one or more _______ or _________ transducers per string.
3. The clipper circuits generally serve to augment the strength and number of partials so as to provide greater possibilities for _________ in the tone-forming filters.
4. Electronic systems to _______ and modify tones of individual musical instruments are being used extensively.
5. A public address system facilitates the __________ of intelligible speech to groups of people.
6. In a public address system the frequency response is typically limited to a range of ___________
7. In a public address system _________ is often a basic requirement.
8. Indoor sound - reinforcement systems generally employ a ________ range of speakers.
9. Speaker matching systems reduce losses in the wiring and increase _______.
10. It is good practice to operate an amplifier ____________ its maximum rated output level.
11. Most PA amplifiers have outputs for a __________ system.
12. The choice of microphones is chiefly determined by _____________.
13. Microphones generally have substantially __________ characteristics than speakers.
14. In speech reinforcements systems the listener hears sound from __________.
15. The maximum tolerable time delay between two arriving wavefronts is _____________

CONVENTIONAL QUESTIONS (C)

1. Explain briefly the working of an electric guitar.
2. How does an electric wind instrument work ?
3. Explain the working of a manual synthesizer.
4. How does a manual synthesizer differ from a programmed synthesizer ?

5. What is the difference between a public address system and a sound reinforcement system ?
6. How will you plan a public address system ?
7. What are the different types of microphones used with public address systems ?
8. Explain the significance of a speaker matching system.
9. What is the difference between a PA system and a megaphone ?
10. Why background music is required ?
11. Explain the working of a simple intercom system with the help of a diagram.
12. Write short noes on :
    (*a*) Formant filter
    (*b*) Haas effect
    (*c*) Paging

21

AUDIO SYSTEMS

For a while hi-fi and stereo were the forgotten components in consumer electronics. All the publicity and spare spending money was going on videos and computers. But now the video boom is levelling off, for the simple reason that most families which want and can afford a video recorder have one. However, there has always remained a hard core of audio enthusiasts who either like playing with hi-fi equipment for the sake of it or like listening to music on hi-fi equipment for the better sound quality it offers.

The traditional aim of a hi-fi system is to reproduce recorded or broadcast sound with the highest possible fidelity to the original. Of course, it can be argued that many modern recordings can exist only as a record. Editing and all the high technology tricks available in a recording studio let the artist create a sound on tape or disc that is wholly artificial, and more perfect than anyone could achieve in live performance. So hi-fi today is concerned with reproducing whatever the artist intended.

21.1 ANATOMY OF A HI-FI SYSTEM

The first essential is to have a *stereo system.* This creates a spread of sound from two loudspeakers, similar to that which we hear naturally with our ears. Virtually all sound reproduction systems are now stereo, but they are not necessarily hi-fi. There are plenty of very hi-fi stereo systems on sale, so *do not confuse stereo with hi-fi.*

There are two basic approaches to hi-fi. The system can be built up from *several separate components,* often from different manufacturers; or it can come as *a single integrated system* in a rack mount or music centre. Both approaches have their good and bad points.

In the late 40's and early 50's, a group of inspired enthusiasts made it their mission to optimize each component of the console's music reproduction system. Usually, this meant keeping those components *separate,* in order to render music with the highest fidelity to the original source. This was the beginning of what we now call *component hi-fi,* Fig. 21.1.

The advantage of separate components is their *flexibility.* Different numbers and combinations of components can be assembled into a system, according to the listener's desire or needs. And *as new technology arrives, only the component affected need be upgraded, not the entire system.* Within each component type, there is a wide variety of features and performance levels so the listener can pick from whatever is available according to appetite or budget, returning to buy a *state-of-art* as needs and circumstances change.

Although component hi-fi has been around since the monaural days, the breadth and depth of choice still puzzles prospective buyers. A brief study of anatomy of a stereo system should put the options in clear focus

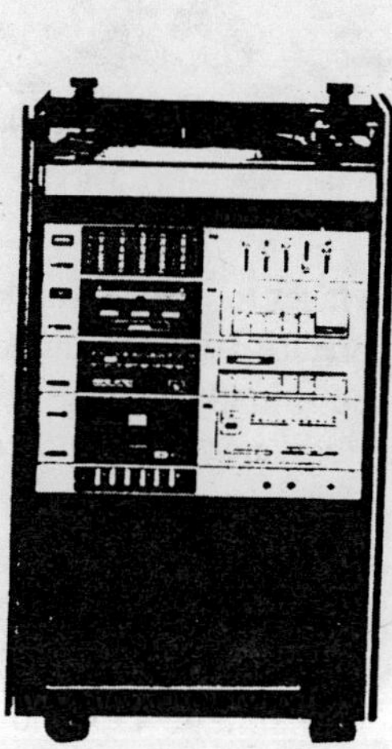

Fig. 21.1 Component System : semi-auto belt-drive record player; graphic equalizer deck; amplifier with pop - up remote control handset; tuner deck and casette deck.

Fig. 21.2 CD's and CD Player.

Basically a *stereo system* comprises three sections, signal source; signal processing and amplification. The first section *extracts* music from a storage medium, such as a disc or tape. The second section *selects and boosts* the weak source signal. The final section *converts* electrical signals into mechanical energy - sound waves through loudspeakers or headphones.

21.2 SOURCE UNITS

Sound is carried by various media. It's the job of a source unit to *extract* the sound from its storage carrier and *convert* it into electrical impulses. Source units are the most numerous among audio components; most system include more than one.

Compact Disc Players, Fig. 21.2, use a laser to extract sound that has been digitally encoded on a disc. Besides the standard 4 3/4 - inch music CD, there are 3-inch minidiscs (dubbed CD-3s) and 4 3/4, 8 and 12-inch CDV-s that add still or motion video to the audio track. Some CD players (dubbed *combiplayers)* handle all disc types, including laser video; others are dedicated to sound reproduction only. *Changer* models house several CDs in a tray or magazine and can be programmed to play discs or individual cuts in any order.

Turntables, Fig. 21.3, also called phonographs, are fitted with a phonocartridge stylus mounted on a tone arm. As the table spins at 33 3/4, 45 or 78 revolutions per minute, the stylus or needle *extracts* sound from the wiggly analog grooves of 7-, 10-, or 12- inch discs.

Tape deck's sound-storage medium is magnetic tape housed either in small cassettes, eight-track cartridges or on larger, open reels (these last two formats are virtually extinct). Tape decks can record sound (on blank tapes) as well as play it back. The term *tape player* refers to a machine that does to have recording ability. Recording may be from live sources, through microphones, or from other source units in a system. Some cassette decks called *editing or dubbing decks,* have transports for two cassettes (see Fig. 21.4).

Tuners, Fig. 21.5, extract radio waves from the atmosphere. Some tuners receive only the high-fidelity, usually stereo, *FM* broadcasts. Most tuners also tune in the *AM* band, to accommodate listeners who lack a wide variety of FM fare. Tuners are available as separate components but can also be part of a popular hybrid component called the *receiver,* which houses *signal processing and amplification circuitry* in the same chassis.

Fig. 21.3 Turntable.

Pre-amplifier, Fig. 21.6, also called a *control amplifier,* is the component to which all source units are connected-*their port of entry into the audio system.* It allows you to *select* the desired source and *route* the signal through the system. It also *processes* or manipulates the source signal with controls for bass and treble for example. Additionally, this component *pre-amplifies* the weak source signals before passing them on to the power amplifier.

Fig. 21.4 A typical dubbing deck with transport for two cassettes.

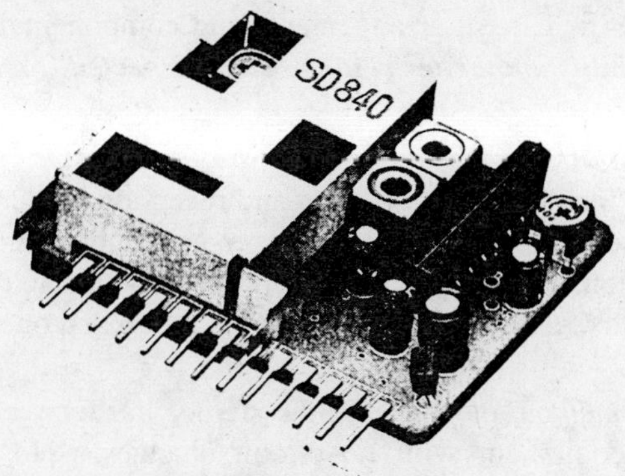

Fig. 21.5 Tuners.

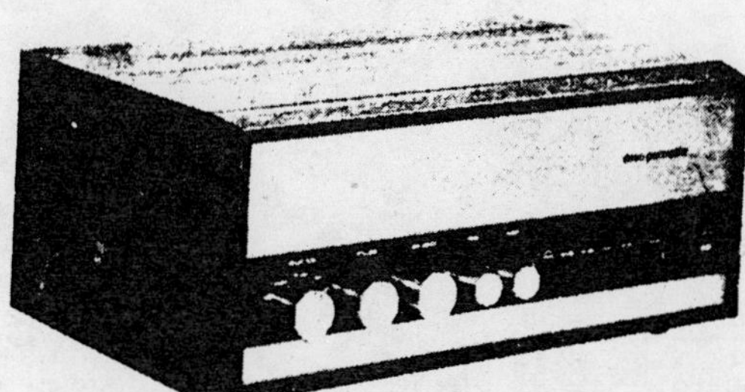

Fig. 21.6 Pre-amplifier - the port of entry into the audio system.

Power amplifier, Fig. 21.7, is *the business end of the system.* The power amplifier boosts all the electrical signals, thereby providing *the impetus that drives the loudspeakers* to literally push air and generate sound waves.

Hi-fi speakers for average living rooms require at least 10-watts of audio output power from the loudspeaker, to handle most conditions. When several rooms, large rooms or outside areas, are to be supplied with the highest fidelity as much as 60-watts of power might be desirable in a home installation.

Audio control centres including pre-amplifiers, equalising networks, boost, droop, loudness, compensation, filtering and switching circuits are more elaborate and usually used instead of the plain pre-amplifier or a simple pre-amplifier equaliser.

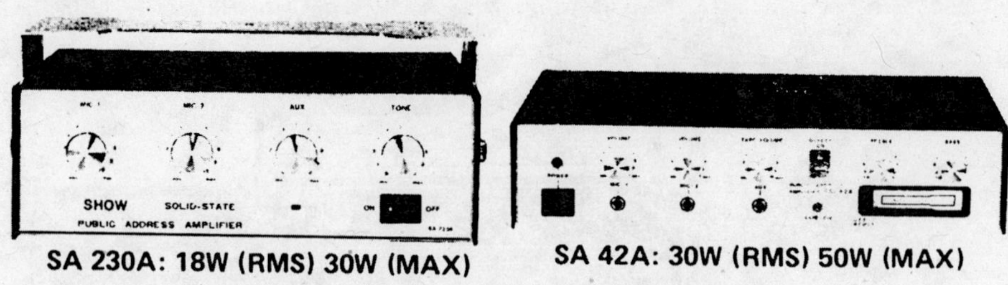

Fig. 21.7 Power amplifier - the business end of the system.

From 50 to 100 dB of *amplification* of very low level signals from pick-ups and other inputs according to the need, plus *compensation and controls* to obtain various emphasis, is provided by the *pre-amplifier-equaliser part of the system* or by a separate *hi-fi control centre unit.*

Hi-fi systems are designed around either of these two types of amplifier arrangements, that is, the *combination* of pre-amplifiers, control, and power amplifier, all in one unit, or the control centre plus basic amplifier units used *separately.* Circuit arrangements and components may be somewhat similar in both arrangements, but *more flexibility and better results can be expected from separate units.*

21.3 SIGNAL PROPAGATION

The first method used for stereo broadcasting and reception is illustrated in Fig. 21.8. Two complete transmitters were used, one sending out the *left* (L) stereo signal, the other the *right* (R) stereo-signal of the same program. All combinations of AM broadcast, FM broadcast, and television sound-channel transmission have been tried. Most popular were the *FM/AM* (one FM station and one AM station) and the *FM/FM* (two FM stations) methods.

The *two transmitter method* of transmitting stereo was convenient because it utilised existing transmitting and receiving equipment with no circuit changes. However, it did have the following *disadvantages:*

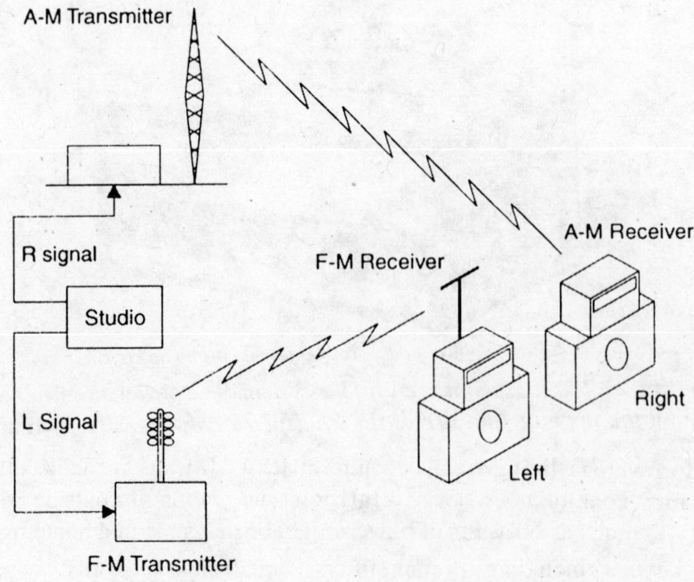

Fig. 21.8 Two transmitter method for radio transmission of stereo-signals.

(*i*) Two channels had to be used for each program, resulting in *wasted spectrum space.*

(*ii*) Differences in propagation characteristics of the waves radiated by the two transmitters led to *variation in signal amplitude and quality within the separate channels.* This was particularly true of FM/AM combinations, where the carrier frequencies are so widely separated.

(*iii*) In some cases, especially with FM/FM, *the duplication of complex receiving equipment represented an excessive expense;* ninety percent of the programs were non-stereo, requiring only one receiver; but to receive stereo, two receivers were needed. Also a number of AM/FM tuners could not receive both bands simultaneously, and so they had to be supplemented with *additional receiving equipment.*

21.4 STEREO MULTIPLEX

The disadvantages of the two-station method made it imperative that a method of transmitting both signals on one carrier be devised. The result is the method called stereo-multiplex. *Multiplex means a method of sending two or more messages simultaneously on one carrier or circuit.* This is exactly what stereo multiplex does - it transmits both channels on the same FM carrier.

The standard FM broadcast system is based on an audio frequency response of approximately 0 to 15 kHz. There is nothing about the basic modulating systems to prevent extension of this audio-frequency range to 75 kHz (further extension would require more bandwidth than is currently allotted by FCC). Therefore, if at some point above the limit of human hearing (but still within the 75-kHz bandwidth) *another signal* is added, it will be amplified and detected by the regular receiver circuits. However, *being above the range of hearing (supersonic) it will not interfere with the regular audio signal.* Thus, the carrier is modulated by the regular audio signal plus the supersonic signal.

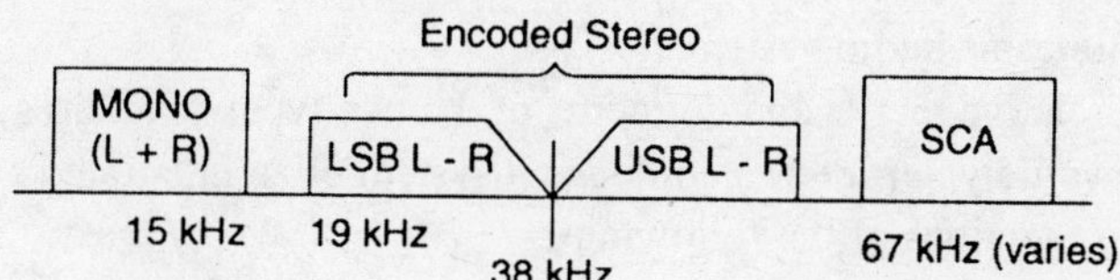

Fig. 21.9 Diagram showing the spectral location of the audio and encoded signals in stereo-multiplex.

The supersonic modulation is known as the subcarrier (SCA in Fig. 21.9). It, in turn is modulated by another audio signal. As long as the bandwidth of the subcarrier is not allowed to extend downward into the range of regular carrier, no interference will occur. This complete modulated subcarrier signal is part of the composite signal which modulates the main carrier. One of the stereo signals is transmitted as modulation of the subcarrier. In receiving, two demodulating circuits are needed. First the main carrier is demodulated to get the first stereo signal and the modulated subcarrier. Then the modulated subcarrier is separated and demodulated to get the second stereo signal.

21.5 COMPATIBILITY

There will always be some listeners who don't care about being equipped for stereo reception. Also, for portable use, and where cost is to be minimised, a conventional single-channel receiver is called for. For this reason, it is important that the stereo system be compatible. *By compatible we mean that while stereo multiplex transmission is in progress a listener with a conventional receiver (one not equipped for stereo) should be able to receive the transmission as a full monophonic signal.*

If one channel of the multiplex system is used for the L signal and the other for the R signal, compatibility is not provided. The owner of the conventional receiver hears only the modulation of the main carrier, which in this case would be just the L signal; the subcarrier frequency is above audibility, so neither the subcarrier nor its modulation is used. *For a truly compatible system, the nonstereo listener should hear the combination of both L and R signals.*

To meet this compatibility requirement, the main carrier can be modulated with a full monophonic signal (L + R), and the subcarrier channel can be modulated by the difference between the two signals (L - R). The (L + R) modulation signal provides the nonstereo listener with his full monophonic reception; his receiver does not respond to and thus ignores the subcarrier.

For stereo reception, the L and R signals are recovered by demodulating the subcarrier signals and adding or subtracting the difference signal from the main carrier (L + R) signals.

Electrical *addition* of the signals can be accomplished simply by applying both signals across a common impedance. *Subtraction* can be accomplished by first inverting one of the signals (changing its phase by 180 degrees), then adding it to the other signal.

One point should be emphasised. *Stereo reproduction is a part of high fidelity, not an "additional feature."* The impression is sometimes wrongly given that "hi-fi" and stereo are two separate stages of development of audio reproduction. This is a fallacy, unfortunately, sometimes encouraged by a statement that some equipment features "high fidelity and stereo."

Stereo is a "fine point" in high fidelity; that is, if you do not have all the good qualities of high-fidelity monophonic system to start with, other forms of distortion will prevent appreciation of the benefits of stereo effect. If the reproduction of your system is clean and clear of harmonic intermodulation and transient distortion, it is likely that improvements in spatial sound effects through stereo will be appreciated.

CONVENTIONAL QUESTIONS (C)

1. What constitutes an audio system ?
2. Discuss the relative merits and de-merits of the two types of audio systems
3. Which one would you prefer, a component system or an integrated system and why ?
4. Describe the functions of the following
   (*a*) source units (*b*) pre-amplifier
   (*c*) audio control centre
5. Explain the two-transmitter method of stereo broadcast. What are its disadvantages ?
6. Discuss the significance of compatibility. How compatibility is achieved by stereo-multiplexing?

22

MONOCHROME TV STANDARDS AND SYSTEMS

The word television means viewing from a distance. In a broader sense, television is the transmission of picture information over an electric communication channel. It is desired that the picture picked up by television receiver be a faithful reproduction of the scene televised from a TV studio. Any scene being televised has a wide range of specific features and qualities. It may contain a great number of colours, gradations of shade, coarse and fine details; motion may be present in a variety of forms and the objects making up the scene are usually in three dimensions.

The more faithful the received TV picture is the more elaborate, expensive and bulky the TV equipment must be. On the otherhand among the important requirements for a TV system, and especially so far a telecasting system, are the utmost in simplicity, reliability and low cost. To simplify the physical facilities employed in telecasts, it is essential to determine which qualitative aspects are of secondary importance in the TV picture. It may be agreed that colour and 3 dimensional effect are second in importance to other aspects more vital to the information transmitted, such as definition and contrast. This is why, in the early days of television, scientists and inventors limited themselves to finding methods for the transmission of two dimensional black-and-white motion pictures with acceptable definition and contrast.

In the early 20th century, TV systems, primitive from the standpoint of the present day were mechanical. Today, TV equipment is all electronic. The replacement of mechanical by electronic television has made it possible to reproduce a high-quality picture approaching the orignal scene in quality

22.1 ELEMENTS OF A TELEVISION SYSTEM

Television is an extension of the science of radio communications, embodying all of its fundamental principles and possessing all of its complexities and, in addition, making use of most of the known techniques of electronic circuitry.

In the case of the *transmission and reproduction of sound,* the fundamental problem is to convert *time variations* of acoustical energy into electrical information, translate this into radio frequency (RF) energy in the form of electromagnetic waves radiated into space and, at some receiving point, *reconvert* part of the resultant electromagnetic energy existing at that point into acoustical energy.

In the case of *television*, there is the parallel problem of converting *space and time variations* of luminosity into electrical information, transmitting and receiving this as in the case of sound, and *reconverting* the electrical information obtained at a receiving location into an optical image.

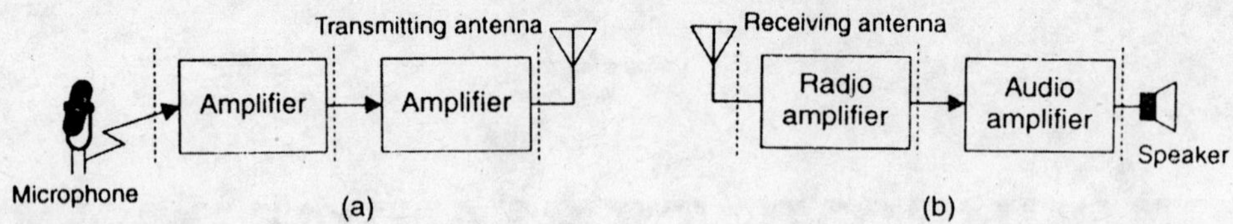

Fig. 22.1 Simplified block diagram of a sound transmitter (a) and a sound receiver (b). Notice that the two systems are virtually mirror images of each other.

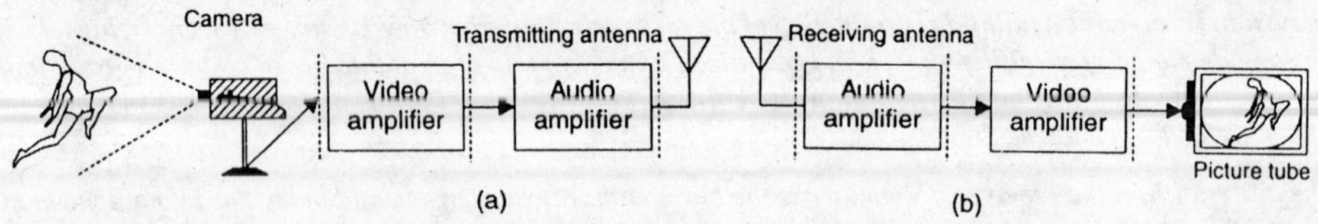

Fig. 22.2 Simplified block diagram of picture transmitter (a) and picture receiver (b).

When the information to be reproduced is *optical* in character, the problem fundamentally is much more complex than it is in the case of *aural* information. In the latter instance, there is dealt with at each instant of time only a single piece of information, since any electrical waveform representing any type of sound is a *single-valued function of time* regardless of the complexity of the waveform. In the corresponding optical case, at any instance there is an *infinite number of pieces of information existing simultaneously* namely, the brightness which exists at each point of the scene to be reproduced. In other words, the information is a *function of two variables, time and space.* Since the practical difficulties of transmitting all this information simultaneously and decoding it at the receiving end at the present time seem insurmountable, *some means must be found whereby this information may be expressed within the form of a single-valued function of time.* In this conversion, the process known as *scanning* plays a fundamental part.

22.2 THE SCANNING PROCESS

Scanning may be defined as *the process which permits the conversion of information expressed in space and time coordinates into time variations only.* Suppose, for extreme simplicity, that an optical image of a scene, perhaps on a photosensitive surface, is scanned by a beam of electrons, i.e. all points on the image are *sequentially* contacted by this beam, and that somehow, as a result of this scanning, *through capacitive, resistive, or photoemissive effects at the surface,* an electrical signal may be obtained that is directly proportional in amplitude to the brightness at the particular point being scanned by the beam. Although the picture content of the scene may be changing with time, if the scanning beam moves at such a rate that any portion of the scene content does not have time to move perceptibly in the time required for one complete scan of the image the resultant electrical information contains the true information existing in the picture during the time of the scan. This *derived information* is now in the form of a signal varying with time.

Consider the image of Fig 22.3(*a*). The light and dark areas represent *variations in brightness* of the original scene. Suppose that a beam of electrons is made to scan this image as shown, starting in the upper left hand corner and moving rapidly across the image in a time t_h, thus forming line 1 as shown, and then made to return instantaneously. This process is repeated until the bottom of the image is reached. The *time variation* of the horizontal component of motion of the scanning beam is that shown in Fig. 22.3(*b*).

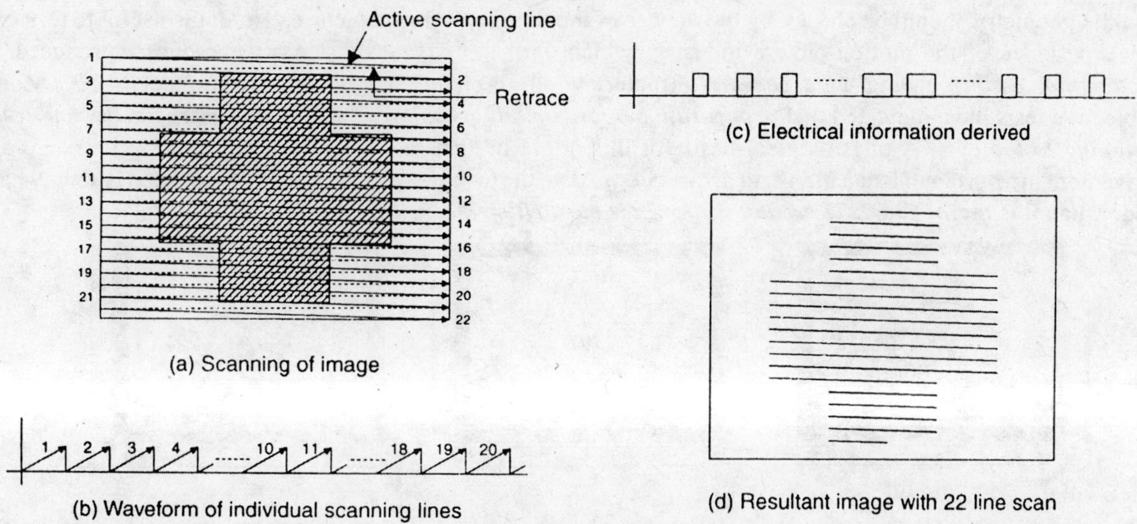

Fig. 22.3. The scanning process.

Means exist whereby an electrical signal corresponding in amplitude to the illumination on the point being scanned may be derived. The electrical output as a function of time corresponding to individual scanning lines is shown in Fig. 22.3(*c*). This electrical information corresponding to the light intensity of a large sampling of points in the original scene is now obtained, with the limitation that *the detail which can be reproduced depends on the completeness of coverage by the scanning beam.* This coverage is determined directly by the total number of scanning lines. If the scanning beam is now made to return to the top of the picture and this process is *repeated,* another sampling of information identical with the first will be obtained unless the scene has changed in the meantime, in which case the new sampling of information will be in accordance with this change.

At the receiving end, the fundamental problem is that of *recombining* the information , which has been broken down into a time function, into an optical image. The image which is reproduced, with the number of lines shown is indicated in Fig. 22.3(*d*)

22.3. SCANNING METHODS AND ASPECT RATIO

The *scanning mechanism* is the portion of the entire television system which deserves the most attention from the standpoint of the necessity of *formulation of standards of methods and performance.* Once the scanning system is standardised, the performance of the system is specifically limited.

The process of scanning makes possible the use of a *single transmission channel.* Otherwise there would be required as many transmission channels as there are simultaneous units of optical information to be reproduced.

In the scanning process the amount of detail actually converted to useful information depends on *the total percentage of picture area actually contacted by the electron beam.* The scanning process is such that the picture area is traversed at repeating intervals, giving in effect *a series of single pictures* much in the same manner that motion pictures are presented. The repetition rate of these successive pictures (referred to as *frame rate*) determines the apparent *continuity* of a moving scene. This rate must be chosen sufficiently high so that neither *discontinuity of motion* nor *flicker* is apparent to the eye.

The scanning method chosen might depend on the *geometry* of the image to be reproduced. It is logical that the geometry should be chosen by the viewer of the final reproduced picture. Here it is useful to rely on the experience of the motion-picture industry, which until the advent of stereo techniques reproduced a *rectangular pattern* having an *aspect ratio* (picture width to picture height) of approximately 4 : 3. Most subjective tests have indicated that *aspect ratios approximating 4 : 3 are most pleasing artistically and less fatiguing to the eye.* A physiological basis for this might be that the eye is less restricted in its range of movement in a horizontal than in a vertical direction. Also, the fovea, or area of greatest resolution is somewhat wider than it is high. Thus, *the area of the fovea is most efficiently utilised.*

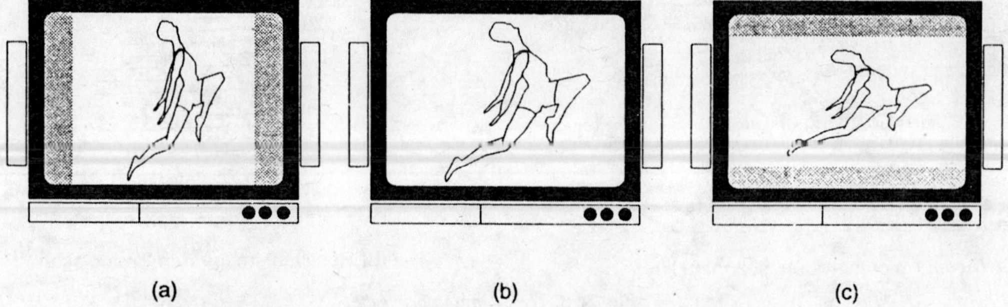

Fig. 22.4 Aspect ratio (a) A picture in which the width-to-height proportion is less than 4:3. (b) A picture in which the width-to-height proportion is NORMAL. Width is to height as 4:3. (c) A picture in which the width to height proportion is greater than 4:3.

Once the geometry of the image has been specified, it remains to determine what *sequence* may be used by the scanning beam to cover the entire area. For a picture of rectangular geometry, any form of spiral or circular scanning would be wasteful of scanned area. Also, scanning of this type would result in *nonuniform coverage* of the area or in *varying scanned velocities. A varying scanning velocity* for the pick up and reproducing devices results in *varying output for constant scene brightness.* Correction for this would be necessary.

Most practical television systems, therefore, resort to *linear scanning* of the image. Also, *it is common practice to scan the image from left to right, starting at the top and tracing successive lines until the bottom of the picture is reached, then returning the beam to the top and repeating the process.* This is shown in Fig. 22.5(a). The direction of the arrows on the heavy lines indicates the forward (*trace*) or useful scanning time. The dashed lines indicate the *retrace* time, which is not utilised in converting picture information into useful electrical information. The movement during the return of the scanning spot to the top of the image is shown in Fig. 22.5(b).

This is the type of scan which would result in a cathode ray device when the scanning-spot position is determined by the horizontal and vertical components of electric or magnetic fields where these components are *repeating linear functions of time* as shown in Fig. 22.5(c). If the repetition rate of the horizontal component is related to that of vertical component by a factor n, then n lines are formed during a complete vertical period. *The retrace times, both horizontal and vertical, are not utilised for transmitting a video signal but may be employed for the transmission of auxiliary information* such as teletext for example.

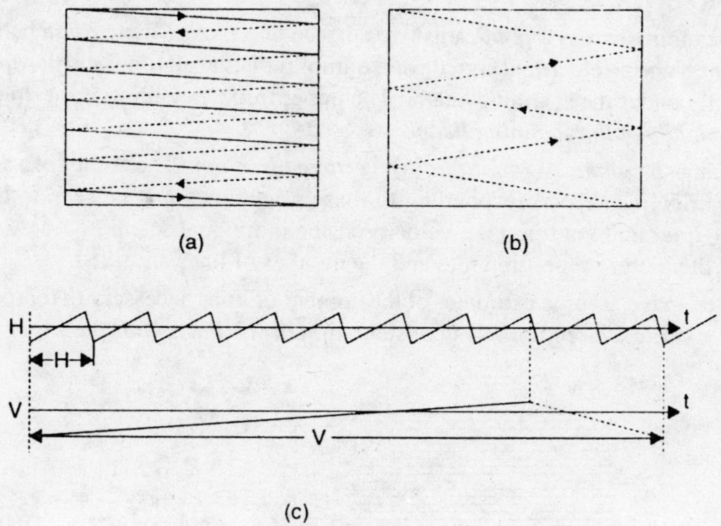

Fig. 22.5 Linear Scanning.
(a) Top to bottom scanning path of the beam
(b) Vertical retrace path of the beam.
(c) Deflecting field waveforms.

22.4. PERSISTENCE OF VISION AND FLICKER

The image formed on the retina is retained for about 20ms even after optical excitation has ceased. This property of the eye is called *persistence of vision,* an essential factor in cinematography and TV for obtaining the illusion of continuity by means of rapidly flashing picture frames. If the flashing is fast enough, the *flicker* is not observed and the flashes appear continuous. The repetition rate of flashes at and above which the flicker effect disappears is called the *critical flicker frequency* (CFF). This is dependent on the brightness level and the colour spectrum of the light source.

In cinema, a film speed of 16 frames per second was used in earlier films to obtain the *illusion of movement.* Lack of smooth movement was noticeable in these films. The present day standard for movie film speed is 24 frames per second and at this speed, flicker effects are very much reduced. This problem is further reduced in modern projectors by causing each frame to be illuminated *twice* during the interval it is shown by means of fan blades. The resulting flicker rate is quite acceptable for cine screen projection, because it is viewed in subdued light and a wide display area.

In television, the field rate is concerned with (*i*) large area flicker, (*ii*) smoothness of motion, and (*iii*) motion blur in the reproduced picture. As the field rate is increased, these parameters show improvements but tend to saturate beyond 60Hz. Further increase does not pay off, and increases the bandwidth. Hence the picture field scanning is generally done at the same rate as the mains power supply frequency, which conveniently happens to be 50 or 60 Hz for the same reasons of reducing illumination flicker from electric lamps. At 60Hz the flicker is practically absent. At 50Hz, a certain amount of *borderline flicker* may be noticed at high brightness levels used to overcome surrounding ambient light conditions. Use of commercial power frequency for vertical scanning also reduces possible effects, like supply ripple and 50Hz magnetic fields, in the reproduced picture.

22.5. VERTICAL RESOLUTION

An electron beam scanning across a photosensitive surface of a pickup device can be made to produce in an external circuit a time-varying electrical signal whose amplitude is at all times *proportional* to the illumination on the surface directly under the scanning beam. The upper frequency limit of the amplifier for such signals depends upon the *velocity of the scanning beam.*

Most scenes have brightness gradations in a vertical direction. The ability of a scanning beam to allow reproduction of electrical signals corresponding to these gradations may be seen to be dependent upon the *number of scanning lines* and not upon the velocity of the individual scanning lines. However, the two are integrally related if the picture-repetition rate and the number of lines are fixed.

It is possible to arrive at some estimates of the number of lines necessary to reproduce known variations in light intensity in a vertical direction from consideration of the *line pattern* in Fig. 22.6

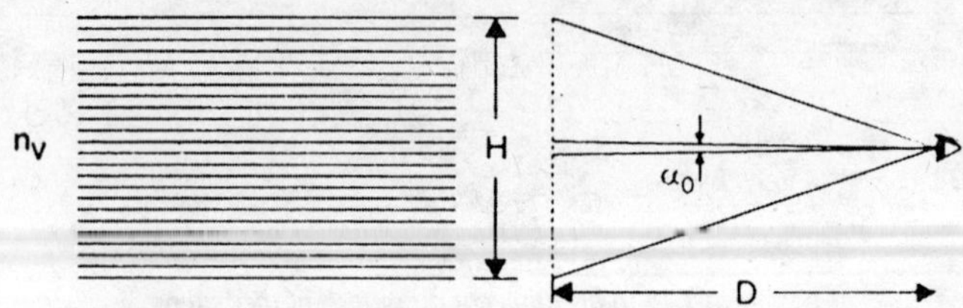

Fig. 22.6. Determination of the maximum number of black and white lines for vertical resolution.

The scanning lines used per frame is referred to as the *scan ratio* of a television system. The realistic limit to the number of lines is set by the *resolving capability* of the human eye, viz. about one minute of visual angle. For comfortable viewing an angle of about 10 to 15° can be taken as the *optimum visual angle.* Hence the best *viewing distance* for watching television is about 4 to 8 times the height of the picture, i.e. a visual angle of about 10° as shown in Fig. 22.6

The maximum number of dark and white elements which can be resolved by the human eye in the vertical direction in a screen of height H is given by n_v according to the relation.

$$H/D = n_v \times \alpha_o \qquad \ldots(22.1)$$

where n_v , is the number of black and white lines of vertical resolution,

α_o is the minimum angle of resolution in radians and D the distance of the viewer.

Problem 22.1. Calculate the number of black and white lines of vertical resolution for a visual angle of about 10°.

Solution: When α_o equals 10°, D/H = 6

$$n_v = \frac{H}{D\,\alpha_o} = \frac{1}{6} \times \frac{60}{1} \times \frac{180}{\pi} = 600 \text{ lines}$$

Problem 22.2. Calculate the number of black and white lines of vertical resolution for a visual angle of about 15°.

Solution: When α_o equals 15° $D/H = 4$

$$n_v = \frac{H}{D\,\alpha_o} = \frac{1}{4} \times \frac{60}{1} \times \frac{180}{\pi} = \mathbf{900 \text{ lines}}$$

The number of alternate dark and white elementary horizontal lines which can be resolved by the eye are thus 600 for 10° visual angle and 900 for a 15° visual angle. There has been a large difference of opinion about what constitutes the maximum number of lines because of the subjective assessment involved. This has been made further difficult to assess because of the effects of the *finite size* of the scanning beam spot.

22.6 PICTURE ELEMENTS

A picture element (pixel) is the smallest area in an image that can be reproduced by the video signal. The size of a pixel depends upon the size of the scanning spot, the number of scanning lines and the highest frequency utilised in the video signal. Its *vertical dimension* is equal to the distance between two scanning units , or centre to centre distance between two adjacent spots. Its *horizontal dimension* will also be the same.

With an aspect ratio of 4:3 the number of pixels on a horizontal scanning line will be 4/3 times the number of pixels on a vertical line. *The total number of pixels activated on the screen reproduce the picture.*

22.7. THE KELL FACTOR

In practical scanning systems, the maximum vertical resolution realised is less than the actual number of lines available for scanning. This is because of the finite beam size and its *alignment* not coinciding with the elementary resolution lines.

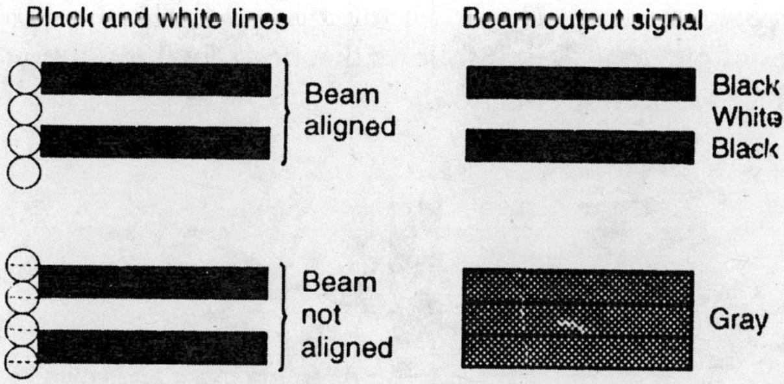

Fig. 22.7 Effect of beam spot size on vertical resolution.

Consider a finite size of a beam spot scanning a series of closely spaced horizontal black and white lines of minimum resolvable thickness, when the beam spot size is compatible with the thickness of the line, as shown in Fig. 22.7. If the beam is in *perfect alignment* the output will exactly follow the lines as *black or white levels.* If, however, the beam spot is *misaligned,* it senses both black and white areas simultaneously. Hence , it integrates the effects of both areas to give a resultant *gray level* output in between the black and white levels. This happens for all scanning positions and the output as well as the reproduced picture is a continuous gray without any vertical resolution at all. In positions of intermediate alignment of the beam, it will be more on one line than on the adjacent line and the output will be reproduced with *diminished contrast.* This indicates that *there is a degradation in vertical resolution due to finite beam size.*

The factor indicating the reduction in effective number of lines is called the *Kell factor.* It is not a precisely determined quantity and its value varies from 0.64 to 0.85. What the Kell factor indicates is that *it is unrealistic to consider that the vertical resolution is equal to the number of active lines.*

The flyback time also reduces the number of active picture elements as depicted in Fig. 22.8. The *horizontal flyback interval* occupies 17% of the line period and *the vertical flyback interval* occupies 6% of the frame period.

In the 625 - line system, the number of active lines lost in vertical blanking (referred to as *inactive lines*) are (625 - 40 =) 585 lines with a Kell factor of 0.7 (average value) the vertical resolutions is (0.7 × 585 =) 409.50 lines. The *horizontal resolution may not exceed this value multiplied by the aspect ratio.*

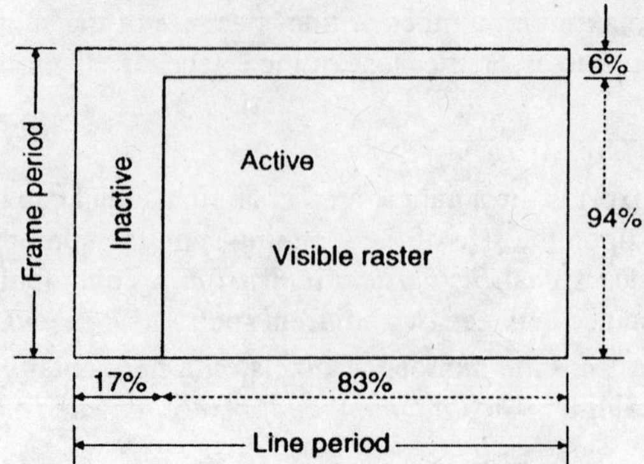

Fig. 22.8 Active and inactive scanning periods.

22.8. HORIZONTAL RESOLUTION AND VIDEO BANDWIDTH

The horizontal resolution of a television system is the ability of the scanning system to resolve the horizontal details i.e., *changes in brightness levels of elements along a horizontal scanning line.* Since such changes represent vertical edges of picture detail, it follows that horizontal resolution can be expressed as a measure of the ability to reproduce vertical information.

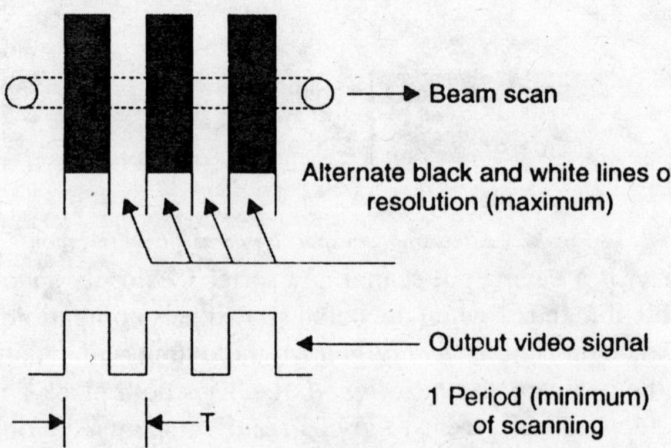

Fig. 22.9 Horizontal resolution and video bandwidth requirements.

The horizontal resolutions in a scanning system depends on the rate at which the scanning spot is able to change brightness level as it passes through a horizontal line across the vertical lines of resolution shown in Fig. 22.9.

In the 625 line system, there are about 410 active lines of resolution. With an aspect ratio of 4:3 *the number of vertical lines for equivalent horizontal resolution* will be ($410 \times 4/3 =$) 546 black and white alternate lines which corresponds to ($546 \times 1/2 =$)273 cycles of black and white alternations of elementary areas. For the 625 line system, the *horizontal scan frequency* (line frequency) is given by:

f_H = number of lines per picture × picture scan rate

$$= 625 \times 25 = 15,625 \text{ Hz} \qquad \ldots(22.2)$$

as each picture line is scanned 25 times per second. The *total line period* is thus

$$T_H = \frac{1}{f_H} = \frac{1}{15,625} \text{ sec} = 64 \text{ µs} \qquad \ldots(22.3)$$

Out of this period, 12 µs are used for the *blanking of flyback retrace.* Thus 546 black and white alternations (273 cycles of complete square waves) are scanned along a horizontal line during the forward scan time of (64-12=) 52 µs. The period corresponding to this square wave is 52/273 or 0.2µs approximately, giving the highest fundamental video frequency of 5MHz. Thus

$$f_{max} = \frac{\text{active lines} \times \text{Kell factor} \times \text{aspect ratio}}{2 \times \text{line forward scan period}} \qquad \ldots(22.4)$$

$$f_{max} = \frac{N_a \times K \times a}{2 \times t_{fH}}$$

where t_{fH} is the horizontal line forward scan period.

The fundamental law of communications, the *interchangeability of time and bandwidth* states that "if a total number of units of information is to be transmitted over a channel in a given time by specified methods, a specific minimum bandwidth is required". If the time available for transmission is reduced by a factor, or if the total information to be transmitted is increased by the same factor, the required bandwidth is increased by this factor. Suppose for example, the picture detail in both horizontal and vertical directions is to be doubled without changing the frame rate. This represents an increase in total picture elements by a factor of 4.

22.9 INTERLACING OF SCANNING LINES

The *vertical resolution* of a system depends on the total number of active scanning lines, whereas the *critical flicker frequency* (CFF) is the lowest possible apparent picture repetition rate which may be used without visible flicker. If during one vertical scanning field *alternate* scanning (odd) lines are formed and during the second scanning field the *remaining* (even) lines are formed, at the end of one frame period *all* the lines are formed. This process, called *interlaced scanning* gives a vertical scanning frequency of twice the true picture-repetition, or frame frequency. The situation is illustrated in Fig. 22.10 and further elaborated in Fig. 22.11

The upper frequency requirement of the video system is based on the true picture-repetition frequency, whereas the large area flicker effects are based on twice this frequency, or the vertical scanning frequency. The principle of *interlacing* the scanning lines, therefore, by cutting the complete picture repetition rate in half, *allows a system to maintain the same resolution and flicker characteristics as a noninterlaced system with a bandwidth requirement only half that of the noninterlaced system.*

Further improvement might result from more complicated interlacing systems. The ratio of field to frame frequency cannot be too great, however, since flicker would begin to show up for small picture areas which would not be covered by all the successive fields making up a complete frame.

The television picture is different from the picture projected in a movie theatre in that it consists of many lines. Another difference is the fact that a projected movie frame appears as a complete picture during each instant it is flashed on the screen. *The television picture is traced by a spot of light.* When the spot traces a television picture, it does so in the following manner;

(*a*) The spot traces line 1 of the picture.

(*b*) The spot traces line 3 of the picture.

(*c*) The spot traces line 5 of the picture ......and so on.

The spot traces all the *odd lines* of the picture. When the spot reaches the bottom of the picture, it has traced all the odd lines i.e a total of 312.5 lines in the 625 line system. This is called the *odd-line field.* Fig. 22.11(*a*).

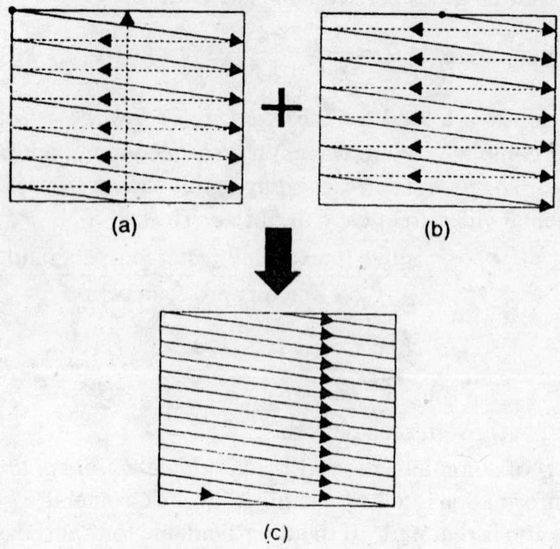

Fig. 22.10 One complete frame in interlaced scanning.
(a) Odd line field in interlaced scanning.
(b) Even line field in interlaced scanning.
(c) Odd-line field in (a) and even line field in (b) interlaced to form a complete frame.

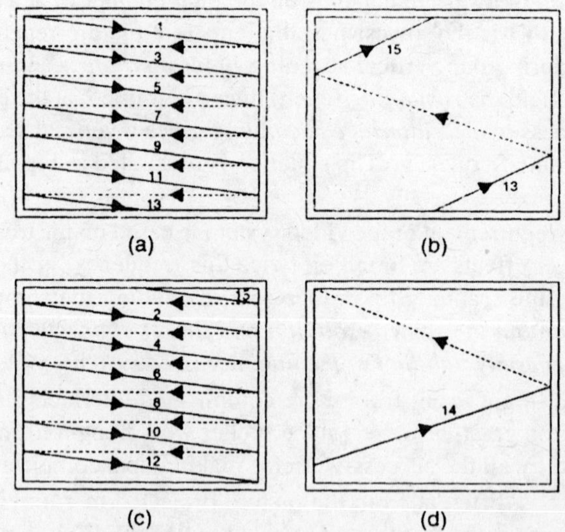

Fig. 22.11 Interlaced scanning: (a) odd line field, (b) transition period; (c) even line field; and (d) transition period.

After the spot has traced all the odd lines, the screen in blanked and the spot is returned quickly to the top of the screen once again. The time during which the spot travels back to the top of the screen is called the *vertical retrace period.* The timing of this period is such that the spot reaches the *top centre* of the screen when it again starts tracing lines in the picture as shown in Fig. 22.11(*b*). On this second tracing, the spot traces,

(*a*) line 2 of the picture
(*b*) line 4 of the picture
(*c*) line 6 of the picture... and so on.

When the spot completes the tracing of all *even lines i.e.* a total of 312.5 lines in the 625 line system and is at the bottom of the picture, it has traced the *even-line field*, depicted on Fig. 22.11(*c*).

The screen is then blanked once again, and the spot is returned to the *top left-hand corner* of the screen, Fig. 22.11(d) and the tracing sequence repeats. *The method of scanning, where the odd lines and even lines are alternately traced, is called interlaced scanning.*

After both the odd-and-even -fields have been scanned, the spot has given us *one frame* of the television picture. The word frame is borrowed from motion pictures, to indicate one complete picture that appears in a split second. *In television, the odd-line field and the even-line field make one frame.*

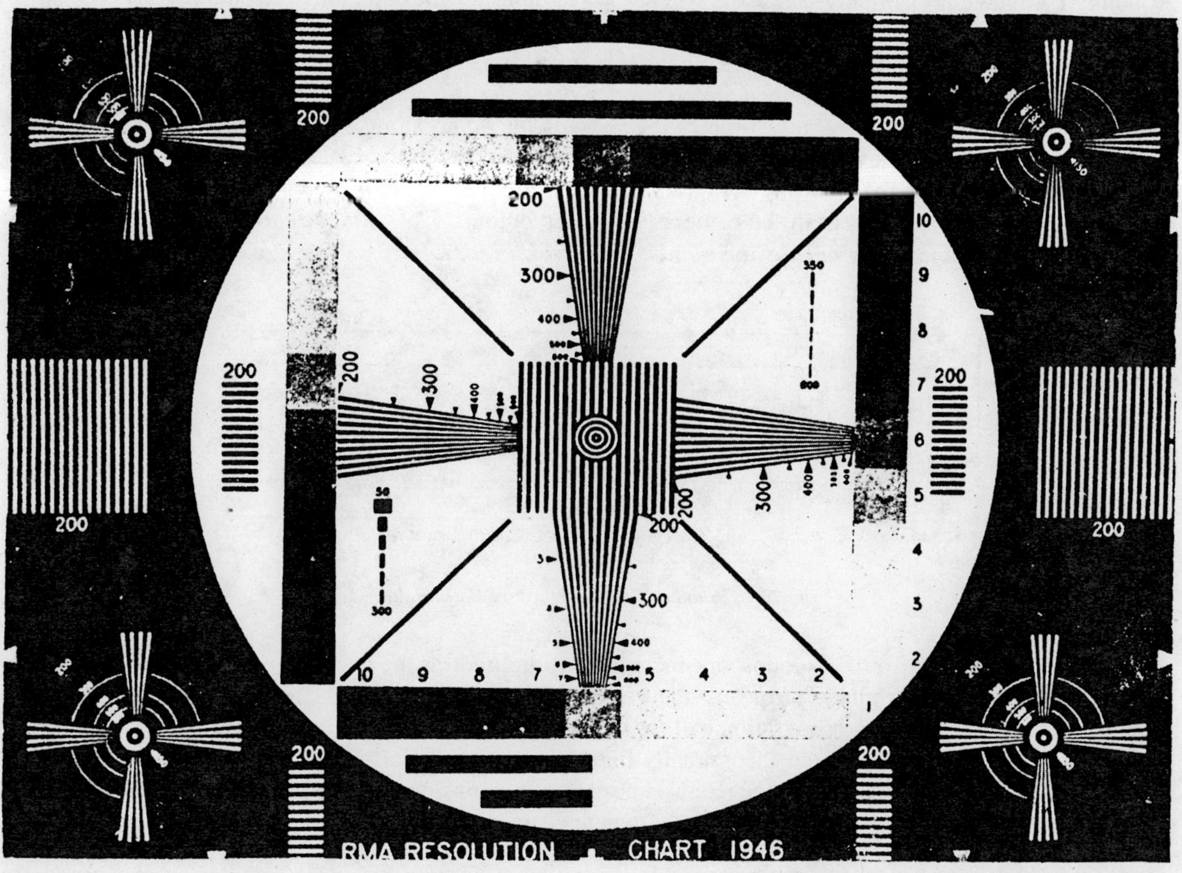

Fig. 22.12 Test Card

22.10. TEST CARD

All TV stations transmit a *tuning signal* for some time before the start of the program proper, *to facilitate the tuning in of receivers.* The signal takes the form of a *Test Card* which, besides other geometrical figures, such as checker board pattern, always includes one or more circles and both horizontal and vertical patterns of lines. The circles and other geometrical figures aid in the correction, where necessary, of frame linearity and of aspect ratio. The vertical pattern of lines serves for the fine tuning of channel selector, and the horizontal one for the accurate adjustment of the spacing. Fig 22.12 shows a test card offering a particularly wide range of possibilities.

You can receive the correct aspect ratio by adjusting for a pattern which resemebles, Fig. 22.12 as closely as possible. The wedges in the test pattern should be as close to equal length as possible; the circle should be as round as possible.

22.11. THE VIDEO SIGNAL

The voltage equivalent of any horizontal line that is scanned *begins* where the electron beam starts scanning the image horizontally, and *ends* where the scanning finishes and the horizontal retrace blanking begins. If we assume that the scanning action continues line after line, as happens in television transmitters, we have delivered to the video pre-amplifier in the camera, a train of picture voltages with intervals of blanking in between. Although the lines scanned are positioned *one below the other*, the voltage representation of the these lines is a *train* of voltage fluctuations, *one following the other*, along a time base. A blanking voltage is applied to the camera tube to extinguish the beam during horizontal retrace, although the signal output from the camera is zero during the blanking interval. In other words, the blanking signal fed to the electron gun of the camera tube does not appear in the camera tube video output. *The voltage representation of every line scanned is separated from the next by the blanking interval, Fig. 22.13.*

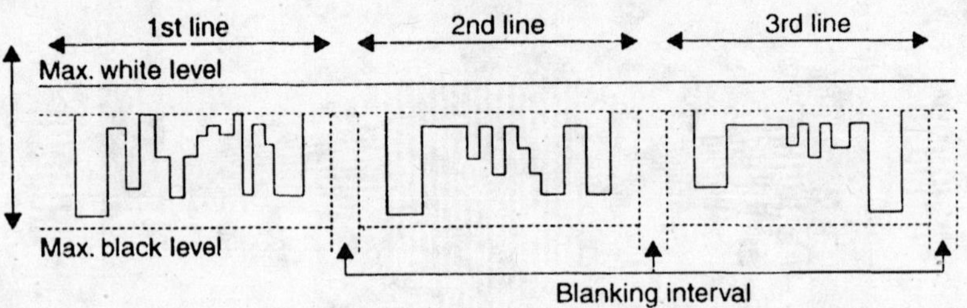

Fig. 22.13. Blanking Interval (Positive Video Polarity)

The video output from the iconoscope increases in amplitude as the image changes from black to white and white in the image produces stronger video signal. This is called *positive video polarity.* In negative video polarity the conditions of operation call for white in the image to produce the theoretically zero signal and black in the image to produce theoretically maximum video signal. The condition of *negative video polarity* can be met with by applying the iconoscope camera output signal to a negative clamping circuit. This is shown in Fig. 22.14. The whole of the signal then appears below the zero-voltage baseline and is entirely in the zone of negative polarity voltage

When the fixed negative bias is added to the positive-polarity camera signal, the result is a voltage variation wherein white in the image becomes the *least negative* voltage and black in the image becomes the *most negative* voltage.

Before it is broadcast, the video signal is passed through a number of amplifying stages employing transistors. *Each amplifying stage inverts the direction of video signal.* It is quite obvious, that a positive polarity video signal can also be converted into a negative-polarity video signal, simply by employing an odd number of amplifying stages.

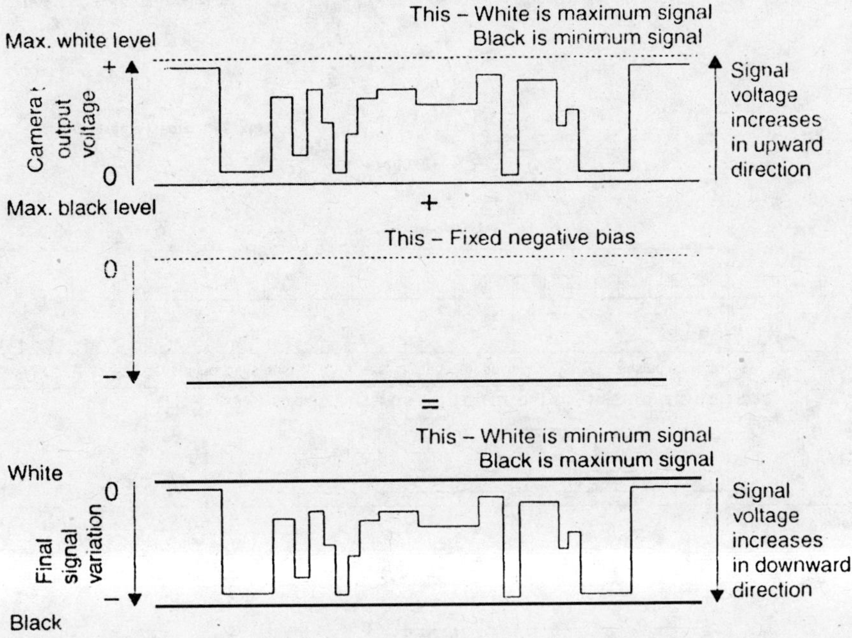

Fig. 22.14 Negative Video Polarity

22.12. CONTROL PULSES

TV receivers work precisely to give you a rock steady, crisp and clear picture, accompanied by scintillating melodious sound. You might think that all this calls for a precise, sophisticated and expensive, TV receiver. *No doubt, precision is essential in television but it is in the transmitting equipment and not in the receiving equipment.* Whether this precision is in the transmitting equipment, or in the receiving equipment, it is immaterial, because anyway it serves the same purpose. To maintain this precision in the transmitting equipment is, of course, convenient, practical, and economical. Were this precision maintained in the receiving equipment, TV receivers would have become complicated, costly, and beyond the repair capability of each and every technician.

In addition to the normal video and sound, the transmitted signal also contains the following *control signals:*

(1) line blanking pulses; (2) frame blanking pulses; (3) line synchronising (sync) pulses; (4) frame synchronising (sync) pulses; (5) pre-equalising pulses; and (6) post-equalising pulses.

The video, sound, and control signals combine to make up the composite video signal. Within the TV receiver, the circuits are slaved to this signal. For example, the frame oscillator, if left free, would drift in frequency, resulting in vertical instability. Similarly, the line oscillator, if left free, would also drift in frequency, resulting in horizontal instability. *The composite video signal locks in the frequency and phase of the frame and line oscillators.*

22.13. COMPOSITE VIDEO SIGNAL

The simple video signal we started with has turned into a video signal which is really complex. The variations of current strength conveying the picture content are here interwoven with a dense system of pulses, Fig. 22.12(a), all working towards a common aim. *If this train of pulses has to be transmitted, a method that suggests itself is to modulate the amplitude of the carrier wave in the rhythm of the pulses.*

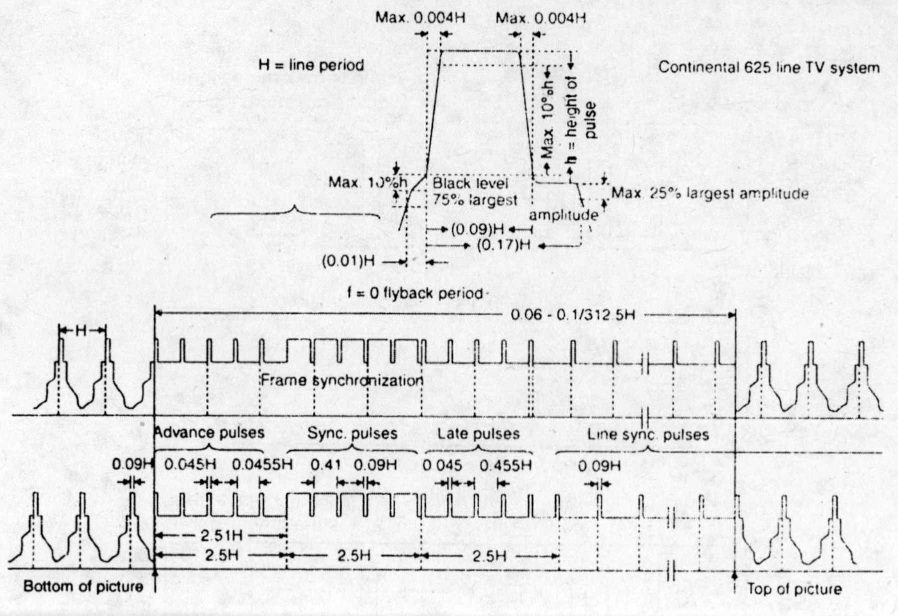
Max. 0.004H
Max. 0.004H
H = line period
Continental 625 line TV system
Max. 10%h
h = height of pulse
Max. 10%h
Black level 75% largest amplitude
Max. 25% largest amplitude
(0.09)H
(0.17)H
(0.01)H
f = 0 flyback period
0.06 - 0.1/312.5H
H
Frame synchronization
Advance pulses
Sync. pulses
Late pulses
Line sync. pulses
0.09H
0.045H
0.0455H
0.41
0.09H
0.045
0.455H
0.09H
2.51H
2.5H
2.5H
2.5H
Bottom of picture
Top of picture

(a)

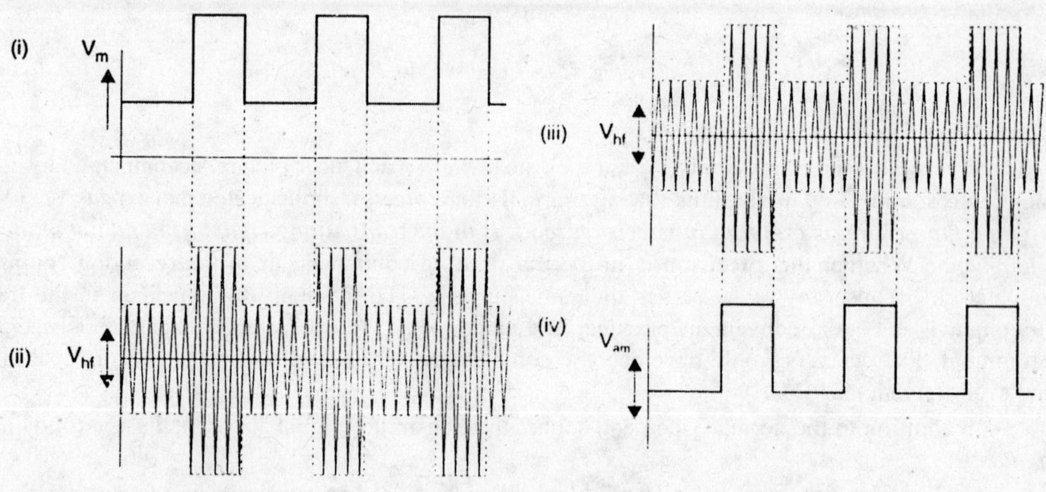
(i)
V_m
(ii)
V_{hf}
(iii)
V_{hf}
(iv)
V_{am}

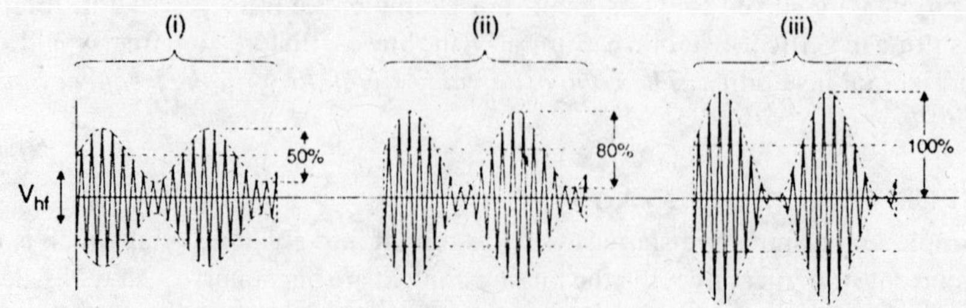
(i)
(ii)
(iii)
V_{hf}
50%
80%
100%

(c)

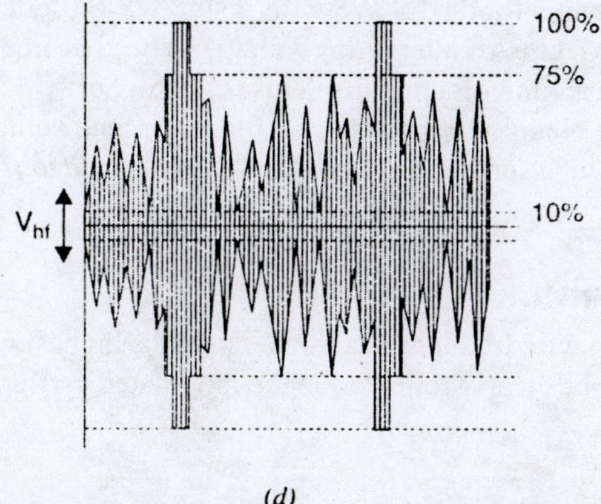

(d)

Fig. 22.15 Composite video signal (a) The train of pulses in a composite video signal. (b) (i) The pulse train as modulation voltage ; (ii) the amplitude modulated carrier wave. (iii) carrier wave with pulse-shaped modulation ; (iv) pulse train as restored by the demodulation process in the receiver. (c) Depth of modulation in AM a carrier wave modulated to depths of (i) 50% (ii) 80% and (iii) 100%. (d) The vision carrier as modulated with the video signal(negative modulation). The levels are indicated as percentages of the maximum amplitude of the carrier: white level 10%, black level 75% "blacker-than-black" region 75-100%.

There will have to be an amplitude detector in the receiver to recover this pulse train from the modulated carrier wave. The detector consists of a simple network that transforms high frequency electrical waves into a direct voltage. The magnitude of the direct voltage produced always corresponds exactly to the strength of the high frequency wave, and thus it will give an accurate representation of the changes in the amplitude of the wave. The output voltage of the detector will thus have the same pulse like character as the voltage modulating the carrier at the transmitter, Fig. 22.15(*b*). *This method of modulation, whereby the amplitude of the carrier wave is made to vary between a maximum and a minimum is called amplitude modulation (AM).*

An important characteristic of amplitude modulation is the *depth of modulation;* it is the difference between maximum and minimum amplitudes of the carrier expressed as percentage of its average amplitude. Sinusoidal modulation with 50%, 80% and 100% depth of modulation is illustrated in Fig. 22.15(*c*). Full modulation of a carrier wave, to a depth of 100%, is obviously the most efficient, but it is avoided in radio and TV transmitters because it makes accurate demodulation in the receiver a very difficult matter. The picture carrier as modulated with the video signal (negative modulation) is shown in Fig. 22.15(d).

22.14 THE HIGHEST VIDEO FREQUENCY

For the picture to be reproduced with sufficient sharpness and wealth of detail, its area must be split up into about 400,000 elements. At the rate of 25 pictures a second, a total of 10,000,000 of these elements will have to be scanned per second. Let us assume an extreme case in which all the elements are alternately bright and dark. The picture signal-the voltage derived from the camera tube-would then be a toothed wave, with five million positive teeth and five million negative gaps, part of which is shown in Fig. 22.16.

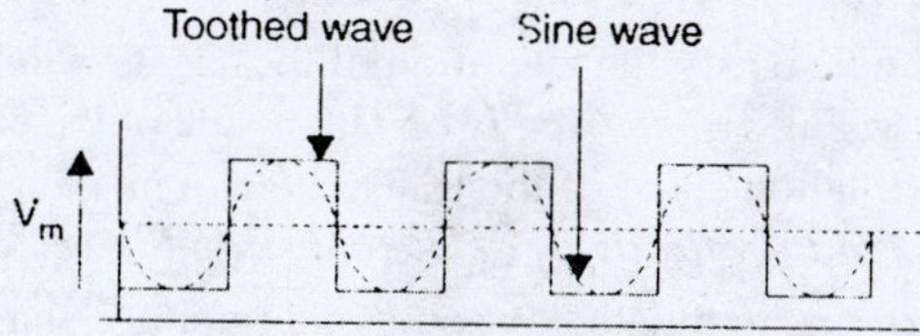

Fig. 22.16 The highest video frequency. A succession of alternately bright and dark picture elements produces a picture signal voltage of toothed shape. As an approximation, the tooth formation can be replaced by a sine wave.

Let us further replace this toothed wave with a sine wave (dotted line). This sinusoidal oscillation represents a frequency of 5MHz. An alternating voltage, with this frequency is the most rapidly changing one that can arise from the scanning of a picture. This is the highest video frequency of the system and plays a dominant role in the whole chain of transmission, from the scene being televised in the studio to the scene being viewed on the picture tube screen. *Obviously, the highest video frequency of a system depends on the number of lines.*

22.15. THE LOWEST CARRIER FREQUENCY

The carrier wave itself must have a frequency far higher than the highest video frequency, otherwise, the shape of the modulation would not be clearly reflected in the modulated carrier. This will be clear from Fig. 22.17.

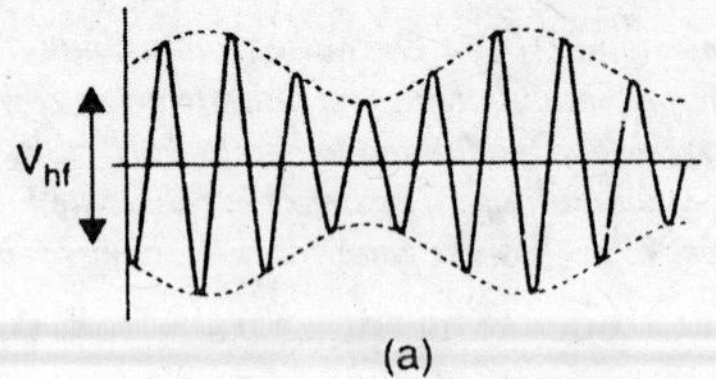

(a)

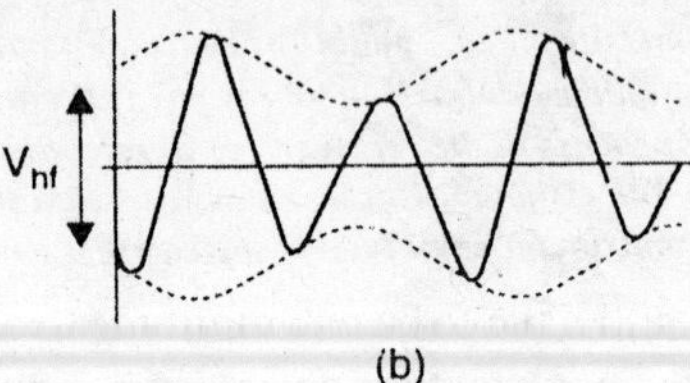

(b)

Fig. 22.17. Relationship between the highest frequency of modulation and the lowest carrier frequency. (a) Here the carrier frequency is five times that of the modulation and the shape of the latter is clearly expressed in the modulated carrier. (b) Here the carrier frequency is only twice that of the modulation and the modulated carrier more or less fails to reflect the shape of the modulation.

With a carrier frequency of at least five times as great as the modulation frequency, the modulation shape finds clear expression in the modulated carrier, Fig. 22.17(*a*). On the other hand, when carrier frequency is only twice modulation frequency, it is scarcely possible to recognise the shape of the modulation, as shown in Fig. 22.17(*b*).

The carrier frequency chosen is, in practice, at least five times as great as the modulation frequency. For television, carrier waves must be used with a frequency higher than about 30MHz. Colour television transmission only becomes possible in the VHF and higher ranges-that is, on wavelengths under 10 metres. In fact, *the lowest frequency officially allocated to television by international agreement is over 40MHz.*

22.16. SIDE BAND FREQUENCIES

The basic form of any process, that repeats itself periodically, is a pure sinewave of constant amplitude and frequency. Processes of a complicated nature, with amplitude or frequency changing in a haphazard manner, are composed of several elementary sinusoidal oscillations. If a 20 kHz wave is amplitude modulated with a 4 kHz wave, there exist in reality three oscillations having constant amplitude: a stronger oscillation of 20 kHz, and two equally weak oscillations of (20 - 4) = 16 kHz, and (20 + 4) = 24 kHz. The three of them appear one above the other. Adding together the amplitudes of the three waves, point by point, results in the amplitude modulated wave Fig. 22.18.

Amplitude modulation does not signify the alteration of the carrier wave amplitude, but what happens is that two new waves are formed, arising *symmetrically* on either side of the carrier. The two new waves are generated automatically in the modulation stage of the transmitter. *The new waves arising in this way, are called sideband waves or are referred to collectively as the upper and lower sidebands of the carrier wave.*

What happens when the carrier is modulated by some other voltage shape, for example, a video signal which, of course, is not sinusoidal. *Completely irregular, as the video information is, and dependent only on differences of brightness between individual picture elements, the video signal is, in reality, made up of a great number of pure sinusoidal oscillations, having various amplitudes and frequencies. This mixture of*

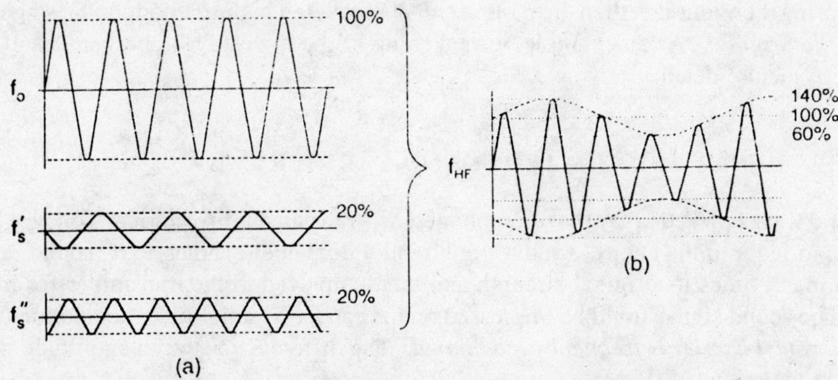

Fig. 22.18 Side waves arising when a 20 kHz carrier is amplitude modulated with a wave of frequency 4 kHz.

frequencies consists of two distinct groups of waves. The *first group* includes all waves with a frequency that is a *multiple of the line frequency.* In the 625 line system, the line frequency is 25 × 625 = 15,625 Hz, and this group, therefore, includes the following frequencies: 15,625 Hz; 2 × 15625 = 3125 Hz; 3 × 15625 = 46,875 Hz, 4 × 15,625 = 62500 Hz and so on. The wave of frequency 15,625Hz is called the *fundamental.* The waves having twice, three times, and four times the fundamental frequency are called the second, third and fourth *harmonics of the fundamental.*

The *second group* of waves, contained in the video signal covers the *picture frequency of 25Hz, and its harmonics,* i.e, 50,75,100 Hz and so on. In the composite video signal, the two groups of waves are so combined that an upper and lower sideband each consisting of a complete group of waves, made up of the picture frequency and its harmonics, is *arranged symmetrically* around each harmonic of the line frequency i.e, at 15,625 31,250, 46,875 Hz etc. This is illustrated in Fig. 22.19.

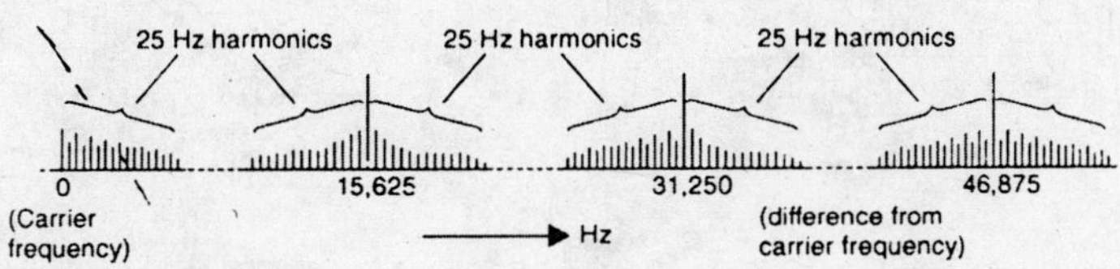

Fig. 22.19. The two groups of sideband frequencies of amplitude modulated video signal.

The waves of the first group are the predominant ones, while the sidebands of the 25Hz group play only a subordinate part and quickly tail off in amplitude. This frequency spectrum, as it is called, exhibits the enormous number of pure sinusoidal oscillations, that go to make up the composite video signal, resulting from the scanning of a picture. *We need take into account only frequencies upto 5MHz.* The finest picture detail is carried by still higher frequencies, but these are beyond the power of resolution possessed by the human eye.

A whole band of frequencies is produced in the modulation stage of a transmitter. The whole of this band has to be *radiated* by the transmitting antenna, *taken up* by the receiving antenna, and *amplified* in the receiver, in order that it can be finally changed back to its original form, the video signal-in the demodulator stage.

This emphasizes the necessity for all the links in the chain of transmission to have sufficient bandwidth. This means that the transmitting and receiving antennas must respectively radiate and take up all the waves from the lowest to the highest sideband frequency, with the same efficiency; it also means that the receiver must amplify all these frequencies to the same extent. If this is not so, and some of the waves are somehow handicapped during their journey, then the picture signal recovered in the demodulation stage will have *a form different from the original.* As an example, a weakening of the lowest or highest sideband frequencies will result in a loss of picture detail.

22.17 THE FREQUENCY MODULATED SOUND CARRIER

The sound waves picked up by the microphones, are translated into electrical voltages by these devices, Fig. 22.20(a), and fed into the control room through individual audio cables. The sound modulated current is amplified to many times its original strength and then conveyed to the transmitter for impressing on the carrier wave. The sound signal could be impressed on the carrier by amplitude modulation (AM). *In the 625 line system, the sound carrier is frequency modulated.* The difference between amplitude modulation (AM) and frequency modulation (FM) is shown in Fig. 22.20

In frequency modulation the amplitude of the carrier remains unchanged, while its frequency is altered in the rhythm of the modulating wave. In the absence of any modulation, the radiated sound carrier has the constant or resting frequency, Fig. 22.20(*d*). *The change in carrier frequency during frequency modulation is known as frequency deviation.*

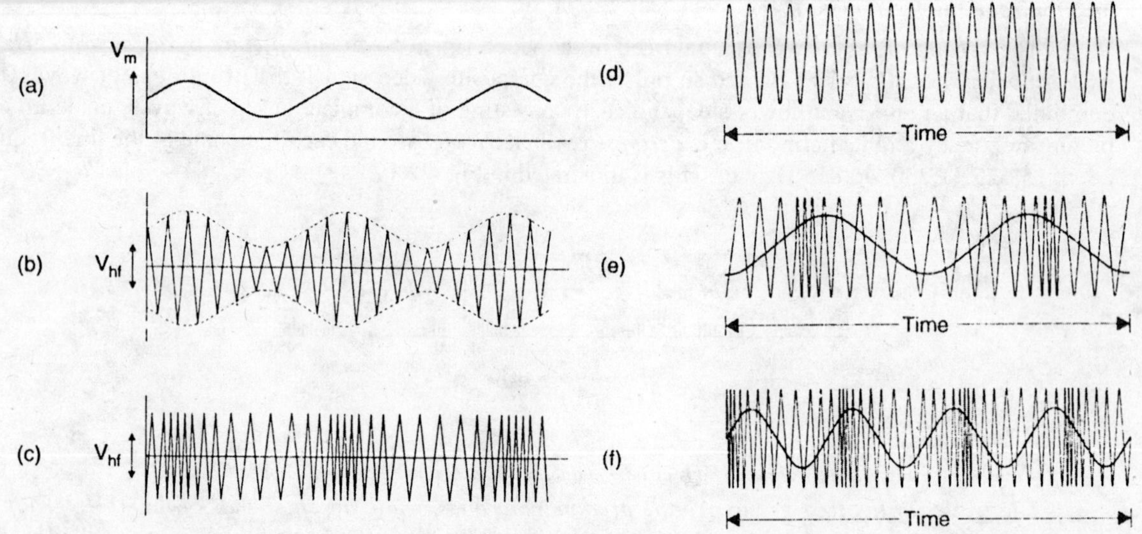

Fig. 22.20 Amplitude and frequency modulation. (a) Vm - A sinusoidal modulation voltage. (b) A carrier wave, Vhf. amplitude modulated by Vm. (c) A carrier wave, Vhf. frequency modulated by Vm. (d) Resting frequency (e) Deviation produced by a whisper. (f) Deviation produced by a bang.

When the frequency *increases above* the resting frequency, it is known as *positive deviation*; when the frequency *decreases below* the resting frequency, it is called *negative deviation.* The change in frequency is called *swing.* This is illustrated in Fig. 22.20(*e*) and (*f*).

In practice it is necessary to lay down a maximum frequency swing for all FM systems just as it is necessary to lay down a maximum percentage modulation for all AM systems. This serves as a basis for the design of FM detectors in receivers. The detectors must work in such a way that, with maximum swing, it delivers the maximum permissible voltage to the loudspeaker or the output stage, whatever the case may be. *The maximum deviation laid down for television sound is* ± *50KHz.*

In general, frequency modulation in the VHF range is an easier matter than the accurate and satisfactory amplitude modulation of the same high frequencies. The shorter the wavelength used, the stronger the undesirable effect that the modulating signal has on the frequency of the carrier; in the end it becomes simpler to incur the slight expense of turning these almost unavoidable fluctuations of the carrier frequency into proper frequency modulation, rather than spend a great deal more on suppressing them. For this reason, *transmitters in the centimetre wave range work almost exclusively on frequency modulation.*

The sidebands arising in FM, constitute a very important problem. In FM, an infinite number of side band frequencies arise *symmetrically* around the carrier frequency even when it is only a single sine wave that is modulating the carrier. The sidebands are separated by a frequency which is always equal to the modulating frequency; at first, with increasing distance from the carrier, their amplitudes vary irregularly, then they fall off regularly and rapidly, as shown in Fig. 22.21 (a). The modulating voltage consists of the sum of several sine waves. Each one of these sine waves will generate a sideband spectrum of its own. The complete frequency spectrum will be extraordinarily extensive and complex. Theoretically, all of these sidebands must reach the detector stage in the receiver, in order that they may be translated into a faithful reproduction of the original sound.

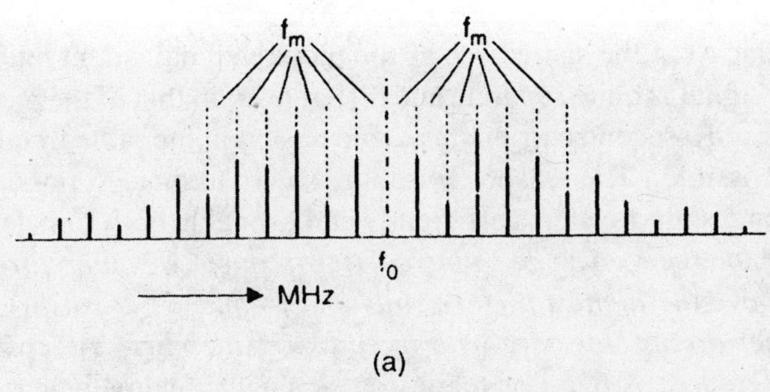

(a)

Fig. 22.21. The side bands of an FM carrier.
(a) f_o = Carrier wave (central frequency) ; fm = separation between the frequencies of individual side waves. The separation is always equal to the modulation frequency.

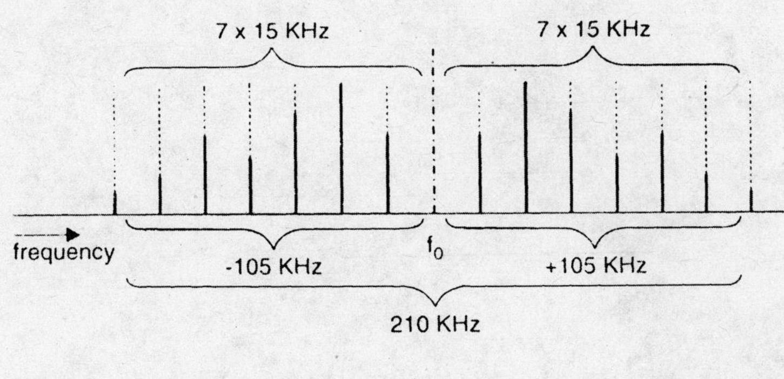

(b)

Fig. 22.21.(b) For sound frequencies upto 15 kHz to be reproduced without distortion it is necessary to transmit sidebands extending to about 105 kHz on either side of the carrier frequency.

The question now arises, how many sidebands are essential for a sufficiently faithful reproduction of the original sound, and how many of them will have to reach the detector. It depends on the ratio between the modulating frequency and the frequency swing. As a rule of thumb, one may say that *at least first five to seven paired waves of each sideband must be transmitted.* Because the frequency separating sideband waves is always equal to the modulation frequency, each group of seven sidebands will extend over a frequency band equal to seven times the modulation frequency, the two bands lying on either side of the carrier. If the modulation frequency has the highest value that has to be transmitted, i.e. 15 kHz (the highest audio frequency), the distance of both the highest and the lowest sideband frequencies from the carrier will be $(7 \times 15) = 105$ kHz, Fig. 22.21(*b*).

We conclude that *the bandwidth of every link in the chain of FM transmission must be at least 200 kHz.* This is a great disadvantage in comparison with any AM system, in which only one upper and one lower sideband wave has to be transmitted and which, for conveying the maximum audio frequency of 15 kHz, only requires a bandwidth of $(2 \times 15) = 30$ kHz. In actual practice, the bandwidth of AM transmitters is often limited to 9 kHz, making the highest audio frequency transmissible 4.5 kHz.

In consequence of the great bandwidth they require, FM systems can only be employed in very high frequency (VHF) ranges, in which a broad frequency band can be allocated to each transmitter without danger of its breaking into the sideband of some other transmitter. It is impossible to do this in the long, medium or short wave bands, it may become possible in the VHF range, that is, with wavelengths under 10 metres and frequencies over 30MHz.

A sound signal is broadcast at the same time as the picture signal; silent television has never existed. The information in the sound signal is quite separate and different from that of the picture signal. The simplest course is to convey it by means of a second carrier wave, and have it in the same frequency range as the picture carrier; and as close to it as possible. The reasons for doing so are to make it possible for a single receiving aerial to take up both picture and sound systems efficiently and to enable both signals to be amplified together, at least in the earlier stages of the receiver. *The sound carrier frequency is, therefore, given a frequency such that it lies just far enough above the highest picture sideband frequency, to ensure that the much narrower sidebands of the sound carrier do not interfere with the picture sideband.* A separation of about 1MHz is sufficient to avoid such interference. A 625 line television channel showing the pass bands of the vision and sound transmitters is shown in Fig. 22.22. The term *sound channel* is frequently used for the narrow frequency band taken up by the sound transmitter within the television channel.

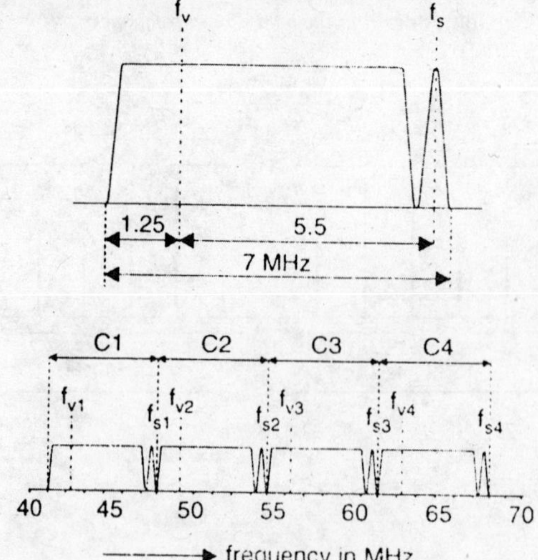

Fig. 22.22. A television channel as showing the frequency characteristics of the vision and sound transmitters. f_v = *vision carrier* ; f_s = *sound carrier.*

22.18 MONOCHROME TV CAMERA

Simplified diagram of a typical monochrome television camera is given in Fig. 22.23. An *optical system* focuses light reflected from the scene onto the faceplate of the camera tube. A *photoelectric process* transforms the light image into a virtual electronic replica in which *each picture element is represented by a voltage.* A *scanning beam* in the pick up tube next converts the picture, element by element into electrical impulses. At the output an *electrical sequence* developes that represents the original scene. The output of the camera tube is then amplified to provide the video signal for the transmitter. A sample of the video signal is also provided for observation in a CRT viewfinder mounted on the camera housing.

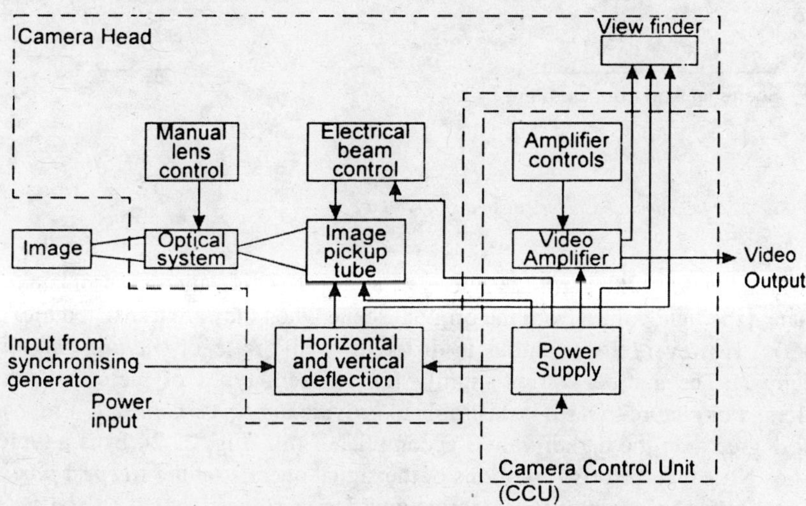

Fig. 22.23 Simplified diagram of a typical monochrome television camera.

Electronic circuits that provide the necessary control synchronisation and power supply voltages operate the TV camera tube. A *deflection system* is included in the TV camera to control the movement of the camera tube scanning beam. Many TV camera receive synchronising pulses from a *studio control unit.* This unit also provides the sync pulses that synchronise the receiver with the camera.

Some TV cameras however generate their own *control signals.* In turn, they provide output pulses to *synchronise* the control unit. Manual controls are also provided at the rear of the camera for setting the optical lens and for zooming. Because of the complex electronic circuits and controls, early TV cameras were rather large and awkward to handle. Recent innovations have revolutionised the cameras' construction.

22.19 CAMERA TUBE CHARACTERISTICS

One important characteristic of all camera tubes is their *light transfer characteristic.* This is *the ratio of the faceplate illumination in foot candles to the output signal current in nano amperes (nA).* This characteristic may be considered as a measure of the efficiency of a camera tube. Typical values of output current range from 200nA to 400 nA.

The term *gamma,* in a television system is applied to camera tubes and to picture tubes. *Gamma is a number which expresses the compression or expansion of original light values.* Such variations, if present are inherent in the operation of the camera tube or picture tube.

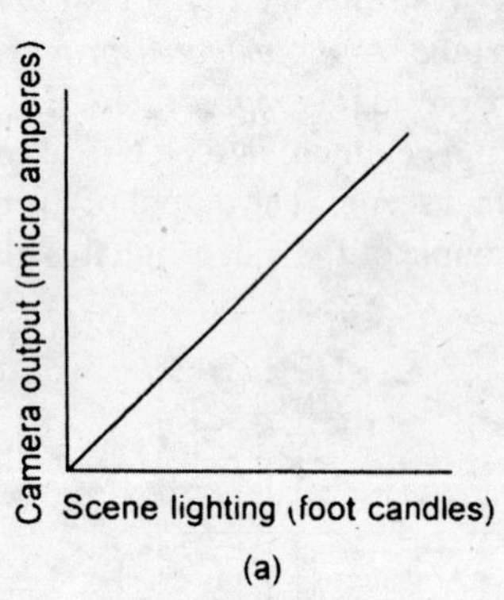

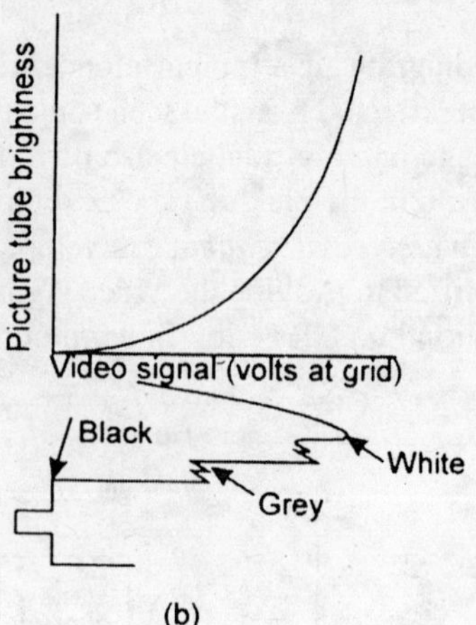

Fig. 22.24 Gamma characteristic of a camera tube

With *camera tubes* the gamma value is generally 1. This value represents a *linear characteristic* that does not change the light values from the original scene when they are translated into electronic impulses, see Fig. 22.24(a). However, the situation is different in the case of *picture tubes* which have gamma of approximately 3. The number varies slightly for different types of picture tubes and also for different manufacturers. For picture tubes it is desirable to provide *improved contrast.* Emphasizing the bright values to a greater degree than the darker values accomplishes this. Fig. 22.24(b) is a typical picture tube gamma characteristic. Note that the bright portions of the signal operate on the *steepest portion* of the gamma curve. Conversely, the darker signal portions operate on a *lesser slope.* Thus it can be seen that the bright picture portions will be emphasized to a greater degree than the darker picture portions.

For proper colour reproduction the *spectral response* of a camera tube is an important parameter. As nearly as possible the tube should have the same spectral response as the human eye. This is necessary to render colours in their *proper tones.* It is also important in reproducing black and white pictures, thereby producing the *proper gray scale.* Tubes designed to operate in a colour camera have a greater response to each of the primary colours. Today, spectral response distribution has made possible the manufacture of camera tubes that are sensitive to the infrared, the ultraviolet and even the X-rays. But variations in spectral response have had little effect on the other operating characteristics of the tube.

If the photosensitive material in a camera tube was able to emit an electron for each photon of light focused upon the material, the quantum efficiency of the material would be 100 percent. The formula for quantum efficiency is:

$$Q_{eff} = \frac{\text{electron}}{\text{photons}} \qquad \ldots(22.5)$$

A quantum efficiency of 100% is almost impossible. However *quantum efficiency is a practical way to compare photosensitive surfaces in a camera tube.* In this comparison, photocurrent per lumen is measured using a standard light source. *A lumen* is the amount of light that produces an illumination of one foot candle over an area of one square foot. The source adopted for measurement is a tungsten-filament light operating at a colour temperature of 2870°K. Since the lumen is actually a measure of brightness stimulation to the human eye, *quantum efficiency is a convenient way to express the sensitivity of the image pickup tube.*

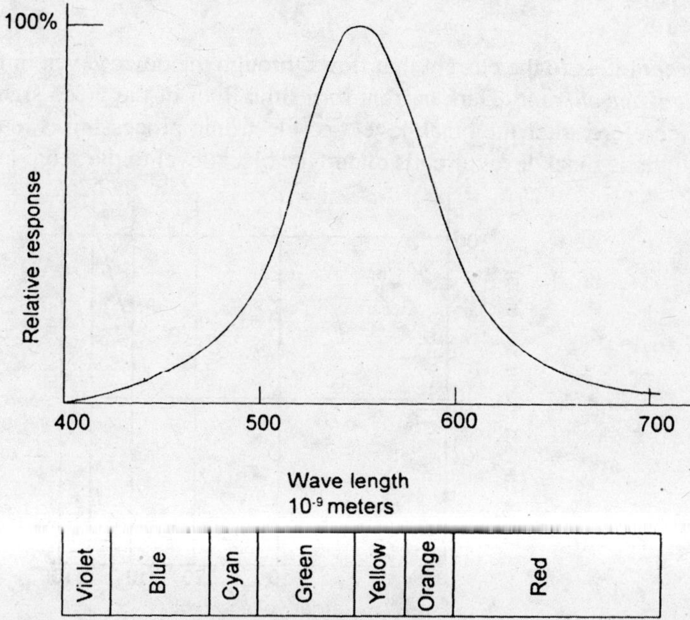

Fig. 22.25 Human spectral response.

Our eyes operate as a frequency - selective receiver that peaks in the yellow-green region at a wave length of about 560 nano metres (560×10^{-9} metres) as shown in Fig. 22.25. Spectral response of Vidicon type 7262A is illustrated in Fig. 22.26.

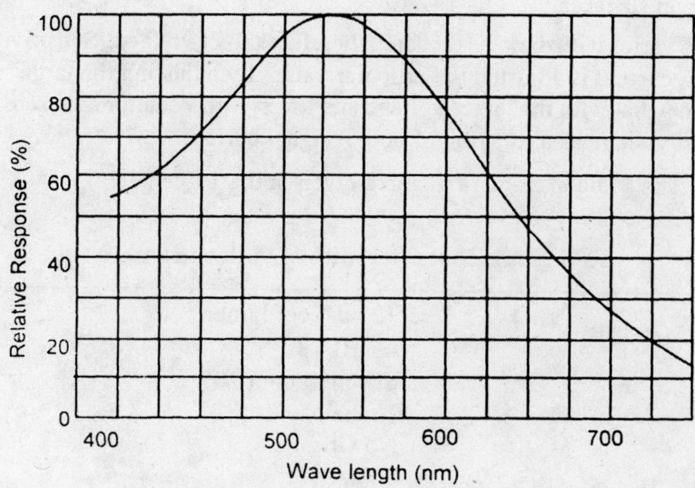

Fig. 22.26 Spectral response of Vidicon type 7262A.

The term *lag* refers to the time lag during which the image on the camera tube decays to an innoticeable value. All camera tubes have a tendency to *retain* images for short periods after the image is removed. Some types do it more than others. *Lag on a television picture causes smear (comet tails) to appear following rapidly moving objects.*

Lag may be expressed as a percentage of the initial value of the signal current remaining 1/20s after the illumination is removed. Typical lag values for vidicon range from 1.5% to 5% Plumbicons have lag values as low as 1.5%.

Dark current refers to the current that flows through the device even in total darkness. It is a form of semiconductor *leakage current.* Dark current forms the floor of the video signal or black level. Variations in *temperature* therefore, alter the black level and electronic processing is required to maintain a *constant black level.* Automatic black level circuits clamp the black level to the setup level of the video signal.

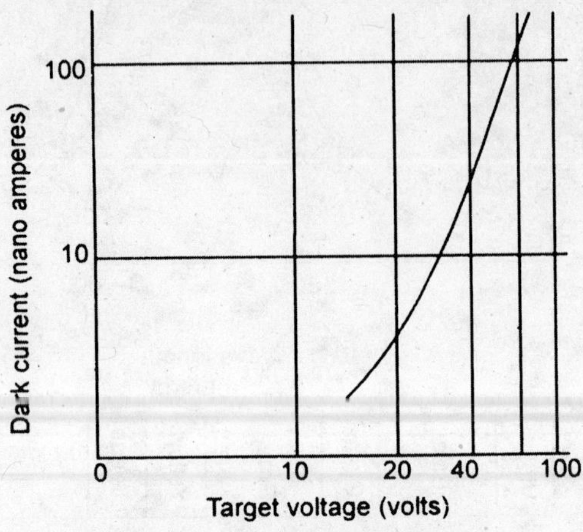

Fig. 22.27 Dark current versus targe voltage.

Dark current also varies with applied *target voltage* in vidicons. One of the factors that lowers sensitivity is the *recombination* of electrons and holes where both cease to act as current carriers. By increasing the *voltage field gradient* in the target the likelihood of recombination decreases and more of the carriers liberated by light reach the target surfaces.

Sensitivity also varies with dark current and, therefore target voltage. Sensitivity rises for higher values of dark current. Dark current is adjusted to particular values by adjusting the target voltage. This means that sensitivity varies somewhat with the target voltage and a very simple automatic sensitivity control can be set up which resets target voltage as a function of output signal current.

Typical characteristics of vidicon camera tube are given in the Table 22.1

Table 22.1 Characteristics of vidicon camera tube

Sensitivity	125 μA per lumen
Signal to noise ratio	45 dB
Image lag	Severe (about 20%)
Halo	No halo
Ghost	No Ghost
Dark Current	20 nA
Resolution	55% at 400 lines (5MHz)
Spectral response	Close to eye's response
Infra-red sensitivity	Moderate
Gamma	0.7
Size	Small, compact (25 mm and 15mm size target)
Applications	Outdoor recording, domestic recording.

22.20 VIDICON CAMERA TUBE

Vidicon is a simple, compact TV camera tube that is widely used in education, medicine, industry, aerospace and oceanography. It is, perhaps, the most popular camera tube in television industry. During the past several years, much effort has been spent in developing new *photoconductive materials* for use in its internal construction. Today with these new materials, some vidicons can operate in exposure to *direct sunlight* or in near-total darkness. Also, these tubes are available in diameters ranging from 1/2 to 4 1/2 inches, and some of the larger ones even incorporate *multiplier sections* similar to those in the image orthicon. The main difference between the vidicon and the image orthicon is physical size and the photosensitive material that converts incident light rays into electrons.

The *image orthicon* depends on the principle of *photoemission* wherein electrons are emitted by a substance when it is exposed to light. The *vidiocon,* on the other hand employees *photoconductivity,* that is, a substance is used for the target whose resistance shows a marked decrease when exposed to light.

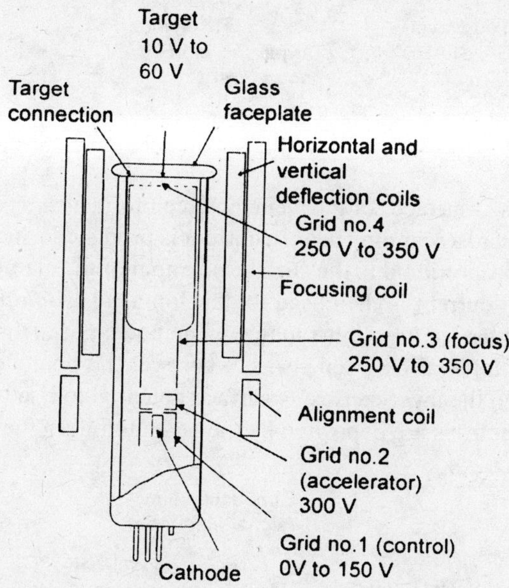

Fig. 22.28 Internal construction and external components of a vidicon camera tube. Voltages shown are typical.

In Fig. 22.28 the *target* consists of a transparent conducting film (the *signal electrode*) on the inner surface of the *faceplate* and a thin photoconductive layer deposited on the film. Each cross sectional element of the photoconductive layer is an *insulator in the dark* but becomes *slightly conductive where it is illuminated.* Such an element acts like a leaky capacitor having one plate at the positive potential of the signal electrode and the other one floating. When the light from the scene being televised is focused onto the surface of the photoconductive layer, next to the face plate, each illuminated element conducts slightly the current depending on the amount of light reaching the element. This causes the potential of the opposite surface (towards the gun side) to rise towards the signal electrode potential. Hence, there appears on the gun side of the entire layer surface a *positive-potential replica* of the scene composed of various element potentials corresponding to the pattern of light *which is focused onto the photoconductive* layer.

When the gun side of the photoconductive layer, with its positive potential replica, is *scanned* by the electron beam, electrons are deposited from the beam until the surface potential is reduced to that of the cathode in the gun. This action produces a change in the difference of potential between the two surfaces of the element

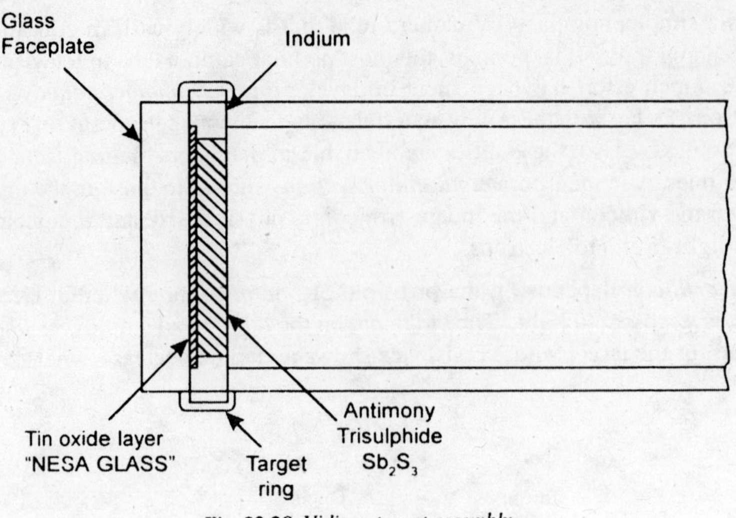

Fig. 22.29. Vidicon target assembly.

being scanned. When the two surfaces of the element which form a *charged capacitor* are connected through the signal electrode circuit and a scanning beam a current is produced which constitutes the *video signal.* The amount of current flow is proportional to the surface potential of the element being scanned and to the rate of the scan. The video signal current is then used to develop a *signal-output voltage* across the load resistor. The signal polarity is such that for highlights in the image, the input to the first video amplifier swings in the negative direction. In the interval between scans, whenever the photoconductive layer is exposed to light migration of charge through the layer causes its surface potential to rise towards that of the signal plate. On the next scan, sufficient electrons are deposited by the beam to return the surface to the cathode potential.

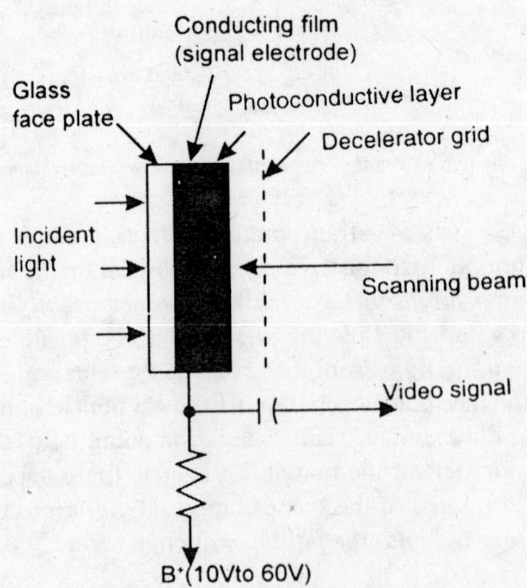

Fig. 22.30 Simplified diagram of the standard vidicon target plate. Output current is taken from the signal electrode.

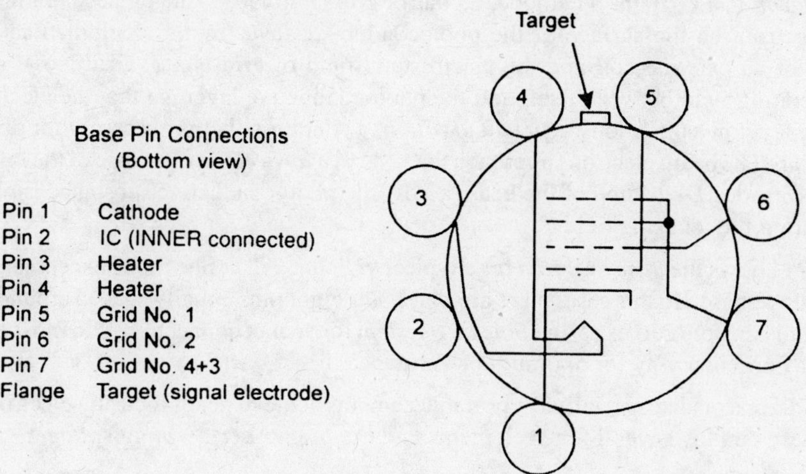

Fig. 22.31 Vidicon base diagram

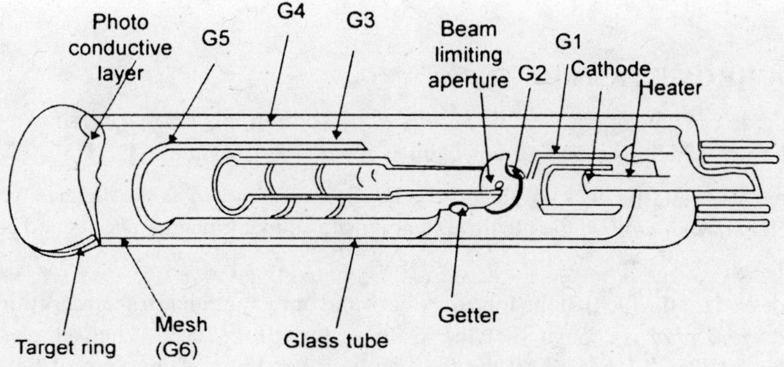

(a) Construction details of H-V tube

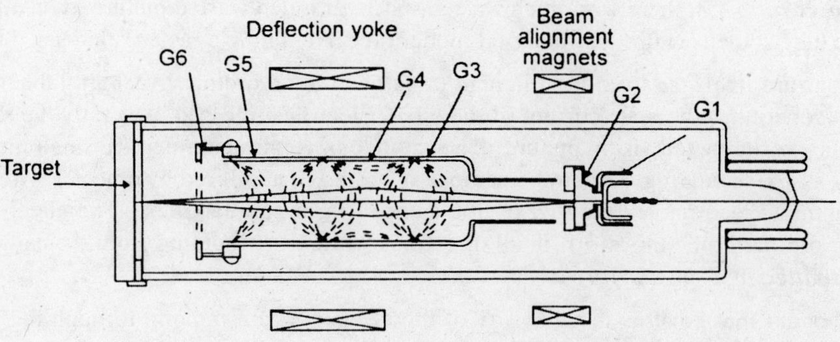

(b) Electroststic focus lens and electrode details

Fig. 22.32 High voltage electrostatic gun uses a unipotential type lens with G3 and G5 at different potentials.

The *electron gun* contains a cathode, a control grid (grid No. 1)and an accelerating grid (grid No. 2). The beam is focused on the surface of the photoconductive layer by the combined action of the uniform magnetic field of an external coil and the electrostatic field of grid No. 3. Grid No. 4 serves to provide a uniform decelerating field between itself and the photoconductive layer, so that the electron beam will tend to approach the layer in a direction *perpendiculor to it,* a condition that is necessary for driving the surface to cathode potential. The beam electrons approach the layer at a *low velocity* because of the low operating voltage of the signal electrode. *Deflection* of the beam across the photoconductive substance is obtained by external coils placed within the focusing field.

Fig. 22.31 shows the *base diagram* for a typical vidicon. All connections except that made to the target ring are made at the base. In this case, there are places for eight pins equally spaced around the sealed exhaust tip. A short or missing pin serves as the *indexing system* for socket connections. In most cases, the tube must be oriented in a particular way for optimum performance.

Some vidicons made especially for portable cameras achieve beam focus by electrostatic means alone, Fig. 22.32. These employ several focus elements and use *higher accelerating voltages.*

Electrostatic lenses achieve a savings in electrical power in several ways. First, the current needed for magnetic focus coils is eliminated, as well as the power consumed by the series regulators that control focus current. Next, because the focus coil is eliminated, the space it occupies in the yoke assembly is eliminated. Magnetic reluctance in the gap between deflection coil poles is thereby reduced and less power is needed for deflection.

22.21 MONOCHROME PICTURE TUBE

The cathode ray tube or CRT-as the *picture tube* is commonly known as (shown in Fig. 22.33) is a large glass structure from which as much air has been evacuated as possible.

At the narrow end (the neck) of the tube is an assembly known as the *electron gun.* This comprises the elements which *emit and control* the electron beam just as a machine gun shoots and controls the direction of a stream of bullets.

The wide end (or the face) of the tube may be round or rectangular in shape. A thin coating or fluorescent material called a *phosphor* is deposited on the inside surface of the face. When a high speed beam, fired from the electron gun, strikes the phosphor, the face of the tube shows a tiny spot of light at the point where it strikes. Thus, *the screen converts the energy of moving electrons into light energy.*

All early television picture tubes had round screens. Since the TV picture is transmitted as a rectangle having an *aspect ratio* 4:3, there was much wasted space, or much wasted picture, if an attempt was made to use the entire tube screen-width. Virtually all modern receivers utilise rectangular-faced tubes.

Electrons are *liberated* from the filament or cathode of an ordinary vacuum tube. It is this stream of electrons that constitutes the plate current of the tube. Electrons released in exactly the same manner in the cathode ray tube form the television picture. The cathode of electron emitter is a small metal cylinder which is covered by an oxide coating. When the cathode is heated to a dull red by a heater wire located inside the cathode mounting, electrons are emitted by the cathode in large quantities. The electrons emitted by the cathode of an ordinary tube move off in all directions while those coming from the cathode of the cathode ray tube are *emitted in a specific forward direction.*

The element that governs the *intensity* of the electron beam moving through the picture tube is the *control grid.* Unlike its counterpart in the common vacuum tube, it is cylindrical in shape and resembles a metal cap rather than a wire screen. This metal cap is called *Wehnelt cylinder,* Fig. 22.34. The front of the Wehnelt cylinder contains a small opening or *aperture* which acts as an outlet for the electron beam. This arrangement is necessary to achieve a *thin* electron beam.

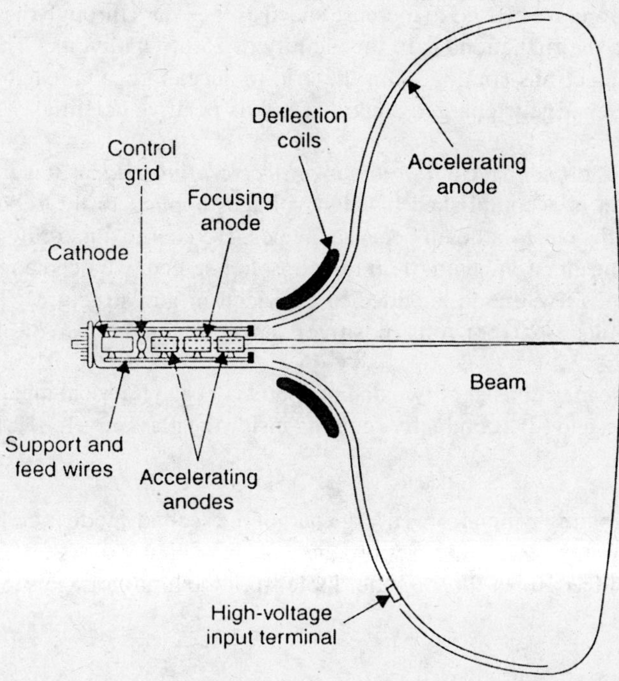

Fig. 22.33 The Picture Tube

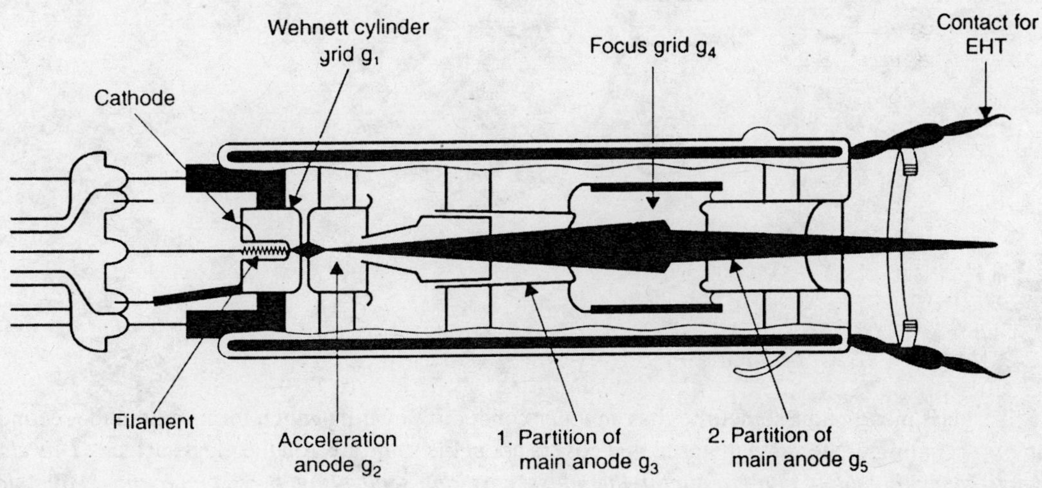

Fig. 22.34 Cathode ray tube construction.

Most picture tubes have a second cylindrical structure, placed adjacent to the control grid, as can be seen in Fig. 22.34. The principal function of the first anode is that of *accelerating* the electrons that emerge from the pinhole in the control grid structure. This accelerating action also causes some degree of beam forming, since the electrons are caused to increase speed as they pass through the apertures in the *first anode.* The potential applied to the first anode is in the vicinity of 250 to 450 volts. The position of this element is such that it causes the electrons coming from the grid to accelerate as a result of the electrostatic force of attraction between the negatively charged electrons and its positive potential.

Some picture tubes do not have a *first anode or accelerating grid*, as it is more appropriately called. In those tubes, acceleration is accomplished by high voltages applied to the *second anode.* The first anode initiates the forming of the electron beam because the positive charge on it draws electrons from the cathode. But, because we want the electron beam to *hit* the phosphor screen, we need another anode located closer to the screen. A second positive anode is added to the electron gun structure. This element usually carries anywhere from 8,000 volts to 20,000 volts (positive) depending on the *type* of picture tube.

The *second anode* may consist of two distinct parts : (1) a cylindrical metallic structure adjacent to and following the first anode, and (2) a conductive coating inside the glass envelope which covers almost the entire area of the glass bell.

The conductive coating which forms a large part of the second anode is colloidal graphite deposit called *aquadag.* In addition to its accelerating action, the aquadag coating plays an important part in maintaining a thin electron beam, and assists in *filtering* the ripple from the high-voltage supply.

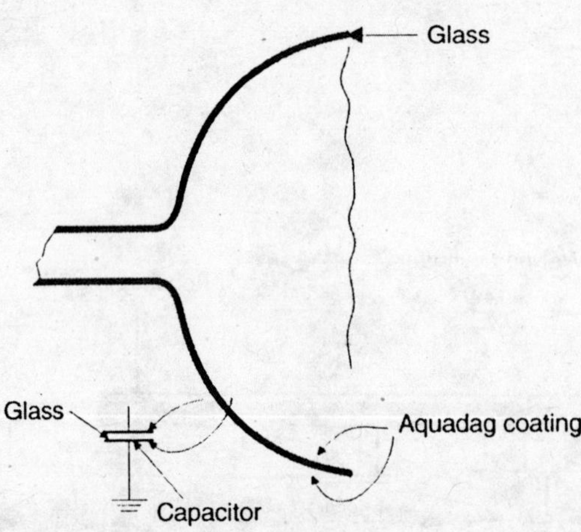

Fig. 22.35 High Voltage Filter Capacitor

The glass picture tube sometimes has an *inner* conductive coating which forms part of the second anode and an *outer* coating of the same material that covers about the same area on the glass surface. *This electrical sandwich consisting of a non-conductor-the glass envelope-separating two conducting layers, forms a capacitor.* Although its capacitance is not very high (approximately 500 pf), it is large enough to serve as a filter-capacitor for the high voltage supply. The outer aquadag coating is connected to the chassis (ground) of the receiver as illustrated in Fig. 22.35. The inner coating is joined to the second anode and is connected to the high voltage terminal inside the receiver. *Thus, this capacitor is connected across the high-voltage supply which provides the potential needed for the second anode.*

22.22 TELEVISION AS A SYSTEM

As shown in Fig. 22.36, television operates as a *system*, the station, and the numerous receivers within range of its signals. The station produces two kinds of signals, *video* originating from the camera or tape recording, and *sound* from a microphone or other source. Each of the two signals is generated, processed, transmitted separately, video as an *amplitude - modulated* signal, and sound as a *frequency-modulated* signal.

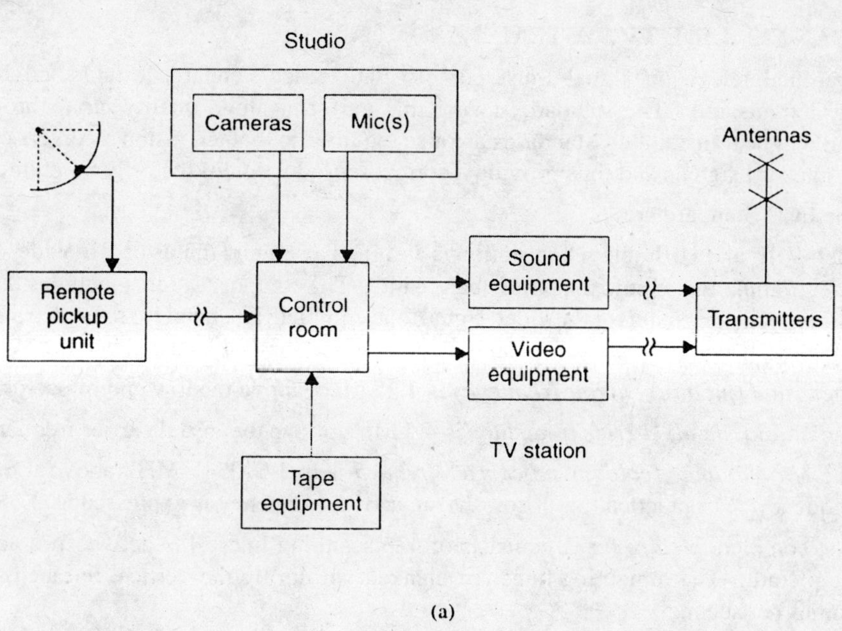

(a)

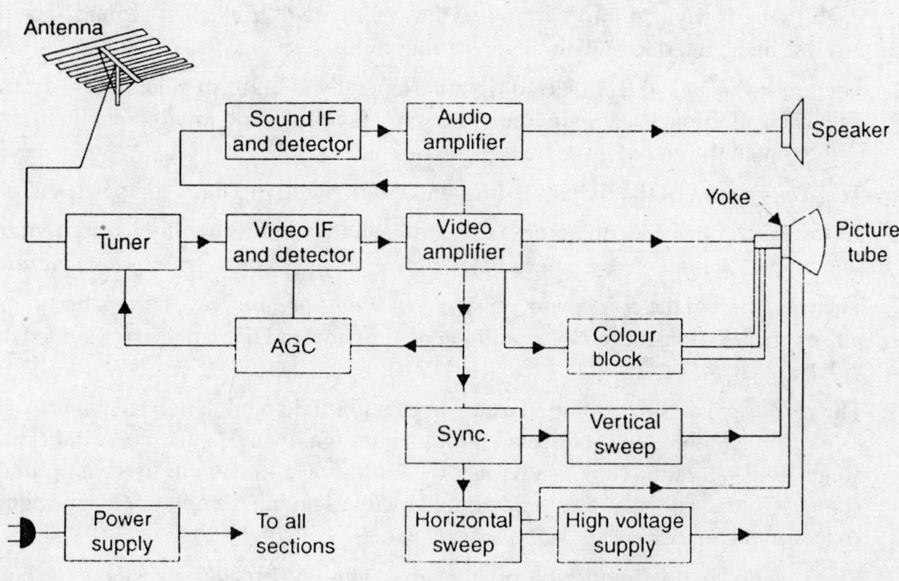

(b)

Fig. 22.36 Television as a system (a) transmitter and (b) receiver.

Considering such things as distance between stations, radiated power, geographic location, and topography each station is assigned a particular *channel* to reduce the possibility of inter-channel interference. A station may operate in either of two *bands.* VHF (very high frequency) channels 2 through 13, extending from 54 to 216 MHz, or UHF (ultra high frequency) channels 14 through 83, extending from 470 to 890 MHz. Each channel occupies a *very wide bandpass* which can best be appreciated by realising that nearly seven radio broadcast bands would be required to accomodate even one TV station. The frequency allocations are in *numerical sequence* except for a couple of *gaps* where frequencies were previously assigned to other services.

22.23 AMERICAN 525- LINE TV SYSTEM

The transmitted television signal must comply with strict standards established by the Federal Communication Commission. The standards insure that uniformly high quality monochrome and colour picture and sound will be transmitted. Standards are also required so that television receivers can be designed to receive these standard signals and thus provide a correct reproduction of the original picture and sound.

The details of the FCC Standard are:

1. Each VHF and UHF television station is assigned a *channel* that is 6MHz wide. The *composite video signal* must fit into the 6MHz bandwidth. The composite signal includes the video carrier, one vestigial sideband for the video signal, one complete sideband for the video signal, the colour signals, the synchronizing and blanking pulses, and the FM sound signal.
2. The *visual (picture) carrier frequency* is 1.25 MHz above the low end of each channel.
3. The *aural (sound) carrier frequency* is 4.5 MHz above the visual carrier frequency.
4. The *chrominance (colour) subcarrier frequency* is 3.579545 MHz above the picture carrier frequency. For practical purposes, the subcarrier frequency is expressed as 3.58 MHz.
5. One complete *picture frame* consists of 525 scanning lines (483 active lines actually produce the picture. The remaining lines are blanked out during the vertical retraced). There are 30 frames per second.
6. Each frame is divided into two *fields*; thus there are 60 fields per second. Each field contains 262 1/2 lines, interlaced with the preceding field.
7. The scene is scanned from left to right horizontally at uniform velocity, progressing downward with each additional scanning line. The scene is retraced rapidly (blanked out) from the bottom to the top, at the end of each field.
8. The *aspect ratio* of the picture is four units horizontally to three units vertically.
9. At the transmitter the equipment is so arranged that a decrease in picture light intensity during scanning causes an increase in radiated power. This is known as *negative picture transmission*.
10. The *video* part of the composite signal is *amplitude modulated.* The synchronizing and blanking pulses are also transmitted by amplitude modulation and these pulses are *added* to the composite video signal.
11. The *colour signal* is transmitted as a pair of amplitude modulation sidebands. These sidebands effectively combine to produce a chrominance signal varying in hue or tint (phase angle of the signal) and saturation (colour vividness). Saturation corresponds to the amplitude of the colour signal. Colour information is transmitted by interleaving the colour signal frequencies between spaces in the monochrome video signals.
12. The *sound signal* is frequency modulated. As in FM broadcast systems the sound signal may also be produced by *indirect FM method.* The maximum deviation of TV is ± 25KHz.
13. Table 22.2 shows the *frequencies assigned* to the United States VHF and UHF channels.

14. According to FCC Rules and Regulations, line 19 of the vertical blanking interval of each field is reserved for a special signal. This is the *vertical interval reference (VIR) signal.* The VIR signal is known as the chrominance reference. While originally intended for station use, it is also utilised by television receivers. At the station, the VIR signal is used as a reference to assure the correct transmitted hue and saturation of the colour signal. At the receiver, the signal and automatic circuits help to provide accurate colour reproduction. When this method is used it eliminates the need for manual colour control adjustment.

22.24 THE 625-LINE SYSTEM

The CCIR-B standard is adopted by our country. The details of the CCIR-B standard are:

1. Each VHF and UHF television station is assigned a *channel* that is 7MHz wide. The *composite video signal* must fit into the 7MHz bandwidth.
2. The *visual (picture) carrier frequency* is 1.25 MHz above the low end of each channel.
3. The *aural (sound) carrier frequency* is 5.5 MHz above the visual carrier frequency.
4. *The chrominance (colour) subcarrier frequency* is 4.43361875MHz above the picture carrier frequency. For practical purposes, the subcarrier frequency is expressed as 4.43 MHz.
5. One complete *picture frame* consists of 625 scanning lines (585 active lines actually produce the picture. The remaining lines are blanked out during the vertical retrace). There are 25 frames per second.
6. Each frame is divided into two *fields;* thus, there are 50 fields per second. Each field contains 312 1/2 lines, interlaced with the preceding field.
7. The scene is *scanned* from left to right horizontally at uniform velocity, progressing downward with each additional scanning line. The scene is retraced rapidly (blanked out) from the bottom to the top, at the end of each field.
8. The *aspect ratio* of the picture is four units horizontally to three units vertically.
9. At the transmitter, the equipment is so arranged that a decrease in picture light intensity during scanning causes an increase in radiated power. This is known as *negative picture transmission.*
10. The *video* part of the composite signal is *amplitude modulated.* The synchronizing and blanking pulses are also transmitted by amplitude modulation and these pulses are *added* to the composite video signal.
11. The *colour signal* is transmitted as a pair of amplitude modulation side bands. These sidebands effectively combine to produce a *chrominance signal* varying in hue and tint (phase angle of the signal) and saturation (colour vividness). Saturation corresponds to the amplitude of the colour signal. Colour information is transmitted by *interleaving* the colour signal frequencies between spaces in the monochrome video signal.
12. The *sound signal* is frequency modulated. As in FM broadcast systems, the sound signal may also be produced by indirect FM method. The maximum deviation of TV is ± 25 kHz.
13. Table 22.3 shows the frequencies assigned to the VHF and UHF channels in India.

22.25 VESTIGIAL SIDEBAND TRANSMISSION

Because of the extensive bandwidth requirements of the video signal, it is desirable to make use of a *bandwidth saving technique.* The signal information is fully contained in each of the two sidebands of the modulated carrier; and *provided the carrier is present, one sideband may be suppressed altogether.* The single sideband transmission technique can reduce the bandwidth requirement to half, viz. 5 MHz. However, it is not possible to do this in the case of a television signal because a television signal also contains very low frequencies including even the DC information. It is impossible to design a filter which will cut off the unwanted band while passing the carrier frequencies and low frequency components of the other sideband, without objectionable phase distortion.

As a compromise, a part of only one sideband is suppressed. The transmitted signal consists of one complete sideband together with the carrier, and a vestige of the partially suppressed sideband as shown in Fig. 22.37. In the vestigial sideband transmission system, *there is a saving in the bandwidth required*, and the filtering required is not so difficult to achieve.

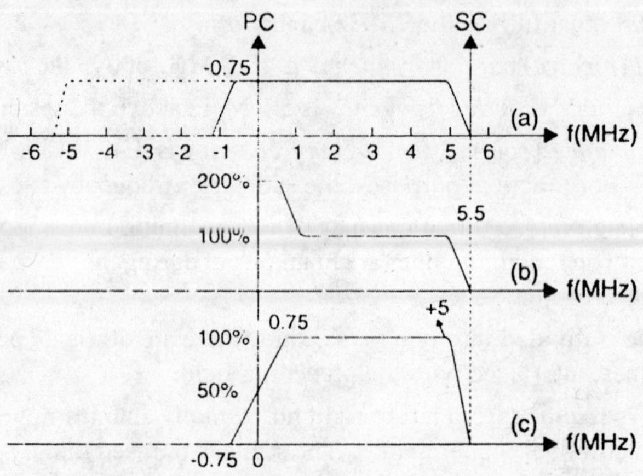

Fig. 22.37 Vestigial sideband channel characteristics. (a) VSB channel, (b) Videopower content and (c) Receiver response.

Table 22.2 Television Channel Frequency Allocations

Channel Number	Frequency Limits (MHz) VHF Television	Picture Carrier (MHz) Station Frequencies	Sound Carrier (MHz)
2	54-60	55.25	59.75
3	60-66	61.25	65.75
4	66-72	67.25	71.75
5	76-82	77.25	81.75
6	82-88	83.25	87.75
7	174-180	175.25	179.75
8	180-186	181.25	185.75
9	186-192	187.25	191.75
10	192-198	193.25	197.75
11	198-204	199.25	203.75
12	204-210	205.25	209.75
13	210-216	211.25	215.75

UHF Television Station Frequencies

Channel Number	Frequency Range MHz	Picture Carrier MHz	Sound Carrier MHz
14	470-476	471.25	475.75
15	476-482	477.25	481.75
16	482-488	483.25	487.75
17	488-494	489.25	493.75
18	494-500	495.25	499.75
19	500-506	501.25	505.75
20	506-512	507.25	511.75
21	512-518	513.25	517.75
22	518-524	519.25	523.75
23	524-530	525.25	529.75
24	530-536	531.25	535.75
25	536-542	537.25	541.75
26	542-548	543.25	547.75
27	548-554	549.25	553.75
28	554-560	555.25	559.75
29	560-566	561.25	565.75
30	566-572	567.25	571.75
31	572-578	573.25	577.75
32	578-584	579.25	583.75
33	584-590	585.25	589.75
34	590-596	591.25	595.75
35	596-602	597.25	601.75
36	602-608	603.25	607.75
37	608-614	609.25	613.75
38	614-620	615.25	619.75
39	620-626	621.25	625.75
40	626-632	627.25	631.75
41	632-638	633.25	637.75
42	638-644	639.25	643.75
43	644-650	645.25	649.75
44	650-656	651.25	655.75
45	656-662	657.25	661.75
46	662-668	663.25	667.75
47	668-674	669.25	673.75

48	674-680	675.25	679.75
49	680-686	681.25	685.75
50	686-692	687.25	691.75
51	692-698	693.25	697.75
52	698-704	699.25	703.75
53	704-710	705.25	709.75
54	710-716	711.25	715.75
55	716-722	717.25	721.75
56	722-728	723.25	727.75
57	728-734	729.25	733.75
58	734-740	735.25	739.75
59	740-746	741.25	745.75
60	746-752	747.25	751.75
61	752-758	753.25	757.75
62	758-764	759.25	763.75
63	764-770	765.25	769.75
64	770-776	771.25	775.75
65	776-782	777.25	781.75
66	782-788	783.25	787.75
67	788-794	789.25	793.75
68	794-800	795.25	799.75
69	800-806	801.25	805.75
70	806-812	807.25	811.75
71	812-818	813.25	817.75
72	818-824	819.25	823.75
73	824-830	825.25	829.75
74	830-836	831.25	835.75
75	836-842	837.25	841.75
76	842-848	843.25	847.75
77	848-854	849.25	853.75
78	854-860	855.25	859.75
79	860-866	861.25	865.75
80	866-872	867.25	871.75
81	872-878	873.25	877.75
82	878-884	879.25	883.75
83	884-890	885.25	889.75

Table 22.3. Television Channel Frequency Allocations

Channel No.	Frequency Range (MHz)	Picture Carrier (MHz)	Sound Carrier (MHz)
1	40-47	41.25	46.75
2	47-54	48.25	53.75
3	54-61	55.25	60.75
4	61-68	62.25	67.75
5	81-88	82.25	87.75
6	174-181	175.25	180.75
7	181-188	182.25	187.75
8	188-195	189.25	194.75
9	195-202	196.25	201.75
10	202-209	203.25	208.75
11	209-216	210.25	215.75
12	216-223	217.25	222.75
13	223-230	224.25	229.75
14	470-477	471.25	476.75
15	477-484	478.25	483.75
16	484-491	485.25	490.75
17	491-498	492.25	497.75
18	498-505	499.25	504.75
19	505-512	506.25	511.75
20	512-519	513.25	518.75
21	519-526	520.25	525.75
22	526-533	527.25	532.75
23	533-540	534.25	539.75
24	540-547	541.25	546.75
25	547-554	548.25	553.75
26	554-561	555.25	560.75
27	561-568	562.25	567.75
28	568-575	569.25	574.75
29	575-582	576.25	581.75
30	582-589	583.25	588.75
31	589-596	590.25	595.75
32	596-603	597.25	602.75
33	603-610	604.25	609.75

34	610-617	611.25	616.75
35	617-624	618.25	623.75
36	624-631	625.25	630.75
37	631-638	632.25	637.75
38	638-645	639.25	644.75
39	645-652	646.25	651.75
40	652-659	653.25	658.75
41	659-666	660.25	665.75
42	666-673	667.25	672.75
43	673-680	674.25	679.75
44	680-687	681.25	686.75
45	687-694	688.25	693.75
46	694-701	695.25	700.75
47	701-708	702.25	707.75
48	708-715	709.25	714.75
49	715-722	716.25	721.75
50	722-729	723.25	727.75
51	729-736	730.25	735.75
52	736-743	737.25	742.75
53	743-750	744.25	749.75
54	750-757	751.25	756.75
55	757-764	758.25	763.75
56	764-771	771.25	770.75
57	771-778	778.25	777.75
58	778-785	785.25	784.75
59	785-792	786.25	791.75
60	792-799	793.25	798.75
61	799-806	800.25	805.75
62	806-813	807.25	812.75
63	813-820	814.25	819.75
64	820-827	821.25	826.75
65	827-834	828.25	833.75
66	834-841	835.25	840.75
67	841-848	842.25	847.75
68	848-855	849.25	854.75
69	855-862	856.25	861.75

70	862-869	863.25	868.75
71	869-876	870.25	875.75
72	876-883	877.25	882.75
73	883-890	884.25	889.75
74	890-897	891.25	896.75
75	897-904	898.25	903.75
76	904-911	905.25	910.75
77	911-918	912.25	917.75
78	918-925	919.25	924.75
79	925-932	926.25	931.75
80	932-939	933.25	938.75
81	939-946	940.25	945.75
82	946-853	947.25	952.75
83	953-960	954.25	959.75

OBJECTIVE TYPE QUESTIONS (A)

1. In the case of transmission of picture and sound (television) the fundamental problem is that of converting.
   (a) time variations into electrical information.
   (b) space variations into electrical information
   (c) time and space variation into electrical information.

2. Scanning may be defined as the process which permits the conversion of information expressed in
   (a) space coordinates into time variations
   (b) space and time coordinates into time variations only
   (c) time coordinates into time variations

3. Aspect ratios of 4:3 are
   (a) most pleasing to the eye (b) least pleasing to the eye
   (c) most fatiguing to the eye (d) least fatiguing to the eye

4. Most practical television systems resort to
   (a) linear scanning (b)circular scanning
   (c) nonlinear scanning

5. It is common practice to scan the image from
   (a) left to right (b) right to left
   (c) top to bottom (d) bottom to top

6. Most scenes have brightness gradations in a
   (a) vertical direction (b) horizontal direction
   (c) diagonal direction

7. The resolving capability of the human eye is about
   (a) half-a-minute of visual angle
   (b) one minute of visual angle
   (c) one and a half minute of visual angle
   (d) two minutes of visual angle
8. The television picture is traced by
   (a) a spot of light (b) a beam of light
9. Precision is essential in television but it is in the
   (a) transmitting equipment and not receiving equipment
   (b) receiving equipment and not transmitting equipment
10. The composite video signal locks in the frequency and phase of the
   (a) frame oscillator (b) line oscillator
   (c) frame and line oscillators
11. The highest video frequency of a system depends on the
   (a) number of lines (b) rate of scanning
   (c) both (a) and (b)
12. In the 625 - line system, the sound carrier is
   (a) amplitude modulated (b) frequency modulated
13. In the 625 - line system the picture carrier is
   (a) amplitude modulated (b) frequency modulated
14. Transmitters in the centimetre wave range work almost exclusively on
   (a) amplitude modulation (b) frequency modulation

FILL IN THE BLANKS (B)

1. Scanning may be defined as the process which permits the conversion of information expressed in.......and ....coordinates into time variations only.
2. The detail which can be reproduced depends on the.................. of coverage by the scanning beam.
3. The process of scanning makes possible the use of .... transmission channel.
4. A varying scanning velocity for the pickup and reproducing devices results in........output for......scene brightness.
5. It is common practice to scan the image from........to........
6. The repetition rate of flashes at and above which the flicker effect.........is called the critical flicker frequency.
7. The total number of pixels activated on the screen.........the picture
8. There is a degradation in vertical resolution due to.....beam size.
9. The horizontal fly back interval occupies......of the line period.
10. The vertical flyback interval occupies......of the frame period
11. The vertical resolution of a system depends on the total number of.....scanning lines.
12. The time during which the spot travels back to the top of the screen is called........period.
13. The method of scanning, where the odd lines and even lines are alternately scanned is called......scanning.
14. In television, the odd line field and the even line field make one................
15. The voltage representation of every line scanned is separated from the next by................
16. Each amplifying stage...........the direction of video signal.

17. The video, sound, and control signals combine to make up ....video signal.
18. The lowest frequency officially allocated to television by international agreement is.......
19. The maximum deviation laid down for television sound is.......
20. At least first ...... to ...... paired waves of each sideback must be transmitted.
21. The bandwidth of every link in the chain of FM transmission must be at least..........
22. The ratio of the faceplate illumination in foot candles to the output signal current in nanoamperes is called the.....................characteristic.
23. With camera tubes, the gamma value is generally.
24. With picture tubes, the gamma value is generally.
25. As nearly as possible, the camera tube should have the same spectral response as the.............
26. The formula for quantum efficiency is.................
27. Quantum efficiency is a convenient way to express the ....of the image picture tube.
28. Lag on a television picture causes.......to appear following rapidly moving objects.
29. Dark current refers to the current that flows through the device even when in total...............
30. A short or missing pin in the base serves as the..................system for socket connections.
31. All electrostatic lenses achieve a saving in...............in several ways.

CONVENTIONAL QUESTIONS (C)

1. What are the elements of a television system?
2. Explain the significance of scanning.
3. What are the different scanning methods?
4. Differentiate between linear scanning, sequential scanning and interlaced scanning.
5. Define the following terms: (a) Frame (b) Critical flicker frequency (c) Odd and even field (d) Trace and retrace (e) Active and inactive lines (f) Aspect ratio (g) Kell factor (h) Flicker (i) Vertical resolution (j) Horizontal resolution (k) Picture element (l) Video bandwidth (m) Test Card (n) Negative video polarity (o) Gamma
6. What are the constituents of composite video signal?
7. Explain the working of a monochrome video camera with the help of a simplified diagram.
8. What are the characteristics of a camera tube?
9. Explain the working of a vidicon a camera tube.
10. What is the difference between a camera tube and a picture tube?
11. With the help of a suitable sketch, explain the working of a picture tube.
12. Compare the 525 lines American system with CCIR-B system. Which system is adopted by our country?

23

COLOUR TV STANDARDS AND SYSTEMS

Colour television has considerably complicated the variables between different television systems. The systems used for television have been thoroughly standardised. There are three different incompatible systems for encoding signals. These three systems are used in different parts of the world. There are various steps in the processing of the video signal. These steps do not necessarily have to be the same in different countries, or not even in the same country, but the end result would always be the same.

23.1. DISPERSION AND RECOMBINATION OF LIGHT

If a stream of sunlight is made to pass through a small circular hole in one of the shutters, it will make a small circular patch on the opposite wall. On placing a triangular prism before the hole an elongated coloured patch of light will be formed on the wall. This is called the *spectrum of light.* The colours formed are in the order shown in Fig 23.1

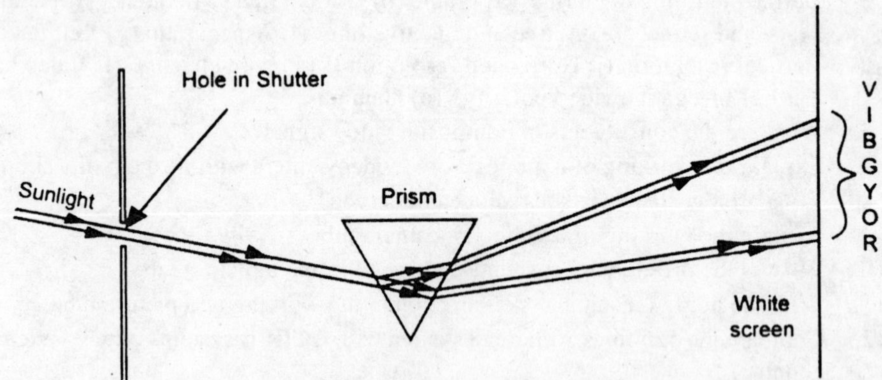

Fig 23.1 Dispersion of light

White light consists of a mixture of seven different colours. The refractive index of glass is different for each colour, so that when white light falls on the prism, each colour in it is refracted at a different angle, with the result that *the colours are spread out to form a spectrum.*

When white light is incident on the prism it is refracted towards the base of the prism, *the violet being deviated the most and the red least.* The separation of white light into its component colours by a prism is called dispersion. Strictly speaking there are many shades of each colour in the spectrum, each shade gradually merging into the next.

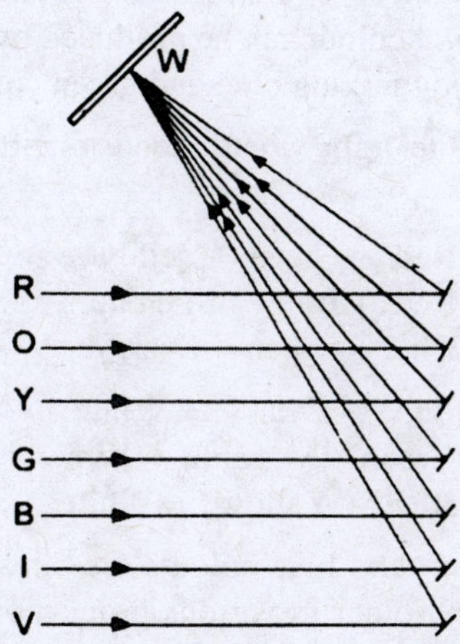

Fig. 23.2 Recombination of spectrum colours by mirrors

The colours of the spectrum may be recombined to form white light by allowing the spectrum to be formed on a row of small rectangular plane mirrors, Fig 23.2. On adjusting the angle which the mirrors make with the incident light so that all the mirrors reflect the light to the same place on the screen, the constituent colours of white light are recombined to form a white patch of light.

When white light falls on any particular body, then either all of the colours in the white light may be *reflected* from the body, when it appears white, or only some of them may be reflected while the others are *absorbed.* In the latter case, the body appears coloured. The energy of the light absorbed is generally converted into *internal energy* so that the body becomes slightly warmer. The *colour which the body presents to the eye is the colour of the light which it reflects.* Thus the leaves of plants appear green since they reflect green light and absorb the other colours. White paper reflects all the colours of the spectrum, while black absorbs all of them. *Blackness is thus due to the absence of light of any colour.*

23.2 PRIMARY AND SECONDARY COLOURS

Colours which cannot be produced by mixing other colours are called primary colours. It is not found possible to produce either red, blue or green colours by mixing two other colours. For this reason red, green and blue are called primary colours.

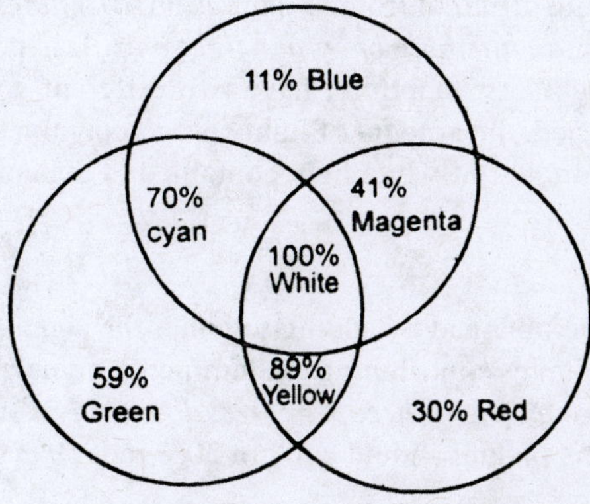

Fig. 23.3 Mixing of the three primary colours.

A secondary colour can be produced by mixing other colours. Thus *yellow* colour can be produced by mixing red and green colours. *Magenta* colour can be produced by mixing red and blue colours. By the same token, *cyan* colour can be produced by mixing blue and green colours. This is shown in Fig 23.3.

Two colours which give white light when added together are called *complementary colours*. The following facts can be verified.

	Red + Green = Yellow		
Primary colours	Red + Blue = Magenta	Secondary Colours	...(23.1)
	Blue + Green = Cyan		
Primary and	Red + Cyan = White		
complementary colours	Green + Magenta = White		...(23.2)
	Blue + Yellow = White		

In colour television, red , green and blue colours are chosen, and are called primary colours. When these colours are combined with each other in various proportions, a wide range of *hues and tints* (colour shades) are produced. These hues and tints are sufficient for presenting any colour picture. Also the *range of colours* produced by combining red, green and blue colours is *wider* than the range produced by combining any other colours.

The colours red, green, and blue are called *additive primaries* and are used when coloured light sources are blended to produce the required colour. Red, yellow and blue are called *subtractive primaries* and are used when a picture on print is viewed by reflected light from a white source.

23.3 ATTRIBUTES OF COLOUR

All primary colours as well as those produced by adding the primary colours *in suitable proportions* are characterised by three main features; hue, brightness and saturation.

Hue : The colour itself is called the hue and depends on the *dominant wavelength* of the light. By *adding* two or more of the primary colours many hues are produced.

Brightness : Each colour produces a certain amount of brightness. Brightness of a colour is determined by the *amount of light energy* contained in it. Light energy is measured in *lumens.*

Saturation : *The amount of white light* contained in a colour determines its saturation level. For instance, a highly saturated or deep red light will contain a much less amount of white light than a dull red light. *Saturation, thus denotes the degree of dilution of colour by white light.* More white light makes the colour dull or less saturated or vice versa.

Brightness of a colour is quite different from its saturation. *Brightness is an attribute of white light and colour light, whereas saturation is an attribute of colour light only.* It is possible to vary one while keeping the other constant. Colour TV receivers, therefore, have two different controls: brightness and saturation. When the *brightness control* is varied, the amount of light energy contained in the colour is varied. When the *saturation control* is varied, the amount of white light contained in the colour is varied.

23.4 LUMINANCE SIGNAL

The luminance signal can be obtained by adding together the signals representing the three primaries, R, G and B. However, the three voltages contributing the luminance signal must be taken in different amounts because *the human eye responds to each of the three primaries differently.* Calculations show that the luminance signal associated with whites of the picture should contain 30% red, 59% green and 11% blue.

$$E_Y = 0.30\ E_R + 0.59\ E_G + 0.11\ E_B \qquad ...(23.3)$$

This equation, pivotal to colour television, needs some explanation. Suppose that a simple pattern made of a white stripe against a black background is projected onto photocathodes of the three pick-up tubes. The dichroic mirrors, Fig. 23.4, will split light from the white stripe into three colour components *R*, *G* and *B*. The gain of the video amplifiers can then be adjusted so that their output voltage is the same i.e;

$$E_R = E_G = E_B$$

This is the *relative sensitivity* of a three tube camera.

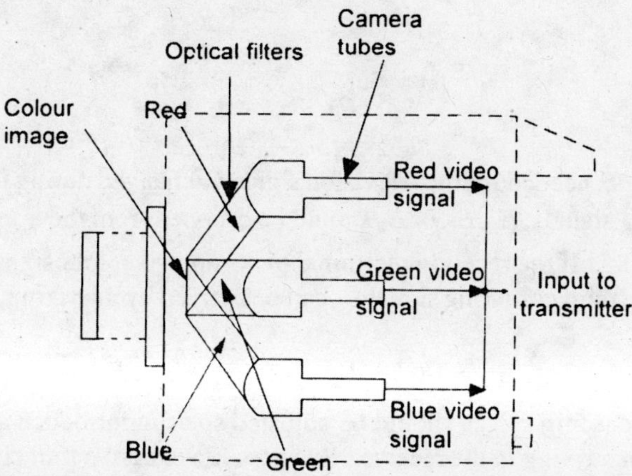

Fig. 23.4. The optics in a colour camera separate the scene into red, green and blue elements.

The *luminance signal* essential to operation of a black-and-white TV, is produced by means of a *matrix*. The circuit of a simple matrix composed of four resistors (three *voltage dividers*) is shown in Fig. 23.5. If the values of R1, R2, and R3 are chosen to be sufficiently high in comparison with R_{out}, the voltage dividers are *mutually isolated* so that the following voltages are developed across the resistor R_{out} :

$$E_{R_{out}} = E_R\,(R_{out}/R_1)$$

$$E_{G_{out}} = E_G\,(R_{out}/R_2) \qquad \ldots(23.4)$$

$$E_{B_{out}} = E_B\,(R_{out}/R_3)$$

By setting the scale factors $R_{out}/R_1 = 0.30$, $R_{out}/R_2 = 0.59$ and $R_{out}/R_3 = 0.11$ the following luminance signal will be secured at the matrix output:

$$E_Y = E_{R_{out}} + E_{G_{out}} + E_{B_{out}}$$

$$E_Y = 0.30\,E_R + 0.59\,E_G + 0.11\,E_B \qquad \ldots(23.5)$$

The following example will provide a better insight into the significance of Equation 23.3 to colour television. Suppose two *blue and white* objects having the same intensity of radiation are to be televised. To an observer watching the two areas on a *colour reproducer*, both of them will appear different in luminance. In accordance with the above equation the luminance of the blue area will be equal to 11% of that of the white area.

On a *black and white reproducer* the areas will not appear coloured. However due to matrixing, the luminance of the blue area will still be equal to 11% of that of the white area because of the voltages E_R and E_G entering the equation for E_Y and coming from the red and green pick-up tubes, disappear and $E_Y = 0.11\,E_B$.

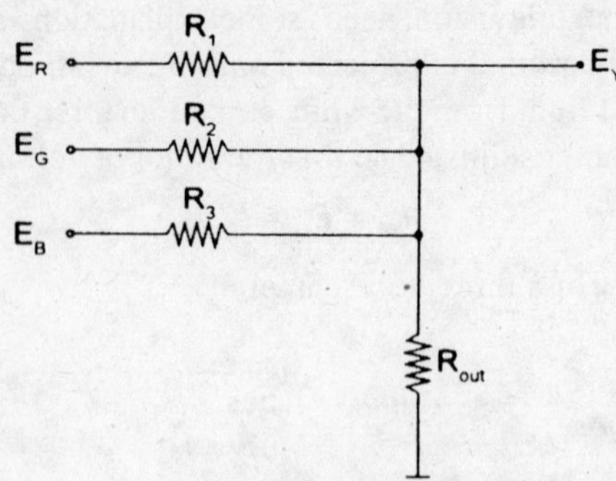

Fig. 23.5 Luminance signal

To keep the *bandwidth* needed for the television signal unchanged during the insertion of the luminance signal, one of the colouring signals, E_R, E_G or E_B should be dropped from the transmission. This may be done on the basis of Equation 23.3. If the receiving terminal picks up two colour signals, say E_R and E_B in addition to the luminance signal, the third colouring signal E_G can be derived by matrixing. It follows from the equation for E_Y that

$$E_G = 1/0.59\,(E_Y - 0.11\,E_B - 0.30\,E_R) \qquad \ldots(23.5)$$

The gain of the succeeding stages should be adjusted so as to introduce a *correction factor* of 1/0.59. In this way , by the use of matrixing in the receiver, it is possible to derive four signals, E_Y, E_R, E_G and E_B from three input signals, E_Y, E_R and E_B.

23.5 CHROMINANCE SIGNAL

The generation of chrominance signal involves the *separation* of the luminance signal, E_Y from the natural colour signals, E_R, E_G and E_B. This is shown in Fig. 23.6. The natural colour signals E_R, E_G and E_B are developed across resisters R_1, R_2 and R_3 respectively. *Part of the voltage* across these resistors (0.3 V across R_1 0.59 V across R_2, and 0.11 V across R_3) is tapped off. The full potential difference across R_1, R_2 and R_3 is simultaneously fed to the adders in stages 2, 3 and 4.

When a white scene is being televised, the three pick-up tubes will extract the three primary colours from the white light and develop *corresponding voltages* across R_1, R_2 and R_3. Since light falls on all the three pick-up tubes with the same intensity, these three voltages would be equal, *i.e*,

$$E_R = E_G = E_B = (\text{say})1\,V$$

Adder stage 1 will produce an output of:

$$0.30\,E_R + 0.59\,E_G + 0.11\,E_B$$

The sum of these voltages is the voltage E_Y corresponding to white light. As such, the output from adder stage 1, will be E_Y.

Thus $$E_Y = 1\,V$$

This output is routed through an L.P. filter (low pass filter to an amplifier stage which is a conventional common-emitter amplifier, so that at its output we obtain the signal ($-E_Y$), whose value in this example will be - 1V.

The ($-E_Y$) signal is *applied simultaneously* to stage 2, 3 and 4. To these stages the full voltages E_R, E_G and E_B are also applied. In the case of white light each of these three voltages are each equal to 1V. Thus each of the 2,3 and 4 stages has two inputs:

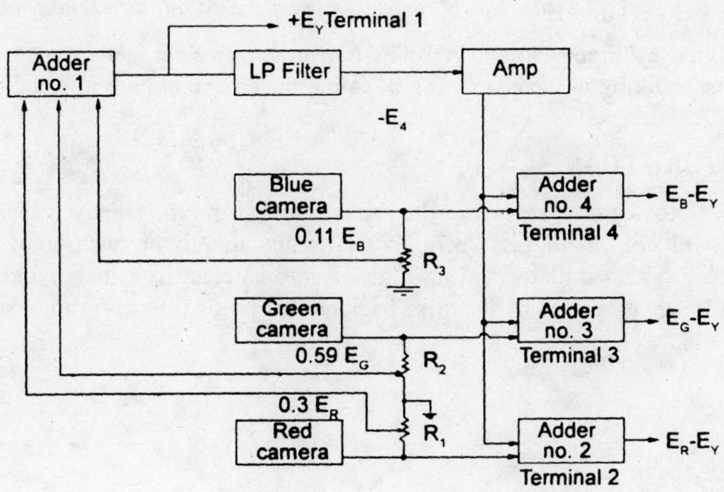

Fig. 23.6 Chrominance signal

$$-E_Y = -1\text{V} \quad \text{and} \quad E_R = E_G = E_B = 1\text{V}$$

Obviously the output of each one of the adder stages 2, 3 and 4 will be zero. These *four outputs* are brought out from terminals 1 to 4 and routed to the TV transmitter after due processing.

This, then, is the process of televising a *white scene*. The luminance signal is available at terminal 1 for transmission and no signal is available at terminals 2, 3 and 4.

Let us next consider a *colour scene*, say a red scene. The red camera will generate the signal E_R. Let E_R, the potential drop across R_1 be 1V. Part of the voltage (0.30V) is tapped off and applied to terminal 1. Since only a red scene is being televised, there will obviously be no output from the blue and green cameras. Hence the input to terminal 1, will be:

$$E_R = 0.30\text{ V}; \qquad E_G = 0\text{V} \qquad \text{and } E_V = 0\text{V}$$

Thus, the output designated as E_Y will be:

$$E_Y = 0.30\text{ V}$$

This voltage is available at terminal 1 for being routed to the transmitter. Also, this voltage is *inverted* after passing through the LP filter and a stage of amplification. After inversion, it is applied to stages 2, 3and 4. This inverted voltage will be

$$-E_Y = -0.30\text{ V}$$

Now the input to stages 2, 3 and 4 will be:

At terminal 2 $\quad E_R = 1\text{V}$ and $-E_Y = -0.30\text{V}$

At terminal 3 $\quad E_G = 0\text{V}$ and $-E_Y = -0.30\text{V}$

At terminal 4 $\quad E_B = 0\text{V}$ and $-E_Y = -0.30\text{V}$

The output of the adder stages available at terminals 1 to 4 will be

At terminal 1 $\quad E_Y = 0.30\text{V}$

At terminal 2 $\quad E_R - E_Y = 0.70\text{V}$

At terminal 3 $\quad E_G - E_Y = -0.30\text{V}$ ...(23.6)

At terminal 4 $\quad E_B - E_Y = -0.30\text{V}$

The voltage available at terminal 1 is the *luminance signal* E_Y. The voltages available at terminals 2, 3 and 4, designated as $E_R - E_Y$, $E_G - E_Y$ and $E_B - E_Y$ respectively are the *colour difference* or *chrominance signals.*

This, in brief, is the technique adopted in the TV transmitter for electrically separating the natural colour and luminance signals, resulting in the generating of colour difference or chrominance signals.

23.6 COLOUR PICTURE TUBE

The colour television camera *separates* the primary hues from the televised scene, Fig 23.4, and the colour picture tube *recombines* them, Fig. 23.7. This is comparable to the function of a mike (which picks up the sound from the program initiator and translates it into its electrical image) and loudspeaker (which retranslates electrical image produced by the mike back into its original form for the listeners)

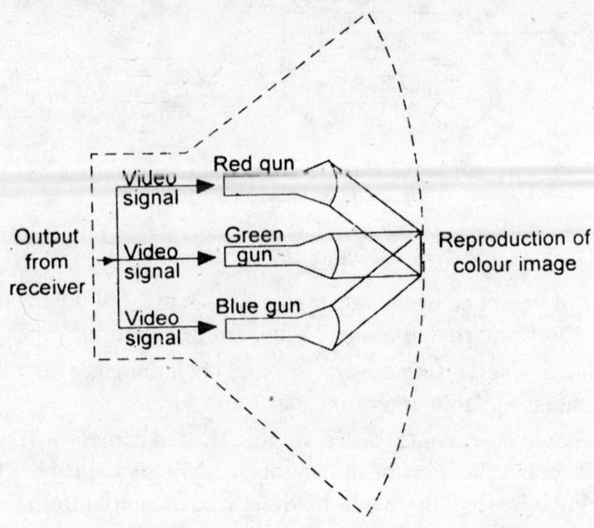

Fig. 23.7 The colour picture tube recombines the primary hues.

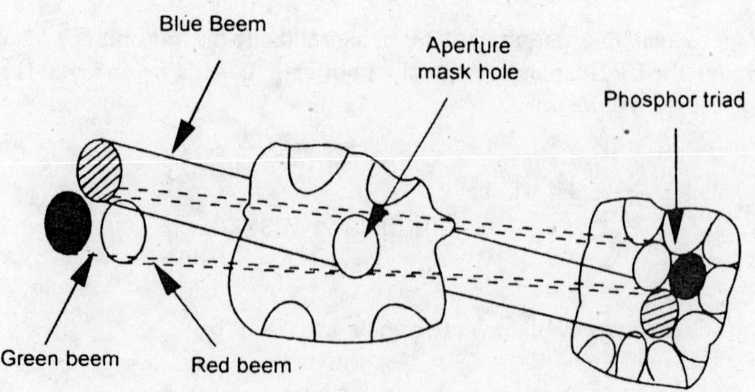

Fig. 23.8.(a)

The colour picture tube is basically a tube which has *three electron guns* in one picture tube envelope. The three guns are placed in the neck of the tube in a triangle (*delta*). One gun is called the *red gun*, the second the *blue gun* and the third the green gun (see Fig. 23.8). The combinations of these primary colours produce all the other colours, including white.

The *phosphor* on the screen or face plate is considerably different from the phosphor of monochorome picture tubes. Instead of a solid coating of one phosphor on the screen, each of the three phosphors is placed in dots along a horizontal axis, that is in one horizontal line, the three basic colour dots of phosphors will be placed in a triangle again and again, until the other side of the screen is reached. There are rows after rows of these triangles on the screen. They make up what is called a *phosphor dot pattern*. The colour tube, called the *trigun tricolcur tube* also has a *shadow mask*. It ensures that the beams from the three electron guns hit their respective phosphor dots. This is shown in Fig. 23.8

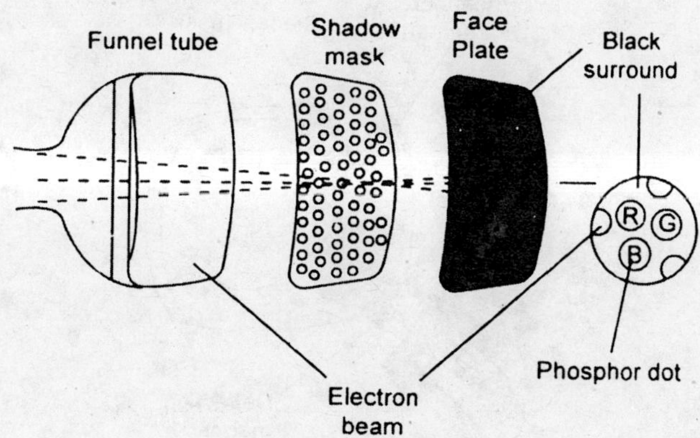

(*b*)

Fig. 23.8 Delta gun colour picture tube.
(a) Delta gun and (b) the shadow mask

The colour tube is *scanned* in the same way as a black-and-white picture tube. If the picture being received is a *black- and white* picture all the three guns will be operating. If however, the picture being televised is *in colour*, the red gun will operate for red objects in the picture, green for green objects in the picture, and so on. If some other colour is required, then the *proper guns* will operate to mix the basic colours and produce the desired colour. The quantity of electrons hitting the phosphor dots is controlled by their respective control grids so as to produce any desired colour.

23.7 DIFFERENCE BETWEEN A MONOCHROME AND A COLOUR PICTURE TUBE

The colour picture tube differs from the monochrome picture tube in the following four aspects:

1. It has *three guns* which provide *three electron beams*, one each for the three primary colours (*trigun, tricolour*).
2. The screen of the colour picture tube is coated with *three different types of phosphor droplets* which separately emit red, green and blue light when the phosphors are bambarded by the high velocity electrons from the three guns.
3. The phosphors are embedded on the screen in a triangular dot pattern (*triad*). Each triad has a group of three phosphor dots.

4. The *shadow mask* which is accurately mounted on the face about 1/2 inch behind the screen is the fourth distinguishing feature between a colour picture tube and a black-and-white picture tube

23.8 DELTA GUN AND IN-LINE GUN

Shadow mask colour picture tubes are trigun, tricolour picture tubes with delta guns, Fig. 23.9. In picture tubes incorporating *delta guns*, each gun is positioned at an angle of 120° along the vertices of an equilateral triangle inscribed in a circle as shown in Fig. 23.10(*a*). *Each gun can have the maximum possible size in the tube because of equal spacing between them. However, convergence adjustments must correct for the guns in different planes.*

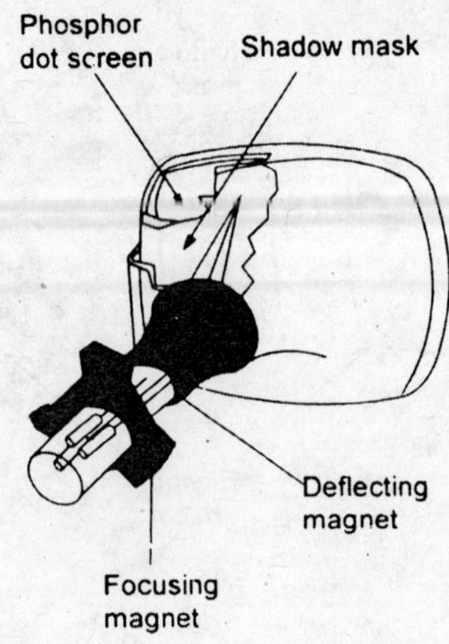

Fig. 23.9 Shadow mask colour picture tube.

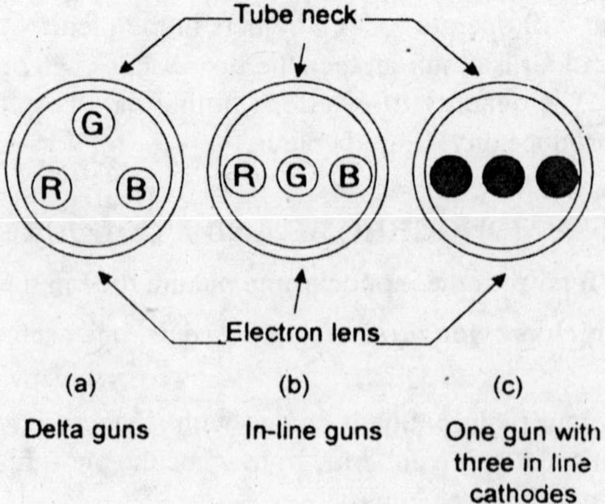

Fig. 23.10 Delta gun and In-line gun. (a) Delta gun (b) In-line gun (c) One gun with three in-line cathodes.

In *in-line guns* all the three guns are in one horizontal plane angled in toward the centre, Fig. 23.10(b). *For the same neck diameter, each in line gun is smaller than each delta gun. With smaller guns, it is more difficult to obtain a high-intensity spot with sharp focus, but convergence is easier with in-line guns.*

There are three types of in-line gun tubes and they have completely replaced the delta gun tubes in colour TV receivers. Some picture tubes employ *one gun with three in-line cathodes,* as shown in Fig. 23.10(c). These cathodes produce three electron beams that are *focused and accelerated by a common gun.*

23.9 COLOUR TV CAMERAS

There are two general categories of colour TV cameras: one group uses a *single* (special) camera tube and the other group uses *three* camera tubes.

In the single-tube colour camera, Fig. 23.11, the tube faceplate has a vertical stripe filter on the front of the tube and a *frequency separation technique* to extract the luminance and chorminance signals.

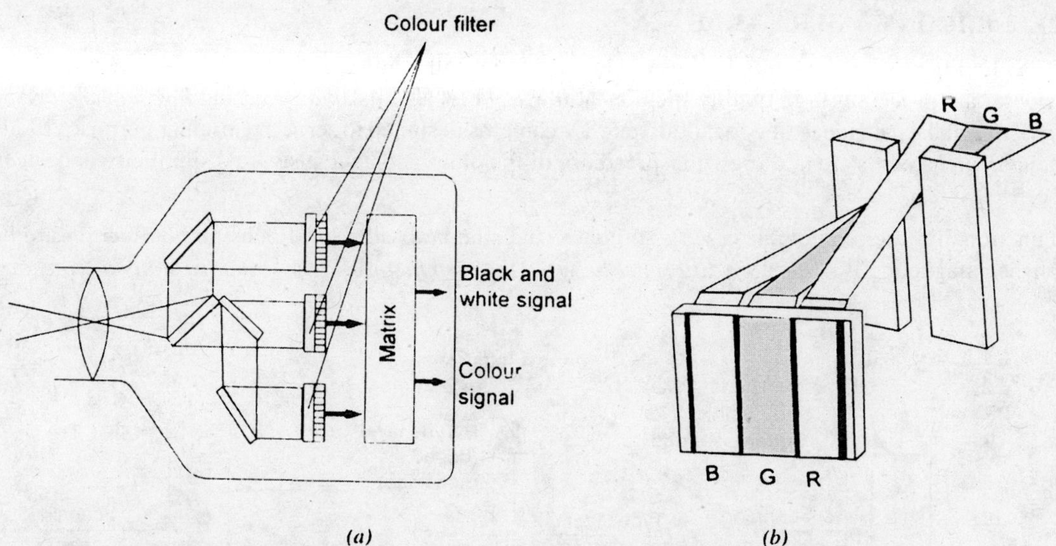

(a) (b)

Fig. 23.11 Single tube colour camera (a) and vertical stripe filter (b)

The basic camera tube is normally a vidicon with 17 mm diameter target. Unlike monochrome tubes, the colour tube has a striped colour filter *bonded* to its face plate just in front of the photosensitive target. In a typical tube the colour stripes may be green, cyan and clear, repeated in sequence across the face of the tube. The stripes themselves run vertically down the face of the tube and are very narrow, with typically some 300 sets of stripes across the picture area.

When three tubes are used, a *separate* tube is employed for each of the three primary colours: red, green, and blue. A colour filter system separates the incoming light from the image into the three colours and focuses each colour onto the face plate of an appropriate camera tube. *The three camera tubes are identical except the photosensitive material in each tube is more responsive to the particular colour for which the tube is used.*

Each colour is first determined by the filter system. The output from each tube consists of voltages that represent the elements of the particular primary colour reflected by the scene. These outputs are then amplified to provide *three channels* of video signals, one channel for each colour.

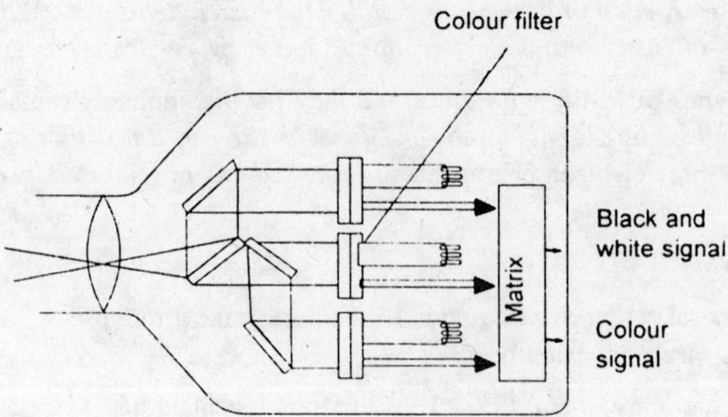

Fig. 23.12 Three-tube colour camera.

23.10 PLUMBICON CAMERA TUBE

The plumbicon, developed by Philips of Holland, is a small lightweight television camera tube that has *fast response* and processes high quality pictures at *low light levels.* Its *small size* and *low power operating characteristics* make it an ideal tube for solid-state TV cameras designed to serve a particular purpose. Modern colour television cameras are making widespread use of the plumbicon because of its simplicity and spectral response.

Functionally, the plumbicon is very similar to the standard vidicon. Focus and deflection are both accomplished magnetically. *The main difference between the plumbicon and the standard vidicon is the target.*

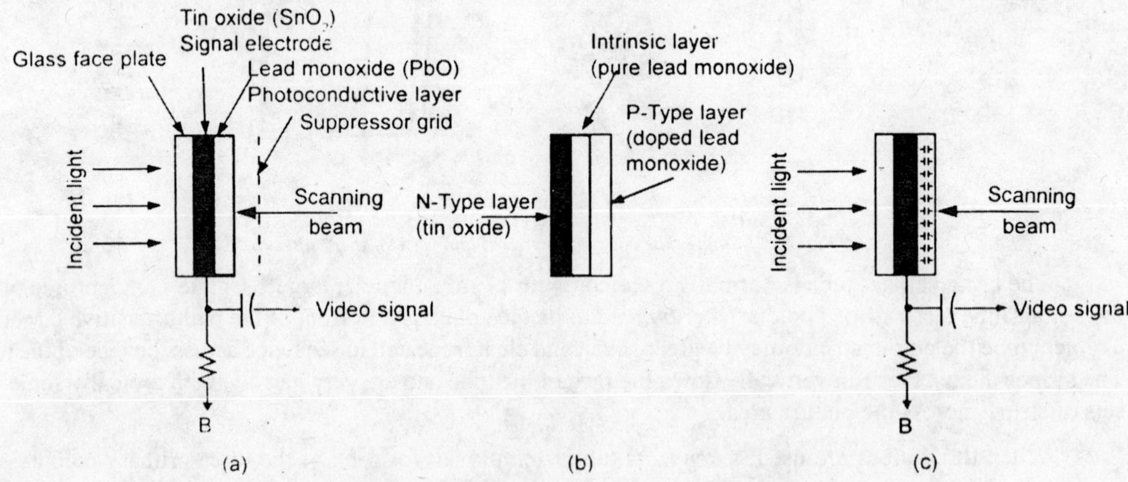

Fig. 23.13. A simplified diagram of the plumbicon target.

As shown in part (*a*) of Fig. 23.13 the inner surface of the glass faceplate is coated with a thin transparent conductive layer of tin oxide (SnO_2). This layer forms the *signal plate* of the target. A photoconductive layer of lead monoxide (PbO) is deposited on the scanning side of the signal plate. These layers are specially prepared to function as three sub-layers. *Each layer has a different conduction mode.*

The tin oxide layer on the inner side of the faceplate is a strong N-type semiconductor, commonly found in transistors. Next to this N-type region is a layer consisting of almost pure lead monoxide. This is an intrinsic semiconductor. The scanning side of the lead monoxide is doped to form a P-type semiconductor. Together these three layer's form a P-I-N. junction diode.

Table 23.1 Typical Plumbicon Operating Conditions and Performance

Operating Conditions (using coil unit AT 1105)	
Cathode Voltage	0 V
Grid No. 2 Voltage	300 V
Signal Electrode Voltage	45 V
Grid No. 4 Voltage	500 V
Grid No. 3 voltage	300 V
Beam Current	150-300 nA
Focusing and Deflection Coil Current	
Monochrome coil assembly KV12	
Focus Current	120 mA
Line Current (P-P)	160 mA
Frame Current (P-P)	25 mA
Colour coil assembly At 1105	
Focus Current	40 mA
Line Current (P-P)	320 mA
Frame Current (P-P)	120 mA
Faceplate Temperature	20 to 45°C (68 to 113°F)
Blank voltage, peak-to-peak, grid No. 1	50V
PERFORMANCE	
Dark Current	-1.5 nA
Gamma of Transfer Characteristic (Gamma stretching circuitry is recommended)	0.95 ± 0.05
Spectral Response, max	500nm
Cut-off XQ1427 G, B	650 to 850 nm
XQ1427 R	850 nm
Limiting Resolution	600TV Lines

The *photoconductive target* of the plumbicon functions much like the photoconductive target in the standard vidicon. Light from the scene being televised is focused through the transparent layer of tin oxide onto the photoconductive lead monoxide. Each picture element charge takes the form of a small capacitor with its positive plate towards the scanning beam. The target signal plate becomes the negative side of the capacitor. When the low-velocity scanning beam lands on the charge element it releases enough electrons to neutralise the charge built up on the element capacitor. The scanning beam current through the external signal plate load resistor develops the video-signal output.

The spectral response of the plumbicon can be varied while it is being manufactured to suit almost any application. Since the tube has gained wide popularity in colour television cameras, it is available with a spectral response suitable for any of the *primary colours.* The particular colour response of the tube is designated by the letter R (red), G (green) and B (blue) following its type number. When this type of tube is intended for *monochrome use*, no letter follows the type number.

23.11 COLOUR TV SYSTEMS

Colour television transmission and reception can be described as a system that reproduces colour images as depicted in Fig. 23.14, that is, a colour scene is *scanned* and *converted* into a corresponding video signal by a colour TV camera. In turn this signal is *processed* through the colour TV system Fig. 23.14(b) and is finally applied to a colour picture tube. *A three-colour picture tube is used to reproduce the colour image.*

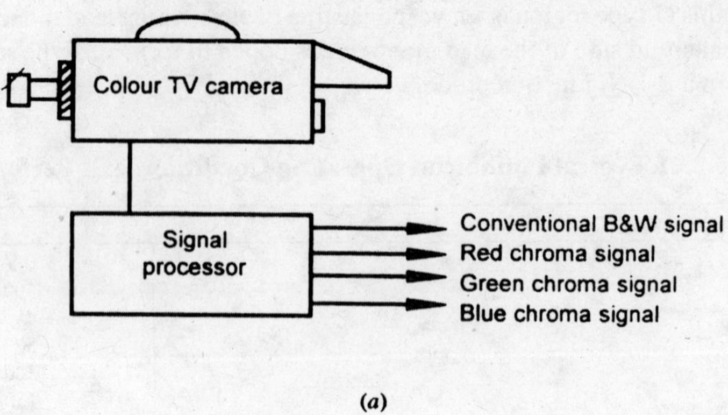

(*a*)

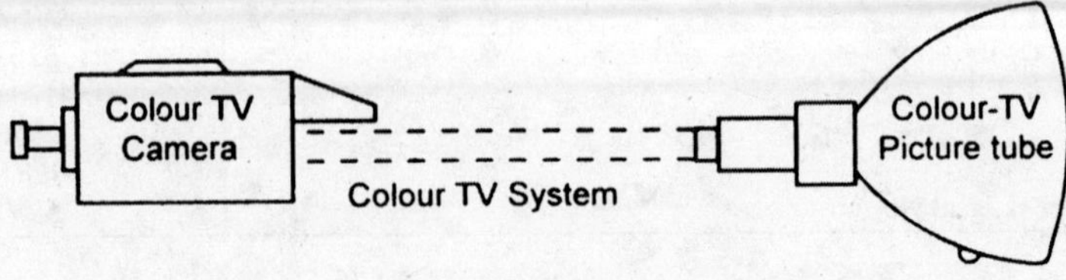

(*b*)

Fig. 23.14 (a) Complete colour signal comprises conventional black-and-white signal plus three chroma signals. (b) Fundamental plan of the colour television system.

In its most general aspect a colour TV receiver can be regarded *as a conventional black-and-white receiver plus a chroma section and a colour picture tube,* as shown in Fig. 23.15. One of the prominent advantages of any of colour television system is its *compatibility.*

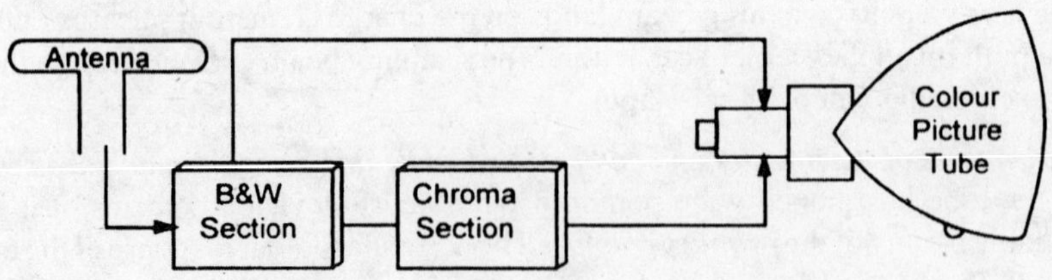

Fig. 23.15 Basic arrangement of a colour television receiver.

23.12 COMPATIBILITY CONSIDERATIONS

An *unmodified* black-and-white TV receiver cannot reproduce the image carried by a colour video signal in monochrome because the signal does not carry the component corresponding to a black-and-white picture. But it is possible by suitably adjusting a monochrome receiver, to reproduce one of the colour images (red, green or blue) in black-and-white. However, these cannot replace a black-and-white picture. For *compatibility* it is, therefore, necessary that in addition to information about colour, the transmitter of a colour television system should send a signal corresponding to black-and-white. This is usually called the *luminance signal* because the various areas of a black-and-white picture only differ in luminance.

The problem of making the colour and black-and-white programs compatible, now can be better understood. All that is required is to combine the two signals and transmit them together. One signal would contain the *luminance* or brightness information and the other the *chrominance* or colour information (hue and saturation). This is shown in Fig. 23.16

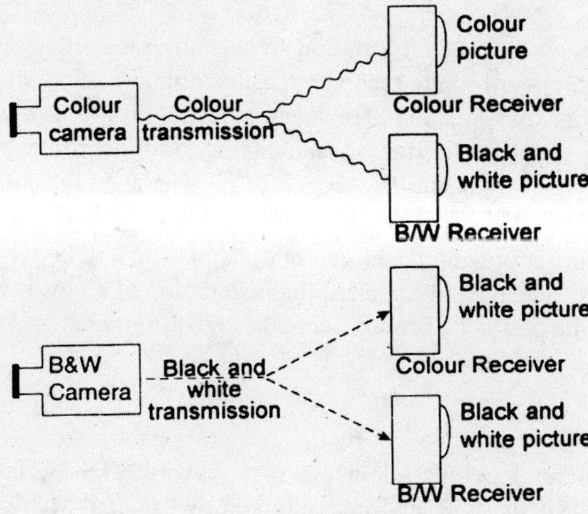

Fig. 23.16 Compatible system.

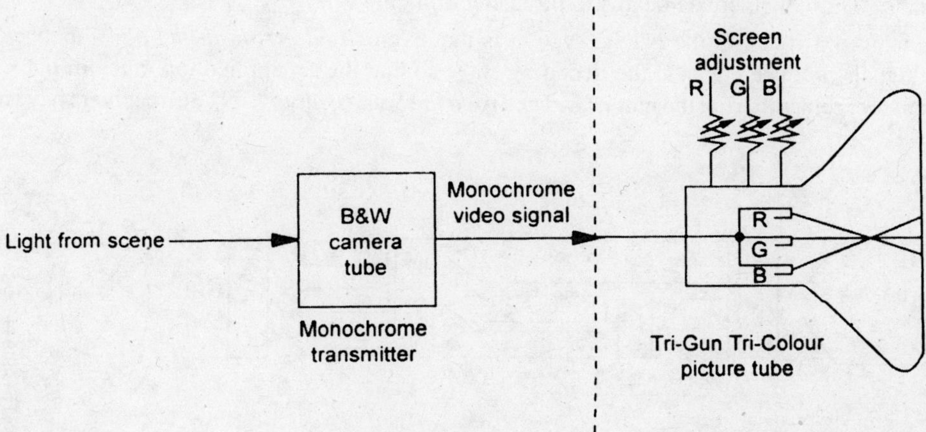

Fig. 23.17 A monochrome Compatible system

For any two systems to be compatible, they must have the same scanning rate, colour subcarrier frequency and colour encoding technique. Presently, we have five different video systems: NTSC, PAL, PAL-M, PAL-N and SECAM. Some of these are *monochrome compatible* i.e ; they can produce black-and-white pictures in the other compatible system. According to compatibility, the systems are divided into two groups with NTSC and PAL-M on one side and PAL, PAL-N and SECAM on the other.

Table 23.2 Monochrome Compatible Systems

525/60	625/50
NTSC PAL-M	PAL PAL-N SECAM

When both the brightness and colour information signals are transmitted a colour picture is produced on a colour receiver. But as a black-and-white receiver does not contain colour circuits, it can display only a *monochrome version* of the colour transmission. However, since all programs are not broadcast in colour, it is also essential that a colour receiver produce a satisfactory monochrome picture on a monochrome transmission. This is called, *reverse compatibility* (Fig. 23.17). The need for compatibility dictates that the colour system operate within the *same bandwidth* as the monochrome system and use the *same standards*.

If a compatible system is not adopted then the colour transmission will be available to only those viewers who are having colour receivers. The cost involved in the installation of a new set of transmitters *just* for the sake of colour programs would, no doubt, be prohibitive. These requirements are kept in view by all countries while switching over to colour transmission.

23.13 THE NTSC SYSTEM

The world's first commercial colour television system was NTSC (National Television Systems Committee) introduced in the USA in 1953, and later adopted by Canada, Mexico and Japan. In the *NTSC system* the two colour difference signals, created by the subtraction of red and blue from the total signal, are not transmitted together, but with one a quarter of a cycle behind the other (*in quadrature*). The signals are then *added* together to form a single chrominance signal. When it reaches the receiver, the decoding circuits inside the TV break-down the chrominance signal and *separate* it from its carrier wave. The two signals are then fed into a matrix which then *combines* them with the luminance signal to *re-create* the three original colour signals, These then create beams in three electron guns.

The main drawback in the NTSC system is that even *slight errors in the phase* between the colour difference signals produce errors at the decoding stage so that the set applied too much of one colour; hence the mnemonic for remembering the name "*Never Twice the Same Colour*". NTSC receivers have a hue control.

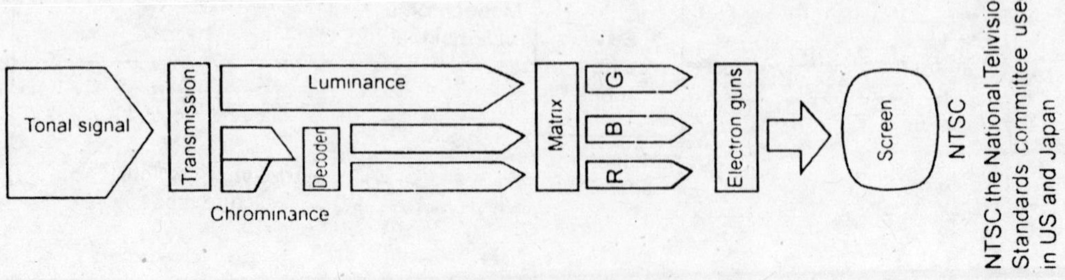

Fig. 23.18 The NTSC colour TV System.

The NTSC system is a simultaneous system which uses quadrature modulation and is compatible with monochrome systems.

23.14 THE PAL SYSTEM

The PAL system, which is a *refinement* of the NTSC system, has been adopted in our country. *The PAL system aims to improve the colour picture.* The signals are transmitted in the same way, but the receiver *delays* the information on every line by means of an ultrasonic device for the exact time needed to compare it with the signal for the next line. If, for example, a certain line of the picture, as received, contains too strong a red signal, the system ensures the next line will be too low on red by reversing the polarity of alternate lines. *The final information passed to the picture tube is the average of the delayed first line and the corrected second line, so cancelling out the error. No hue control is necessary.*

The PAL (Phase Alternate Line) colour TV system may be said to stand midway between the NTSC and SECAM system. The PAL system is explained by the engineers in USA as "*Pay for Added Luxury*".

The PAL system is not a standard in its own right since the name refers only to the *technique* used to encode the colour information. *PAL standards* exist in both the *625/50* (erstwhile German Democratic Republic, UK, Australia and many other all using a 4.43 MHz colour subcarrier) and the *525/60* (the PAL-M system used exclusively in Brazil which has a 3.58 MHz colour subcarrier) scanning frequency standards. There is also a PAL-N version used exclusively in Argentina which is a *cross between the 625/50 scanning and 3.58MHz* subcarrier.

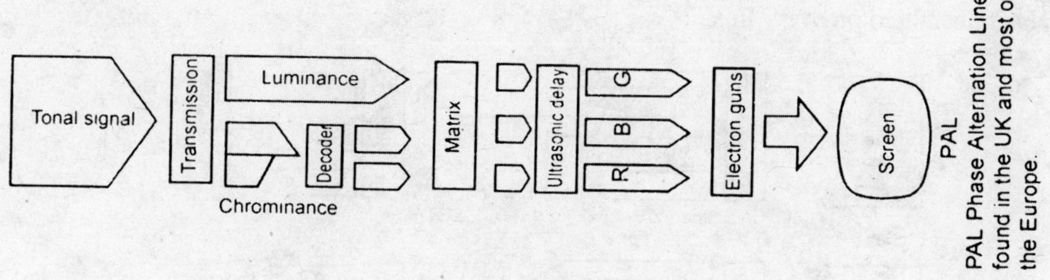

Fig. 23.19 The PAL Colour TV system.

PAL is a simultaneous compatible TV system using quadrature modulation. In contrast to SECAM, colour information in the PAL system is transmitted on every line.

Details of the different colour TV broadcasting systems are given in Table 23.3.

23.15 THE SECAM SYSTEM

In 1954, Henri De France, a French engineer, published the description of a new colour television system called the "*Henri De France System*". However, due to a number of limitations this system did not find practical application. In 1956-1957, it gave way to a new design from which a whole family of systems has grown under the common name of SECAM (Sequential couleur a memoire). It is used in the erstwhile USSR and France.

The *SECAM system* has a different mode of transmission. The colour difference signals are not arranged a quarter of a cycle apart, but are kept separate by transmitting them on *alternate lines* of the picture. *Delay lines*, inside the receiver then *hold up* one set of signals so that they can be *recombined* to build a picture from alternate lines of signals from the original scan within the camera. The SECAM system does not, however, give good picture on a black-and-white receiver because it is difficult to separate the signals from their carrier.

In 1960-1961, the SECAM system was further modified to enable the colour difference signals to be frequency modulated onto a subcarrier, which considerably improved the performance of the system. *The use of frequency modulation, and the sequential transmission of chrominance signals are the major features of SECAM as distinct from the NTSC system.* In a SECAM receiver the chrominance signals are recovered on a time rather than a phase basis, thus rendering any synchoronous detectors (as required in the NTSC system) unnecessary. *Frequency modulation has made the SECAM system insensitive to amplitude, frequency and phase distortions in the transmission circuit.*

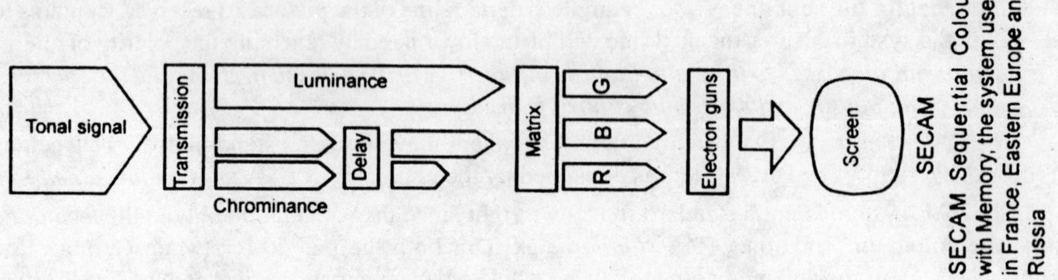

Fig. 23.20 The SECAM colour TV system

SECAM is a compatible colour TV system. Its distinguishing feature is the fact that two colour-difference signals are transmitted sequentially on alternate lines, frequency modulated on a subcarrier while the luminance signal is transmitted on every line.

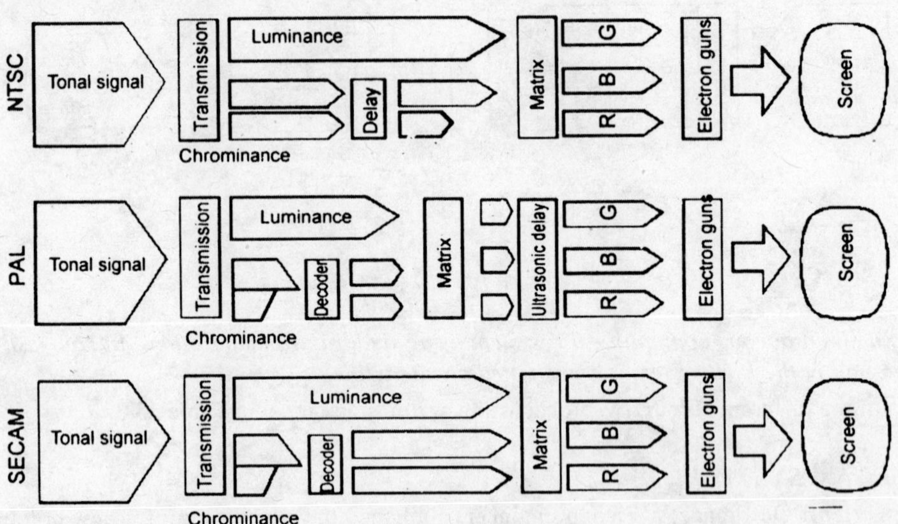

Fig. 23.21 Diagrammatic plans comparing and contrasting the three principal television broadcasting systems.

23.16 CONSOLIDATED CCIR SYSTEM-B STANDARDS

Based on Appendix A of IS : 4545 - 1968.

The main characteristics of the CCIR system-B for monochrome television, adopted in India are given below.

Video Characteristics

Number of lines per picture	625
Interlace ratio	2 : 1
Field frequency	50 fields/sec
Picture frequency	25 pictures/sec
Line frequency tolerance	15625 lines/sec ± 0.1%
Aspect ratio	4 : 3
Scanning sequence for lines	Left to right
Scanning sequence for fields	Top to bottom
Video bandwidth	5 MHz
Approximate gamma of picture signal	0.5

Composite Video Signal Characteristics

Line period H	64 µs
Line blanking period HB	18.5 to 19.2% of *H*, 11.8 to 12.3 µs
Front porch FP	2 to 2.8% of *H*, 1.3. to 1.8 µs
Line sync pulse HS	7 to 7.7% of *H*, 4.5 to 4.9 µs
Build-up time of line sync pulse	0.31 to 0.62% of *H*, 0.2 to 0.4 µs
Field period V	20 ms
Field blanking period FB	[(18 to 22) *H* +12] µs, 1164 to 1420 µs
Duration of pre-equalising pulse sequence	2.5 *H*, 160 µs
Duration of vertical pulse sequence	2.5 *H*, 160 µs
Duration of post equalising pulse sequence	2.5 *H*, 160 µs
Interval between field sync pulse (serrations)	7 to 7.7% of *H*, 4.5 to 4.9 µs
Duration of equalising pulse	3.4 to 3.75% of *H*, 2.2 to 2.4 µs
Build-up time field sync pulse	0.31 to 0.62% of *H*, 0.2 to 0.4 µs
Build-up time of field blanking pulse	6 µs

Note: Build-up time-rise-time (10 to 90%) of edges

Radio Frequency Characteristics

Channel bandwidth	7MHz
Sound carrier relative to picture carrier	+5.5 MHz
Sound carrier relative to the nearest edge of channel	-0.25 MHz
Vestigial sideband lower	-0.75 MHz
Minimum attenuation of vestige beyond - 1.25 MHz	20dB
Type and polarity of video modulation	A5, AM negative
Sync level as percentage of peak carrier	100%
Blanking level as percentage of peak carrier	72.5 to 77.5%
Difference between black level and blanking levels as percentage of peak carrier	0 to 2%
Peak white level as percentage of peak carrier	10 to 12.5%

Sound Modulation

Type	F3 FM
Carrier deviation	± 50 kHz at 100% modulation
Pre-emphasis	50 µs
Ratio of radiated powers of vision and sound	5:1

23.17 BROAD CASTING OF TV PROGRAMS

The public television service is operated by broadcasting picture and sound from picture transmitters and associated sound transmitters in three main frequency ranges in the VHF and UHF bands. By international ruling of the ITU, these ranges are exclusively allocated to television broadcasting. Subdivision into operating channels and their assignment by location are also ruled by international regional agreement. The continental standards are valid as per the CCIR 1961 Stockholm plan. The details of the various system parameters are as follows.

Band	Frequency	Channel	Bandwidth
I	(41) 47 to 68 MHz	2 to 4	7 MHz
II	87.5 (88) to 108 MHz	VHF FM Sound	
III	174 to 223 (230) MHz	5 to 11 (12)	7 MHz
IV	470 to 582 MHz	21 to 27	8 MHz
V	582 to 790 (860) MHz	28 to 60 (69)	8 MHz
VI	11.7 to 12.5 GHz	superseded by satellite	
Special	68 to 82 (89) MHz	2(3) S	7 MHz
Channels	104 to 174 and	Channels	
Cable TV	230 to 300 MHz	S1 to S20	7 MHz

Types of Modulation

Vision : C3F (vestigial sideband AM)

Vestigial sideband ratios:

0.75 MHz/4.2 MHz = 1:5.6 for system M 525/60.6, MHz

0.75 MHz/5.0 MHz = 1:6.7 for system B 625/50, 7 MHz

1.25 MHz/5.5 MHz = 1:4.4 for system I 625/50, 8 MHz

The saving of frequency band is about 40%; the polarity is negative because of the susceptibility to interference of the synchronizing circuits of early TV receivers (exception : positive modulation); residual carrier with negative modulation 10%(exception 20%).

Sound : F3E ; FM for better separation from vision signal in the receiver (exception : AM).

Sound carrier above vision carrier within RF channel, inversion at IF ; (exception : standards A, E and, in part, L).

Intermodulation distortion in TV transposers and TV transmitter with common vision-sound amplification and in Cable Television ; 20 : 1 : 0.2 for dual sound broadcast in the B/G standard *Channel bandwidth* are 5/6/7/8/14 MHz depending on standard, conventional values are 6/7/8 MHz and 14 MHz are still valid for a certain transition period.

23.18. SYSTEM PARAMETERS (FOR STANDARDS B/G)

	Channel - 1	*Channel - 2*
IRF sound carriers		
IF Frequencies	fvision + 5.5 MHz (+/-500 Hz), eqvt. to 353 fh	fvision + 5.7421875 MHz (+/-500 Hz).
Vision/sound power ratio	13 dB	20 dB
Modulation	FM	FM
Frequency deviation Max.	M +/- 50 kHz	< +/-50 kHz
nominal value	+/- 30 kHz	+/- 30 kHz

Pre-emphasis	50 μs	50 μs
AF bandwith	40 to 15000 Hz	40 to 15000 Hz
Sound Modulation		
Mono	mono	mono
Stereo	(L + R)/2 = M	R
Dual sound mono	channel A	Channel B
Identification		
Pilot carrier frequency	—	54.6875 kHz (+/- 5 Hz) Eqvt. to 3.5 fh)
Modulation	—	AM (with identification frequency)
Modulation depth	—	50%
Identification signal		
Mono		none
Stereo		(fh/133 =) 117.5 Hz
Dual sound		(fh/57 =) 274.1 Hz
Frequency deviation of transmitter carrier (due to pilot tone)		+/- (2.5 kHz +/- 0.5 kHz)
Synchronization		pilot carrier and identification frequencies phase-locked with fh

The two *sound channels* arrive from the studio via radio link with 15 kHz bandwidth at the TV transmitter, where matrixing is performed for compatibility ; (L + R)/2 for channel-1, R for channel-2. An additional sound modulator is used to modulate the second sound carrier with sound channel-2 and with the AM modulated pilot carrier.

The mode *identification* is transmitted in (data) line 16 (329) of a normal TV picture from the studio to the dual sound coder of the TV transmitter via the conventional TV lines (*i.e.* not the sound lines). From the 13 usable words of this data line the first two bits of word 5 are provided for mode identification in bi-phase code as follows :

Identification	Bit 1	Bit 2
Stereo	1	0
Mono	0	1
Dual sound	1	1
Fault	0	0

Vision/sound power ratio is 3 : 1/4 : 1/10 : 1/20 : 1, depending on the standard. Ratios of 5 : 1 and 10 : 1 are conventionally used ; 20 : 1 is used in Germany, *its advantage being energy saving and low*

Table 23.3. Colour TV Broadcasting Systems

Colour Systems	PAL				
Standard Broadcast Systems Items;	B CCIR	D	G CCIR	I	N
Channel bandwidth (MHz)	7	8	8	8	6
Video bandwidth (MHz)	5	6	5	5.5	4.2
Frequency difference between audio and video	+ 5.5	+ 6.5	+ 5.5	+ 6	+ 4.5
Vestigial sideband (MHz)	0.75	0.75	0.75	1.25	0.75
Scanning lines	625	625	625	625	625
Field frequency (Hz)	50	50	50	50	50
Line frequency (Hz)	15625	15625	15625	15625	15625
Audio modulation method	FM	FM	FM	FM	FM
Video modulation method	AM (Negative)	AM (Negative)	AM (Negative)	AM (Negative)	AM (Negative)
Colour subcarrier frequency (MHz)	4.332618	4.332618	4.332618	4.332618	4.332618
Intermediate frequency (MHz)					
Audio Video	33.4 38.9	30.5 37.0	33.4 38.9	32.9 38.9	41.25 45.75
	West Germany, Italy, Kuwait, Bahrain, Jordan, Algeria, Thailand, Indonesia, Australia, New Zealand, Holland Norway, Switzerland, Sweden, Qatar, Malaysia, Pakistan, Denmark, India	China	West Germany, Austria, Other European and African nations. *In Belgium and Yugoslavia vestigial sidebands are 1.25 MHz.	England (UHF) Hongkong (UHF) Ireland, South Africa	Argentina

OBJECTIVE TYPE QUESTIONS (A)

1. The colour which the body presents to the eye is the colour of the light which it
   (a) absorbs (b) reflects
2. Primary colours
   (a) can be produced by mixing other colours
   (b) cannot be produced by mixing other colours
3. The colours red, green and blue are called
   (a) subtractive primaries (b) additive primaries
4. The attributes of colour are
   (a) Phase (b) hue
   (c) amplitude (d) saturation
   (e) brightness
5. The signal essential to the operation of a black-and-white TV is
   (a) chrominance signal (b) luminance signal
   (c) composite video signal (d) control signal
6. In the NTSC system, the two colour difference signals are
   (a) transmitted together (b) not transmitted together
   (c) transmitted in quadrature (d) transmitted in anti-phase
7. Hue control is necessary in the
   (a) PAL system (b) NTSC system
   (c) SECAM System (d) none of these
8. Colour information in which system is transmitted on every line
   (a) PAL (b) SECAM
9. Which television system is used in India
   (a) FCC (b) CCIR-B
   (c) PAL-B (d) PAL-G

FILL IN THE BLANKS (B)

1. White light consists of a mixture of .... different colours.
2. A .... colour can be produced by mixing other colours.
3. Hue depends on the .... wave length of light.
4. The amount of white light contained in a colour determines its .... level.
5. Brightness of a colour is determined by the amount of .... contained in it.
6. Brightness is an attribute of ....
7. Saturation is an attribute of .... only.
8. The human eye responds to each of the three primaries ....
9. The receiving terminal picks up two colour signals the third colouring signal .... is produced by ....
10. The generation of the chrominance signal involves the .... of the luminance signal from the .... signals.
11. The colour picture tube .... the primary hues from the televised scene.
12. The colour picture tube has .... electron guns in one picture tube envelope.

13. The delta gun picture tube has a .... mask.
14. The quantity of electrons hitting the .... dots is controlled by their .... control grids.
15. The screen of the colour picture tube is coated with .... different phosphor dots.
16. Some colour picture tubes employ one gun with three .... cathodes.
17. In the plumbicon camera tube, colours and deflection are both accomplished ....
18. The main difference between the plumbicon and the standard vidicon is the ....
19. Plumbicon camera tubes are available with a spectral response suitable for any of the .... colours.
20. For any two systems to be compatible, they must have the same..........., colour subcarrier frequency and colour............techniques.
21. Monochrome compatible systems can produce .... pictures in the other compatible system.
22. The main drawback in the NTSC systems is that even slight errors in the .... between the colour difference signals produce errors at the decoding stage.
23. Colour information in the PAL system is transmitted on ....
24. PAL standards exist in the .... system and .... system.

CONVENTIONAL QUESTIONS (C)

1. What are the constituents of white light ?
2. Explain the difference in additive primaries and subtractive primaries.
3. What is the difference between a primary colour, secondary colour and complementary colour?
4. Differentiate between hue, brightness and saturation.
5. What is the need for a separate luminance signal ?
6. Discuss the relative sensitivity of a three tube camera.
7. How are chrominance signals obtained ?
8. What is the difference between a television camera tube and a picture tube ?
9. How does the operation of a colour picture tube differ from that of a monochrome picture tube?
10. What are the differences between a delta-gun colour picture tube and an in-line colour picture tube ?
11. Briefly explain the working of a colour camera tube.
12. How does a plumbicon differ from a vidicon ? Discuss their relative merits and demerits.
13. What are the requirements of a compatible system?
14. How the different television systems have been evolved ?
15. Give the salient features of the television system used in our country

24

MONOCHROME AND COLOUR TELEVISION CONTROLS

There are three types of controls or adjustments in a TV receiver. In the first group come those controls which are accessible to and used by the viewer. These are called user or customer controls. The second group consists of those controls which the service technician adjusts while setting up the receiver in the customer's house. These service controls are usually located on the rear apron of the receiver. The third group includes the alignment adjustments for tuning the frequency response and bandwidth of RF, IF, sound IF and colour circuits. Proper alignment of these adjustments, which requires specialised knowledge and equipment, is usually done only in the service shop. The service adjustments made on the colour receivers are mainly concerned with the proper setup of the colour picture tube and also require special equipment.

Many colour receivers include automatic circuitry. This circuitry is usually activated by pushing a special button that is often designed to light up when it is engaged. These include automatic fine tuning, automatic tint, and preset controls for colour, tint, brightness and contrast. The preset controls are a duplicate set of controls for the same functions as the front-panel controls. They are preset at the factory or by a service technician to give an adequate picture in normal reception conditions. A discerning viewer will often be able to obtain a better picture by careful adjustment of the manual controls than with the automatic preset system. The automatic button permits the less critical viewer to obtain a usable picture quality.

24.1 OPERATING CONTROLS COMMON TO BLACK-AND-WHITE AND COLOUR RECEIVERS

So as to become familiar with the common controls, *first turn the colour or chroma control fully counter clockwise, switching it off.* This would eliminate all colour information from the screen. The adjustments now would be the same as of a black-and-white receiver.

The television viewer must be able to adjust the *operating conditions* of the receiver. The names of most *user controls* describe what they do and this simplifies understanding their adjustment. Table 24.1 lists the operating controls a viewer may have to adjust in order to set up the television receiver for proper operation.

The operating controls are set up in the following way. The *ON-OFF switch* is turned to the ON position and the receiver is allowed enough time to warm up. The *automatic control* is set to off. The *volume control* is adjusted for the presence of some sound. The *brightness control* is adjusted to show a raster on the picture tube, Fig. 24.1. *The channel selector* is then switched ON to bring in the desired station and the *fine tuning* is adjusted to bring in the best picture and sound. In colour receivers, the setting of the fine tuning is critical. Many receivers now include *automatic fine tuning* (AFT) to eliminate this problem. If multiple lines or an unsynchronised picture appears, the *vertical hold* is adjusted to stop any up and down motion. Similarly, the *horizontal hold* is adjusted to eliminate horizontal or diagonal bars. The *level of the operating controls* such as brightness and contrast should then be readjusted to give the desired picture and sound. The initial and final touch-up adjustments may be made in any sequence.

Fig. 24.1 Raster-the scanned or illuminated area of the TV picture tube.

Table 24.1 Operating Controls

Control	Function	Location	Circuit Affected
1. On-off	To turn power ON and OFF	Front panel, sometimes part of volume	Power supply
2. Volume	Control sound level	Front panel	Audio amplifiers
3. Brightness	Adjust CRT light output	Usually front panel	CRT bias
4. Contrast	Adjust ratio of black to white on CRT	Front or rear panel	Video amplifiers
5. Channel selector	Tune in station	Front panel	Front end
6. Fine tuning	Exact station setting	Front panel	RF oscillator
7. Horizontal hold	Lock picture horizontally	Front or rear panel	Horizontal oscillator frequency
8. Vertical hold	Lock picture vertically	Front or rear panel	Vertical oscillator frequency
9. Tone	Audio-frequency response	Front panel	Audio amplifiers
10. Automatic	Simple adjustment (see text)	Front panel	Many (see text)
11. Video peaking	Sharpen picture detail	Front or rear panel	Video amplifiers
12. Colour	Adjust colour intensity	Front panel	Chroma amplifier
13. Tint	Adjust to correct shade of colour	Front panel	4.43-MHz phase.

In some colour receivers, there are *additional user controls* to adjust picture and sound. These may include *tone control* and *video peaking*. The tone control varies the sound frequency response, emphasizing either high (treble) or low (bass) frequencies. The video peaking control emphasizes or reduces the high-frequency response of the video circuits in the receiver. For *strong signals*, emphasized high frequency results in crisp sharp pictures. For weak or *fringe signals*, mimimum high frequency response, within the range of the video peaking control, reduces noise in the picture.

Fig. 24.2 The level of operating controls should be adjusted to give the desired picture and sound.

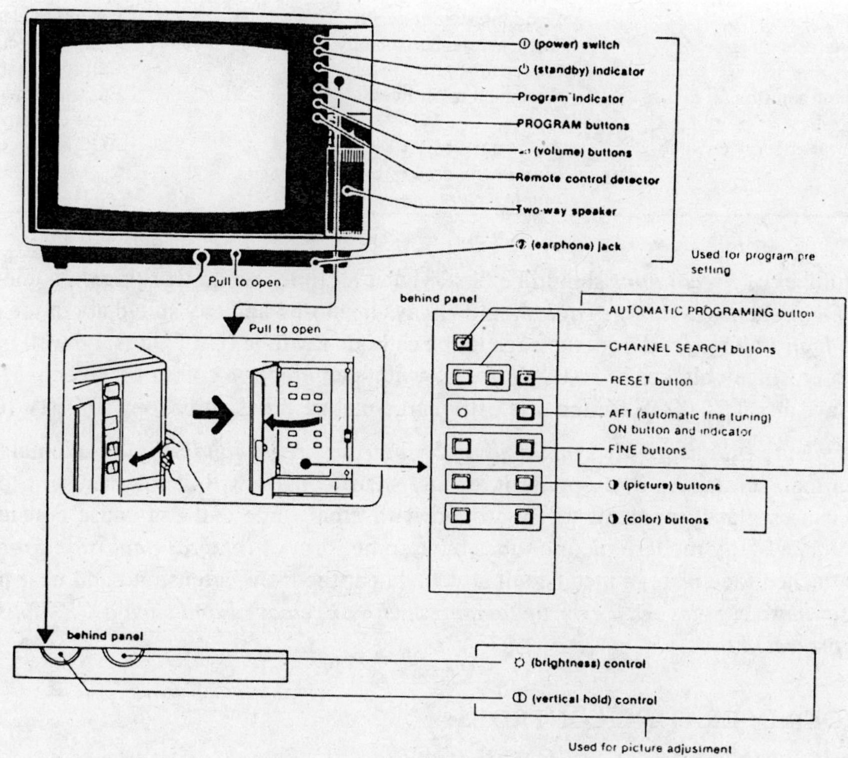

Fig. 24.3 Location and function of controls SONY KV-2062 E(RM - 622) Colour TV Receiver.

24.2 COMMON SERVICE CONTROLS FOR BLACK-AND-WHITE AND COLOUR RECEIVERS

Television service technicians must adjust many different types of television receiver controls. Understanding the *circuits* and the *functions* of each control will help technicians to make proper adjustments. Since you are not expected to adjust these controls properly or to be familiar with receiver circuitry at this point. Table 24.2 is to be referred to for information on the effects of service controls. This table lists most service controls with a general description of their *function* and of the *circuits affected.*

Table 24.2 Service Controls common to black-and-white and Colour Receivers

Control	Function	Circuit Affected
1. Circuit breaker	Reconnect *B* + after overload	Power supply
2. Automatic gain control	Set level of detector output	RF and IF bias
3. Vertical Size (height)	Height of raster (bottom)	Vertical sweep output
4. Vertical Linearity	Equalize height near top and bottom of raster	Vertical-output stage bias
5. Width	Horizontal raster size	Horizontal output transformer
6. Horizontal linearity	Equalize width of left and right side (adjust efficiency)	Damper waveform controls output current
7. Brightness range	Adjusts operation of front-panel control	Voltage applied to brightness control
8. Noise inversion	Optimize sync for available signal strength	Bias on noise inverter
9. Vertical centering	Position raster vertically	Control of electron beam (vertically)
10. Horizontal centering	Position raster horizontally	Control of electron beam (horizontally)
11. Focus	Sharpen the raster lines	Control of narrow electron beam
12. Pincushion adjustment	Straighten raster lines	Electric or magnetic control of raster edges (usually top and bottom)
13. Normal-raster-service switch	Aid in colour CRT setup *Must be in normal position to get a usable picture*	AGC and vertical sweep circuit

A number of *precautions* should be observed at all times while working on a television receiver with the power ON. The *high-voltage cage* should always be closed and any shield about the regulator must be in place. The *high-voltage lead* from the cage to the cathode ray tube (CRT) must be well insulated and dressed out of the way, since voltages of 30,000V are now quite common in colour receivers. Most black-and-white receivers have about 20,000V for the sets using large picture tubes and about 10,000V for smaller ones.

The picture tube is highly evacuated and requires careful handling. Accidentally hitting the picture tube with a tool, scratching it, dropping it, or any sharp contact with it may cause it to implode. Such an implosion causes glass fragments to fly around with great force and can cause considerable damage and physical injury. Many modern picture tubes have some form of *integral implosion protection.* This can be an epoxy attached face plate, a metal shell attached near the front, a tension band near the front of the tube, etc. *None of these systems must ever be tampered with or removed and such a CRT must never be replaced with an unprotected tube.*

24.3 COLOUR OPERATING CONTROLS

In addition to the controls normally found in a black-and-white television receiver (see Fig. 24.5), the colour receiver has two operating controls for the *colour rendition* of the picture. The colour control, sometimes called *chorma or saturation control* adjusts the amount of colour information in the scene. With the control turned fully counter-clockwise, the picture turns black-and-white. As the control is rotated clock-wise the picture goes from delicately tinted shades, through stronger pastel shades, to bright colours and finally to *colour overload* manifest as an indescent appearance and streaking.

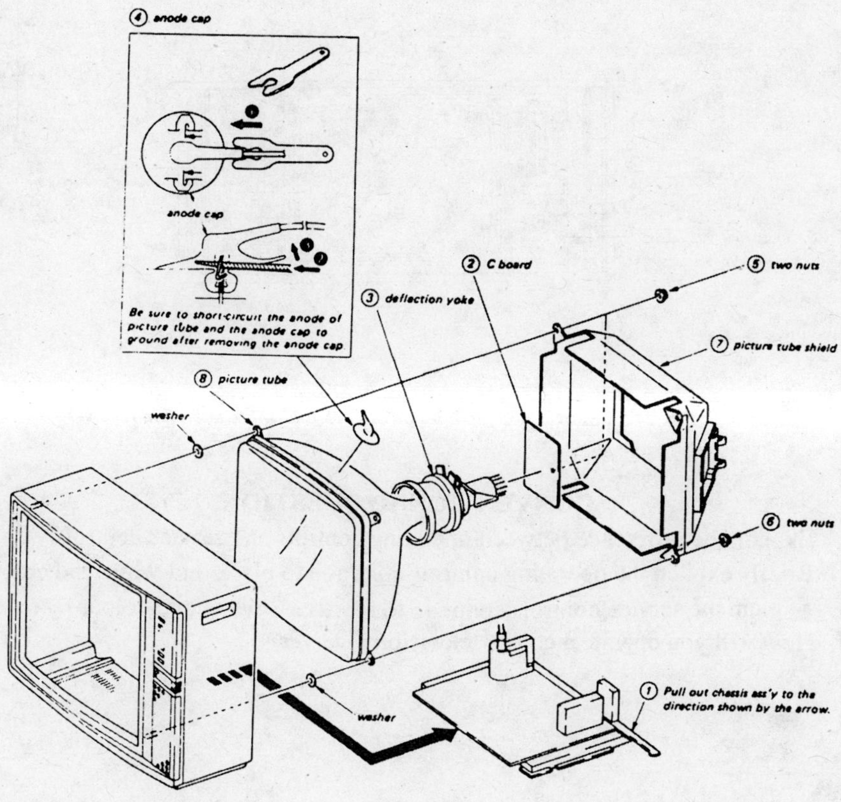

Fig. 24.4 Various steps involved in the removal of picture tube of the colour television receiver shown in the Fig. 24.3.

The second control is the *tint or hue control* (see Fig. 24.3). This control changes the actual hue or colour of the scene.

The most readily available reference colour on the screen is the flesh colour. In most receivers, turning the tint control *counter clock-wise* from its centre position will make the flesh tones magenta and turning *clockwise* from the centre will make the flesh tones green. To adjust for the proper flesh tones, the colour control must be set for near normal colour levels. Flesh colour is not the only *reference colour* the viewer may use. *In many cases, the colour of some well known, advertised product or package can be used to adjust the tint control correctly.* Red is generally the easiest colour for this purpose.

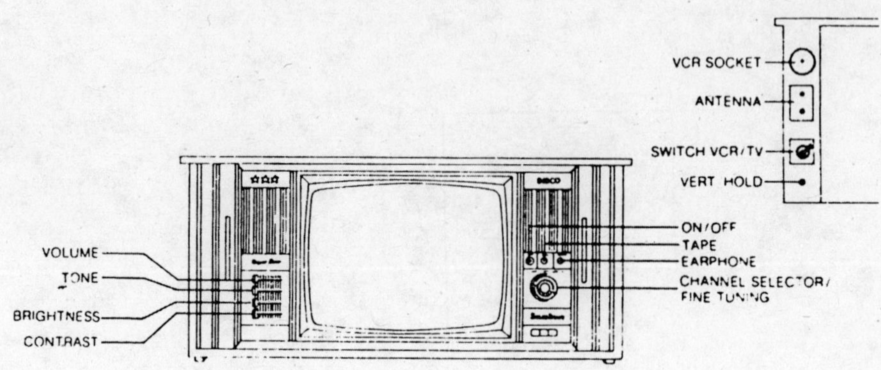

Fig. 24.5 Controls and facilities on a black-and-white television receiver.

CONVENTIONAL QUESTIONS (C)

1. Explain the difference between operating controls and service controls.
2. Briefly explain the operating controls common to black-and-white and colour receivers.
3. Explain the service controls common to black-and-white and colour receivers.
4. How will you operate a colour television receiver?

25

VIDEO TAPE RECORDING AND REPRODUCTION

Video is a word which has become so familiar that it is all too easy to forget how recent is its impact on consumer electronics. The idea of recording and playing back television pictures is as old as television; a primitive video disk was on sale in Selfridges of London in 1938.

Yet the first post-World War II experiments in recording television pictures and playing them back were not made with the domestic market in mind. The intention was to remove the stranglehold live television broadcasting imposed on the medium. Before video recording sequences had to be shot on film which was expensive or they had to be transmitted live. Since broadcasting was closely tied to the evenings, this overloaded the studios and, moreover, any mistakes in the production went out on the air immediately, giving the producer no chance to reshoot or edit them out afterwards. Repeats were impossible unless the production was shot all over again, or transferred to film.

Recording the signals which went over the air on magnetic tape seemed an obvious solution, but it proved amazingly difficult to achieve. To preserve a picture, vastly more complex information has to be recorded than with the highest quality sound reproduction. Simply recording the finer details of a picture meant a huge increase in the area of tape used to store the information. This meant, in turn, that the speed of the tape over the recording and playback heads had to be much higher than in audio recording.

After years of development, the American company Ampex produced the first working video recorders for broadcast use. Ever since then, more and more television has been recorded on tape, so that now-a-days it is rare to see live broadcasts. The main exception is sports and even these events are often recorded as they are transmitted for later repeats or instant action replays.

25.1 VIDEO TAPE

Video tape comes in various widths to fit various tape machines. Every tape machine can work with *only one* width of tape. There's 1/4 inch, 8mm, 1/2 inch, 3/4 inch, 1 inch and 2 inch tape. *Format* is a term for generic design of a range of hardware. There is more to format than just size; *tape speed* and other *electronic differences* which make *not all* 1/2 inch tapes playable on *all* 1/2 inch machines. To simplify matters, various manufacturers and video-associations have agreed to *standardise* on several formats. So if a tape is recorded on a certain format machine, it should playback on the same format machine, regardless of the manufacturer. The most popular formats today are 3/4 U, VHS, Beta and 8 mm. About 90 percent of homes with video cassette recorders have *Video Home System* (VHS). Most others have *Beta* or *8 mm*. If you look closely on a video cassette, you may see printed or embossed on the cassette itself the trademark U-matic or the number 8 or the Greek letter β beta to indicate the *format of the cassette*. This is shown in Fig. 25.1.

Fig. 25.1. The trademark embossed on the cassette indicates the format of the cassette.

Video tape for consumer use is available in *cassette* only and works in basically the same way as an audio tape. *Audio tape* is used for recording and playback of a rather narrow band of frequencies when compared with *video tape.* The audio tape recorder handles signals that correspond to *the range of human hearing.* Video tape is used for recording *the picture and any sound that accompanies the picture.* Video tape moves *more slowly* than the one that's in a typical audio-cassette recorder. In this connection it is imperative that a video tape should contain *superior properties* in order to provide acceptable performance.

In any cassette, whether audio or video, it is the *size* of the cassette and the *thickness* of the tape that determine the amount of tape that can be loaded into the cassette. *An increase in running time can be obtained by making the tape thinner and running it at slow speed.*

The width of a video tape is generally 1/8 inch to 1/4 inch. The width of video tape for domestic use is 1/2 inch. *Although attempts are being made to encourage the use of 1/4 inch tape, the tape most commonly used today by non-professionals has a width of 1/2 inch (12.7 mm)*

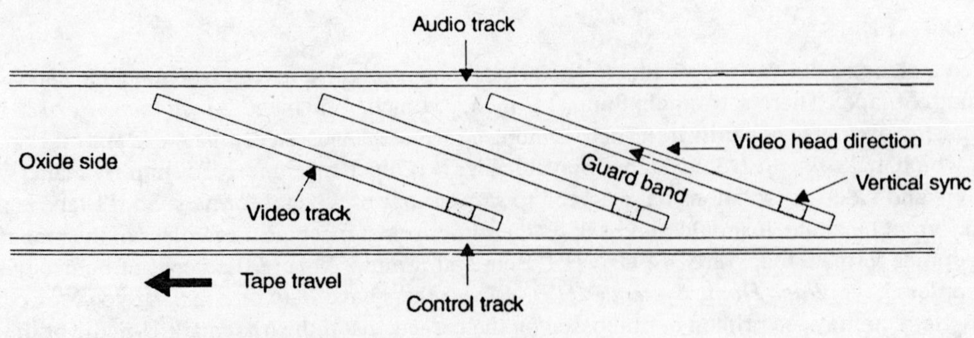

Fig. 25.2. (a)

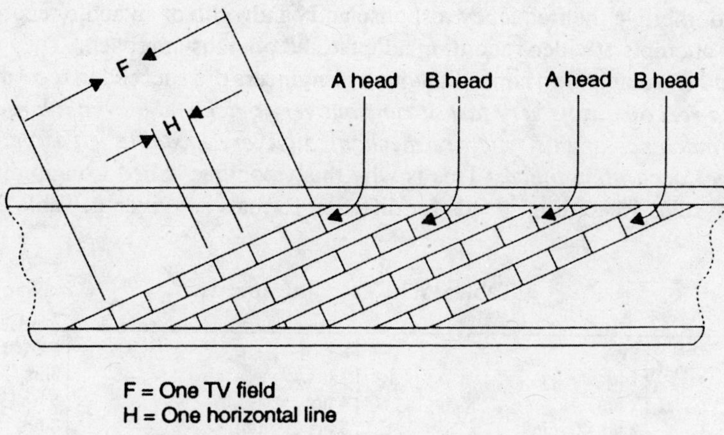

(*b*)

Fig. 25.2 Enough information can be crammed on a video tape by helical scan.

Early *professional video recorders* recorded video tracks across a 2 inch (5 mm) wide tape. Since then tapes have grown narrower and narrower, down to Compact Video Cassette (CVC) format 1/4 inch (6 mm) tape. This has been made possible by *helical scan,* the process of recording tracks at an angle across the tape, done by a pair of recording heads on a high speed rotating drum. *If the heads were fixed as in audio, the tape would have to move impossibly fast for enough information to be crammed on to it,* Fig 25.2.

25.2 PROBLEMS IN VIDEO RECORDING ON TAPE

At first, it might seem very easy to record colour television pictures on *magnetic tape*. After all, it is only necessary to feed the composite video signal into a tape deck, record it just like audio, and then replay the output of the tape deck into the antenna socket of a television receiver. However, there are many problems. The first problem is that the *frequency range* needed to record video is much wider than that for audio. A hi-fi audio signal has a bandwidth, or frequency range of around 20 kHz. But a colour television signal, as broadcast or relayed down a cable, has a bandwidth of around 5 MHz. Frequencies of this magnitude are necessary to control the electron beams as they scan several hundreo lines of the picture, several dozen times a second.

Remember too, that the scanning electron beam is continually changing in strength, so that the spot of light it makes on the television screen is *continually changing in brilliance.* The three beams of a colour tube need to be *controlled* to an accuracy of literally millionths (10^{-6}) of a second hence the need to handle frequencies of millions of cycles a second. If this kind of high frequency signal is fed into an ordinary audio tape recorder, the high frequencies which have very short wave lengths-just merge into one another to make an indecipherable mush.

There is also a problem concerning *synchronisation.* The electron beams have to be kept very accurately *in step* and for this some very precisely timed synchronisation pulses are necessary. Any slight deviation in the speed of the recorded tape will alter the timing of the pulses and upset the picture synchronisation, making it *tear sideways or roll vertically.*

25.3 EARLY VIDEO

It is possible to improve the frequency response and bandwidth of an audio recorder by *increasing* the tape speed. So early attempts at video recording adopted the obvious approach. They ran the tape very fast to widen the bandwidth sufficiently to cope with video. Engineers did succeed in recording television signals this way. *But when a reel tape runs very fast, it runs out very quickly. So even a large reel only lasts for a few minutes. Also, machines of this type are mechanically very unstable. It is very difficult to get the synchronisation pulses accurately timed.* This is why this technique called Longitudinal Video Recording (LVR) fell by the wayside. Recording speeds for different formats are given in Table 25.1.

Table 25.1 Recording speeds

VHS	**SP**	**LP**	**EP (SLP)**
T-60	1 hr	2 hrs.	3hrs
T-120	2 hrs	4 hrs.	6 hrs
T-160	2 hrs 40 min	5 hrs 20 mim	8 hrs
Beta	**BI**	**BII**	**BIII**
L-125	15 min	30 mm	45 min
L-250	30 min	1 hr	1 hr 30 min
L-500	1 hr	2 hrs	3 hrs
L-750	1 hr 30 mim	3 hrs	4 hrs 30 min
L-830	N/A	30 min.	5 hrs
8 MM	**SP**	**LP**	
MP-30	30 min	1 hr	
MP-60	1 hr	2 hrs	

The concept of LVR was investigated again by BASF in Germany and Toshiba in Japan. Both firms built prototype longitudinal video recorders intended for domestic use. But they were never launched on the market because by then another techniuqes called *helical scan* had completely taken over. There was just no point in offering an alternative approach that was neither cheaper nor better.

Helical scan was the brainchild of an American Company called Ampex. The big breakthrough came in 1956, when the Ampex Corporation demonstrated the world's first practical video tape recorder at a conference of the National Association of Broadcasters in Chicago. A team of Ampex engineers (including Ray Dolby who was later to become famous for his noise-reduction systems) had finally succeeded in recording television pictures on *slow-running tape.* They used *frequency modulation* (FM), as standard for high-quality stereo radio transmissions.

Previous attempts at video recording had relied on *amplitude modulation (AM),* as used for low-quality medium and long wave-radio. Ampex also abandoned the idea of trying to run the tape very *fast* through a machine. Instead they used *wide tape* (50 mm. or 2 inch wide) and ran it *slowly* (38 cm or 15 ips) past a *rapidly rotating drum* which carried four video heads. As the drum rotated over the *slowly moving tape*, the video heads laid as *series of recording tracks* across the width of the tape, like the rungs of a ladder. Because the Ampex recorder used four video heads on the rotating drum it was called a quadplex or *Quad machine.*

Early Quad machines were bulky and inexpensive, and could initially only record black-and-white pictures. Before long they were modified to record colour, and *the Quad format became standard for bradcasters around the world.* Machines of this type are still used in many television studios. Soon afterwards the technique was modified for use with narrower tape. Ampex in the United States and Toshiba in Japan, who were working in parallel, both came up with the idea of helical scan recording. *For helical scan the rotating video drum carries two heads and rotates at an angle to the slowly moving tape. In this way the heads lay a series of oblique recording tracks across the tape width, like the helical threads of a screw.*

Toshiba built a laboratory prototype helical scan recorder in 1958 and demonstrated it in 1959. In the 1960s many different kinds of helical scan recorders were produced. for professional, industrial and educational use. A few even found their way into homes for domestic use. But *they were difficult to use and offered limited recording time.* The tape was one half or one quarter the width of Quad tape, but it was still on *open spools* that had to be *threaded* through a tortuous path around the video head drum. The video heads are very fragile, and clumsy threading could easily cause expensive damage. Even the largest reel of tape could offer less than an hour's playing time. Compared to what video-home offers today, with a playing time in excess of 4 hours, the early helical scan machines were very primitive.

25.4. FM MODULATION

The techniques involved to record and replay video signals are not so straight forward as for audio signals. The main problem is the *relatively large bandwidth* of the video signals which for a broadcast-standard signal extends from DC (about 25 Hz in practice) to 5.5 MHz. For domestic video recorders a more limited response is adequate the HF roll-off occurs at about 2.5 MHz, permitting a more sparing use of the tape.

A magnetic tape system's output level is proportional to the *rate of change* of the magnetic flux, so that output is directly geared to frequency. Thus *each halving of frequency (octave) halves the output signal, giving the tape/head interface a charateristic 6 dB/octave curve*, Fig 25.3. Because the difference in levels between magnetic saturation of the tape's coating and the inherent noise of the system is about 1000 : 1, corresponding to 60 dB, it is plain that the 6 dB/octave will permit a maximum of *ten octaves* to be fitted between the *noise floor* and the *overload point*, so long as massive compensation is provided to equalise replay levels across the frequency spectrum.

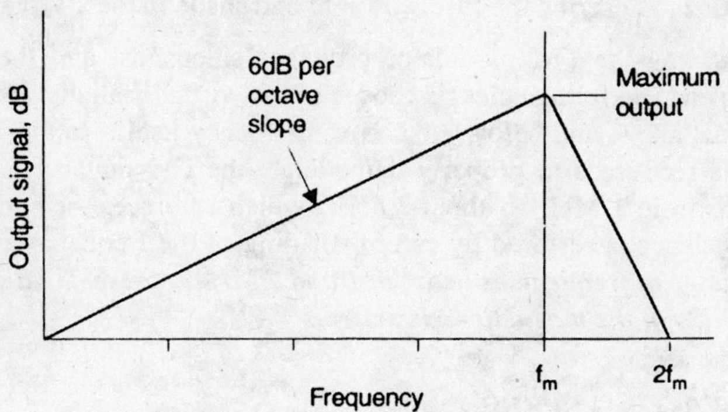

Fig. 25.3. The voltage induced in the playback head increases at a 6 dB/octave rate as the frequency of the recorded signal increases. When the wavelength equals the head gap, the induced voltage is zero.

A television signal, even the bandwidth-restricted one described above, occupies fifteen or sixteen octaves and so cannot be *directly* recorded on tape by any means. *Indirect methods* of recording are possible, however, and they involve modulation of the video signal onto a carrier. Invariably an *FM carrier* is chosen for this purpose. The use of FM increases noise immunity, masks shortfall in signal strength stemming from slight tracking errors and imperfect head/tape contact, and permits either (*a*) its use as recording bias for a second signal carrying the chroma information or (*b*) the facility to drive the tape coating into magnetic saturation on each FM carrier cycle to further improve S/N ratio.

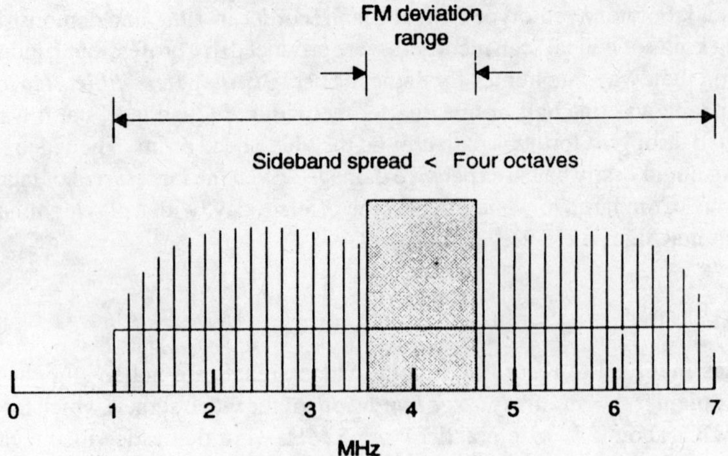

Fig. 25.4. Reduction of octave range of signal by FM modulation : the solid block represents carrier deviation range, and the lined-areas the first order sidebands.

The way in which the FM carrier technique ***reduces*** the octave range is shown in Fig. 25.4. Carrier frequencies are assigned for both extremes of the luminance signal wavefrom, typically 3.8 MHz for *Sync tip* and 4.8 MHz for *peak white*. This actually permits the recording of DC (zero frequency) video signals since a *constant level of white or grey will give a constant FM carrier frequency*. During each line sync pulse the carrier falls to 3.8 MHz for its 4.7 μs duration, and during 52 μs active line period the carrier frequency rapidly deviates from 4.1 to 4.8 MHz to describe the levels of light and shade in the TV picture.

In deviating in this way the FM modulator produces sidebands, and the *modulation index* (the relationship between video and FM frequencies) is chosen so that virtually all the sideband energy is confined to the *first pair* of sidebands above and below the carrier frequency itself. In the VHS system, for instant, enough sideband energy is recovered to properly demodulate the FM signal when the record and replay frequency repsonse extends from 1 MHz to about 7 MHz, which embraces the entire lower sideband and a portion of the upper one-balance is restored by careful shaping of the frequency response of the playback amplifier. *An operating range of frequencies between 1 and 7 MHz represents an octave range of less than four-well within the capability of the magnetic tape system.*

25.5. HEAD GAP AND WRITING SPEED

Although the octave range has necessarily been reduced using an FM carrier system the maximum frequency required to be recorded has greatly increased. *Peak white* occurs at 4.8 MHz, and the upper sideband signal- most extensive when sharply-defined detail is being recorded-extends towards 8 MHz. *The head/tape transfer system must be capable of passing such frequencies and the magnetic surface of the tape capable of retaining them.* The period of an 8 MHz signal is 125 μs and during this short time enough tape must traverse the video head gap to adequately imprint the entire cycle as a magnetic pattern in the tape coating. With a typical head gap of 0.5 microns (5×10^{-7} m) the tape-to-head speed needs to be 5 m/s ; this is called *writing speed.*

The achievement of such a high writing speed is very difficult in a *direct transport system,* where extremely high spool and capstan speeds would be required. A solution is *to rotate the heads* themselves against the tape ; the heads are mounted on a spinning head-drum, and protrude beyond its surface to make intimate contact with the tape ribbon which itself is wrapped around the drum. For domestic (and many professional) video formats the arrangement takes the form of helical-scan setup illustrated in basic form in Fig. 25.5

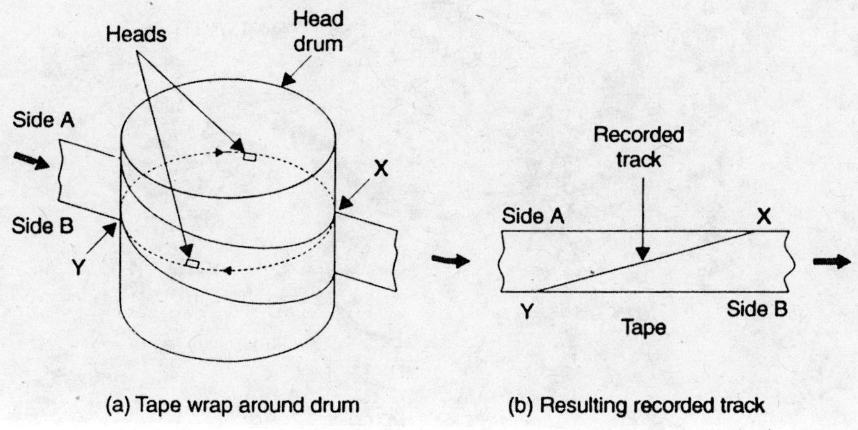

Fig. 25.5. Basic principle of helical tape scanning.

25.6. HELICAL TAPE SCAN

The tape is wrapped around approximately 180° of the drum's periphery and takes a *helical path* due to the *tilting* of the head drum and a precision-machined guide rabbet on the lower (stationary) part of the head-drum. The head assembly spins anticlockwise at 1500 r.p.m. which confers the required high writing speed. *All that is required of the tape transport system now is that it moves the tape along by one track-width per head scan in order that successive tracks are laid down side by side and just abutting each other during record.* At playback the same tape transport system ensures that as each new head sweep presents itself to the tape wrap a fresh track is lined up in its path for readout. In a typical system using 49 μ m-wide video tracks the tape progresses through the deck at 2.34 cm/s, pulled around the head drum by a downstream capstan/pinchwheel assembly. At the entry and exit points of the tape head-drum wrap are positioned guides which precisely *aligns the path of the tape around the drum.*

The head-tips must maintain intimate contact with the magnetic surface of the tape. In audio-recorder practice this is achieved by *pressure-pads* which hold the tape tight against the heads. That is obviously impractical in a helical-scan system, where a degree of *lateral tension* is imparted to the tape in its head-wrap by (*a*) a back-tension brake on the feed spool ; and (*b*) the friction of the tape against the surfaces of the upper and lower head drum's sections. Against the taut tape ribbon the rotating video head-tips, whcih project about 50 μm from the drum face surface, push out a *moving stretch-spot* in which the required head-tape pressure is set up.

Because the tape wrap only covers *half* the circumference of the head drum, and signal transfer must take place *all* the time, conventional domestic head drums are fitted with two videoheads. As one leaves the tape wrap at the end of a scan, another (diametrically opposite) begins a new scan of the tape. The signals from each head are routed in turn into the replay amplifiers by a switch during *playback*-the swtich is *synchronized* by a head-position sensor associated with the head drum. During *record* both heads are driven with write signals, though *only the one traversing the tape is recording at any one moment.*

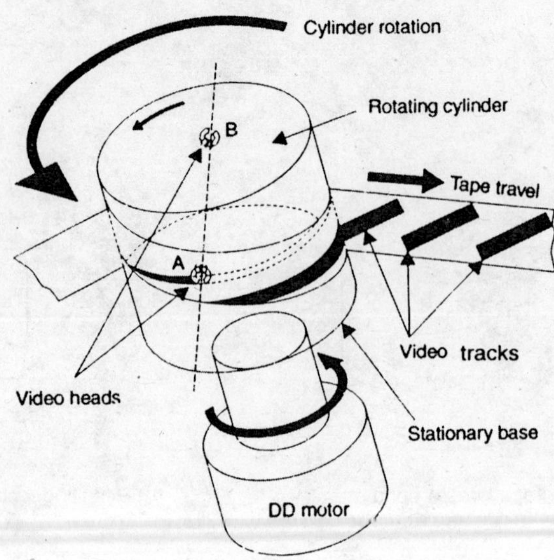

Fig. 25.6. Head Scan and Tape Tracks

The *transfer* of video signals to and from the spinning video heads is accomplished by a *rotating transformer* whose ferrite core is arranged as two shallow discs concentric with the head drum itself. One half is *stationary* and the other half *rotates* with the head disc, magnetic coupling between the two taking place via a very small air gap. One pair of windings is provided for each head ; in multi-head machines three or four pairs of windings are required. *Separate rotatry heads are employed for hi-fi sound signal transfer. For these a separate rotary transformer assembly is usually provided.*

25.7. DOMESTIC VIDEO

The first big breakthrough that led to today's boom in domestic video came from Japan and Holland. In 1970 Sony, JVC and Matsushita of Japan agreed to adopt a uniform standard, the *U-format*, for industrial and semi-professional video. U-format machines went on sale two years later, and have now become the *industrial standard,* worldwide. They are also used for education and broadcasting. The U-format recorder uses tape which is 19 mm or 3/4 in wide and is contained in a cassette. *The user does not need to thread the tape round the videodrum. All this is done automatically as the cassette is loaded in the machine.*

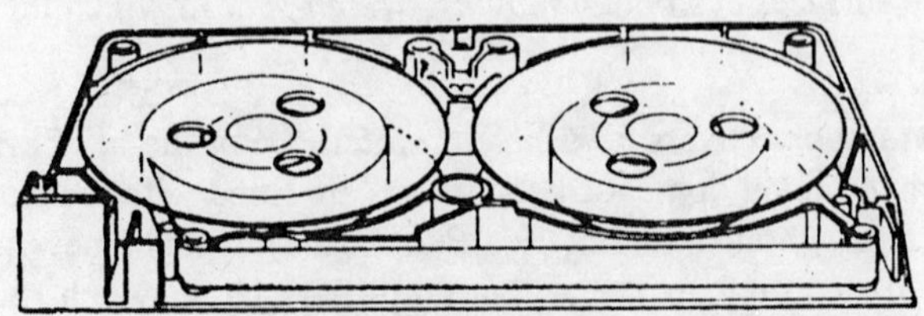

Fig. 25.7. An exploded view of a standard VHS tape cassette.

At the same time in Europe, Philips of Holland also unveiled a cassettee video recorder, the N-1500 series of VCR machines. The first Philips VCRs, shown in 1972, were intended for industrial and educational use. Two years later Philips produced a domestic version. The Philips VCR used 12.5 mm or 1/2 in. tape in a cassette, *automatically threaded by the machine mechanics.* Both the U-matic and Philips VCR could only offer upto one hour's playing time from a single cassette. This is more than enough for industrial use but inadequate for the home where most people want to tape programs off air that last for more than an hour.

The U-format proved to be far more reliable as an industrial machine than the Philips VCR. The U-format was never intended as a domestic format. It required clumsy add-on units to make it suitable for off-air taping and replay through a domestic television set. But the Philips VCR had these features built in. For a few years, in the mid-1970s, the Philips VCR sold fairly well on the doemstic market in Europe *even though maximum playing time remained at one hour per cassette, and it cost a relatively large amount of money to feed it with tape.*

25.8. SLANT AZIMUTH SYSTEM

Meanwhile in Japan, all the major electronics companies were striving to produce a domestic video unit that would record for longer than 1 hour, and would be relatively cheap to feed it with tape. In Europe, Philips was also carrying out similar research. Many differnet formats were developed and several were lanuched on the market. The next breakthrough which opened the door to *longer playing time and cheaper feeding costs,* was the rediscovery of an idea patented in 1964 by Prof. Okamura of Japan. This is the so called *slant azimuth system,* which is used on every modern domestic video recorder, illustrated in Fig. 25.8.

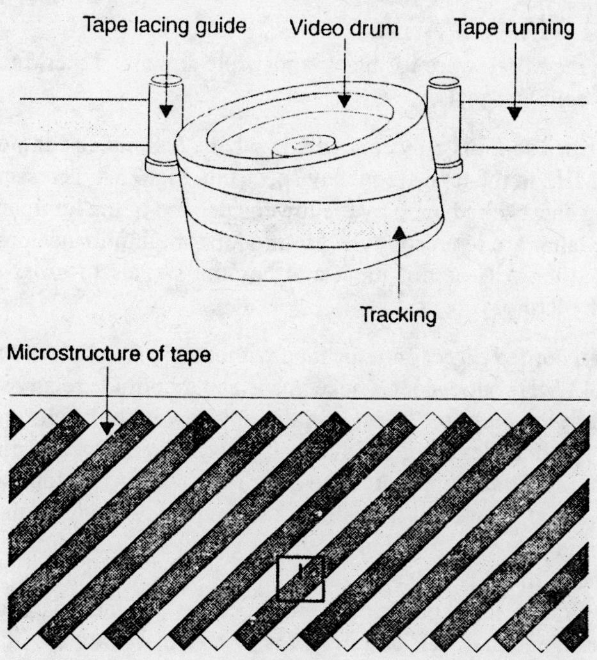

Fig. 25.8. Microstructure of tape in slant azimuth.

To get more playing time on a given length of tape, all you need to do is *run the tape slowly and make the individual tracks narrow,* so that the individual helical tracks are packed more closely together. The problem is that if the tracks are too close they interfere with each other and cause *crosstalk.* The result of Okamura's idea was *to cancel crosstalk on adjacent recordings.* It was ideally suited to modern helical scan recorders, which use two heads on the video drum, plus perhaps some *extra heads for special effects.* If the two heads are set at *slightly different angles* they will lay down tracks across the tape which have a slightly different characteristic. It is not difficult, then, to ensure that *each head can only read its own track and ignore those on either side.* To achieve this the azimuth alignment of the recording heads is *slightly skewed.* The effect is very similar to the loss of high frequency or treble performance obtained from an audio tape recorder if the playback head is not accurately aligned with the tape. The use of slant azimuth enabled video designers to position the video tracks very tightly together. This is why it is now possible to obtain many hours of recording from the relatively short length of magnetic tape in a modern domestic video cassette.

In Europe Philips modified the orginal VCR-N1500 format to use slant azimuth. At the same time the tape speed was slowed. This more than doubled the playing time available from a Philips cassette. The new machines, the VCR-N1700 series went into the market in late 1977 and early 1978. Soon afterwards Grundig of Germany further modified the tape format, slowing the tape speed even more so that playing time was further extended. But this format, the Grundig Super Video Recorder (SVR) soon became obsolete. So did the Philips N1700. Both formats were replaced by the *Video 2000 format* jointly developed by Philips and Grundig.

Philips and Grunding V2000 machines first appreaed on the British market in 1980. However, it was not until 1981 that the format started to produce good results and present any serious competiton for the two Japanese formats, VHS and Beta, which had by then gained a very firm foothold in the market place. This is why the V2000 format, although *technically superior,* has captured only a small share of the market.

25.9. COLOUR RECORDING

The first video tape recorders were for black-and-white signals. Later, methods were developed that permitted the recording of colour signals.

What has been described so far is only concerned with the *luminance* component of the TV signal. The space between zero and 1 MHz in the tape-frequency spectrum, Fig. 25.4, is reserved for the *chroma* signals, and to prevent it from being encroached upon by the lower sideband of the luminance FM signal, the rise time and frequency range of the latter are reduced in the pre-modulator lumminance processing stages ; any short comings in this bandpass filter will permit luma and chroma signals to *cross-modulate* with consequent *patterning* in the play-back picture.

The colour signal is recorded *direct* onto the tape without any form of modulation system. It is already encoded and based on a 4.43 MHz subcarrier whose phase and amplitude relative to the accompanying burst signal describe the colours in the picture. To interfere in any way with the chroma signal would be to invite hue and saturation errors ; to decode and demodulate it to baseband for recording and encode/modulate it during replay would be a complex and unpredictable business. The PAL colour signal cannot be recorded in its 4.43 MHz form as broadcast, however, it is above the frequency range of the baseband signals that the system can handle ; the timing jitter introduced by the mechanics of the recording/playback process would play havoc with the phasing of the subcarrier signal and hence hue ; and the bandwidth demand would be too great to accomodate in the limited tape spectrum. Two processes are carried out on the chroma signal to slot it into the 0-1 MHz space assigned to it. It is *bandwidth restricted* in a bandpass filter which limits its sideband excursions to ± 500 kHz on each side of the carrier frequency ; and it is *frequency shifted* by a heterodyne process to a new base frequency around 600 kHz. Thus is formed a *lower-definition, lower-frequency* chrominance signal with all charactersitics of phase amplitude and burst features intact. *The new frequency allocation and clipping of wings tailors it for the position it occupies in the tape-frequency specturm.* It is added to the FM luminance signal and passed to the heads for recording. Fig. 25.9 (*a*) shows the complete signal spectrum on tape.

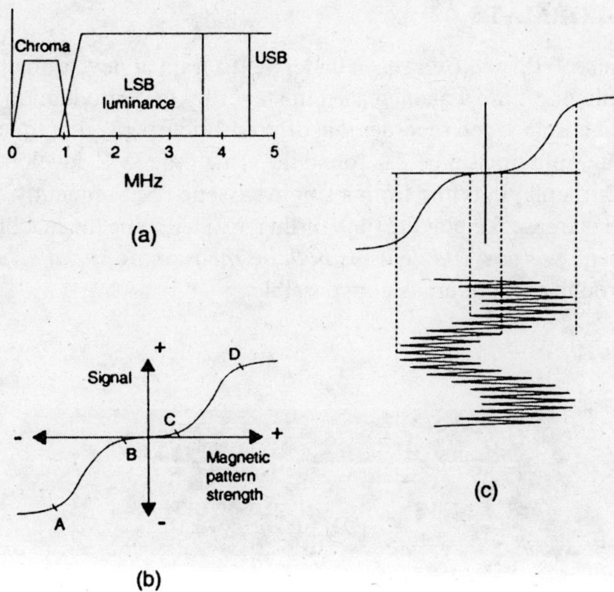

Fig. 25.9. Video signals on tape: (a) frequency spectrum; (b) transfer curve; (c) effect of recording (HF) bias.

Because the transfer characteristic of magnetic tape is not linear, Fig. 25.9 (*b*), a *recording bias* must be provided and added to the signal to be recorded if severe distortion is not to take place. It is normal practice in audio tape recorders to add a relatively high frequency (say 40 kHz) switching signal to the audio waveform before it is applied to the recording head ; the locally generated bias signal is *very carefully controlled in amplitude* so that its peaks sit in the centres of the linear section A-B and C-D in Fig. 25.9 (b). As the record signal is rapidly switched between the two centre points by the bias signal, it is printed *in linear fashion* as magnetic patterns in the tape, as Fig. 25.9 (c) shows. *The ability to work on both sections of the transfer curve increases the replay level and improves S/N ratio.*

In the *colour-under recording system* used in domestic video recorders the FM luminance carrier acts as AC bias source for the low-frequency chrominance signal. The two are added in the recording amplifier, and the variations in bias frequency do not matter at all-during replay a low-pass filter in the chrominance amplifier removes all signal components above 1MHz.

For the scheme to work the *levels* of both FM luminance carrier and chrominance signal must be *closely controlled* and their pre-set amplitudes depend very much on the *characteristics of the tape* to be used. The level of the FM carrier--*luminance writing current* - must be set to take the tape's magnetic coating half way to saturation in each direction. The *chrominance record level* then must be trimmed so that the largest expected colour signal amplitude (fully saturated cyan or red) drives the magnetic surface of the tape to just short of saturation in the positive direction, and to just short of non-linearity in the negative direction. Correct setting of these levels ensures low signal distortion and maximum S/N ratio.

In the audio world, different *tape materials* (oxide, metal, chrome, etc.) are available, and need different bias and drive settings for optimum performance. *For video recorders as yet, the levels are pre-set to the type of tape available for that format.* The metal and metal-evaporated tapes used with Video 8 format have very different bias, coercivity and saturation characteristics to conventional tape as used in other formats. When colour is not present in the signal to be recorded (mainly vintage films) a useful gain in noise performance can be achieved by increasing the gain of the luminance recording amplifier; a 6 dB boost in writing current drives the tape into magnetic saturation in both directions on each cycle of the FM carrier.

25.10. VHS AND BETA FORMATS

The Japanese formats, VHS and Beta, each underwent a train of developments in Japan and the United States. For each format, playing time was increased in stages by the introduction of slower tape speeds. In May 1975 Sony announced a Beta video recorder that offered 1 hour's playing time. In early 1977 the format was modified to give 2 hour's playing time. At around the same time JVC developed the VHS system, which from the start offered 2 hour's playing time from a single cassette . Subsequently, options were built in both VHS and Beta machines to increase the playing time further, by switching the machine to run at slower speeds. Thus *the actual playing time per cassette depends both on the running speed selected and the length of the tape in the cassette*, but around eight hours is now possible.

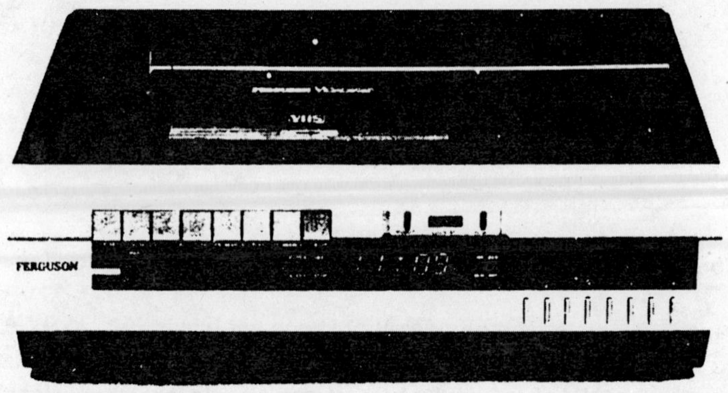

Fig. 25.10. VHS Format : (a) Ferguson 3V29 VHS Video Cassette Recorder :

Beta machines on the American market can be switched between Beta II speed which is 2 cm (0.787 in) per second and Beta III which is 1.3 cm (0.525 in) per second. VHS machines on the American market can be switched between SP mode which is 3.3 cm (1.3 in) per seconds, LP mode which is 1.67 cm (0.65 in) per second and EP mode which is 1.1 cm (0.43 in) per second. In Britain a VHS recorder runs the tape at 2.34 cm (0.9 in) per second, and a Beta machine runs the tape at 1.87 cm (0.73 in) per second. With a normal cassette this gives 3 hours for VHS or 3 hours and 15 minutes for Beta. But 4 hour cassettes, with more tape, are available.

VHS now has *stereo capability*, though it is nowhere near high fidetity. However, with introduction of *FM sound* the sound quality on VHS has improved dramatically. Engineers have found a way to encode the audio signal *inside* the video signal, resulting in a very high quality audio reproduction. As far as picture is concerned it is now possible to obtain professional, high definition pictures from the VHS format, though the equipment is not compatible with domestic VHS, as the machines run at much higher speeds.

One or two manufacturers have now adopted LP or *half speed* record and play modes. *The capacity of a tape is effectively doubled, though there is a noticeable degradation of the picture and sound quality* - in practice the reduction in quality shouldn't matter too much, being quite acceptable.

There is, of course, no possibility of playing a Beta cassette on a VHS machine or vice versa. The cassettes are of different size, and the only feature they have in common is the use of tape which is 12.5 mm (1/2 in) wide. The picture quality on screen, the sound quality and all the features and facilities which VHS and Beta formats offer are virtually identical.

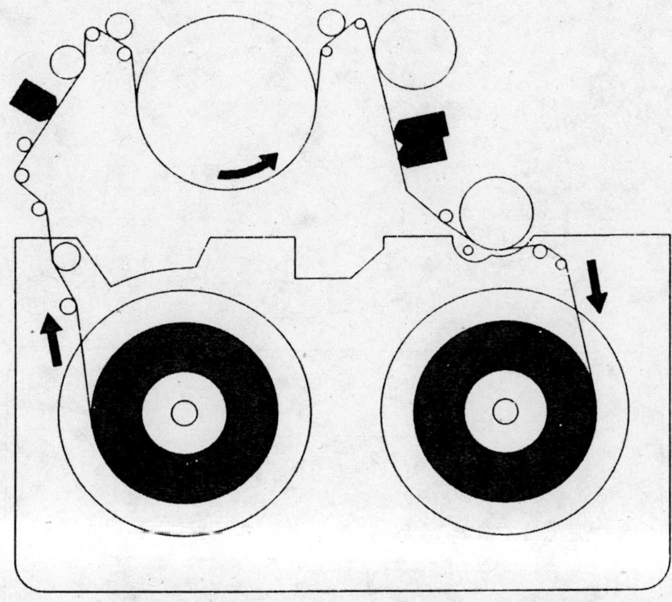

(b)

Fig. 25.10 VHS Format: (b) An explanation of how the tape is picked up from cassette to video drum when placed in the ON mode.

Fig. 25.11. (a)

Although VHS and Beta tapes recorded in one country cannot be replayed in another, the cassettes themselves are physically identical. In many respects they are both miniature versions of the industrial U-Matic. The two tape spools in the cassette are *coplanar*, that is *the spools are side by side and the tape runs direct from one spool to the other.* One reason for the eventual failure of the Philips VCR formats was that the cassette used *coaxial* spools - in other words, spools stacked one on top of the other. This means that the tape had to run a *skew path* which made it much more difficult to control. The tape would jam, and even break, especially in ageing cassettes. With modern video recorders it is virtually unheard of for the tape to jam or break in a machine that is working properly.

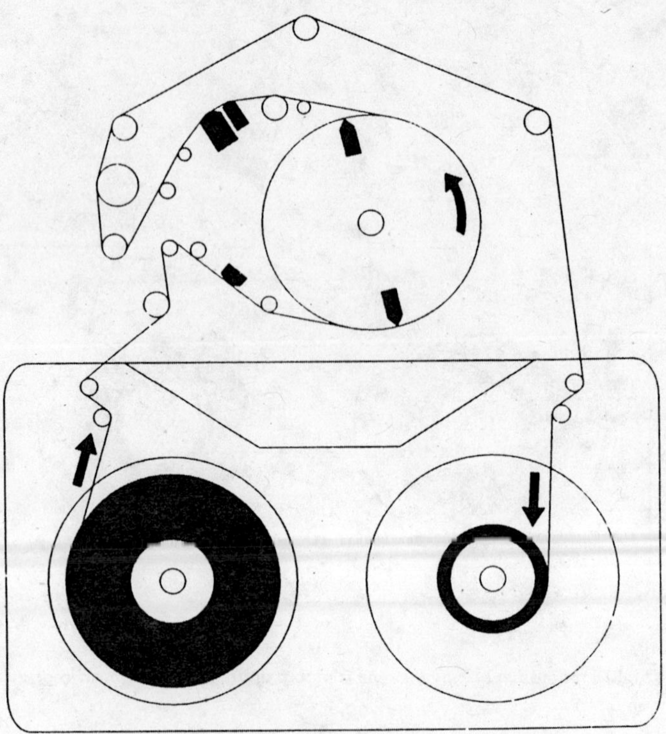

(b)

Fig. 25.11 Beta Format: (a) Sanyo VTC 5400 P Beta Video Cassette Recorder; (b) The tape pick up pattern described by a Beta video tape cassette when loaded into a recorder.

25.11. VIDEO 2000 FORMAT

The Philips V 2000 cassette also uses 12.5 mm (1/2 in) tape and coplanar spools. It looks very similar to a VHS cassette. The tape runs at a speed of 2.44 cm (0.95 in) per second. The major difference between V 2000 and the VHS and Beta formats is that *only half of a V 2000 tape width is used on each pass through the recorder; the cassette is then taken out, flipped over and put back in the recorder; the tape then passes through the machine in the opposite direction with the other half of its width used.* This exactly parallels the situation for audio cassettes where only half the tape width is used on each pass. This approach doubles the playing time but it also means that twice as much video information has to be put onto that tape, because only half its width is being used. It is as if the V 2000 recorder is using 6.25 mm (1/4 in) tape.

The information is packed on by making helical tape tracks *very narrow* and *very close* together. To ensure that the rotating video heads of a V 2000 machine are kept in accurate alignment with the video tracks across the tape *the individual heads are mounted on tiny piezo electric crystals.* All the time that the heads of a V 2000 video recorder are scanning across the tape, they are continually being fed by a *control signal* that keeps them accurately in line with the tape tracks. This is one reason why the V 2000 format has great potential, technically. It was its late appearance on the market and the poor performance of early V 2000 recorder that handicapped the recorder commercially. All around the world VHS is a clear market leader with Beta second and V 2000 usually well behind.

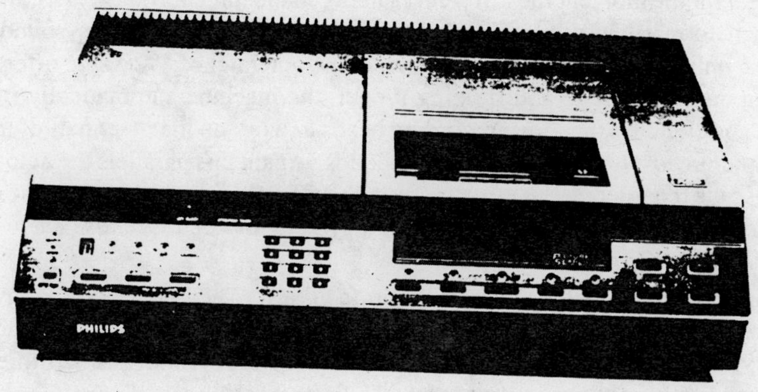

(a)

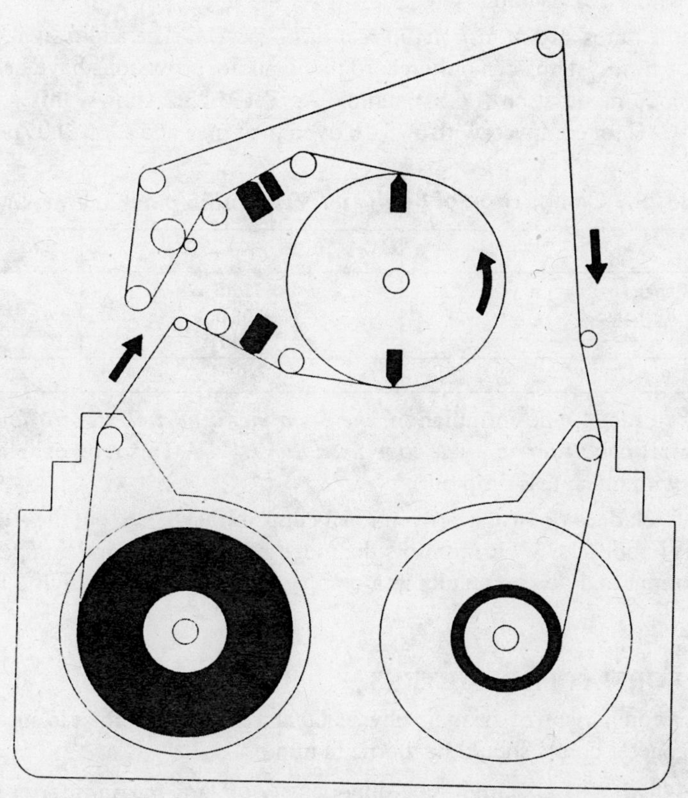

(b)

Fig. 25.12 Video 2000 Format : (a) Philips VR 2022-V2000 Video Cassette Recorder: (b) The pattern traced by V 2000 tape when engaged in the recorder.

25.12 COMMON FEATURES

Let us forget for the moment the differences between the three formats: VHS, Beta and V 2000. Suffice it to say that each format can now offer virtually identical picture quality, sound quality and all-round performance. The only major remaining difference is the potential of V 2000 to offer *auto - reverse*. Instead of the user taking out the cassette and flipping it over, the machine automatically reverses the direction of travel of the tape and switches the video recording heads to work on the other half of the tape width. This can be done with the picture lost for only a couple of seconds. Again this parallels the auto-reverse facility offered by some audio cassette recorders and it effectively doubles the continuous playing time available from a similar cassette. Obviously *auto - reverse cannot be offered by VHS and Beta because each of these formats utilises the full tape width on the first pass.*

All the formats offer exactly the same basic facilities. They all record and replay colour television pictures and sound. Originally they could record only mono sound, but modern machines record in stereo. Every domestic video recorder since the original Philips VCR-N1500 has had three features which were available only as expensive optional extras for the U. Format machines. *Every domestic recorder has a built in tuner, a built-in modulator and a built in timer.*

25.13. 8mm FORMAT

The 8mm video format has been designed to meet the requirements of *smaller size and lighter weight.* The 8mm video tape is only 8mm wide as compared to the VHS video tape, which is only 12.6 mm wide. As a result the mechanism is much smaller. In the 8mm video system, the tape runs at 14.3 mm per second and utilises a 40 mm drum with a 221° wrap.

The 8 mm format features an *FM hi-fi audio recording system*. The audio signal is recorded as an FM carrier. Although most 8 mm systems can only record mono audio, provisions have been made in the format to record stereo pulse code modulation (PCM) audio. As Table 25.2 shows, this gives 8mm a far better performance than VHS-C (C for compact) with a 75 dB dynamic range and a low 0.05 percent wow and flutter rating.

Table 25.2 Comparison of 8mm and VHS audio playback performance

	8mm	VHS-C (SP)
Dynamic Range Frequency Response Line-in:	75 dB 20 Hz 20 kHz	45dB 50Hz 11kHz(SP)
Wow & Flutter	0.05%	0.5%

Using an advanced metal tape formulation, *the 8mm metal particles boast four times the magnetic energy of the conventional cobalt ferric-oxide tape used in VHS-C.* This tape enables 8mm to record and playback video signals with much less drop out.

The 8mm video track does not use a conventional control track head, but the *Automatic Track Finder* (ATF) circuit. With this circuit, the VCR monitors the video head position and compensates for error in the tape path. This new format and design results in a compact and low weight high-quality camera recorder (camcorder).

25.14. VIDEO TAPE FORMAT PARAMETERS

Table 25.3 gives a comparison of the main physical features of each format in current use. In comparing the data the following general points should be borne in mind.

(*a*) Narrow video tracks give high recording density on tape but an inferior noise performance; at track widths below 25 microns good tracking is difficult to achieve with a Dynamic Track Following (DTF) or Automatic Track Finder (ATF) system.

(*b*) Head drum diameter is directly related to video-track writing speed; the larger the drum (for a given wrap angle, i.e 180 degree) the greater the writing speed and the wider the frequency response.

(*c*) Linear tape speed determines the sound-channel frequency response where longitudinal tracks are used-the higher the speed the greater the frequency range.

(*d*) Longitudinal track audio S/N ratio depends on track-width; the half-width stereo tracks for use with stationary audio heads have the worst S/N ratio.

(*e*) All half-wrap head drums for use with 50 fields and TV systems rotate at 1500 r.p.m.in order that one-half revolution occupies exactly the period of one TV field; in 60Hz systems the head-drum speed is 1800 r.p.m.

Table 25.3. Physical characteristics of the four main videotape formats

Format	VHS	Beta	V2000	Video 8
Tape	12.65	12.65	12.65/2+	8
SP linear tape speed, cm/s	2.339	1.873	2.442	2.051
Standard drum diameter, mm	62	74.5	65	40
Audio track width, mono mm	1.0	1.05	0.65	0.5*
SP video track width μ m	49	32.8	22.5	34.4
Video writing speed, m/s	4.86	5.83	5.08	3.12
Video track angle to tape	5°57	5°01	2°39	4°54
Video head azimuth offset	±6°	±7°	±15°	±10°

* Only used for auxiliary purposes
\+ Flip-over cassette

25.15 VIDEOTAPE TRACK CONFIGURATION

The standard VHS track pattern is shown in Fig. 25.13 The tape advances at 2.34 cm/s to give tracks 49 microns wide. They are scanned from bottom to top of the tape. The upper tape edge is reserved for a conventional sound track 1mm wide, which for stereo is split into two 0.35 mm tracks separated by a 0.3 mm guard band. The lower tape edge carries the control track, which provides a positioning reference for the video tracks themselves, and is used to guide the head sweeps during replay.

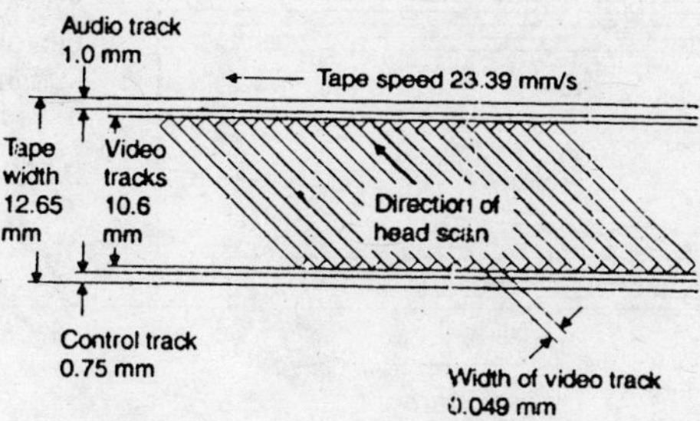

Fig. 25.13 Standard VHS track pattern

Fig. 25.14 illustrates the Betamax track pattern. Tape width is 12.65 mm again, but *video tracks are narrower* at 33 microns with a corresponding lower linear tape speed of 1.87 cm/s. The larger Beta head drum of 74.5 cm diameter is responsible for this format's *high writing speed* of 5.83 m/s compared with VHS format's 62 mm drum and 4.86 m/s writing speed.

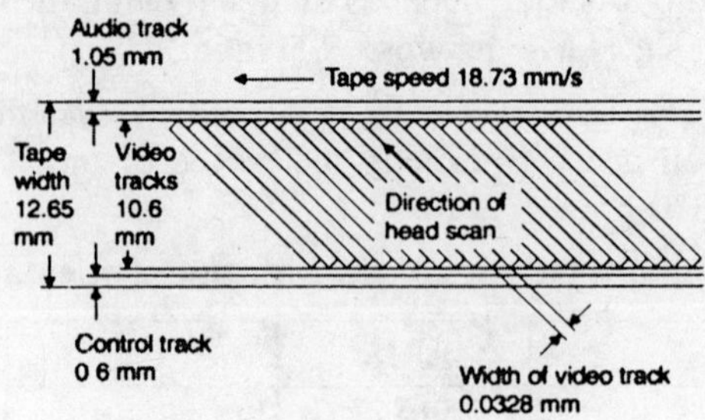

Fig. 25.14 Betamax track pattern

The third format using 12.65 mm tape is the V2000 system shown in Fig. 25.15. This is a *double-sided arrangement* with only one-half of the tape width used; in effect it is a 6.3 mm (1/4 in) system sharing with the compact audio cassette the *turnover feature* whereby both sides can be used to effectively double playing time. *V2000 video tracks are the narrowest of all SP domestic format ones at 23.5 microns* calling for a linear tape speed of 2.442 cm/s with the specified head-drum diameter of 65 mm. In laying down these narrow tracks, and particularly in tracing them during replay, head positioning is governed by an *electronic steering system*, a variant of which is also used in the most recently developed format, Video 8.

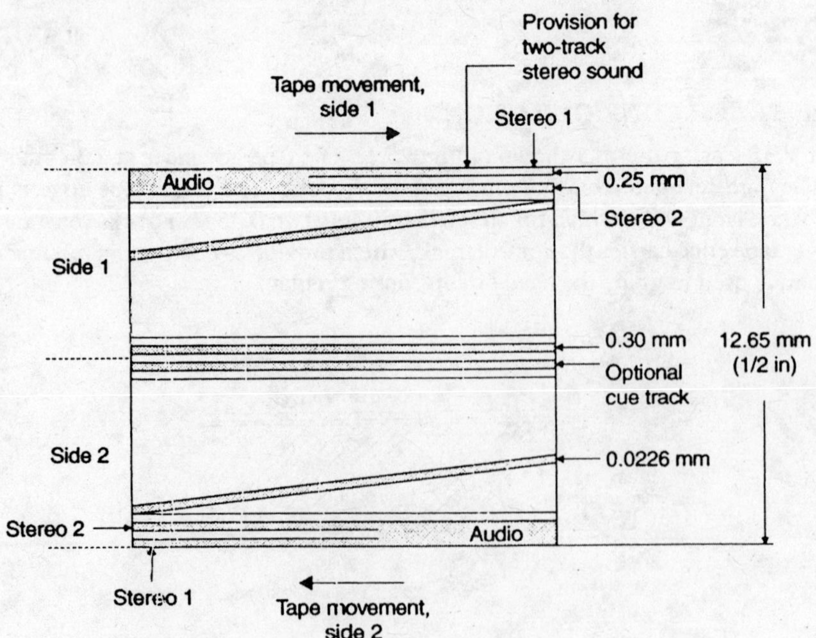

Fig. 25.15 V2000 track pattern.

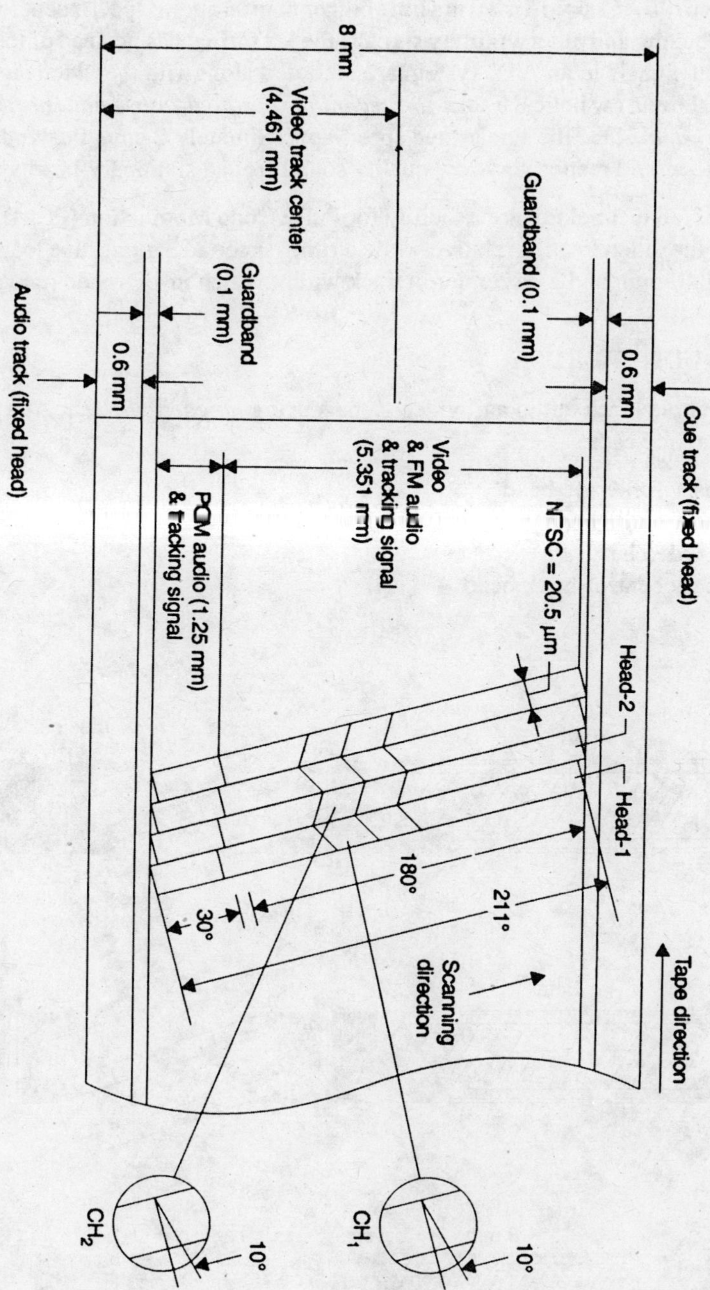

Fig. 25.16 The 8 mm. track pattern.

Tape parameters of Video 8 are given in Fig. 25.16. *A major departure from previous practice is the tape width. This narrow tape lends itself well to portable mobile and light weight video applications*, especially since its cassette package (95 × 62.5 × 15 mm) is a little bigger than an audio tape. Except for special applications like program indexing/cueing and other auxiliary signals, the V8 format has no need of longitudinally recorded tracks. Tracking control signals in an ATF system are recorded alongwith the video signals themselves, and sound is also carried in the narrow helical tracks-*this technique of sound with vision (hi-fi) record and playback is essential in the V8 system* since the linear tape speed-approximately 2 cm/s in standard-play mode and 1 cm/s in long-play mode-would render very low quality sound from a stationary-head system.

Other features of V8 technology are a facility for Pulse Code Modulation (PCM) audio recording on a 30 degree extension of the video track; a relatively low writing speed of 3.1 m/s, due to the conveniently small head drum diameter of 40 mm; and a 34.5 micron track width, which in LP mode reduces to 17.2 micron.

25.16 VIDEO AND AUDIO HEADS

VCRs record and playback audio and video signals using *magnetic heads*. At a maximum, all VCRs use the following magnetic heads:

(*a*) A stationary full-erase head
(*b*) A stationary audio head
(*c*) Rotating video heads
(*d*) A stationary control-track head

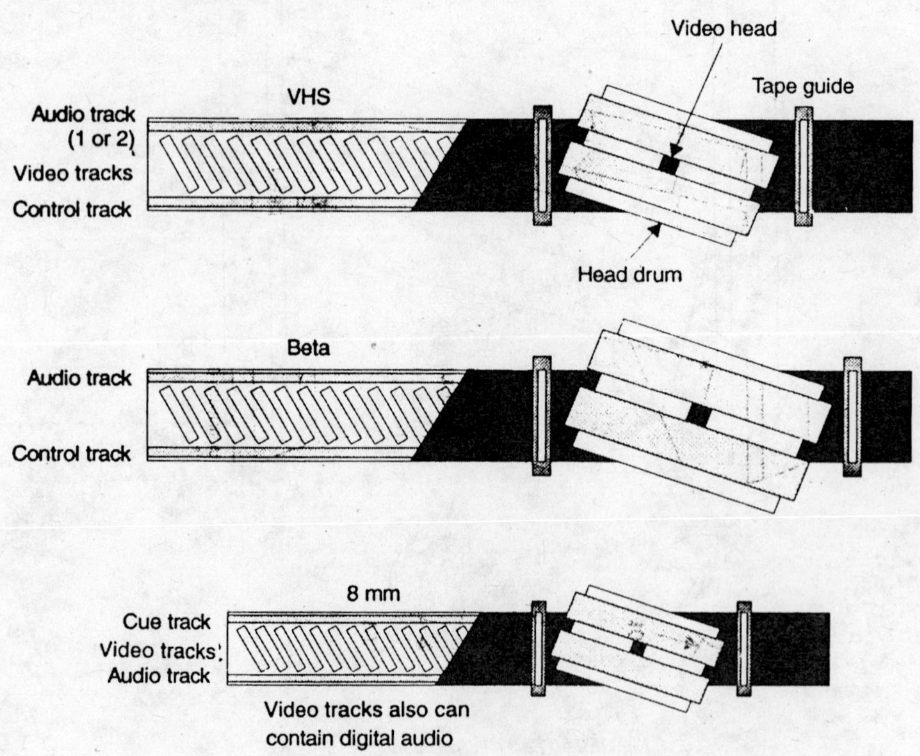

Fig. 25.17 The three home video formats and how the information tracks are recorded on each one.

The *way* these heads record information on the tape is similar for the three formats, though the exact *track widths* and *specifications* vary. The recording formates for VHS, Beta, and 8 mm are shown in Fig. 25.17. *The relative location* of the heads in the three types of VCRs is shown in Fig. 25.18

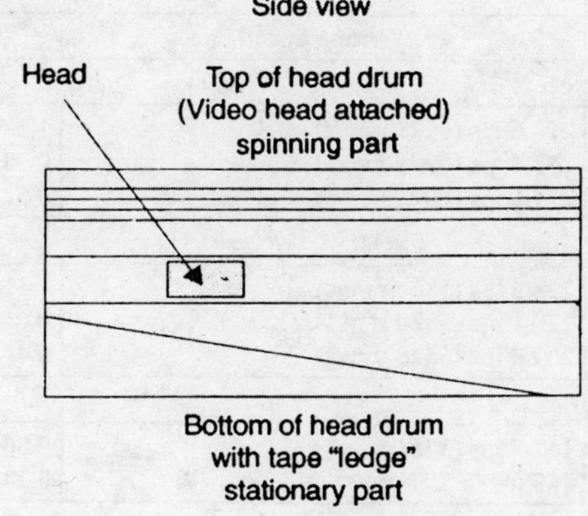

Fig 25.18 A simplified view of a video head. The bottom of the drum is stationary; the top spins with the head.

(*a*) *Full-erase head*: The full-erase head erases the entire tape *and is on only during recording*. The erase head blanks out any previous recordings that might be on the tape. In some VCRs a *secondary audio erase head* can be used. This head erases just the audio track.

(*b*) *Audio head*: In the *record mode* the tape moves past the audio head, where the head impresses the sound signal on a thin track on the edge of the tape. In the *playback mode* the head picks up the signal recorded on tape.

(*c*) *Video heads*: VCRs use *rotating heads* to record and playback video, Fig. 25.19. The heads are mounted in a polished metal cylinder called the *head drum* (other names are used as well including *scanner head cylinder* and *head wheel*).

The heads are at an *angle* so they record a series of long, diagonal tracks on the tape, as shown in Fig. 25.20. The tape might *creep* slowly through the VCR, but the rotating video heads *spin* at 1500 revolutions per minute, effectively covering about 250 inches of tape per second.

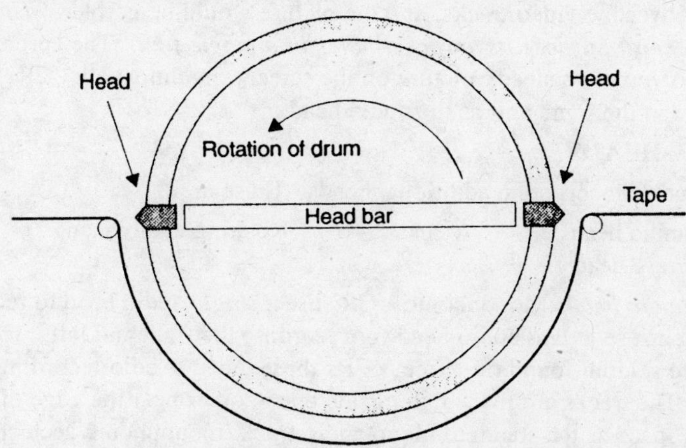

Fig. 25.19 Top of a video head drum, showing both heads and the 180 degree path of the tape.

Table 25.4 shows the *relative speeds* of the tape traveling through a VCR versus writing speed, which is the amount of tape covered by the video heads in one second.

Table 25.4 Writing speeds

Format and Speed	Writing Speed	Linear Tape Speed
Beta		
B-I B-II B-III	274.6 ips (6975 mm/s) 275.4 ips (6995 mm/s) 275.6 ips (7002 mm/s)	1.57 ips (40 mm/s) 0.78 ips (20 mm/s) 0.52 ips (13.3 mm/s)
VHS		
SP LP EP	228.5 ips (5804 mm/s) 229.1 ips (5820 mm/s) 229.3 ips (5826 mm/s)	1.31 ips (33.35 mm/s) 0.66 ips (16.7 mm/s) 0.44 ips (11.12 mm/s)
8 mm		
SP LP	147.7 ips (3751 mm/s) 148 ips (3758 mm/s)	0.56 ips (14.3 mm/s) 0.28 ips (7.2 mm/s)

Note: Writing speed increases slightly at slower linear tape traveling speeds.

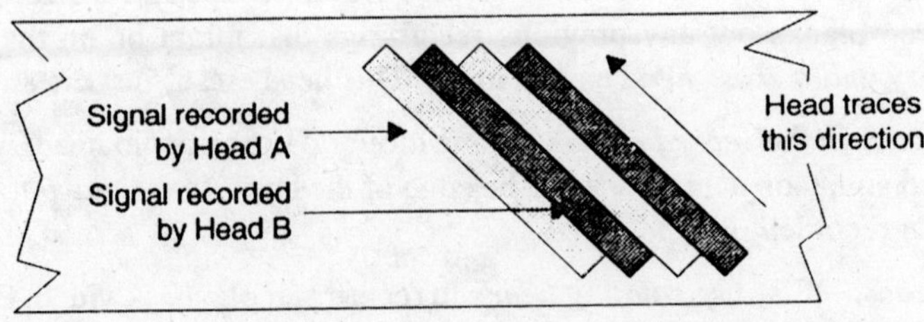

Fig. 25.20 The heads on the head drum record adjacent tracks on the tape.

(*c*) *Control-track head*: During recording, *the control track records a series of 25 Hz pulses*. These pulses are used to synchronise the video heads during playback so they pass directly over the tracks that were previously recorded. Without the control track the video heads might not scan directly over the video tracks, and the picture would be garbled. *The control tracks serve the same general purpose as sprocket holes in a movie film.* The sprocket holes help align each frame so you see a steady picture on the screen. In almost all VCRs, the control-track head is mounted in the same unit as the audio head.

25.17 ADDITIONAL HEADS

Some VCR models incorporate additional heads. These are:

(*a*) Stereo audio heads
(*b*) Rotating audio heads
(*c*) Flying erase head

(*a*) *Stereo audio heads*:Monophonic VCRs use a *single* audio head to record and playback sound. Older stereo VHS decks use a *dual* audio head for recording the right and left tracks separately. Stationary stereo audio heads are seldom found on home VCRs these days. Audio recording with two audio heads is called *linear stereo*. The tracks are placed alongside one another near the edge of the tape (see Fig. 25.21) and occupy the same space as the standard mono audio track (monophonic decks pick up both the right and left channels together). *Noise reduction circuitry (such as Dolby) is often used with linear stereo decks to improve the sound quality.*

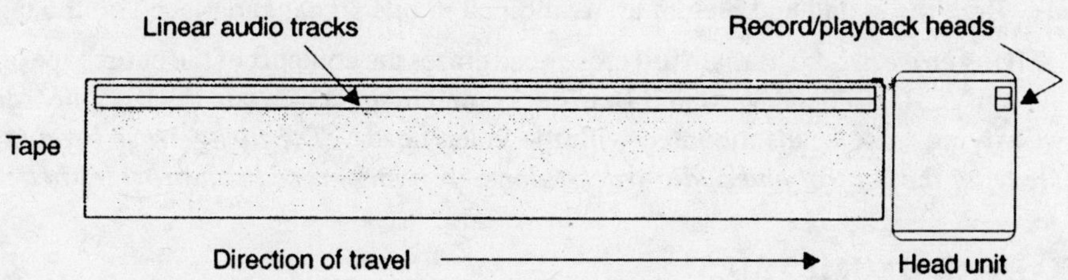

Fig. 25.21 A linear stereo head records the right and left sound tracks separately at the top of the tape.

On a number of linear stereo VCRs, one or both of the channels can be *independently erased or activated* for audio dubbing. The audio erase head is built into the audio head as shown in Fig. 25.22. With audio dubbing, you record only the audio segment, but leave the video and control tracks intact. Most linear stereo decks let you record on either the left or right stereo channel, leaving the other channel untouched. Beta VCRs record stereo audio mixed with the picture information.

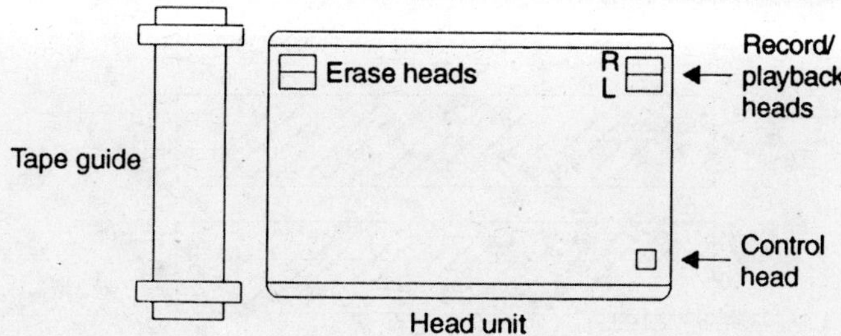

Fig. 25.22 Close up of the control/audio head unit. The unit contains both the control and audio heads and can also contain one or more audio erase heads.

(*b*) *Rotating audio heads*: VHS hi-fi decks - those that record stereo sound in high fidelity - have two *additional heads* mounted with the rotating video heads. These heads record audio information in an FM carrier signal. Even though the audio and video tracks are intermixed, they do not interfere with one another because the azimuths of the heads are different ,as depicted in Fig. 25.23. *Magnetic heads do not pick up information that is recorded on substantially different angles*. The video heads themselves record at different azimuth angles.

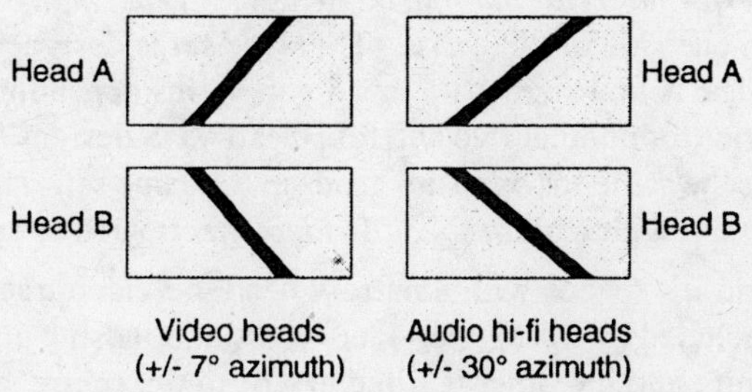

Fig. 25.23 The video heads are slanted at an angle to prevent interference from adjacent signal tracks.

Beta stereo hi-fi decks record picture and high-fidelity sound using the machine's standard complement of video heads. Both the sound and picture are combined in one signal and placed on the tape together.

(*c*) *Flying erase head*: The main full erase head erases the contents of the entire tape prior to recording. The 8mm table top VCRs, top-end VHS and Beta decks, and many 8 mm and VHS camcorders use one (and sometimes two) flying erase heads mounted with the video heads. *The flying erase head (s) erases just the video tracks, leaving the control and audio-tracks alone. A flying erase head provides better, cleaner edits.*

25.18 VIDEO HEAD GEOMETRIES

The size of the gap found in all magnetic recording heads varies from 28 to 90 microns wide, depending on the application. The video heads in typical *VHS* decks have a head gap of only 38 microns. The gap in conventional *Beta* video heads is even smaller, 30 microns, and in *8 mm* decks, the gap is 20.5 microns.

VCRs use head gaps of various sizes to improve performance at slower record and playback speeds and to accomplish noise-free special effects. Wider heads record a stronger signal and are desired over narrow head gaps. A problem occurs when using wide heads to record and playback video at the slower LP, EP or BIII speeds.

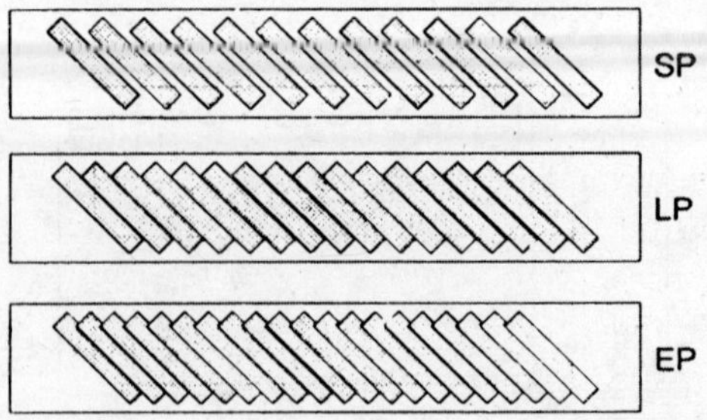

Fig. 25.24 Track spacing at the three VHS speeds of SP, LP, and EP (SLP)

Fig. 25.24 shows the geometry of the video tracks laid down by the video heads with a conventional VHS-VCR at SP, LP and EP speeds VHS design parameters allow for 58 microns between each video track at SP speed. *The spacing decreases proportionately at the slower speeds* - 29 microns and 19 microns for LP and EP speeds, respectively. At the fast speed there are *guard bands* as wide as 28 microns between the tracks. As the tape slows down, the tracks get closer together and they overlap by as much as 11 microns.

Obviously, if the deck is equipped with wide heads, playback and recording will be satisfactory at the SP speeds because of the guard bands between the tracks. *But operation will be severely limited at slower speed* because the tracks will almost stack on top of one another during recording.

To provide optimum performance at both fast and slow speeds, many VCRs use four heads instead of just two. The gap is wider on one side of the heads. The *wide set* is designed for SP speeds and the *narrow set* is for LP and EP speed. The actual size of the head gap varies depending on the make and model. Fig. 25.25 shows the head geometries for popular two and four-head VHS designs from JVC and Matsushita. Note that the head gap is not always the same for opposite heads in the same set. This provides the deck with better performance during playback and with certain *special video effects* such as still-frame and slow motion.

Even at the slow LP and EP speeds with a narrow head the video tracks still *overlap*. The effects of this inevitable overlap are precluded by tilting the video heads at opposing angles. In *VHS system* one video head is tilted at +6° to the centre and the other is tilted at - 6° to the centre, as shown in Fig. 25.26. The 12 degree difference in azimuth reduces or eliminates the *crosstalk* that would otherwise occur if the heads were on the same plane. In *Beta* decks, the azimuth is +7° and -7°, and in *8 mm* the heads are angled ± 10°.

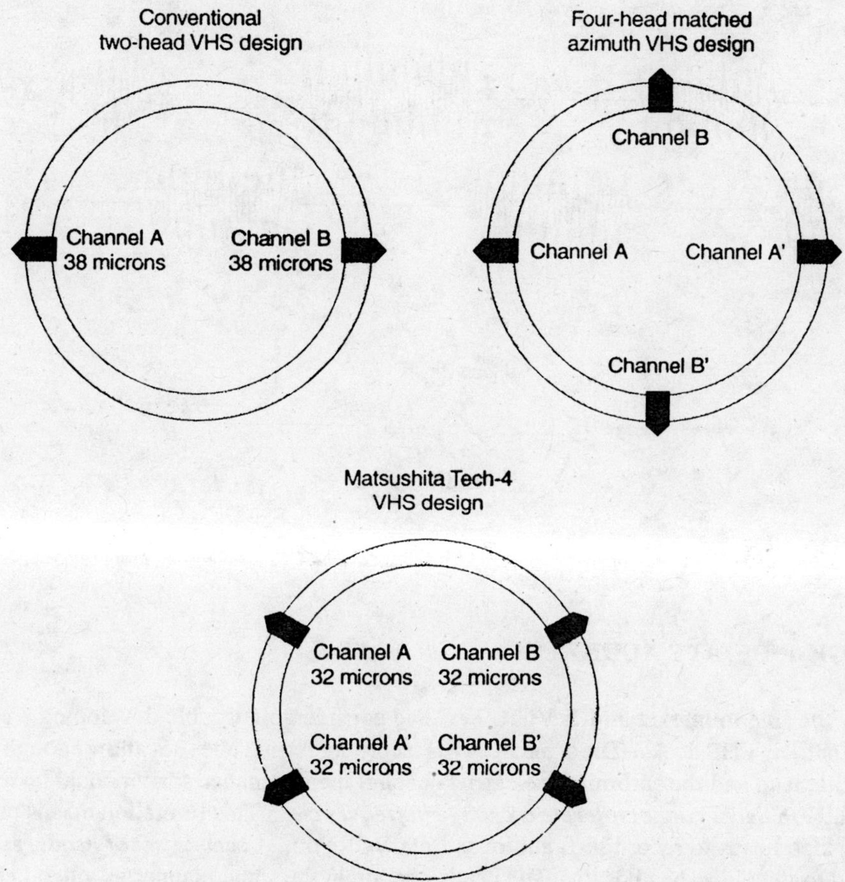

Fig. 25.25 A sampling of head geometries for typical VHS machines.

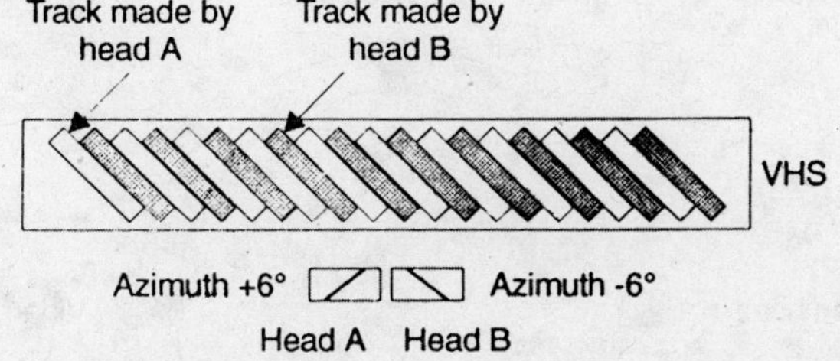

Fig. 25.26 Alternating video tracks are recorded at different angles to minimise cross-channel interference.

25.19 HEAD SWITCHING

During recording, the signal feeds alternately to the heads via a head switching circuit. The head is turned ON as it touches the tape and makes its half-circle pass around the head drum. Just before the head is about to leave the tape it turns OFF and the signal is applied to the opposite head. The video waveform for head switching is shown in Fig. 25.27.

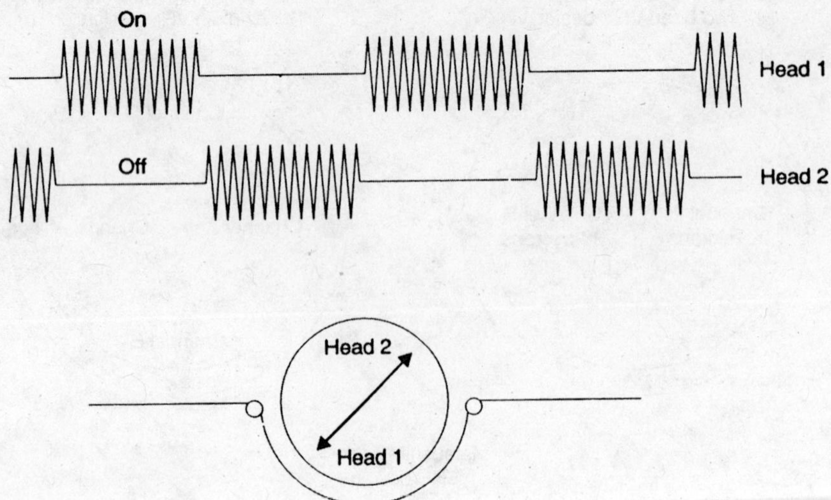

Fig. 25.27 The action of switching the video heads ON and OFF as the video drum rotates.

25.20 VHS HIGH-FIDELITY AUDIO

Beta was the first to market a hi-fi VCR. JVC had considerable trouble developing a competing hi-fi audio standard for their VHS decks. The standard VHS recording format does not allow enough space between the luminance sideband and the chrominance carriers even if the luminance carrier could be moved up a few notches. In fact, *VHS decks cannot tolerate a change in frequencies*. This limitation means that VHS-VCRs cannot use the video heads to record hi-fi audio, as Beta VCRs do. *A second set of heads*, as shown in Fig. 25.28, are added to the video-head drum. The heads operate in the same manner as video heads. One head in the set records a field's audio information, the next head records another, and so forth. The gap of the hi-fi heads can vary from about 6 microns to 42 microns.

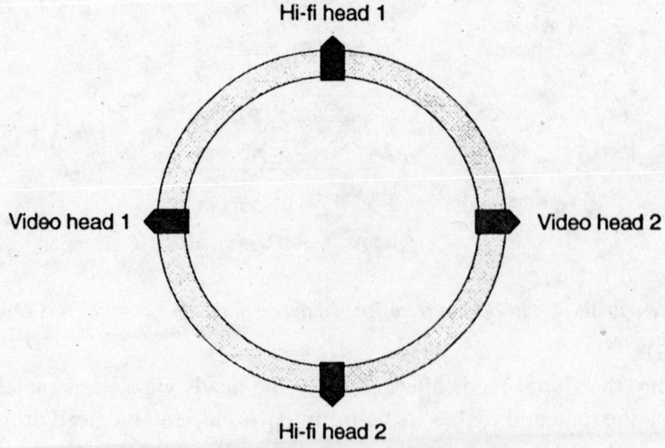

Fig. 25.28 The position of the video and audio hi-fi heads on a typical VCR.

In the VHS hi-fi system, the audio signals are recorded first, then the video signals are recorded. Normally *overrecording* with the video heads would lead to complete or partial erasure of the audio track but in practice this does not happen-thanks to an old recording technique.

It has been known for many years that *low-frequency signals penetrate into the tape further than high-frequency ones*. Because the hi-fi audio portion is lower in frequency than the 3.4 to 4.4 MHz luminance carrier, it travels deeper into the tape. JVC calls this technique *depth multiplex recording*.

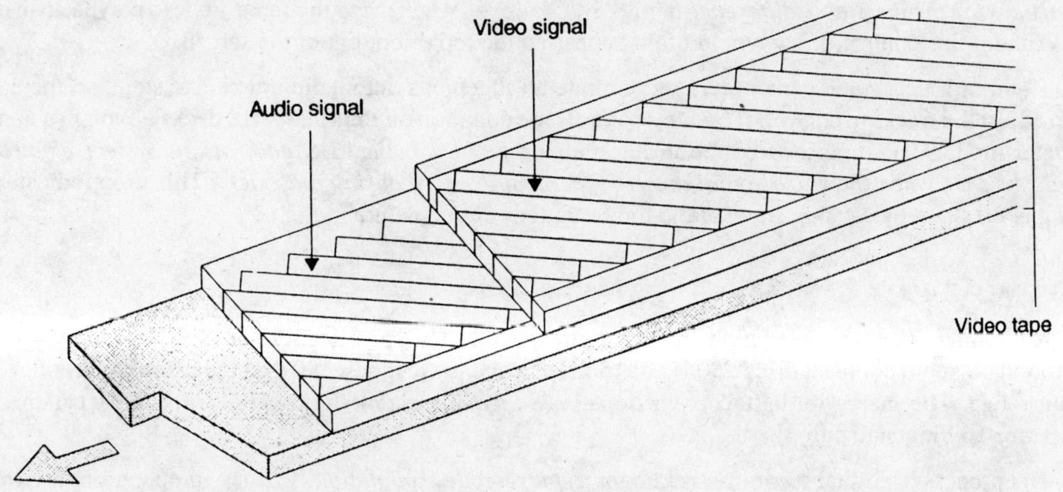

Fig. 25.29 In hi-fi recording sound is recorded deep inside the tape, with the video signal recorded on top.

During recording the hi-fi heads record their signal a *split second before* the video heads record theirs. The audio signal goes *deeper* into the tape, so when the video heads pass by, most of the already-recorded video information remains. *To avoid crosstalk between the audio and video carriers, the hi-fi audio heads are tilted* at ± 30°. *The extreme azimuth also reduces crosstalk between right and left stereo channels.*

The hi-fi heads record the audio on FM carriers, as shown graphically in Fig. 25.30. The frequencies are 1.3 MHz for the *left* audio channel and 1.7 MHz for the *right* audio channel. *Compatibility with decks that do not feature hi-fi is retained because the sound also is recorded on the linear audio tracks.*

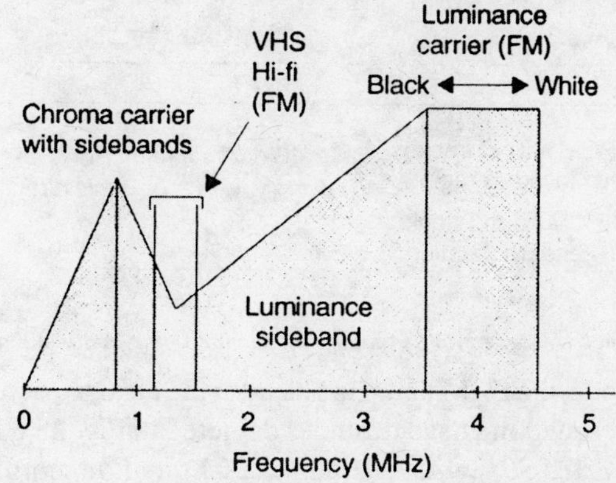

Fig. 25.30 Signal layout for the VHS hi-fi recording format.

Because the hi-fi audio is recorded with the video track *effects such as audio dubbing are not possible.* There is no way to *selectively erase* the audio portion of the program and replace it with another track. However, many VHS hi-fi machines also record stereo audio onto one or two linear audio tracks. *These tracks can be selectively erased and recorded.*

25.21 AUTOTRACKING

One of the banes of using a VCR is *fine-tuning* the picture using the *tracking control.* This control *compensates for the inevitable difference between VCRs,* even when using the same VCR to playback an older tape. Without this control, snow would often appear on the top or bottom of the screen.

The 8 mm decks and camcorders incoroprate an ingenious automatic tracking system, so there's no need to adjust the tracking control. Tracking controls are common on Beta and VHS decks although a number of the later-model VHS units incorporate an autotracking feature. *In the VHS autotracking system, electronics in the deck sense when the video signal drops 15 percent or more during playback.* This drop indicates the tracking is off, thereby causing snow at the top or bottom of the picture.

25.22 PCM AUDIO

All 8 mm decks can record high-fidelity digital sound. The specifications are almost as impressive as those found on audio compact discs : 50 Hz to 15 kHz frequency response; 90 dB signal-to-noise ratio, 88 dB dynamic range. The analog-to-digital conversion uses *8-bit quantization* (by comparison, compact audio discs use superior 16 bit quantization).

The process of digital recording is known as *pulse-code modulation* (PCM). In operation, a series of pulses that represents digital data is recorded on the tape. The pulses are derived by converting the original analog audio signal to digital form. For playback the pulses are processed in just the reverse to reconstitute analog audio from the digital. There are two digital audio modes: video-plus-audio and all-audio (the latter is sometimes called *multi-PCM*). In the *video-plus-audio mode*, the digital sound is recorded with the video picture (the track also can be added after the picture is recorded). In *all-audio mode* the deck records only sound. Depending on the recording speed, you can store up to 24 hours of digital sound on a 2 hour tape (12 hours at SP speed) Fig. 25.31 shows the track layout for video-plus-audio and all-audio recording.

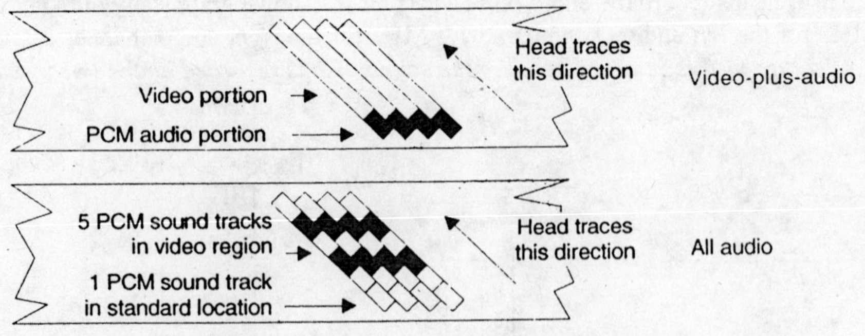

Fig. 25.31 Tracks recorded on 8 mm tape in the video-plus-audio and all-audio recording.

In video-plus-audio mode, the PCM audio is placed at the head of each video track. This is accomplished by wrapping the tape further around the head drum, as depicted in Fig. 25.32. The standard *180° wrap* is used for video recording, and the *extra 30° wrap* is used for PCM audio recording. PCM audio is also found on some high-end VHS decks. Most use higher-quality 14-bit quantization.

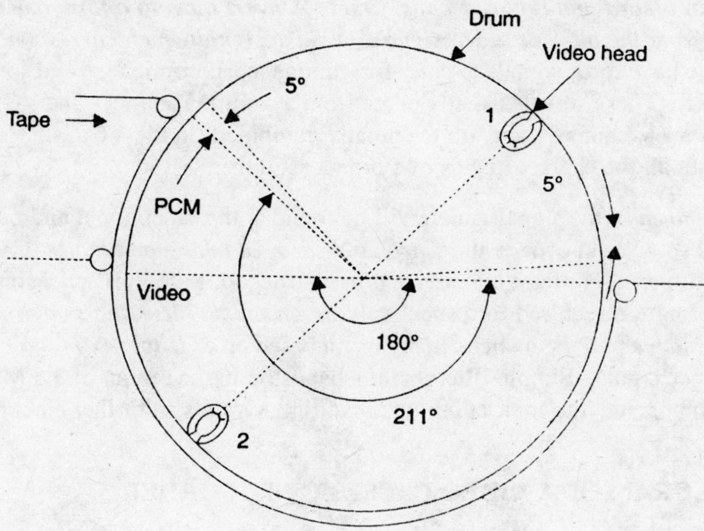

Fig. 25.32 The tape is wrapped around the head in the 8mm system an extra 30° to accommodate the PCM audio track.

25.23 THE CVBS SIGNAL

The TV video signal cannot be recorded directly on tape in its basic form, due to bandwidth and noise problems. This is true of both luminance and chrominance components and *each has to undergo a great deal*

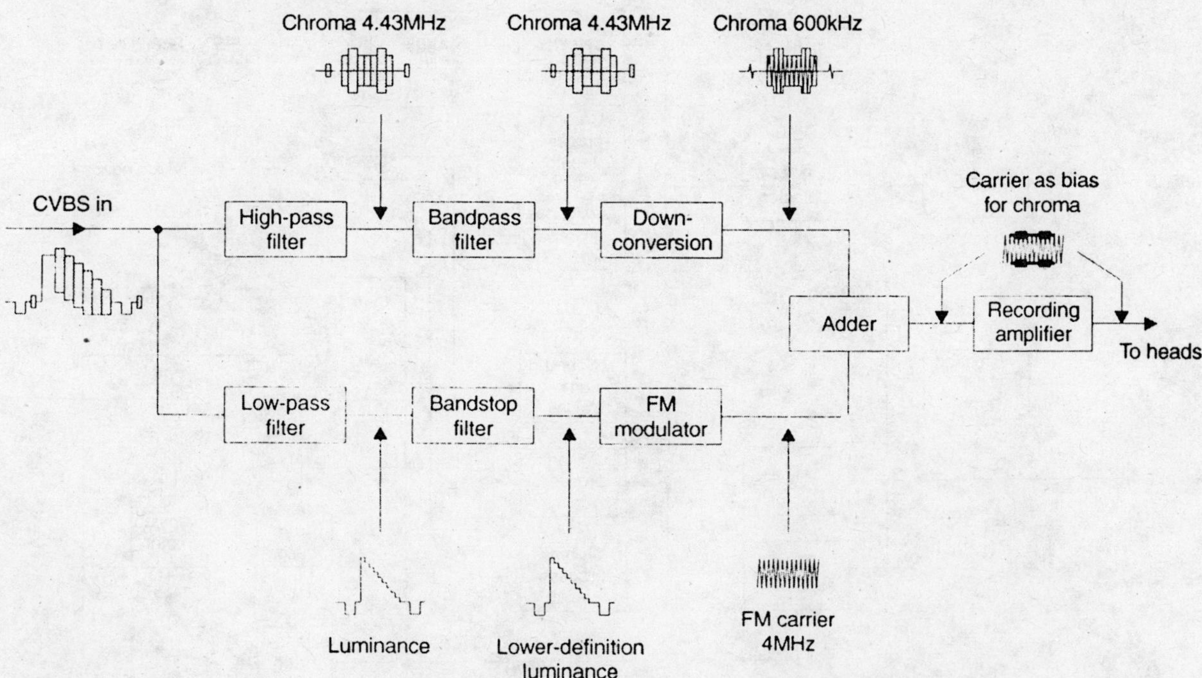

Fig. 25.33 The route and nature of video signals in record mode. The waveform on the right corresponds to that at the top of Fig. 25.9.

of processing in both record and replay modes, first to convert them to a form amenable to the tape system and then to restore from the off-tape video signals, a video signal which corresponds as closely as possible to that recorded. We have three signals to cater for: luminance, chrominance and sound, and each is handled separately in the VCR. These signals usually come from a common source - the VCR's detector working on broadcast signals or a TV camera - and are eventually combined (in the VCR's RF modulator) into *a single composite signal* with all the characteristics of a broadcast transmission.

The *standard form* of the signal which will be found at the video input and output sockets of cameras VCRs monitors and so on is known as the *Chroma, Video, Blanking and Syncs* (CVBS) signal. Its level is invariably 1V peak-to-peak. The term *luminance* is used to describe the V, B and S components which together form the signal information required to respectively *describe, stabilise and synchronise* a monochorme TV picture. The *chrominance* part, is in *encoded form* and based on a frequency of 4.43 MHz. This is above the range of the luminance signal. Simple filters with a bandstop in the region of 3.5 MHz *intercept and divert* these two vision components into their appointed recording circuits as further elaborated in Fig. 25.33.

25.24 BLOCK DIAGRAM OF A VIDEO CASSETTE, RECORDER

Apart from the *deck mechanism* the recorder consists of an ensemble of printed circuit panels, each dealing with one or more aspects of the machine's operation. A block diagram of a typical *basic machine*, showing signal paths is given in Fig. 25.34.

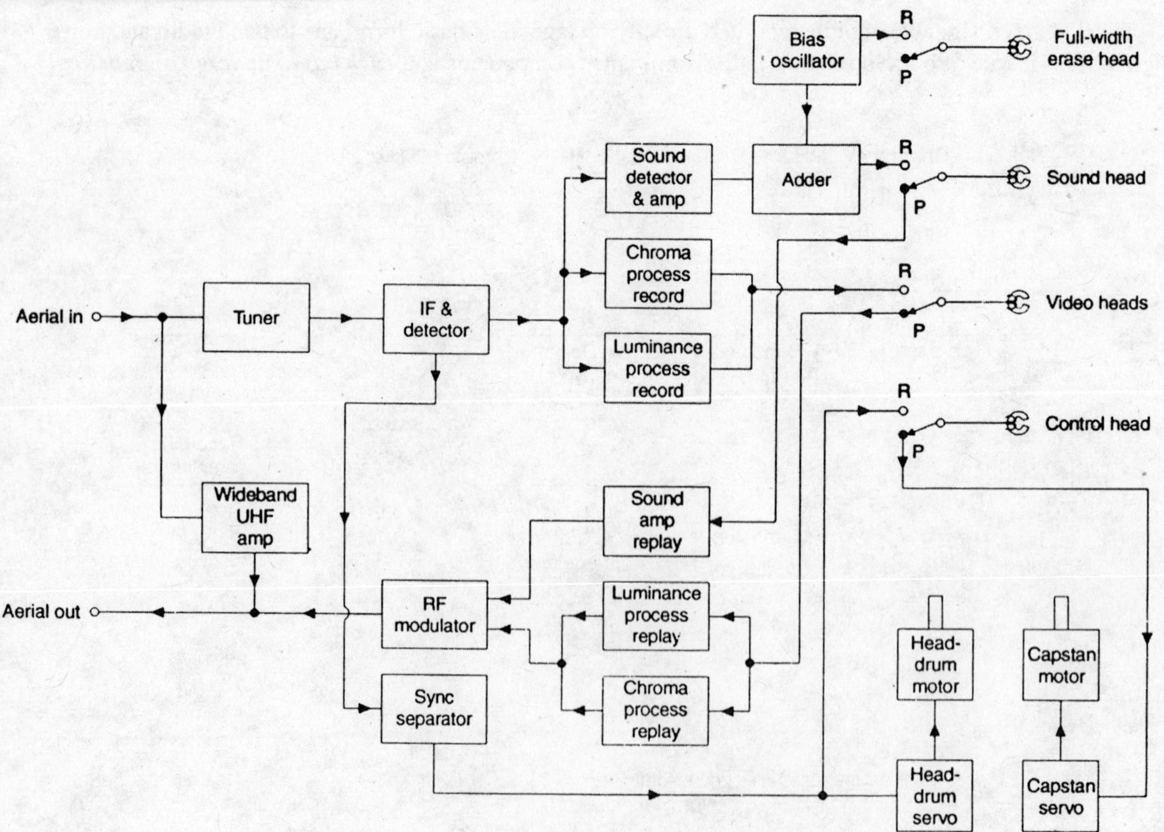

Fig. 25.34 Block diagram of a typical basic machine.

During *record* the VCR's receiver section (tuner, IF and detector) converts broadcast transmissions into baseband sound and vision signals. *The sound signal passes to the audio recording head after the addition of a bias signal, which is also fed at high power to the full-width erase head to wipe existing tracks off the tape.* The vision signal is split by filters into *luminance* (black-and-white) and *chrominance* (colour) components which are dealt with separately before being added together for application to the *rotating video heads.* A sync separator strips the *timing pulses* (part of the video waveform) from the video signal for processing in the *video-head-drum motor servo*-they are used to keep the head sweeps synchronised to the picture signal. These timing pulses are also recorded on the tape by the *control track head* for use as reference during replay.

During playback mode, the tiny signals from the tape heads are switched to replay processing circuits for sound, luminance and chrominance, all three again being separately dealt with. The re-created sound and vision signals come together in the RF modulator, where they are impressed on a UHF carrier, then passed out of the machine en route for the TV set. The control track is replayed into the head drum or capstan servo via a tracking control to enable the video heads to accurately scan the existing tape tracks.

25.25 CONTROLS ON VIDEO CASSETTE RECORDERS

VCPs (video cassette players) *lack* a recording capability—they only playback tapes and lack the means to record. This means VCPs dispense with a lot of extraneons electronics, as well as the programmable record timer. As a result VCPs are a little less expensive than full-fledged recording VCRs. However they require the same level of mechanical maintenance as full-fledged recording VCRs. The controls on a typical VCP are depicted in Fig. 25.35.

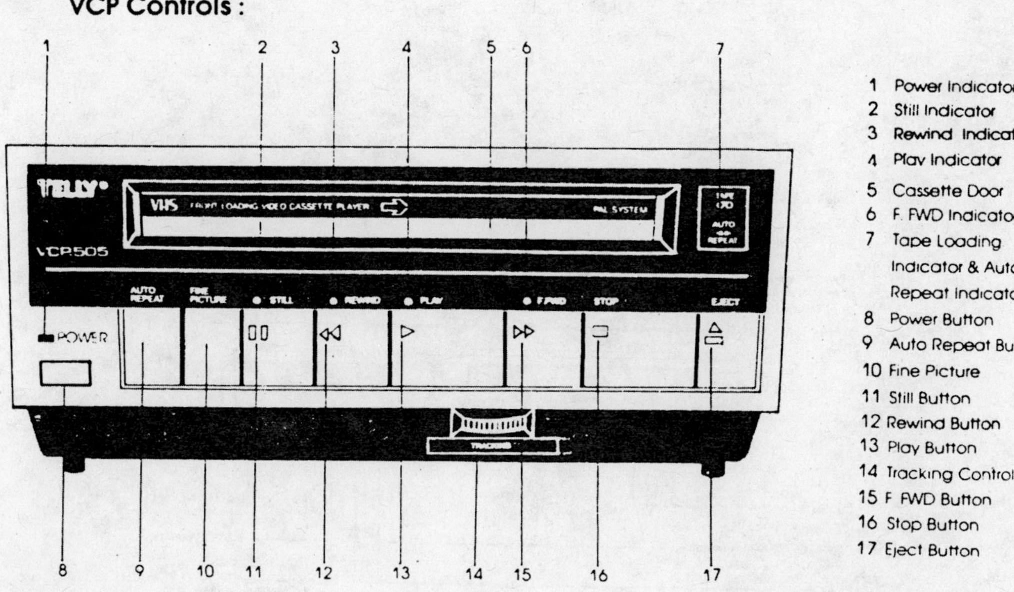

Fig. 25.35 Controls on a typical video cassette player.

Most VCRs have the following typical controls:

POWER: Turns the power On and OFF

EJECT: Ejects the cassette from the VCR.

PLAY : Places the deck in the PLAY mode.

FAST FORWARD: Fast forwards the tape; *no picture on screen.*

REWIND : Rewinds the tape; *no picture on screen*

CUE (or fast forward scan) : Scans the tape at 3 to 30 times the normal speed in forward; *high-speed picture on screen*

REVIEW (or reverse fast scan): Scans the tape 3 to 30 times the normal speed in reverse; *high-speed picture on screen.*

PAUSE: Stops the tape during playback; a *still frame* appears on the screen (*the deck stops automatically after about five minutes to prevent excessive tape and video-head wear*).

RECORD: When used with the PLAY button, places deck into RECORD mode.

AUDIO DUB : Records *just audio* over previously recorded program (*video portion must be previously recorded*)

VIDEO DUB : Records *just video* over previously recorded program.

TRACKING : Adjusts tracking to reduce or eliminate *noise bars* in picture (*not found on 8mm decks*).

OTR : One-touch recording; starts deck in RECORD mode and *records for preset period.* On many decks, each press of the OTR (or EXPRESS) button increases the recording time by 30 minutes.

TIME buttons: Set the record timer.

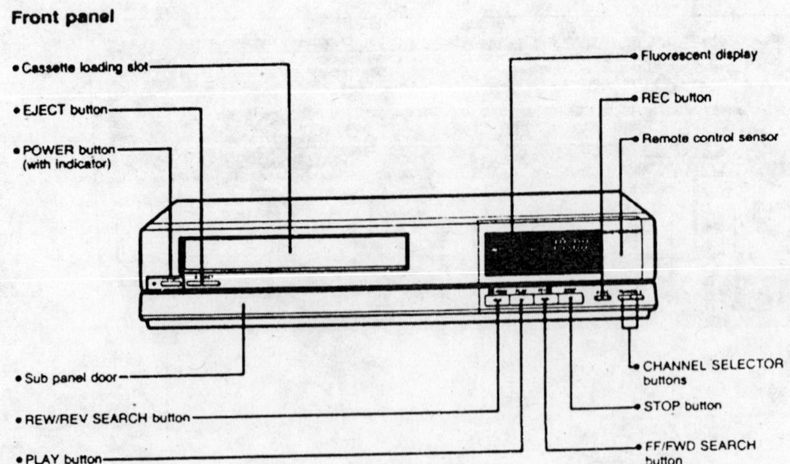

Fig. 25.36(a) Front Panel

CLOCK buttons: Set the clock (many decks set the time and day of week).

CHANNEL 3/4 output (back of deck): Switches RF modulator output to channel 3 or 4.

***LINE : (With** A/V terminals)*: Switches between RF or direct AUDIO/VIDEO input terminals (*might be automatic on some models*).

Most VCRs have a *counter* and *counter* RESET. The counter is operated by the *take-up spindle (either mechanically or electronically) or the control-track pulses* and indicates *relative tape position.* The counter is reset to "0000"by pressing the RESET button.

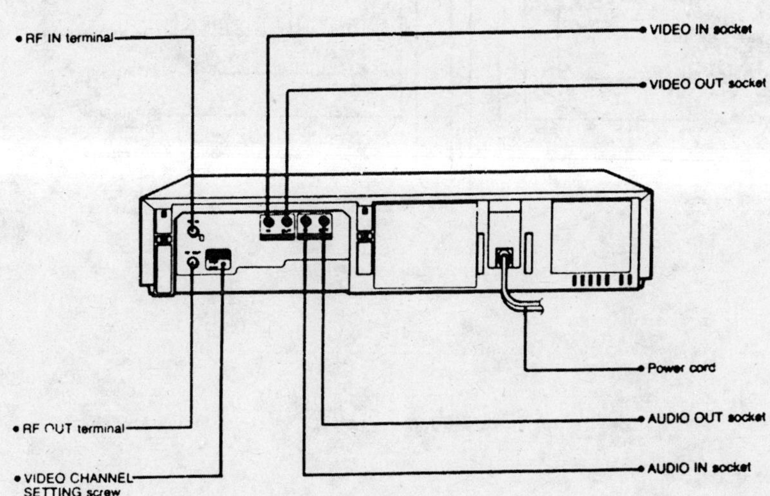

Fig. 25.36 (b) Rear Panel

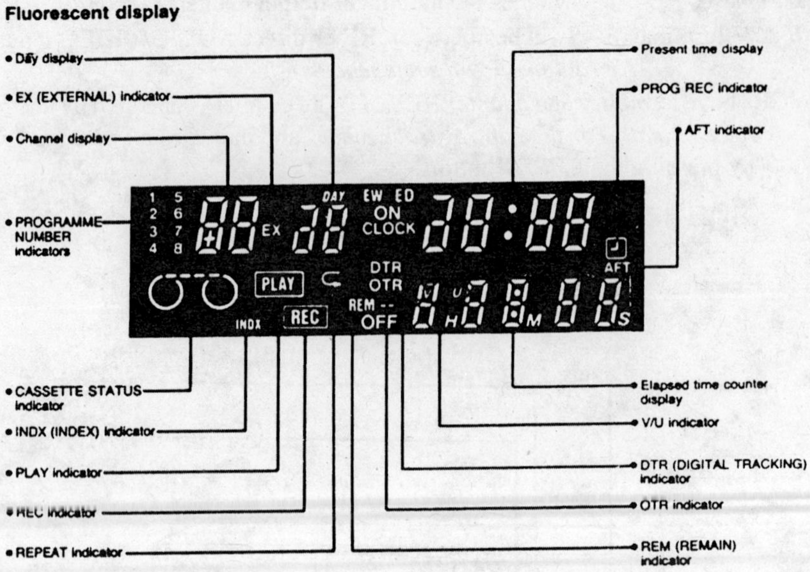

Fig. 25.36 (c) Flourescent Display

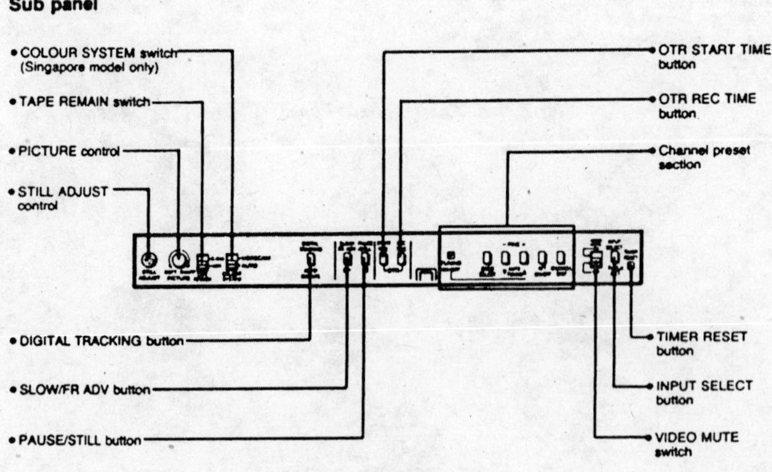

Fig. 25.36 (d) Sub Panel

Fig. 25.36. Controls on a typical VCR

OBJECTIVE TYPE QUESTIONS (A)

1. The most popular formats today are
   - (*a*) 3/4 U
   - (*b*) VHS
   - (*c*) V2000
   - (*d*) Beta
   - (*e*) Video 8
2. For helical scan the rotating video drum carries
   - (*a*) one head
   - (*b*) two heads
   - (*c*) three heads
   - (*d*) four heads
3. As the frequency of the recorded signal increases the voltage induced in the playback head
   - (*a*) increases at a 3dB/octave rate
   - (*b*) decreases at a 3dB/octave rate
   - (*c*) increases at a 6dB/octave rate
   - (*d*) decreases at a 6dB/octave rate
4. A bandwidth restricted television signal
   - (*a*) can be directly recorded on tape
   - (*b*) cannot be directly recorded on tape
   - (*c*) can be recorded on the tape by indirect methods
   - (*d*) none of the above
5. The FM carrier technique
   - (*a*) enhances the octave range
   - (*b*) restricts the octave range
   - (*c*) reduces the octave range
   - (*d*) none of the above
6. The modulation index is chosen so that virtually all the sideband energy is confined to the
   - (*a*) first pair of sidebands
   - (*b*) first two pairs of sidebands
   - (*c*) first three pairs of sidebands
   - (*d*) none of the above
7. In slant azimuth system, the two heads are set at
   - (*a*) the same angle
   - (*b*) slightly different angles
8. The locally generated bias signal is very carefully controlled in
   - (*a*) phase
   - (*b*) amplitude
9. The actual playing time per cassette depends on
   - (*a*) running speed of tape
   - (*b*) thickness of tape
   - (*c*) length of tape in the cassette
   - (*d*) width of tape in the cassette
10. With LP record and playback modes the capacity of a tape is
    - (*a*) effectively halved
    - (*b*) not affected
    - (*c*) effectively doubled
    - (*d*) increased insignificantly
11. Beta tapes
    - (*a*) cannot be played on VHS machines
    - (*b*) can be played on VHS machines
    - (*c*) can be played on VHS machines with a Beta to VHS adopter.
12. On which format half of the tape width is used in a single pass?
    - (*a*) VHS
    - (*b*) Beta
    - (*c*) V2000
    - (*d*) 8 mm
13. In which format video heads are mounted on tiny piezo electric crystals?
    - (*a*) Beta
    - (*b*) 8 mm
    - (*c*) VHS
    - (*d*) V2000
14. Which format offers the facility of auto-reverse ?
    - (*a*) V2000
    - (*b*) VHS
    - (*c*) Beta
    - (*d*) 8mm

15. Which format meets the requirements of smaller size and lighter weight?
    (*a*) 8 mm (*b*) V2000
    (*c*) Beta (*d*) VHS
16. To provide optimum performance at both fast and slow speeds, many VCRs use
    (*a*) two heads (*b*) four heads
    (*c*) six heads (*d*) eight heads

FILL IN THE BLANKS (B)

1. Video tape for customer use is available in ....only.
2. Video tape is used for recording the............and any............that accompanies the picture.
3. An increase in running time can be obtained by making the tape............and running it at a............speed.
4. The tape most commonly used by non professionals has a width of ....
5. If the heads are fixed, the tape will have to move incredibly .... for ....information to be crammed on to it.
6. When a reel tape runs very fast it.............out quickly.
7. The.................format has become standard for broadcasters around the world.
8. For helical scan the rotating video drum carries...........heads.
9. For helical scan the rotating video drum rotates at an ....to the slow moving tape.
10. Each halving of frequency.............the output signal.
11. A constant level of white or grey will give a ....FM carrier frequency.
12. An operating range of frequencies between 1 and 7 MHz represents an octave range of less than....................
13. The transfer of video signals to and from the spinning video heads is accomplished by a ....transformer.
14. In the U-format recorder threading of the video tape is done.............as the cassette is loaded in the machine.
15. If the tracks are too close together, they interfere with each other and cause...........
16. Extra heads are used for ....effects.
17. Two processes are carried out on the chroma signal to fit it into the 0 to 1 MHz slot. It is ....and......................
18. The ability to work on both sections of the transfer curve........the replay level and.................S/N ratio.
19. For video recorders the levels of bias are..........to the type of tape available for the format.
20. The actual playing time depends both on the running.................. selected and the.............of the tape in the cassette.
21. The only feature common in VHS and Beta cassettes is the ....of the tape which is..................wide.
22. Auto-reverse cannot be offered by VHS and Beta formats because each of these formats utilises the full...........of the tape on the first..........
23. The 8mm format features an FM hi-fi audio.............system.
24. The 8mm metal particles boast four times the............energy of conventional cobalt oxide tape used in VHS.C.
25. Narrow video tracks give high..............density on tape but an inferior................performance.

26. Head drum diameter is................related to video track writing speed.
27. Longitudinal track audio S/N ratio depends on..................
28. All half-wrap head drums for use with 50 fields/TV systems rotate at .... r.p.m.
29. V2000 format is a ....sided arrangement.
30. Video heads are mounted at an angle, so they record a ....of long diagonal..................on the tape.
31. The control track records a series of .... pulses.
32. The ....tracks serve the same general purpose as sprocket ....in a movie film.
33. Audio recording with two video heads is called............ stereo.
34. Magnetic heads do not pickup information that is recorded on substantially different................
35. The flying erase head (s) erases just the ....tracks.
36. A flying erase heads provides better, cleaner...........
37. VHS decks cannot tolerate a change in..................
38. In VHS hi-fi systems, the ....signals are recorded first, then the ....signals are recorded.
39. Low-frequency signals..................into the tape further than high-frequency signals.
40. The extreme azimuth (± 30 degree) reduces................between right and left stereo channels.
41. Tracking control...............for the inevitable difference between VCRs.
42. In PCM video-plus-audio mode, the standard 180° wrap is used for video recording and an extra..........wrap is used for PCM audio recording.

CONVENTIONAL QUESTIONS (C)

1. What are the problems in video recording and how are these overcome?
2. Describe the basic principle of video recording.
3. Compare audio cassette recording with video cassette recording.
4. What is a format? Compare the three most popular video formats. Enumerate the relative similarities of the three formats.
5. Explain why signals are attenuated as the gap size increases beyond $\lambda/2$.
6. What is the relation between gap width and writing speed?
7. What techniques are adopted to record and replay video signals?
8. With the help of a suitable sketch, explain the slant aziumuth system of video recording.
9. How the colour signal is recorded on the tape?
10. What is the significance of recording bias ?
11. Briefly explain the colour-under recording system.
12. Compare video and audio heads.
13. What is the function of a flying-erase head?
14. Explain why head switching is required? How is it done?
15. With the help of a block diagram, explain the working of a basic video cassette recorder on record and replay.
16. Differentiate between a VCP and a VCR. Briefly explain the most-often-used controls on a VCR.

26

VIDEO DISC RECORDING AND PLAYBACK

The enormous popularity of the gramophone record and its playback equipment and the rapid growth of television made it an obvious idea to look at the possibility of recording video signals on a disc. The use of a disc, as an information carrier, has the advantage that one has immediate access to any program part, but of even far greater importance is its low price by using a method of production which could be similar to that of the gramophone disc.

Research on this subject had started in Philips laboratories by 1969. Earlier investigations in the same field were made, but it was only the availability of the laser as a light source with very high brightness and manufactured in an industrial way that offered the possibility of a practical realisation. It was in September, 1972 that the first demonstration for the International trade press was given with a laboratory set-up and using a master disc as the information carrier.

Subsequent demonstrations in Berlin in 1973, Tokyo in 1974 and New York in 1975 proved the feasibility of the system and created great interest. The object of these demonstrations was to prove the superiority of the optical system over the alternative systems and thus promote the creation of one standard.

26.1 OPTICAL RECORDING MEDIUMS

The storage of information by optical methods has many advantages over the conventional method of magnetic recording. As optical recording had obviously much to offer, an intense research began for materials on which information could be stored with the aid of a laser. Three classes of materials that seem suitable for the optical recording of information are: tellurium-selenium alloys, organic compounds and magneto-optical materials.

26.2 TELLURIUM-SELENIUM ALLOYS

One of the new materials for the storage of information is a *polycrystalline tellurium-selenium alloy to which small quantities of other alloys (arsenic, for example) are added to give better control of the melting point and the stability of the material.* A thin layer of the alloy is applied to a substrate. A narrow laser beam is used to melt this material locally so that holes are created with the same depth as layer. *During the readout process, with a less intense laser beam, the presence or absence of holes produces differences in the reflection of laser light. These differences in reflection represent the information in coded form.*

The disc with a tellurium selenium alloy is ideally suitable for use as a *storage medium* for both digital data (alpha-numeric information or digital audio) and video recording.

The use of tellurium-selenium alloys also makes it possible to *record* information on a disc, *erase* it and then use the disc again to *record new information*. By choosing the energy output of the laser appropriately (compared with the level necessary for the 'hole' disc) the polycrystalline material is *melted* locally, but no holes are formed. After the laser pulse, the molten areas cool down so quickly that they solidify in a metastable amorphous phase. These *amorphous domains* reflect differently from the crystalline surroundings on readout. *Erasure* takes place when a laser with sufficiently high energy level transforms the amorphous domains in the crystalline phase.

In most applications, the disc can be *used and erased many times*. In principal, storage of both digital data and video recording is possible because of the high S/N ratio.

26.3 ORGANIC COMPOUNDS

Organic dyes exist that absorb a great deal of light and *have a high reflectance even when applied in thin layers*. These thin layers of organic compounds seem to be a promising alternative to tellurium-selenium alloys. The *memory effect* is again obtained by melting the material locally with a laser to create small *pits*. The difference from the tellurium -selemium alloy is that *these pits do not normally penetrate through to the substrate. The reflectance varies with the depth of the pit. The difference in reflection created by the pattern of pits is used when the information is being read. The melting process is irreversible, so the disc can only be written once.*

It has been found that these organic compounds retain the information just as well as the hole discs with tellurium-selenium alloys. These compounds have also been found to be highly resistant to heat and moisture. Another attractive feature is the simple *spin-coating process* for applying organic compound to the disc.

This type of disc has many applications. *The S/N ratio obtained experimentally is high enough for the storage of both digital and video information.*

26.4 MAGNETO-OPTICAL MATERIALS

Amorphous magnetic gadolinium-iron-cobalt compounds have been known for a long time. *A laser beam can be used to heat the material locally, reverse the polarity of small areas and freeze it in this state.* This technique makes it possible to *write* on a magnetised layer in a pattern of areas of opposite magnetisation directions. *This type of pattern can then be read out with polarised laser light.* The direction of polarisation of the reflected light is rotated slightly with respect to the polarisation of the original laser beam as a result of the Kerr effect. The written areas on the disc can, therefore, be distinguished from the unwritten areas and information can be read out. *The information can be erased just as easily as it is written.* The areas to be erased are heated by the laser, while an external magnetic field is applied with the same direction as the original magnetisation of the layer; the magnetisation of the heated area reverts to its original direction after cooling. *The information can be written, erased, and rewritten as often as required.*

At present the S/N ratio is only moderate, so this information storage method is suitable for digital data only (alpha-numeric information and digital audio signals). It could very well be possible to improve the S/N ratio sufficiently for recording of video signals.

26.5 EVOLUTION OF VIDEO DISC

The concept of a gramophone record that produces pictures first became a reality in the 1920 s, long before the development of the first video cassette. John Logie Baird successfully created a mechanical television system by recording simple visual signals on an ordinary 78 r.p.m. gramophone record and in 1935 his invention went on sales in Selfridges. Purchasers of Baird's video disc were able to view various British public figures in profile. The invention failed to capture the imagination of the public and was eventually withdrawn.

Fig. 26.1 A picture produced on a receiver by one of Baird's video discs in 1938. It lasted several seconds.

It was 35 years before the first video disc appeared but it was never actually launched on the British market. It was due to have been marketed by a company called Teldec, jointly by the German company *Tele*funken and *Dec*ca in the U.K. Made of plastic and with a playing time of only ten minutes the disc was named *TeD*; however due to a mixture of technical problems and disagreement between Teldec partners, it never became commercially available. The Teldec debacle, however, set the video disc ball rolling, prompting a burst of activity from the electronics companies around the world as they attempted to improve upon the original idea.

Philips was the first off the mark unveiling its video disc at a press conference in Eindhoven in 1972. It has taken the company another ten years to introduce the first commercially available video disc and video disc player in the U.K.

In the United States, RCA introduced its *Selecta Vision system*, and in Japan government pressure for technical standardisation finally produced some results with JVC developing its *VHD video disc system* (VHD is the short form of *Video High Density*).

The use of disc, as an information carrier, has the advantage that one has immediate access to any program part, but of even greater importance is its *low price* by using a method of production which could be similar to that of the gramophone disc.

It was only the availability of laser as a light source with very high brightness and manufactured in an industrial way that offered the possibility of a practical realisation.

Subsequent demonstrations in Berlin in 1973, Tokyo in 1974, and New York in 1975 proved the feasibility of the system. The object of the demonstrations was to prove the *superiority* of the optical system over the alternative systems and thus create the promotion of one standard.

In accordance with this endeavor, an arrangement was reached with Pioneer in 1974 to combine efforts in further developing the system. At the same time there was the numerous advantage of the availability of a wealth of software for the Pioneer and Philips video disc systems, since known as *Pioneer Disco-Vision System* and *Philips VLP system* (Video Long Play). The Pioneer and Philips video disc system is much more than just an entertainment, item. *It is a new communication system offering the consumer information of his own choice at a time that suits him.* Pioneer Laser Disc Player is shown in Fig. 26.2.

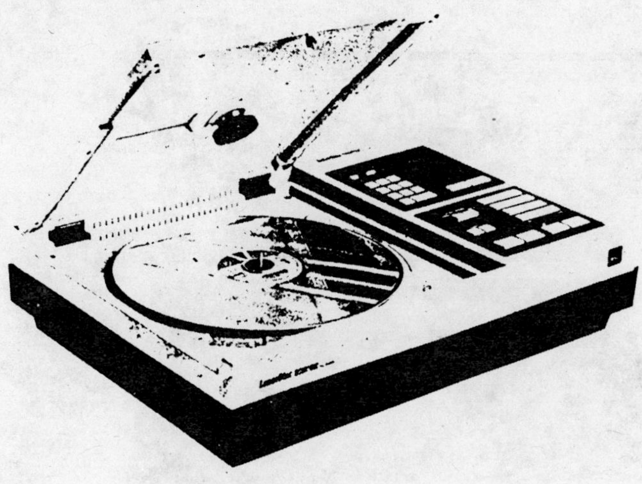

Fig. 26.2. Pioneer Laser Disc Player

The superiority of the system originates from the application of optics for reading out the information on the disc. Due to the absence of a physical contact between the disc and reading out device, *no wear occurs.* There is an effective *protection of information* on the disc against finger prints, dust and scratches. In this respect the video disc is superior to the gramophone disc! Finally, the optical system offers a number of *playing modes* to be utilised that are inconceivable with mechanical systems.

26.6 VIDEO DISC

The Pioneer and Philips video disc resembles very much in its physical shape a gramophone record Fig. 26.3. It consists of a transparent plastic material with *standardized diameters* of 30 cm and 20 cm and a *thickness* of 1.1 mm. The most striking differences with an audio record are *its mirror like appearance due to a reflective coating* and the fact that *it can be played on one side only.* The basic difference however, lies in the *structure of information.* Whereas the audio disc has grooves, the walls of which are modulated with an audio signal, the video disc, in view of the requirements of a much higher information density, has *tracks with a much finer information* and with a *spacing that is about 60 times smaller than that of the audio disc.*

Two basic types of disc are available; one rotating at 1800 r.p.m. for areas with *NTSC* television system and another rotating at 1500 r.p.m for areas with *PAL or SECAM* television system. For obvious reasons, the two types are not interchangeable.

The information on the disc exists as a *spiral track* starting at the inside of a fixed diameter and moving to the outside. Average *track pitch* is 1.6 μm, resulting in a playing time of approximately 30 minutes for a 30 cm diameter disc. The information in the track consists of small depressions called *pits* with variable distance and length. *Pit width* is 0.4 μm, where as *pit depth* is approximately 0.1 μm. Instead of depressions, other elements suitable for absorption or dispersion of light can also be used and do exist mainly as a result of the mastering system used.

The *total track length* on a 34 cm disc is about 34 km and the *linear speed of tracking* varies from 37 km/h at the innermost track to 100 km/h at the outermost track. One complete television picture (two fields) requires a *surface area* on the disc of 0.5 mm at the innermost track.

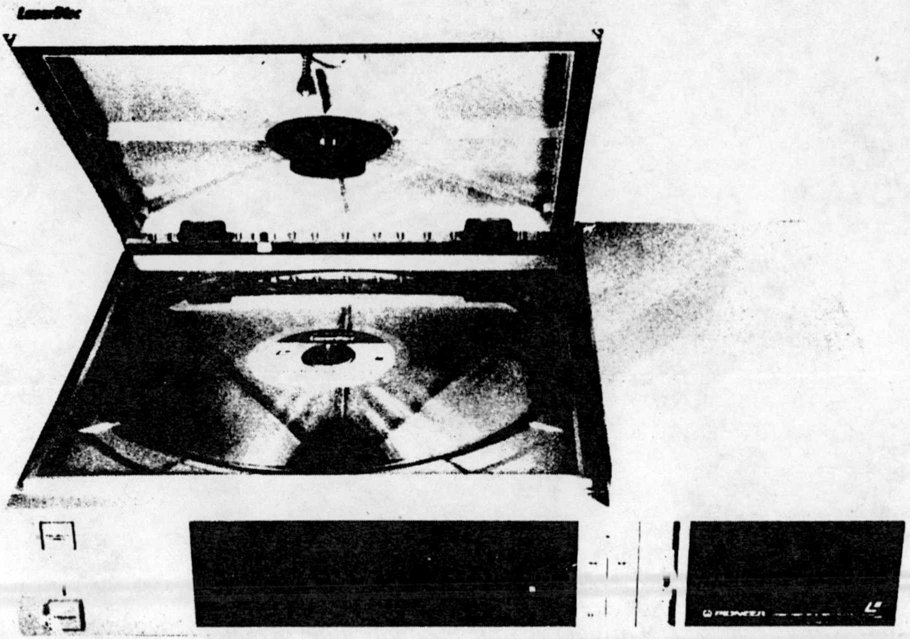

Fig. 26.3 The video disc resembles a gramophone record but has mirror-like appearance. The player is only slightly larger than a normal hi-fi turntable.

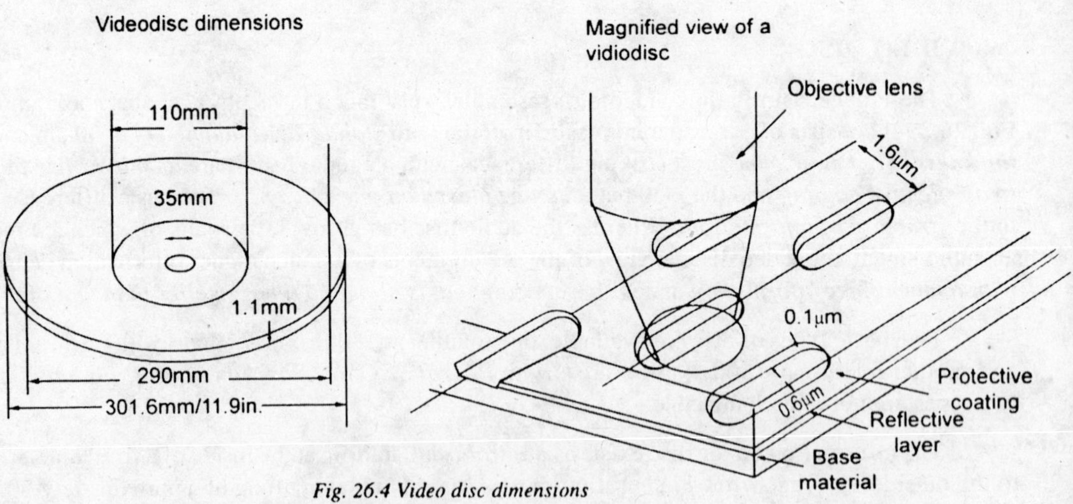

Fig. 26.4 Video disc dimensions

26.7 VIDEODISC MASTERING AND REPLICATION

The production process of a video disc is more or less comparable with that used for conventional gramophone record. First a *master recording* is made. It consists of a glass plate with a photosensitive layer deposited on one side. The coded signal of the information to be stored *modulates* the beam of a 1 mm laser which writes the information in the surface of the disc. Cutting is done on *real time basis;* that is it requires only as much as the program lasts and recording takes place at the disc's *rotational speed* of 30 r.p.s. for NTSC and 25 r.p.s. (revolutions per second) respectively for PAL and SECAM. In principle, every normal type of TV signal source can be connected to the cutting device. *In practice however, 50 mm magnetic recording tape is used as a program carrier.*

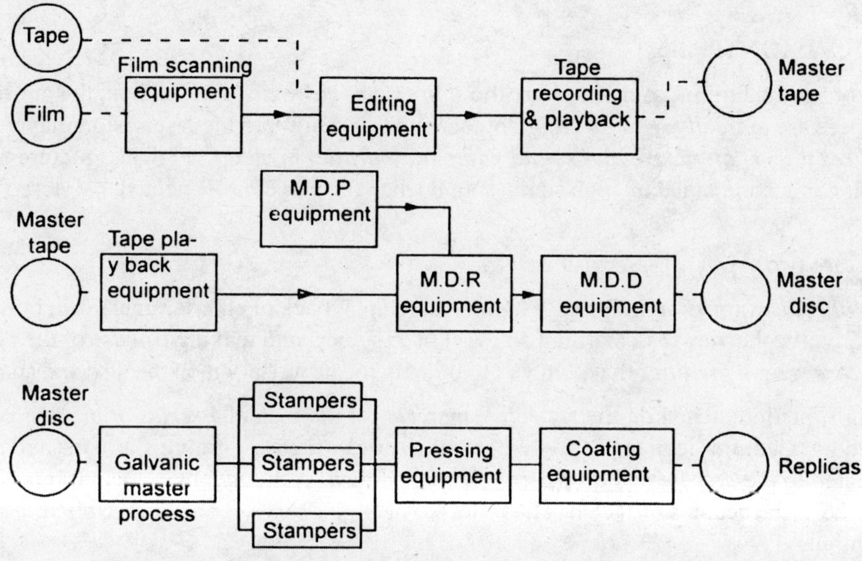

Fig. 26.5 Video disc mastering and replication.

Exposure to the laser beam is followed by a development process which leaves a pattern of pits on the master from which, via a galvanic process, *stampers* are made which are used for disc production in a way similar to processing of gramophone records.

After processing, an extremely thin metal coating, not more that 0.04 μm thick is deposited on the information side which is then sealed with a *protective layer*, as shown in Fig. 26.6.

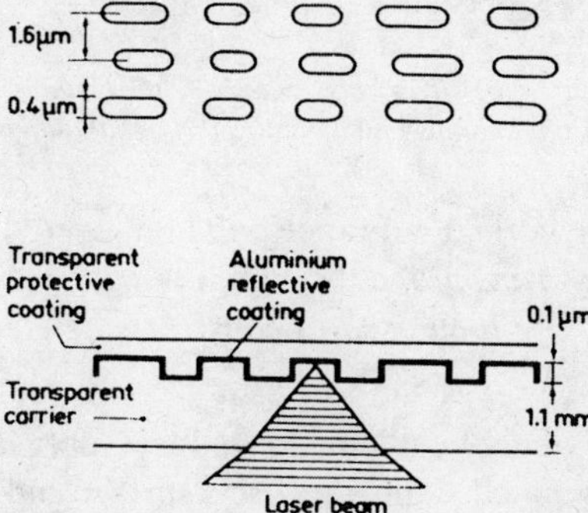

Fig. 26.6 The upper half of this diagram shows the tracks of pits, while the lower half shows a cross section of the disc indicating how the information pits are protected by a transparent protective coating.

26.8 VIDEO DISC FORMATS

The three *non-interchangeable* video disc formats fall into two basic categories ; optical and capacitance. The *laser optical system* (also called VLP) which is employed by Philips, uses a laser beam to recover the electronically encoded information stored on the disc. *The capacitance system (also called capacitance electronic disc or CED), employed by both JVC and RCA,* uses a stylus and tracking arm similar to that of a conventional record player to recover the information recorded on the grooves of the disc. There are two variations of the laser optical system *reflective* and *transmissive*. There are also two variations of the capacitance system, the *video/audio high density system* (VHD) and the capacitance electronic disc system (CED).

26.9 LASER VISION (Philips)

The core of the Philips laser vision are the *shiny silver video discs* which hold the pre-recorded program material. These are *optically read* by a high precision laser, built into the laservision player. *The laservision system features a level of picture and sound without a parallel in video.* Flawless picture reproduction, full of rich colours, is accompanied by high stereo sound when the player is connected to a stereo TV or hi-fi audio system.

There are two types of Laservision discs

Long play discs for straight forward uninterrupted playback of entertainment programs and *Active play discs* which involve the viewer in learning a new skill or subject and making full use of the player's many and versatile *picture search facilities* that help you to quickly locate any point on the disc and study parts in detail.

The information etched on the *reflective material* of the disc in a *series of pits* is read off with a fine beam created by a helium-neon *laser*. By means of a series of lenses, gratings, prisms and mirrors, the laser beam is directed to the *disc underside* where it is moved by a scanning lens. When it hits a pit in the silvery surface, Fig. 26.7, the beam is reflected. The reflected beam then passes back to the photodiode which creates the signal output.

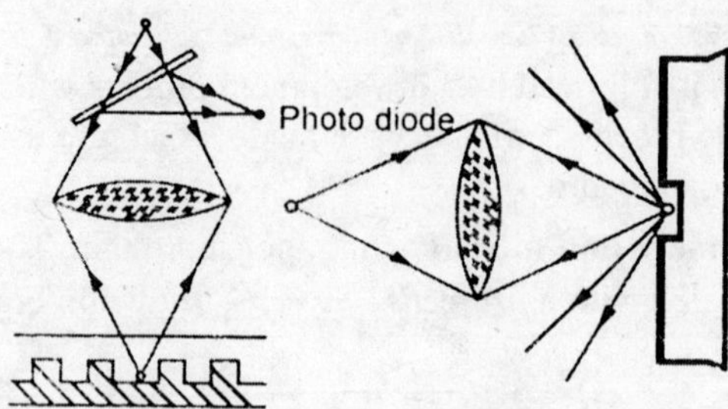

Fig. 26.7 When the scanning beam hits a pit in the silvery disc underside the beam is reflected. The beam then passes back to the photodiode which creates the signal output.

In a similar systems, which is worked in France by Thomson CSF, the laser beam is not reflected from the surface of the disc, but is actually *transmitted* through it. *Variations in the quality of the beam, created by the information etched on the disc, are then used to construct the electrical signals needed to create a TV picture.*

Any Laservision disc may be freely handled, cleaned and replayed as often as you wish. The program material is held in billions of tiny pits punched into the disc's surface and then covered by a transparent material which protects it against dust scratches and finger marks. Because the disc is scanned by a laser light beam, *its life is virtually infinite*. Laservision discs are marketed by a steadily increasing number of leading program distributors under their own *labels*.

26.10 SELECTAVISION (RCA)

The *cheapest and the most basic* of the three video disc formats, with the current models offering only mono sound and picture search, Selectavision discs, based on the capacitance system, have a maximum *running time* of one hour per side.

The RCA stylus recovers information form the disc surface by *direct electrical means*. The Selectavision disc is either formed from or coated with electrically conductive material and the stylus serves as a conductive electrode. Although the groove appears to be smooth, it does have *tiny pits along the bottom* which produce changes in electrical capacitance between the disc surface and the stylus electrode. These *changes in capacitance* are sensed by a tuned circuit to produce an output of video signals.

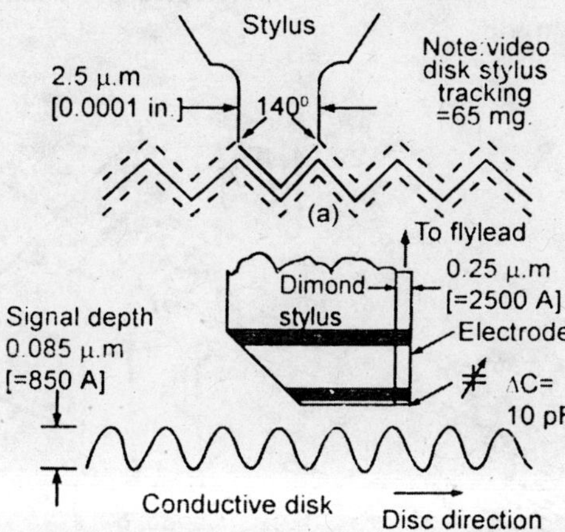

Fig. 26.8 The RCA Selectavision system employs a capacitance sensing stylus electrode, which is guided along shallow grooves and detects capacitance changes as it passes over shallow pits. (a) Front view (b) Side view.

The TeD and RCA systems, rely on a stylus following a minute surface groove and *fail to function if the disc surface is damaged*. For this reason, the floppy TeD disc is always stored in a *sleeve* which is itself loaded into the player for automatic extraction of the vital foil, the sleeve remaining inside the player until the disc has played. RCA discs, which must also be warp free and are rendered useless by finger marks, which confuse the capacitance, effect are stored in a *caddy* from which the player automatically extracts the disc for playing.

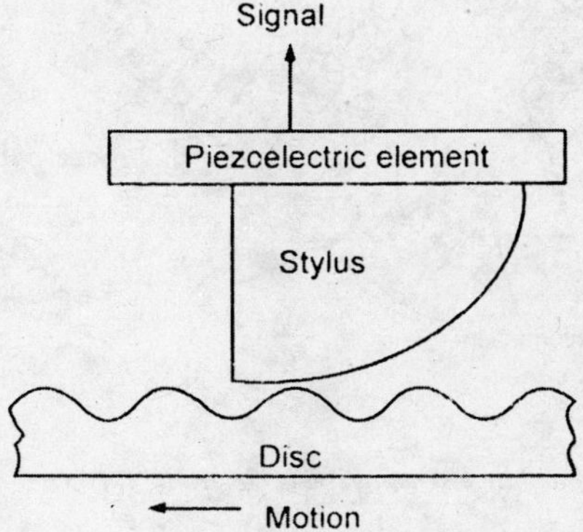

Fig. 26.9. The Ted employs a stylus shaped like the prow of a boat which senses mechanically the hill-and-dale modulations that contain the information.

26.11 VIDEO HIGH DENSITY (JVC/Thorn-EMI)

Technically and price wise, VHD falls between Laservision and Selectavision. The JVC system resembles the RCA approach in its use of *capacitance pickup* from an electrically conductive disc with a spiral of pits on its surface. But the 25cm (10 inch) JVC disc has a *smooth surface with no groove*. The stylus electrode is constrained to follow the spiral pits by a servo system. The spiral of *program pits* (which the electrode stylus tracks to produce pictures on TV screen) is *interlaced* with a spiral of *tracking pits* which are sensed to control the servo system to guide the electrode.

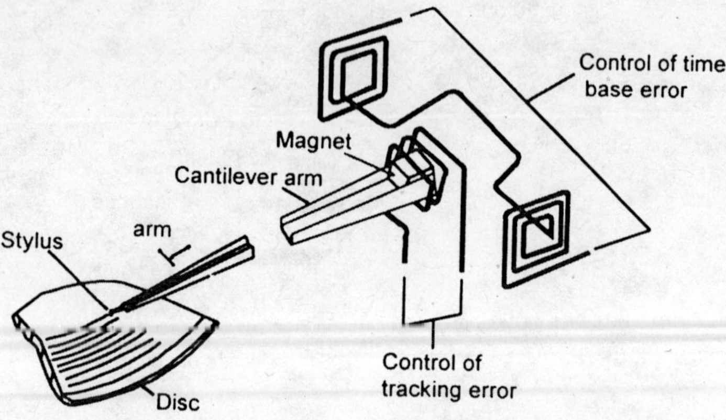

Fig. 26.10 Video signal pickup in VHD system uses a cantilever arm.

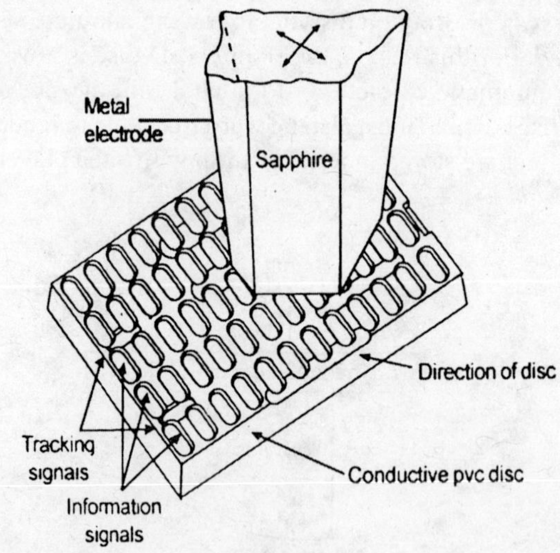

Fig. 26.11 The JVC system also employs a capacitance sensing stylus electrode. It is guided along the correct path by servo pits.

Fig. 26.10 illustrates the method of controlling the stylus. The stylus is mounted at the end of a cantilever pickup arm opposite that end on which a magnet is attached. Fixed coils are positioned near the magnet: a single coil is wound around but not in contact with the magnet, and a pair of vertical coils are positioned, one each on either side of the coil, and in phase opposition to each other. Thus, *the stylus can move transversely and longitudinally in response to the particular current flowing in these coils. The current is varied by a tracking error signal*, or by a command to move the stylus to a desired track, permitting various functions during playback.

The JVC disc rotates at twice the speed of the RCA disc, and gives a playing time of one hour each side. It can also provide *instant replay* of selected short passages. *Freeze frame* is possible but not easy, for example, it requires a memory to store one of the two frames recorded for each revolution of the disc

Because there are no *grooves* and *no tracking stylus to follow them*, the JVC electrode is required only to *slide over* a smooth disc surface and will doubtless have a *longer life* than any other system employing a stylus in groove. However, no one yet knows how long a disc will last, especially when used to display *still frames*. With any contact system there is inevitably some abrasion.

Although the JVC system is relatively simple to construct, it requires *complex servo control system* which may put the price higher than the mechanically very simple RCA system.

26.12 VIDEO DISC SYSTEMS—A COMPARISON

Common to all video disc systems is the process in which a *program* (originally recorded on magnetic tape) is recorded on to a *master metal disc*. The metal master is then used to mass produce *plastic discs* which are played on the video dics player (VDP).

Plastic discs for the *laser video (LV) optical pickup system* are coated with metal on one side (the recorded surface) and then *bonded* with the metal inside for protection. Carbon is added to discs for the two *capacitive pickup systems*. CED and VHD, to make the disc conductive. A lubricant for smoothing the pickup and reducing wear is added to the CED discs. This *lubricant* is necessary because the CED system has grooves on the disc for stylus tracking. This makes the CED system simpler (no servo tracking is required) but does produce some wear. However, the wear on a CED disc in no way compares with the wear of a conventional audio record. RCA demonstrated this by playing a single groove of a CED disc 9000 times without noticeable deterioration in the video display.

The three video disc systems also have similarities. *All three systems use a plastic disc rotating on a turntable. In all systems, the player picks up information represented by changes in the disc surface and converts the information into signals for playback on a television set. All systems use frequency modulation (FM) for both video and audio signals. Each disk has a spiral track to carry the information rather than a series of circular tracks.*

In spite of all the basic similarities, *the systems differ not only in the pickup technique (optical versus capacitive) but also in the format in which the information is encoded and in the method by which information is tracked. Other differences include size, material, rotation speed and signal-protection schemes.*

26.13 RECORDING SYSTEM

In the optical video disc there is a *single information track* in which all the information is stored for the reproduction of a colour television program with two sound channels and data signals.

The *non-linearity* of the master recording process limits the choice of possible encoding techniques and a *two-level signal recording* was found to be the most attractive solution. On this track *the information is encoded in the length and the spacing of the pits* or, in other words, for a rotating disc in the repetition frequency, determined by the average length of the pits, and a pulse width modulation of the frequency, determined by the modulation of the length of the pits, Fig. 26.12

The *composite video signal* employed in the video disc system is frequency modulated on a carrier at 8 MHz which is pulse width modulated by two hi-fi audio channels at 2.3 MHz and 2.8 MHz.

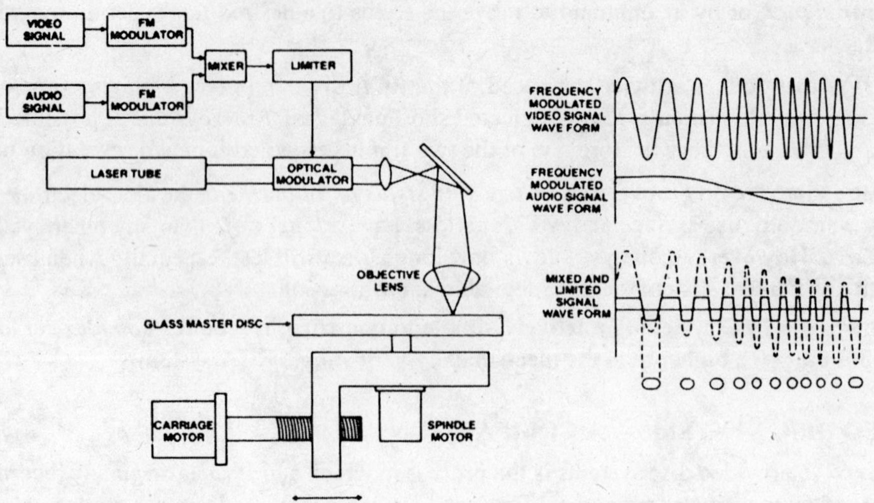

Fig. 26.12 Video disc recording system

Fig. 26.13 shows the block diagram of the signal processing for coding the video and audio system. Before FM modulation of the *video signal* pre-emphasis time constants of 50 μs and 12.5 μs are employed. The *audio signals* are FM modulated on carriers of 2.3 MHz and 2.8 MHz with a frequency deviation of ± 100 kHz and a pre-emphasis of 75 μs. The two audio carriers are summed with the FM carrier and after limiting, the output signal is used to modulate the intensity of a laser beam passing through an electro-optical modulator in the master recording machine.

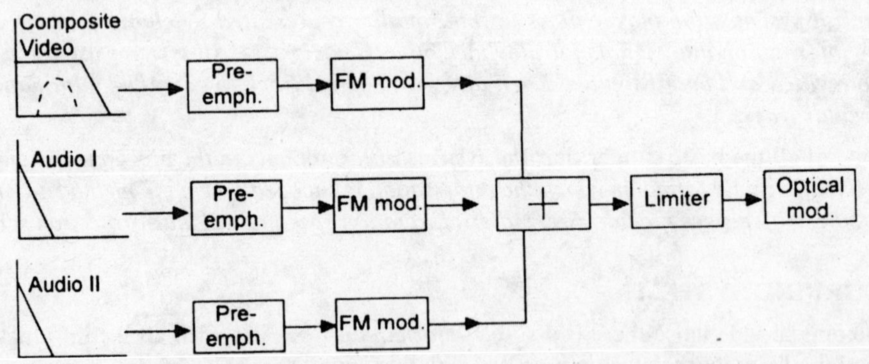

Fig. 26.13 Signal processing—Encoding.

The *spectrum* of video and audio signals is given in Fig. 26.14. The *master* is used as a starting point for the production of discs.

26.14 PLAYBACK SYSTEM

In reading back the information, the reflected light returning from the disc falls on a photodiode and its output is *amplified and corrected* according to the frequency characteristic of the player. A high-pass filter separates the video information and the filters have a crossover frequency at 3.5 MHz. The separated FM signals are then demodulated and a *de-emphasis* is applied to compensate for the *pre-emphasis* employed before recording, in order to achieve a better S/N ratio and a more uniform frequency response. The playback system is shown in Fig. 26.15.

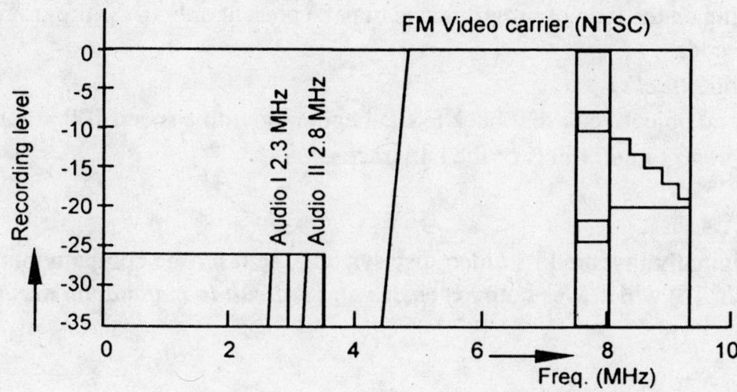

Fig. 26.14 The spectrum of video and audio signals. Black level is at 8.0 MHz white level at 9.2 MHz.

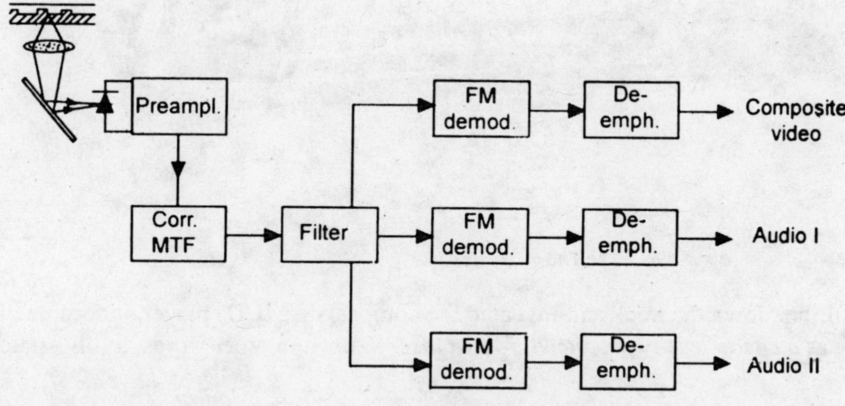

Fig. 26.15 Signal processing—Decoding.

26.15 DIGITAL ADDRESS SIGNALS

Apart from video and audio information, the disc contains a number of signals, *inserted in the non-visible lines of the blanking periods*. These are in the first place *Vertical Interval Reference (VIR) and Vertical Interval Test (VIT) signals for testing purposes (lines 19,20,282,283) and secondly a* number of digital address signals for various purposes (lines 17,18,280, 281).

The *address signals* have the following functions:

A. *Lead-in tracks (min 600 tracks = 20 seconds)*

Start code: The "read" objective is sent with normal speed forward to the program area on the disc.

The " start" code is present in all fields of the lead-in tracks.

B. *Program area*:

1. *Picture Code*: Consists of a picture number and a key for either continuous play in the normal forward mode or a stop key for automatic stopping in the "freeze frame" mode.
2. *Chapter Code*: Consists of a chapter number and a key for either continuous search during depression of the search button or a stop key for automatic stopping of the search action. The chapter code is optional, depending on the type of program, and may be present only in the program area in those fields without a picture code.

C. *End tracks (min 600 tracks)*

End Code: The read objective is sent back to the beginning with a speed 100 × normal speed.

The end code is present in all fields of the end tracks.

26.16 SOLID-STATE LASER

When Philips originally invented its video disc system, the laser the company planned to use was an older, bulky type (Fig. 26.16) which was both *expensive and difficult to manufacture and align*.

Fig. 26.16 The original 'bulky' laser

As Philips went further down the track and invented the compact disc (CD) player, it became obvious that what was required was *a cheap, easily produced, robust laser system*; in other words, a solid-state laser.

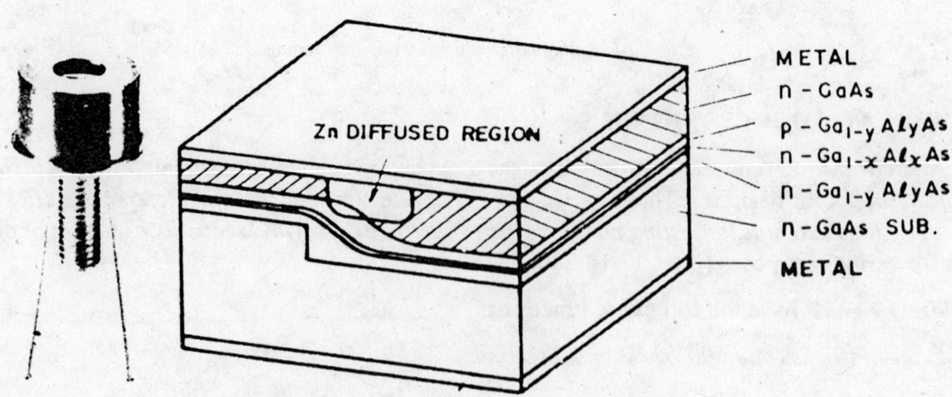

Fig. 26.17 Matsushita's terraced substrate (TS) semicondutor laser.

The first commercially viable, and commercially available solid-state laser came from Matsushita, Japan's largest electronics concern. Called the *MEL 4745 laser*, it was developed by the specialist semiconductor division of Matsushita, Matsushita Electronics Corporation. It is shown in Fig. 26.17.

In order for a semiconductor laser to be ideal for use in optical information processing systems, the following factors are essential:

1. Low threshold current
2. High power-output
3. Stable single transverse mode oscillation
4. Circular radiation pattern of laser beam
5. Long life

Conventional semiconductor lasers, however, could not satisfy all of these conditions. Matsushita succeeded in satisfying all of the above key factors by adopting a unique terraced substrate(TS) construction. *The TS laser can confine laser beams in a far narrower wave guide - resulting in higher power output with lower threshold current.*

In the fabrication of TS lasers, the face of the n-GaAs *substrate* is *terraced* by chemical etching, on which a double heterostructure is formed by successively growing an n-GaAlAs *cladding layer*, a GaAl. As *active layer*, a p-GaAlAs cladding layer and an n-GaAlAs *isolation layer*.

The active layer is grown thick at the terraced part, forming a modified rib-waveguide structure. *Such a structure is very effective for confinement of a laser beam and for a stable single transverse mode oscillation.* The TS structure also improves the concentration of injection current in the active region resulting in a remarkable decrease in *non-effective current*. The ratio of the *effective current* flowing through the terraced active region to the total injection current is in excess of 80 percent.

Through these improvements, the threshold current has been reduced to a level of lower than 30 percent of conventional lasers while still keeping a high power output.

The *TS laser MEL 4745* has a built-in PIN photodiode and can be easily connected with an APC circuit to keep the light output constant.

Characteristic Features-TS Laser MEL 4745	
Threshold current	30 mA
Light output	5mW
Laser wavelength	810 nm
Radiation angle	13 degree (parallel) 30 degree (perpendicular)
Mode	Single transverse mode oscillation
Built-in PIN photodiode.	

26.17 FOCUSING

A thick video disc with a thickness of 1.1 mm rotating in free air at 30 Hz is generally *stable but not critically damped*. In order to restrict the action of transverse waves in the frequency range 0-15Hz and also to compensate for the "umbrella" form deflection of the disc under its own weight, a *stabilisation system* has been incorporated. *Stabilisation is achieved in the player by means of an air bearing, which is created by the centrifugal action of the rotating disc on the surrounding air.*

A plate with a central hole is positioned in the vicinity of the rotating disc's surface. This creates an air bearing having both the desired damping and stiffness characteristic. Disc stabilisation obtained in this way is by no means sufficient to guarantee the required accuracy on the *distance* between the disc and read objective.

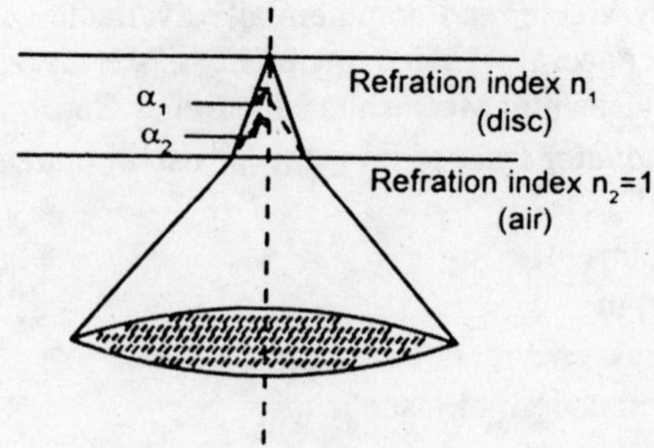

Fig. 26.18 Numerical aperture NA $= n_1 \sin \alpha_1 = n_2 \sin \alpha_2$

In order to read the microscopically small information details on the disc, an objective lens with a large numerical aperture is required, which thus has a small depth of focus. *Numerical aperture (N.A) is defined as the product of the refraction index and the sine of the angle between the optical axis and the outermost light ray contributing to the imaging* (Fig. 26.18). With N.A = 0.4 a maximum *out-of-focus* allowed is 2 µm. In view of tolerances in the disc and player construction this accuracy can only be realised by means of a *servo control system* including a *moving read objective* that can follow the undulations of the disc.

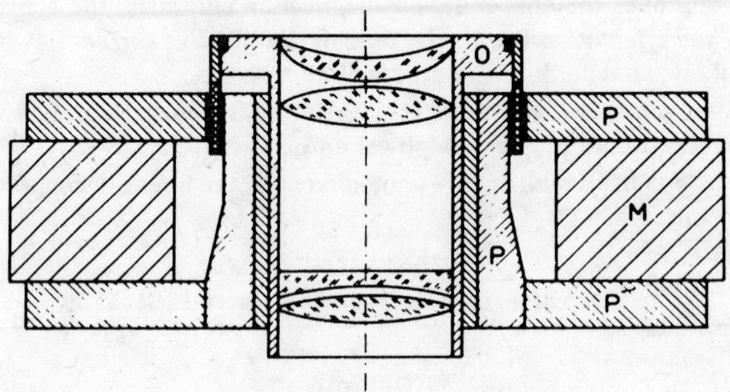

Fig. 26.19 Principle sketch of moving objective.

As can be seen in Fig. 26.19 the objective lens is mounted in a system similar to that of a loudspeaker voice coil and thus operates according to the *electrodynamic principle.* Depending on the direction and magnitude of current through the coil it executes *controlled vertical movements* in order to follow the irregularities.

To understand the way in which the *control signal* is obtained, it is necessary first to have a closer look at the photodiode which detects the light beam reflected by the disc. In fact this diode consists of two segments E and F and a third one between the two which is composed of four quadrants A, B C and D (Fig. 26.20). The light beam focused on the central segment will normally create a *circular spot* and thus all four segments will receive an equal amount of light. The sum of the electrical signals over these four quadrants is the RF signal or video-information.

However, on its way to the diode the reflected beam passes a *cylindrical (astigmatic) lens.* As a result, the light spot on the diode which normally has a circular shape obtains an *elliptical deformation* if the information plane in the disc changes its distance from the objective. If this occurs each quadrant no longer

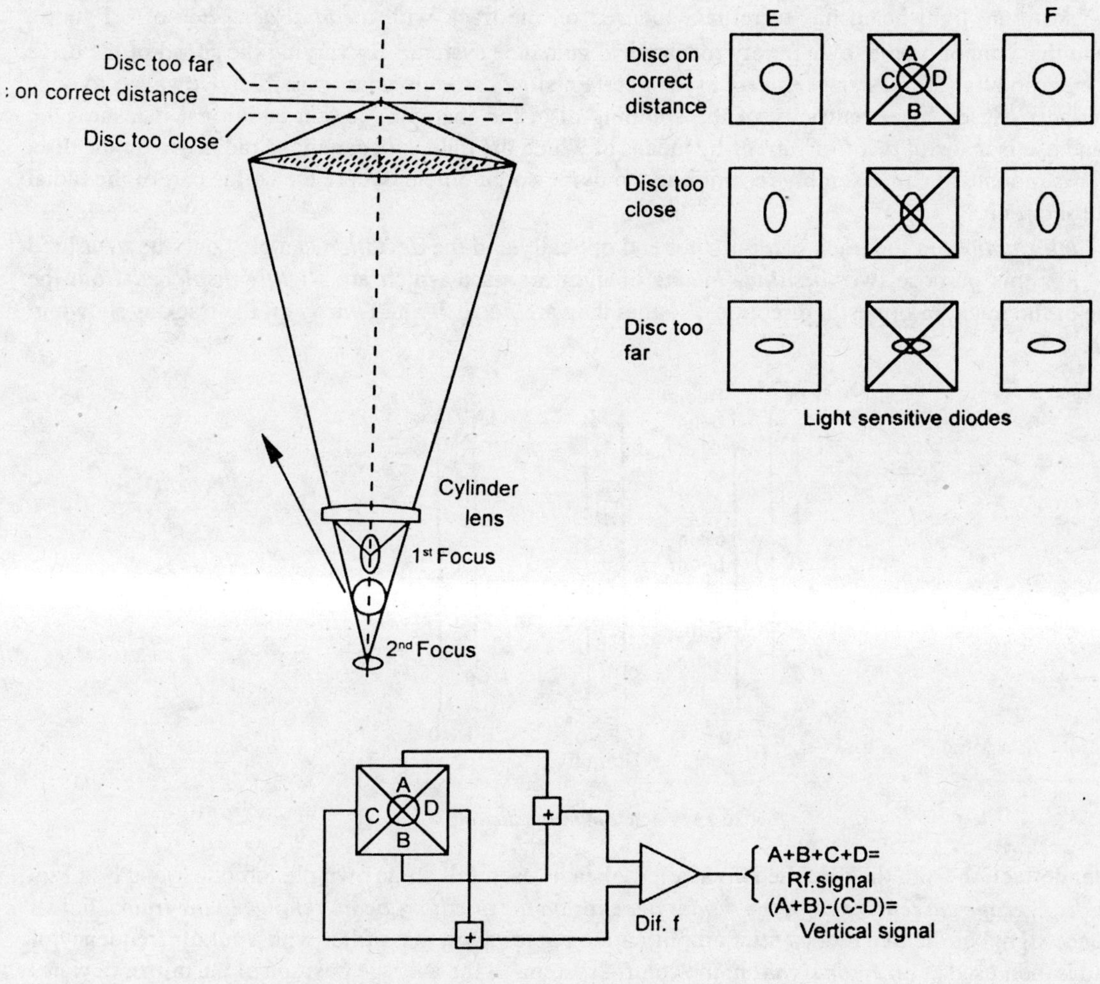

Fig. 26.20 Principle of astigmatic focusing

receives an equal amount of light and so a difference signal can be derived. *This difference signal, once it has been amplified and processed, can be used to change the position of the objective lens with respect to the disc.*

The largest vertical movements of the disc occur at the speed of rotation and decrease rapidly at higher frequencies. Disc specification sets the maximum value at 10 g but the player can cope with accelerations as high as 14g and still remain below an out-of-focus value of 2 μm.

26.18 RADIAL AND TANGENTIAL TRACKING

The information on the disc is contained in a *spiral that is read from the inside to the outside.* For this purpose the 'read' objective and other optical elements are mounted on a *sledge* driven by a small DC motor, and moving radially under the disc. With an average track pitch of 1.6 μm and the disc rotating at 1800 r.p.m. this means an average linear speed of the sledge of 3 mm/m.

The scanning light beam has to remain focused on the track with a *radial accuracy* of 0.1 μm, a requirement that cannot be met by a purely mechanical guidance system. By varying the speed of the drive motor by incorporating it in a *servo control system* certain slow corrections are possible. However, to cope with the effects caused by eccentricity of the spinning disc, for example, certain additional measures are required and use is made of *pivoting mirror* by means of which the light spot can move radially over the disc. This mirror is mounted in an assembly resembling a moving coil ammeter where the coil is part of the radial servo control circuit.

The *information* on the track can only be read optically and the *deviation* can also only be measured optically. For this purpose two *auxiliary beams* of light are used which are *slightly displaced* from the centre-line of the track, in opposite directions, so that they are *partly on* and *partly off* the track as shown in Fig. 26.21.

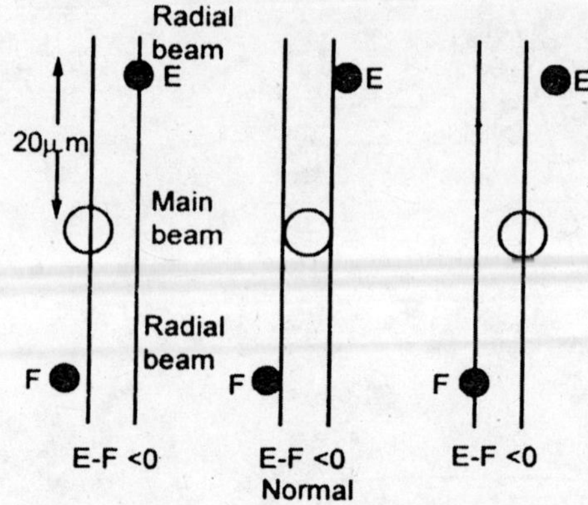

Fig. 26.21 Principle of track following.

After deflection from the disc the two availiary beams each fall on its own photodiode E and F in Fig. 26.21 and *the average current through the diodes depends on the auxiliary beam relative to the track.* In fact the difference signal of the two diodes after amplification passes a low-pass filter with a cut off frequency of 20 kHz and is then used as an *error signal* in the control system. If the average position of the mirror deviates from its zero position, the average current is used to contiol the slide mirror for correction of the position of the sledge.

26.19 BLOCK DIAGRAM OF VLP PLAYER

From the discussion so far, the larger part of the block diagram will be self-explanatory. A few items, however, require an explanation.

In the event of a *drop-out* (*an interruption of the signal due to any imperfection in the disc*) a drop-out detector operates as a switch by means of which *a parallel channel, fitted with a line scan delay, can provide information of the previous line*. Of the two available audio channels, either one of the two or both (in the case of stereo), can be connected to the VHF modulator. Since normal television receivers are monophonic, the two demodulated audio signals are available at output sockets at the back of the VLP player. These can be connected to a stereo amplifier.

26.20 VLP PLAYER CONTROLS

The player Fig. 26.23, is housed in a plastic case of modern design. All operating controls but one are arranged on a sloping panel at the front, together with a number of *LEDs indicating which function has been switched on.*

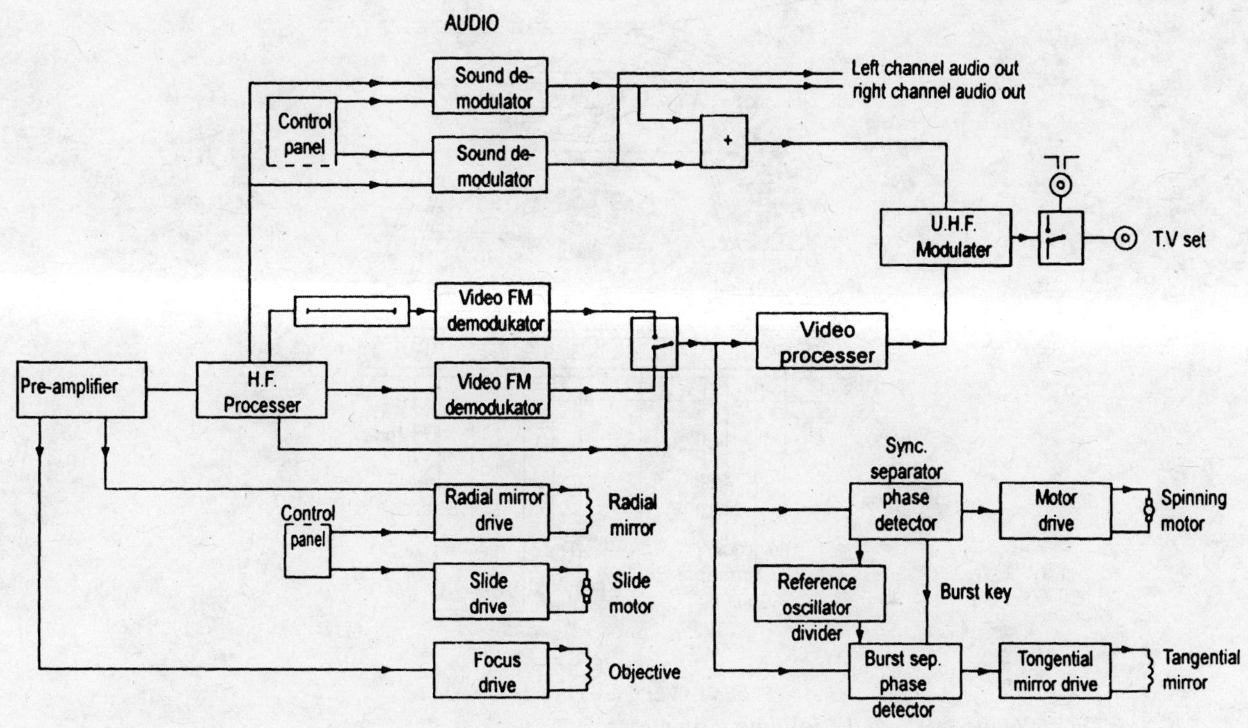

Fig. 26.22 Block diagram of VLP player.

Measures have been taken to prevent opening the lid when the disc is spinning and or the laser is ON. In order to satisfy the most stringent safety requirements a *double protection against laser radiation is provided for when the lid is open*. Not only is the laser supply switched OFF automatically, but also a small metal shield is placed in the light path. Below the lid and fixed to it is the stabilisation disc and a magnetie device for clamping the video disc on its turntable. The Pioneer system uses a spring clamp to secure the disc.

The spinning motor is a standard DC cordless direct-drive rotor type.

The electronic circuit is divided into *functional units* of the same type as used in Philips colour TV sets. They can be *plugged in and out* easily, thus simplifying fault finding and repair. All optical components, the laser with its ignition unit, the sledge drive mechanism and the preamplifier are situated on the sledge.

In order to take advantage of the versatility of the system, the VLP player has been equipped with a number of operating controls for a variety of functions. Apart from the latch for releasing the lid, all other controls are of the push button type and arranged in a logical sequence.

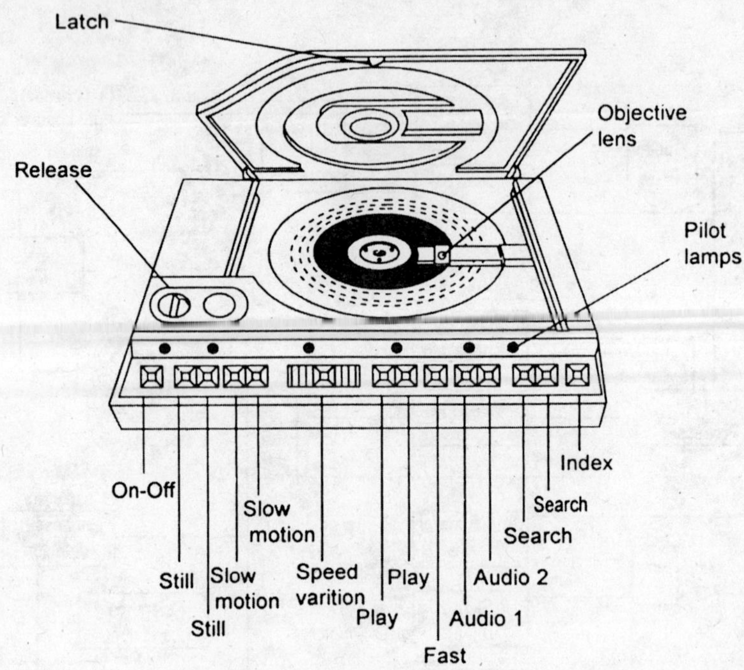

Fig. 26.23 VLP Player operating controls.

On the player we find the following functions:

Mains switch: ON/OFF

Freeze frame: Two push buttons in order to obtain a *still picture*. Repeated depressing of the button *still forward* produces next picture. Repeated depressing of the button *still reverse* gives the preceding picture.

Slow motion: Separate push buttons for *forward* and *reverse* action in combination with a sliding potentiometer *vary the speed* between 30 frames (normal speed) and 4 seconds per picture.

Normal speed: Separate push buttons for *forward* and *reverse* at normal speed.

Fast forward: Forward motion at a speed three times normal. *Fast forward action is obtained only while keeping the button depressed.* After release the apparatus is in the still picture mode.

Sound: The player has *two separate sound channels*, each of which can be switched ON or OFF. This not only gives the player stereo capability, it also means that *dual-language discs* could be used. Using this feature, the sound would, of course, be mono.

Search: Separate push buttons for forward and reverse search through the program with a 100 × normal speed. This means that it takes 18 seconds from start to end of a 30 cm diameter disc. The search action takes place while keeping the button depressed. Release of the button brings back the apparatus in the playing mode it had before activating search.

Index: Picture number and chapter number can be made to appear on the screen by pressing the 'index' button. *When the button is pressed once the chapter number appears, when it is pressed again the picture number appears, and when it is pressed a third time the picture number disappears.*

26.21 AUTOMATIC ACTIONS

A great many automatic functions have been incorporated in the VLP player both to ensure a *flawless performance* of the apparatus under all circumstances and in order to make it *foolproof.* A number of other automatic functions are particularly intended for making it *user friendly*. These are:

Priority mode: After switching ON the apparatus will always be in the normal play forward mode with both sound channels connected.

Program end: At the end of a disc the 'read' objective will automatically move with 'search speed' to the beginning of the disc. The same will happen after interruption of the playing of a disc by releasing the lid.

Program start: If the 'read' objective lens arrives at the beginning of a disc and the lid is closed, the player will start again in the normal 'play froward' mode.

Stop during normal play: If an *appropriate picture code* is on the disc the player stops immediately at that picture and remains in the 'freeze frame' position.

Stop during search: If an *appropriate chapter code* is on the disc, the player stops the search action at the beginning of a chapter. Search is resumed after releasing and subsequent depressing of the 'search' button.

26.22 OPTICAL MEMORY DISCS

Magnetic discs are based on ideas which were current 30 years ago in the recording industry. With a normal record you start playing on the outside run and the track goes round and round in a spiral to the centre. One problem with the cassettes is that you have to record all the information in a *serial form.* So if you have recorded say 8 programs, and want the information stored in the last one, you have got to start all the way from the beginning until you get to the information you want. One way of overcoming this setback is to take a *normal LP-type record approach*, where although you have got serial recording format, you have designated points called *tracks*. As there is a certain amount of random access, it is possible to 'drop in' at any point.

A magnetic disc, Fig. 26.24, is divided up into *discrete tracks* in the same way as a normal LP record. But instead of the tracks being in a spiral *they are all concentric*. A small magnetic recording head, just like a recording head on a small cassette recorder, is then moved from say track 0 to, say, track 35 (Fig. 26.25). The disc unit is instructed that the computer wants to record the next chunk of information on track 5, and it moves the head to track 5 and dumps the information in a similar way to how the information was recorded on a cassette tape, i.e by modulating the 1's and 0's pulses with a frequency that can be recorded magnetically.

A few years ago IBM came along with the *floppy disc*, which is now being made in a whole series of different formats, 8", 5.25" and now 3.5" in diameter. If a sheet of plastic covered with a thin layer of magnetic material is spun at 300 r.p.m. it keeps relatively rigid

The disc is split into a *series of tracks*, which vary from about 35 to 80. Each track is itself split into a number of *sectors*.

The number of sectors vary, but it is usually about 16. In other wards the disc is made of a *series of blocks*. So to store information it must be possible to keep track of which block or group of blocks the information is stored. One of these tracks is designated the *index tracks*, and on that track you store an index of what information is on the disc and where it is. It's like looking up something in a book; there is an index which tells the recording head exactly *where* all the information is stored and usually *how much* information is there. There are many more sophisticated methods where you might have various check digits or other information stored on the index tracks.

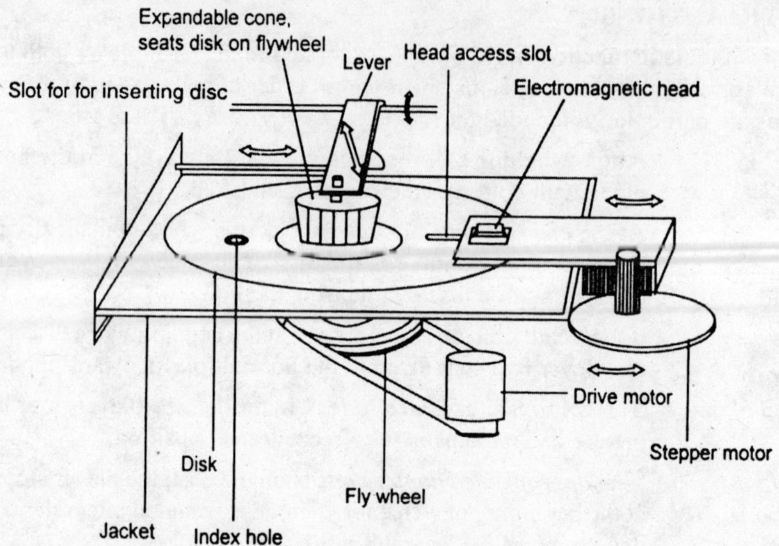

Fig. 26.24 Floppy disc and floppy disc drive.

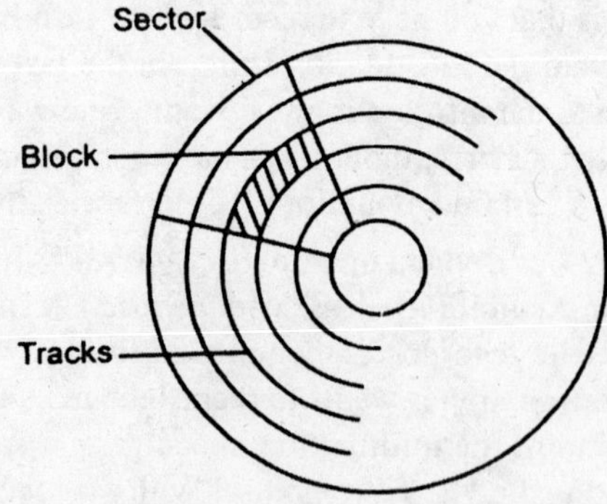

Fig. 26.25 How data is recorded on a floppy disc.

Any one who has used *disc drives* will have heard it *whirring away* making clunking noises. This is the head trying to find the appropriate track. The *size* of each block can vary from between 256 bytes to 1024 bytes. *Unlike a cassette recorder the information can be spread all over the place.*

OBJECTIVE TYPE QUESTIONS (A)

1. The three video disc formats are
   (*a*) interchangeable (*b*) non-interchangeable
2. With the video disc as information carrier one
   (*a*) does not have immediate access to any program part
   (*b*) has immediate access to any program part
3. The Pioneer and Philips video disc can be played on
   (*a*) both sides (*b*) one side only
4. The information on the video disc exists as a
   (*a*) spiral track starting from outside (*b*)groove starting from inside
   (*c*) spiral track starting from inside (*d*)groove starting from outside
5. Pioneer and Philips video discs have standardised diameters of
   (*a*) 10 cm (*b*) 20 cm
   (*c*) 30 cm (*d*) 40 cm
6. There are two types of laservision discs
   (*a*) Standard play (*b*) Long play
   (*c*) Extended play (*d*) Active play
7. The RCA stylus recovers information from the disc surface by
   (*a*) indirect electrical means
   (*b*) direct electrical means
8. The TeD and RCA systems
   (*a*) function even if the disc surface is damaged
   (*b*) fail to function if the disc surface is damaged
9. In response to a tracking error signal the stylus in a VHD system can move
   (*a*) longitudinally (*b*) transversely
   (*c*) longitudinally and transversely
10. With a single information track, the information is encoded in
    (*a*) the length of the pits (*b*) the spacing of the pits
    (*c*) the length and spacing of the pits
11. The old bulky lasers were
    (*a*) expensive (*b*) difficult to manufacture
    (*c*) difficult to align (*d*) (a), (b), (c)
12. The TS laser has
    (*a*) higher power output with lower thereshold current
    (*b*) lower power output with higher threshold current

FILL IN THE BLANKS (B)

1. The superiority of the Laser Disc Player originates from the application of............for reading out the information on the disc.
2. There is an effective.........of information on the disc against finger prints, dust and scratches.
3. Average track pitch is.............
4. Pit width is................whereas pit depth is........

5. In practice, however, ...magnetic recording tape is used as a....... carrier.
6. The laser optical system is also called............
7. There are two variations of the laser optical system, ...and..........
8. Long play discs are for straight forward.............playback of entertainment program.
9. Active play discs involve the viewer in learning a new............
10. ....is the cheapest and the most basic of the three video disc formats.

CONVENTIONAL QUESTIONS (C)

1. Compare a video disc with a gramophone record.
2. What are the similarities between the three video disc formats?
3. Briefly explain the different optical recording mediums.
4. With the help of a suitable sketch explain the construction of a video disc along with the dimensions of
   - (*a*) diameter
   - (*b*) track pitch
   - (*c*) pit width
   - (*d*) pit depth
   - (*e*) thickness
   - (*f*) protection layer
5. Briefly explain the recording and playback system of an optical video disc.
6. Describe the digital address signals.
7. Explain the working of a TS laser.
8. What is the significance of radial tracking ?
9. What are the different controls on a VLP player?
10. What is the diffeence between a video disc and an optical memory disc ?
11. How data is recorded on a floppy disk ?
12. Explain the following :
    - (*a*) serial form
    - (*b*) discrete tracks
    - (*c*) blocks
    - (*d*) index tracks

27

THEATRE SOUND SYSTEM

Theater sound systems differ considerably from the audio systems. The chief distinction is in the method of sound recording and reproduction. For theater systems, the audio signal is recorded on film and is reproduced by means of a light beam and photoelectric cell. The sound may be recorded on the same film as the picture or it may be recorded on a separate film. In most cases the recording consists of an optical band from 0.1 to 0.2 inch in width, running at the right-hand side of the picture and next to the sprocket holes in the film. When wide screens were introduced in 1952, the traditional set of speakers behind the screen were supplemented by a speaker system placed at various location is the audience area so that a sound passage could always be radiated from a realistic source. Thus, the new technique anticipated quadraphonic sound. The Reeves system employed seven sound channels. The Cinemascope system, introduced in 1953, employed a four channel sound track utilising a magnetic strip on the film. In 1954, MGM introduced an optical three-channel sound system with subsonic signals to switch various groups of speakers on and off.

27.1. SOUND TRACK

A sound track is *a narrow strip of exposed film beside the image on a motion-picture film,* as illustrated in Fig. 27.1. The density of the track varies in accordance with the amplitude of the recorded audio signal. Although persistance of vision is adequate at an exposure rate of 16 frames per second, a film with a sound track is exposed at the rate of 24 frames per second. *This increase in film speed permits recording of.*

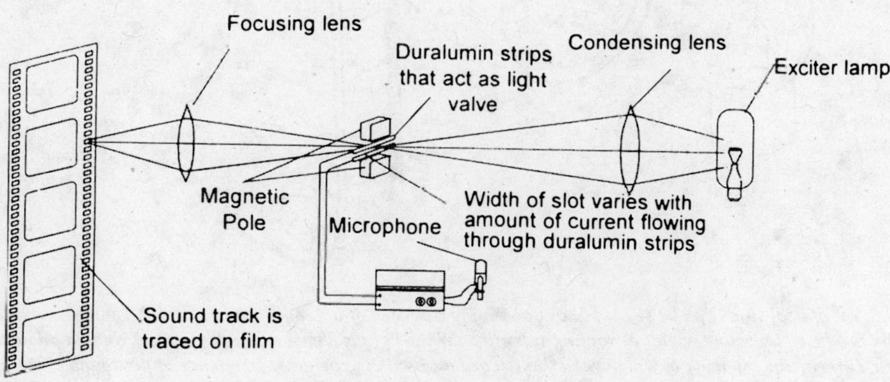

Fig. 27.1. Basic method of recording a sound track on film.

comparatively high audio frequencies. Original picture and sound *takes* are generally recorded on separate strips of film to facilitate *editing* of both the picture and sound material. In the final stage of production, the sound track with its *changes and dubs* is exposed on a film strip along with the edited picture framers. *It is positioned 15 inches ahead of the corresponding picture image.* In turn, a *free loop* with constant velocity passes through the sound transducer while the picture images proceed in *sudden jerks* through the projection lenses.

Fig. 27.2. A variable density sound track.

Basically sound is recorded in film as depicted in Fig. 27.1. Audio signals from the microphone are stepped up by an amplifier and are passed through a pair of *metal strips* that vibrate between the poles of a permanent magnet. These metal strips operate as a light valve, with a slit approximately 0.001 inch wide. When the audio-frequency currents flow through the strips, the magnetic filed exerts a force on the strips and causes the width of the slit to vary accordingly.

The beam of light passing through this slit varies in intensity thereby exposing light and dark bands along the film track. These bands appear as shown in Fig. 27.2. and form a *variable-density sound track.*

Next, in the reproduction of the signal by the projector an arrangement is employed as shown in Fig. 27.3. A beam of light is passed through a narrow slit and focused on the sound track and then into a *photoelectric cell.* Variations of light intensity in the photocell cause corresponding variations of current through the cell and reproduce the audio signal.

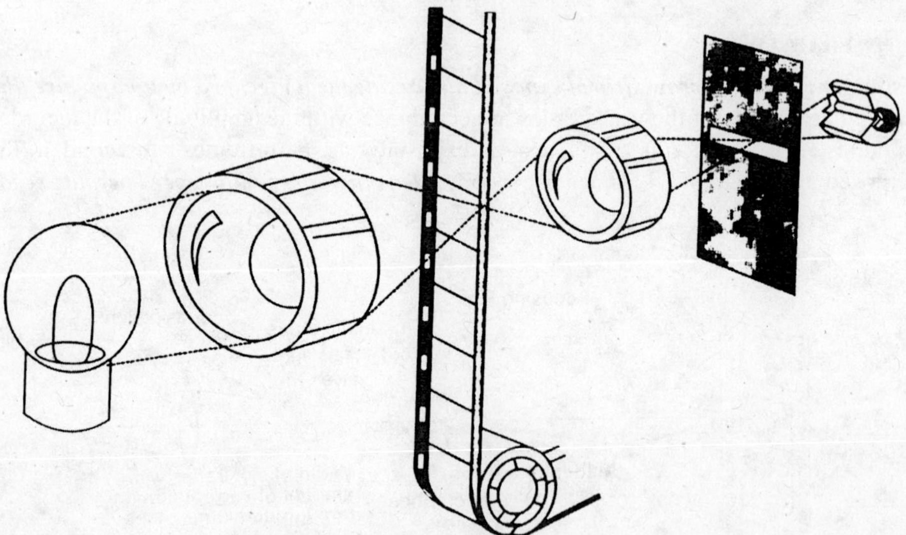

Fig. 27.3. An optical track can be regarded as containing a record of a lamp's variations in brightness, where these variations were originally produced by sound waves of varying intensity. When light is shone through the track on to a photocell, the cell produces an electric current, the intensity of which varies in accordance with the original frequency of the sound. This current is played back through loudspeakers.

Instead of employing a variable-density sound track, a *variable-area sound track* is often used, as shown in Fig. 27.4. These are two basic types of variable-area sound tracks, call unilateral and bilateral sound track.

Fig. 27.4. Variable area sound tracks, (a) unilateral type (b) bilateral type

A *unilateral track* has an unsymmetrical form with tooth like projections of various sizes, Fig. 27.4.(*a*). On the other hand, a *bilateral track* has a symmetrical form that varies in area from one point to another Fig. 27.4(*b*)

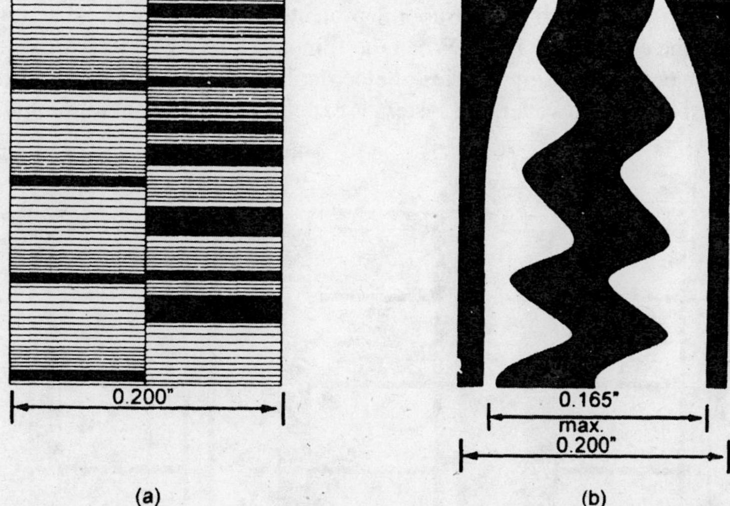

Fig. 27 5. Push pull sound tracks used in original recordings. (a) Variable-density type (b) Variable-area type.

Original recordings are often made with *push pull sound tracks* of variable-density or variable-area type, Fig. 27.5. This type of sound track is *twice as wide* as a single-ended sound track. The chief advantage of push-pull sound tracks is the cancellation of even harmonics, allowing higher fidelity to be obtained. In the final stage of production, the push-pull track is recorded as a single-ended track on the films that are released for exhibition.

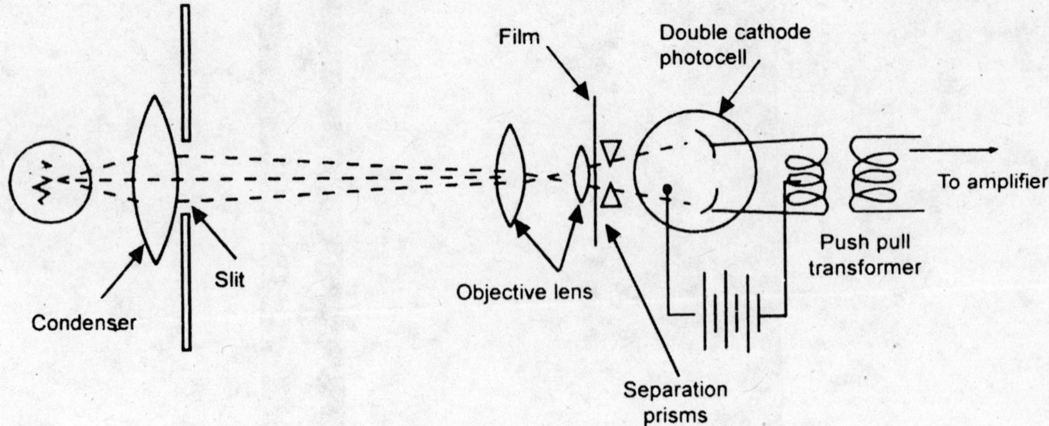

Fig. 27.5. Sound reproduction from a push-pull sound track.

Push pull sound tracks are reproduced by means of a *double-cathode photoelectric cell* as shown in Fig. 27.6. The output from the push-pull transformer is a *single-ended audio signal.*

27.2. TYPES OF SOUND FILM

There are two types of sound film, optical and magnetic. *Optical film* is used exclusively in the 16 mm cameras in the amateur field, though some Super 8 projectors can replay it. The sound track is recorded photographically along the edge of the film. When the film is projected, the track passes between an exciter lamp and a photoelectric cell, generating a series of electrical impulses which are amplified and converted to sound waves *via* a loudspeaker. However, the system is expensive and the photographic track has to be added by a commercial laboratory.

Fig. 27.7. The two types of sound film-magnetic (near right) and optical (centre and far right). Magnetic striped film is coated with two strips of iron oxide, one on each edge. The wider stripe on the left side records sound, while the narrower strips on the right hand edge is a balance stripe to help the film wind evenly on the spool. It, too, can be used for recording music or special effects. In optical film the sound track is recorded photographically along the edge of the film.

With the *magnetic film* the sound track is recorded on a narrow stripe of magnetic oxide, applied to one side of the film. This can be easily added at home provided that the laminate variety is used though it is usually more convenient to buy a sound cartridge with the stripe already in place. A second narrower stripe along the other edge ensures that the film spools evenly. This is known as the *balancing stripe*. The magnetic stripe itself works in exactly the same way as an ordinary magnetic tape.

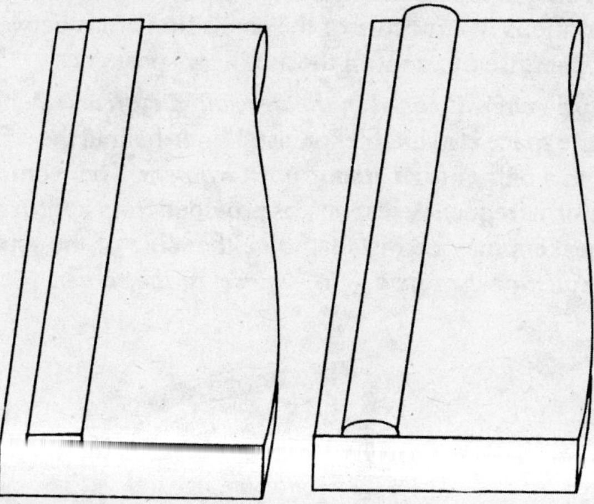

Fig. 27.8. A magnetic stripe to record sound can either be laminated right or applied as magnetic paste left by a professional laboratory. Professional striping is inexpensive and is often superior to home striping, as the stripe has better contact with the sound heads.

27.3. THEATER SOUND REPRODUCTION SYSTEMS

A sound track is played in a theater by means of a *sound head*. Fig. 27.8 shows two sound heads are employed so that little or no interruption of a program occurs when changing from one head to another. The

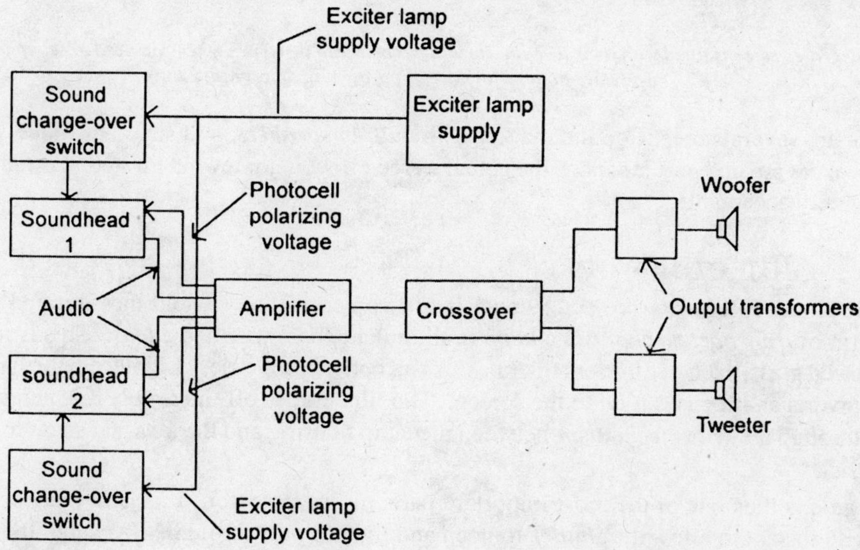

Fig. 27.9. Block diagram of a theater sound system

audio outputs from the sound heads are fed to an amplifier and then to speakers that are customarily mounted *behind* the screen. In addition, speakers may be mounted on *either side* of the screen. The amplifiers are essentially the same as those used in high-fidelty systems and their power output depends on the size of the theater. It is standard practice to utilise at least one woofer and one tweeter *behind* the screen. In addition a woofer and a tweeter may be installed on *either side*.of the screen to enhance the volume of dramatic sound passages. These *auxiliary speakers* are switched into or out of the circuit by means of a control track. This *control track* consists of variations in exposure of the small film areas between adjacent sprocket holes. An electronic switch is thereby energised to control the auxiliary speakers.

Horn-type speakers are generally used in theater sound systems. A horn speakers provides increased operating efficiency and ample space is available for installation behind the screen. *Multicellular horn tweeters* are commonly employed with a pair of *logarithmic horn woofers.* The woofers are much larger than those in high-fidelity systems. A uniform frequency response is provided from approximately 70 to 7000 MHz. Similar woofer tweeter auxiliary speakers may be installed on either side of the screen. It is the general concensus that *theater sound is most realistic when its origin is close to the screen.*

Fig. 27.10. Integrated screens fitted with their own loudspeakers, come into their own when you need to set up the equipment quickly and have no time for careful positioning of extension units.

There are several screens on the market with *built in speakers,* and stereo or mono sound. Naturally, since the speakers are not very far apart, the actual stereo effect is somewhat limited, though the sound quality can be extremely acceptable.

27.4. HOW A PROJECTOR WORKS

In the simplest terms, a projector consists of two *spools*-one to hold the film and another to take it up-some form of *transport mechanism* a *lamp* to illuminate the image and a *lens*. This is made from several pieces of curved glass, each of different thicknesses and optical properties. Light from the lamp shines through the image, inverts and focuses it on to the screen. The film comes off the *supply reel* and is threaded on to a guide path through the *gate mechanism*-between the lamp housing and the lens and so on to the *take-up reels*, Fig. 27.11.

The *gate* is thus one of the most important parts of the projector. It is here that each frame in turn is stopped for an instant to allow the *shutter* to open and light to pass through. Basically, the gate consists of a fixed plate with a rectangle cutout it ; as the film is guided through the gate, a spring pressure plate presses it against this *aperture.*

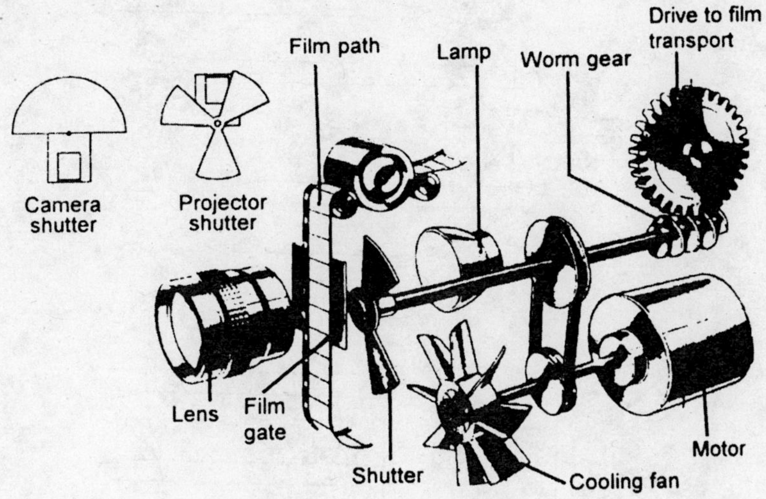

Fig. 27.11. Camera and projector mechanisms are essentially very similar. Both work on the principle of advancing film frame by frame, holding each frame steady in the optical section for a fraction of a second, and exposing or projecting it. In each case a shutter passes between light source and film during the film advance phase. However, the camera has one shutter blade, the projector three. One of the projector shutter blades prevents exposure of the moving film. The other two each interrupt the projected frame, so that each frame is shown three times. If you have filmed at 18 frames per second, then you will view at 54 frames per second. Film is taken from the feed spool by the rotating feed sprocket wheel. It is received by the gate assembly which consists of a spring loaded pressure plate and a fixed aperture plate with spring-loaded side guides. The pressure plate pushes against the film and, with the side guides on the aperture plate, helps to keep the path of the film straight and steady. The claw pushes the film through the gate one frame at a time. A take-up sprocket wheel then takes the film from the gate area to the take-up reel. Because the movement of the sprocket wheels is continuous and that of the claw is not, there is a loop between each wheel and the gate.

The *pressure* has to be enough to ensure the film lies flat ; at the same time, the spring controls this, so that there is no danger of the film being scratched.

The film is moved to the next frame by a *claw*. This is a small pin, which, when the shutter closes, enters a perforation on the film, pulls it down quickly and withdraws as the shutter opens again. The whole process happens 18 or 24 times every second, depending on the running speed of the projector.

Projector lamps are specially designed to be small and to give a very bright light. On most modern projectors, the light source is a small filament, low voltage lamp allowing light to be *concentrated* efficiently onto the gate. *These lamps have built in mirorrs, with a dichroic coating.* This reflects the light well but allows most of the heat generated by the lamp to escape rearwards, thus keeping the film as cool are possible. A projector will take only the type of lamp designed for it.

Most modern projectors have an *automatic threading mechanism,* which considerably simplifies the business of actually setting up the film. All the projectionist has to do is to place the reel of film on to its *extended arm,* set the projector in motion and push the trimmed film into a slot. Many projectors, too, use one or more *sprockets,* whose teeth engage in the perforation and help transport the film through the mechanism. These sprockets are at constant speed, so that *loose loops of film* are formed between them and the gate. This counters the intermittent motion of the claw.

Smoothing out this action is all important as far as sound projectors are concerned. This must be done before the film reaches the sound heads, which are used to record on, or playback, the magnetic stripe at the edge of the film. *In some ways, the sound projector is like a tape recorder and silent projector all in one.* The projector contains an amplifier and usually a small loudspeaker. Thus, it can be operated as a *completely self-contained unit,* though better results are usually achieved if an extension speaker is used.

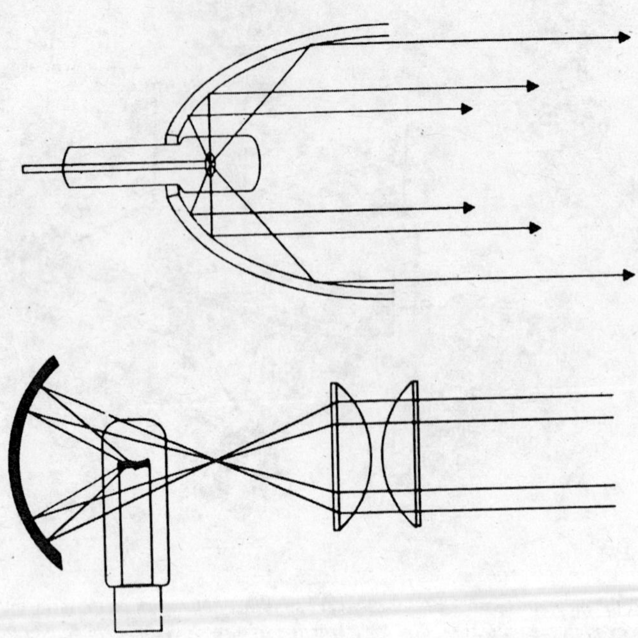

Fig. 27.12. A projector lamp must combine the highest possible degree of illumination with the lowest possible heat. The condenser lenses in the lamps concentrate the light to a pin point. New designs combine a reflector with a lower voltage lamp and so need no condenser.

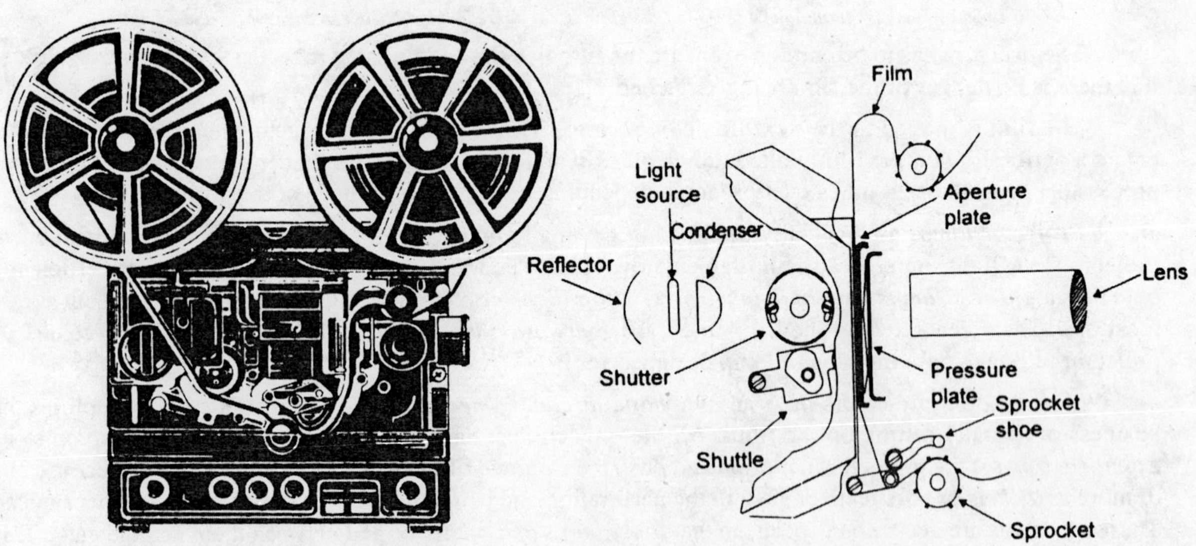

Fig. 27.13(a) A typical projector. (b) Simple motion-picture projection system.

Lenses are of two kinds *fixed focal length* and *zoom*. In the first case, the projector or screen has to be moved physically to alter the size of the projected image ; in the second case, this can be done by simply adjusting the zoom.. However it is important to check the quality of image this produces. Some low priced models can give poor image definition, especially round the edges of the picture.

27.5. SOUND PICK-UP

The sound track pick-up is either advanced or retarded in relation to the picture to compensate for the physical displacement of the photocell or magnetic head relative to the picture heads as shown in Fig. 27.14 and 27.15.

When *magnetic sound tracks* are used on 35 mm or 70 mm film, the sound track is *retarded* or behind the picture. *Optical sound tracks* are *advanced* relative to the picture. In the first instance, it is termed *pull-up*, and in the second, *pull-down*, with reference to the picture aperture. For *8-mm* projection, an optical sound track is advanced 22 frames, and for a magnetic track it is advanced 18 frames.

The actual displacement of the sound track will depend on whether the track is magnetic or optical. For reproducing a *35 mm* optical sound track, the photo tube and the associated equipment is placed in the sound head below the picture head, and the magnetic reproducing head is housed in a penthouse above the picture head.

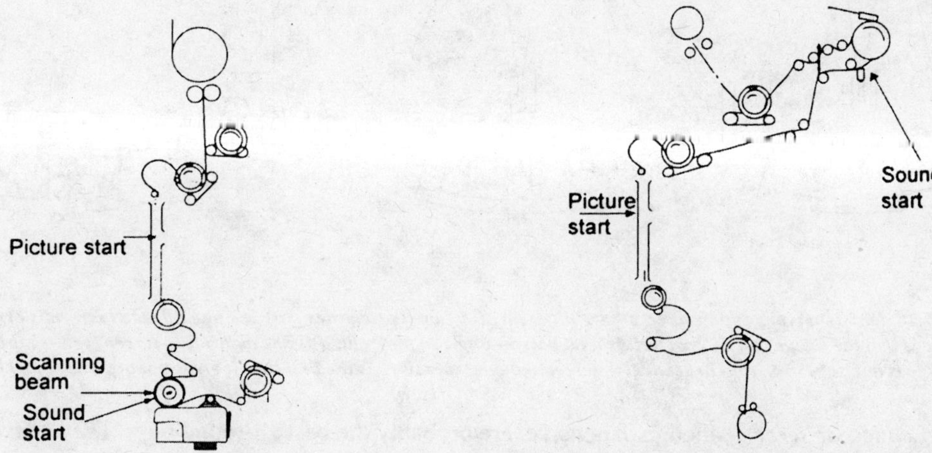

Fig. 27.14. Film threading path for 35mm and 70mm film using optical soundtrack reproduction.

Fig. 27.15. Film threading path for 35mm and 70mm film using magnetic sound track reproduction.

In large theaters where the projection throw is greater than 100 ft, it may be necessary to advance or retard the sound track a few sprockets to bring the sound into proper synchronization at audience distances of 100 ft or more from the screen. This may be accomplished for an optical sound track by threading the sound start mark at the 19th rather than the 20th, frame, *In the average theater it is necessary to emit the sound before the corresponding picture frame is projected,* since sound travels at approximately 1100 ft/s or about 50 ft/frame for a normal projection rate of 24 frames/sec. The projectionist can place the sound in synchronization in the theater by varying the length of the threading path between the projector and the sound head.

The length of film between the aperture opening and the scanning beam must be 26 frames. The film has a loop in it as it leaves the aperture channel. *The size of this loop determines the length of film that will exist between the two points.* When the film is badly damaged, worn, or poorly spliced the lower loop is usually lost. Depressing the *loop setter* until it hits the stop then releasing it resets the loop. The lower loop will react automatically to the precise size required for synchronization of the picture and sound.

27.6. CINE SCREENS

Glass-beaded screens are fairly expensive and need to be handled carefully as the surface is delicate. They are very directional and give a very bright image to those sitting near the projection beam. However, the brightness appears to diminish rapidly at wider angles, so *this kind of screen is best suited to long rectangular rooms.* There are two drawbacks ; with some screens the glass beads are visible if viewed closely, and others can give a slightly unnatural, grainy look to colour film. The optimum viewing distance is about eight feet.

Matt white screens are not directional but the image is not particularly bright. Given the good light output of most modern projectors, they are perfectly satisfactory for home use, but *it is essential to ensure that the room is blacked out carefully.* This is because any stray light falling on a matt screen can seriously affect the quality of the image.

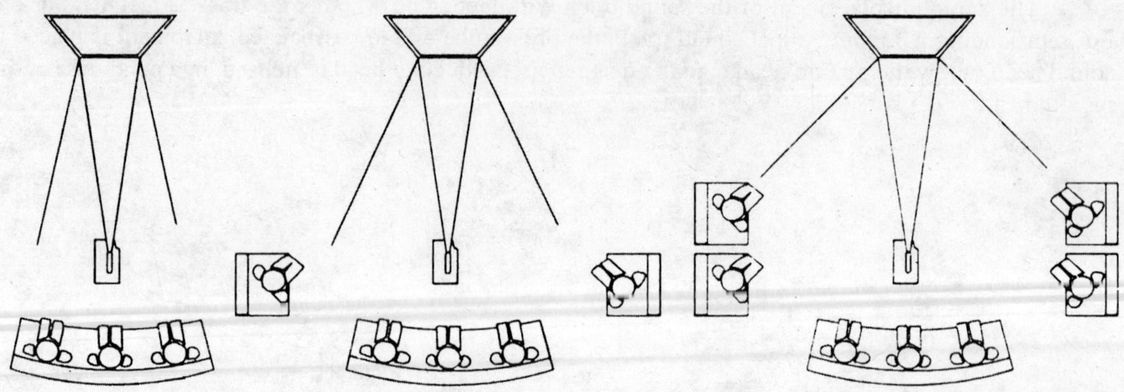

Fig. 27.16. Different types of film screens have different directional properties. (a) The glass beaded screen is extremly directional, this means that the image can be seen well from a narrow angle in the front. (b) The matt white screen (centre) is less directional, but the image is not as bright. (c) The lenticular screen though expensive, is the best buy. It gives a sharp image and has a wide angle of view.

Lenticular screens, though expensive, are probably the best all-round buy. *They offer a sharp, bright image over a fairly wide area and are not so sensitive to handling as other screens.* They can also be cleaned easily Their characteristic ribbed appearance comes from the ridges and indentations on the surface ; these ensure that the light is well reflected. As with glass beaded screens; the best viewing position is therefore a few metres away.

OBJECTIVES TYPE QUESTIONS (A)

1. Increase in film speed permits recording of
   (*a*) low frequencies (*b*) high frequencies
   (*c*) medium frequencies (*d*) none of the above.
2. In film reproducing system, reproduction of sound is done by using
   (*a*) photoconductive cell (*b*) photoelectric cell
   (*c*) photoresistive cell (*d*) photoemissive cell
3. The sound track is recorded photographically on
   (*a*) magnetic film (*b*) optical film
4. Speakers are mounted customorily
   (*a*) on either side of the screen (*b*) behind the screen
   (*c*) (*a*) and (*b*) (*d*) (*a*) or (*b*)

5. The auxiliary speakers are switched into or out of circuit by means of a
   (*a*) mechanical switch (*b*) control track
   (*c*) electronic switch (*d*) none of the above.

6. The sound track pickup is
   (*a*) advanced with respect to the picture
   (*b*) retarded with respect to the picture
   (*c*) (*a*) or (*b*)
   (*d*) none of the above

7. In the average theater it is necessary to emit the sound
   (*a*) along with the corresponding picture
   (*b*) after the corresponding picture
   (*c*) before the corresponding picture

8. Glass beaded screens are
   (*a*) directional (*b*) non-directional

9. It is essential to ensure that the room is blacked out carefully where using
   (*a*) glass beaded screens (*b*) matt screens
   (*c*) lenticular screens.

10. Cine screens which are not sensitive to handling are
   (*a*) lenticular screens (*b*) matt screens
   (*c*) glass beaded screens

FILL IN THE BLANKS (B)

1. A unitlateral track has an .... form
2. Push-pull tracks are .... as wide as a single-ended sound track.
3. The chief advantage of a push-pull track is the cancellation of .... harmonics.
4. Push-pull sound tracks are reproduced by means of a .... photoeletric cell.
5. The sound track is recorded .... along the edge of optical film.
6. With the magnetic film the sound track is recorded on a .... stripe of magnetic material, applied to .... side of the film.
7. A woofer and a tweeter may be installed on either side of the screen to .... the volume of dramatic sound passages.
8. .... type speakers are generally used in the theater sound systems.
9. Theater sound is most realistic when its .... is close to the screen.
10. The .... is one of the most important parts of the shutter.
11. Modern projection lamps have .... mirrors with a dichroic coating.
12. A projector lamp will take only the type of lamp .... for it.
13. Most modern projectors have an .... threading system.
14. Magnetic sound tracks are .... relative to the picture.
15. Lenticular screens offer a sharp bright image over a .... area.

CONVENTIONAL QUESTIONS (C)

1. What is a sound track and how is it recorded ?
2. Differentiate between unilateral, bilateral and push-pull sound tracks.
3. What are the advantages of push-pull sound tracks ?
4. Explain the variable density method of sound recording.
5. Which film is preferred in sound recording and why ?
6. Explain theater sound reproduction system with the help of a block diagram.
7. Explain briefly how a projector works
8. Briefly explain the following :
   (*a*) free loop ; (*b*) gate ;
   (*c*) shutter (*d*) pull-up ;
   (*e*) pull down and (*f*) zoom lens.
9. Compare the different types of screens used in image projection.

28

VIDEO SYSTEMS

Video systems have become an essential part of the household entertainment. There are two main types of video systems. Paralleling their audio counterparts, there are tape-based systems and disc-based systems. With the tape-based video systems you can, in your own home, both record and playback television programs and your own electronically produced home movies. The disc-based systems, by and large, are for replay of pre-recorded material only.

28.1 CANON PORTABLE VIDEO SYSTEM

The Canon Portable Video System (*tape-based*) is shown in Fig. 28.1. The *VC-20 camera* is capable of operating in almost any conditions and gives the user almost *total creative control* over the end result. The VR-10 *recorder* turns out to be the perfect partner for both the camera and tuner/timer, producing good pictures from either source. Features such as *insert edit* give this recorder the capability of producing truly professional home recordings. This facility, used with audio dubbing, gives the user the opportunity of achieving with one machine the kind of editing normally possible only in multi-machine studios.

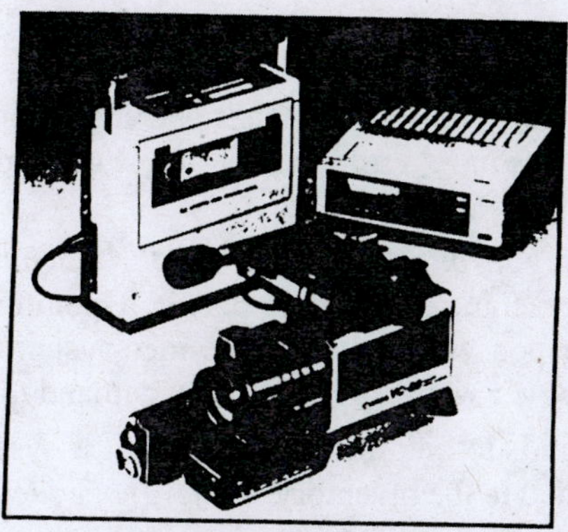

Fig. 28.1 Canon Portable Video System

The *tuner/timer* has all of the features of much larger domestic video recorders. Its electronic channel selection is an innovative idea and its programmable capabilities are more than adequate. *One touch recording* (OTR) will start recording a program at the touch of a button.

As a complete unit, you have a very impressive combination which does far more than a home video recorder. The system offers the scope for truly creative moviemaking.

28.2 LASERVISION-VIDEO DISC SYSTEM

Laservision, is without doubt, the most sophisticated video disc system. *Its future lies in the industrial/educational market* and not the mass consumer market. The disc itself is the same size as an audio LP but is *silver covered* and *reflective* like a mirror.

Stereo sound is available when the player is linked to a hi-fi and Laservision, Fig. 28.2 offers a full range of *special features* including slow motion, freeze frame and picture search.

Fig. 28.2 Laservision - The Philips video disc system.

The Philips discs also offer numerous *interactive possibilities* once they have been specially pre-programmed. For example, they could be used to test students by posing questions and then providing the correct answers.

28.3 INTERACTIVE VIDEO SYSTEMS

When video can be made to respond to the wishes and instructions of the operator, it is said to be *interactive*. Both video tapes and discs have a role to play in this new application of video, which promises to introduce new methods of learning.

The aim of interactive learning through video is to let the machine take over from the teacher. If it incorporates a programmed tape, a machine on which to play it, a monitor and a device, linked to a print-out, into which the student can punch responses, an interactive video system can check a student's progress and understanding step by step, and discover which areas of study demand the teacher's personal attention.

The next step in sophistication is to link the tape player of the system to a computer such as Apple II. With the potential of a computer added to the inventory, the *level of education* can be made more flexible and more technical.

Whenever interactive programs are used with VCR systems, they always meet the same drawback: the *accesstime*. This is the time it takes to spin the tape from the *end* of one section to the *start* of the next. The solution is to use video discs instead of tapes. *Because of the way in which they are constructed and played, video discs allow virtually immediate access to any other part of the disc.* All the machine has to do is move its playback head a few centimetres and the next section of the recording appears on the screen. The player

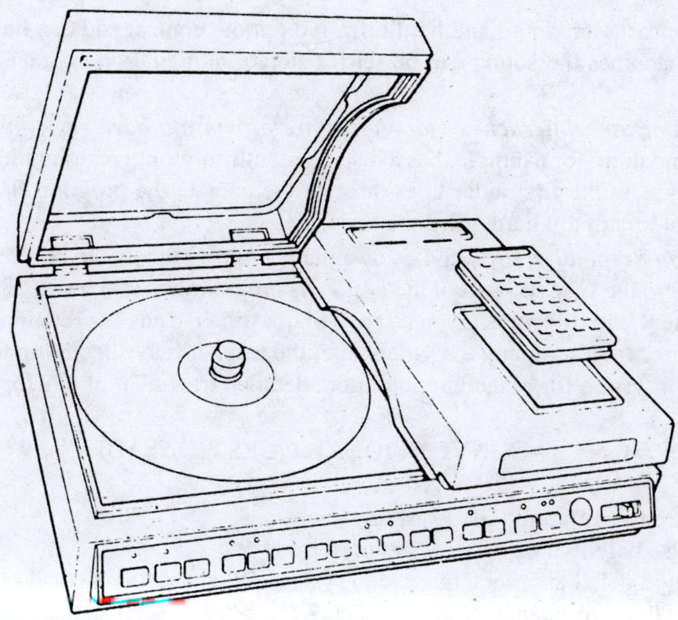

Fig. 28.3 Attached to a TV monitor, the Pioneer Disco Vision Player becomes an interactive video system.

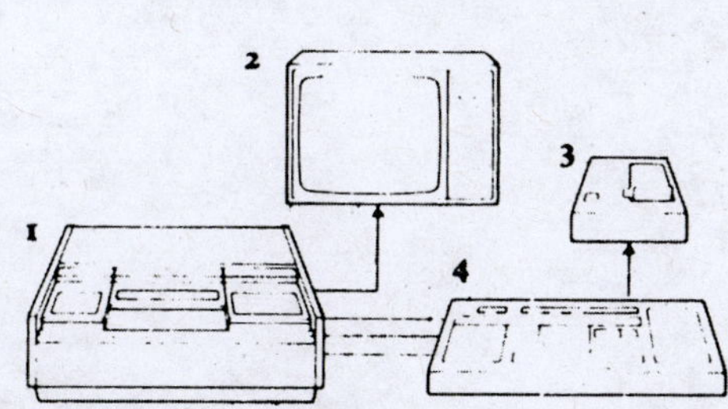

To make a program for the Sony Video Responder system, the teacher first buys or records a video tape for play in the Random Access U-matic, 1, connected to a TV monitor, 2. The Cue Programmer, 4, is then inserted into the circuit and used to write a series of questions to accompany each segment of the tape. The printer, 3 confirms the data entered at every stage.

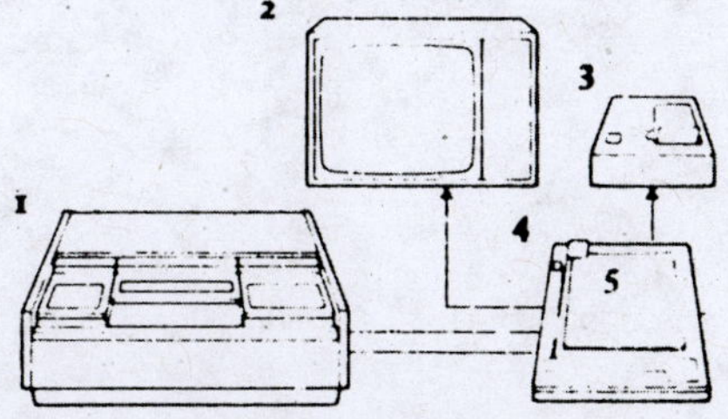

In use during a training course, the Student's Responder repaces the Cue Programmer, 4. After inserting the tape, and following the guidelines given in the work book, 5, the student presses appropriate buttons along the bottom edge of the Responder. Answers may be recorded on the printer, 3, to give a permanent record of a student's progress at every stage.

Fig. 28.4 Sony Video Responder System

incorporates a microprocessor, so a hand held infra-red remote control pad can be used to interact with the disc. On advanced machines the sound can be set for stereo, bilingual or a choice of commentaries at *two levels of difficulty.*

Linked to a computer, interactive video discs are proving to have some fascinating examples. The American Heart Foundation, for example, has a disc on mouth-to-mouth resuscitation, in which the computer is linked to a dummy. As the first aider goes through the course, the program flashes back comments on progress such as 'first breath too hard, third too weak',

The *Sony Video Responder*, Fig. 28.4, is designed to allow students to progress at their own pace. The tape that is inserted into the U-matic part of the system is programmed in a series of up to seven segments. At the end of each segment, the student is given a series of questions to answer relating to what has been shown. In writing the questions to accompany the video tape, the teacher may specify up to three attempts for each question, and make provision for branching into more detailed treatment of any topic.

CONVENTIONAL QUESTIONS (C)

1. What are the different video systems?
2. Why disc based video systems are preferred?
3. Explain in detail an interactive video system. What are its practical applications?
4. What is the significance of Sony Video Responder?

29

THE PHILOSOPHY OF TROUBLESHOOTING

Troubleshooting is basically the process of determining why a device or circuit does not operate as it should so that it can be remedied within the shortest possible time. To do this satisfactorily, it is necessary to have a broad knowledge of how things work and to be able to recognise the symptoms of a defective device or circuit so as to form a strategy of action for repair. Troubleshooting is the most important aspect of audio video systems.

29.1 BASIC PHILOSOPHY OF TROUBLESHOOTING

Troubleshooting is the process of determining why a device or circuit does not operate as it should. Troubleshooting has but one objective, *finding as quickly as possible what is wrong so that it can be remedied.* To do this efficiently, it is necessary to have a broad knowledge of how things work. To be efficient in troubleshooting and servicing the technician must also know what test *instruments and procedures* are necessary to repair it. Technicians must be able to diagnose i.e. they must be able to recognise *the symptoms of a defective device or circuit, and then able to form a plan of action for repair.*

Almost three quarters of the servicing time is devoted to *finding the trouble* and one quarter to *remedying it.* Obviously, then, it is in the search for the trouble, which we call troubleshooting that we can make the greatest gains through logical organisation of the thought and by efficient execution.

In troubleshooting, the technician becomes a *detective,* and the defective circuit or component is the *culprit* which must be *exposed and eliminated* in the shortest possible time. The troubleshooting process involves the following steps:

1. Gathering all possible *evidence.*
2. *Interpreting* the evidence and putting it together to narrow down the trouble-causing possibilities to as small a number as possible.
3. Making *routine tests* that eliminate each of these possibilities in turn until the actual trouble source alone remains.

The range of time and effort required to accomplish the above three steps is different for different symptoms. At one extreme is the case in which a *particularly significant evidence* immediately tells the technician what is wrong and enables him to remedy it immediately. At the other extreme is the case of an *intermittent condition* which occurs only occasionally. The source of this trouble can be determined only by operating the equipment for long periods of time and thoroughly checking all possibilities each time the intermittent condition occurs. Even though the amount of time in the two extreme cases mentioned above is drastically different, *the logical organisation of thought is the same,* with the three main steps listed above necessary in each case. In the general troubleshooting of devices and circuits, there are usually no hard and fast rules to be followed. The technician must gather all possible evidence and then apply his knowledge of theory, test instruments, tools, backed by experience, to the solution of a particular trouble symptom.

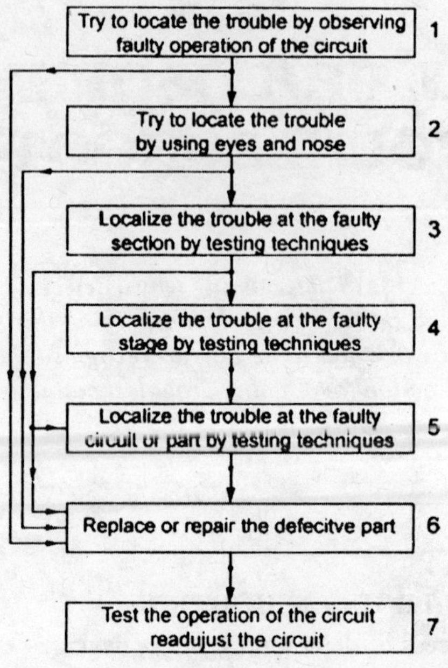

Fig. 29.1 A good time saving troubleshooting approach.

For troubleshooting the technician sets up a set of procedures and tests. What he does in a particular case depends on the nature of the trouble symptom and its indication. The efficient technician prepares a *mental check list.* As he goes down the mental check list he eliminates many of the items which do not apply to the specific symptom. Thus he makes sure that each possibility has been considered.

Many technicians have done an excellent job of technical service work, yet failed in their business because they did not have the *proper respect for the customer.*

If anything out of the ordinary is to be done, such as relocating wires, drilling etc, the technician should always obtain the customer's permission. The whole problem is one of *attitude* and is not at all difficult if one starts off on the right foot. Once the technician has the customer's confidence, the rest is easy and just a matter of common sense.

Obtain all possible information from the customer himself. The customer can, no doubt, impart bits of information about the failure which may ultimately save hours of work. Did the failure take place gradually or suddenly? Was any trouble noticed before complete failure took place? Such information will usually not be extracted from the customer without a certain amount of patience. The technician will frequently come across the customer who would like to solve the problem by *wishful thinking,* which he does but with 'I know it must be only a loose wire, it went off so suddenly', or 'It requires only a minor adjustment? In fact he knows practically nothing about it at all. All this is part of the game. The ultimate objectives of repairing the equipment and also keeping the customer's good will are best obtained by *patience and equanimity* and never by telling the customer off. Argument and controversy should be avoided by placing the responsibility for and choice of any debatable repairs squarely on the shoulders of the customer.

Examination of the equipment is usually made after the customer has been questioned about it. The technician may prefer to examine the equipment and make his own observations first. There is no hard and fast rule.

Try to determine by *visual inspection* if any components overheat, and then try to interpret and find the cause of trouble. It helps us to approach the actual source of trouble more quickly.

The gathered evidence can now be used to localise the trouble to one particular section of the equipment. Component tests in that section will then lead to one or more guilty components. We make use of two well known methods of detective work:

1. Deduction
2. Elimination

We cannot always deduce that the trouble must be in just one thing or in just one location. We can ordinarily place it in *one general section.* Then by the process of elimination we eliminate all of the possibilities except the real trouble cause.

When the trouble is fairly well *localised*, transistors are checked, if necessary, and other elimination tests made to determine the actual component(s) at fault so that the necessary adjustment can be made.

29.2 ROUTINE TESTS

• Visual Inspection:

Unless the *probable source of trouble* is known, troubleshooting should usually begin with a *thorough visual inspection.* Loose connections, broken wires, poor solder joints, burned out resistors, or even a burned out fuse can very often be located in this way.

Let us take an example of a *dead equipment.* You might adopt a lengthy procedure to come to a conclusion. Perhaps if you had inspected the attachment plug before replacing anything, you would have discovered that a wire had become unfastened from a plug terminal screw. With this corrected, the equipment might be restored to its normal operation. In this case, a *thorough visual inspection* would have saved time and possibly some expenses. In complicated troubleshooting and repair more time and more money can be wasted by procedures which would have been quite unnecessary if a visual inspection had been made first.

• Smell, Hearing and Touch

The senses of smell, hearing and touch can be *valuable aids* in troubleshooting. Overheated resistors, transformers, choke coils, and insulation materials can very often be detected by their *peculiar odor.* This can help the technician in *locating* certain defective components. When the *overheating* of any component or part in a circuit is noticed *the circuit should be turned off immediately,* since this may add further troubles.

Overheated transformers and choke coils will sometimes produce a cooking, or frying *noise,* that can easily be heard, the source of noise can thus be located.

The sense of hearing can be exploited by using the loudspeaker as a test instrument. Lack of sound or distortion of sound heard through it can indicate specific circuit defects which the experienced technician can quickly locate.

Overheated components can, in many cases, be located by touch. This is especially true of transformers, choke coils, electrolytic capacitors, transistors, and integrated circuits. If anyone of these components feels excessively warm, there is reason to believe that a defect exists in the *component or the associated circuit.*

• Continuity Tests

Following the preliminary visual inspection and the smell, hearing, and touch tests, troubleshooting involves continuity tests made with an ohmmeter. Such a test should be made whenever an *open circuit*

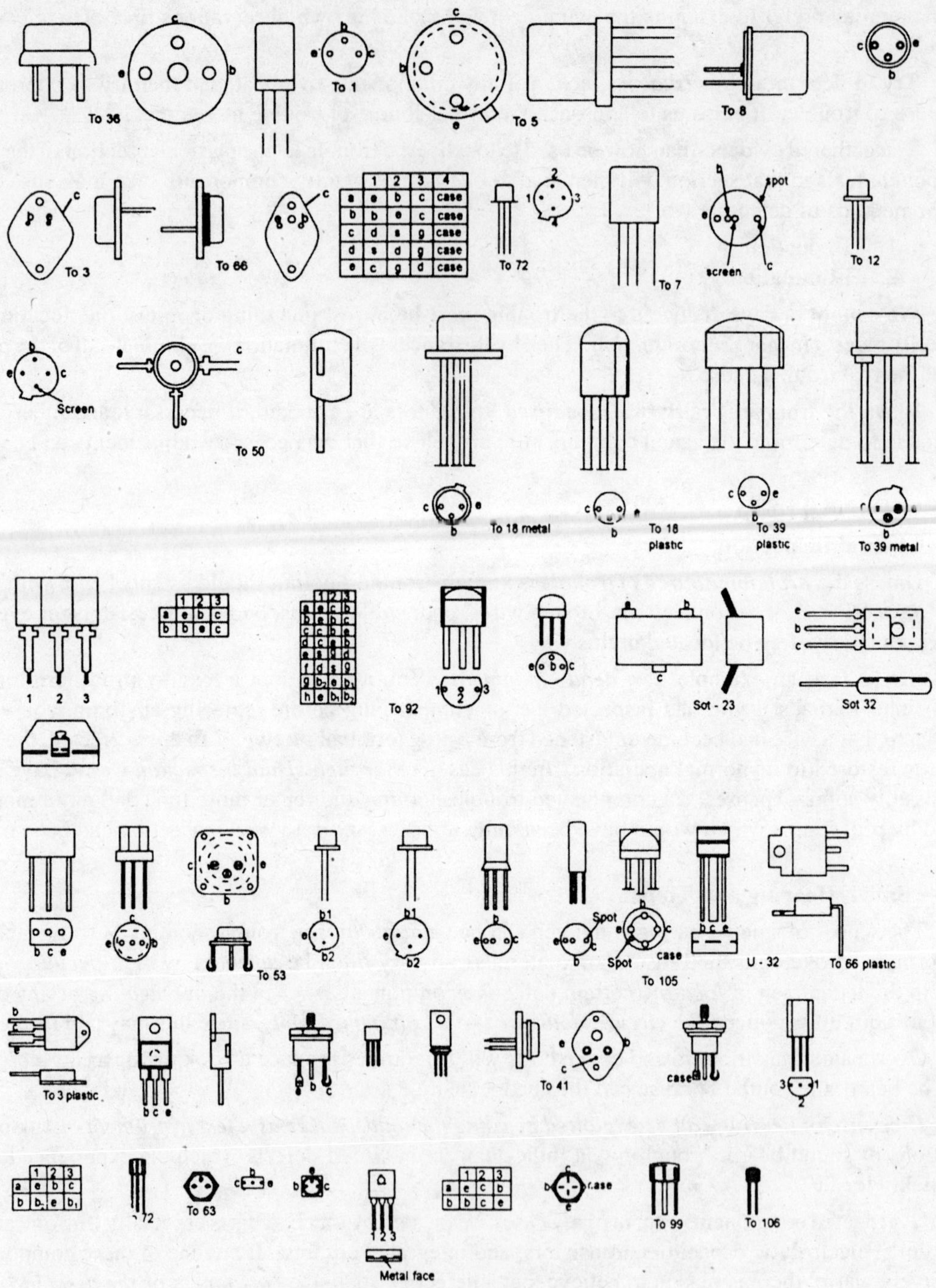

Fig. 29.2 Transistor lead and base identification chart.

condition is suspected. This test can also be used to test *short circuits.* An ohmmeter should never be used in a circuit to which a voltage is applied. Also, the component should be isolated by disconnecting one of its leads before continuity tests are performed.

- **Transistor Testing**

The transistors in a circuit are generally not tested, until there is definite reason to believe that at least one of the transistors is defective. Transistors can be tested *in-circuit,* as well as *out of circuit.* In-circuit testing is more difficult, but it is possible, in most instances, if you apply your knowledge of electronics in a logical way.

- **Voltage Tests**

Regardless of the method you use to find a defective stage, after you have located the stage you must find the defective part within it. *Measuring the voltages is one of the best methods of testing within a stage, because with this method the circuit is tested with the power applied.* Some defective parts check good when the power is not applied. For example, a coil in the collector circuit of a transistor can show continuity when measured with an ohmmeter. However, the continuity of the coil might be lost entirely when collector current flows through it. The loss of continuity occurs because the collector current is several times the current used to test the coil with an ohmmeter.

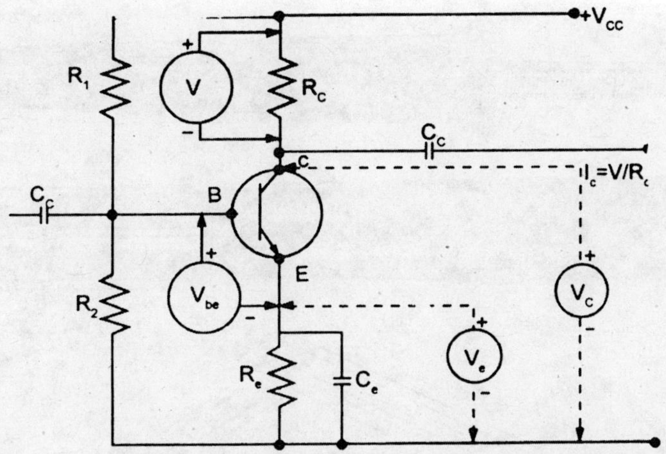

Fig. 29.3 Calculating the current from a voltage reading.

The voltmeter test usually can be made without disconnecting any portion of the circuit. By applying Ohm's law, ($V = I \times R$), you can also calculate the current from a voltage reading. The current in milliamperes is equal to the voltage in volts, divided by the resistance in kilohms.

- **Testing Integrated Circuits**

To isolate the fault to an IC stage, start by *signal checks.* Give the required signal at the *input* pins and measure at the *output* pins. The required signal can be taken from an AF signal generator or from another equipment in perfect working order. The output signal may be checked by an oscilloscope. If an oscilloscope is not available the signal *supposed to be present* at the output pin, may be taken from an AF signal generator or a working equipment of a similar model and injected at the output pin to check whether the stage/stages *beyond the IC* are working. If the signal is passing from the output pin of the IC, but does not pass from the input pin of the IC, the IC stage is obviously defective. The fault might be in the IC or in the circuit associated with the IC (external components).

As easy method of testing the IC is by testing the voltage on IC pins. Standard voltages at different IC pins are given in the schematics of the equipment you are working on. The voltages are first *measured* and then *compared* with those in the schematic. In case a change of voltage at any pin of the IC is observed, check the components connected to that pin. If these components are found OK and if all other voltages on the IC are correct, *remove the IC from the circuit to give it a resistance check.*

In modern equipment, ICs are either soldered into the circuit, or fixed in an *IC socket,* Fig. 29.4(*b*) which in turn, is soldered into the circuit. In the first case, removal of the IC demands a *desoldering operation.* A desoldering pump, illustrated in Fig. 29.4(a) is most suitable for the job. In case a desoldering pump is not available, a small brush can be used for removing the solder from the IC pins during desoldering operation, to make it easier to remove the IC. Both the methods work equally well. If the IC is fixed in a socket, removal of the IC requires a bit of care, patience and, of course, common sense. *Lift the IC from the base equally from both the sides with the help of a poker or tweezer and then remove it straight off.*

Resistance check of an IC involves the measurement of the resistance of the IC Pin No. 1 with respect to the remaining pins *in both directions* (with the multimeter leads reversed). Tally both the forward and reverse resistance of the suspected IC with those of a good IC. Standard resistance readings of the IC in both directions, can be kept as a record for *future guidance.* If the IC is found to be defective *replacement* of the IC is the only remedy and this should be done with utmost care.

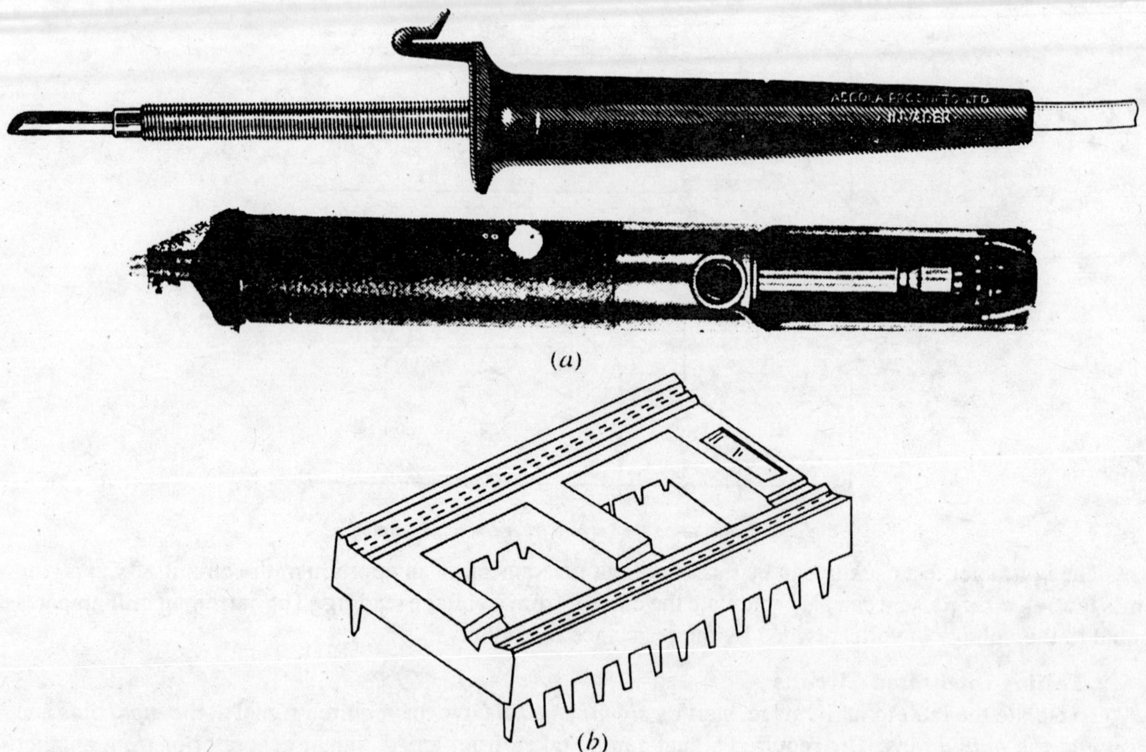

Fig. 29.4 Desoldering integrated circuits. (a) Desoldering pump and low wattage soldering iron. (b) IC socket

• **Stage Substitution**

Although stage substitution seems to be a rather *drastic troubleshooting measure* it is sometimes justified. *Stage substitution is a useful technique, however, when normal troubleshooting methods fail in circuits which have considerable interaction.*

Some manufacturers provide an entire circuit board (*module*), often containing several stages mounted on cards. These boards can be *replaced* in a matter of seconds. This prevents the customer having to do without the service of the equipment while only one small part is 'in for repair'. In these days of mass production, it is often more *economical* to replace an entire mass-produced circuit than to repair it. *In fact, many of the newer integrated circuits are 'unrepairable', and they are replaced as an entire unit.* Some of these integrated circuits are equivalent to three or more conventional stages with numerous transistors, doides, resistors, capacitors etc.

• Brute Force Technique

One typical brute force method of trying to remove an intermittnet is *soldering every suspected connection* on a printed circuit board. Another is *deliberately overloading a circuit component* to determine whether it becomes defective under additional strain. *The idea is that a defective part can withstand considerable temporary overloading without damage, but a defective (or almost defective) part will fail when overloaded.* By this means the intermittent (or a possible intermittent) is eliminated.

Extreme caution is necessary with any overload test. Brute force technique, as the name suggests, is to be applied with utmost care.

• Duplicating The Trouble

Many defects in electronic circuits are subject to the *healing effects of the test procedure*. Either an electrical or a mechanical disturbance can cause the circuit to develop *temporary good health.* Replacing a part that is suspected to be defective can cause enough circuit disturbance to make the trouble to disappear *temporarily but not permanently.*

If you are uncertain of your diagnosis, and if the trouble does disappear, *return the original part or adjustment to the circuit to determine whether you can duplicate the trouble.* Again replace the part or change the adjustment, and note whether the trouble again is removed. If the exact trouble *can be duplicated and again removed*, you can be reasonably sure that you have found the cause. This procedure should be used when the trouble is intermittent, or when you are not absolutely sure of your diagnosis and what you did to eliminate the trouble.

• Conclusion

Nothing in this world is perfect. It should be understood that the indications discussed in this chapter are relative. A certain amount of *deviation from the ideal* can be tolerated in most of the factors, without interference with normal performance. The technician will learn by experience *what degree of perfection will be worth while*, and will thus avoid hours of unnecessary effort to obtain slight improvements which would not be noticed by the customer.

29.3 COMPONENT BEHAVIOUR AND TESTING

(1) Resistors

Although manufacturing mistakes could result in defective resistors, most failures in resistors are caused by defects in the circuit. Burned resistors are usually the result of three causes:

- A shorted transistor.
- A shorted or leaky capacitor.
- Shorts or arc in the wiring or component.

A burned carbon resistor, especially if it is overheated slowly, may decrease in resistance. A sudden or drastic short usually burns the resistor open. Intermittent resistors are rare, but they are possible. *The type that is most likely to be intermittent is the wire-wound resistor.* Unfortunately, intermittent resistors usually look perfectly normal.

A resistor *identical* to the resistor you are replacing is always your ideal choice; however, this can be impossible or inconvenient. Rules for replacing resistors in a circuit are:

- The *wattage rating* of the replacement should be *at least as high as* the original rating, but can be higher if mounting is not a problem.
- In some high frequency circuits, some resistors cause an *inductive effect* that results in poor performance. Carbon resistors do not cause these inductive effects.
- If necessary a resistor with a higher wattage rating can be replaced by connecting carbon resistors *in series or parallel combinations.*

  In most circuits, the resistance of the replacement resistor can vary as much as 20 to 25 percent with no problems.

R1 through R17, resistors, carbon-composition or film, 1W, having the following resistance values:

3.3Ω	680Ω	47kΩ
10Ω	1,000Ω	100kΩ
27Ω	1,800Ω	220kΩ
56Ω	4,700Ω	560kΩ
100 Ω	10kΩ	1MΩ
470 Ω	27kΩ	

S1, switch, single-gang nonshorting rotary, 1-pole, 17-position, with pointer-knob. Switch plate, dial-type, shopmade. Dial should have resistance values printed, typed, or lettered on its face to show which resistor is connected to the substitution box text leads.

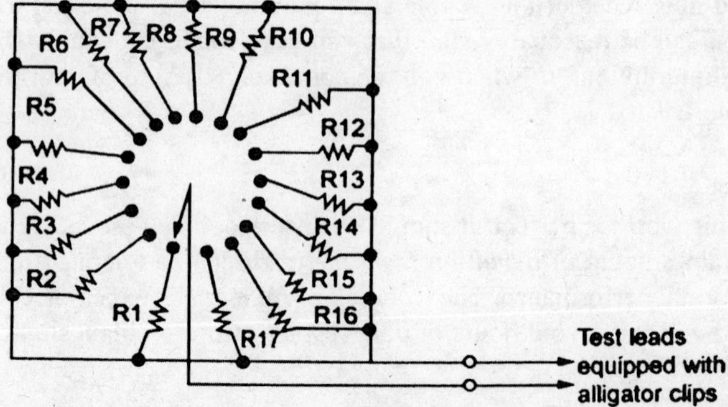

The circuit diagram and component details of a resistor substitution box, which, no doubt, is an indispensable tool in troubleshooting electronic equipment, in general

Fig. 29.5 Resistor substitution Box

(2) Capacitors

Essentially only two troubles occur in capacitors

(*i*) A capacitor may become *leaky or shorted.*

(*ii*) A capacitor may *open*.

The technique of capacitor troubleshooting can be divided into three categories.

These include finding : (*a*) a leaky capacitor, (*b*) a shorted capacitor, and (*c*) an open capacitor.

These techniques also can be divided further into finding troubles in;

- non-electrolytic capacitors; and
- electrolytic capacitors.

The dielectrics of non-electrolytic capacitors are air, wax or oiled paper, polystyrene, mica, ceramic, etc. *Mica and air dielectric capacitors are least likely to develop leakage*. The relatively small 'natural' leakage in electrolytic capacitors can be ignored within limits.

In non-electrolytic capacitors, *leakage* is considered to be any DC resistance between the capacitor plates, no matter how small the current it passes. A tiny amount of leakage current can be measured in nearly all capacitors at their normal working voltage or above, if the leakage is measured with an extremely sensitive test set. These leakage resistances which can be 500 MΩ or more, generally can be ignored; normally they do not harm circuit performance.

The *size of the leakage* that can be tolerated depends on the circuit itself. In *high impedance* (or high resistance circuits) a leakage of 50MΩ can cause noticeable distortion by changing the bias and the quiescent current. On the other hand, leakage of 100kΩ or even less, might not cause trouble in *low impedance* solid, state circuits.

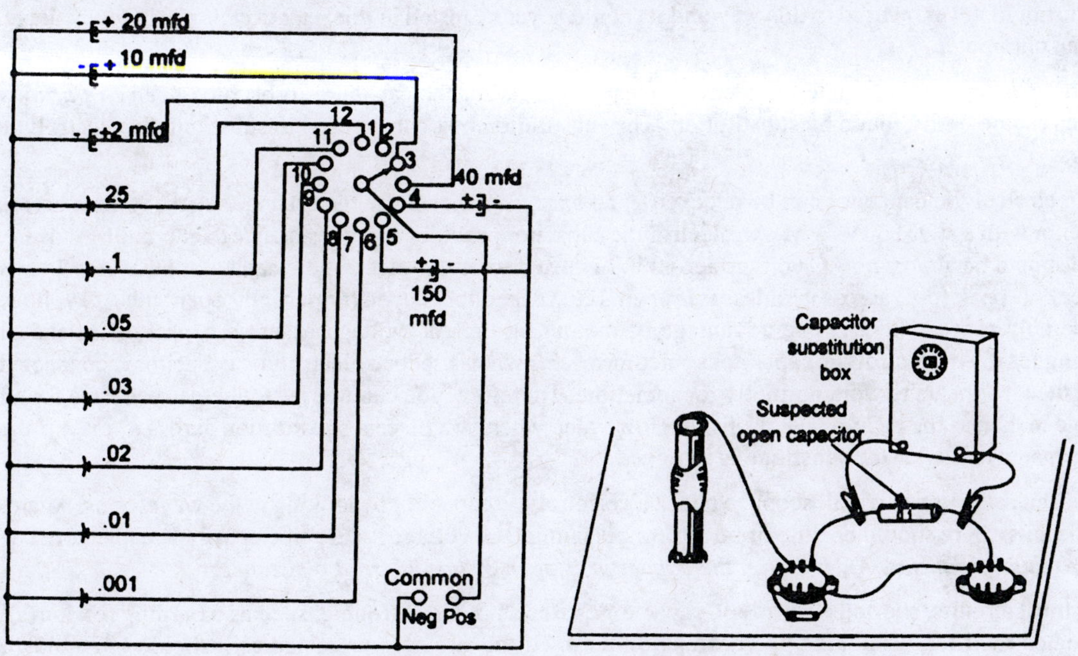

Fig. 29.6(a) Schematic diagram of capacitor substitution unit, (b) Testing a suspected open capacitor

A leakage of less than 50MΩ in a non-electrolytic capacitor is an indication of failure of the capacitor dielectic. Therefore, the capacitor should be replaced even though it is not affecting the performance of the circuit.

Several methods can be used to detect leakage in capacitors. Special instruments are available for testing leakage or other quantities. An ohmmeter is not the best instrument for checking capacitor leakage unless the leakage is a relatively low resistance. The voltage in the ohmmeter circuit usually is not high enough to break through the barrier resistance of the dielectric. In practice, capacitors often show *no leakage* when tested with an ohmmeter, but *considerable leakage* is indicated when about 56 to 100 volts is placed across them.

One method of checking capacitors is the *brute force method.* Another method of checking capacitors, the one most commonly used, is the *method of substitution* i.e. substituting a known good capacitor for a suspected capacitor. Substitution boxes, Fig. 29.6(*a*), are available for this purpose. If it is suspected that a capacitor is *open*, connect the substitution box across the capacitor, as shown in Fig. 29.6(*b*). Leave the suspected faulty capacitor connected into the circuit. Then note if the performance of the circuit returns to normal. For a suspected *leaky or shorted* capacitor, disconnect one lead of the capacitor, and connect the substitution box. Set the substitution box to a value that is the same, or nearly the same, as that of the suspected capacitor and again check the performance of the circuit.

Capacitor substitution boxes cannot be used in high frequency circuits, or very high frequency circuits, in which the long test leads would affect the circuit performance. However, in most cases, you can use the substitution box to test circuits in power amplifiers, audio amplifiers and broadcast (AM) radios.

Some *natural leakage*, usually not less than 10kΩ as measured with an ohmmeter, occurs in electrolytic capacitors. Since electrolytic capacitors are nearly always used in low impedance or low resistance circuits, this leakage is not a problem, and it can be ignored. The leakage resistance in the *reverse direction*-that is, if a voltage source is connected backward across the capacitor-is low. This also is normal; however, an electrolytic capacitor should never be used in this way and it should never be tested in this way except with a low voltage, as from an ohmmeter.

The most common trouble in electrolytic capacitors is for the capacitor to become *open or partially open*. This can be easily found by substitution. The substitution box can be used to substitute for electrolytic capacitors.

An electrolytic capacitor can be checked in an *energised circuit* by measuring the AC voltage across the capacitor with a VOM or VTVM. An electrolytic capacitor provides a low resistance to alternating current, and there should be almost no AC voltage across it. In most circuits, an electrolytic capacitor measures 2 volts or less peak-to-peak (p-p) across it, unless it is open. If the capacitor is open (or partially open) the AC voltage is considerably more. There is one advantage of measuring the AC voltage across a capacitor instead of substituting for it. An electrolytic capacitor sometimes *heals* when a second electrolytic capacitor is connected across it; then it appears to work normally for sometime. Therefore, you can't be sure whether you have found the trouble maker, if the AC voltage drops to a low value when you place a substitution across it, even if the voltage remains low after the substitute is removed.

You can also use an oscilloscope to check an electrolytic capacitor by looking at the waveforms. Again, in most circuits, there should be little or no AC (or pulsating DC) voltage across an electrolytic capacitor. The *presence of high frequency AC is a sign the capacitor is open or has internal resistance*.

In most circuits, shorted electrolytic capacitor causes an obvious trouble, such as a burning resistor(s), an overheated rectifier, a smoking power transformer or (if the circuit is protected as it should be) a blown fuse. You can detect a shorted electrolytic capacitor with an ohmmeter. Set the ohmmeter to the "Rx1" or "R x 100" scale, and measure across the capacitor. *Two readings should be taken-one with the leads reversed.* If a low resistance reading results in each direction, the capacitor is shorted.

You can get an indication of whether an electrolytic capacitor is 'good' or 'bad' with an ohmmeter. Set the ohmmeter to 'R x10' or 'R x 100' scale and connect its leads across the capacitor. The meter needle should first *deflect almost fully*, and then fall back toward infinite resistance as the capacitor becomes fully charged. Reverse the leads from the ohmmeter to the capacitor, and you should observe the same needle deflection as before.

Yet another method of checking electrolytic capacitors is with a DC power supply. Connect the capacitor to be tested (using the correct polarity) and permit the capacitor to charge fully. Remove the capacitor, and after several seconds, use a screw driver or jumper wire to discharge the capacitor by connecting its leads together. A good electrolytic capacitor can hold a charge for several minutes, however, if a full charge is held for more than 15 seconds, the capacitor is normally good. *Be careful not to exceed the voltage rating of the capacitor in this test.*

Depending on the *purpose of a capacitor in the circuit*, a capacitor within 10 percent, or even within 20 percent, of the same capacitance can be substituted without a noticeable difference in circuit performance.

Bypass capacitors are selected to provide ample bypassing. There is no distinct dividing line between satisfactory and unsatisfactory capacitance values. Of course, *replacement with the exact size is a logical choice, but it is not always entirely necessary.*

If a choice is possible, it is usually practical to replace a capacitor with a *larger* size, rather than with a smaller size. In most circuits, you can use replacement capacitors which are 200 to 300 percent larger than the original; however, *a capacitor more than about 25 percent smaller should not be used.*

In some circuits, the exact capacitor size is essential for proper circuit performance. *If the capacitor size is less that about 470pF, an exact replacement should be used*, since the capacitor is more likely to be used in a critical circuit.

One essential precaution is to *replace a ceramic or mica capacitor with an identical capacitor type, unless its use in the circuit is known.* Paper dielectric capacitors do not operate satisfactorily at frequencies above about 10MHz, because the rolling of the foil causes the capacitor to have some built-in inductance. However, *ceramic capacitors of the correct size can be used as satisfactory replacements for paper capacitors in nearly all circuits.*

Only temperature sensitive ceramic capacitors should be used in timing circuits. If you are in doubt, use an exact replacement for the capacitor, or you can use paper or polystyrene dielectric capacitors in low frequency timing circuits. Many ceramic capacitors are extremely stable, exceeding the stability of paper types by a satisfactory margin.

Ceramic capacitors designed especially for bypass application, in which the capacitance value is not critical, usually increase in capacitance as the surrounding temperature increases. These capacitors are referred to as *guaranteed minimum value* (GMV) capacitors-the size stamped on the capacitor is its smallest value of capacitance. For example, capacitor stamped 0.002 μF might have an actual capacitance of perhaps 0.005 μF or more; however, it cannot have less than 0.002 μF capacitance.

Mica capacitors were used as high frequency capacitors before ceramic dielectrics were developed. *Mica capacitors are stable, seldom develop internal leakage or shorts, and work well in high frequency and high power circuits.*

When you replace capacitors in high frequency circuits, be sure to keep the leads for the capacitors *as short as possible.*

In the case of electrolytic capacitors if the capacitance of the replacement capacitor is the same as or larger than the original capacitance, and if its voltage rating is the same as or larger than the original rating, use the capacitor if it can be mounted satisfactorily.

(3) Inductors and Transformers

Inductors and transformers are employed extensively in any electronic equipment. The manner of showing the inductors on the schematic gives the symbol *special meaning*. These pertinent facts are detailed here.

A *single arrow indicates* that a single adjustment accounts for the variation of the inductance of the winding(s). If the arrow is shown *above* the coil, it indicates that the adjustment is accessible from the *top* of the unit, if the arrow is shown *below* the coil it indicates that the adjustment is accessible from the *bottom* of the unit. A *dotted or dashed line square* surrounding the symbol(s) indicates that the entire unit is a *single assembly* and the housing is a *shield* that can house all the components shown inside the border. Openings in the metal housing on top or bottom, or both allow *access* to the adjustable component.

In general inductors and transformers do not become defective as often as other components such as resistors, capacitors etc. When a defect does occur it is usually an *open* (broken wire) or a *short* (insulation breakdown). A coil may be open because of excessive current or because chemical action (corrosion) has eaten away the copper wire. Other causes are *thermal expansion* of the coil form, or *mechanical damage* caused by careless handling.

Often, the break occurs at the solder terminal. In this event the coil can be easily repaired by cleaning the tip of the wire and resoldering it to the terminal. *Open coils* can easily be located by checking for continuity with an ohmmeter. *Shorted coils* are more difficult to locate because only a few of the turns (or even a single turn) may be shorted. An ohmmeter will not reveal this defect because the short does not change the total resistance of the coil appreciably.

(4) Ohmmeter Tests

The ohmmeter used as a *continuity checker*, Fig. 29.7 is useful for locating open coils. Depending on the associated circuitry, it may be necessary to *disconnect one end* of the coil from the circuit before making the measurement. A burned out secondary winding in an audio output transformer, for example, will still show a continuity reading because it is shunted by the voice coil of the speaker.

Manufacturer's *schematics* and *service manual* frequently indicate the resistance of coils and transformer windings. When such information is available, the actual ohmmeter reading should be *compared* with these values. Even when such values are not available, an experienced technician will know the approximate values of various types of inductors and transformers.

When an *open circuit* is located, some thought should be given to the probable cause of this defect. An open caused by excessive current, for example, is distinguishable from an open caused by corrosion or mechanical damage.

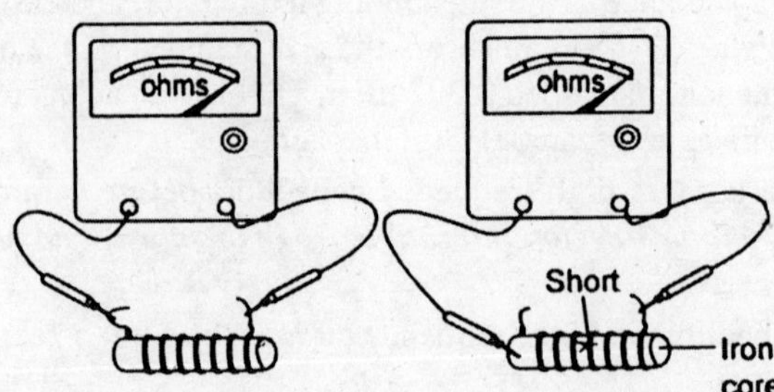

Fig. 29.7 Making a continuity test with an ohmmeter

One of the first tell tale symptoms of a faulty power transformer is a waxy material flowing from the transformer. This is accompanied by an *odor* that is easy to remember. Once you have smelled this odor, it is relatively easy to determine whether a transformer is burned or is burning, merely by *sniffing*. When burning is the problem it is likely that the trouble has not originated in the transformer but in some other component in the circuit. If you replace the transformer without correcting the troubles, the replacement may also burn out for the same reason as the original. Possible shorts in other circuits connected to B supply should also be taken into consideration. If suspected, these circuits can be checked, in turn, *by disabling one circuit at a time.*

In addition to checking the resistance of coils and transformer windings, the ohmmeter is also useful for checking *leakage and shorts. A primary-to-secondary short* in an interstage transformer causes the B voltage to be grounded through the core. This short increases the current drain from the B supply, and may damage the filter choke, rectifier, or power transfomer *Winding-to-winding shorts* in a transformer are located by connecting the two ohmmeter leads to the two windings. The ohmmeter should show an extremely high reading-corresponding to the resistance of the insulating material in the transformer. In a good transformer this resistance is many kilo-megohms. If the transformer has more than two windings, similar tests should be made between *all pairs of windings*, primary to secondary 1, primary to secondary 2, secondary 1 to secondary 2, and so on. In addition, each winding should be checked for possible shorts to the core. *All ohmmeter tests should, of course, be made with the power off.*

Shorts between *adjacent turns* or *adjacent layers* are often difficult to detect with an ohmmeter, because they produce only a small change in the total resistance of the winding. This does not imply that the short is not important if it involves a few turns. Even a single shorted turn will prevent normal operation of the circuit and can cause excessive current. A shorted turn, in the secondary of a power transformer, for example, can burn out the primary winding. Since a heavy current circulates through the shorted turn, and since this power must come from the primary circuit, primary current rises above its normal value. Similarly a short in the B+ circuit will often burn out the primary winding of the transfomer.

Turn-to-turn, or *layer-to-layer* shorts can sometimes be identified. All secondary leads are disconnected, and a light bulb is wired in series with the primary. If the transformer is not defective, only a small magnetising current will circulate in the primary circuit-not enough to light the bulb. If the transformer has shorted turns, however, the primary current will be *high enough to light the lamp.*

(5) Voltmeter Tests

Defective inductors and transformers can often by located with voltage checks. A power transformer, for example, can be tested simply by measuring the voltage across each winding. *The voltage outputs of the two sections of a centre-tapped secondary should be measured separately and compared.* These voltages should be equal within manufacturing tolerances. *Unbalance* in the two halves may indicate shorted turns, particularly if the transformer has been running hotter than normal.

If a power transformer appears to be overheating this would lead to the assumption that it is either *shorted or overloaded.* If the transformer is cold, it can be assumed to be either open or unloaded. In either case the results obtained by an ohmmeter will verify the assumption.

In power transformers the *primary* usually is indicated by black wires. The *high voltage secondary* is indicated by red wires (with a red-and-black wire where a centre tap is used) and the *low voltage leads* are green, yellow, or slate in most cases.

The auto-transformer is an exception to the rule that *there is no direct metallic connection between the primary and secondary.* In the autotransformer, if the whole winding is used as the primary, the autotransformer has *step-down* ratio. If part of the winding is used as the primary the autotransformer has a *step-up*-ratio. The autotransformer is sometime constructed with several *fixed taps*, to permit variation of the secondary voltage.

29.4 COLOUR CODE FOR RESISTORS

Carbon resistors are made in power ratings of 1/8W, 1/4W, 1/2W, 1W, and 2W,. Their sizes generally differ and are identified by their diameter and length. Standard tolerances are ±5% and ±10% of the nominal value although for special applications, such as, in measuring instruments, precision instruments etc, carbon resistors are sometimes made within *closer tolerance*. The highest permissible ambient temperature for the normal carbon resistor is 70°C.

Carbon resistors are manufactured in a *standard range of values*. For Philips resistors, this range is as follows:

10	15	22	33	47	68	100
11	16	24	36	51	75	110
12	18	27	39	56	82	120
13	20	30	43	62	91	etc.

In the range given here we have indicated the *units*: it will be understood, of course that the *same sequence* has also been adopted for hundreds and thousands. The resistance value of Philips resistors is always given in the *last part of the code number*. In the circuit diagram of Philips Service Documentation, this value is indicated by figures and letters E, K or M, For example:

Code No.	Indication in Philips Circuit diagram	Value
3115 211 03101	100E	100
3115 211 03123	12k	12,000
3115 211 03392	3k9	3,900
3115 211 03225	2M2	2,200,000

The 10th and the 11th digits, from left to right, of the code number, represent numerical values which are to be multiplied by a factor denoted by the 12th (i.e. last) digit. The last digit denotes the *multiplication factor* as specified below:

8 = 0.1

9 = 1

1 = 10 (*i.e.* followed by 1 zero)

2 = 100 (*i.e.* followed by 2 zeros)

3 = 1,000 etc.

4 = 10,000 etc.

5 = 100,000 etc.

6 = 1,000,000 etc.

Most of the manufactures identify their resistance value by means of a *colour code*.

Colour	Tens	Units	Multiplier	Tolerance
Black	0	0	1	20%(M)
Brown	1	1	10	1%(F)
Red	2	2	100	2%(G)
Orange	3	3	1000	-
Yellow	4	4	10000	-
Green	5	5	100000	-
Blue	6	6	1000000	-
Violet	7	7	-	-
Grey	8	8	-	-
White	9	9	-	-
Gold	-	-	0.1	5%(J)
Silver	-	-	0.01	10%(K)

Close Tolerance Resistors

1% resistors have five coloured bands instead of four. The colour coding is basically the same except that they have three bands indicating the significant values, the fourth band is the multiplier and the fifth is the tolerance.

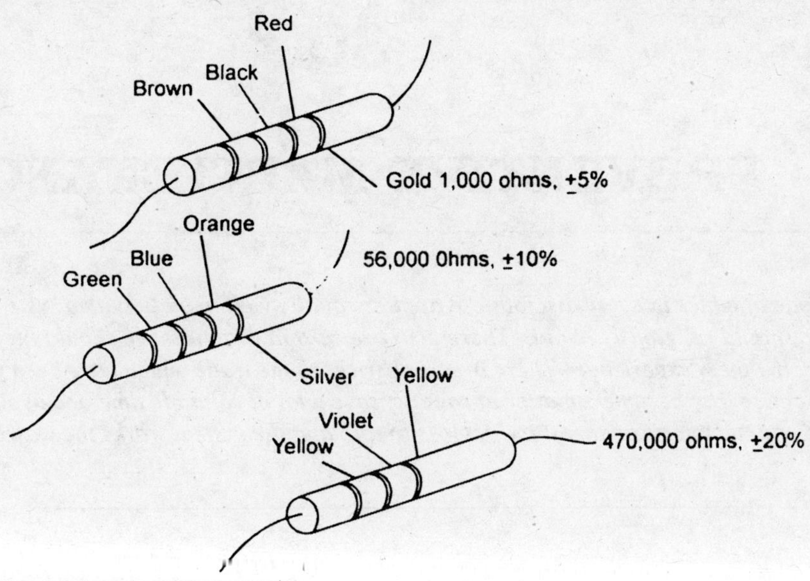

Fig. 29.8 Illustrating colour coding pattern for resistors

CONVENTIONAL QUESTIONS

1. Discuss the basic philosophy of trouble shooting.
2. How will you perform routine tests?
3. When, why and how will you replace a transistor?
4. How will you check an IC (*a*) in-circuit and (*b*) out-of-circuit?
5. What precautions will you take when applying brute force technique?
6. How will you test a transformer by (*a*) resistance measurement and (*b*) voltage measurement?

30

PREVENTIVE MAINTENANCE

There is a code of ethics in every discipline-it applies equally well to audio/video systems. There are certain practices we should religiously avoid. There also are certain practices we should literally follow. We also learn a lot from others experience. There are some tricks of the trade which have been passed on to us. After sustained practice they become a matter of routine, save a lot of valuable time and avoidable headache. The end result is optimum performance - the devices breath and the system works the way it should.

30.1 TAPE RECORDERS

Preventive maintenance is basically concerned with the *correction of a potential trouble* and thereby avoids repair costs and down time. Properly performed, it ensures top performance from a recorder at all times in much the same way that periodic tune-ups and lubrication ensure top performance from an automobile.

Carry out a *through visual inspection.* Check tape guides; *they should be clean and polished* with no dirty deposits. Check the mechanical action of the play, fast forward and rewind keys; *they should operate smoothly and interlock properly.* Any faults here should be dealt with by a skilled technician because the mechanical adjustments needed are not simple, and have to be carried out in *proper sequence.*

Maintenance work which is required to keep a stereo cassette system in perfect condition can be classified into three types:

1. Mechanical maintenance of the deck.
2. Electrical maintenance of the deck and amplifier.
3. Cassette maintenance.

Of these three, the untrained operator can tackle some of the mechanical maintenance but *electrical maintenance is seldom required and needs special knowledge and instruments.*

30.2 MECHANICAL MAINTENANCE

The only mechanical maintenance that is normally required on a deck is the cleaning of the heads and capstan spindle.

Tape heads, and many of the mechanical parts of a tape player, tend to collect dust and dirt. Additionally, *oxide particles* from the tape itself can cause a gradual buildup on the tape head which, just as gradually, results in deterioration of the units playback performance. For this reason, tape players require periodic maintenance to keep them in top operating conditions and to prevent the possibility of an unnecessary malfunction.

Periodic maintenance for some users may be once a day, and for others, once a week. It all depends, of course, on the *operating conditions* and *amount of usage.* Generally speaking, most tape player customers complain only when the sound becomes distorted or operation is erratic. It would be of advantage if a *periodic maintenance schedule,* based on the use of the machine, is set up. Operating conditions, too, must be taken into account.

Cleaning Tape Heads: Deposits on the tape head are equally detrimental to the performance of the machine. The magnetic flux pattern at the gap of a recorder tape head *spreads out* so that its width is very much greater at a distance of a few thousands of a millimeter from the head surface that is close to the head. A greater width of the flux pattern on the tape, see Fig. 30.1, means that fewer reversals of flux can be recorded per millimetre of tape, which in turn means that the higher frequencies are recorded at lower amplitude. A similar state of affair's exists on playback; *any spacing between the tape and the head causes a loss of high frequencies.*

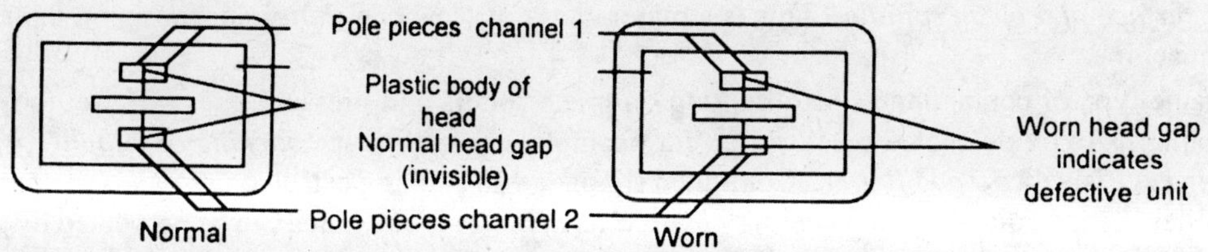

Fig. 30.1. Normal and worn stereoheads.

The *erase head* is less critical, but a bad buildup of dust will degrade the erase performance to the extent that a faint remaining signal can be heard when a previously recorded tape is used to make a new recording. This is undesirable.

Extreme care must be taken when cleaning tape heads, for they are easily damaged, mechanically as well as magnetically. Tape heads are made either of nickel alloys or of glassy ferrite materials. The metal nickel alloy materials are easily chipped. No hard metallic object should ever be held near the tape heads. No abrasive material (apart from tape) should ever be rubbed against them. In addition, because the heads detect magnetic flux, no maintenance object or object made of material which can be magnetised (iron, steel, cobalt, nickel) must be allowed near the heads

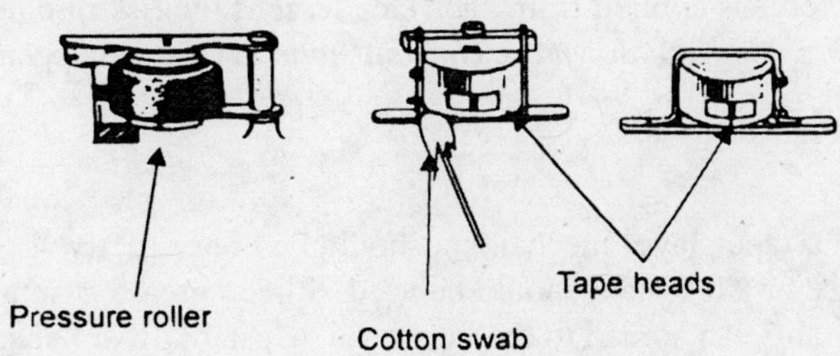

Fig. 30.2. Cleaning the heads.

When heads need cleaning, the safest material to use is *cotton wool* held in plastic tweezers or on the end of wooden splints. This is illustrated in Fig. 30.2. The cotton wool should be moistened with *industrial alcohol* and rubbed gently against the heads until no trace of brown oxide deposits can be seen.

Some tape players have a convenient removable cover or flap which allows access to the tape head.

Do not use cleaning fluids which are not specifically recommended for tape head cleaning, since they may damage the head as well as other components in the tape player. And never attempt to clean the heads with anything metal. Some tape player manufacturers (for example, Philips) also produce their own *head-cleaning solvent* which they specify to be used. However, isopropyl alcohol is recognised by most manufacturers as being a suitable cleaner.

Cleaning Capstans, Rollers and Pads:

Deposits of dirt on the capstan spindle are undesirable for two reasons. Firstly, a sticky deposit on the spindle will encourage the tape to *wrap around the spindle*, causing jamming and usually leaving the capstan spindle in even worse condition, because the large amount of friction between the spindle and the tape just before jamming takes place will rub off even more of the tape on the spindle. The other problem is that dirt on the spindle will not be perfectly evenly distributed, *so that the dirty surface is not so perfectly circular as the polished surface of a clean spindle.* This is a major cause of *wow and flutter* appearing on a previously satisfactory machine.

The same type of contaminants also buildup on pinch rollers and pressure pads. These, too, require periodic cleaning. An oxide coated pinch roller, for example, can cause tape *sticking or slippage.* This, in turn, results in excessive *wow and flutter.* Continued slippage and sticking can also produce a permanent flat spot on the roller. This frequent and unnecessary *byproduct of poor maintenance* can be prevented by periodic cleaning.

In all cases, continued operation of dirty, oxide coated rollers and pads increases the possibility of clogging the surface pores of these components. Eventually, *rubber rollers become hard, pads become stiff and brittle, and both loose their effectiveness.*

Normally, the only sure way to clean the capstan, rollers, etc, is to remove the player and take off the cabinet cover. In this way, cotton swab or cotton pad moistened with isopropyl alcohol can be used effectively. In the case of the capstan and pinch rollers, they can be more easily cleaned by applying power to the player and activating the motor switch to operate the moving parts.

Great care must be taken to ensure that no force is applied which might bend the spindle or damage the bearing, nor must even the mildest abrasive cleaning powders or glass paper be used on the spindle, whose surface is machined, polished and plated to an extremely fine finish. Cleaning must be confined to gentle treatment with solvents and the best method known is the use of cotton wool moistened with alcohol. The cotton wool is held in plastic tweezers and wrapped in a half-turn round the capstan spindle, which should be revolving. Very gentle pressure on the tweezers will then be enough to clear sticky deposits from the spindle, and the fact that force is applied from each side lessens the risk of damage. *The amount of alcohol used should not be so great as to run down the capstan spindle.* If this happens, stop the machine and mop up the excess liquid.

30.3 LUBRICATION

Any lubrication of a tape player mechanism should be done strictly according to the *manufacturer's instructions.* A light grease or a lubricant should be used. Silicon grease is sometimes required. When using it, silicon and other lubricants can spread to the tape head, capstan, drive belts, pulleys and flywheel if care is not taken during its application.

Generally, the drive surfaces of a cam and ratchet assembly, the solenoid plunger, the pivot points of parts in an eight-track player are the only areas requiring lubrication.

Many of the auto-stereo tape players now use special bearing for capstans, motors, and pinch rollers that require lubrication. However, the *metal-to metal moving parts* sometimes require a small amount of lubricant, especially if that particular part has been replaced.

No general rules on lubrication can be given here, because machines vary so much, except to note that *no lubrication of any sort should ever be used unless advised in the machine manual.* The reasons that many bearings are sealed for life, and others consist of low friction can cause damage by swelling the plastic materials. *If any form of lubrication is advised, it must be carried out using the recommended oil in the smallest possible quantity.* Remember that tape dust will settle on any oily surface; transforming the oil into a grinding paste which is capable of damaging small bearings quite rapidly. *Oil can, thus, cause considerable damage if wrongly applied,* so that unless a manual is available with advise on lubrication, no oil should be used on the capstan bearing.

30.4 ELECTRICAL MAINTENANCE

Very little electrical maintenance is needed or can be carried out on a stereo cassette recorder. One quality, however, can contribute greatly to better quality of recordings, and that is *head demagnetisation.*

In use, of course, recording heads are being magnetised in alternate directions by the bias frequency and also by the signals being recorded. This magnetisation is *alternating*, and should not level any permanent magnetisation in any direction. It can happen, however, that because of the way in which the bias signal is turned off when the stop key is pressed, or the shape of the bias signal itself, the head can be left magnetised in one direction or the other. This is usually prevented by making the stop key turn off the signal frequency first, then the supply to the bias oscillator so that the bias signal dies away comparatively gradually instead off being switched off abruptly, before the head is disconnected. Even when this is done, differences in the speed of operating the stop key may catch the mechanism out, and sometimes disconnect the head before the bias signal has completely died away, so leaving the head slightly magnetised. In addition, the weak magnetic field of the earth, and the stronger field around iron and steel objects can cause weak magnetisation of the record/play head.

The effect of a permanently magnetised head is to add a signal to the tapes which pass the head; not a signal of any particular frequency but a noise signal, because some particles of tape are affected by the weak permanent field and some are not, so that a random rearrangement of molecular magnets takes place, producing noise on playback. In addition, some erasure of very weak signals takes place at random.

Since demagnetisation is fairly simple, *its better to damagnetise as an item of routine maintenance; perhaps once a month.* There are many good commercially available demagnetisers. These demagnetisers come in the form of AC powered hand operated units, while others are built into cassettes and cartridges for DC operation in the vehicle (in the case of auto-stereos).

Regardless of the method used to demagnetise a tape head and surrounding components, *always have the demagnetiser at least to face away from the tape player before turning it on and off.* Otherwise, you may defeat the demagnetising process.

To demagnetise a tape head, slowly move the demagnetiser, see Fig. 30.3, tips over the tape head, tape guides, and capstan. *Then slowly withdraw the unit to a safe distance (two feet or more) before unplugging it.*

Fig. 30.3. Head demagnetiser

It is quite easy for many oxide particles to become lodged in the head gap and stay there even though the head is cleaned and demagnetised. The pores in rubber pinch rollers and pressure pads can also become digged with these tiny particles, which are difficult to remove.

The *wet demagnetising process* means demagnetising while the surface of the tape to be cleaned is still wet with cleaning solution. *It is desirable to have the tape head facing up, which allows the alcohol to soak in for a few minutes before demagnetising.* The alcohol tends to loosen the oxide particles and helps the demagnetiser in removing them.

The same procedure can be very effective in cleaning rubber pinch rollers and felt pressure pads when surface pores have been clogged with oxide from continued use.

30.5 TAPE SPLICING

There are a number of tape splicers on the market which help to make repair of a valuable tape relatively quick and simple. *The nature of the magnetic tape allows it to be readily cut and replaced.* However, there are some conditions which you should be aware of regarding the splicing itself.

Many pressure-sensitive tapes available today are not the best thing to use for splicing magnetic tape. Splices in a reel of magnetic tape are subject to a great deal of pressure. These pressures cause the adhesive in some tapes to creep out or *bleed* around the splice. This not only weakens the splice, but also introduces the possibility of the adhesive contamination of the tape head and other parts. It could even cause adjacent layers of tape to bond together. So the best and safest approach is *to select a product specifically made for splicing magnetic tape.*

A broken tape in a cassette can normally be brought out of the case without too much difficulty. Many tape cases can also be taken apart if necessary. Some tape cases are sealed together, while others use plastic looking tabs which can be released by inserting a small blade screw driver through an access hole provided to each tab. Others are simply pressed together. In any event, *the broken ends of the tape can usually be retrieved.*

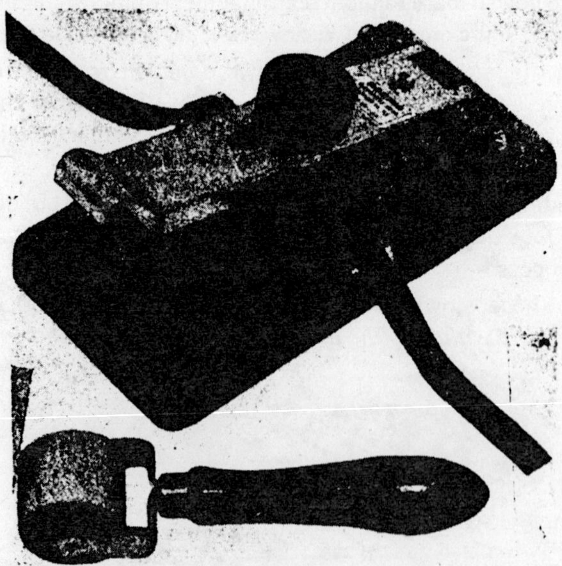

Fig. 30.4. A typical tape splicer.

A *splicing block*, Fig. 30.4 is generally the best type of splicing jig to use. To make the tape splice, lay the two ends of the tape in the splicing block tape groove so that they overlap. Lay the tape in the block with its oxide side down. *Magnetic tape is always spliced on the opposite side of the oxide.*

Cut the two tapes in the slot provided. Normally, a *diagonal cut* is desirable. Remove the cut off pieces, but leave the two ends of the tape butted together. Centre the splicing tape over the junction of the tape ends and press the splicing tape into position. Rub the entire spliced area of the tape to be sure it adheres evenly and firmly. Remove the splicing tape carrier and the splice is completed.

If a splice is properly made with good *splicing tape*, it will probably last as long as the tape. Rewind the tape onto the reels and *give it a quick check before returning it to your customer.*

30.6 CARE OF CASSETTES

Cassettes require more care and maintenance than open reel tapes, because part of the mechanism, which in the open reel recorder is contained in the deck, is incorporated in the cassette itself. In a cassette, the tape is *wound* on small reels, with no moving side checks, and is *guided* by grooved posts to the heads. The tape is *pressed* against the heads by a pressure pad which is part of the cassette itself.

Avoid at all costs unbranded or unknown brand casettes sold at low prices. Some of the standard Indian brands are: EMI, MIL, Philips, T-series, Venus...... *Substandard brands may contain tape of reasonably good quality, but are seldom mechanically satisfactory.*

Fig. 30.5. Cassette Storage.

Good cassette care starts with suitable *cassette housing*. Cassette *storage*, Fig. 30.5, is not such a problem as disc storage, but some care should be taken against the two main enemies of cassette tapes; dust and magnetic fields. *Avoid keeping cassettes in the recorder or lying about uncased.*

Apart from care over storage, cassette maintenance consists of care of pressure pads, Fig. 30.6 ensuring free movement of reels. *Pressure pads should be inspected at intervals of say about two to six months.* The pad should have a *soft surface*. Maintenance consists of brushing the pad with an old tooth brush.

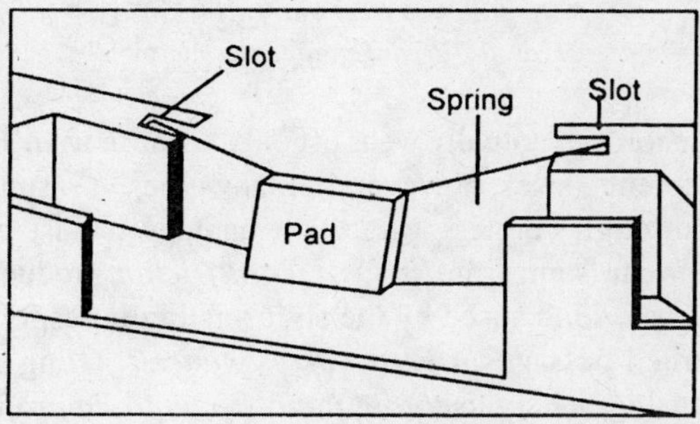

Fig. 30.6. Pressure pad

A pressure pad spring which is bent inwards can be straightened by holding the cassette with one finger and thumb placed approximately over the spring mounting slots and using plastic tweezers to bend the spring gently outwards in a direction opposite to that of the permanent bow.

When the tape is eased away from the pressure pad, the movement of the reels can be checked. Each reel should move easily. If there is any trace of jamming, place the cassette in the recorder, switch on, and use the *fast forward* to wind the tape to the end. Then use *rewind* to place the whole length of the tape on the other reel. The reels should now be free and the tape evenly wound.

Before the cassette is inserted into the cassette recorder, check whether *loose tape* is hanging from it as this could get entangled in the mechanism. Tighten it with a pencil.

30.7 RECORD CARE

Records should be stored in a *cool, dry place,* out of the sun and away from any radiators during the winter. They should never be left in an enclosed area exposed to sun's rays. Temperatures above 30°C can cause serious warping and other damage. Low temperatures make records brittle, but they will become normal again when returned to room temperature.

Records should be stored in albums or in sleeves, and filed on edge, snugly, so as to stand straight and not lean over. Shelving should be sufficiently wide to keep the record from bending over the edge, Fig. 30.7. Flat storage is satisfactory if each record is well supported over its entire area, and the stack comprises records of one size only. *Records should never be left on the support of a changer mechanism* they can warp at the edges or middle as a consequence. Nor should they be stored so tightly that the sleeves or albums stick together; trying to pry the latter apart may cause breakage. *The record storage area should be reasonably clean and free of dust and maintained at a normal room temperature.*

Fig. 30.7 Record storage mechanism

Even with the best care, records eventually wear off only because *wear from playing is inevitable*. Wear can result from a number of different causes, however. In many cases, it results from *sheer abuse*, sometimes unknowingly and sometimes through carelessness. Such damage is just as harmful as if it were done deliberately. In other cases wear can stem from *stylus-record friction* (reducible by means of spray-on wet and dry lubricants), *groove deformation* caused by the stylus acting in response to excessive tracking force, stylus response to heavily recorded passages, *stylus self-resonance* reacting with recorded frequencies at or near the resonant point and causing the stylus to *dig* into the record, *improperly set antiskating forces and electrostatic charges between record and stylus.*

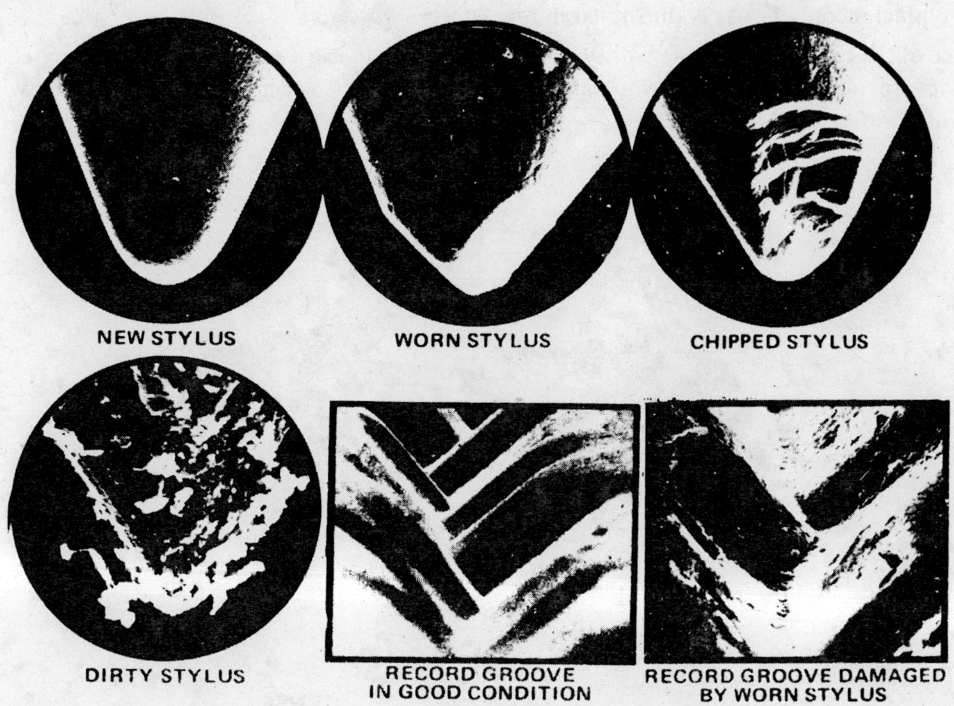

Fig. 30.8. Record wear can result from a number of different causes.

Records soon become noisy and scratched if handled with greasy, gritty fingers and gripped at the grooved portions. If a record is handled in this fashion, *scratches cause the noise level to increase, and the grease attracts dust.* This, in turn, causes more scratches and still more noise.

Many of the worst types of scratches are caused in removing and replacing the records in albums or sleeves. *Always spread the sleeving at the edge before removing or replacing.* A good way is to *squeeze* the sleeve against your body and then push the record out slightly with a pencil inserted into the centre hole. Then you can grab the edge and slowly ease out the record, keeping the sleeve bellied out so that its contact with the record is just the reverse process when replacing the record. *Once the record is out, always handle it by the edges, or by edge and centre, never by the grooved area.*

Warping is another problem that is sometimes owner-caused. Fortunately, minor warps can be cured, although the record will never be as good as new again. If the warp is not too severe, it can be straightened by placing the record on a flat piece of plate glass in a room just a little warmer than usual. If necessary you can make a sandwich with another sheet of plate glass on top.

Another wrong habit is that of placing the pickup arm on the record in some spot other than the beginning. Whenever this is done, no matter how carefully, the record sustains a little dig or scratch. Each of these little *potholes* brings up the noise level, besides causing stylus wear and vibration, and in a short time you can throw away both record and stylus. Fig. 30.8 illustrates the different causes of record wear.

Records must be kept as clean as possible. Dirt, in addition to causing surface noise, may collect around the stylus point, introducing severe distortion. There are *cleaning devices* that can be used, such as a brush attached to the tone arm. It picks up the dust as the record turns. It, and the dust it continues to collect add weight to the tone arm, requiring that the arm be rebalanced. One solution to this problem is a *separate arm.* These can be mounted on the player case anywhere, temporarily or permanently, so that the records are kept clean via the brush and round plush pad., Other means of keeping records clean can be obtained in kit form.

A typical record cleaner is illustrated in Fig. 30.9

One of the best ways to clean a dirty or greasy record is to use a soft cloth, lightly dampened in a tepid solution of a soapless detergent, Or you can use plain tepid water, not hot; then immediately wipe dry with another soft, lintfree cloth.

Fig. 30.9 Record cleaners.

30.8 CARTRIDGES

Because the tonearn is supported by the stylus/cantilever during record/play, and because the cantitever has a spring-like action, they form a *vibrating system.* Like all vibrating systems, this one also has a *resonance frequency.* It's important that the resonant frequency of the vibrating system falls in what is called the *safe range* - above the frequencies of record warp and below the frequencies where the music takes place.

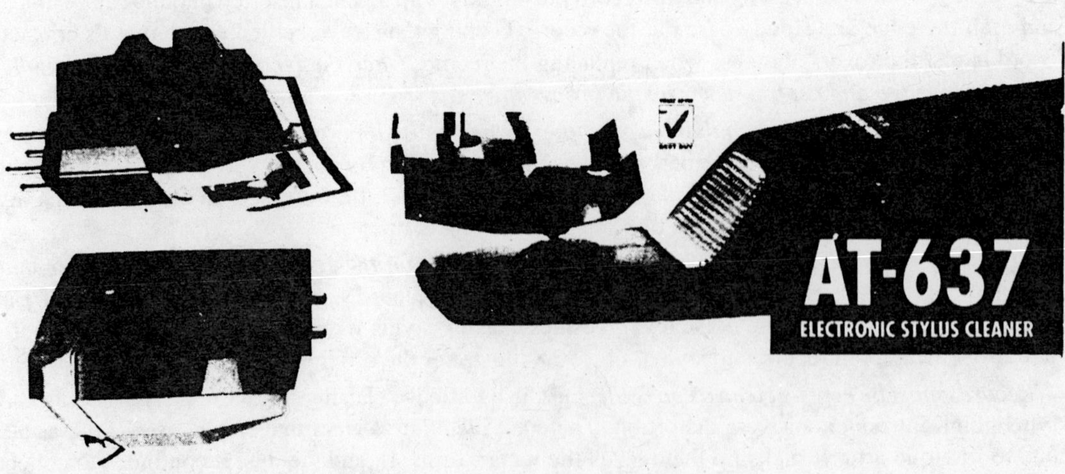

Fig. 30.10 Audio Technica Cartridges

The only way to achieve this is to carefully match the compliance or springiness of the cartridge's cantilever with the weight of the cartridge, head shell and tonearm. Here, the advice of a knowledgeable dealer is required.

Because of the many variables involved with catridges, the only consistent guide is *sonic performance.*

Listen to the hi-fi system using high quality turntable and tonearm and relatively new records. Use the same amplifier and speakers so you can judge the *overall system impact of the sound produced by the cartridge.*

The following golden rules should be kept in mind :

- It always pays to change to a better cartridge. *It is the component in a hi-fi system that makes the first and only contact with record grooves, so it can never be too good.*
- Be sure to *choose a cartridge with a compliance that matches your type of tonearm.* Your dealer will be able to help you in this matter.
- *Never play old or damaged records with a new cartridge.* This may ruin the diamond stylus which in turn will ruin your new records
- *Remove dust from the records and the stylus each time you play.* When necessary, use a record cleaner or a cleaning fluid recommended by your dealer.
- *Have your cartridge checked regularly* (at leaste once a year) by a good hi-fi dealer who has the necessary equipment.

30.9 MICROPHONE MAINTENANCE

Severe *machanical shock* should be avoided for all types of microphones, even the robust moving-coil units, so *when in transit they should be adequately packed.* Some makers supply *foam-lined cases* and they are ideal for storage and transport.

Care must also be taken when connecting to an electrical circuit. *No current from an external source should ever be allowed to pass through the microphone.* Moving-coils are tolerant of small currents such as from an ohmmeter, built ribbons are especially vulnerable and no attempt should be made to measure their resistance.

Dampness should be avoided with all types, as should *excessive heat.* Electret and capacitor units are more prone to heat damage than the dynamics. *Dust* has little effect on a diaphragm or cone, but can cause problems with a ribbon by building up a deposit between the edge of the ribbon and the magnet pole-piece. *Especially to be avoided are iron filings which are attracted by the powerful magnet and are almost impossible to afterward dislodge.*

Fig. 30.11 The microphone.

With electret microphones, the *battery*, although long-lasting, does need occasional replacement. Here, as with all other battery equipment, the battery should be removed if the microphone is to be stored for sometime, unless it is a mercury cell. Ordinary zinc battery cells corrode from the inside and when finally the battery case is holed, the corrosive fluid can then cause irreparable damage to the microphone. *A smear of silicon grease on the battery terminals helps to prevent oxidization and noise.*

Apart from these points microphones need very little maintenance or attention, and if not misused should remain active for many years with little if any deterioration of performance

30.10 CD PLAYERS

There is considerable disagreement among CD player manufacturers concerning the need for *periodic maintenance* or routine checks. At one extreme, a certain manufacturer recommends *replacement of a few parts after a given number of playing hours or playing times*. (They recommend that the *optical pick up* be replaced at 5000 hours of playing time, and *all motors* be replaced after 10000 plays). At the other extreme, another manufacturer recommends no periodic replacement cleaning, lubrication or adjustment of any kind. "*Fix it if it breaks down is the rule*" (It is fair to say that this rule will probably be observed religiously).

Somewhere between these two-extremes, other manufacturers recommend adjustment (electrical and/or mechanical) *only as needed* to put the player back in service, or when certain parts or assemblies have been replaced. However, most CD player manufacturers recommend a *complete checkout* using the *recommended test disc*, after any service.

Follow the manufacturer's recommendations. Regard a CD player in the same way as you do an audio/stereo player or TV (that is, *bring it in for service when it breaks down*.) However, there are some *maintenance procedures* that apply to users of CD players.

Cleaning the CD player

Use a soft, clean cloth to wipe off dust and dirt accumulated on the player. If necessary, moisten a soft cloth with diluted neutral detergent to remove heavy dirt. *Never use paint thinner, benzene, or other solvents*, since any of these solvents react with the surface and cause colour changes and melting.

Objective lens care

The objective lens is a key part of the CD player. Note that the lens surface must be clean (and free of moisture) in order to maintain the best performance. *Never try to touch the lens surface. Keep the disc compartment closed to keep out dust and dirt. Open the disc compartment only when inserting and removing the disc.* If too much dust or dirt accumulates on the objective lens, sound quality can be degraded. Dust can be removed from the objective lens with an air blower for a camera lens.

30.11 CD CARE

Since compact discs are *played without physical contact with the playing surface*, the discs do not wear out or degrade in quality. Also, light scratches or small amounts of dust or finger-prints have almost no effect on the sound. A heavy accumulation of dust and dirt, however, can reduce sound quality and should be removed by wiping both sides of the disc with a clean soft, dry cloth. *As a matter of routine cleanliness, all discs should be handled by their edges the same as a phonograph record.*

There are a variety of playability problems associated with *warped CDs*. Such problems include hanging (or sticking), skipping, and loss of focus. A severely warped disc can even hit the player lid or objective lens. To determine if a CD is warped, lay the disc on a flat surface and press down on the centre. If no movement is detected, turn the disc over and again press down in the centre. If movement is detected on either side, measure the amount of movement. The distance should be no more than 3 mm. (just under 1/8 in). *If the amount of warpage is over 3 mm, the disc is defective and should be replaced.*

- Use the following procedure *to remove scratches* from the surface of a compact disc
- Use a polishing compound. *Do not use a rubbing compound.* Place a small amount of polishing compound on a dampened soft cloth. With a *light, circular motion*, rub the surface of the disc until the scratches or marred area is restored to a luster.

- If the compound begins to dry up on the disc surface before the scratch is removed, *add a small amount of water* to the cloth and continue rubbing.
- After the marred area has been restored, remove any residue from the disc surface with a clean, dry cloth.

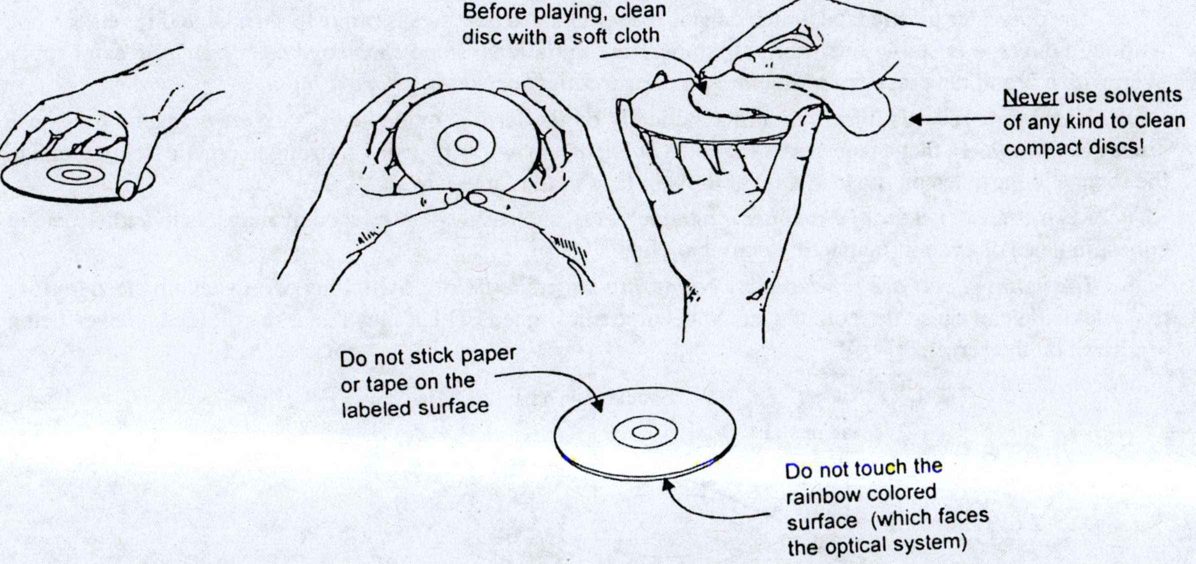

Fig. 30.12 Handling compact discs.

If this procedure does not remove the scratch or mar, the disc is probably beyond hope.

Try to *avoid cold temperatures*. When CDs are subjected to extremely cold temperatures, allow 45 minutes to 1 hour for the disc to return to room temperature before playing.

To prevent warping, *always store CDs on edge*. A standard phonograph record rack is a good storage device for CDs. Do not stack CDs, if at all possible.

CD cleaners

A soft, dry cloth will remove most dirt found on a CD. However if a CD is exposed to very dirty fingers, liquids, tobacco smoke, fire and brimstone, and so on, you may need help in restoring the disc. Cleaners can be of great help in these *extreme cases*.

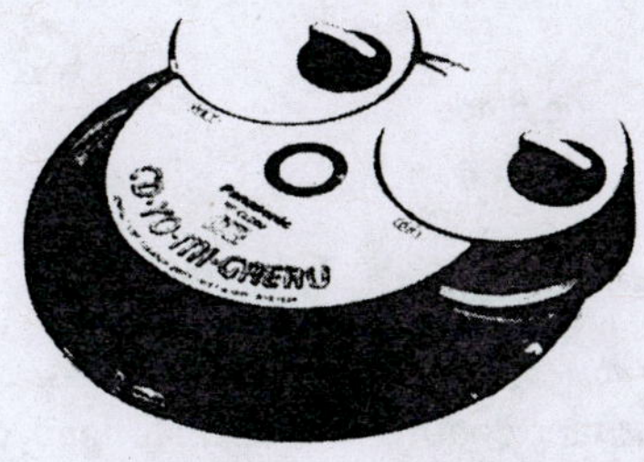

Fig. 30.13 Panasonic's RP. CC 200 disc cleaner.

30.12 VIDEO CASSETTES

Video tape is more than 20 microns thick. It has to stand up to *winding on and off* high-speed rotating drums and *threading and unthreading* hundreds of time during its lifetime, *without breaking or distorting* as much as a thousandth of an inch.

The polyester plastic tape which carries the specialised coatings is prone to shrinkage if it gets too hot. Although the tape is stable over normal temperature variations, some *shrinkage and stretching* can happen even with a brand new tape, so *try winding it forward and backward before using it.*

Video tape records information magnetically on the ferrous oxide layer; so *stray magnetic fields can erase recordings.* In fact a tape eraser (called a *bulk eraser*) works by using a strong magnetic field to realign the magnetic particles on the tape so that they all face in the same way.

Keep cassettes clear of any other magnetic fields, such as those produced by magnets in audio speaker cones, in tapes or even in magnetic document clips.

The worst effects are produced by collapsing magnetic fields. Switching off a vacuum cleaner close to a television can cause the colour guns to become misaligned, which may result in coloured patches being produced on the screen.

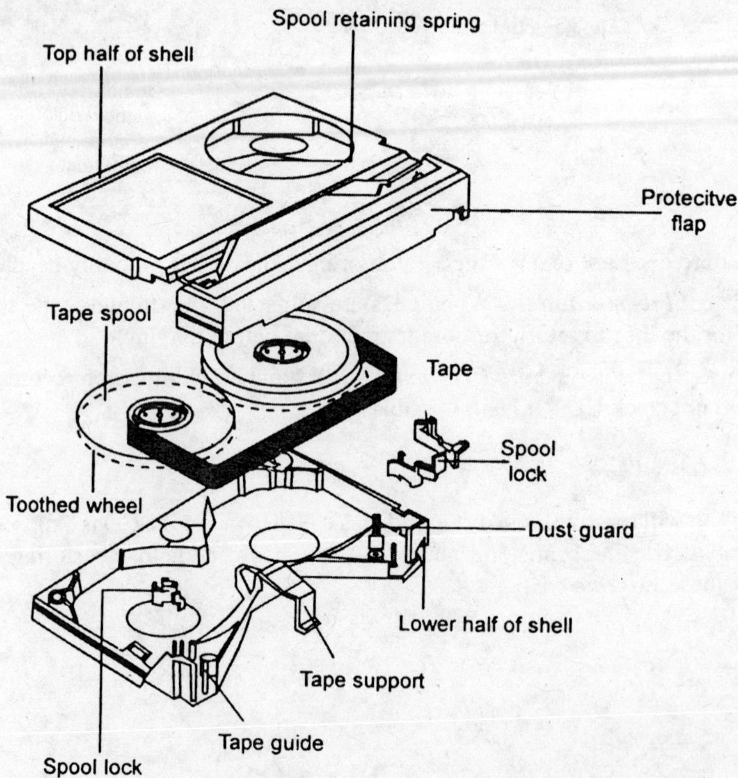

Fig. 30.14 A video cassette might look like an audio cassette, but inside it is very different. The tape spools are mounted on toothed wheels, which are released only when the tape is inside the recorder, to prevent it from slackening. The two halves of the plastic shells are held together by screws (five for domestic VHS and six in a stronger arrangement for industrial machines).

In time, the *magnetic particles* of the tape coating will be dislodged, causing *white streaks* on the picture. This is called *dropout*, and it is usually more common in the first few minutes of a tape. One way of postponing this problem is to *begin new recordings at different times over the first five minutes of tape to reduce head wear at any given point.*

Break off in the direction of the arrow

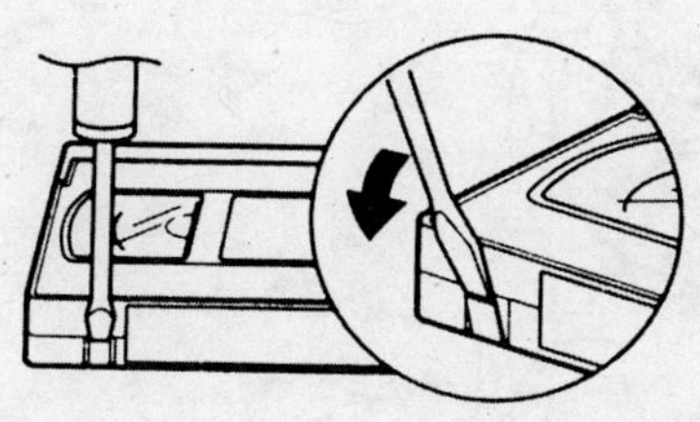

To record on the video cassette again

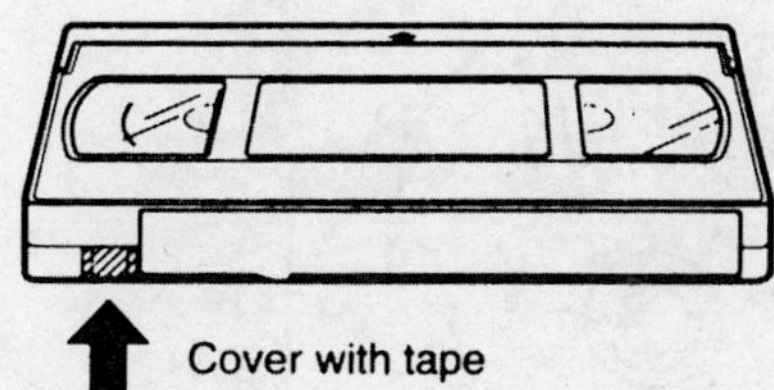

Fig. 30.15 Safety tab-protecting valuable recordings from erasure.

Video cassettes have a *safety tab to prevent accidental erasure of recordings.* When a cassette is loaded into a VCR, a *probe* detects the presence of the tab. *If it has been removed the tape cannot be erased.* If it is necessary to record on the tape at some future date, the tab hole can be covered by a piece of adhesive tape. VHS, Betamax and U-matic cassettes each have *one* tab. Video 2000 cassettes have *two* tabs.

30.13 LOADING A CASSETTE

Most VCRs are *top loaders.* The detailed procedure for loading a cassette in top-loading VCR is illustrated in Fig. 30.16. The cassette drawer opens for loading and unloading by rising about 3 in.(7.5 cm) above a top panel. *Allow for this when fitting a VCR into shelving.*

Loading a cassette

Before a cassette can be inserted into a top-loading VCR, the cassette drawer must be opened by pressing the 'Eject' control. Hold the cassette with the protective flap farthest away from you and with the loose spools facing downward. Push the cassette into the open cassette drawer. The cassette drawer must then be closed by pushing it down until it locks into place. The recorder is now ready for playback or recording.

It is not necessary to prepare a front-loading model before inserting a cassette. The cassette is held as before and pushed into the loading slot. When it is about half way in, the motorized mechanism takes over, pulling the cassette away from you and down into the machine.

When loaded, a cassette can be seen through a window in the top loader's cassette drawer. This is not so with an all-enclosed front loader, and in some machines it is possible to insert a second cassette on top of the first. VHS, Betamax and U-matic cassettes can be inserted only if they are the right way up. Video 2000 cassettes can be inserted either way up, but the protective flap must always be farthest away from you.

Fig. 30.16 Loading a cassette in a top-loading VCR.

Differences in tape lacing in the three VCR formats are illustrated in Fig. 30.17.

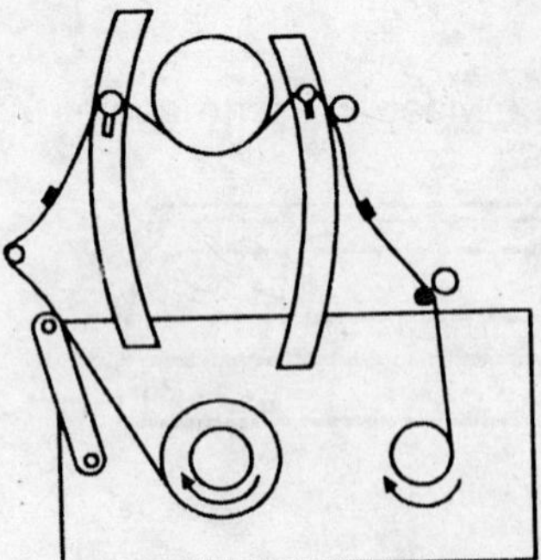

When a VHS tape is inserted in a recorder , the spool locks are released and two guide poles pull a length of tape out of the cassette. Because the tortuous tape path would place severe strain on the tape during fast wind , it is only laced up during normal - speed playback . M - load (or M - wrap) takes its name from the shape of the tape when laced up , above.

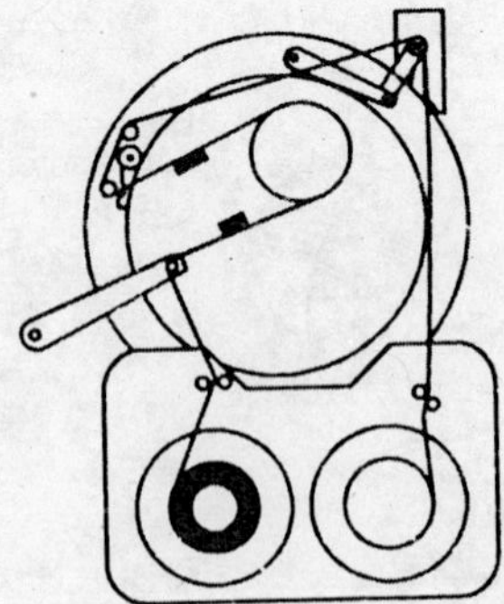

In the semi - professinal U-matic format developed by Sony , the tape is paced up as soon as the cassette is inserted in the recorder . Capstans on a circular guide track catch the tape and pull it round the videodrum . Tension arms take up any slack in the 3/4 in (19 - mm)tape . U - matic format has become the standard for the first cable TV stations in the UK.

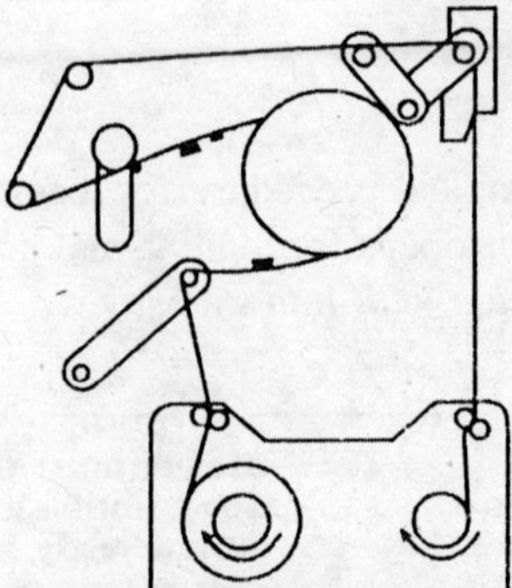

Betamax , developed by Sony , is a smaller version of U - matic , using narrower 1/2 in (12.5 - mm) tape.Lacing operates in the same way as U - matic , although a smaller mechanism is used . The tape remains in contact with the video heads for all playback and winding functions until the ' Eject ' button is pressed . Despite this , there is no excess head wear .

Fig. 30.17 Tape lacing in the three VCR formats.

30.14 VIDEO CASSETTE RECORDERS

What keeps a VCR working well? It needs the *right environment* to work in: not too cold, nor too hot, humid, dusty or prone to vibrations or magnetic fields. It needs what many engineers call *mechanical sympathy* on the part of its owner: do not hit it, drop it, shake it, stand heavy objects on it, or spill things into it. *Use the controls gently and do not confuse it by trying to give it too many contradictory commands at once because you are unsure about what to do next.*

The only repair attention it will need is *general cleaning and demagnetizing* every so often. It is difficult to be specific about how often chores like these need to be carried out. Most experts say that preventive maintenance (*doing the job before the need becomes obvious*) is to be avoided.

It is better to watch out for picture distortion or signs of partial erasure on some of your tapes, which are indications that the recording heads are retaining some of their magnetism, before calling a service engineer to *demagnetize (or de-gauss)* them.

Much more common is *the need to clean the heads* of particles of magnetic oxide which flake off the tapes in regular use. It is difficult to give an accurate figure of how often this is needed. Manufacturers' and testers' estimates vary from 'up to every 1,000 hours of use' to 'as few as every 400 hours or so' depending on the *cleanlines of the atmosphere* in which the machine is kept, and the *quality of the tapes* being used. Poor-quality tapes may shed more oxide particles than the latest, best-quality tapes do.

Once again the answer is to *watch for the symptoms.* Progressively noisier pictures(that is, grainy and speckled with snow) on all your tapes during playback will indicate that the heads are becoming clogged and that it is time to clean them.

Opinions differ sharply too on whether you should do the job yourself, or call in a service engineer. The problem is that the heads are extremely delicate, and the slightest slip in cleaning them could easily cause damage which would cost far more to repair than the price of having the heads cleaned professionally in the first place. Add to that the fact that most manufacturers' guarantees are *rendered invalid* by opening up the machine to perform an operation like this, and *it hardly seems worth while trying to undertake the work oneself.*

Fig. 30.18 Do not use a video head cleaner unless you suspect that video heads are clogged, and thus use it sparingily.

Care and maintenance check list

1. Make sure your VCR is *horizontal* before you operate it.
2. Never stand a VCR on the floor, or on a carpet.
3. Keep your VCR *well ventilated* at all times. Never cover the ventilation grilles.
4. Do not *expose* your VCR to direct sunlight or heat; keep it in a cool, dry place.
5. If you have disconnected the machine for some time, do not use it immediately after reconnecting it. *Give it time to warm up before playing a tape.*

6. Do not *expose* your VCR to moisture from spilled liquids, sudden changes of temperature, or the damp air near a window.
7. After playing a cassette, *remove it* from a VHS or Betamax machine, and *never transport the machine with a cassette in place.*
8. If a fault occurs, *do not try to repair it, and do not continue to use the machine.* Call a service engineer.
9. Keep the VCR's *original packing* in which to transport it.
10. Do not use solvent cleaners on a plastic VCR case. Use a little *isopropyl alcohol* (white spirit) to remove stubborn or oily marks.
11. Wipe the VCR over regularly with a damp cloth, using clean water, or an anti-static cleaning fluid obtained from a video dealer. Dust or vacuum carefully around and behind the VCR to prevent dust and fluff accumulating. *Keep the cassette compartment closed when cleaning.*
12. Do not use a *tape-head cleaner* unless you suspect that your video heads are clogged, and then use it sparingly. If in doubt, call a service engineer.
13. Run your VCR in record mode for an hour every couple of weeks, *even if you do not wish to record.* If you use it only for playback, the heads will gradually become magnetized, causing a loss of picture quality which is symptom of permanent damage to the recordings. *The strong RF signal used for recording reverses this process.* If you suspect your tape heads need demagnetizing, call your service engineer.
14. Take care *not to drop or knock* your VCR. This could cause the video heads to fall out of alignment, a condition which can only be repaired in a video laboratory.
15. Run *trailing leads* round the edge of the room, or cover them with a mat or rubber electrical cord protector, so that people cannot *trip over* them or damage them by treading on them.

CONVENTIONAL QUESTIONS (C)

1. What is the significance of preventive maintenance?
2. How will you carry out the electrical and mechanical maintenance of a tape recorder?
3. Describe the dry and wet procedures of cleaning heads.
4. List the steps you will follow in splicing tapes.
5. Record wear can result from a number of different causes. List them.
6. How well can you ensure top class sonic performance from a cartridge?
7. How will you carry out a maintenance check of microphones?
8. Explain in details the procedure to remove scratches from the surface of a compact disc.
9. What is the correct procedure for handling CDs?
10. Explain the following
    (*a*) Safety tab — (*b*) Head demagnetiser
    (*c*) Pressure pad — (*d*) Record cleaner
    (*e*) Dropout — (*f*) Bulk eraser
11. What steps will you take to avoid shrinkage and stretching in a brand new video-tape?
12. How will you carry out the maintenance check of a VCR?

31

ACOUSTICAL MEASUREMENTS

"I often say that when you measure what you are speaking about, you know something about it, but when you cannot express it in numbers, your knowledge is of a meagre and unsatisfactory kind. It may be the beginning of knowledge, but you have scarcely, in your thought advanced to the stage of Science whatever the matter may be."

-- -Lord Kelvin

A scientific field advances at a speed determined by the availability of adequate measuring instruments. "To measure is to know" is a truism that applies to the audio-video field as well as to others.

31.1 THE MICROBAR

Sound creates pressure. For that pressure to be *sensed*, it must fall on some measuring instrument whether it be a laboratory instrument or the human ear-drum.

One of the more common units of pressure is the *atmospheric-pressure* or the pressure from the weight of the air which surrounds the earth. At sea level normal atmosphere pressure is 14.7 pounds per square inch (see Fig 31.1).

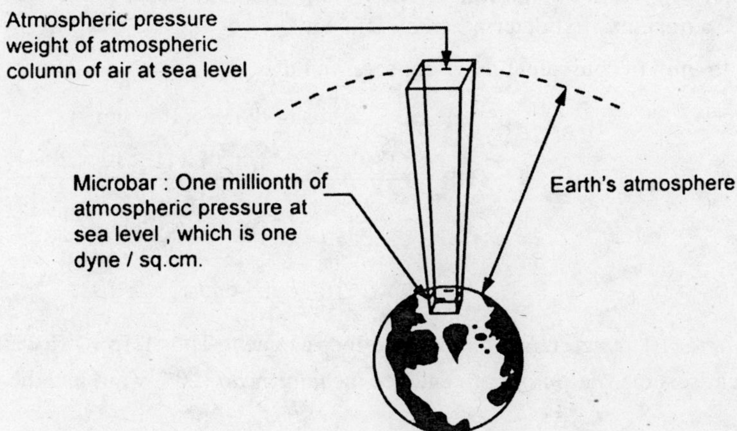

Fig. 31.1 A microbar is a millionth (10^{6}) of the barometric pressure on 1 sq.cm at sea level and is equal to 1 dyne/sq.cm.

We will *convert* 14.7 lbs/sq in to a unit easily measured in the acoustic laboratory. Pounds are converted to ounces, ounces to grams and finally grams to dynes.

Pounds	×	ounces	×	grams/ounce	×	dynes/gram	=	dynes
1	×	16	×	28.35	×	980.7	=	444,800

$$14.7 \text{ pounds of force} = 14.7 \times 444,800 = 6.59 \times 10^6 \text{ dynes}$$

$$6.59 \times 10^6 \text{ dynes/sq.in} = 659.10^6/6.46 = 1.025 \times 10^6 \text{ dynes/sq.cm}$$

If we accept a two percent accuracy in our measurements, we arrive at a figure of 1×10^6 *dynes / sq. cm. as the pressure applied by the atmosphere.*

However it is *far too large* for the minute pressures met in acoustic measurements. Since atmospheric pressure is often referred to as *barometric pressure*, one *micro* atmosphere of pressure, equal to 1 dyne/sq.cm. has been named a microbarometer or *microbar,* a quantity which forms the *basis* of all acoustic measurements.

To give an idea of the actual sound level of this unit, 1μ*bar is the approximate pressure produced by an average human voice speaking at a distance of 2 feet.*

31.2 THE MICROWATT

We have converted a common unit of *pressure* to a useful acoustic unit, we can proceed in a similar manner and convert a common unit of *work* to a useful acoustic unit.

In nature *unless a force accomplishes some useful work, it is wasted.* Work is accomplished only when an applied force moves something in the direction of the force. The simplest way to measure the quantity of work is to relate the weight being moved to the distance it is moved. These factors *weight and distance*—are combined to form a unit called foot-pounds. For example if you carry a 1-lb weight over a 1 ft distance, you have worked an equivalent of 1 ft-lb.

In moving a weight over a distance, the time it takes to move that weight over a given distance must also be considered. Now *horse power* has a very definite value -33,000 ft. lb. of work per minute ; that is, a standard horse can be expected to pull a load of 100 lb over a distance of 330 ft in one minute or a load of 1000 lb over a distance of 33 ft in one minute *or any combination of pounds and distance whose product is 33,000 in one minute.* a quantity first determined by Dr. Watt.

Converting ft.lb /min to equivalent dyne-cm/sec, we have

$$1\text{HP} = \frac{\overbrace{(33,000 \times 444,800) \times (1 \times 12 \times 2.54)}^{\text{lbs to dynes} \times \text{ft to cm}}}{\underbrace{60}_{\text{Min. to seconds}}}$$

$$= 746 \times 10^7 \text{ dyne-cm/s}$$

The *constant factor* 10^7 dyne-cm/sec has been defined as a *watt.* Thus 1Hp = 746 watts. For the acoustical engineer, 10^7 dyne-cm/s. is the watt. For convenience the *microwatt* (10^{-6} watt) and the *milliwatt* (10^{-3} watt) are used.

31.3. ACOUSTIC RESISTANCE

Properties such as acoustic resistance and acoustic current are *analogous* to electrical resistance and electrical current. Just as electrical resistance varies from substance to substance, *acoustic resistance also varies from substance to substance.* Our objective is *to determine the acoustic resistance of the air around us through which sound waves travel to our ears.*

A travelling sound wave is actually a wave of energy. Vibratory energy is imparted to the air molecules by the vibrating sound source and this energy is transmitted through the air to the ears. If it is a *pin-point source,* the sound field has a circular cross-section while at the same time it is *unidirectional*-always away from the source. The sound energy is thus spread out over a large area, the size of the area depending upon the *distance* from the source. The practical way to specify the strength of this energy is to specify it as it flows through given size *window* of the space around it: 1 sq.cm (see Fig. 31.2).

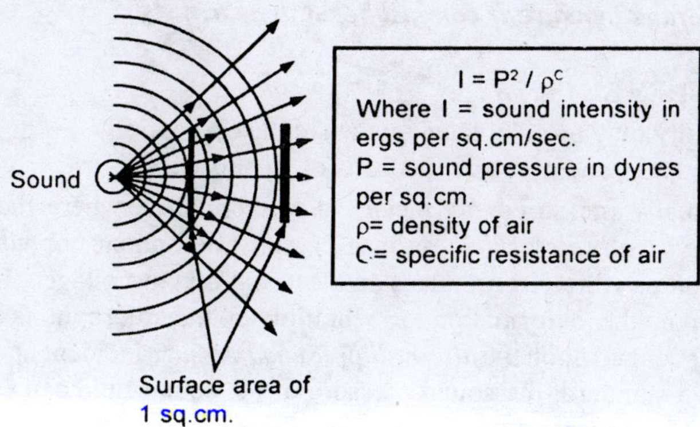

Fig. 31.2 The intensity, I, of a sound wave is the flow of energy through a square centimetre. The intensity of flow through a square centimetre close to the source is greater than through the same area removed from the source.

The amount of energy flowing through a square centimetre is called the *sound wave intensity, I.* The intensity of the sound wave will deerease with distance. *Acoustic power* is measured in dyne cm/s; the intensity of the sound wave passing through a 1 sq. cm of space is given in dyne- per square centimetre per second. The dyne-cm is know as the *erg.*

The *acoustic resistance* of air depends first on the *density* of air and *velocity* with which these vibrations are passed. The *characteristic impedance* of air is proportional to the *number* of air particles available per cubic centimeter, *p*, and the *speed of sound, c*, in centimetres per second - or proportional then to the product *pc* (see Fig. 31.3). Then *pc* the factor that determines how much sound will be radiated through the air, is :

$pc =$	0.00120	×	34,400	=	41.4 acoustic ohms
	p		c	=	42-acoustic ohms (appr.).

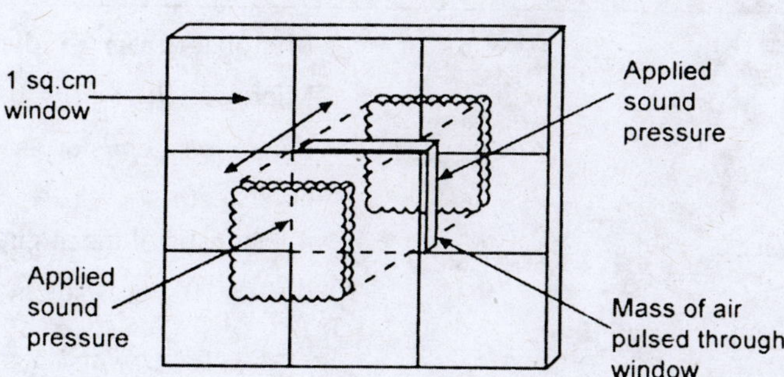

Fig 31.3 The specific acoustic resistance of air is 42 acoustic ohms. The units are grams/sq.cm/sec.

31.4. EAR SENSITIVITY

Fig 1.11 presents a graph of the sensitivity of the ear for various levels of intensity of sound. The curves are the statistical results of a large number of measurements. We are primarily interested in *the minimum intensity of sound that the human ear can hear,* represented by the lowest curve on the chart. On this curve we choose *for convenient reference purposes* the 1000 Hz area. The point of minimum audibility, or threshold of hearing is 10^{-9} ergs/sq.cm-sec. Converting ergs/sec to watts we obtain for the *threshold of hearing* : $10^{-9}/10^{7}$ or 10^{-16} watts/sq.cm, often referred to as the *zero reference level,* the basis for sound level scale. *It is the intensity which the average human ear can just begin to hear.*

31.5. SOUND PRESSURE LEVEL (SPL)

Acoustic power intensity may be calculated by the simple expression $I = p^2_{rms} / 42$ which relates sound pressure, sound intensity, and acoustic ohms on the basis of sq.cm/sec.

Sound pressure is usually measured, by means of microphones, where the *active elements* of the microphone are either *vibrationally deformed,* as in a crystal microphone, or *vibrated,* as in a dynamic microphone. The deformation or vibration produces an electrical output voltage. The *voltage output* of the microphone will depend upon the deformation or vibration of the microphone element. The extent of deformation or vibration will depend upon the *alternating sound pressure* incident upon it. Once a microphone is calibrated against a known standard, the sound pressure of the sound field can easily be measured by its electrical equivalent output of the measuring microphone.

To develop a reference zero level pressure relation, that is one based on zero level of sound intensity, let's rearrange the equation to $p^2_{rms} = 42I$. Thus,

$$I = 10^{-16} \text{ watts/sq.cm}$$

$$p^2_{rms} = 42 \times 10^{-16} \text{ (on a watt basis)}$$

$$p^2_{rms} = 42 \times 10^{-8} \text{ (on a dyne -cm basis)}$$

and

$$p^2_{rms} = 2.04 \times 10^{-4}$$

Since the unit of pressure is dynes/sq.cm, the zero level is 0.000204 dynes/sq.cm. For convenience, this figure is rounded off to 200 microdynes /sq.cm and has been standardised as such and is *usually given as* 2×10^{-4} *dynes /sq.cm.* The constant is quite useful in comparing all loudspeaker output measurements.

Table 31.1 Tabulation of Acoustic Measurement Terms

Term	Symbol	Unit
Sound Pressure	p	Microbar (μbar) = 1 dyne /sq.cm
Sound Power	P	Microwatt (μ watt) = 10 dyne-cm/s = 10 ergs/s
Sound Intensity	$I\ (= p^2_{rms}/pc)$	dyne-cm/sq.cm/s or ergs/sq.cm/s.
Specific Acoustic Resistance	ρc	42 acoustic ohms in gm/sq.cm/s
Zero Power Level	dB	0dB = 10^{-9} ergs/sq.cm-s = 10^{-16} watts/sq.cm
Sound Pressure Level	dB	0dB = 0.000200 dynes/sq.cm = 200 microdynes/sq.cm
Decibels	dB	dB = $20 \log_{10} p + 74$

31.6 SPL MEASUREMENTS

The *decibel notation* is used to compare two quantities of power on a logarithmic basis. It is also a means of converting a large numerical ratio into a small, easily handled number.

Table 31.2 The decibel scale is a means of compressing large changes in physical quantities into small numerical changes

	Power	dB Gain	Volts	
Power ratio of 2 to 1 represents a 3-dB change	.1 watts	Reference 0	.01 volts	A voltage ratio of 2 to 1 represents a 6 dB change
	.2	3		
	.4	6	.2	
	.8	9		
	1.6	12	.4	
	3.2	15		
	6.4	18	.8	
	12.8	21		
	25.6	24	.6	
	51.2	27		
A power increase of 1000 to 1 is equivalent to a 30-dB gain	102.4	30	3.2	
		33		
		36	6.4	
		39		
		42	12.8	
		45		
		48	25.6	
		51		
		54	51.2	
		57		
A voltage increase of 1000 to 1 is equivalent to a 60-dB gain		60	102.4	

On a power basis, $\text{dB} = 10\log_{10} p_2/p_1$

On a voltage basis, $\text{dB} = 20\log_{10} p_2/p_1$

where p_2 is the larger power (voltage) and p_1 is the smaller power (voltage).

Sound pressure is analogous to sound pressure (voltage).

Thus, to compare two sound pressures, we will employ the relationship:

$$\text{dB} = 20\log_{10} p_2/p_1$$

To specify the zero reference level of sound pressure 200 micro dynes/sq.cm, we use this figure as the lower or reference sound pressure ; substituting the value of 2×10^{-4} for p_1

$$dB = 20[\log_{10} P_{rms} - \log_{10}(2 \times 10^{-4})]$$

$$= 20[\log_{10} P_{rms}(-0.3010 + 4)]$$

$$= 20\log_{10} P_{rms} + (20 \times 3.7)$$

$$dB = 20\log_{10} P_{rms} + 74$$

If we measure the pressure in a sound field by means of a microphone and voltmeter, which are calibrated in terms of voltage for a given pressure in the sound field, we may quickly calculate *pressure level* of the sound field. To simplify matters the measuring system is calibrated in voltages referenced to that voltage output of the measuring system when the sound field is 200 microdynes/sq.cm, and the meter is calibrated in decibels (based on 200 microdynes / sq.cm). Such a meter is known as *sound level meter,* and it measures the *level* of the sound pressure above that of the reference 200 microdynes/sq.m.

The dB ratings of sound pressure represent a *ratio,* whose *smaller value* is the level of reference, placing the measured sound pressure at some higher level. The expression Sound Pressure Level is abbreviated as SPL.

The sound output of a loudspeaker is determined by :

- the weight of the moving system comprising of the cone and the voice coil
- the magnetic flux in the gap
- the frequency being radiated
- the baffling conditions of the loudspeaker.

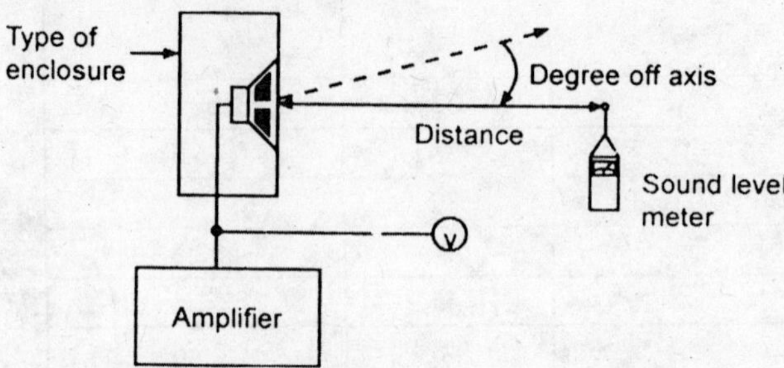

Fig. 31.4 Sound pressure level must state all conditions of measurement : power input to speaker, type of enclosure, distance to point of measurement, and angle to axis at which measurement is made, along with frequency or band of frequencies being measured.

The loudspeaker manufacturer usually supplies the SPL for the speaker mounted in an infinite baffle if it is a *cone type unit.* If it is *horn-loaded,* the SPL is relatively independent of this type of baffling. But once the SPL is given, the user can evaluate the unit for his particular needs.

The power output of the loudspeaker depends upon the amount of power fed into the loudspeaker. So we must specify that power. Stating the *rated power* serves a double purpose; not only is the reference power given for the SPL stated, but the rated power is also specified as a guide to the application to which the speaker is to be put.

A loudspeaker seems much louder especially for middle and high frequencies when you are *directly in front of it* than when you are off to one side. Speakers have a tendency to *beam* their higher frequency output directly out in front of them, on their 0 degree axis, and so this *placement* must also be specified.

32

PERFORMANCE TESTS

It is important that no pre-set controls are adjusted until the performance of the machine has been established and, if possible, compared with the specifications. That distortion you are trying to cure could be within the manufacturer's tolerance and efforts to improve upon it might adversely affect other parameters such as signal-to-noise ratio.

If a specific fault exists it will be found out readily by working through the signal chain and isolating the fault to a specific mode. With a stereo-machine suspect voltages can be compared with those of the working channel. If a fault is common to both channels it is likely to be due to the power supply or the oscillator or to diode, relay or solenoid switching and their associated supplies.

Where a specific fault cannot be pin-pointed but the machine seems to be giving inferior results, the best approach is to check the main parameters of performance and it will probably be found that a series of minor adjustments will produce a major improvement in overall quality. The first step, of course, is to clean and demagnetise the heads.

32.1 PERFORMANCE TESTS MICROPHONES

(1) *Microphone Sensitivity:* It is necessary to know what electrical output can be expected from a microphone, because this determines what amplification would be needed, so *governs the required specification of the following amplifier.* Conversely, if the amplifying equipment has already been chosen, *a microphone with the required output can be selected.*

Sensitivity figures specify *a particular electrical output for a specific sound pressure input.* The latter is obviously necessary because a loud sound will generate or modulate a larger electrical signal than a quite one. Unfortunately, there are a number of ways by which the *electrical output* is expressed, and to further complicate matters there are also several ways whereby the *sound pressure level* is described. Thus, it can be very confusing when comparing microphones of different makes when *different modes of expression* are used. They can be related, however, and fortunately the relationship is arithmetically simple

Electrical output is most often expressed in *millivolts* (mV) but it is also frequently described in dB. *In this case, the reference level is 1 V which thereby becomes 0 dB.* So 1 mV will be expressed as - 60 dB. As this is below the reference voltage, *the smaller the figure, the higher the output;* thus - 54 dB is 6 dB, or twice as much i.e. 2 mV. Strictly speaking, the reference level should be given as dBV. So 1 mV would be written as - 60 dBV. *The suffix is not always used as it is supposed to be understood.*

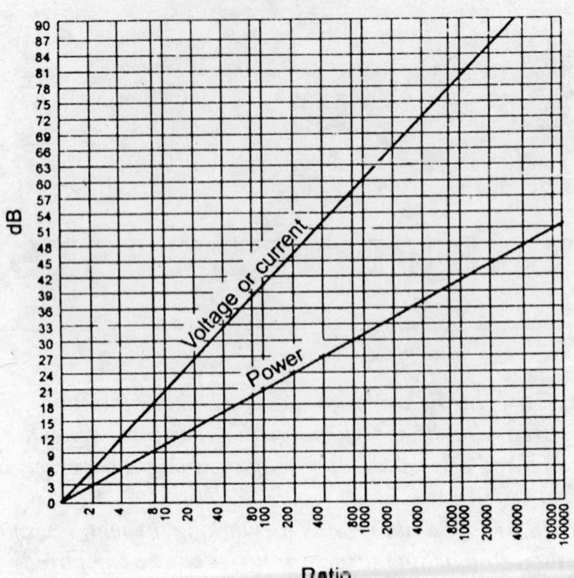

Fig. 32.1 Chart of decibel ratios.

Effect of impedance: The output voltage must vary according to the impedance. A microphone with a transformer giving a *high impedance* will also give a high output voltage, while one with a *low impedance* will give a low output voltage. The instruments may otherwise be identical and so deliver the same amount of electrical energy, yet the sensitivity figure will be different for each. For this reason, sensitivity voltages when expressed in decibels or directly in voltages should be considered in conjunction with the impedance of the instrument. *Any comparison is only valid between microphones of the same impedance.*

If the voltage increases with a rise in impedance the current drops, and vice-versa, hence, the power which is the product of voltage and current remains the same provided it is *correctly matched* into the load circuit. *So although voltage rather than power generation is the prime concern in most microphone circuits, a power rating does give a more unambiguous picture of the sensitivity.*

The *milliwatt* is used as the standard of reference. In the case of a *voltage rating*, the sensitivity is usually measured *unterminated*, that is, without the connection of a load circuit. With a *power rating*, though, current must flow, and so there must be an external load circuit. *600 Ω is, at present, the standard impedance circuit used for professional equipment.*

One milliwatt in a circuit of 600 Ω impedance is the power resulting from an applied voltage of 0.773 V or 773 mV. As the microphone output will be far less than this, the milliwatt, like the volt, is too large a unit and is therefore used as a *reference level.* Actual values are given in minus decibels. To distinguish these from dBV, the suffix *m* is used, so the sensitivity specification would read : 58 dBm (0dB = 1 mW into 600 Ω).

Pressure Level : 1 μ bar is the approximate pressure produced by an average human voice speaking at a distance of 2 feet. At 8 inches which is about the usual distance a microphone should be held from a speakers lips, the pressure will be in the region of 9 μ bars. If then the microphone sensitivity is quoted as 0.1 mV per μ bar, an output of about 0.9 mV could be anticipated.

An alternative unit often used is the dyne. This is a unit of force, but we need a unit of pressure. The unit employed is the *dyne-per-square centimetre* written as dyne/cm^2. Fortunately this unit is identical to the μ bar, so a direct comparison can be made between microphones rated in μ bars or dyne /cm^2. Either of these expressions can be used when the electrical output is given in mV or dBV, but when power ratings are used which are much smaller, the pressure unit is increased tenfold. Thus

0dBm = 1mV/10 μ bars, or 0dBm = 1 mW/10 dynes/cm^2

Yet another unit of force is the Newton (10^5 dynes). The related unit of pressure is obtained by spreading this force, over a square metre to give the newton-per-square metre, abbreviated to N/m^2. *This unit is just ten times larger than the μ bar or dyne/cm^2.* It enables a comparison to be made to the other ratings with just a movement of the decimal point. Thus a rating of 0.1 mV per μ bar becomes 1.0 m V per N/m^2. In the case of dBV ratings an increase of 10 times is indicated by adding 20 dB, so in our example, this would be - 80 dBV /μ bar = -60 dBV / N/m^2.

As with the dBm power rating the basic unit is already increased by ten, there is no change; so

0dB m = 1 mW/10 dynes/cm^2 is equivalent to 0dB m = 1mW /N/m^2.

Table 32.1 shows *the various ways in which the same microphone can be rated.*

Table 32.1.
Various ways of expressing the sensitivity of the same microphone

0.1 mV per μ bar
0.1 mV per dyne/cm^2
80 dbV/μbar (0dB 1V)
58 dBm (0dB 1 mW into 600Ω per 10 dyne cm^2)
1.0mV per N/m^2
60 dBV/Nm^2 (0dB 1V)
58 dBm (0dBm 1mV into 600Ω per N/m^2)
1.0 mV/Pa

(2) *Microphone Frequency Response* : Ideally, the electrical output of the microphone should be the *same* for all frequencies within the audio spectrum. In practice, this is difficult to achieve due to resonances, interference, pressure-doubling at the diaphragm and other factors. *The way in which the output varies according to frequency is termed the frequency response.*

The most informative manner of giving the frequency response is by means of a graph. This is plotted on a *logarithmic horizontal scale* which denotes frequency and a *linear vertical scale* which shows the output level in dB. The *reference level* is set at 1 kHz, and if the response curve goes up or down it is said to vary by the indicated number of dB from that level at other frequencies.

The *dB scale markings* are somewhat arbitrary; they may give 0dB as the 1 kHz level and then show plus or minus figures above and below that, or they may start at 0dB at the bottom of the graph and work up. In this case the 1 kHz level will always be at a round number, say 20 or 30 dB. The actual numbers given matter little; *it is the number of dB between 1 kHz level and other portions of the curve that is important.*

All we need to know then, is the number of dB from one horizontal line to the next so that we can ascertain the extent of curve *deviations.* Most charts use *heavy lines* to denote the 5 or to 10 dB intervals, with four *fine lines* giving five spaces in between. Thus, each space indicates either 1 or 2 dB depending on the calibration of the heavy lines. Some charts have 4 or even 5 dB to each space and thereby the vertical curve excursions are greatly *compressed.* This has the obvious effect of making the curve look a lot *flatter* than it really is, and so *at a quick glance the curve of a poor microphone can look quite resonable.* Compare Figs 32.2. (*a*) and (*b*).

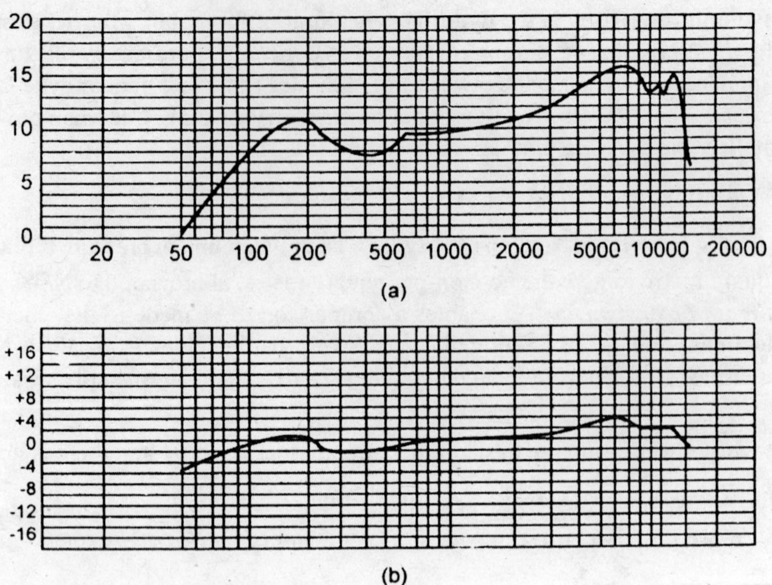

Fig. 32.2 Compare charts (a) and (b). Frequency response of (b) looks smooth with gradual roll-off at the limits, whereas (a) looks peak, uneven and with rapid cut-off. Actually they are the same: vertical calibration of (a) is in 1 dB divisions, while that of (b) is in 4 dB units. Always check vertical calibrations carefully.

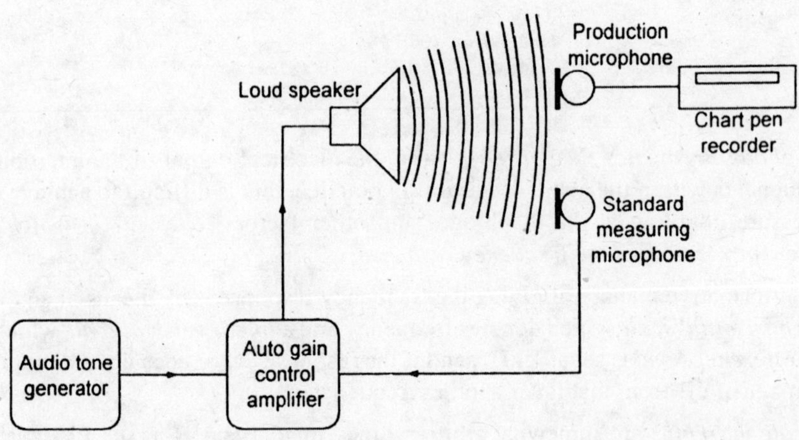

Fig. 32.3. Arrangement for producing frequency response chart with pre-recorder. Any variations due to the loudspeaker are detected by the standard measuring microphone and used as a control signal to alter the gain of the amplifier. Thus a constant level is maintained over the whole frequency range.

The frequency response curve given in maker's brochures cannot be expected to apply exactly to every instrument of that type. Deviations of dimensions, diaphragm elasticity and other factors all within the normal manufacturing tolerances can and do produce modifications of the response curve in individual microphones. The curve should be regarded only as a *rough average,* and inclined to be *somewhat idealised and smoothed out* by the firm's advertising department.

For most medium-quality applications this is near enough, but the better instruments are supplied complete with their own frequency response graphs *which have been made with, and apply only to, that particular microphone.* Comparing several of these for the same model will reveal that the overall curve is similar, especially in the middle and lower registers, but some minor differences may be seen at higher frequencies where small dimensional variations are related to the shorter wavelengths.

These *individual charts* are made on a perforated strip of paper printed with the graph lines by means of a *pen recording-machine* which is driven from an amplifier fed by the microphone under test. The sound is provided by a loudspeaker which is supplied from an *electronic tone-generator* via an amplifier with automatic gain control. This control is operated by a *standard laboratory measuring microphone,* which is mounted alongside the production model. Thus, any variations due to resonances in the loudspeaker produce a *correction signal* from the measuring microphone which alters the amplifier gain to compensate.

As reflections from the walls and ceiling of an ordinary room cause *interference and reinforcement* effects at various frequencies, the whole operation must be carried out in an *anechoic room, i.e.* one that is completely deadened and non-reflective to sound. Such rooms are large and expensive to fit out, so only the larger manufacturers usually have this facility. However, the above discussion gives a deep insight into the method adopted to plot frequency -response curves for individual microphones by the manufacturers.

(3) *Microphone Directional Response* : The behaviour of microphones to sounds coming from different directions, and the polar diagrams that are plotted to indicate this, have already been discussed in Chapter 5. In addition, a *directivity factor* is often quoted by makers and this can be one of two characteristics.

In one case it is simply *the difference between the sensitivity of the maximum and minimum points of pickup, the minimum point varying from 180° of the cardioid to 90° of the figure-of-eight.* This is expressed in dB and values of 15 to 20 dB are typical. The figure can in fact be obtained by examining the polar diagram if it is expressed in dB. Other terms also used for this factor are : rejection ratio ; discrimination ; cancellation; and front-to-back attenuation.

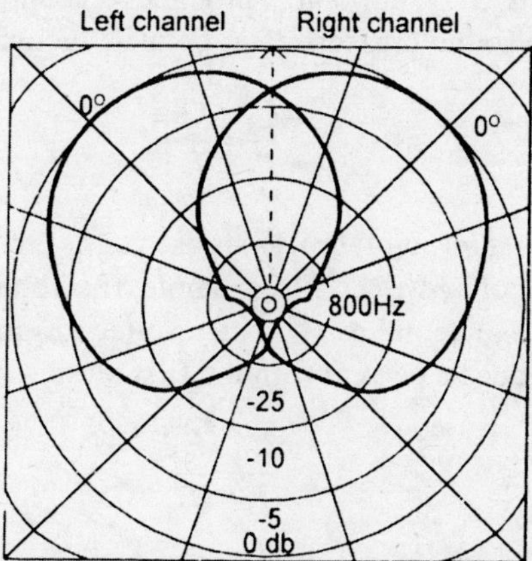

Fig. 32.4 A directional microphone designed for stereo pick up.

In practice, this figure is of limited usefulness. *It is not always possible to arrange for unwanted sounds to come from the angle of minimum pickup;* in fact, more often than not they come from many directions at the same time. A more realistic specification, then, is one that gives *the ratio of sound picked up in the forward direction compared with sound picked up in all other directions.* It is to be expected that this figure would be less than the previous one.

If a directional and an omnidirectional microphone are placed in a reverberant room facing a sound source, the acoustic power received by the *omni* will be greater than that picked up by the *directional* unit because it will include reverberation from all directions. The difference is the *diffused sound field*, and the ratio between this and the direct or *plane sound field* as received by the directional microphone is squared to give the sound power concentration (symbol: gamma, γ) or directivity index. Expressed mathematically it is:

$$\gamma = \left[\frac{\text{transm factor plane sound field}}{\text{transm factor diffused sound field}} \right]^2 \qquad \ldots(32.1)$$

For a *cardioid* microphone the index is about 3 while for a *hypercardioid* it is 6. In the case of *gun microphones* it can vary from about 6 to 10.

This figure can be very useful. It is possible to workout simply how much further away a *directional microphone* can be placed from the sound source before the diffused sound reaches the same level as for *omni*. *The further away from the source the greater the diffused or reflected sound in proportion.*

As acoustic power decreases in proportion to the square of the distance, *the relative distance is the square root of the directivity index.* Thus, a *cardioid* can be placed $\sqrt{3} = 1.73$ or nearly 1 3/4 times the distance from the source as an omnidirectional instrument. A *hypercardioid* can be positioned $\sqrt{4} = 2$ times the distance, while the *gun microphones* can be mounted from 2 1/2 to 3 times the distance away.

As the directivity index is a ratio, it can also be expressed in dB but the power logarithm ratios apply. Thus the *cardioid* has a power ratio of 4.8 dB, the *hypercardioid* 6 dB and the *gun microphone* 7.8 - 10 dB. If these various microphones were placed at the same distance from the source then the diffused sound would be reduced by the dB values here noted in comparison with an omnidirectional instrument.

(4) *Reciprocity Calibration of A Microphone* : The statement after Helmholtz of the acoustical reciprocity theorem may be stated in the form :

"If a simple source of strength Q_1 at point *A* produces a sound pressure P_2 at *B*, then a simple source of strength Q_2 at *B* will produce a sound pressure P_1 at point A such that

$$\frac{Q_1}{P_2} = \frac{Q_2}{P_1} \qquad \ldots(32.2)$$

In this form the *acoustical reciprocity theorem* is seen to be analogous to a commonly used *reciprocity theorem for linear passive electrical networks.* For example, if a voltage E_1 in the first mesh of the circuit of Fig. 32.5 produces a current I_2 in the second mesh, then a voltage E_2 inserted into the second mesh, where the current I_2 was measured, will produce a current I_1 in the first mesh such that

$$\frac{E_1}{I_2} = \frac{E_2}{I_1} \qquad \ldots(32.3)$$

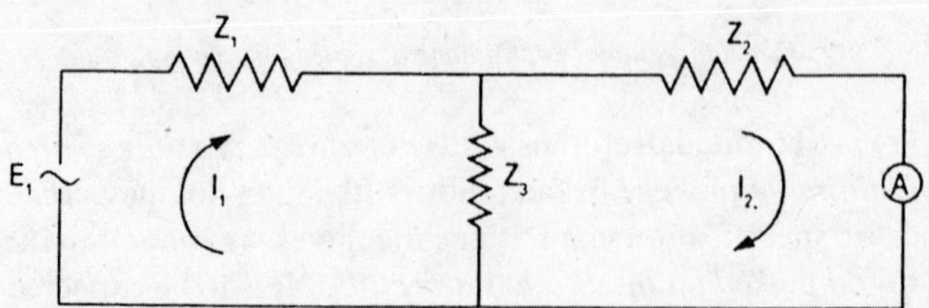

Fig. 32.5. Illustrating the reciprocity principle for a linear passive electrical network.

Since the principle of reciprocity applies separately to both electrical and acoustical systems and since acoustical and mechanical elements of an electroacoustic transducer may be transferred to equivalent electrical elements, we may anticipate that linear electroacoustic transducers will also satisfy an electroacoustic version of the reciprocity theorem.

The experimental calibration of microphones is greatly facilitated by the application of the electroacoustic reciprocity theorem. In one method three transducers are employed - the microphone A *to be calibrated*, a *reversible transducer B* which obeys the reciprocity principle and a *loudspeaker C*. A convenient type of reversible transducer is a small back-enclosed, direct radiator loudspeaker. In *step one* of the calibration, the reversible transducer *B* is placed at a specific position in the sound field of the speaker *C* as shown in Fig. 32.6 (*a*) and the open-circuit voltage generated in *B* by *C* is measured. This voltage is

$$E_B = M_B\, P_C \qquad \text{...(32.4)}$$

where M_B is the microphone response of *B*, and P_c is the sound pressure produced at *B* by speaker *C*. In *step two* the microphone *A* is substituted for the transducer *B*, as shown in Fig. 32.6 (*b*), and its open-circuit voltage E_A is measured for the same sound pressure P_C employed in step one. Then

$$E_A = M_A\, P_c \qquad \text{...(32.5)}$$

where M_A is the desired microphone response of *A*. These two equations may be combined to give

$$M_A = M_B\, \frac{E_A}{E_B} \qquad \text{...(32.6)}$$

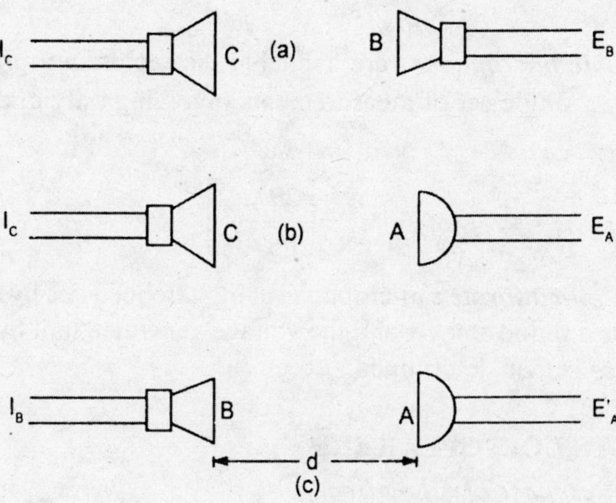

Fig. 32.6 Schematic representation of the three steps used in the reciprocity calibration of microphone A.

In the *final step* of the calibration, the reversible transducer *B* is substituted for the speaker *C*, as indicated in Fig. 32.6 (*c*), and the output E'_A of microphone *A* is measured for a current I_B in transducer *B*. Then

$$E_A = M_A\, P_B \qquad \text{...(32.7)}$$

Combining this with the expression

$$P_B = S_B\, I_B \qquad \text{...(32.8)}$$

which gives the pressure P_B produced by the transducer *B* as a function of its speaker response S_B, we obtain

$$E'A = M_A\, S_B\, I_B \qquad \text{...(32.9)}$$

Then from equations 32.6 and 32.9

$$M_A^2 = \frac{M_B}{S_B} \cdot \frac{E_A E'_A}{E_B I_B} \qquad \text{...(32.10)}$$

The reciprocity constant is given by $\frac{2d\lambda}{\rho_0 C}$ where p_0 is the specific acoustic resistance of the medium, and λ is the wavelength being transmitted. *The reciprocity constant is independent of the type or constructional details of any particular transducer.* In mathematical form the reciprocity principle may be stated as

$$\frac{M}{S} = \frac{2d\lambda}{\rho_0 C} \qquad \text{...(32.11)}$$

where M is the *free-field microphone response* of the transducer, expressed in volts per N/m^2, and S is its *speaker response*, expressed in N/m^2 per ampere.

Substituting the reciprocity constant of equation 32.10 for M_B / S_B,

$$M_A = \left[\frac{2d\lambda E_A E'}{\rho_0 C E_B I_B}\right]^{1/2} \qquad \text{...(32.12)}$$

This method has the advantage of avoiding the necessity of attempting to produce measurable or calculable sound pressures, since *all the basic measurements, other than the distanced are electrical in nature*. These measurements should be made either in a *sound deadened room* such as an anechoic chamber, in large rooms where reflections are negligible, or outdoors at a considerable distance from reflecting surfaces. It is customary to convert the results obtained by equation 32.12 to volts per microbar by multiplying the factor by 0.1.

If two *identical reversible microphones* are available the reciprocity principle makes it possible to calibrate both microphones by a single set of measurements involving only one step. Here $E_A = E_B$, so that equation 32.12 reduces to

$$M_A = M_B = \left[\frac{2d\lambda E_A}{\rho_0 c I_B}\right]^{1/2} \qquad \text{...(32.13)}$$

Finally, it is possible to *self-calibrate* a microphone at high frequencies by using short pulse techniques. The microphone is first used as a sound source, and the voltage generated in it by reflection of the pulse from a *100 percent reflecting surface* is then determined.

32.2 PERFORMANCE TESTS LOUDSPEAKERS

(1) *Measurement of Pressure Response of A Loudspeaker* :

Loudspeakers are commonly rated both as to the total *acoustic power* radiated and as to the *acoustic pressure* produced. *The pressure response of a loudspeaker is a measure of the sound pressure produced at designated positions in the medium with the electrical input, the frequency, and the acoustic conditions specified.*

One method of specifying pressure response is to give the effective or RMS sound pressure measured at a specified position *for one volt input* to the terminals of the voice - coil. It is given by

$$S_E = \frac{P}{E} \qquad \text{...(32.14)}$$

where P is the measured effective sound pressure in microbars and E is the voltage applied to the voice-coil. The resulting behaviour with frequency, as calculated from equation 32.14 represents the sound pressure that would be obtained from the speaker if it were fed from an amplifier that automatically delivers an input of 1 volt *irrespective of the loading impedance of the voice coil.*

In the second method, a *constant current of one ampere* or a *constant power of one watt* is supplied to the voice-coil. Significant responses are obtained in the second method.

The pressure response may also be *expressed in decibels,* relative to an arbitrary response of 1 microbar per volt of electrical input, by the equation

$$n_S = 20 \log \frac{P}{E} \quad \text{...(32.15)}$$

If it is desired to express the level of response in terms of the commonly used *acoustic reference pressure of 0.002 microbar* the response levels as computed by equation 32.15 must be increased by the addition of + 74 dB.

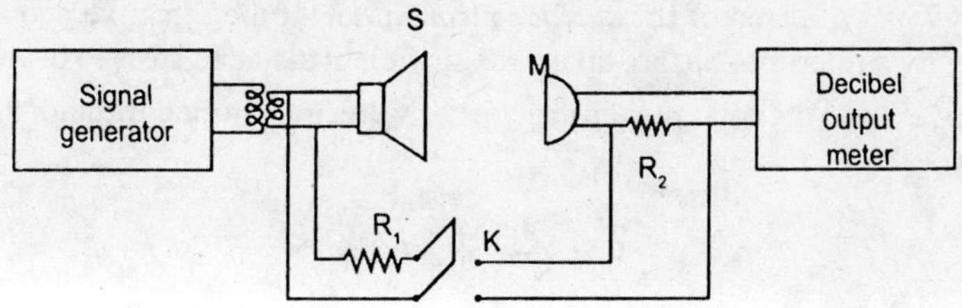

Fig. 32.7 Measurement of output pressure level of a loudspeaker.

When a source of sound is present in a room *the resultant sound at any point is a combination of the direct and reflected sound.* The *direct sound* is independent of the room and is characteristic only of the source. The *reflected sound,* however, depends not only on the source but also on the geometrical configuration of the room and the absorption characteristics of its walls, and on the coordinates of the source and point of observation. It is obvious, therefore, that, unless the measuring conditions are *identical,* the response characteristics obtained in any ordinary room will vary considerably. Hence the outdoor or free-space measurement of the frequency-response characteristic offers the simplest method of loudspeaker measurements that may be duplicated and compared among various laboratories. For this reason, *outdoor measurements are considered as a standard test for showing the performance of a loudspeaker.* In practice it is observed that measurements made in large sound deadened or anechoic chambers are equivalent to outdoor measurements at all frequencies except those below about 100 Hz.

A general schematic arrangement showing one specific way to obtain the factor P/E in equation 32.14 is given in Fig. 32.7. This particular arrangement has the desirable feature that *it does not require an absolute, calibration of the measuring system. However it necessitates the use of a calibrated microphone having a known open-circuit response* n_m *in decibels relative to a reference level of 1 volt produced by a pressure of 1 microbar in a free progressive wave.* It can be shown that the pressure response of the speaker in decibels is

$$n_S = A - B - n_m - 20 \log \frac{R_1 + R_2}{R_2} \quad \text{...(32.16)}$$

where A is the output of the measuring system in decibels, with K open and the microphone M picking up sound from the loudspeaker S, and B is the output of the measuring system with K closed and the microphone shielded from sound. The resistance R_2 should be small enough compared to the internal impedance of the microphone so that the measured output M does not change when R_2 is short-circuited. *In order to reduce errors resulting from in accuracies in the relative readings of the measuring system,* R_1 *should be so selected as to obtain a value of B in the range of values obtained for A.* The loudspeaker should be mounted either in the cabinet with which it is to be used or, if the speaker unit alone is to be tested, in a rigid baffle of adequate size. The microphone should be placed at a distance of *at least* 5 ft. (150 cm) from the speaker, or at some designated distance.

Acoustic pressure responses measured on the axis of a loudspeaker are in general higher than those measured in other directions.

(2) *Measurement of Acoustic Power Output of A Loudspeaker* : The sound power output from a loudspeaker driven at any particular frequency may be obtained *by measuring the rate of flow of acoustic energy through any arbitrary surface that entirely encloses the sound source.* When a loudspeaker is mounted in a wall, this surface is normally that of a hemisphere centered on the loudspeaker. For speakers mounted in cabinets or horns, it is normally a sphere centered on the loudspeaker. The surface of the sphere (or hemisphere) is divided into *incremental areas,* and the power transmitted through each is determined by the product of the *intensity I* and the area of the *surface element dS.* If the effective acoustic pressure response measurements, as made by a microphone, are taken at a sufficient distance so as to be external to the so-called near-field, then $I = P^2/\rho_o c$, and the total power in acoustic watts transmitted through the surrounding surface is given by

$$W = \frac{1}{\rho_o c} \iint P^2 dS \qquad \text{...(32.17)}$$

The acoustic output characteristics of a loudspeaker may also be determined by measurement of the *resistance component* of the input electrical impedance under various load conditions. Then equation 32.18 could be used to compute electro-acoustice efficiency of the speaker

$$\eta = 1 - \frac{R_B}{R_A} \qquad \text{...(32.18)}$$

The first step in such a procedure is to block the voice-coil so that it cannot move. The resistance R_B measured in this condition is a direct measurement of the *resistance of the voice coil.* The second step is to measure the *input resistance* R_A under normal conditions of *air* loading. Equation 32.18 is used to compute the efficiency of a loudspeaker *at frequencies well above that of mechanical resonance.*

(3) *Loudspeaker Phasing* : When groups of speakers are mounted at one point or near together so that they cover the same area, it is necessary that they be *correctly phased.* If the speakers are not phased properly, *the sound waves from the improperly phased speakers will tend to cancel out, reducing the effective output of the system.*

There are a number of ways to check the phasing of speakers. Two speakers may be connected to an amplifier in the same manner. They will be connected when installed, and placed close together *facing each other.* A low frequency signal from an audio generator or a record should be fed through the amplifier. By listening to the speakers, it is possible to determine, whether or not the *low frequencies* are being cancelled. If the low frequencies are absent when the speakers are facing each other, then the phasing is correct if the, speakers, are to be mounted so that they face *in the same direction.* If the low frequencies can be heard, when the test is made, then the low frequencies are not being cancelled, and the phasing is correct if the speakers are to be mounted *facing away from each other.* If the phasing is incorrect in either of the two cases just described all that is necessary is that *the connections to one of the voice coils be reversed.*

The phasing of speakers may also be checked by using the circuit shown in Fig. 32.8. The apparatus consists of a pair of headphones connected to the input of an amplifier through two long cords and a double-pole, double-throw switch. The amplifier is equipped with an *output indicator.* The sound system whose speakers are to be checked for correct phasing should be turned ON and a constant tone fed into its input. The double-pole, double-throw switch should be marked to show *in phase* and *out of phase.* This may be done by holding both phones in front of one speaker and noting the position of the switch which gives the greatest indication on the output meter. This position should be marked in phase and the other position out of phase. The equipment may now be used to check the phasing of two speakers. One phone is held in front of each speaker. *If the greatest output is indicated when the switch is in the in phase position, then the speakers are phased properly.*

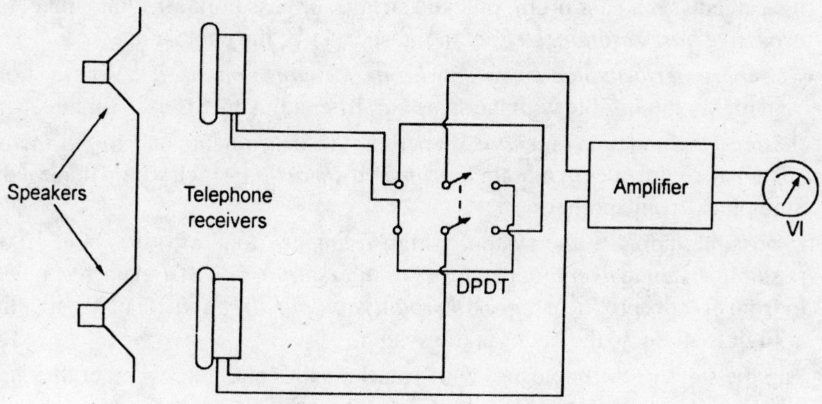

Fig. 32.8 Setup to determine loudspeaker phasing.

In *stereo systems* phase checking and correcting are made relatively simple. In most stereo systems, a *phase-reversing control* is included, connected either in the speaker output circuits or earlier in the amplifier chain. However, this control does not provide for phasing of the centre speaker. If a single tone or other signal is applied to the input of both channels at the same time, the relative outputs from the right and left speakers will indicate the phasing. If the phasing is *wrong* the two signals will differ by 180 degrees and try to cancel each other. If the phasing is *correct* the signals reinforce each other. Thus, *the correct phasing should produce much greater volume.*

There is a difference between *speaker phasing* and *system phasing*. The final output phase depends on the windings of the output transformers, connections between pre-amplifiers and amplifiers, connections of cartridges and input devices, and other things besides speaker connections. Therefore, checking the *relative phasing* of the speakers themselves is not always enough.

(4) *Loudspeaker Listening Tests* : The proof of the pudding is in the eating, the saying goes and it goes for speakers that, specifications and measurements notwithstanding, *the real assessment is a listening test.* At every stage from design, through manufacturing and sales down to the final consumer, *ears are the final arbiters of quality.*

The final test of a speaker system is *how it sounds.* The listener who does not know what to look for in such tests is at a disadvantage. He may overlook factors which later, after he has purchased the system, may assume new importance. Experience has shown that proper attention to certain features will go a long way toward assuring continued satisfaction.

There are no standards of performance. Most prospective purchasers of speaker systems do not have equipment to make the tests necessary to check a system against proper standards any way. *So listener tests must be made as personal comparisons between different available systems.*

The following suggestions should tell the reader *what to look for in making listener tests on a speaker system.* Each feature should be *compared* from one system to the next in an attempt to evaluate the one most pleasing. In any tests, every effort should be made to *eliminate other components of the whole high-fidelity equipment by using the same or similar auxiliary equipment with each different speaker system.* If the reader has not had experience in distinguishing between audio-frequency tones of different frequencies, it is suggested that he obtain an *audio signal generator* or *test record* made for the purpose, connect it to a speaker, and run through the audio range, *familiarising himself with the sound of audio tones of low, middle-range and high frequencies.* He will then be better prepared for the listening tests which follow.

(*i*) Look for *distinctions* between various kinds of bass notes, such as those from different musical instruments. If a bass drum, plucked strings, or bass horn all sound alike, then there is probably *excessive bass resonance effect* in the speaker or the enclosure.

(*ii*) *Compare performance on voice with performance on music.* Voices should be crisp, natural, and highly intelligible with the same amplifier adjustments used for music from the same source.

(*iii*) Listen for sudden *crescendos* of percussion instruments and to *pizzicato*-passages of music. Check for evidences of *hangover transient distortion* which would make it difficult to distinguish one pluck from another.

(*iv*) If possible, apply to the system a high-frequency tone of, say, about 10,000 Hz. If this is not possible, listen to *needle scratchings* or *interstation noise* from an FM tuner. Then walk around in front of the reproducer, *checking relative output directivity.* The broader the output distribution at high frequency, the better is the system.

(*v*) Tap the sides of the enclosure with your knuckles and note whether they are deadened properly or respond with undesirable vibration. *All parts of an enclosure structure should be solid and dead.*

(*vi*) Be sure that, for each speaker system checked, the amplifier *boost controls* are adjusted for the same pleasing performance. Also check *speaker balance control* (if any) for best balance between woofer and tweeter output. This control is sometimes referred to as the *treble adjustment* of the speaker system. *It is unfair to compare systems in which all controls are not adjusted for optimum performance.*

32.3 PERFORMANCE TESTS - TAPE MACHINES

1. Head Alignment : High frequency response and interchannel corsstalk are extermely dependent on head alignment. In most tape transports, the heads are *adjustable* and can be aligned as required to establish correct *head-to-tape interface.* The adjustment is very precise and must be accomplished in accordnace with the instructions given in the manufacturers manual. *There are five basic adjustments involved in correctly positioning a head as shown in Fig. 32.9.*

(*i*) **Tilt :** The first basic adjustment is establishing a true vertical position for the head face, Fig 32.9 (a), with reference to its contact with the tape. *The correct altitude is when the head neither tilts toward nor away from the tape surface. Establishing the correct vertical altitude is important to maintain uniform tension across the entire tape width in contact with the head.* If the tape is under more tension at one edge than at the other, total intimate contact between the tape and head will be disrupted. The difference in tension can also cause the tape to slew away from the centre line of the tape path.

(*ii*) **Height :** The next basic head adjustment it head height, Fig 32.9(*b*). *Improper height manifests itself in tracking and crosstalk.* On multitrack recordings this particular adjustment is very *critical* in that loss of output, noise and inter-channel crosstalk result if the playback head gap is not tracking the recorded path on the tape perfectly. If recording with a head that is not properly adjusted in height, it may be impossible to play that tape back on another machine.

(*iii*) **Tangency :** Tangency Fig 32.9(*c*), is simply the *squaring of the record/playback gap(s) to the tape surface.* Correct tangency is important to ensure proper head to tape contact at the gap. *If the contact is incorrect, high frequency response will be degraded; and more important the system may become oversensitive to dropouts.* Dropouts are usually caused by debris or contamination which separates the tape's oxide from the gap. Thus, any interruption of head to tape contact will result in the stated degradation. If the separation is severe, complete signal loss can result.

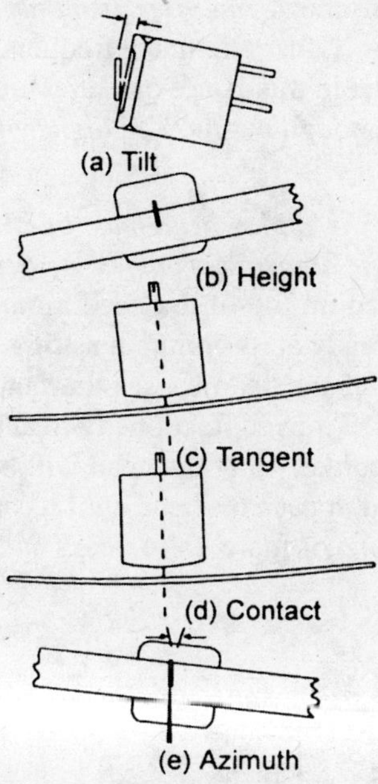

Fig. 32.9. Head adjustment planes.

(*iv*) **Contact :** Contact is the head position with respect to tape wrap around Fig. 32.9(d).

Correct head -to-tape contact is ensured by the slight bending in "wrap" in the tape path as it passes over the head. *Poor contact can result in poor high frequency response and oversensitivity to dropouts.*

All cassettes and cartridges are equipped with *pressure pads,* which force the tape against the head by applying pressure to the tape backing. When inspecting the head position these pressure pads must also be checked. If the pad is deeply worn, the cassette/ cartridge must either be re-recorded and the old one thrown away or the pad be replaced. Also, because of the intimate contact between the pad and the tape backing, tape surface contamination will tend to stick to the pad. Such buildup may create hard spots on the pad, forming an uneven surface and leading to tape squeal, loss of contact, and excessive tape and head wear.

(*v*) **Azimuth :** The most important head adjustment is that of azimuth Fig 32.9(e). If the playback head is not parallel to the recorded poles on the tape, serious loss in high frequency response results. To be sure of compatibility and interchangeability, *the record and playback gaps must be adjusted so that they are exactly perpendicular to the centreline of the tape.* This adjustment is best accomplished by using an alignment test tape that is recommended by the manufacturer of the specific machine. The test tape has a carefully recorded high frequency signal which, when played back, is used to determine the output level. Any head misalignment is readily apparent in the loss of output.

When using a *test tape* to check azimuth, a variety of methods can be employed, the best being that recommended by the manufacturer. However, in the event of non-availability of a service manual, there's another way to check the azimuth. With the machine in the play mode, the volume at a normal listening level, and the tape running, deliberately skew the tape (gently, with the sharpened wooden part of a pencil) across the head. If the volume (or the signal level if the machine has a meter) *increases,* then the existing azimuth is incorrect and must be adjusted. If the volume or level *drops*, then the existing aziumth can be assumed to be correct. Azimuth can also be adjusted with the help of test equipment set up shown in Fig. 32.10

To adjust the azimuth a millivoltmeter should be connected to the line output and adjusted initially for a midscale reading from the calibration tape. *Azimuth is commonly set at 10 or 15kHz with reel-to reel recorders, and at 6.3 kHz with cassette recorders.* Other alignment frequencies are specified by some manufacturers but if the recommended tape is not available this is not of critical importance and any azimuth calibration tape can be used provided it gives a high enough output. *The higher the frequency used the more accurate will be the azimuth alignment.*

With a three-headed recorder the replay head must be aligned before the record head, whereas with a combined record/replay head a single adjustment only is required. Whilst the azimuth tape is being replayed firm finger pressure should be applied on top of the head, in a rocking motion. The movement should be sufficient to indicate if the head is already at its optimum setting: if it is, any movement will cause the output reading to *fall*. If the *slight rocking* of the head causes the output to *rise* when the head is pushed in one direction, the adjusting screw should be moved fractionally to allow the head to go in that direction. Azimuth is a very critical operation and it is unlikely that the head will be far from true unless it has been tampered with. If a major adjustment is required, loosen the head until it can be rocked easily, allowing the output peak to be found. There will probably be one or more false peaks and it is important that the *correct maximum* is found.

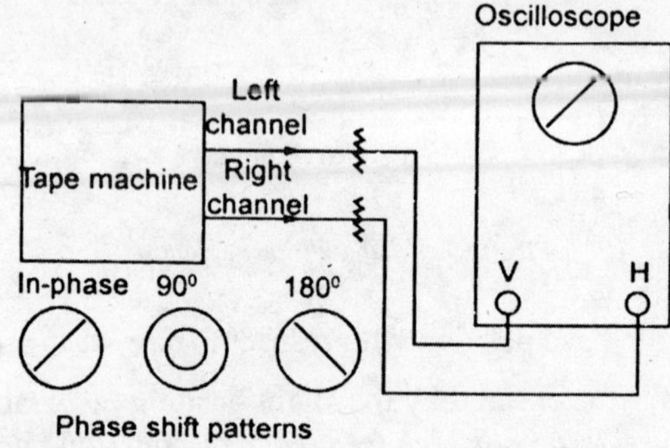

Fig 32.10 Azimuth adjustment

With a *stereo machine* it is unlikely that any one azimuth setting will give the maximum output from both channels and a *compromise setting* will probably have to be adopted. *The best approach for stereo, or multitrack machines, is to adjust for minimum phase error having established the near optimum setting.* The easiest way to do this is shown in Fig. 32.10. With a single-beam scope one channel should be fed to the horizontal input and the other to the vertical input : the head should then be adjusted, fractionally, to give an *in-phase display.* With a double beam scope, the two channels can be individually displayed and the relative phases compared.

To adjust the *record head,* the output from the replay head should be *monitored* as previously whilst a high frequency tone is recorded simultaneously on both tracks. The record head should then be adjusted following the method outlined for the replay head.

2. Replay Response : A datum should be established with a millivolmeter connected to the recorders line output, as in Fig. 32.11 then the tape should be played through and readings noted. *The response should be substantially flat over the range of the machine* and any significant deviation compensated for with the appropriate preset control where fitted,

In the event of no adjustment being provided or if adjustment does not give a satisfactory response, then either the *head* must be worn or the *associated amplifier* must be faulty. If component failure is expected a likely culprit is one of those forming *the equalisation feedback loop.* Fig. 32.12, particularly if the response is satisfactory at other speeds. Other possibilities are high resistance switch contacts, dry joints or mechanical damage in cases where the switch bank is mounted directly to a printed circuit board.

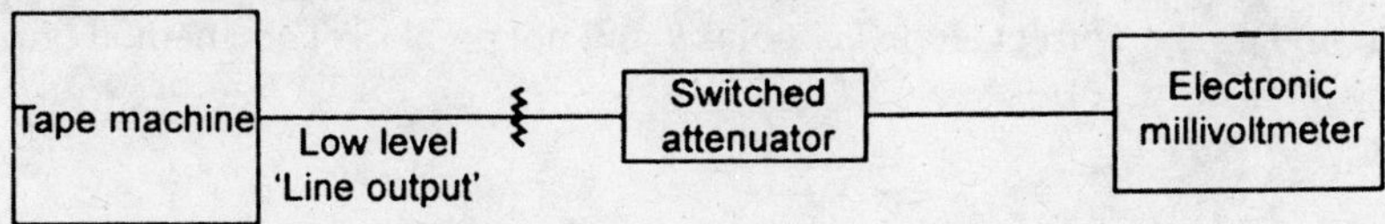

Fig 32.11 Test set up for replay response.

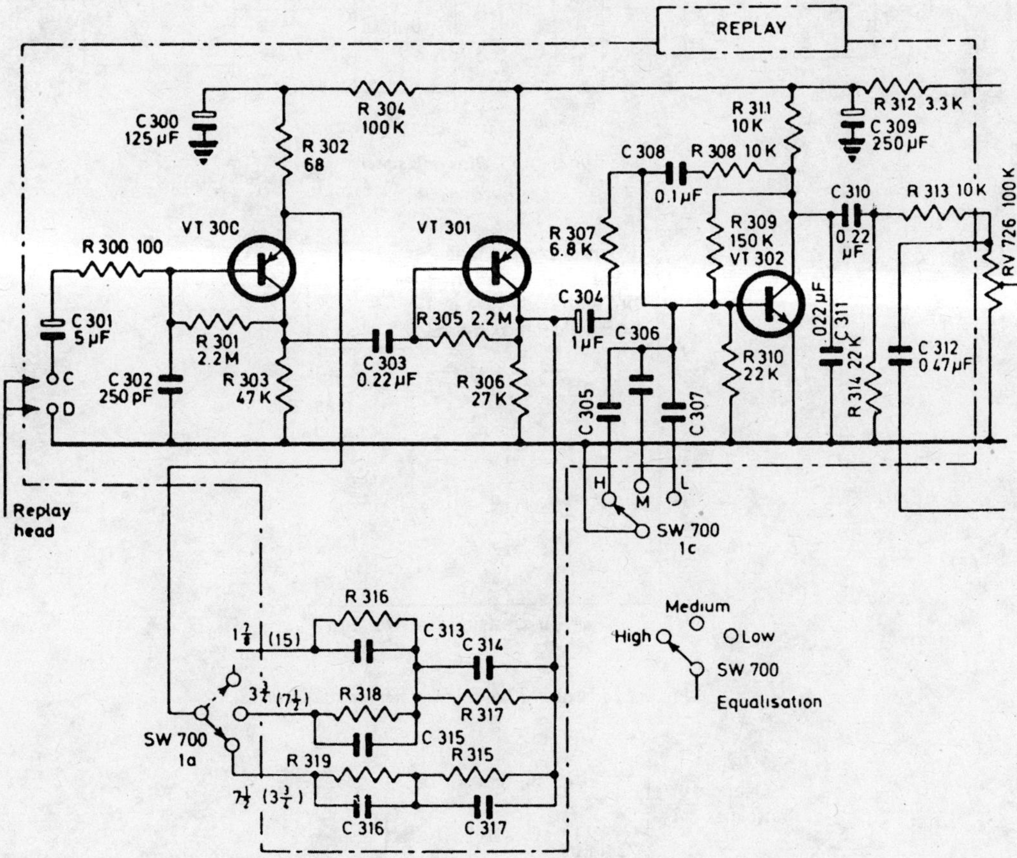

Fig. 32.12. Equalisation feedback loop.

3. Bias Adjustment. In order to optimise the bias a number of *conflicting requirements* have to be taken into account, and the resultant setting will be a *compromise* between distortion, frequency response and signal-to-noise (S/N) ratio.

With a *three-headed reel-to-reel recorder* biasing is a relatively straight forward operation and the result of any adjustment is readily apparent. If the machine is well designed, when the bias has been optimised for the highest speed it will be correct for the others also. Fig. 32.13 shows the test setting required. A 1 kHz sine wave is fed to the recorder and the input level is adjusted to give a deflection of 0-VU on the record level meter. The millivoltmeter is then set to give a mid-scale reading when monitoring the off-tape signals. If the bias level is now reduced, the meter reading will fall considerably and this reduction should be continued until a minimum output is obtained. Then progressively increase whilst thus metering the off-tape signal : continue to increase the bias until firstly a maximum reading is obtained and secondly the output begins to fall. *The optimum bias will be in the region of 1 to 1.5 dB below maximum output.* (see Fig. 32.14). It is important that the *-1dB point* is found on the correct slope *i.e.* point *b* and not point *a*. This method of biasing is often referred to as *over-drop.*

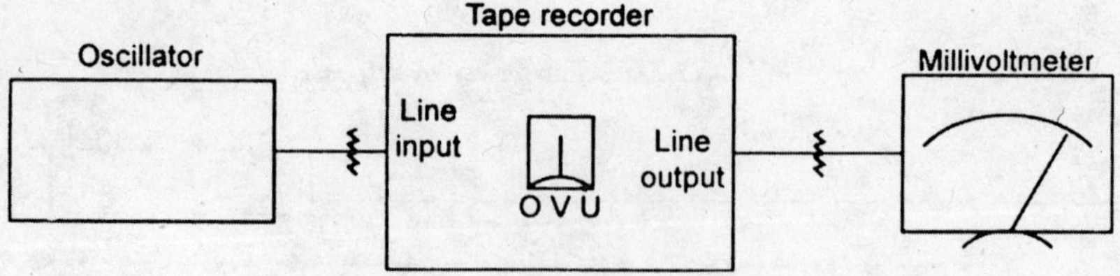

Fig. 32.13. Bias adjustment

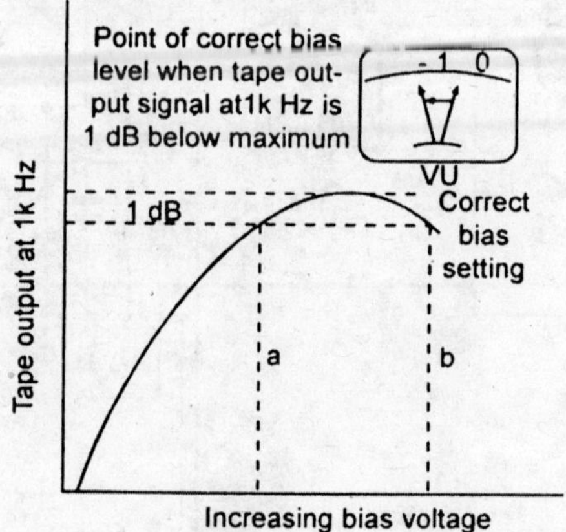

Fig. 32.14. The 1 dB over-drop method of biasing.

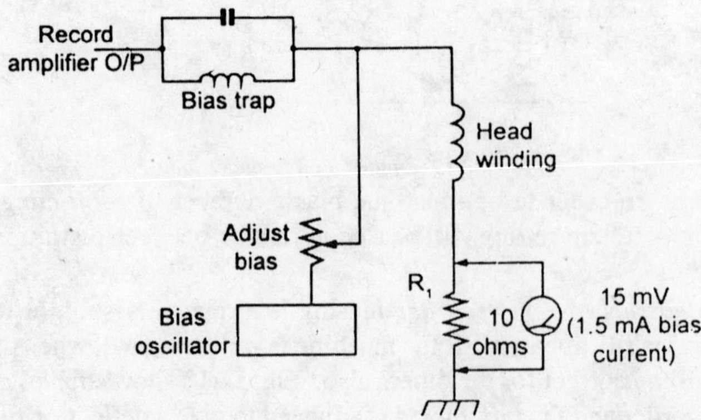

Fig. 32.15. Measurement of bias current in terms of voltage across R_1 (10 ohms)

Frequently the service data for recorder will recommend a value for the *bias current* and the above procedure can be short-circuited. If the recommended bias current is say 1.5 mA, this is equivalent to 15 mV across R_1. This setting may have to be modified slightly according to the tape used in order to arrive at the *specified balance* between frequency response, signal-to-noise ratio and distortion.

The biasing of a cassette recorder is *extremely critical* and in the absence of detailed service information, the *1 dB over-drop method* can be tried but the *reference frequency* should be 6.3 kHz rather than 1 kHz. Some *trial and error* may be required to bring the machine within the published specification.

4. Overall Response. Use the test set up as shown in Fig. 32.16. Inject a *reference signal* of 333 Hz or 1 kHz and adjust the input level control to obtain meter reading of 0 VU. Now reduce the oscillator output by 20 dB and record the reference frequency. ***It is necessary to record at a level of - 20 dB to prevent tape saturation at high frequencies due to the record pre-empahsis.***

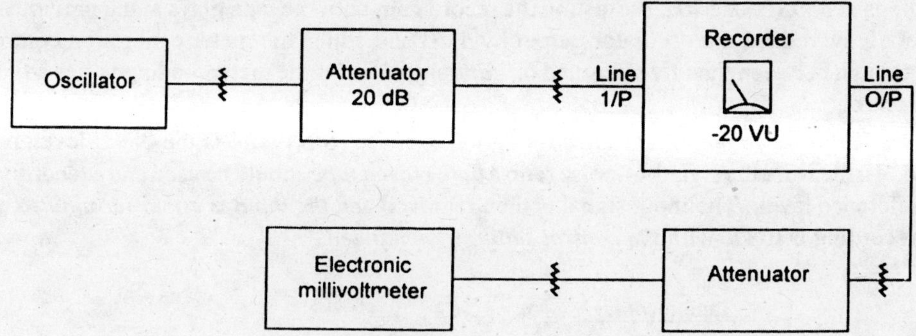

Fig. 32.16. Measurement of overall response

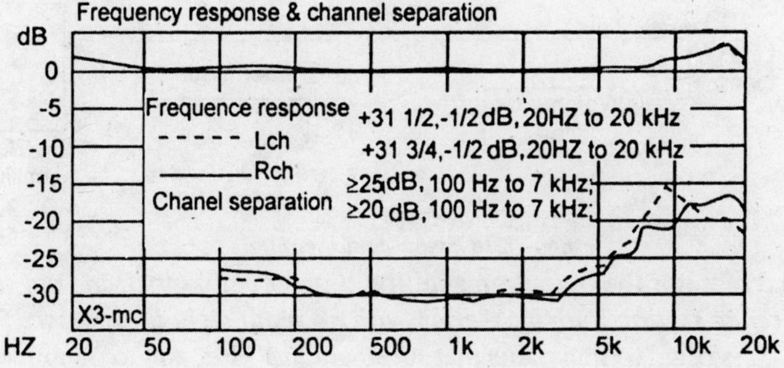

Fig. 32.17. Typical overall frequency response curves at speeds of 7 1/2 and 3 3/4 ips (top and immediately above). The curves are essentially flat from 50 to 20,000 Hz at 7 1/2 ips and from 50 to 10,000 Hz at 3 3/4 ips.

Having recorded a frequency run, replay the tape and note the deviation from the datum set by the reference frequency. If the response is uneven, the record head and amplifier should be checked and, where possible, the equalisation adjustment to give a response within the specified limits. The overall response of a typical tape machine is given in Fig. 32.17.

5. Meter Sensitivity. On domestic recorders the *0 VU level* is usually established with reference to a *given level of distortion* frequently 3% at 333 Hz or 1 kHz. Alternatively it may be referred to a *given tape flux level.*

In the first instance the meter is adjusted by making a series of recordings at the reference frequency at progressively increasing input levels. The *distortion factor* of these recordings is measured to find the input level which produces 3% distortion. A recording is then made at the *determined input level* and the meter calibration control (*not the zero adjustment*) is set to give a meter reading of 0 VU.

To calibrate the meter against a specific flux level obtained from a fully modulated tape at the *Line output.* Replay a reference level tape (320 nW/m or 250 nW/m according to tape speed) and measure the replay level at *Line out* socket. Adjust the replay gain to obtain the specified output. Load a blank tape and make a series of recordings at 333 Hz or 1 kHz, adjusting the record gain until the tape plays at the previously established output level. Now reduce the oscillator output by 4 dB and adjust the meter calibration control for a 0-VU indication. Switch between the *incoming* and *off-tape* signal and if the meter reading varies adjust the replay calibration control.

6. Signal-to-Noise Ratio. A *common reference* for signal-to-noise (S/N) ratio is the 0-VU level, but +3 VU may also be used. To measure the signal-to-noise ratio a *bulk erased tape* should be used and a recording made to the specified reference level. The input signal is then removed and *the input is either terminated or shorted* and a further recording is made *with the control settings unchanged.*

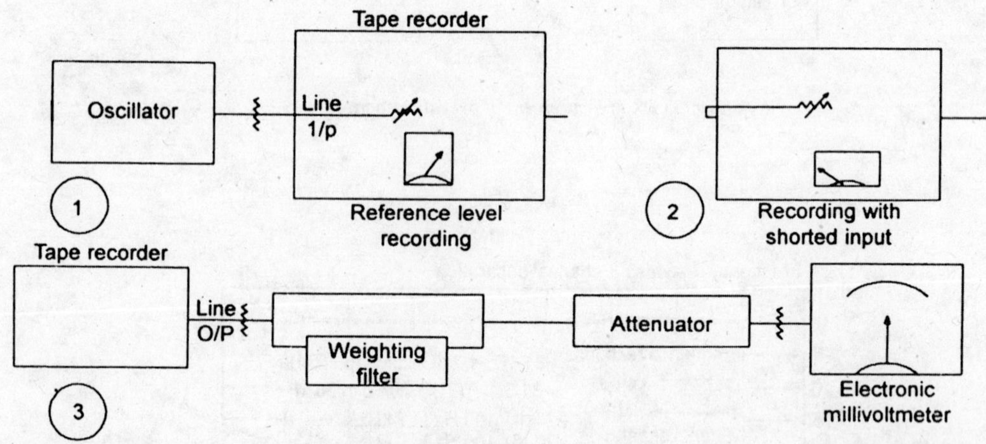

Fig. 32.18. Signal-to-noise ratio.

Replay the tape and monitor the output on a millivoltmeter across the *Line out* socket. *The difference in level between the reference section and the section with no input signal, expressed in dB, is a measure of the signal-to-noise ratio.* It is usual to make this measurement at 1 kHz, and to obtain a *weighted noise figure* a filter of the required characteristic *having unity gain at 1 kHz* should be inserted before the measuring instrument (See Fig. 32.18). If the measuring instrument incorporates a *switched attenuator* the ratio can be read off directly in decibel.

7. Tape Speed. The easiest way to check a machine's speed is to replay a *calibration tape of known frequency* and to connect a frequency meter to the output of the machine. If a frequency meter is not available a *calibrated audio osicallator* should be connected to the horizontal input of an oscilloscope. A tape of the same frequency is then replayed and the output of the recorder connected to the vertical input of the oscilloscope. *If the frequencies match an ellipse will be produced.* If an accurate test tape is not available it is no use recording one on the machine under test, as it will replay the test frequency accurately, although at the wrong speed.

8. Cassette Tapes. To realise the best performance from magnetic tapes with respect to frequency response, distortion and noise needs a *careful balance* among recording bias, recording and playback equalisation and recording level. Although the design of the tape recorder's heads and electronic systems are also vital factors *for any given recorder and tape formulation the key parameters are the bias and equalisation characteristics.*

The frequency response of any magnetic tape is *far from flat* and *considerable equalisation is required at both high and low audio frequencies to meet audio recording standards.* The equalisation is divided between the recording and playback amplifiers, but *playback frequency response is standardised to permit tapes made on one machine to be played on another.* This still leaves the recorder designer considerable latitude in setting bias levels and recording equalisation for the best results with his heads and available tapes.

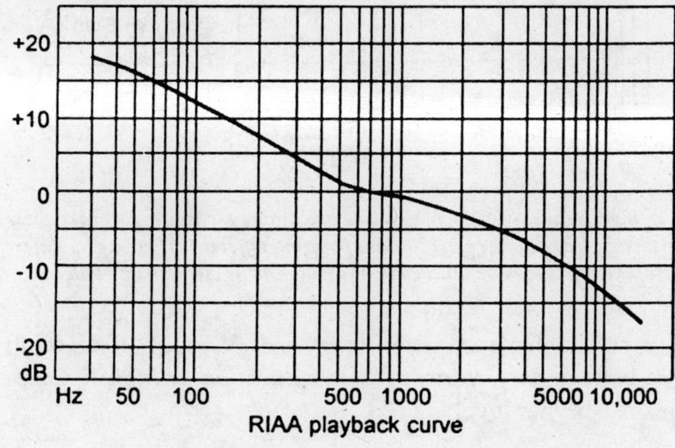

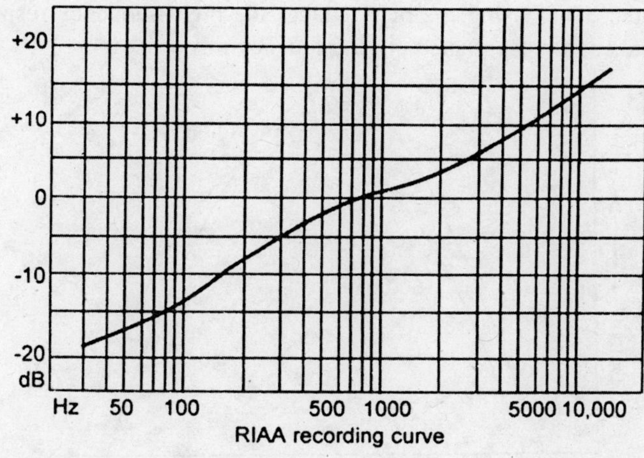

Fig. 32.19. RIAA recording and playback curves.

Except in the higher price brackets, most home tape recorders have no adjustment accessible to the user for optimising performance with any particular tape. They are generally designed to operate satisfactorily with a variety of tapes, although *the manufacturer's specifications may be realised only with a certain recommended tape formulation.*

Cassette recorders which must extract the last bit of performance from the tape in order to meet reasonable high-fidelity standards at their 1 7/8 ips operating speed are especially critical in their adjustment for the specific tape used. In particular, the bias level is extremely important if a reasonably uniform response in the 8000 Hz to 16000 Hz octave is to be obtained (see Fig. 32.20). Although the user normally has no means of optimising bias for his tape, the recorder manufacturer recommends specific tape brands and types which will enable his specifications to be met, and qualified service technicians can adjust almost any machine for use with any tape.

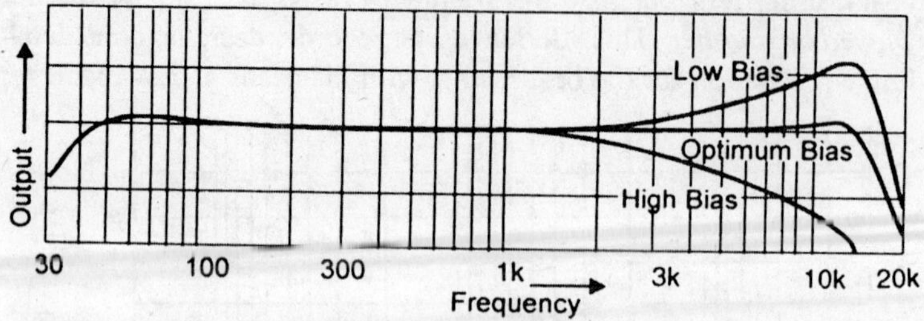

Fig. 32.20 Curve shows the effect on high-frequency response caused by varying a cassette recorders bias adjustment. With a given bias setting, you can encounter almost as great a range of variations by using different tapes having different bias requirements. But none of the tapes may be inherently any better or worse than any other.

All frequency response measurements are made at a *recorded level* of -30 dB relative to the 0-dB reading of the recorder's meter. This is necessary *to avoid tape saturation at high frequencies,* which can give erroneous response indications if a higher level is used.

Recordings are made over the full frequency range with different tapes. If the playback response in the 10000 Hz to 15000 Hz range deviates more than 2 dB from the mid-range level, the bias is adjusted slightly and the measurement repeated. Increasing the bias reduces the high frequency response, while less than optimum bias results in a peaked, or accentuated, high frequency response.

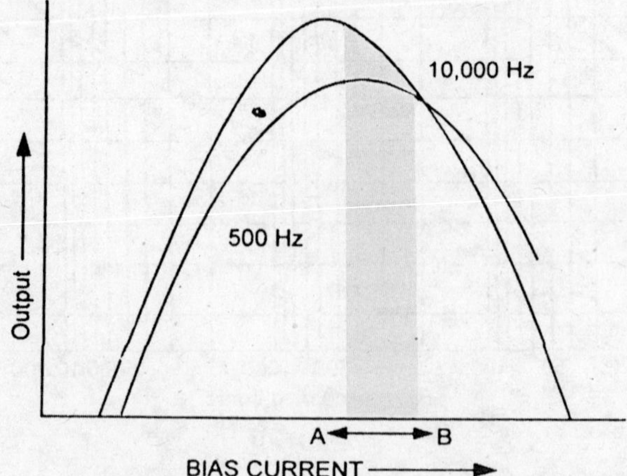

Fig. 32.21. How bias current affects mid-and high frequency signals output from a cassette tape. Note that for the current range between A and B 500 Hz output hardly varies at all but 10 kHz output varies substantially.

Using the optimum setting make 10000 Hz recordings at different levels around the 0 dB point observing the playback distortion for signs of overload. Determine the level (relative to 0-dB) which results in 3% distortion (predominantly third harmonic) in playback. *The 1000 Hz output from this 3% distortion is considered to be the maximum usable output of the recorder with each tape.*

Operate the recorder in RECORD mode, with no input signal and during playback measure the noise level (so called *bias noise*). This is expressed in decibels below the maximum recording level.

One of the major weaknesses of cassette tapes is the *output fluctuations due to tape coating in homogeneity (dropouts) and erratic tape-head contact.* The latter is a function of the mechanical design and assembly of the cassette. Unevenness of tape hub friction and tape widening can cause an irregular output amplitude similar to that resulting from actual tape loading defects.

Some cassette assemblies use *screw-assembled(S) cases.* If the tape breaks it is usually possible to open the cassette and retrieve both ends for splicing. On the other hand, the *welded (W) case* mould on most cassettes cannot be opened non-destructively, so that a tape breakage means the loss of the cassette.

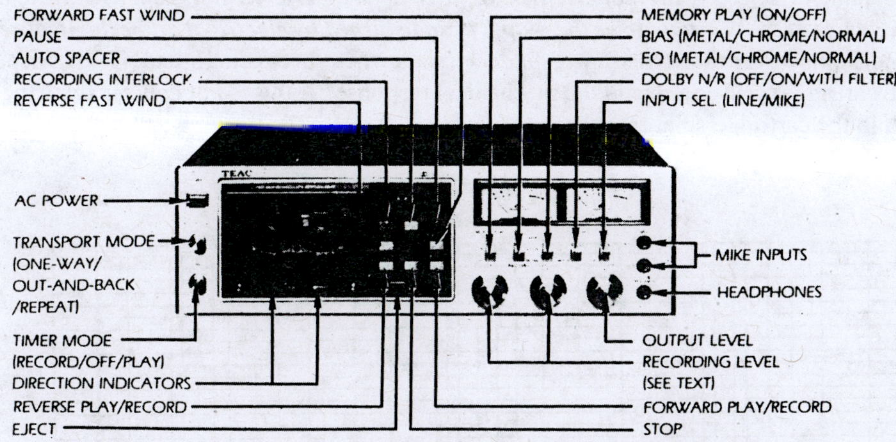

Fig. 32.22 Teac Model CX-650 R bidirectional cassette deck.

One of the significant factors in cassette sound improvement has been the development of chromium dioxide (CrO_2) tape. *Properly used, it can slightly extend the high frequency response and significantly reduce noise levels.* However, it requires somewhat different bias, equalisation and operating levels for fully effective results. Many recorders are now provided, with a *switch* marked "CrO_2" and 'Normal' (see Fig. 32.22).

32.4 PERFORMANCE TESTS-PHONO EQUIPMENT

(1) *Cartridge's Sensitivity :* Cartridges can sound fairly different from one another - a fact that makes *listening* unusually important (more so than for any other component category except *loudspeakers* the electromechanical transducers at the *opposite end* of the playback chain).

The most basic measurement is the one that relates groove velocity to cartridge output voltage. *The higher the output for a given groove velocity the higher the sensitivity.* This is much like a loudspeaker's sensitivity rating turned on its head. Whereas a speaker's sensitivity depends in large measure on its efficiency as an *electrical motor,* a cartridge's sensitivity mostly reflects its efficiency as an *electric generator.* Reciprocal functions breed reciprocal specifications. Sensitivity is usually given in millivolts per centimetre per second (mV/cm/s) of groove velocity. Typical ratings for *fixed -coil and high - output moving coil cartridges* range from 0.5 to 1.5 mV/cm/s, with most clustered around 0.75 and 1.25 mV/cm/s. *Phono pre-amplifiers are designed with these sensitivities in mind so that they will not overload on strong signals from very high output cartridges or be excessively noisy on weak signals from pickups of only moderate sensitivity.* However, the lower the sensitivity of the cartridge being used, the further you will have to advance the volume control to obtain the desired sound level and, all else remaining equal, the poorer the signal-to-noise ratio will be. This is a serious concern with most *moving coil* cartridges, which typically have sensitivities an order of magnitude lower than those of *fixed-coil* pickups. As this type of cartridge has become more popular, many manufacturers have begun building *head amplifiers* into their high-end electronics.

It is important that *the sensitivities of the two channels be as nearly the same as possible,* irrespective of what the output of a particular channel happens to be. If one channel is significantly more sensitive than the other, the stereo image will *pull* to that side. *Channel balance can be expressed in dB relative to the mean of the sensitivities of the two channels at 1 kHz* (channel balance often varies slightly with frequency - hence the *1 kHz reference*). Therefore, a *channel - balance specification* of less than or equal to ±1/2 dB indicates that there is no more than a 1 - dB difference between the sensitivities of the two channels. *A rating of ± 1 dB is resonable, and any imbalance can be corrected with balance control.*

2. *Cartridges's Frequency Response :* A cartridge's frequency response is one of the most important indices of its quality. *Virtually all modern pickups have almost perfectly flat response through the bass and mid range, because at those frequencies the output is undisturbed by electrical or mechancial resonances.* At high frequencies, however the *stylus-tip resonance* (somewhere between 15 and 40 kHz depending on the design of the cartridge) causes a rise in the mechanical response. If this is not *compensated* the result will be a treble peak in the cartridge's net frequency response.

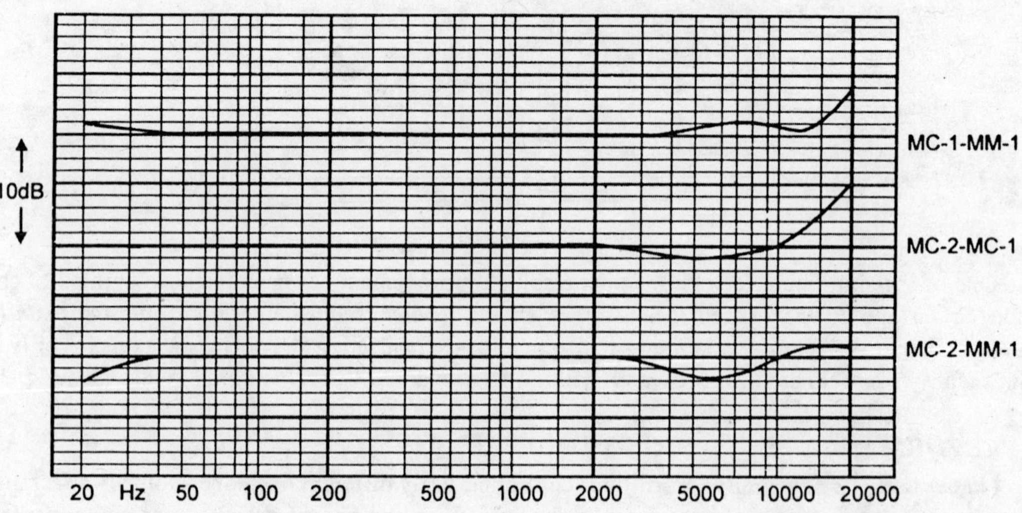

Fig. 32.23 Differences in frequency response between three typical cartridges.

Fortunately, a magnetic cartridge's electrical circuit consists of a set of coils; *these form a low-pass filter that progressively attenuates signals as they increase in frequency (rolls off the high end,* in audio jargon). The bigger the coils, the higher the inductance and, therefore, the bigger the bite out of the treble. *Designers of fixed coil (moving-iron and moving magnet) pickups routinely balance this filtering action of the coils against the rising response of the mechanical system to achieve, with varying degrees of success, flat overall response.*

Moving coil cartridges must use *small low-inductance coils* to prevent the moving mass of the stylus assembly from becoming too high. Besides extracting a penalty in sensitivity, this practice makes the electrical response of the moving-coil pickup assembly essentially flat to well beyond the upper limit of the audible band leaving the designers with the uneviable task of trying to make the cartridge's mechanical response flat to 15 kHz or beyond. However, whatever response they achieve *will be almost independent of the load* presented by the resistance or capacitance of the phono input combined with the capacitance of the connecting cables. This is not true of most *fixed-coil* cartridges, whose high inductances *interact* with the load, altering the electrical response. Manufactueres have to anticipate this effect and design their cartridges for flat response into a *standard resistance* of 47,000 ohms (which all phono inputs are supposed to provide) in parallel with some capacitance. The *recommended capacitance* varies according to the design of the cartridge, but rarely exceeds 450 picofarads (pF). *Although you cannot normally remove capacitance from a system that inherently has more than a particular cartridge demands, it is fairly easy to add what's necessary to one with less.*

Advances in cartridge technology are reducing the *susceptibility* of fixed coil cartridges to variations in loading and making moving coil pickups *flatter and more sensitive.* Given the current state of the art, *response* within ± 1 dB from 30 Hz to 15 kHz, and a *variation* of ± 2 dB over the range is very good.

(3) *Channel Separation:* Channel separation is the ability to keep information for one channel from *leaking* into the other. It is tested by measuring each channel's output in response to a signal that should ideally *produce output from only one channel.* The ratio of the desired output to the unwanted output, normally expressed in decibels (dB) is the separation.

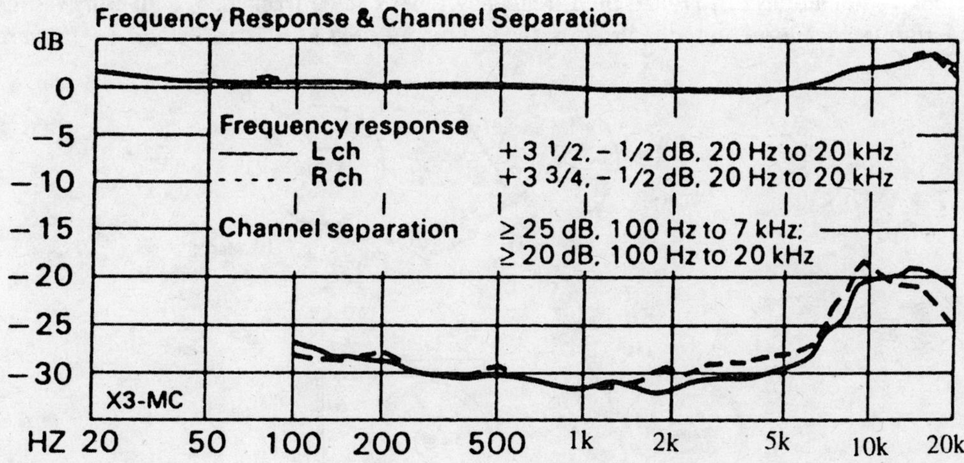

Fig. 32.24 Frequency response and channel separation.

Fig. 32.24 includes a graph with curves for frequency response and channel separation. The line across the top is the *frequency response plot;* the roughly *bowl-shaped* curve below indicates how much lower the *unwanted output* from the lower channel is than the *desired output* from the driven channel. About 15 dB of separation at midband is usually sufficient for good stereo reproduction; *20 to 25 dB is excellent.* Separation normally deteriorates somewhat at the frequency extremes, especially at the high end, but this usually is not audible.

(4) *Tracking Ability:* Regardless of how good a pickup is in other respects, it will be worthless if the stylus can't maintain intimate contact with the record groove. The degree to which it is able to do this in the face of high groove velocities is its tracking ability. *Tracking ability is a function of cartridge's design and of how hard its stylus is pushed down against the record.* The force that works to hold the stylus in the groove is called the *tracking force*; it is applied by the turntable's tone arm and usually ranges from 3/4 to 3 grams, depending on the pickup.

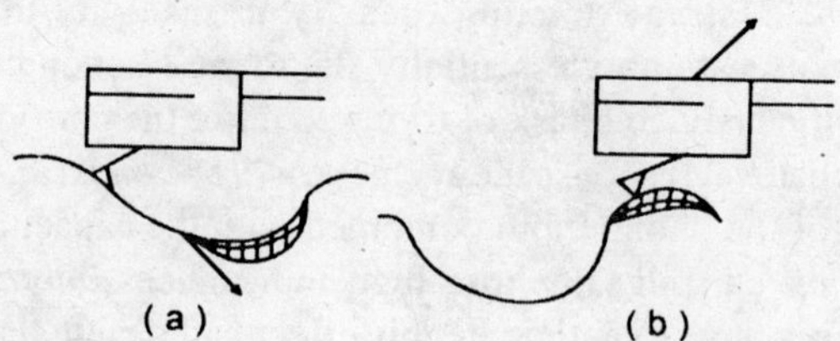

Fig. 32.25 In (a), when the stylus reaches the bottom of the "valley" it tends to try and continue in the direction of the arrow. In (b) the stylus also tends to continue in the direction of the arrow, thus losing contact with the record's surface. The two shaded areas show areas of increased record wear due to excessive force (a) and too light a tracking force (b).

All else remaining equal, the higher the tracking force, the better the tracking. But there are limits. One is that *too much force will collapse the cantilever suspension, making the cartridge unusable*. Another, more subtle problem is that *increasing the tracking force also tends to accelerate record wear*. This difficulty can be mitigated somewhat but not a great deal by careful attention to the *shape* of the stylus contact areas. Consequently, *most manufacturers design their pickups to track at forces from 1 to 1 1/2 grams.*

Tracking ability can be measured in a number of ways. In one method we measure the maximum levels at which various combinations of low-and high-frequency tones can be tracked. We give the results in dB above a standard reference level, upto the limits of the test record : + *18 dB laterally and + 9 dB vertically.*

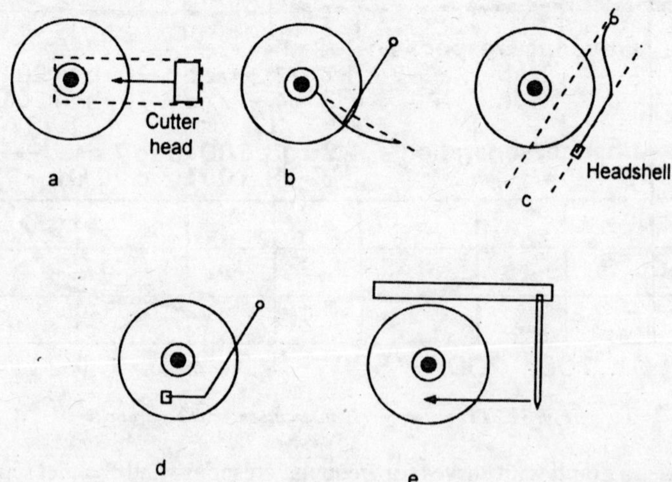

Fig. 32.26 The elimination of lateral tracking error. (a) The cutter head on the lathe moves tangentially across the lacquer. The cutter is therefore always at right angles to the grooves (b) If a straight pivoted arm were used for replay, the stylus would follow a curved path. (c) If a curved arm is used, with the head shell overhanging the centre of the record, the lateral tracking error is minimised. (d) An alternative is to use a straight arm but to offset the headshell. (e) Some manufacturers have produced elaborate 'parallel tracking' turntables. These have no tracking error but this does not automatically give an audible advantage.

In fact, even at their best, *cartridges are probably the highest distortion link in the audio playback chain.* Distortion on the order of several percent routinely occurs just because *it is impossible to use a playback stylus that exactly matches the shape of the disc cutting stylus.* However, designers have worked hard to come as close as possible to that ideal shape within the realm of practicality. Hence the current prevalence of various sorts of *line-contact style* especially on premium pickups.

Additional distortion arises from misalignment of the stylus to the groove. Unless you use a linear-tracking tone arm, which keeps the stylus constantly tangent to the groove, some slight *lateral misalignment* will almost always exist. However, correct tonearm geometry and careful installation can minimise such misalignment.

Vertical alignment, particularly as defined by the stylus rake angle (SRA) and the vertical tracking angle (VTA), also affects distortion. *Stylus rake angle* is the angle between the axis of the stylus tip and a line perpendicular to the record surface. The *vertical tracking angle* is formed by the stylus cantilever's pivot point, the point at which the stylus contacts the record, and the record's surface. Ideally, a cartridge's SRA and VTA should *match* the corresponding angles of the cutting stylus during disc mastering. (The nominal standard of VTA is 20 degrees). *Unfortunately, not all records are cut at the same angles, so it is nearly impossible to get everything right all the time on playback.* However, the distortion caused by even fairly severe stylus misalignment (especially vertical) can often pass unnoticed.

(5) *Turntable* : The turntable performs two basic tasks. One is *to turn the record being played at a constant, correct speed.* The other is *to hold a phono cartridge in the proper position with the correct tracking force.* (We're assuming here that the turntable comes with its own tone arm; some high-end models don't). And it must perform both tasks as self-effacingly as possible, to prevent any unwanted additions to the signal picked up from the groove.

Any mechanical system with moving parts generates vibrations, inconsistencies, and wobble, and in a turntable, these cause wow, flutter and rumble. *Wow and flutter are short term variations.* Wow, which usually is the more audible of the two is just low-frequency flutter. Such fluctuations can add an unpleasant sourness or coarseness to the sound, especially on sustained notes. Maximum allowable wow or flutter percentage for high fidelity performance from a turntable is about 0.15 percent, from a changer about 0.2 percent.

Rumble is low-frequency motor noise transmitted through the platter to the pickup stylus. It is reported in dB below a specified reference level, much like *signal-to-noise ratio,* which is essentially what it is. Wow and rumble can be checked with a test record. *Acceptable rumble* can be checked by setting the volume and bass controls to their normal positions, setting the treble at full ''cut,'' and setting the stereo controls ON (if it's stereo player). Then play a *test record.* If you can't hear any rumble, the player is OK.

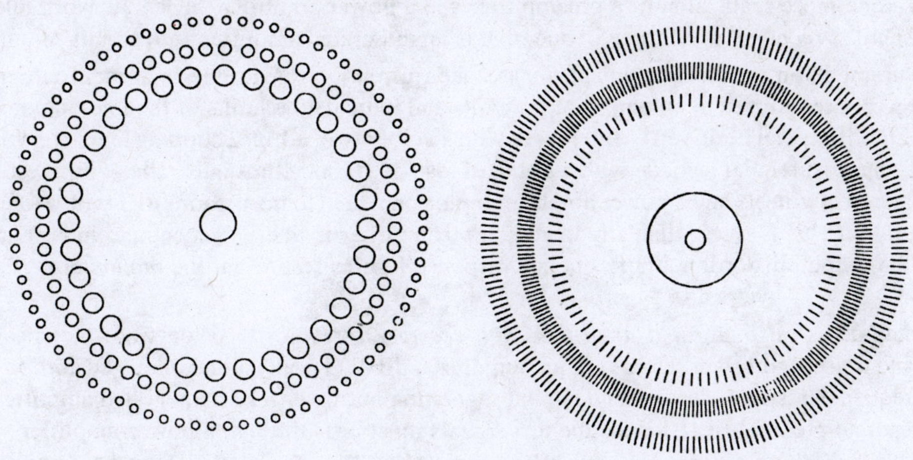

Fig. 32.27 Stroboscope

Accuracy of speed can be checked with a *stroboscope*—a circular card with a series of dots or bars printed in concentric circles. Each circle is marked for a specific speed. The card must be used with a light, preferably a fluorescent one. A neon lamp is perfect and lamps are made just for this purpose. The card is slipped on the turntable just like a record and the turnable is then started at the speed to be checked. If the dots or bars for that frequency seem to be standing still, fine; if they begin to drift, not so good. You can check for drift by holding a pointed object over the spot where the dots or bars coalesce into a single spot or line (Fig. 32.27). Then, if the single spot or bar drifts away from under the point, try to count the number of dots or bars that it drifts in one minute. If the number is *less than 35,* forget it; your accuracy is only about 1/2 percent. If it is more, *check your line voltage;* it may need regulation.

(6) *Tonearm :* Provided that pivot friction is negligible, a tonearm's most important mechanical characteristic is its *effective mass.* This is not the tonearm's physical weight, but the inertia a stylus feels as it moves the arm while playing a record. Effective mass has no inherent significance, it matters only because of the role it plays in determining the frequency of the *maximum / cartridge resonance.* The other factors are the *cartridge's weight* (which adds directly to the arm's effective mass) and *low-frequency dynamic compliance.* Increasing any of these tends to lower the resonance frequency, while decreasing any of them tends to raise it.

Ideally, the resonance should be about 10 Hz. If it is too much *lower* than that, the system will overrespond to record warps; if it is too much *higher* the system's low-frequency response and tracking ability will be adversely affected.

Normally anything within the range from 8 to 12 Hz is acceptable. The arm's effective mass is measured using a calibrated cartridge. The results appear in *tonearm/cartridge matching graph* and can be used with the weight and compliance figures for any phono cartridge if the two are a good match.

A tonearm's most important electrical characteristic is its *lead capacitance.* It matters only because the frequency responses of many fixed-coil pickups are affected by capacitive loading. You therefore need to know the sum of the phono-input and arm-lead capacitances to determine whether they make a good match for a particular pickup.

32.5 PERFORMANCE TESTS-RECEIVERS, AMPLIFIERS AND TUNERS

(1) *Receivers :* A *preamplifier* is the part that you probably use most to adjust the volume, switch sources and so forth. Every integrated amplifier and receiver has a preamplifier section built into it. An *integrated amplifier* is a one-piece realistaion of a preamplifier and a power amplifier (hence the word integrated) and is tested as such. A *receiver* is all three in one; that is an integrated amplifier with a built-in-tuner.

Measurements on a *separate preamplifier* are taken primarily from its line-level main output. Sensitivity and noise figures are referred to an output of 0.5 volts and indicate the limits of the line-output stage. Since virtually all amplifiers can be driven to full power with a volt or two, a 1 kHz clipping level of 3 volts is usually sufficient. This measurement is made with a standard load of 10,000 ohms in parallel with 10,000 picofarads which simulates a low-impedance power amplifier input connected to the preamplifier with a long run of high capacitance cable. This is worse than any loading a preamplifier is likely to encounter in normal use, which means that the output, distortion and frequency response figures respresent the *minimum performance you can expect.*

Measurements on *integrated amplifiers and receivers* are mostly taken from the speaker outputs. Sensitivity and noise figures are referred to an output of 0 dBW (1 watt) into 8 ohms. Distortion usually will be higher, frequency response less extended, and separation narrower for an integrated amplifier or receiver than for a separate preamplifier because the test signals must pass through a power-amplifier section. The outputs-at-clipping measurements are not at all comparable. Those for amplifiers and receivers are in dBW and reflect the capabilities of the power amplifier, while those for preamplifiers proper are in volts and indicate the limits of the line-output stage.

Total harmonic distortion (THD) is measured from 20 Hz to 20 kHz at an output of 2 volts. This is done twice; once through the phono input (to include the effects of phono-preamplifier circuitry) and once through a high -level input. The results are reported separately. In the case of a single result, distortion through either input is less than 0.1 percent.

Frequency response is measured through the auxiliary input on the preamplifier. Deviations are referred to the output at 1 kHz and are reported over two ranges; one representing the maximum deviation in dB over the entire audio band, the other the frequency range over which response drops no more than 3 dB below the 1 kHz response.

The Recording Industry Association of America (RIAA) *playback-equalistion curve* is applied to signals passing through the phono-preamplifier section to *reverse* the equalisation recorded into all phonograph records and thereby *restore* flat response. We show the accuracy of this equalisation from 20Hz to 20kHz between the frequencies at which response is down 1 dB relative to the output at 1 kHz, whichever comes first.

Sensitivity is the amount of input signal required to obtain a 0.5V (or in integrated amplifiers and receivers-0dBW) output with the volume control all the way up. It is measured through the auxiliary input, the fixed-coil phono input, and (if there is one) the moving-coil phono input.

To measure the *signal-to-noise ratio,* set the volume control for a 0.5 volt (or 0.dBW) output from a 1 kHz input of 0.5 volt (500 millivolts) at the aux input, of 5 millivolts at the fixed coil phono input, or of 0.5 millivolt (500 microvolts) at the moving coil phono input, The auxiliary measurements are made with a 1,000 ohm resistive source, and the moving-coil measurements ar taken with a 10 - ohm source impedance. You should expect to see figures of 80 dB or higher for the aux input and of 75 dB or higher for the fixed-coil phono input. The moving coil input will likely have signal-to-noise ratio several dB lower than that of the regular phono input.

Phono overload is the input in millivolts required to cause clipping at 1 kHz through the phono input. For a fixed-coil input 100 millivolts is adequate. The moving-coil overload point is usually a function of the additional gain provided by the preamplifier stage and the fixed coil inputs overload margin.

Separation is checked at 100 Hz, 1 kHz and 10 kHz. About 30 dB is all that is necessary for proper stereo reproduction.

(2) Amplifiers : Every stereo system includes a *power amplifier* whether as separate unit, as half of integrated amplifier (along with a preamplifier section), or as the final stage in a receiver (following tuner and preamplifier circuitry). Its sole job is to provide the voltage and current necessary to drive loudspeakers. Whereas the design of a preamplifier is relatively straight forward, that of a power amplifier involves some important trade offs and compromises.

The most important thing to know about a power amplifier is *how much power it will deliver into your speakers.* Two amplifiers with identical continuous power ratings into an 8 ohm load may have very different maximum outputs into a 4 ohm load. For this reason measurements are made on the continuous clipping at 1 kHz with both channels driven into 8-and 4-ohm loads.

Although the law requires that components be rated according to their continuous power output, this measurement does not give a true indication of how an amplifier will perform on most music. *Music consists primarily of short term transient information not continuous tones.* The EIA amplifier testing standard therefore provides for a measurement called *dynamic headroom.* This is the amount in dB by which an amplifier can exceed its rated continuous power into 8 ohms on a series of tone bursts of a specified duration. *All else being equal, the greater the dynamic headroom, the better.*

Dynamic headroom is based on the manufacturer's rated power which is usually given only for 8-ohm loads. Many loudspeakers have impedances that are *lower* (sometimes much lower) than that over much of the audible range, and almost all of their impedances include *reactive* (non resistive) components that increase the amount of current the amplifier's output stage must handle. It is therefore highly desirable to know an amplifier's transient power output capability into impedances that are difficult to drive than the 8-ohm resistive load traditionally used for specification and testing. Consequently dynamic power is reported not just into 8 ohms, but also into 4 ohms and unless the manufacturer indicates that the amplifier will not operate safely driving such a low impedance into 2 ohms.

When an amplifier is driven beyond its power limits (that is into clipping) it will generate large amounts of distortion. But below that point, its output should be very clean. *Total harmonic distortion* (THD) is measured from 20 Hz to 20 kHz at the amplifiers rated power and at 0 dBW (1 watt) into 8 ohms. Although the ear is remarkably tolerant of distortion in the presence of music (several percent can easily go unnoticed in complex passages) a THD figure of 0.1 percent or below is preferred.

It is vital that *frequency response* be flat over the audible range (from between 20 and 30 Hz to about 15 kHz) and that noise be low enough not to cause any audible degradation of the signal. Today's amplifiers are almost invariably flat within 1/2 dB from 20 Hz to 20 kHz.

Sensitivity (the input voltage required to produce a 0 dBW output into 8 ohms with any level or volume controls turned all the way up) is almost never something you have to worry about. Ratings from 100 to 200 millivolts are typical.

Input impedance is never a problem. 20,000 ohms or more is preferred, but in most cases 10,000 ohms is adequate.

Damping factor is an expression for an amplifier's output impedance at low frequencies. The lower the impedance, the higher the damping factor. It is unlikely to make much difference so long as it's higher than 20. Output impedance is measured at a number of frequencies up to 20 kHz.

3. Tuners. Tuner specifications fall into two categories : radio frequency (RF) and audio frequencies (AF). The former pertains to how well a tuner functions as a radio receiver while the latter defines the quality of its audio output under good signal conditions.

Of the *RF specifications* the most basic is sensitivity ; how much signal is needed to get *listenable reception*. Signal strength is expressed in dBf referenced to 1.0 femtowatt (10^{-15} watt), where 0 dBf equals 1.0 femtowatt and every doubling of the signal strength (in femtowatts) amounts to an increase of 3 dB and the current EIA standard pegs listenable at a single-to-noise ratio of 50 dB.

Sensitivity ratings of 16 dBf in mono and 38 dBf in stereo are about average for high quality home units. To supplement the raw sensitivity number, the tests include mono and stereo *quieting curves,* with signal-to-noise ratio plotted on the vertical axis and signal strength on the horizontal axis. The farther to the left a curve begins and the steeper its descent, the better. Most *stereo quieting curves* flatten out somewhere between 65 and 75 dB down at a signal strength of 65 to 85 dBf. The *mono curves* level off at much lower signal strength, usually somewhere between 70 to 80 dB down.

Signal-to-noise ratio is measured at a signal strength of 65 dBf, which is about where most tuners approach full quieting in stereo. Signal-to-noise ratios are typically between 65 and 75 dB in stereo and as much as 10 dB better in mono.

Just as important as a tuner's ability to *pull* in desired signals is its ability to *reject* unwanted ones. *Selectivity* is a measure of how well a tuner discriminates against broadcast channels close to the one you have tuned. The number normally reported is *alternate channel selectivity*, indicating how completely broadcasts spaced two channels apart (400 kHz on FM band) are rejected. Tuners average about 60 dB of alternate channel selectivity.

Capture ratio is a measure of how well a tuner discriminates between signals on the same channel. In most areas, a tuner will receive a multitude of *reflections* in addition to (or even instead of) the *direct* signal from the transmitter, a condition that will create noise and distortion in the tuner *unless it can latch onto just one of the signals and ignore the others.* The amount by which one signal must be stronger than a competing one for the former to be demodulated and the latter to be suppressed is the tuner's capture ratio. Most of today's home units have a capture ratio of 1/2 dB or less. The very best tuners have a capture ratio of 1 dB or less.

AM suppression is a measure of how immune the FM detector is to amplitude modulation of the carrier. This is important for removing noise from the signal, especially any that may be generated by mutlipath. An AM suppression ratio of 50 dB is reasonable, a rating of 60 dB is excellent.

Even after the radio signal has been demodulated into audio, there still remain certain elements that must be removed if the highest possible fidelity is to be maintained. Most important is the *19 kHz pilot tone* that turns on the stereo demodulator. 60 dB or more of *pilot suppression* is desirable to reduce the likelihood of Dolby mistracking when you record off the air. It is likewise desirable to attenuate the *38 kHz stereo sub-carrier* by as much as possible - 60 dB or better.

A tuner's principal *audio only specifications* are frequency response, channel separation and noise. The *frequency responses* of good tuners are usually within ± 1 1/2 dB from 50 Hz to 15 kHz. Since 15 kHz is the upper limit allowed in FM broadcasting there is no reason for a tuner's response to reach higher.

Separation between signals in the two stereo channels is essential to maintain *correct stereo imaging*. About 20 dB between 300 Hz and 3 kHz is desirable. Any decent tuner should be far better than that.

The EIA standard calls for *harmonic distortion* to be measured at 100 percent modulation at each of three frequencies : 100 Hz, 1 kHz and 6 kHz. It also provides for a mono intermodulation (IM) distortion measurement and a stereo test for intermodulation with a 19 kHz pilot. No good home tuner will exhibit more than 1 percent distortion on any of these test signals in either mono or stereo, and most will show less than 0.5 per cent.

33

TROUBLESHOOTING PROCEDURE

Servicing without service data is like fighting with one hand tied behind your back. Circuit diagrams, component layout diagrams and block diagrams are invaluable service aids in restoring unhealthy equipment to normal health. Troubleshooting charts create the right mental attitude for tackling block-specific and circuit specific problems. Equipped with the right mental attitude and, last but not the least, tools and test equipment relevant to the job at hand, the technician should be able to accept and tackle any challenging service problem. Service aids save mental torture and valuable time and time saved is money saved.

33.1 SERVICING WITHOUT SERVICE DATA

Servicing without service data is like *fighting with one hand tied behind your back.* Servicing without service literature can be *tedious and time consuming.* Many smart technicians refuse to touch one of these Made in Japan, Hong Kong, Taiwan etc, electronic devices. They figure out they will invest more time in *flying-blind servicing* than they will be able to recover in service charges.

The trick is to find ways to nullify the *unattractive qualities* of the work by using your knowledge and experience to work out *a diagnostic procedure that keeps unnecessary moves to a minimum.* Make maximum use of *generalised service data,* and employ different techniques than the ones used with service data and stick to that procedure religiously.

Check the easy things first. Nothing is more frustrating than to remove a hard-to-get-out chassis for a lot of fruitless voltage testing to discover the trouble was a broken antenna lead you could have repaired in a minute. Always check easily repaired possibilities first. Battery voltage and current drawn should be checked routinely, even if the customer says the battery is brand new. It could have grown old on the shelf, or the set could have a high current drain that exhausts the battery quickly. The millimeter can be inserted in a snap connected battery lead easily by removing one snap and using the meter leads to restore the connection. In case individual cells in series furnish the current, use a special jig made up of a couple of pieces of brass shim stock cemented to opposite sides of a piece of thin fish paper. When this is inserted between the end connector of one of the batteries and its contact, milliammeter leads touched to the two pieces of shim stock will read the total current being consumed. Your experience will serve as a guide to the *normal current* for the number of transistors used.

Carefully look over the chassis with a pair of magnifying glasses for loose connections, broken wires or charred resistors. Look especially for dangling fine wires going to the antenna. These leads are often broken by the customer in changing batteries or in attempts at do it yourself servicing.

Gently flex the printed circuit board, especially along the edges. What you are hoping for, of course, is *a sudden restoration of normal functioning,* indicating a broken or loose connection has been temporarily corrected.

The next thing to try is *signal injection and/or signal tracing.* With the proper equipment, even voltage checks can be made from the top of the printed circuit board. Just listening for a *click* when the set is switched on will tell you if the output transistor is drawing collector current and if the output transformer and speaker voice coil are OK. If you hear the click they are; if not, something is wrong with one of these components.

Another *crude signal-injection test* is to touch the HOT or centre lead of the volume control with the volume full on. If this produces a quite *noticeable hum,* the signal can pass from the volume control clear through the audio amplifier.

But it is usually better to use, *signal generators* for injection and a *signal tracer* to pick off the signal at various points downward from where you put the test signal in.

Look at Fig. 33.1. If you know the audio amplifier is working you can start right in by putting an RF signal into the antenna and picking it off at the detector diode or as far forward as you need to go to pick up the signal. *You can zero in on the defective component that is losing the signal from two directions.*

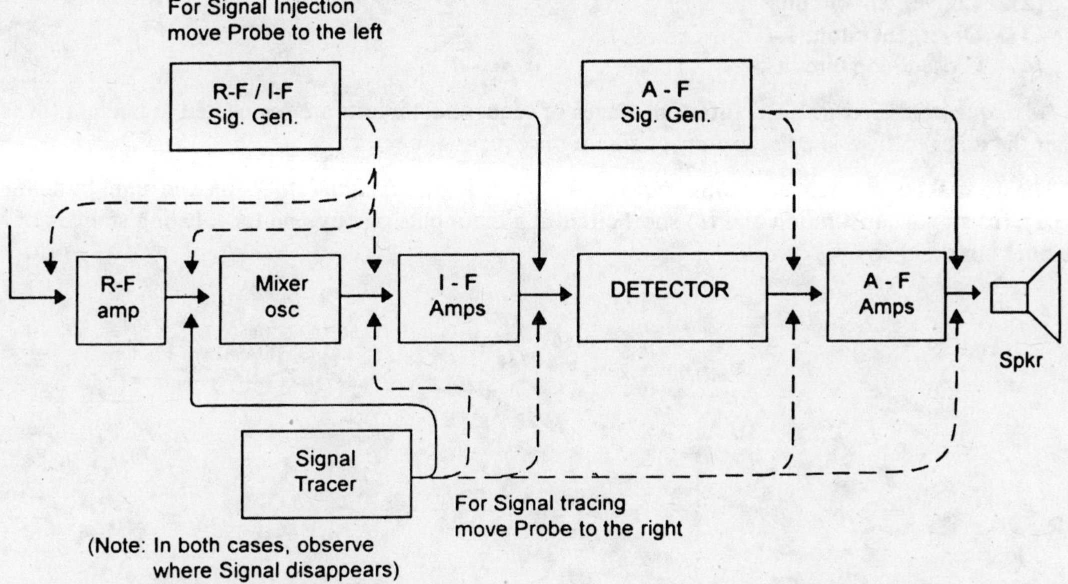

Fig. 33.1 How to use signal generators and a signal tracer to check out an unfamiliar receiver.

It's pretty tricky trying to work a signal-injection or signal tracing probe into the *proper points* on top of the printed circuit board without shorting out a couple of leads and possibly blowing a transistor. Special *extensions* (available as accessories or locally fabricated) are available that will slip over the probe tips of VOM, VTVM, signal tracer, or signal generator. When not needed, they can be quickly removed.

Once the *loss of signal* has been cornered in a small portion of the circuit, you fall back on your technical knowledge and your *generalised service data* to pinpoint the faulty component. If you have got to change a component or repair a broken connection you have to decide whether it will be quicker to remove the board or improvise a repair from the top.

Here are some helpful suggestions - *tricks of the trade.*

- Don't try transistor replacement-especially if they are *soldered in* - as a form of first aid. Neither should you blindly assume the fault can't be in a transistor. *Probably, operating the equipment for long periods of time overheated the output transistors in spite of their being heat sinked.*
- Keep a *card file* by make and model number on data less equipment. *List the symptoms, the service procedure by which the fault was found and how it was repaired.*

- If you are really convinced servicing can't be done without diagrams and voltage data and if you are going to waste your time, *you'd better not tackle it.*
- If you consider this kind of servicing a *challenge* to your technical proficiency and if you are willing to invest some *low-return time* in working out a procedure that best suits you and your instruments, *you'll end up making money.* Bear in mind that *it takes an experienced and shrewd navigator to make port by dead reckoning.*

33.2 CIRCUIT CONFIGURATIONS

Electronic circuits can be broadly grouped into four types of circuit configurations for identifying faults. These are :

(1) Sequential circuits
(2) Convergent circuits
(3) Divergent circuits
(4) Closed loop circuits.

(1) Sequential Circuits : The different stages of sequential circuits are connected in tandem (in series, one after the other). There is a *linear flow of signal* in sequential circuits.

The normal approach to servicing sequential circuits is by (*a*) signal injection and signal tracing (see Fig. 33.1), (*b*) signal substitution and (*c*) short circuiting the inputs one by one for locating sources of hum, transformer buzz and so on.

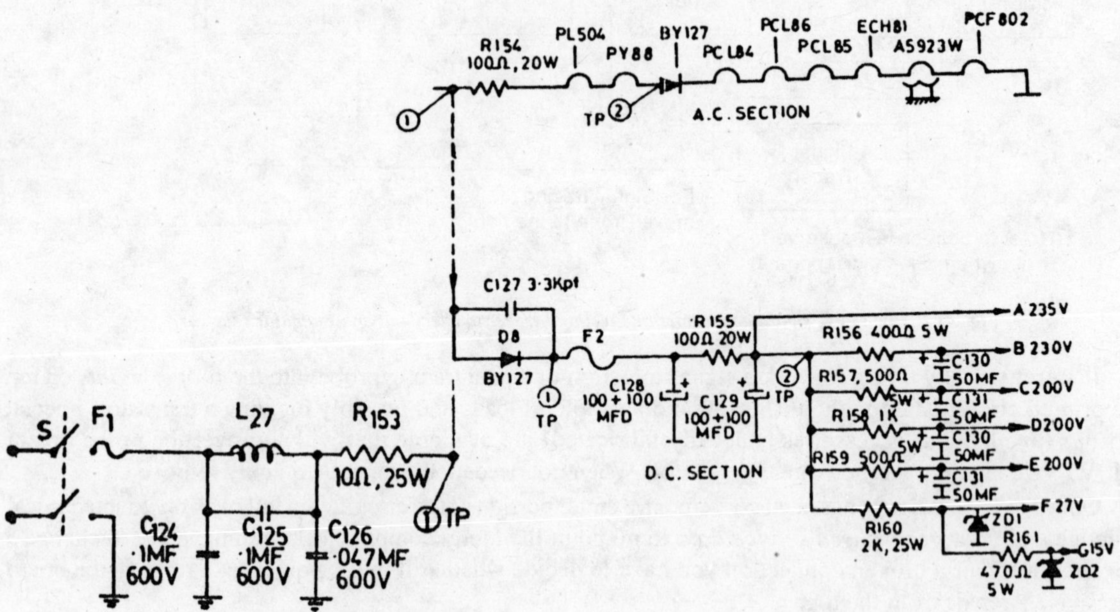

Fig. 33.2 Power supply system of a hybrid television receiver (Divide and Rule).

The troubleshooting approach is based on '*Divide and Rule*' policy. The stage under investigation is split in the middle (depending on the type of circuit) to ascertain whether the fault is beyond the split or ahead of it (see Fig. 33.2 for example).

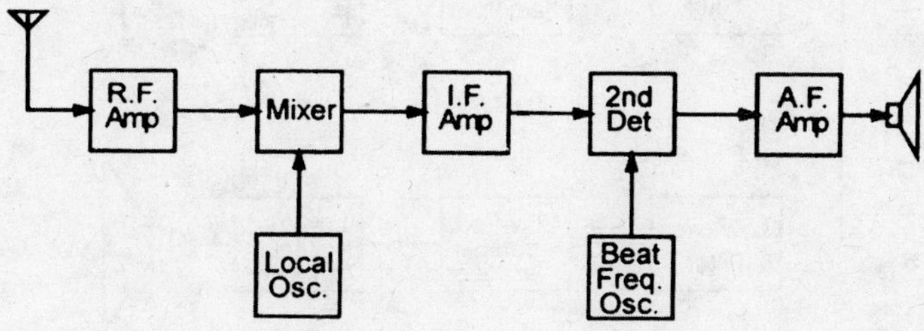

Fig. 33.3 Block diagram of a superheterodyne receiver

A unique example of a sequential circuit is the superheterodyne receiver shown in block schematic in Fig. 33.3. *The point of spilt could be the detector stage.* The faulty stage can by localised by the *click test,* followed by further tests to localise the source of trouble.

(2) *Convergent circuits* : Convergent circuits have more than one input and deliver a single output. An example of a convergent circuit is given in Fig. 33.4. The different program sources feed into the control centre. The output of the control centre feeds into a power amplifier which provides the drive for the loudspeaker.

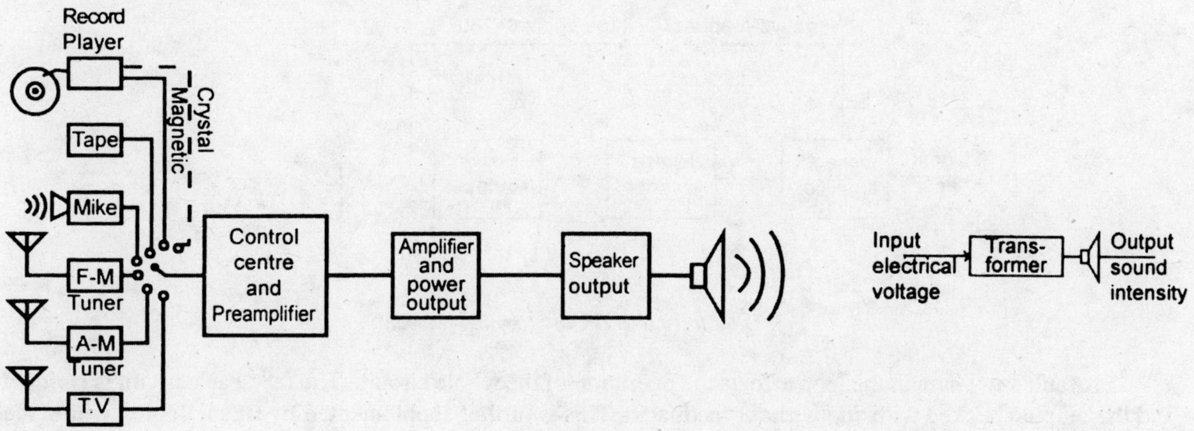

Fig. 33.4 Block diagram of a high fidelity system.

A convergent circuit is infact a *multiplexer* - with more than one input and a single output. *Each input circuit is tackled individually to localise the source of trouble.*

(3) *Divergent Circuits* : A divergent circuit is in fact a *demultiplexer.* It has one input and several outputs. The power supply system shown in Fig. 33.2 is a divergent circuit. It derives its input from commercial AC mains and provides outputs at different voltages and current ratings.

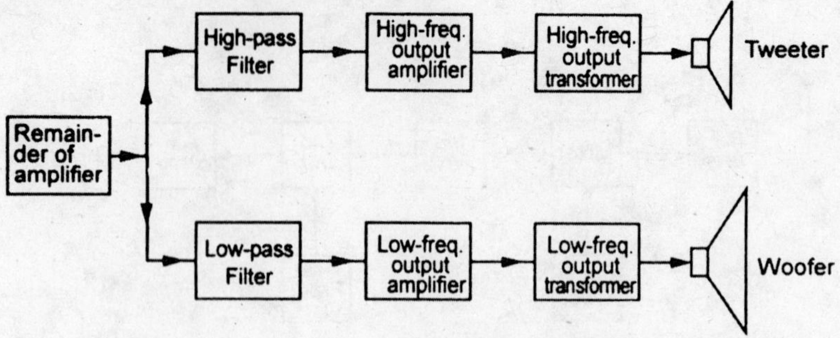

Fig. 33.5 A dual-speaker system.

The dual-speaker system shown in Fig. 33.5 is yet another example of a divergent circuit. The amplifier feeds its output, through a frequency selective network, to woofer and tweeter. The troubleshooting approach follows *isolation technique.*

Each stage is isolated in turn from a *strategic point*. If the fault is removed by isolating a stage, the isolated stage is faulty. Further tests are then carried out to localise the faulty component.

(4) *Closed Loop Circuits.* All automatic control circuits (AGC, ACC, ABC, AVC and so on) form a closed loop and are classified as closed loop circuits (see Fig. 33.6).

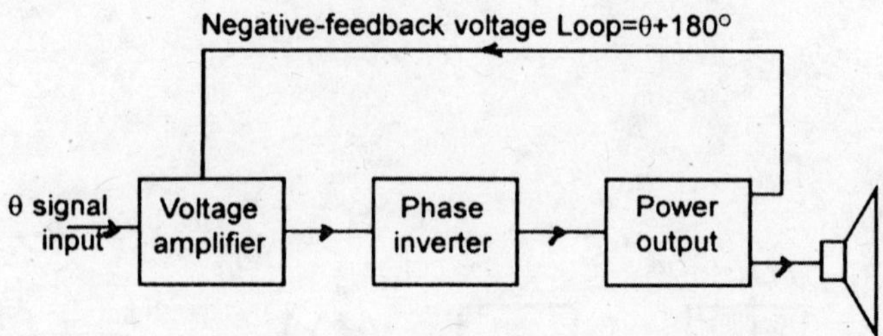

Fig. 33.6 Closed loop circuit - Negative feedback.

A fault anywhere in the loop affects the operation of the whole circuit. The feed back circuit is *isolated* and the output checked with and without feedback. This is further supplemented by signal tracing within the stages other than the feedback circuit. Thus the fault is localised either *in the feedback loop,* or *outside the feedback loop.*

33.3 TESTING SOLID-STATE DIODES AND TRANSISTORS

Aside from ICs, which contain a multiplicity of circuits, the two major semiconductor devices are diodes and transistors, (In monolithic ICs. diodes are often formed by laying down a transistor pattern in which the collector and base are shorted together to form the cathode of the diodes, and the emitter forms the anode.) Since it is simpler to test diodes, we will begin with them. But whether you are testing a discrete or an IC diode, the procedures are the same.

Solid-State Diodes

Diodes have three basic requirements: (1) they must be able to pass current in one direction (*forward current*) while preventing or limiting current flow in the opposite direction (*reverse current*): (2) for a given forward current *the voltage drop across the diode should not exceed a specific value*: and (3) for a specific reverse voltage, *the reverse current should not exceed a given value.*

All of the foregoing are best determined *out of circuit*, using an oscilloscope and appropriate data sheets or a diode tester. However an ohmmeter can be used to good effect in checking a diode's current-passing/limiting capabilities. *A good diode will exhibit high resistance in the reverse direction and low resistance in the forward direction.*

If the resistance in the reverse direction is low, the diode is probably *leaking.* If it is low in both directions the diode is most likely *shorted.* If high resistance is found in both directions the diode is probably *open.* The basic factor in determining a good diode in addition to high-reverse/low-forward resistance, is the *ratio* between the two resistances. The actual ratio depends upon the type of diode: for *small signal diodes* the ratio should be at least 1:100, forward-to-reverse (*front-to-back* is the more common name). For *power diodes*, a satisfactory ratio is at least 1 : 10. For *IC diodes,* the ratio should be at least 1 : 30. Incidentally to check IC diodes you must have a schematic of the IC itself. Since most equipment schematics show an IC simplify as a sideways triangle.

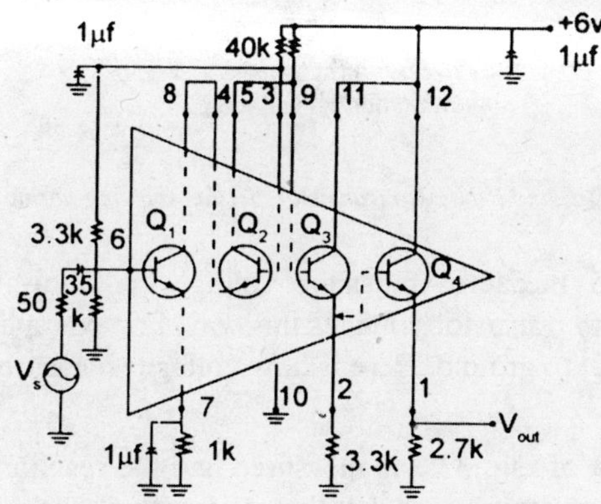

Fig. 33.7. Schematic diagram for a CA 3018 cascode video amplifier.

Transistors

The two most common types of transistors, be they discrete or integrated, are the *bipolar* (base, emitter, and collector) and the *unipolar* or field-effect (gate, source, and drain) types.

Bipolar Transistors. Basically, linear circuits consist of resistors, capacitors, etc., and bipolar transistors. The following paragraphs describe how to trace troubles to these transistors, and then how to check the transistors themselves. Again, remember that on a schematic diagram the symbol for an IC is usually a sideways triangle, as shown in Fig 33.7. The dotted lines have been inserted by the author and are *not* normally given in the schematic. So, as with IC diodes, you must have the IC schematic if you want to check out an IC transistor.

Bipolar transistor circuits are best tested via *voltage checks*, using the applicable schematic diagram and a sensitive VOM.

Fig. 33.8 shows the basic connections for both p-n-p and n-p-n transistor circuits: the coupling and bypass capacitors have been omitted for simplicity. *In practically all transistor circuits, the emitter-base junction must be forward-biased to obtain current flow through a transistor.* In a *p-n-p,* this means that the base must be made more negative than the emitter. The emitter base junction will then draw current, causing heavy electron flow from collector to emitter. In an *n-p-n,* the base must be made more positive than the emitter for emitter-to-collector current flow.

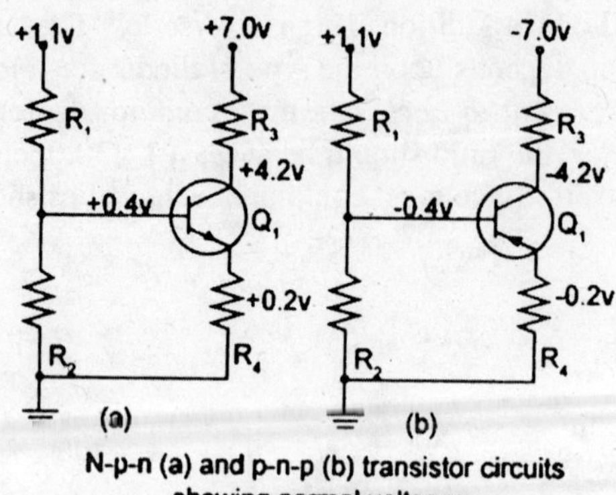

Fig. 33.8. N-P-N (a) and P-N-P (b) transistor circuits showing normal voltages.

The most common way to measure transistor voltages is *between ground and the element*: manufacturers' data generally specify transistor voltages this way. For example, all the voltages for the p-n-p of Fig. 33.8 are negative with respect to ground. Here is how voltages measured at the transistor element can be used to analyze failure.

Assume that the p-n-p circuit of Fig. 33.9 is measured and the resultant voltages are those shown in Fig 33.9. The first clue that something is wrong is that the collector voltage is almost the same as the source voltage at R_3 ; indicating that very little current is flowing through R_3 . The resistor could be defective, but the trouble is more likely caused by a bias problem. The emitter voltage depends mostly upon the current flowing through R_4 thus unless the value of R_4 has changed drastically, the problem is one of base bias.

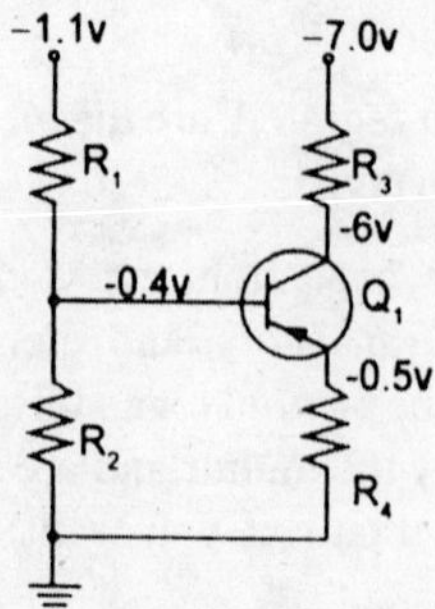

Fig. 33.9. P-N-P transistor circuit of Fig. 33.8 with abnormal voltages.

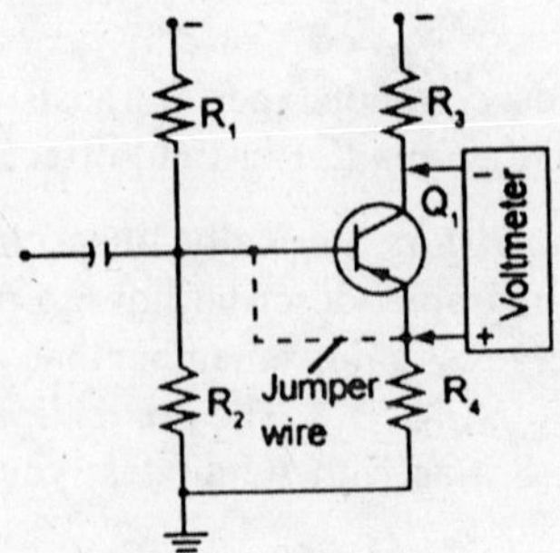

Fig. 33.10. In-circuit transistor test setup.

The next step, then is to measure the voltage at R_1 . If the bias source voltage is say, -0.9 volt instead of the required -1.1 volts, the problem is obvious: there is probably a defect in the power supply. If the bias source voltage is correct then either R_1, R_2 or Q_1 is defective.

The next logical step is to power-down the equipment and measure R_1 and R_2. If either is incorrect, there's your solution. If both values are correct, check R_4 , just to be sure. However, it is more likely that Q_1 is faulty. (A short note on the foregoing : the capacitors are not shown in Figs. 33.8 and 33.9. So, before you pull Q_1 or the IC it belongs to, check out any capacitors that might be across R_1 and R_4.)

Having reasonably assumed that Q_1 is faulty, there is a way to make sure . Fig. 33.10 illustrates the test setup if you do not have a transistor tester. With the voltmeter connected and the circuit operating, measure and note the *difference potential*, emitter-to-collector. Now shut off the power and short-circuit the emitter-base junction, as shown by the dotted-line jumper. Reapply power, and remeasure the emitter-collector potential: if it is not much higher than the original measurement, the transistor is defective.

Transistors (discrete) can also be checked *out of circuit* via simple resistance measurements (again assuming you do not have a transistor checker). Bipolar transistors are either p-n-p or n-p-n types. This means that there are two p-n junctions in each type. And, since a p-n junction is, in effect, a semiconductor diode, it can be tested in the same way. So, to check an n-p-n type, the ohmmeter leads are first connected as shown in Fig. 33.11(*a*) then with the leads reversed so that the meter COM lead is connected to the base and the + lead to the emitter.

The ratio of front-to-back resistance should be at least 1:30. If this step checks out, the process is then repeated for Figs. 33.11 (*b*) and 33.11 (*c*), each time reversing the leads to determine the front-to-back ratios.

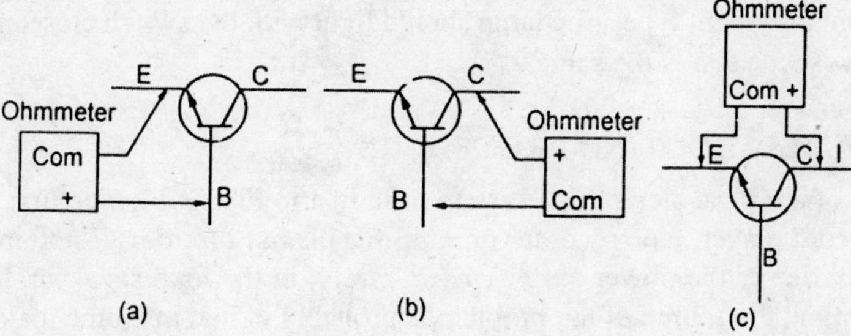

Fig. 33.11. Resistance testing an n-p-n transistor. (a) Testing emitter-base forward resistance. (b) Testing collector-base forward resistance. (c) Testing collector-emitter forward resistance.

As for p-n-p types, the procedure is exactly the reverse. That is, if it were a p-n-p in Fig 33.11 (*a*) the meter COM lead would be connected to the base and the + lead to the emitter, and then reversed.

Unipolar (Field-Effect) Transistors. Unipolar (field-effect) transistors can be checked in a variety of ways. A good VOM will tell you whether the proper source and output voltages exist. Or, you can use an ohmmeter to check front-to-back ratios. *But with IC field-effect transistors (FETs) you must be extra careful.* Their performance depends on the relative perfection of the *insulating layer,* if any (some IC FETs do not have an insulating layer) between the gate (equivalent to the base of a biopolar transistor) and the active channel.

Should this layer become punctured by inadvertent application of excess voltage to the external gate connection, like the battery voltage across the leads of an ordinary 1.000 ohm/volt VOM, the damage done is irreversible.

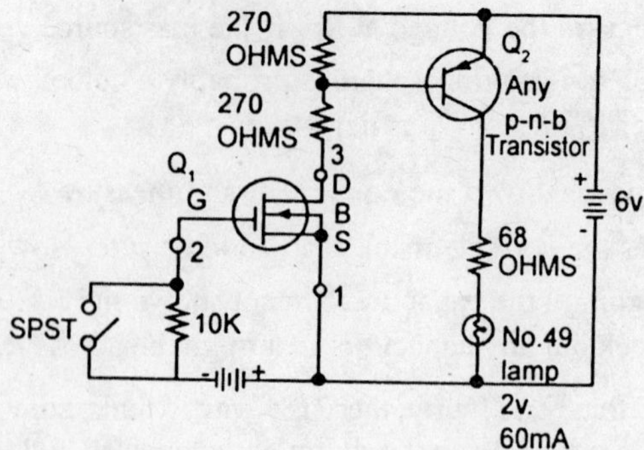

Fig. 33.12. Simple go/no go test circuit for MOS transistor. Q_1 is an n-channel depletion type ; p-channel would have the arrow head pointing outward.

The circuit shown in Fig. 33.12 will enable you to determine whether your 1C MOSFET is open or shorted. It is a simple go/no-go circuit that will test out-of-circuit, n-channel depletion type or p-channel enhancement types.

The substrate (B) and source (equivalent to the emitter of a bipolar) are connected to terminal 1, the gate to terminal 2, and the drain (equivalent to the collector of a bipolar) to terminal 3. If the MOSFET is a *dual-gate* type, test each separately.

For n-channel depletion types, if the lamp lights with the switch open, then goes off when the switch is closed, the transistor is "*good*," i.e. neither shorted nor open. If the lamp lights whether the switch is open or closed the transistor is *shorted*. If the lamp does not light at all, regardless of switch position, the transistor is *open*. For p-channel enhancement types, the lamp should light with the switch closed and go out when the switch is opened, otherwise the test is the same.

33.4. GENERAL TROUBLESHOOTING PROCEDURE

A block diagram of a typical stereo hi-fi system is shown in Fig 33.13. The first step is to determine which section (tuner, record player, tape recorder or amplifier) is out of order, If nothing works the trouble lies either in the preamplifier or the power amplifier (or, rarely, in the loud speaker). If the malfunction is confined to only one section, the source of the problem is probably in that instrument alone.

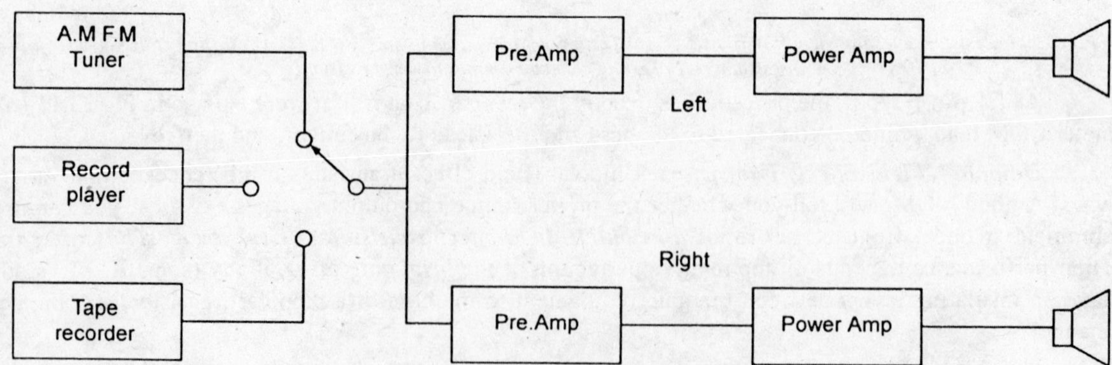

Fig. 33.13. Block diagram of a typical stereo hi-fi system.

Since there are *two identical amplifier systems* in a stereo amplifier you can use the amplifier itself to determine which section-the preamp or the power amp-is not functioning. If, for instance, the left channel has no output but the right channel operates normally, connect the right-channel preamplifier to the left-channel power amplifier. If the left channel then functions normally this indicates that the left-channel preamplifier is not functioning. If the unit still does not function, presumably the power amplifier is malfunctioning. If simple voltage/resistance measurements and new transistors/tubes do not repair the faulty unit it's time to go to Table 33.1.

33.5. TROUBLESHOOTING TUNERS, RECEIVERS, AND AMPLIFIERS

Table 33.1 is a chart of typical problems encountered in modern AM/FM tuners and receivers, and in amplifiers Remember *these are hints not cure-alls.* You still must use your head to figure out what's wrong in many instances.

Table 33.1 Troubleshooting Tuners, Receivers, and Amplifiers

	Symptom		Probable Cause		Remedy
1.	No sound mono	a.	NoACsupply	a.	Check plug cord and AC outlet
		b.	Tape monitor switch in wrong position	b.	Reset
		c.	Blown fuse	c.	Replace: if blows again check for shorts, wrong voltages
		d.	Defective position on switch that connects convenience outlet to AC supply	d.	Check by connecting directly to AC wall outlet
		e.	Defective power supply rectifier	e.1	If electron tube, check on tube tester.
				e.2	If solid-state, check forward-to-back resistance; refer to discussion on solid-state diodes
		f.	Open or shorted filter capacitor	f.	Disconnect and "jump" with known good capacitor of same rating.
2.	No sound, one channel	a.	Loose or open speaker connections	a.	Check, using cabling to other speaker
		b.	Defective brilliance or presence switch	b.	Interchange speakers and check
		c.	Defective amplifier output stage	c1.	If electron rube, check on tube tester
				c2.	If solid-state, check voltages against schematic; refer to earlier discussion on transistor testing
3.	Low or poor sound mono	a.	Weak tube(s)	a.	Check transconductance(s) on tube tester
		b.	Open filter capacitor	b.	See Remedy 1f
		c.	Defective power supply rectifier	c.	See Remedies le1 and le2.
		d.	Poor or broken antenna connection	d.	Check and repair

	Symptom		Probable Causes		Remedies
4.	Low or poor sound one channel	a.	Same as Probable Causes 3a, 3b, 3c.	a.	Same as Remedies 3a, 3b, 3c.
		b.	Defective channel in source e.g. cartridge	b.	Interchange source connections: if *same* channel power, amplifier is bad-trouble shoot; if *other* channel bad, source or connecting cable is bad - check with new cables.
5.	Low or poor sound, front or back channels	a.	Incorrect speaker phasing.	a.	Rephase
		b.	Defective amplifier	b.	Interchange front and back amplifiers; if formerly bad channel is now good troubleshoot defective amplifier per Symptoms 1 and 2 for each channel
		c.	Defective preamp	c.	Connect source directly to power amp, if possible; if good now, troubleshoot preamp via voltage checks.
		d.	Defective channel cabling	d.	Interchange channel cabling, first input then output; replace defective cabling.
6.	Flat, distorted sound	a.	Defective FM detector stage(s)	a1.	If tube, check with *new* one.
				a2.	If solid-state, check voltages via schematic; refer to earlier discussion on solid state diode and transistor testing.
				a3.	Check discriminator transformer for short or open ; if set is old, adjust transformer slugs per instruction mannual.
		b.	Lower power supply voltages	b.	See Remedies 1e1, le2 and If
		c.	Defective brilliance or presence switch	c.	Interchange speakers and check
		d.	Speaker cone out of balance	d.	Interchange speakers and check
7.	Static	a.	Loose connections	a.	Check all cable and antenna connections
		b.	Shorted tube (s)	b.	With set playing, *gently* tap all tubes with *insulated* end of screwdriver; replace tube(s) causing noise when tapped.
		c.	Dirty controls or switches	c.	Clean all controls/switches with cleaner spray or with hyponeedle and trichloroethylene, *not* carbon tet
		d.	Filter capacitor arcing internally	d.	See Remedy 1f
		e.	Dirt between tuning capacitor plates	e.	Blow, *brush* gently, or liquid-spray dirt away.

		f.	Tuning capacitor plates touching	f.	Rotate capacitor slowly while checking with magnifying glass; *gently* spread plates apart to avoid short in other direction.
8.	Vibration in sound	a.	Same as Probable Cause 6 c	a.	Same as Remedy 6c.
		b.	Defective audio output stage	b1.	If tubes, check driver and output tubes on tube tester.
				b2.	If solid-state, check voltages, looking for open or shorted capacitor.
9.	Hollow sound	a.	Microphonic tube(s)	a.	Try Remedy 7b; defective tube(s) will cause a hollow, ringing sound in speaker: try mixer first.
10.	Oscillator drift in FM tuner on receiver	a.	Defective automatic frequency control (AFC) diode	a.	See remedy le2
		b.	Defective reactance tube	b.	Check on tube tester
		c.	Component in oscillator circuit changing value with heat	c1.	If tube, replace; defective tube won't warm up enough on tube tester to give correct indication.
				c2.	Let tuner/receiver warm up until drift begins, then shut off and check resistances against schematic; may have to repeat, as components cool quickly
11.	Hum	a.	Cables not grounded properly	a.	Push all cable connectors firmly into sockets
		b.	Broken cable shielding (braiding)	b.	Wiggle cabling; if hum comes and goes, replace cabling; also, check soldering of braid to cable connectors.
		c.	Open or shorted filter capacitors	c.	See Remedy 1f
		d.	Aging component	d.	Adjust hum balance control, if any
12.	''Motor-boating''	a.	Open filter capacitor	a.	See Remedy 1f
		b.	Open cathode or screen bypass capacitor	b.	Check for these and replace; must check with one end unsoldered.
		c.	Open emitter or collector bypass capacitor	c.	Same as Remedy 12b

33.6 TROUBLESHOOTING RECORD PLAYERS

Table 33.2 is a chart of typical problems encountered in modern record playing units. Like Table 33.1 it lists symptoms, probable causes and remedies. It should not be taken as a cure-all either. In fact for some symptoms, the only cure may be a trip to the manufacturer, Remember that you are dealing with *delicate mechanisms,* and you can't be too careful.

In addition to the chart some suggestions on *general checkout and adjustment* are given in the following paragraphs. Read them *before* you try to troubleshoot; you may find that the information is just what you were looking for.

Typical Adjustment Procedures

The information presented below applies in general only. If you want really detailed data, obtain a copy of the *service manual* for the particular model from the manufacturer.

Leveling the Changer

Most changers are equipped with springs to permit them to *float* on the baseboard assembly. *Leveling screws are located at key points on the base plate to facilitate leveling.* The leveling procedure is a follows:

1. Place a small carpenter's level or torpedo level on the turntable so that the level points forward and back.
2. Adjust the leveling screws on the front and back of the base plate until the turntable is level. (*Note: The leveling screws should never be turned so tightly that the changer does not float freely. Adequate spring action between the base plate and the baseboard is essential.)*
3. Place the level parallel to the front of the changer on the turntable.
4. Adjust the leveling screws on the left and right sides of the base plate untill the turntable is level.
5. Repeat steps 2 and 4 until the turntable is completely level.
6. *For changers without leveling screws, place shims under the springs.*

Tone Arm Height Adjustment

The tone arm height adjustment is not critical, but it is important. *Too low* an adjustment can cause the tone arm to hit the side of the record and the stylus to strike the changer base during record play. After record play, the tone arm will not come to rest on the rest post. *Too high* an adjustment will prevent the stylus from setting in the record groove.

Most changers have a *tone arm height -adjustment screw* located under the rear of the tone arm. Some changers have this screw located on the tone arm casting itself, whereas others have it mounted on the tone arm hinge and bracket assembly. No matter where the tone arm height-adjustment screw is located, the operation is basically the same. *The height-adjustment screw will mate with a flat surface, maintaining constant tone arm height throughout the cycling operation.*

Whereas most record changers require a different tone arm height adjustment, a general rule is to *set the tone arm height (during the cycling operation) so that the stylus rides approximately 1/8 inch above the maximum number of records the changer is designed to hold.* The tone arm height adjustment is made by lifting the tone arm, locating the tone arm adjustment screw, and adjusting the screw until the tone arm is at the proper height.

Stylus Set-Down Position

If the stylus set-down position is not adjusted properly, you may hear only part of the record; if the stylus sets down outside the record, the result may be a broken or prematurely worn stylus.

In most record changers, the set-down adjustment screw is located on the tone arm hinge assembly. To perform a typical tone arm set-down adjustment, proceed as follows:

1. Lift the tone arm and locate the set-down adjustment screw.
2. Place a 10-or 12-inch record on the platter.
3. Cycle the changer.
4. If the stylus sets down prematurely, turn the screw *counterclockwise.* This will make the stylus set down closer to the spindle.
5. Repeat steps 3 and 4 until the set-down position is correct.
6. If the stylus sets down after the lead-in grooves turn the adjustment screw *clockwise*. This will move the set-down point away from the spindle.

7. Repeat steps 3 and 6 untill the set-down position is correct.

Stylus Pressure

The pressure of the stylus on the record during play is often the difference between good and poor sound reproduction and between long and short record and stylus life. It is for these reasons that a stylus pressure gauge is an invaluable tool.

There are basically two methods of adjusting for and maintaining good stylus pressure. The first and less common is to use *external balance weights*. By manipulating these weights (they are usually threaded on a long screw at the rear of the tone arm) as one would the weights on a balance scale, the pressure on the record may be adjusted.

The second, and by far the most common method in record changers is to use a *balance spring* under the tone arm, attached to the underside of the arm and the hinge bracket. Moving the spring toward the *front* of the tone arm decreases the stylus pressure. Moving the spring toward the *rear* of the tone arm increases the stylus pressure. In some changers, the spring is moved from hole to hole on a *stylus pressure clip*, whereas in others the spring is moved by turning a *pressure adjusting screw.* Refer to the changer service manual for recommended stylus pressure and method of adjustment.

Pickup Replacement

If local troubleshooting indicates a faulty pickup, the pickup and stylus assembly may be replaced as follows;

1. Remove the electrical connecting lugs from the pins on the pickup (two for *monaural*, four for *stereo* pickups). *Be sure to mark the position of the wires.*
2. Unscrew *attaching hardware* and remove the pickup.
3. *Obtain exact or equivalent replacement.*
4. Secure replacement to tone arm with attaching hardware.
5. Reattach the electrical connectors to pins of the replacement pickup. (*Be sure to get them on the same pins.*)

Stylus Replacement

Each pickup manufacturer has his own stylus positioning technique. *Generally it is a good idea to note how the stylus was removed in order to facilitate easy replacement.* Most styli are mounted on a wire arm shaped to fit a corresponding groove in the cartridge. In most cases all that is required to remove a stylus is to apply pressure to the holding spring and slip out the stylus, being very careful not to damage the contacting mechanism. *Some styli are replaceable by the manufacturer only.*

Checking Platter Speed

The easiest and most economical method of checking platter speed is with a *stroboscopic disc.* This is usually made of paper, cardboard, or light plastic and comes in various sizes and different types of markings. No matter what the size or markings, the method of use and the results are the same.

When a stroboscopic disc is placed on a platter, illuminated with a fluorescent light or neon lamp and rotated, the pattern should appear stationary at the speed being tested. If the platter is rotating *too slowly,* the pattern will appear to be moving counterclockwise. If the platter is rotating *too fast,* the pattern will appear to move clockwise.

To test the speed of a platter, proceed as follows:

1. First test *with only one record* on the platter and the tone arm on its rest.
2. Second, test *with only one record* on the platter and the *stylus on the record.*
3. Third, *add one record at a time* (to the maximum number of records the changer will hold), performing the stroboscopic test each time.

Record-Dropping, Tripping, and Indexing

Unless you are thoroughly familiar with the intricacies of these mechanisms, *don't fool with them.* Adjustments here are mostly for the trained technician.

Final Checkout

Take the following steps to make sure that your changer is performing the way it should:

1. Place a 7-inch record on the spindle and *cycle the changer.* Make sure that the record-drop mechanism is functioning correctly and that the stylus set-down position is correct.
2. Repeat step 1 using a 10 inch record.
3. Repeat step 1 using a 12-inch record.
4. If the changer handles records of *mixed sizes* without difficulty, place several records of each size on the spindle and cycle the changer. Make sure that the indexing mechanism is functioning properly by noting whether the stylus sets down in the proper position for each record size.
5. Place a stroboscopic disk on the platter and *check the platter speeds.*
6. Check whether the *pickup* is seated properly in the tone arm.
7. Check whether the *stylus* is properly seated in the pickup.
8. Check the *condition of the stylus* for wear.
9. Using a *stylus pressure gauge*, check whether the stylus pressure accords with the manufacturer's specification.
10. Check the *tone arm height.*

Table 33.2 Troubleshooting Record Players

	Symptom		Probable Cause Mechanical		Remedy
1.	Stylus lands improperly	a.	Improper setting of landing adjustment.	a.	Refer to manufacturer's manual on exact settings for arm height adjustment and landing adjustment
2.	Mechanism fails to cycle (changer)	a.	Cycling mechanism stuck	a1.	Turn platter manually, power off, through one complete revolution: then retry with power on
				a2.	Remove platter and turn cycling gear in direction opposite to normal rotation until gear teeth mesh with turntable hub teeth when turntable is replaced
		b.	Defective (loose or broken) cycling mechanism springs	b.	Check for proper spring tension or loose, broken, or disconnected springs
		c.	Slippage due to excess lubrication in cycling mechanism	c.	Remove all excess lubrication: relubricate per manufacturer's specifications
3.	Stylus skips record grooves	a.	Improper adjustment of tone arm height	a.	Readjust arm per manufacturer's specification

		b.	Counterblance device improperly adjusted or broken	b.	Readjust or replace per manufacturer's specifications
		c.	Warped record	c.	Try newest record you have: if still skips readjust stylus force per instruction manual
		d.	Player not level	d.	Check and adjust
		e.	Tone arm set down mechanism not set properly	e.	Check and adjust per instruction manual
		f.	Antiskating force not set properly	f.	Check and adjust per instruction manual
		g.	Tone arm/cartridge vertical tracking angle not optimum	g.	Check instruction manual for adjustment; if *none* try remounting cartridge without spacers (if any). or with 1/16 inch spacer (if none) at *front* end of cartridge, then rebalance tone arm
		h.	Tone arm automatic lift or change mechanism binding	h.	Check and clean/lubricate per instruction manual
		i.	Stylus force set too low	i.	Readjust per instruction manual
		j.	Stylus or record dirty	j.	Clean both
		k.	Stylus very worn	k.	Check and replace
		l.	Tone arm wiring under platter is binding	l.	Check and free up.
4.	Platter turns too slowly or not at all.	a.	Oil or grease on inner rim of platter or on drive wheel (s)	a.	Clean throughly
		b.	Poor spring tension or broken springs in drive mechanism	b.	Check and repair or replace
		c.	Defective motor	c.	Check for open or shorted field or armature
		d.	LowAC supply	d.	Check wall outlet this-may be temporary drop, due to high current appliances on same line
		e.	Motor or bearings insufficiently lubricated	e.	Relubricate per manufacturer's specifications

	Symptom		Probable Cause Electrical		Remedy
1.	No output	a.	Poor connection at preamp or amplifier input connector	a.	If all other program sources play through preamp or amplifier, check connections and soldering on connector or pickup leads

		b.	Loose or broken pickup leads	b.	Check leads for continuity between pickup and connector; replace open lead(s) with identical wire
		c.	Defective pickup	c.	Check out with new pickup
2.	No or low output on one channel of stereo setup	a.	Defective pickup	a.	If any other program source plays properly on both channels, one side of stereo pickup is defective; replace pickup
3.	Noisy output	a.	Scratchy records	a.	Try new or slightly used record; if no change
		b.	Poor connection at pickup, preamp or amplifier	b.	Check all lead and connector connections for tightness, soldering etc.
		c.	Cracked or defective pickup	c.	Check out with new pickup
		d.	Pickup lead insulation worn through where it goes through arm	d.	Check all insulation and replace detective pickup lead
		e.	Pickup lead stranding broken under insulation	e.	Check lead continuity and jiggle lead at same time to see if meter pointer oscillates
		f.	Dust or dirt on stylus	f.	Brush off with soft brush, *not a cloth*; brush from back of stylus to front
4.	Hum	a.	Poor connection at preamp or amplifier	a.	Make sure that all connections are tight
		b.	Poor shielding of motor	b.	Try surrounding motor with metal (tinfoil, aluminium foil)
		c.	Preamp or amplifier too close to pickup	c.	Separation must be at least two feet
		d.	Pickup lead shielding broken or "cold-soldered" to connector	d.	Check for breaks and replace; resolder even if joint looks good
		e.	Turntable mechanism and preamp or amplifier not properly grounded to each other ; same for pickup arm	e.	If necessary, put in your own ground lead; if one exists check all connections
		f.	Hum balancing control on preamp or amplifier improperly adjusted	f.	Adjust hum balancing control for minimum hum, with no record playing and volume control slightly above normal listening level

33.7 TAPE RECORDER TROUBLESHOOTING

All tape recorders are similar in principle; *the major differences are in the mechanical arrangements and the complexity of the electronics.* Hence this information is of a basic nature; specific problems can be solved only with the help of the manufacturer's service manuals and your *growing familiarity* with the machines you come across.

General Techniques

Before doing anything else *become familiar* with all the manufacturer's data such as exploded views, mechanical flow diagrams, circuit schematics and any other information. *Who knows, tucked away in a corner of the service manual may be the one little item that will solve your problem.*

Next, be sure that all the necessary tools are handy and clean. The tools generally used for tuner/receiver and record players servicing are applicable also to tape recorder servicing. However for proper servicing you'll need the test equipment listed in Table 33.3.

Routine Maintenance

Preliminary checks of the electrical components are the same as with any electronic circuits. They include *visual checks* for burned resistors, bulging capacitors, cracked tube bases, loose or broken connections, tube transconductance and short tests; *voltage and current checks against manufacturer's specifications* and the like. The fact that these are high-fidelity circuits doesn't change the techniques; it just means that *the tolerances are more critical and must be adhered to more closely.*

Table 33.3 Tape Recorder Servicing Materials and Equipment

Item	*Function*
Head demagnetizer/ degausser	Removes residual magnetism from tape heads
Tape splicer	Patches, edits, or adds to tapes
Splicing tape	Patches, edits or adds to tapes
Leader tape	Splices to tape ends for easier threading; splices between tape sections/programs for easier editing
Speed-check tape	Checks recorder/player speeds for accuracy
Bulk eraser	Removes rapidly all traces of previously recorded material and ensures a "clean" tape
Trichlorethylene	Removes oil and grease
Denatured (isoproply) alcohol*	For cleaning heads, capstan and all rubber surfaces
Hypodermic needle and/or eye dropper	Cleans motors, rotating controls, slide switches, etc.
Lateral adjustment tape	Adjusts heads in azimuth
Magnifying glass	For checking printed-circuit boards, head gaps, etc.
Tape cleaner and lubricant	Cleans and lubricates tapes
Head lubricant	Lubricates heads
Light machine oil	Lubricates motor, sliding parts (use separate hypodermic or eye dropper for oiling), rotating parts
Head cleaner tape	Cleans heads; smooths out wear spots on heads
Clean and test cartridge	Cleans and aligns heads; cleans capstan; balances speakers
Crosstalk adjustment kit	Corrects cartridge recorder/player crosstalk

*Rubbing alcohol

Servicing the *mechanical portions* is another matter: almost any mechanical part can cause trouble. However, as with all equipment some general techniques obtain. These include *visual checks* for dust, excess grease, broken or loose switch contacts, cracked rubber and loose or broken spring (or other) belt drives; checks for wobble or binding of moving parts; and *checks for proper sliding of slide switches.*

Cleaning and Inspection

Following the preliminary checks, a good cleaning of the *transport mechanism* is in order. Oxide loosened from the tape during normal operation builds up on tape heads guides, pressure roller, pressure pads, and capstan. This accumulation can result in poor playback response, weak or distorted recordings and poor erasure. It can also cause an up and down travel of the tape between the pressure roller (pad) and capstan. Remove these oxide deposits and any other dirt, grime, or oil with *denatured alcohol and cotton swabs* which make handy tools for this purpose. (Note. *Cleaning and lubricating kits* containing all needed items are sold by most supply houses, including Philips. In some cases, specific cleaning procedures are recommended by the manufacturer; in others follow the general procedure below :

1. Dip the applicator into the cleaning solution; then squeeze excess solution from it.
2. Apply applicator to the faces of the heads *swabbing back and forth* until all traces of oxide are gone. Use a *dry applicator* to remove any excess solution after cleaning is complete.
3. Repeat the same procedure on the pressure roller.
4. Turn the machine on, and hold the *wet applicator* firmly against the pressure roller. This will cause the solution to contact the capstan, thus wiping off the oxide. Continue the process until all traces of oxide are removed.
5. Using a *cloth tape,* operate the machine as though it were in playback. This will not only clean all the critical parts again but put a high, smooth polish on them.

While cleaning inspect all parts for wear. The oxide coating on the tape is an abrasive, causing wear on the heads. As they wear the gaps widen, causing both lower output level and poorer high-frequency response. *The gaps must be inspected with a magnifying glass; they are too narrow for ordinary visual checks.* The record and playback gap(s) should appear as a fine, uniformly wide line. The erase head has two gaps, each of which is about five times as wide as the record and playback gap(s). *Any increase in gap width, or irregularities in the gaps, such as shown in Fig. 33.14 indicate the need for head replacement.*

Check the *tape guides* for excessive wear. Worn guides can cause misalignment of the tape with the heads.

Check the *pressure pads* to be sure they hold the tape flat- but not too tightly-against the heads. Pads that are too tight wear the heads rapidly and can cause tape speed variations. Pads packed too tight or with glazed or hardened surfaces, can be "fluffed out" again by judicious use of an emery board or a small file. The file used for the points of an automobile distributor will do just fine. Take care, however, not to scratch the heads, Should the pressure pads require recementing due to looseness, apply the adhesive very carefully. Excess cement will harden the pads.

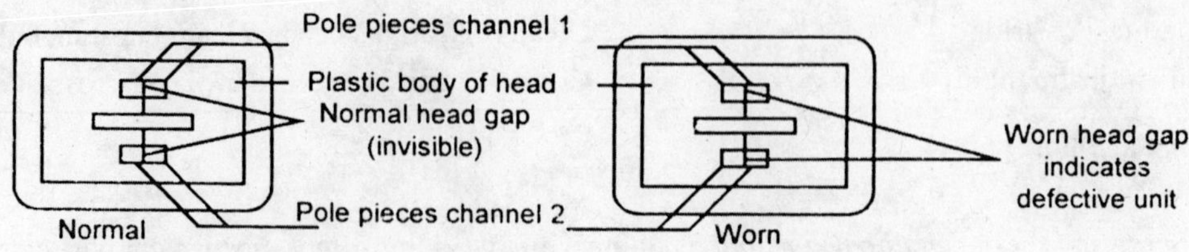

Fig. 33.14 Normal and worn stereo heads.

General Techniques

Before doing anything else *become familiar* with all the manufacturer's data such as exploded views, mechanical flow diagrams, circuit schematics and any other information. *Who knows, tucked away in a corner of the service manual may be the one little item that will solve your problem.*

Next, be sure that all the necessary tools are handy and clean. The tools generally used for tuner/receiver and record players servicing are applicable also to tape recorder servicing. However for proper servicing you'll need the test equipment listed in Table 33.3.

Routine Maintenance

Preliminary checks of the electrical components are the same as with any electronic circuits. They include *visual checks* for burned resistors, bulging capacitors, cracked tube bases, loose or broken connections, tube transconductance and short tests; voltage and current *checks against manufacturer's specifications* and the like. The fact that these are high-fidelity circuits doesn't change the techniques; it just means that *the tolerances are more critical and must be adhered to more closely.*

Table 33.3 Tape Recorder Servicing Materials and Equipment

Item	*Function*
Head demagnetizer/ degausser	Remove residual magnetism from tape heads
Tape splicer	Patches, edits, or adds to tapes
Splicing tape	Patches, edits or adds to tapes
Leader tape	Splices to tape ends for easier threading; splices between tape sections/programs for easier editing
Speed-check tape	Checks recorder/player speeds for accuracy
Bulk eraser	Removes rapidly all traces of previously recorded material and ensures a "clean" tape
Trichlorethylene	Removes oil and grease
Denatured (isoproply) alcohol*	For cleaning heads, capstan and all rubber surfaces
Hypodermic needle and/or eye dropper	Cleans motors, rotating controls, slide switches, etc.
Lateral adjustment tape	Adjusts heads in azimuth
Magnifying glass	For checking printed-circuit boards, head gaps, etc.
Tape cleaner and lubricant	Cleans and lubricates tapes
Head lubricant	Lubricates heads
Light machine oil	Lubricates motor, sliding parts (use separate hypodermic or eye dropper for oiling), rotating parts
Head cleaner tape	Cleans heads; smooths out wear spots on heads
Clean and test cartridge	Cleans and aligns heads; cleans capstan; balances speakers
Crosstalk adjustment kit	Corrects cartridge recorder/player crosstalk

*Rubbing alcohol

Servicing the *mechanical portions* is another matter; almost any mechanical part can cause trouble. However, as with all equipment some general techniques obtain. These include *visual checks* for dust, excess grease, broken or loose switch contacts, cracked rubber and loose or broken spring (or other) belt drives; checks for wobble or binding of moving parts; and *checks for proper sliding of slide switches.*

Lubrication

Most manufactures supply specific instructions about lubricating their machines. What it all boils down to is *not to lubricate them unless absolutely necessary,* such as when a moving part is being replaced. Your machine was probably factory lubricated; this lubrication is good for approximately one year. On the other hand the manufacturer's service manual for your machine may provide special instructions as to *where, when, how much and what to use. In the final analysis, the service manual should be your "bible" on everything.*

However, some general procedures can be observed,

- Never use lubricants that can oxidize and harden.
- *Never use oil or grease on the capstan and pressure roller.*
- When necessary to lubricate, *be sure that no oil or grease gets on any of the rubberized parts or on any parts that might transfer the oil or grease to the tape or drive surfaces.*
- *Heads can be lubricated with a silicone-base lubricant.* This type of lubricant will not affect rubberized surfaces of the drive components.

Head Demagnetization

The more the machine is used the greater the *residual magnetism* picked up by the heads. This residual magnetism *decreases* the machine's high frequency response while *increasing* both noise and harmonic distortion, Therefore, one of the first items on your maintenance agenda should be *head demagnetization* (following cleaning). Heads should also be demagnetized any time they have been checked for continuity with an ohmmeter or if magnetic tools have been used in the vicinity.

To demagnetize the head, proceed as follows;

1. Insert the tool tip between the pressure pad and head.
2. Turn on the current to the tool and press it *gently against* the head.
3. *Slowly* pull the tool tip away from the head. Don't turn off the tool until it is completely out of the vicinity of the head.

Head Alignment

High-frequency response and interchannel crosstalk are extremely dependent on head alignment. *In most transports, the heads are adjustable and can be aligned as required to establish head-to-tape interface.* The adjustment is very precise and must be accomplished in accordance with the instructions given in the manufacture's service manual. *There are five basic adjustments involved in correctly positioning a head.* These adjustment are discussed in detail in Chapter 32.

Worn Heads

If alignment of the head (or heads) doesn't solve the problem, the next step is to replace the offending head. Worn heads can cause any number of discrepancies in a test of frequency response. *Mono* heads are not too expensive, but *stereo* heads are and *quad* heads more so ; all are replaced as you would replace a tube in a radio, except that the heads have to be reconnected to wires. *Head replacement must be followed by head realignment.*

Capstan Adjustment

Unless the capstan and pinch (pressure) roller are perfectly parallel, the tape will tend to ride up and down during its travel and may wrinkle and tear. Fig 33.15 shows *what can happen when the capstan is out of true alignment and what does happen when the capstan is properly adjusted.* All machines are provided with capstan adjustments ; the exact method depends on the manufacturer's mounting arrangements.

Automatic Cutoff Switch Adjustment

The cutoff switch in some machines is usually *a microswitch in series with the motors.* Tape tension during operation is enough to keep the switch off. When the tape runs out, or breaks the switch arm releases, turning off the motor. Switch adjustment usually calls for loosening the mounting screws and repositioning the switch so that normal tape tension during operation is enough to keep the switch off. If the switch itself is immovable, simply bend the switch arm to obtain the desired effect.

Electric Motors

Defective motors can be detected by *speed-timing tests.* Subnormal speeds indicate that motor bearings or directly driven rotating parts, may need cleaning or lubricating. On the other hand, line voltage may be low. A variable line transformer (variac, or the equivalent) may be able to fix that.

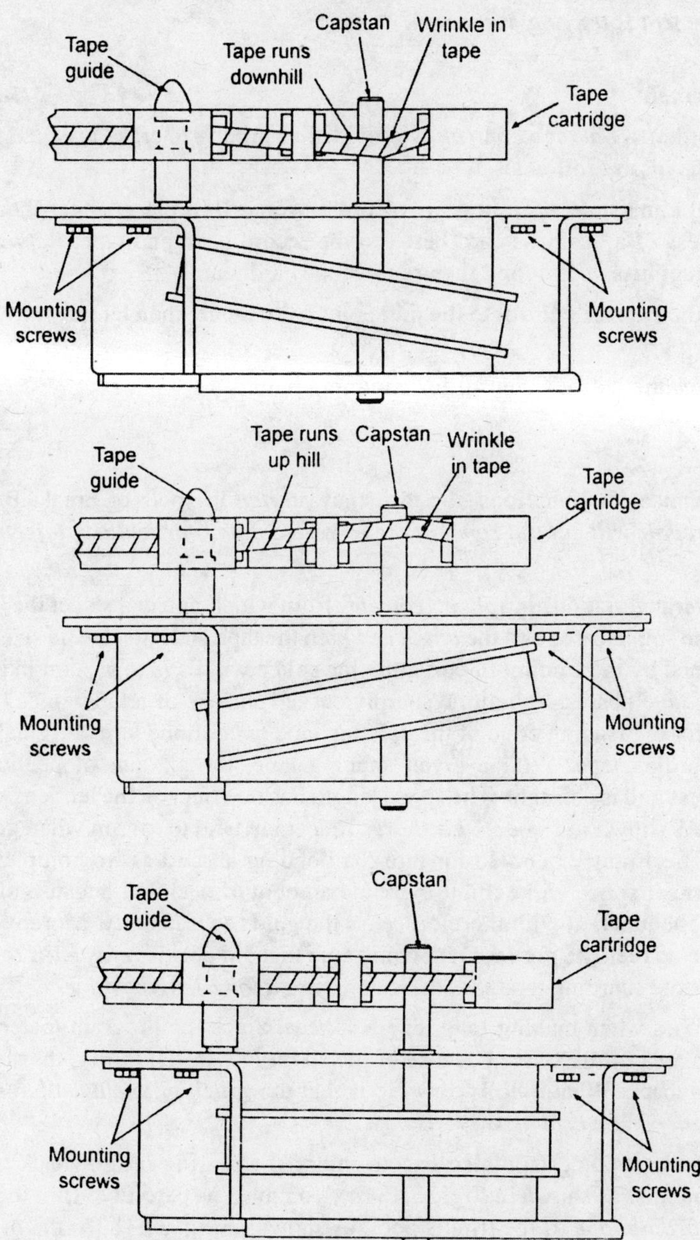

Fig. 33.15 Effects of flywheel position. (a) Flywheel bracket too far to the left; (b) flywheel bracket too far to the right; (c) flywheel bracket properly adjusted.

Rubber Driver Components

One of the causes of *wow* is the slippage of rubber drive surfaces due to age. These surfaces harden with age, tending to become smooth and polished. Trying to rough up the surface as a temporary expedient is useless; not only does the surface quickly become smooth again, but it may lose its roundness because of the roughening process. *The best repair is to replace the entire unit.*

Rubber belts show their age in *looseness* or spots that become hard and tend to *slip* during operation. Here, again, *replacement is the only remedy.*

Erase Head Adjustment

In some recorders a *balance control* is provided to minimize erase hiss and hum. In such machines, you can minimize the hiss as follows:

- With the machine operating in "record" insert a thin piece of cardboard or plastic, about the thickness of a business card between the record head and tape. *Reposition* the erase head for minimum hiss, lock it and then remove the cardboard.
- *Reset* the balance control to the mid point of its range; then let the machine run *until thoroughly warmed up.*
- *Readjust* the balance control for minimum hum.

Tape Splicing

Due to mechanical malfunctions, the tape may *stretch* in spots or break. Before attempting repair, *remember that the repair will cut out some of the recorded data, particularly affecting four-and eight-track tapes.*

There are several variations in *splice geometry* from which you can select the best combination. These include the *size* of the spliced area and the *angle* at which the tape ends meet. The size (actually, splice length) is basically determined by the amount of curvature the splice will have to sustain in its path from reel to reel. When the recording tape passes around the sharply curved surface of a tape guide, as shown in Fig. 33.16, there is a tendency for the leading edge of the splicing tape to continue in its original direction that is to *peel away* from the recording tape. With a given splicing tape, the amount of peeling is determined in the manufacturing process and is constant. The splice length has no effect on the tendency to peel, but is important for another reason. As shown in Fig 33.17 a *short splice* may tend to loosen when going around a tight bend because the area of peel may extend so far into the bonding around as to completely free one end of the recording tape. A longer splice will exhibit the same amount of peel, but because of its length, will not free the recording tape. The junction will, therefore, pass the guide successfully. Moreover, once the spliced area gets back on the take up reel, the portion of splicing tape that did peel is *refastened to the tape* by the pressure of succeeding layers of recording tape. So, in general, *use a long-length splice.*

A problem arises when making long length splices. Since the program material may drop in level as much as 4 dB in the area of the splice because of the slight loss in flexibility, the shorter the splice, the less the effect during playback, What it boils down to is that the *splicing tape must have high adhesive strength so that splicing length can be kept as short as possible.*

The *width* of the splicing tape also has an important bearing on playback quality. When using the *full-width* splicing method as shown in Fig. 33.18 (*a*), you must be careful to trim the splicing tape *exactly at the edges of the recording tape*. If the trim is poor, as shown in Fig. 33.18 (*b*) the *overhanging* splicing tape will adhere to an adjacent layer on the reel. In addition, even though some tape splicers are designed to cut the tape as shown in Fig. 33.18(*c*) to ensure against overhang, there is still a chance that some adhesive will ooze out of the edges. Fig. 33.18(*d*) shows a splicing tape that is *slightly narrower* than the tape to be spliced.

This type offers a number of advantages without any apparent disadvantages. It won't overhang if laid down properly ; a straight forward tape splicer can be used and there isn't any sacrifice in strength, since it covers about the same area as that shown in Fig. 33.18 (*c*).

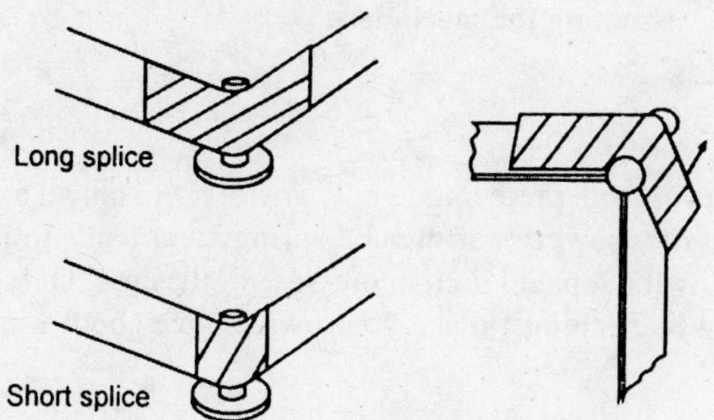

Fig. 33.16 Splice length and bend radius.

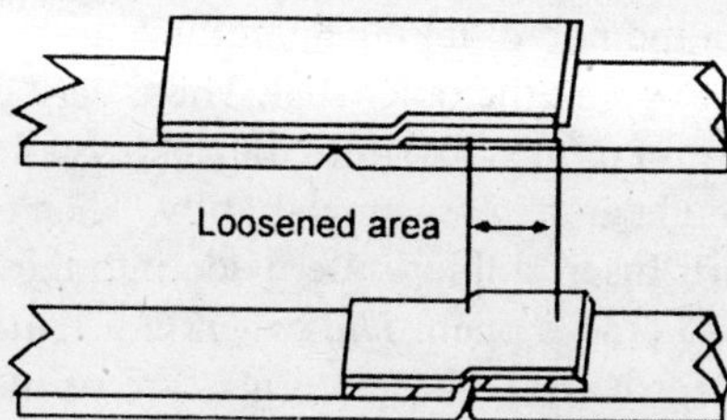

Fig. 33.17 Enlarged view of loosened splice.

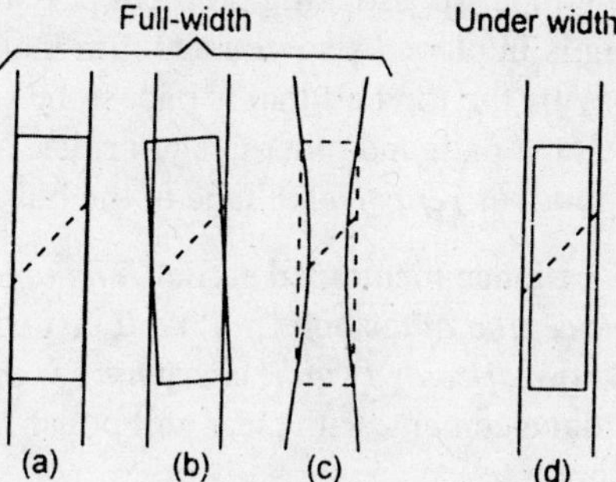

Fig. 33.18. Splicing widths

As for the actual splice the most desirable method is to *cut the recording tape at an angle between 45°and 60°*. Any angle of cut greater than 60° increases the distortion during playback because the head then sees the cut as an abrupt change. However, making a cut shallower than 45°, while lessening the distortion during playback, increases the chances that the points of the recording tape will peel back.

Regardless of the type of splice, *the most important consideration in the process is cleanliness.* Your hands must be as clean as possible ; a fingerprint on the oxide can drop the output level several decibels. Dirt can also weaken the splice. Once the splice is made, rubbing the splicing tape with your fingers will ensure a good bond and eliminate *air pockets*. It is advisable to use gloves for this operation.

33.8 TROUBLESHOOTING CASSETTES

It's very possible that the malfunction lies not in the machine but in the cassette or cartridge itself, so *when a malfunction does occur, always try to play or record a different cassette or cartridge before examining the machine.* If the *new one* works properly, then you know the *old* is at fault: if the new one doesn't work, then you are justified in troubleshooting the machine.

Cassette Troubleshooting

Unless the cassette is a valuable prerecorded one, *do not* try to take it apart; it takes a world of patience and very nimble fingers. The *tape* is very narrow and springy and tends to jump out of its path. The *guide rollers* are held in place only by the top and bottom pressure of the case. Only the tape holds the pressure pad and spring in place. Knowing these things, make your own choice about whether or not to troubleshoot the cassette.

Cassettes are sealed either by screws or by ultrasonic welding. If your particular cassette is screwed together, simply take out the screws. Be careful since they may be *self-tapping* and if you exert too much pressure, you could crack the case or strip the threads in the case.

If the case is welded the best method of taking apart the halves is to *gently slide* a knife or single-edge razor blade around the seam once to break the outer seal. Then, *very gently* rock the blade edge against the seam at the back of the case (be careful of the tabs) until the blade is all the way through the weld. Now gently pry apart the point of entry until you hear the case crack lightly. Then repeat the *blade-and-pry process* at the sides. Now comes the difficult part. Insert a thin knife blade into the *cracked-open seam* at any corner and pry very gently until the corner also cracks open. *The covers are sealed more firmly at the corners because they have greater surface areas.* Repeat until all four corners are-open. The case should be loose now, but be especially careful of the front recesses where the heads and pinch roller enter- the tabs there are very flimsy.

Holding the case *with the back facing you,* gently lift the top half away until you can see inside. If everything looks normal, that is, the tape isn't sticking on the top, remove the top half. Now you can check the interior to make sure everything is in place. Also you can turn the reels by hand and see where anything is binding; if it is, alleviate it. It may be the top half that is bad, so be sure to check it for burrs, nicks, excess plastic (bumps, for example) in the tape path and other irregularities. You can now *replace* the pressure pad if it is stiff, dirty, or otherwise no good, or *refasten* the tape to the hub if that's the trouble.

Resealing the case takes the patience mentioned earlier. *The tape and guide rollers have a tendency to fight each other. When one is in place, the other pops out.* And just when you think you've got it made, the pressure pad and spring fall down. So *work slowly until everything is in properly, then lower the top down on the bottom.* Reseal with model airplane cement, with clear nail polish, or with nail polish remover, The latter melts the plastic, making a good bond.

A *broken tape* is one repair you can make without taking the case apart, *but only if you can see both ends of the break.* Just pull both ends out until you have enough to put into the *splicer,* and make the splice. One note of caution-*make sure you pull the ends out of the recess in which you see the pressure pad* or you will have part of the tape running outside the case when the splice is finished. Once the splice is finished, rewind one of the hubs by hand until all the tape is back inside again and fairly snug against the pressure pad.

33.9 TROUBLESHOOTING MACHINES

Table 33.4 lists some common symptoms of reel type recorders, their possible causes and remedies. Table 33.5 is a troubleshooting chart for cartridge and cassette-type machines.

Table 33.4 Troubleshooting Reel-Type Recorders

	Symptom		Probable cause-Mechanical		Remedy
1.	No tape drive in "Record" or Play	a.	Jammed, slipping or broken drive belt	a.	Clear jam ; replace belt
		b.	Disconnected or broken, pressure roller spring	b.	Reconnect or replace spring
		c.	Oil on pressure roller or capstan.	c.	Clean
2.	No tape drive in fast "Forward" (Wind)	a.	Fast-forward spring (or belt) broken or disconnected	a.	Replace or reconnect
		b.	Oil on take up reel rest or on drive parts	b.	Clean
		c.	Reels binding	c.	Clean and lubricate reel shafts
3.	No tape drive in Rewind"	a.	Rewind spring (or belt) broken or disconnected	a.	Replace or reconnect
		b.	Oil on supply-reel rest idler, or other drive parts	b.	Clean
		c.	Reels binding	c.	Clean and lubricate reel shafts.
4.	No tape drive in ''Rewind'' and "Wind"	a.	Slipping or broken belt	a.	Replace
		b.	Pressure roller stuck against capstan	b.	Check pressure roller spring and replace if necessary
5.	No tape drive at all	a.	Jammed or broken belts.	a.	Check and clear jam or replace belts
		b.	Defective motor	b.	Replace
		c.	On models with automatic cutoff, check for inoperative shut off switch	c.	Repair or replace
6.	Take-up reel doesn't turn (supply OK)	a.	Loose take-up clutch	a.	Readjust or replace clutch pads
		b.	Broken or disconnected take-up spring	b.	Replace or reconnect
		c.	Take-up belt broken or slipping	c.	Replace
7.	Tape spiilage	a.	Loose take-up clutch	a.	Readjust or replace clutch pads
		b.	Defective braking	b.	Check brake springs rollers, etc.
		c.	Oil on reel rests or on brakes	c.	Clean

		d.	Supply-reel drag not operating properly	d.	Check for wear or oil ; clean or replace worn part; check and replace broken spring
8.	Tape breaks continually	a.	Tight take-up clutch	a.	Adjust clutch
		b.	Brakes not releasing sufficiently	b.	Check brake springs for stretch or break and replace
9.	Machine squeals	a.	Tape is brittle and dried out	a.	Try lubricating
		b.	Incorrect tape tension	b.	Entire drive must be checked for binds, lubrication or slippage
		c.	Reels are warped	c.	Rewind tape on new reels
		d.	Tape guides or reels improperly aligned	d.	Check and adjust or repair guides and reel shafts
		e.	Dirt on heads	e.	Clean
		f.	Dry or worn bearings.	f.	Lubricate or replace
		g.	Dirt on pressure pads or pressure pads too tight	g.	Clean or replace pads
10.	Speed selector defective	a.	Loose control knob.	a.	Check and tighten setscrew
		b.	Spring broken or disconnected	b.	Replace or reconnect
11.	Tape counter not operating	a.	Broken counter	a.	Replace
		b.	Broken or slipping drive belt or spring	b.	Replace loose or broken parts
12.	Uneven tape speed flutter wow	a.	Take-up clutch too loose or too tight	a.	Readjust or replace pads
		b.	Bearings are binding	b.	Lubricate
		c.	Oil on drive components.	c.	Clean
		d.	Friction drags worn	d.	Replace
		e.	Defective motor (intermittent)	e.	Check insulation for grounds; replace motor
			Electrical		
13.	No "Playback"	a.	Open head	a.	Check continuity
		b.	Defective function switch	b.	Check for bent or broken contacts
		c.	Defective amplifier	c.	Check and replace defective part (s)
		d.	Open speaker	d.	Check continuity and replace
		e.	Defective output connector	e.	Check for loose or bent contacts and repair

14.	Distorted or weak Record" no "Record";	a.	Defective microphone	a.	Check with new microphone
		b.	Bias-erase oscillator not functioning properly	b.	Check voltages, resistance, etc.
		c.	Defective coupling capacitor,oscillator-to head	c.	Replace defective part
		d.	Function switch defective	d.	Check for dirty, loose or broken contacts
		e.	Dirty record head or worn head	e.	Clean and re-record ; replace worn head
		f.	Loose pressure pads	f.	Recement new pads
		g.	Defective input connector	g.	Check for loose or bent contacts and repair
		h.	Level indicator defective	h.	Check operation with new unit
15.	Improper high frequency response	a.	Dirty, worn or misaligned record playback head	a.	Clean, repair or replace defective head
		b.	Bias too high or too low	b.	Check biasing circuitry and oscillator, and repair or replace as necessary
		c.	Magnetized heads	c.	Demagnetize
		d.	Pressure pads worn or loose	d.	Recement new pads
16.	Incomplete or no "Erase"	a.	Dirty worn, open or misaligned head	a.	Check and repair or replace
		b.	Faulty erase circuitry	b.	Replace oscillator; check surrounding circuits for defective part
		c.	Erase pressure pads dirty, worn, or loose	c.	Clean or recement new pads
		d.	Open coupling capacitor, erase-to-oscillator	d.	Replace defective part
17.	Low output level	a.	Defective bias oscillator	a.	Replace oscillator; check surrounding circuits for defective part
		b.	Worn, dirty or misaligned record-playback head	b.	Check and repair or replace
		c.	Dirty, worn, or loose pressure pads	c.	Clean or recement new pads
		d.	Weak tube or defective component in amplifier	d.	Check and replace defective parts.
		e.	Record-level indicator not functioning	e.	Check operation with new unit
18.	Distorted output	a.	Bias oscillator defective	a.	Replace and check surrounding circuits for defective part

		b.	Record-level indicator functioning improperly	b.	Check operation with new unit
		c.	Heads magnetized	c.	Demagnetize
		d.	Worn or misaligned record-playback head	d.	Check and repair or replace
		e.	Defective equalizing network	e.	Check equalizing network and replace defective parts; entire amplifier response must then be checked
		f.	Defective amplifier	f.	Check by frequency and gain runs and replace weak or defective parts
		g.	Dirty or loose switch contacts	g.	Check and repair or replace switch
		h.	Weak erase	h.	See Symptom 16
19.	Noise; hum	a.	Improper bias	a.	Check oscillator circuit and replace weak or defective parts
		b.	Noisy tubes	b.	Replace tubes
		c.	Noisy controls	c.	Clean all controls; switch contacts
		d	Loose grounds or shields	d.	Check and repair
		e.	Power supply filter defective	e.	Check and replace filter components
		f.	Shorted tube	f.	Check and replace
		g.	Magnetized heads	g.	Demagnetize
		h.	Low record current because of defective level indicator	h.	Check operation with new unit

Table 33.5 Troubleshooting Cassette-Type Recorders

	Symptom		Possible Cause and Remedy
1.	Tape winds unevenly in cassette	a.	Plastic flash-notably around knockout-pin marks on inside of flat surface of case in area where tape winds-interferes with tape winding
		b.	Poor guiding evidenced by "curling" of tape (see Symptom 7)
		c.	Cassette not fully seated
		d.	Tape not falling into guides; check for mold flash on inside of guidepost holes
		e.	Pressure-roller yoke not pivoting freely; this causes a delay in equalization of pressure between top and bottom of pressure roller and capstan, which in turn may cause a stepping in the tape winding ; the binding of this pivot can be relieved by prying the fork of the yoke apart slightly with a screwdriver
		f.	Check for bent pins

2.	Cassette climbs up during operation	a.	Check for bent pins
3.	Groove worn in pressure roller by tape	a.	Machine fails to shut off at end of play (see Symptom 4)
		b.	Excessive torque in take-up clutch (see Symptom 6)
		c.	Excessive drag on tape, causing creepage through capstan; check freedom of turnable bearing and counter
		d.	Insufficient pressure on pressure roller due to stretched spring or bent parts
4.	Machine fails to shut off automatically	a.	Uneven winding of tape in case (see symptom 1)
		b.	Excessive drag in control mechanism (cam or levers)
		c.	Insufficient torque in clutch (see Symptom 6)
		d.	Excessive drag on tape (see Symptom 3c)
		e.	Insufficient pull on tape by capstan (see Symptom 4c)
5.	Tape winds around capstan or pressure roller	a.	Failure of take up because of low torque in take-up clutch (see Symptom 6), or poor winding of tape in cartridge causing excessive drag on take up reel (see Symptoms 1, 2 and 6)
6.	Torque of clutch high or low	a.	Clutch torque outside of limits ; may be due to actual slippage of adjustment (locking screw or nut should be checked)
		b.	Oil or loose particles, such as burrs, may contaminate clutch lining (torque may either increase or decrease)
		c.	Lack of lubrication on end play bearing
		d.	Unlubricated washer between pulley and take up shaft bearing or between turntable spring and shaft bearing
7.	Poor tape guiding	a.	Note: Poor tape guiding is evidenced by tape "curl" as it leaves the capstan; (tape "curl" means that one edge of the tape is tighter than the other as it leaves the capstan; it may require very close observation to detect "curl")
		1.	Cassette must be fully seated and contain new (unstretched) tape
		2.	Tape "cur!" is usually due to misalignment of the capstan.
		3.	Tape "curl" may be due to the tape not operating in the guiding grooves of the tape guide posts (see Symptom la)
		4.	The operator may have failed to seat the cassette properly
		b.	Piles of oxide built up on the heads
8.	Tape won't rewind	a.	Binding of tape in the cassette (see Symptom 1)
		b.	If the capstan is not rotating and motor is stalled drive system is probably overloaded
		1.	Check to see that there is no excessive friction or binding
		2.	Check to see that rewind turntable shaft is free and properly lubricated
		3.	Check to see that idler bearing is properly lubricated

		c.	If the capstan is rotating:
		1.	Check if the idler is pressing the drive belt against the rewind pulley
		2.	Check for friction or binding
		3.	Check all staked or riveted parts
		d.	If capstan is not rotating but motor is running check the belt tension; tension may be low, or the belt may have oil or grease on its surface
9.	Mechanical squeal	a.	Clutch torque high causing belt slippage on take up pulley
		b.	Belt slipping on take up pulley because of oil or grease on belt
		c.	Insufficient rewind idler lubrication
		d.	Tape rubbing over the heads; when this occurs the heads will usually have a pile or coating of oxide built up on their faces, which must be cleaned (a tape which rubs off excessively on the heads should be discarded)
10.	Two tracks heard at once	a.	Check guiding (see Symptom 7)
		b.	Check head height (slight misadjustment of head height may not be observed on playback only)
11.	Difference in level of two heads	a.	Heads should be clean to eliminate possibility of one core making better contact with tape than other core
		b.	May have the same cause as Symptom 10
12.	Azimuth of head wont adjust properly	a.	Adjustment screw may be overtight : bracket may be bent
13.	Pressure pads do not open sufficiently for loading of tape	a.	Excessive drag in pressure-roller levers
		b.	Improper spring tension
		c.	Excessive lubrication under pressure pad mechanism
		d.	Damaged pressure pads

CONVENTIONAL QUESTIONS (C)

1. Is it advisable to service without service data ?
2. What are the tricks of the trade ?
3. Give one example each of each type of circuit configuration.
4. Explain the difference between signal tracing and signal injection.
5. How and where will you apply the circuit isolation technique ?
6. How will you test ; (*a*) unipolar transistors (*b*) bipolar transistors.
7. How will you troubleshoot (tuners, receivers and amplifiers) for the following symptoms ?
   (*a*) No sound, one channel (*b*) Flat, distorted sound
   (*c*) Hum.
8. Give a detailed description of adjustment procedures for record players.
9. How will you replace a pickup ?
10. How will you check for stylus pressure and platter speed ?
11. How will you troubleshoot for the following trouble symptons in record players ?
    (*a*) Stylus skips record grooves, (*b*) Platter turns too slowly or not at all.
    (*c*) Noisy output.

12. Give a detailed description of routine maintenance, both mechanical and electrical, of tape recorders.
13. What is the practical significance of servicing materials and test equipment for tape recorders ?
14. How will you carry out the different adjustments on tape recorders.
15. How will you troubleshoot for the following trouble symptoms in tape machines ?
    (*a*) No tape drive in fast forward
    (*b*) No tape drive at all
    (*c*) Tape spillage
    (*d*) Tape breaks continually
    (*e*) Uneven tape speed
    (*f*) No playback
    (*g*) Incomplete or no erase
    (*h*) Tape winds unevenly in cassette.
    (*i*) Machine fails to shut off automatically
    (*j*) Tape won't rewind
    (*k*) Two tracks heard (crosstalk)
    (*l*) Pressure pads do not open sufficiently for loading of tape.
16. What are the malfunctions in cassette tapes and how will you rectify them ?
17. How will you open a cassette to retrieve damaged tape ?
18. Give the detailed procedure for splicing tape.
19. Explain the procedure for head adjustment.
20. When and how will you replace worn heads ?

Index

A

B

C

D

E

F

G

H

I

K

L

M

N

O

P

Q

R

S

T

U

V

W

ANSWERS

Objective Type Questions—A **Fill in the Blanks—B**

Chapter-1 (A)

1. (c)&(d) 2. (d) 3. (c) 4. (a) 5. (c) 6. (c)

(B)

1. disturbance 2. Energy and momentum 3. directive interference 4. Samephase
5. rapid, louder 6. 343 m/s 7. elasticconstants,density 8. reflected
9. echo,sound 10. threshold, hearing

Chapter-2.

(A)

1.(a) 2.(d) 3.(c) 4.(a) 5.(a) 6.(c)
7.(c) 8.(a)

(B)

1. odd 2. $v/2l$ & $v/4l$ 3. antinodes 4. transverse vibrations 5. 2l
6. no energy

Chapter-3

(A)

1.(a) 2.(b) 3.(c) 4.(a) 5.(b) 6.(b)

(B)

1. funnels,vibrate 2. nerve impulses 3. stimulus, loudness
4. inverse square law 5. decibel scale 6. personal judgement, vary
7. high, low 8. differences. intensity

Chapter-4

(A)

1. (a) 2. (c) 3. (d) 4. (b) 5. (c) 6. (b)
7. (a) 8. (d) 9. (b). 10. (a) 11. (c) 12. (b)

(B)

1. low frequencies 2. 300 to 3400 Hz 3. regular frequency
4. pitch frequency 5. musical interval 6. octave
7. 2:1 8. chord 9. range of binary digits
10. unwanted 11. audible 12. uniformly,audible

Chapter-5 (A)

1.(a) 2.(b) 3.(a) 4.(a) 5.(a) 6.(b)
7.(a) 8.(c) 9.(a) 10.(c) 11.(b) 12.(c)

(B)

1. low 2. high 3. sounds,direction
4. omnidirectional 5. telephone systems 6. hiss
7. bimorph 8. voltage source capacitor 9. a hum bucking coil
10. very low,impedance 11. polarising voltage 12. ribbon microphones

Chapter-6

(A)

1.(a) 2.(a) 3.(a) 4.(b) 5.(a) 6.(a)

(B)

1. Headphones
2. Headsets
3. pull, vibrate
4. mass, stiffuess system resonance
5. capacitive
6. balanced configuration, -harmonic distortion
7. polarised
8. activated
9. 117dB

Chapter-7 (A)

1. (d) 2. (c) 3. (b) 4. (a) 5. (c) 6. (c)
7. (a), (b) & (c) 8. (a) 9. (b) 10. (c) 11. (a)&(b) 12. (a)

(B)

1. reverse
2. vibrate
3. polorise
4. back & forth
5. better
6. ceramic
7. voice coil
8. inertance
9. capacitance
10. damping
11. frequency
12. horn

Chapter-8

(A)

1. (a) 2. (b) 3. (b) 4. (a) 5. (a) 6. (a)
7. (d) 8. (d) 9. (c) 10. (a)

(B)

1. low frequency
2. capacitive
3. open
4. reinforce
5. maximum
6. extensions

Chapter-9

(A)

1. (b) 2. (c) 3. (b) 4.(c)

(B)

1. one channel
2. two seperate channels
3. two channel sound
4. discrete
5. pop music
6. much less

Chapter-10 (A)

1. (b) 2. (b) 3. (a) 4. (b) 5. (a) 6. (b)

(B)

1. persistence
2. overlapping of speech
3. deadness
4. absorption co-efficient
5. frequency
6. m^2-Sabin
7. detrimental
8. irregularities
9. deteriorate
10. less
11. 20 to30 dB
12. time of reverberation
13. absovbers
14. insulators
15. adjoining room
16. same room

Chapter-11 (A)

1. (c) 2. (a) 3. (a). 4. (a) 5. (a) 6. (c)
7. (b) 8. (a) 9. (c) 10. (c) 11. (c) 12. (a)
13. (b) 14. (b) 15. (a)

(B)

1. mixed
2. preamplifiers
3. electrical signals
4. placement
5. contact
6. generating
7. spacing
8. boost
9. 75
10. impedance
11. mone
12. mass
13. wow and fluttter
14. adjustable

Chapter-12

(A)

1. (c)	2. (a)	3. (a)	4.(a)	5.(b)	6.(c)
7(a)	8.(c)				

(B)

1. large
2. speed
3. run
4. bad, too small
5. even
6. odd

Chapter-13 (A)

1.(a)	2.(b)	3.(a)	4.(b)	5.(b)	6.(b)
7.(b).	8.(c)	9.(b)	10.(a)	11.(c)	12.(c)
13.(c)	14.(c)				

(B)

1. reversible processes
2. rate of change of flux
3. iron
4. acicular particles
5. composition, alloy
6. erase head
7. grasps
8. thinner
9. domestic
10. professional
11. recorded simultaneously
12. dual-track recording
13. interlocked
14. small pause
15. digits
16. recording level meters

Chapter 14 (A)

1.(c)	2.(a)	3.(b)	4.(c)	5.(c)	6.(c)
7.(d)	8.(c)	9.(a)	10.(c)	11.(a)	12.(a)
13.(b)	14.(a)	15.(c)	16.(b)	17.(c)	18.(a)
19.(b)	20.(b)				

(B)

1. linear
2. efficient
3. complete
4. strength
5. five, seven
6. separate
7. constant speed
8. top bend
9. tapped
10. intentional
11. comparable
12. treble bass

Chapter-15

(A)

1.(c)	2.(a)	3.(a)	4.(b)	5.(c)

(B)

1. equalised
2. normal output
3. separate
4. compatibility switches
5. relative
6. minimised
7. high stability

Chapter-16

(A)

1.(a) 2.(a) 3.(b) 4.(a) 5.(c) 6.(a)

(B)

1.double-ended
2. compressor, expander
3. compressor
4. quiet
5. 2:1
6. modified

Chapter-18

(A)

1. (a) 2. (b) 3. (a) 4. (b) 5. (c) 6. (a)

(B)

1.smaller
2. voltage
3. sine wave
4.easy
5. low-pass filters
6. radically

Chapter-19

(A)

1. (b) 2. (b) 3. (b) 4. (c) 5. (b) 6. (d)

(B)

1. binary number
2. twice
3. information
4. pivoting arm
5. buffer memory
6 filling degree
7. interlaced
8. main
9. 44.1 kHz
10. CIRC
11. 0.1 μm
12. error burst

Chapter-20

(A)

1.(b) 2.(b) 3.(c) 4.(a) 5.(c) 6.(c)
7.(c) 8.(b) & (d) 9.(b)

(B)

1. modify
2. electromagnetic or piezoelectric
3. variety
4. amplify
5. communication
6. 300 to 5000Hz
7. reliability
8. wide
9. efficiency
10. below
11. constant voltage
12. internal use
13. better
14. two sources
15. 25 ms

Chapter-21

(A)

1. (c) 2. (b) 3. (a) & (d) 4. (a) 5. (a) & (c) 6. (a)
7. (a) 8. (b) 9. (a) 10. (c) 11. (a) 12. (b)
13. (a) 14. (b)

(B)

1. space time
2. completeness
3. single
4. varying, constant
5. left to right
6. disappears
7. reproduce
8. finite
9. 17%
10. 6%
11. active
12. vertical retrace
13. interlaced
14. frame
15. blanking interval

16. inverts
17. composite
18. 40 MHz
19. ± 25 kHz
20. five, seven
21. 200 kHz
22. transfer
23. 1
24. 3
25. human eye
26. Q_{eff} = electrons/photons
27. sensitivity
28. smear
29. darkness
30. indexing
31. electrical power

Chapter -23

(A)

1. (b)
2. (b)
3. (b)
4. (b), (d) & (c)
5. (b)
6. (c)
7. (b)
8. (a)
9. (b) & (d)

(B)

1. seven
2. secondary
3. dominant
4. saturation
5. light energy
6. white light and colour light
7. colour light
8. differently
9. E_G, matrixing
10. separation, natural colour signals
11. recombines
12. three
13. shadow
14. phosphor, respective
15. three
16. in-line
17. magnetically
18. target
19. primary
20. scanning rate, encoding
21. black-and-white
22. phase
23. every line
24. 625/50, 525/60

Chapter -25 (A)

1. (a) through (e)
2. (b)
3. (c)
4. (c)
5. (c)
6. (a)
7. (b)
8. (b)
9. (a) & (c)
10. (d)
11. (a)
12. (c)
13. (d)
14. (a)
15. (a)
16. (b)

(B)

1. cassette
2. picture, sound
3. thinner,show
4: 1/2 in
5. fast, enough
6. runs
7. U
8.two
9.angle
10. halves
11. constant
12.four
13. rotating
14. automatically
15.crosstalk
16. special.
17. bandwidth restricted, frequency shifted
18. increases, improves
19. preset
20. speed, thickness
21. width, 12, 5 mm
22. width, pass
23. recording
24. magnetic
25. recording, infirior
26. directly
27. trackwidth
28. 1500
29. double
30. series, tracks .
31. 25 Hz
32. control, holes
33. linear
34. angles
35. video
36. edits
37. frequencies
38. audio, video
39. penetrate
40. crosstalk
41. compensates
42. 30°

Chapter -26

(A)

1.(b)
2.(b)
3.(b)
4.(c)
5.(b)&(c)
6.(b)
7.(a)
8.(b)
9.(c)
10.(c)
11.(a)&(c)
12.(a)

(B)

1. optics
2. protection
3. 1.6 qm
4. 0.4 μm, 0.1μm
5. 50 mm
6. Philips VLP system
7. reflective,transmissive
8. uninterrupted
9. skill
10. Selectavision(RCA)

Chapter -27

(A)

1.(b)	2.(b)	3.(b)	4.(b)	5.(b)	6.(c)
7.(c)	8.(a)	9.(b)	10.(a)		

(B)

1. unsymmetrical
2. twice
3. even
4. double cathode
5. photographically
6. narrow,one
7. enhance
8. horn
9. origin
10. gate
11. builtin
12. designed
13. automatic
14. retarded
15. wide